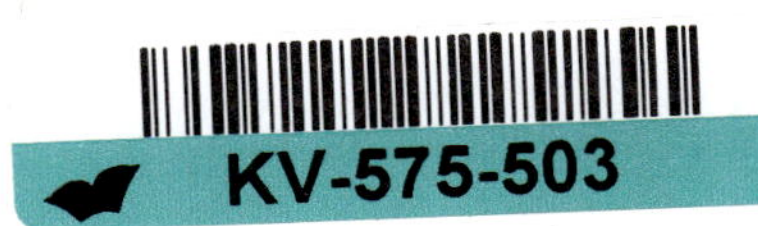

SURFACE MODIFICATION OF METALS BY ION BEAMS

SURFACE MODIFICATION OF METALS BY ION BEAMS

Proceedings of the International Conference on Surface Modification of Metals by Ion Beams, Kingston, Ontario, Canada, July 7–11, 1986

EDITORS: W. A. GRANT, R. P. M. PROCTER
and J. L. WHITTON

Reprinted from the journal
Materials Science and Engineering
Vol. 90, Nos. 1–2

ELSEVIER APPLIED SCIENCE
LONDON and NEW YORK

ELSEVIER APPLIED SCIENCE PUBLISHERS LTD
Crown House, Linton Road, Barking, Essex IG11 8JU, England

Sole Distributor in the USA and Canada
ELSEVIER SCIENCE PUBLISHING CO., INC.
52 Vanderbilt Avenue, New York, NY 10017, USA

British Library Cataloguing in Publication Data
International Conference on Surface
Modification of Metals by Ion Beams
(1986: Kingston, Ont.)
Surface modification of metals by ion beams: proceedings of the International Conference on Surface Modification of Metals by Ion Beams, Kingston, Ontario, Canada, July 7–11, 1986.
1. Metals — Surfaces 2. Ion bombardment
I. Title II. Grant, W. A. III. Procter, R. P. M.
IV. Whitton, J. L. V. Materials science and engineering
671.7 TS213
ISBN 1-85166-141-7

Library of Congress CIP Data Applied For

Printed in The Netherlands by Krips Repro B.V.

Preface

This volume contains the Proceedings of the International Conference on Surface Modification of Metals by Ion Beams held at Queen's University at Kingston, Canada, from July 7–11, 1986. Over 100 participants from 15 countries attended.

This conference was the fourth in the series and the first to be held outside Europe. The aim of the conference, as on the previous occasions, was to bring together physicists, chemists and engineers interested in ion beam modification of metals from both fundamental and applied aspects. Sessions were held on the topics of tribology, oxidation, amorphization, ion beam mixing and morphological effects, aqueous corrosion, enhanced deposition, adhesion, industrial applications and equipment and methods.

Twelve invited speakers addressed areas of special interest. Contributed papers were presented orally or in poster sessions. The format of the conference allowed for much formal and private discussion on the state of the field and on future prospects for research and applications.

As at the previous conference in Heidelberg, the growth of industrial applications of ion beam modified materials was assessed by an international panel of experts. The concensus appeared to be that although industrial acceptance of these materials is growing only slowly it is, nontheless, growing at an accelerated rate each year.

It is a pleasure to thank all the speakers and those who presented poster contributions. We gratefully acknowledge the help and advice given by the various committees, international, programme and local. Queen's University Conference Office provided efficient friendly assistance along with superb accommodation and cuisine. The Physics Department secretaries, Miss Alison Cunninghame and Mrs Phyllis Bentley were invaluable before, during and after the conference week. As may be seen from the list of sponsors, help came from some associated but also from some unexpected sources, all of which was gratefully received. We thank all authors and referees for responding so well in submitting and processing papers which, in no small way, assisted the publisher and editorial staff of *Materials Science and Engineering* in having this volume printed so expeditiously.

Finally, while the scientific part of the conference was appreciated by all, the social programme, aided enormously by glorious weather (the best week of summer) encouraged contact between participants from the various countries and has already spawned many new collaborative research programmes. A lasting memory of SM^{2}IB'86 is the spectacular sunset viewed from the deck of the Island Queen during the cruise among the Thousand Islands.

W. A. GRANT
R. P. M. PROCTER
J. L. WHITTON
Proceedings Editors

Conference Organization

International Committee
H. Bernas (Orsay, France)
G. Dearnaley (Harwell, U.K.)
W. A. Grant (Salford, U.K.)
L. Guzman (Trento, Italy)
H. Herman (Stony Brook, NY, U.S.A.)
J. K. Hirvonen (Bedford, MA, U.S.A.)
G. K. Hubler (Washington, DC, U.S.A.)
M. Iwaki (Saitama, Japan)
S. T. Picraux (Sandia, Albuquerque, NM, U.S.A.)
R. P. M. Procter (Manchester, U.K.)
H. Ryssel (Erlangen, F.R.G.)
J. L. Whitton (Kingston, Ontario, Canada) (Chairman)
J. M. Williams (Oak Ridge, TN, U.S.A.)
G. K. Wolf (Heidelberg, F.R.G.)

Local Committee
J. A. G. Alexander
K.-C. G. Hui
T. Laursen
G. A. Mattiussi
J. D. MacArthur
M. C. Ridgway
P. J. Scanlon
R. W. Smith
J. L. Whitton (Chairman)

Programme Committee
H. Hawthorne (NRC, Vancouver, B.C., Canada)
R. Kelly (IBM, New York, U.S.A.)
G. K. Hubler (NRL, Washington, DC, U.S.A.)
I. V. Mitchell (AECL, Chalk River, U.S.A.)
R. G. Saint Jacques (INRS, Varennes, France)
J. L. Whitton (Kingston, Ontario, Canada) (Chairman)

Supporting Organizations
Bank of Montreal
Calvin Klein Jeans (Canada)
Eaton Ion Beam Systems Corporation
Instituto per la Ricerca Scientifica e Tecnologica (Trento, Italy)
Natural Sciences and Engineering Research Council of Canada
Oak Ridge National Laboratory
Queen's University, Faculty of Arts and Science
Queen's University, Department of Physics
U.S. Office of Naval Research
Universität Heidelberg (Heidelberg, F.R.G.) (*Hosts for SM²IB 1984*)

CONTENTS

Materials Science and Engineering, 90 (1987) 1–8

Formation of Icosahedral Al–Mn and Al–Ru by Solid State Processes*

D. M. FOLLSTAEDT and J. A. KNAPP

Sandia National Laboratories, Albuquerque, NM 87185-5800 (U.S.A.)

(Received July 10, 1986)

ABSTRACT

We have used ion beam mixing and solid state interdiffusion of interdigitated Al/Mn or Al/Ru layers to form icosahedral surface alloys in the solid state. Higher temperatures are required to form the icosahedral phase in Al–Ru than in Al–Mn, both with mixing (300 °C vs. 150 °C) and by interdiffusion (400 °C vs. 300 °C). Annealing to higher temperatures produces a transformation to more stable crystalline phases. Formation by these techniques indicates that these icosahedral phases nucleate and grow in the solid state in preference to crystalline phases of lower free energy.

1. INTRODUCTION

During the past 2 years, considerable attention has been focused on newly discovered metastable phases which exhibit icosahedral symmetry in their diffraction patterns [1, 2]. Icosahedral symmetry is inconsistent with lattice periodicity; none the less, diffraction patterns from these materials have sharp reflections, which indicates long-range order. These phases are now widely believed to be examples of a fundamentally different form of solid matter termed "quasicrystalline" [3], whose atomic order allows symmetries not permitted for periodic structures (crystals). This new order is in some ways intermediate between crystalline and amorphous order [4]. The icosahedral phases thus are being studied to learn more about the fundamental structural and thermodynamic properties associated with such order and to determine whether their unusual symmetry might produce useful properties.

To obtain basic information about this new structure, we have used surface-alloying methods to prepare icosahedral alloys [5, 6]. Formation by these methods gives new insight into their properties not obtainable with the usual preparation methods of melt spinning and splat quenching. Two such methods to be discussed here, ion beam mixing [7, 8] and thermal interdiffusion [9], are solid state transformation processes and do not involve quenching a molten alloy. Producing the icosahedral phase by these methods shows that kinetics can favor formation of the icosahedral phase in the solid state as well as from the liquid state. Furthermore, these studies are yielding important information on the structural and thermodynamic relations between the icosahedral and amorphous phases, as well as other phases of similar composition.

The Al–Mn icosahedral phase has been studied by many workers and is the best characterized of these phases [10]. We have investigated this alloy system with the two methods [7–9] and can place our results into the context of the larger body of available information. Here we summarize the results on Al–Mn and present new results, showing for the first time that icosahedral Al–Ru can be produced by these methods as well. With Al–Ru, we examine icosahedral phase formation with aluminum and a refractory 4d transition element. The high ruthenium melting point makes Al–Ru alloys difficult to produce with melt spinning [11], but this difficulty is of less importance for our techniques. We find that icosahedral Al–Ru requires higher temperatures than Al–Mn to be produced by these solid state processes and is stable to higher temperatures.

*Paper presented at the International Conference on Surface Modification of Metals by Ion Beams, Kingston, Canada, July 7–11, 1986.

0025-5416/87/$3.50

2. ION BEAM MIXING

Alternating layers of Al/Mn or Al/Ru were deposited on clean substrates of aluminum or sapphire (Al_2O_3) in a vacuum of about 5×10^{-7} Torr. Typically, five to eight pairs of layers were used to produce total alloy thicknesses of 50–150 nm. The relative thicknesses of the aluminum and manganese layers were adjusted to obtain average manganese concentrations of 12–45 at.%, while a more limited ruthenium concentration range of 16–23 at.% was examined. The alloy thickness and concentration were examined with Rutherford backscattering analysis using 1.5 MeV helium ions ($^4He^+$) and some Al/Mn layers on aluminum substrates were examined for oxygen and carbon contents by nuclear reaction analysis with a deuterium beam. A typical impurity content was 3.5×10^{16} O cm^{-2} and 1.7×10^{16} C cm^{-2} for eight pairs of layers and 150 nm thickness. The oxygen is believed to be mostly in the form of surface oxides and is equivalent to about 5 nm of Al_2O_3.

Ion mixing was carried out by irradiating the deposited layers with a 400 keV xenon ion (Xe^+) beam. The sample was clamped to a copper block; either the ion beam current was controlled so that the temperature did not rise above 30 °C or the block was heated and maintained at a constant temperature during the irradiation. The projected xenon ion range exceeds the alloy layer thickness and thus most of the xenon ions pass through the layer. Samples for transmission electron microscopy (TEM) analysis were made either by jet electropolishing (aluminum substrates) or by ion milling (Al_2O_3 substrates) from the back side.

2.1. Al–Mn

Our results obtained with ion beam mixing of Al–Mn alloys have been presented in detail elsewhere [8]. The phases which we observe at different manganese concentrations are summarized in Fig. 1, and the composition range for the icosahedral–amorphous phase (14–20 at.% Mn) is indicated. Mixing the Al/Mn layers at 60 °C or lower produced an amorphous phase with three clearly identified diffuse rings in the electron diffraction pattern, while mixing at 100–200 °C or higher produced the icosahedral phase with five or more rings. Thus, temperature plays an important role in determining the degree of

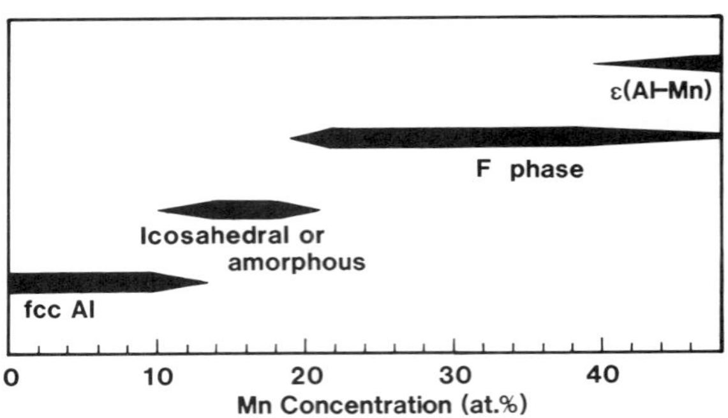

Fig. 1. Diagram indicating the phases observed in Al–Mn alloys formed by ion beam mixing (1×10^{16} Xe$^+$ cm^{-2}) at 30–150 °C for 0–50 at.% Mn.

order in the mixed layers. Our sharpest most complete ring patterns of the icosahedral phase were obtained by ion mixing at 150 °C; the diffraction pattern and associated dark field TEM image of the icosahedral grains (up to 50 nm in diameter) are shown for a 17 at.% Mn sample in Fig. 2. More than 15 rings can be identified, and their lattice spacings [6] accurately match those determined for the icosahedral phase in melt-spun samples by high resolution X-ray diffraction [12]. Similar microstructures have been obtained with ion beam mixing by other workers [13].

The minimum manganese concentration required to form single-phase icosahedral material by ion beam mixing is about 14 at.%; at lower manganese concentrations, excess f.c.c. aluminum remains in the sample. For higher manganese concentrations of 21 at.% to above 30 at.%, a crystalline phase replaces the icosahedral phase. This phase, labeled the "F phase", has not previously been reported [14]. We have determined that the rings from this phase can be fitted to a b.c.c. lattice with $a_0 = 0.519(2)$ nm. An unidentified phase was observed by others in ion-beam-mixed samples at manganese concentrations above those producing the icosahedral phase [15]; we find that its lattice spacings [16] fit the F phase, in agreement with our results. At a still higher managanese concentration of 44 at.%, hexagonal ε-(Al–Mn) is observed; this phase is expected to persist up to 62.5 at.% Mn based on the Al–Mn equilibrium phase diagram [17].

From these ion-beam-mixing studies, several important results have been obtained. First, since the nature of the xenon ion cascades is expected to change very little between sub-

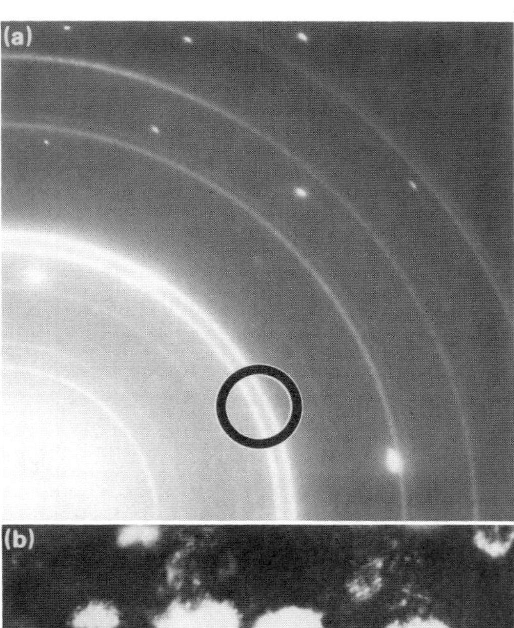

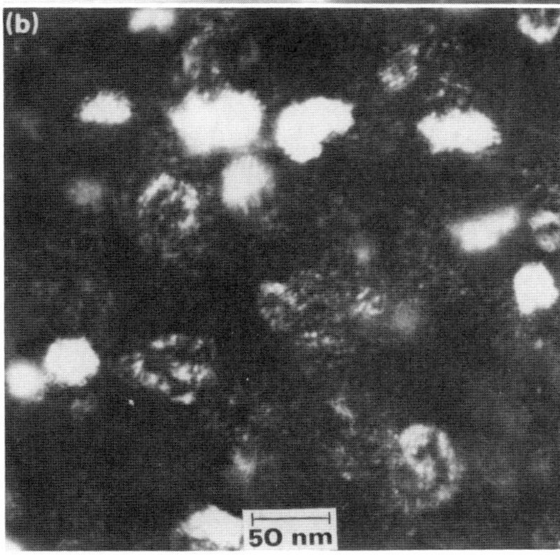

Fig. 2. (a) Diffraction pattern for Al–Mn; (b) dark field image obtained using the indicated rings of the icosahedral phase in an aluminum surface alloy with 17 at.% Mn on Al_2O_3 (spots in (a)) formed by ion beam mixing at 150 °C.

strate temperatures of 60 and 100 °C, formation of the icosahedral phase apparently does not take place within the ion cascade, *i.e.* it does not occur during the "prompt" regime (within about 1 ps or less) of the cascade nor during the "cooling-down period" (about 1–100 ps) in the nomenclature of ref. 18. The increase in sample temperature needed to produce the icosahedral phase is consistent with an expected increase in atomic mobility of atoms in the predominantly aluminum matrix after the cascade, which would place its formation in the "delayed" regime [18]. Atomic mobilities would be further promoted by radiation-enhanced diffusion because of the numerous point defects produced by the xenon ions. From this and other work [7], it is clear that the icosahedral phase is stable against irradiation-induced amorphization above 100 °C, but not below 60 °C.

An alternative to the quasicrystalline structural model of icosahedral phases involves fine-scale twinning of crystals with twin planes making 72° (=360°/5) angles with each other [19]. We have observed icosahedral phase grains of less than 40 nm and perhaps down to 10 nm [5], which would require twinning to occur on this unusually fine scale. These and other experimental results indicate that "microtwinning" is not the correct structural model; quasicrystalline models are generally taken to be the most promising explanation of structures with this novel symmetry [20].

The discovery of amorphous Al–Mn is also relevant to understanding the icosahedral phase, since its diffraction pattern exhibits icosahedral short-range order [7]. Such short-range order is hypothesized for liquids and used with Landau theories to predict that a solid with icosahedral long-range order can be energetically favored over other symmetries [21]. Since amorphous alloys have liquid-like structures, identifying the icosahedral short-range order indicates that such treatments are applicable to Al–Mn.

2.2. *Al–Ru*

We have examined the Al–Ru system to a more limited extent and find both similarities with and differences from the Al–Mn icosahedral alloys. Ion beam mixing an alloy with 18.5 at.% Ru at 150 °C yielded icosahedral material. However, the alloy was not single phase, as was observed for Al–Mn at this concentration. The diffraction pattern from this alloy is shown in Fig. 3(a). It contains five diffuse rings at positions expected for a fine-grained icosahedral alloy, with the qualification that the bright ring corresponds to the bright doublet seen in Fig. 2. Also seen in Fig. 3(a) are superimposed sharp arcs (marked with arrows) indexing to cubic (B2) AlRu. A dark field image obtained with the bright ring is shown in Fig. 3(b). In this figure, large particles are seen in the thinnest material at the edge of the hole; these particles are obscured by the high density of fine particles in the thicker material. The large particles

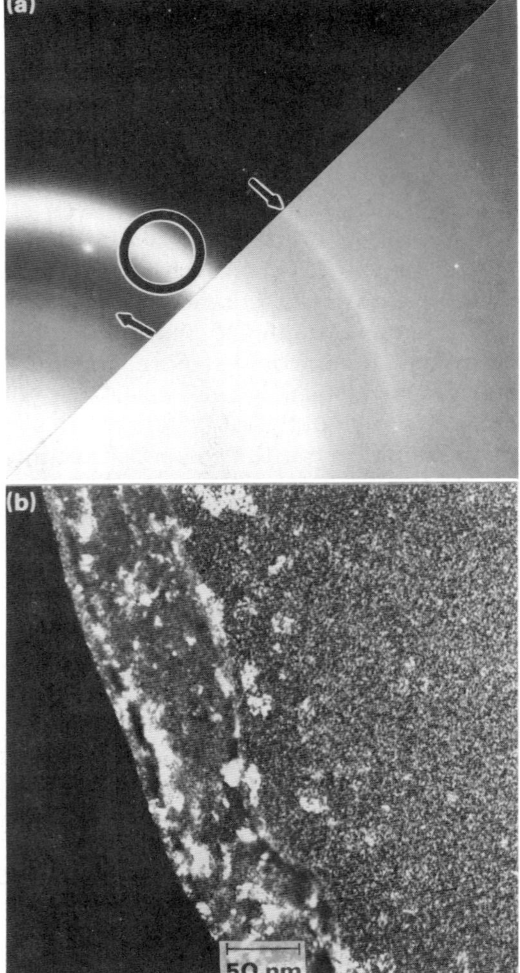

Fig. 3. (a) Diffraction pattern (two exposures) for Al–Ru; (b) dark field image obtained using the indicated rings from an aluminum surface alloy with 18.5 at.% Ru formed by ion beam mixing at 150 °C. The icosahedral phase is identified with the diffuse rings in (a) and the fine grains in (b), while cubic AlRu is identified with the sharp arcs (arrowed) in (a) and the larger particles near the edge in (b).

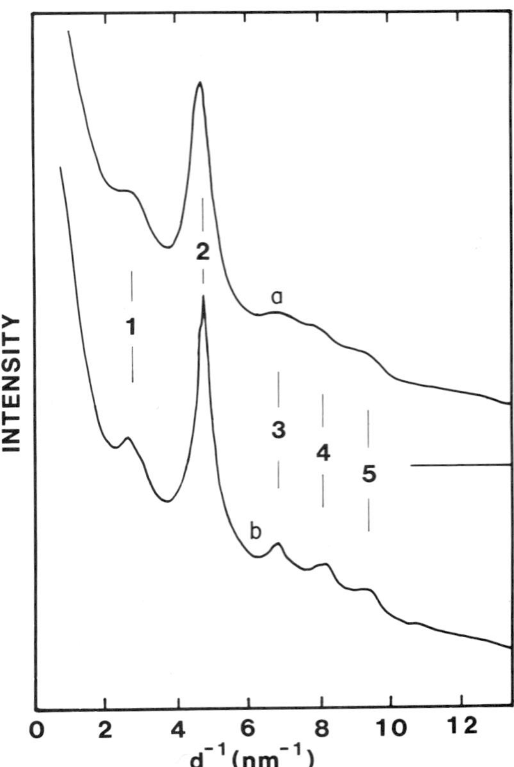

Fig. 4. Densitometer traces of electron diffraction negatives: curve a, from Fig. 3(a) using a scan which avoided the AlRu arcs, ion beam mixed at 150 °C; curve b, from a negative obtained after annealing the Al/Ru layers used in Fig. 6 for 15 min at 500 °C. The linear vertical intensity scale is arbitrary and has been displaced for curve a.

(greater than 10 nm) near the irradiated surface are apparently cubic AlRu and produce the sharp rings, including one obscured by the diffuse ring and included in the aperture used to obtain the dark field image. The fine particles (less than 2 nm) are the icosahedral phase.

The five diffuse rings of the icosahedral phase in Fig. 3(a) are also seen in the densitometer trace in Fig. 4; from their width, perhaps it should be asked whether the alloy is really icosahedral or whether it might be more properly called amorphous. Amorphous metals generally have three rings, those numbered 2, 4 and 5 in Fig. 4. Since we are able to resolve numbers 1 and 3 as well, and since the five rings fall at the positions expected for the most intense rings of the phase, we have chosen to identify the material as the icosahedral phase. Others have suggested that liquid-quenched amorphous Al–Mn alloys are in fact "microquasicrystalline" [22], *i.e.* not truly amorphous but extremely fine-grained icosahedral material. We are currently investigating amorphous alloys formed by ion beam mixing to determine whether they have additional icosahedral order beyond the short-range order mentioned above.

The results with Al–Mn suggest that the icosahedral grain size might be increased above that in Fig. 3(b) by mixing at a higher temperature. However, when we mixed the alloy with 18.5 at.% Ru at 225 °C, the cubic AlRu phase became more dominant, although diffuse rings from the icosahedral phase could still be seen in the diffraction pattern. Increasing the ruthenium concentration to 23 at.% further increased the intensity of AlRu and diminished the icosahedral phase.

We were able to obtain larger grains in surface alloys free of AlRu by decreasing the ruthenium concentration to 16 at.% and performing the ion beam mixing at 300 °C, as shown in Fig. 5. The diffraction pattern in Fig. 5(a) has sharper rings (superimposed on spots from Al_2O_3) than Fig. 3(a), consistent with the larger (about 30 nm) grain size. We can identify each ring in Fig. 5(a) with a peak in the high resolution X-ray powder diffraction pattern for icosahedral Al–Mn [12], with the proviso that the atomic spacings are scaled by $d_{Al-Ru}/d_{Al-Mn} = 0.984(2)$, where we have used the α-Al_2O_3 reflections with $a_0 = 0.5127$ nm as an internal diffraction standard. This excellent agreement with 13 rings indicates that this phase is indeed an icosahedral phase with a slightly smaller atomic spacing than that of Al–Mn. However, although the pattern in Fig. 5(a) contains the brightest rings observed for Al–Mn, a few weak rings (for instance two seen in Fig. 2(a) just inside the bright doublet (circled)) are not observed in Fig. 5(a) for Al–Ru. Those rings were also not observed in X-ray diffraction from melt-spun icosahedral Al–Ru [11]. These considerations suggest either that the structure factor governing the diffracted intensities of icosahedral Al–Ru has changed relative to that of Al–Mn or that the Al–Ru phase has defects which have preferentially reduced the intensity of these rings.

Ion beam mixing the Al–Ru alloys at 150 °C produced an icosahedral phase, but temperatures higher than those used for Al–Mn were required to achieve comparable grain sizes. These results again suggest that the icosahedral grains grow by diffusive processes after an ion cascade has ended, but comparable diffusion rates require higher temperatures for Al–Ru.

The B2 (CsCl) phase AlRu was observed in the samples mixed at 150 and 225 °C. Such phases with transition metals and aluminum are often radiation stable and form during ion beam mixing [23]. Furthermore, they can form with relatively low transition metal concentrations [24]. Since the phases Al_3Ru_2, Al_2Ru, Al_6Ru and $Al_{12}Ru$ are found in the phase diagram [25] but do not form in our alloys, the phase fields with AlRu can be expected to extend to lower concentrations in a metastable phase diagram appropriate for ion-beam-mixed Al–Ru [26]. Ion beam mixing can produce phases found in high temperature regions of phase diagrams [27], and thus the AlRu may have formed in the ion cascade during the "cool-down period" [18] when the available thermodynamic states are being sampled.

3. SOLID STATE INTERDIFFUSION

To investigate forming icosahedral alloys by interdiffusion, the same sequences of layers were deposited on TEM grids with a Formvar layer across the grid bars. These samples were annealed *in situ* in the microscope (in a vacuum of about 10^{-7} Torr) and their microstructures were monitored during annealing. Typically, the anneals were for 15 min intervals at successively higher temperatures.

3.1. Al–Mn

Annealing Al/Mn layers with 14 and 18 at.% Mn transformed them to fine-grained icosahedral alloy layers [9]. The transformation began within 5 min at 250 °C but was not complete after 1 h. The reaction went to completion within 2 min at 300 °C and produced a single-phase alloy for 18 at.% Mn; however, unreacted aluminum remained in the 14 at.% Mn alloy. Continued annealing at 300 or 350 °C produced crystalline phases. With the addition of 4.5 at.% Si to an alloy with 20 at.% Mn, we were able to anneal up to 425 °C without complete crystallization of the layers and we obtained the largest icosahedral grain sizes with this method, 3–14 nm. With this treatment, nine relatively sharp diffraction rings (the bright doublet was resolved) matching the icosahedral phase were detected.

The formation of icosahedral Al–Mn by the more widely used rapid quenching methods indicates that the phase nucleates from the liquid in preference to crystalline phases in spite of their lower free energies. The favorable nucleation might be promoted by the short-range icosahedral order of the liquid alloy; similar arguments can be made for nucleation of the icosahedral phase during

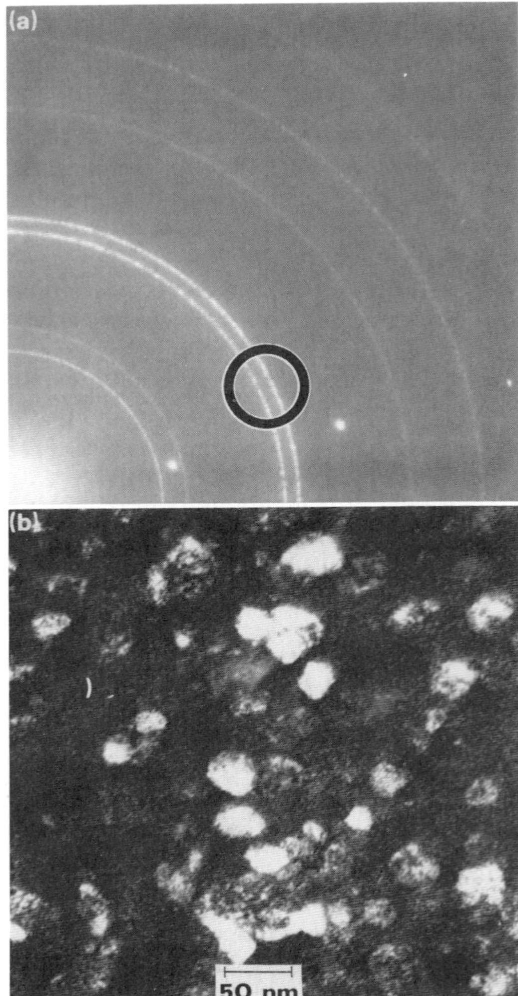

Fig. 5. (a) Diffraction pattern for Al–Ru; (b) dark field image obtained using the indicated rings of the icosahedral phase in an aluminum surface alloy with 16 at.% Ru on Al_2O_3 substrate (spots in (a)) formed by ion beam mixing at 300 °C.

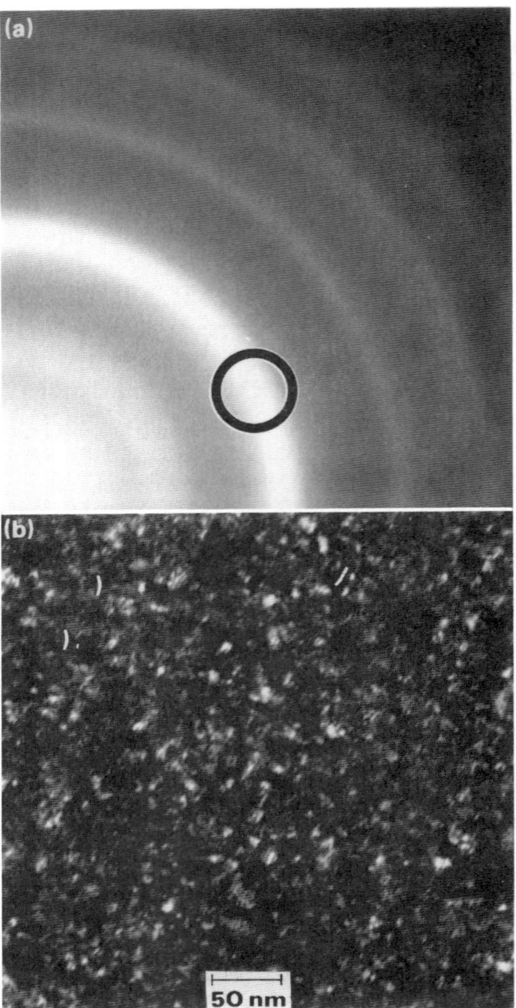

Fig. 6. (a) Diffraction pattern for Al–Ru; (b) dark field image obtained using the indicated rings of the icosahedral phase in an aluminum alloy with 22.5 at.% Ru on a TEM grid, produced by annealing up to 550 °C for 15 min.

the annealing of amorphous phases. However, the alternating elemental layers do not have the icosahedral short-range order throughout their total thickness, although the layer interfaces might locally have such order and it might also be present in the deposited manganese layers. The formation of the icosahedral Al–Mn by solid state interdiffusion therefore demonstrates that uniform icosahedral short-range order is not a requirement for forming icosahedral phases.

3.2. *Al–Ru*

We have examined layered structures with an average of 22.5 ± 1.0 at.% Ru during *in situ* annealing. There was some evidence that a reaction was occurring (an increase in diffuse scattering) at 200 and 250 °C during the 15 min anneals. At 300 °C an inner diffuse ring was clearly identified, and the aluminum rings were weaker. The bright ring corresponding to the unresolved doublet was stronger, and diffuse rings at larger radii were also seen. The f.c.c. aluminum was still detected at 350 °C but disappeared at 400 °C, where the aluminum and ruthenium had apparently fully reacted. Continued annealing up to 550 °C produced a slight sharpening of the rings and gave the ring pattern shown in Fig. 6(a). Dark field imaging with the bright ring illuminated particles 3–8 nm in diameter, as seen in Fig. 6(b). Crystallization was observed to occur in these alloys at temperatures between 500 and 600 °C. We tentatively identify the resulting crystalline phases as $RuAl_{\approx 2.5}$ [28] and Ru_2Al_3 [29].

A densitometer trace from the negative of a diffraction pattern similar to Fig. 6(a) but obtained after annealing at 500 °C is shown in Fig. 4; five rings are clearly identified, and structure can be seen on numbers 1 and 4. On comparison of Figs. 4(a) and 4(b), it is apparent that this anneal produced sharper rings than the ion-beam-mixing treatment at 150 °C. However, neither ring pattern in Fig. 4 is as sharp as those produced by ion beam mixing at 300 °C (see Fig. 5(a)).

We attempted to heat the alloy rapidly to 500 °C, hoping to avoid first nucleating the phase at lower temperatures and to obtain fewer larger grains. Although the temperature went from 20 to 500 °C in only 2 min, the icosahedral phase formed when the temperature reached 400 °C, and a similar microstructure was obtained.

Like Al–Mn, icosahedral Al–Ru nucleates in preference to crystalline compounds and grows during interdiffusion of the elemental layers; however, higher temperatures are needed to complete the reaction for Al–Ru. The driving force for formation of compounds from elementally segregated layers is the heat of mixing which lowers the overall free energy of the set of layers by forming an alloy mixture. The heat of formation has been calculated by Miedema [30] for binary alloys with 1-to-1 composition ratios. For Al(Mn) the value is -43 kJ (g-atom)$^{-1}$; using a graphical method devised by Miedema [31] based on the same analysis, we obtain essentially the same value for Al–Ru, which is consistent with the similar evolution of their alloy layers.

4. DISCUSSION

The formation of icosahedral Al–Mn and Al–Ru phases by two solid state techniques indicates that these phases nucleate and grow more readily than crystalline phases of lower free energy in the solid state (as well as liquid). We have suggested [9] that this is because the icosahedral phase uses a less-ordered arrangement [4] of icosahedron-shaped units of atoms [32, 33], attached to each other, whereas the competing crystalline phases use essentially the same units arranged into a more-ordered crystalline lattice structure [34]. The less-ordered attachment of such units would seem to be more readily accomplished than arrangement into a repeating unit cell, which is consistent with observation. With both techniques discussed here the icosahedral phase forms by diffusion in the solid state, and in both cases a higher temperature was required for Al–Ru than for Al–Mn. Apparently, diffusion is slower at a given temperature in the Al–Ru alloys. Icosahedral Al–Ru was also stable against crystallization to higher temperatures than Al–Mn (500 °C *vs.* 350 °C).

On annealing, amorphous Al–Mn transformed to the icosahedral phase and then to crystalline Al_6Mn. Icosahedral Al–Ru also transformed to crystalline phases at sufficiently high temperatures. These results indicate that these icosahedral phases are metastable with respect to formation of equilibrium crystalline phases but are more stable than amorphous phases.

The formation of metastable alloys by solid state interdiffusion has recently been extended to a "mechanical alloying" process [35, 36]. With this method, powders of two elements are mixed by agitation with steel balls and forced to adhere to each other when the powder is compressed between impinging balls. During this process, solid state interdiffusion produces a powder of metastable material, *e.g.* amorphous Ni–Ti [35]. Because several amorphous alloys can be produced both by simple interdiffusion alone and by mechanical alloying [36], we suggest that mechanical alloying may also be able to produce icosahedral Al–Mn and Al–Ru, perhaps requiring the elevated temperatures used here. This method holds promise for producing larger quantities of these phases which could then be consolidated into bulk form.

We have used both ion beam mixing and interdiffusion during solid state annealing to produce single-phase icosahedral layers of both Al–Mn and Al–Ru on sapphire substrates. Because this substrate is transparent and insulating, such samples will now permit the study of properties such as electrical conductivity and optical transmission. We have recently also used such surface layers on sapphire to determine that the melting point of icosahedral Al–Mn is 910 ± 20 °C for 20 at.% Mn [37]. Since icosahedral Al–Ru is stable to higher temperatures, we speculate that its melting point is above that for Al–Mn.

ACKNOWLEDGMENTS

We wish to thank W. L. Johnson and S. Anlage of the California Institute of Technology for valuable discussions about the Al–Ru alloy system. The assistance of M. Moran and G. Schuh in preparing and characterizing samples is gratefully acknowledged.

This work was performed at Sandia National Laboratories, supported by the U.S. Department of Energy under Contract DE-AC04-76DP00789.

REFERENCES

1 D. Shechtman, I. Blech, D. Gratias and J. W. Cahn, *Phys. Rev. Lett., 53* (1984) 1951.
2 R. D. Field and H. L. Fraser, *Mater. Sci. Eng., 68* (1984) L17.
3 D. Levine and P. J. Steinhardt, *Phys. Rev. Lett., 53* (1984) 2477.
4 P. W. Stephens and A. I. Goldman, *Phys. Rev. Lett., 56* (1986) 1168.
5 J. A. Knapp and D. M. Follstaedt, *Phys. Rev. Lett., 55* (1985) 1591.
6 J. A. Knapp and D. M. Follstaedt, *Materials Research Society Symp. Proc.*, Vol. 51, Materials Research Society, Pittsburgh, PA, 1986, p. 415.
7 D. M. Follstaedt and J. A. Knapp, *J. Appl. Phys., 59* (1986) 1756.
8 J. A. Knapp and D. M. Follstaedt, *Nucl. Instrum. Methods B*, to be published.
9 D. M. Follstaedt and J. A. Knapp, *Phys. Rev. Lett., 56* (1986) 1827.
10 Abstracts in Sessions on Quasicrystals, *Bull. Am. Phys. Soc., 31* (3) (March 1986).
11 P. A. Bancel and P. A. Heiney, *Phys. Rev. B, 33* (1986) 7917.
12 P. A. Bancel, P. A. Heiney, P. W. Stephens, A. I. Goldman and P. M. Horn, *Phys. Rev. Lett., 54* (1985) 2422.
13 D. A. Lilienfeld, M. Nastasi, H. H. Johnson, D. G. Ast and J. W. Mayer, *Phys. Rev. Lett., 55* (1985) 1587.
14 P. Villars and L. D. Calvert, *Pearson's Handbook of Crystallographic Data for Intermetallic Phases*, Vol. 2, American Society for Metals, Metals Park, OH, 1985, p. 1019.
15 D. A. Lilienfeld, M. Nastasi, H. H. Johnson, D. G. Ast and J. W. Mayer, *Materials Research Society Symp. Proc.*, Vol. 51, Materials Research Society, Pittsburgh, PA, 1986, p. 427.
16 D. Lilienfeld, personal communication, 19 .
17 R. P. Elliott, *Constitution of Binary Alloys, First Supplement*, McGraw-Hill, New York, 1965, p. 43.
18 D. M. Follstaedt, R. S. Averback and M.-A. Nicolet, *Rep. SAND85-2465*, January 1986 (Sandia National Laboratories).
19 M. J. Carr, *J. Appl. Phys., 59* (1986) 1063.
20 J. W. Cahn, *Mater. Res. Soc. Bull., 11* (2) (1986) 9.
21 D. R. Nelson and B. I. Halperin, *Science, 229* (1985) 233.
22 L. A. Bendersky and S. D. Ridder, *J. Mater. Res., 1* (1986) 405.
23 M. Nastasi, L. S. Hung and J. W. Mayer, *Appl. Phys. Lett., 43* (1983) 831.
L. S. Hung, M. Nastasi, J. Gyulai and J. W. Mayer, *Appl. Phys. Lett., 42* (1983) 672.
24 D. M. Follstaedt and A. D. Romig, Jr., in J. T. Armstrong (ed.), *Microbeam Analysis — 1985*, San Francisco Press, San Francisco, CA, 1985, p. 173.
25 F. A. Shunk, *Constitution of Binary Alloys, Second Supplement*, McGraw-Hill, New York, 1969, p. 40.
26 D. M. Follstaedt and S. T. Picraux, in L. H. Bennett, T. B. Massalski and B. C. Giessen, *Alloy Phase Diagrams, Materials Research Society Symp. Proc.*, Vol. 19, Elsevier, New York, 1983, p. 94.
27 R. S. Averback, in D. M. Follstaedt, R. S. Averback and M.-A. Nicolet (eds.), *Proc. 2nd Workshop on Ion Mixing and Surface Layer Alloying*, in *Rep. SAND85-2465*, January 1986, p. 27 (Sandia National Laboratories).
28 L.-E. Edshammar, *Acta Chem. Scand., 20* (1966) 427.
29 *Powder Diffraction File*, Joint Committee on Powder Diffraction Standards, International Center for Diffraction Data, Swarthmore, PA, 19 , Card 19-46.
30 A. R. Miedema, *Philips Tech. Rev., 36* (8) (1976) 217.
31 A. R. Miedema, personal communication, 19 .
32 P. Guyot and M. Audier, *Philos. Mag. B, 52* (1985) L15; *53* (1986) L43.
33 V. Elser and C. L. Henley, *Phys. Rev. Lett., 55* (1985) 2883.
34 W. B. Pearson, *The Crystal Chemistry and Physics of Metal Alloys*, Wiley-Interscience, New York, 1972, p. 714.
35 R. B. Schwarz, R. R. Petrich and C. K. Saw, *J. Non-Cryst. Solids, 76* (1985) 281.
36 E. Hellstern and L. Schultz, *Appl. Phys. Lett., 48* (1986) 124.
37 J. A. Knapp and D. M. Follstaedt, to be published.

Materials Science and Engineering, 90 (1987) 9-12

Measurements of Sputtering Yield at the Critical Angle of Incidence*

H. HASUYAMA, Y. KANDA, K. NIIYA and M. KIMURA

Department of Energy Conversion Engineering, Kyushu University, Kasuga, Fukuoka 816 (Japan)

(Received July 10, 1986)

ABSTRACT

A new method is proposed for measuring the values of $S(\hat{\theta})$, the maximum sputtering yield at the critical angle $\hat{\theta}$ of incidence in the present paper. It is based on the fact that a single thin wire evolves to a cone with an apex angle $\pi - 2\hat{\theta}$ at a high irradiation above 5×10^{19} ions cm^{-2}, and then the cone is eroded, conserving its shape. The value of $S(\hat{\theta})$ can be obtained from observation of its steady erosion. The method presented here is useful for avoiding the laborious setting of samples at the angle $\hat{\theta}$ and ensures sputtering yield measurement at the correct incident angle $\hat{\theta}$. It is found in the experimental results that the values of $S(\hat{\theta})$ for the irradiation of 20 keV argon (Ar^+) ions on the polycrystalline copper and silver wires are 23 atoms ion^{-1} and 37 atoms ion^{-1} respectively at $\hat{\theta} = 83°$.

1. INTRODUCTION

A number of investigations have been carried out to study the angular dependences of the sputtering yield $S(\theta)$ of solids by ion bombardment as summarized in ref. 1. It has been demonstrated that $S(\theta)$ reaches its maximum value at the angle $\hat{\theta}$ of incidence which is associated with maximum energy deposition at and near the surface. The measurements of $S(\hat{\theta})$ are few because the setting of the angle of incidence is very difficult in the range of near-glancing angles on planar targets.

We have developed a method to measure the critical angle of incidence using scanning electron microscopy (SEM) after irradiating a thin cone-shaped wire without being influenced by the sputtered atoms and reflected ions from the surrounding target materials. It was confirmed [2, 3] that the slope of the sample surface became stable and converged to be straight at the situation where the incident angle reduced to the critical angle $\hat{\theta}$ with respect to the surface normal, after a high ion fluence of approximately 5×10^{19} ions cm^{-2}. This result enables us to measure the sputtering yield at the critical angle of incidence using SEM observations without complex experimental procedures.

In the present paper the method for measuring the sputtering yield at the critical angle of incidence and the results obtained from 20 keV argon (Ar^+) ion bombardment of polycrystalline copper and silver wires are described. Comparisons of the results with available data are also made.

2. EXPERIMENTAL TECHNIQUES AND SCANNING ELECTRON MICROSCOPY ANALYSIS

Thin wires of polycrystalline copper (purity, 99.9%) and silver (purity, 99.99%) samples with a diameter of 30–50 μm, whose tips were conically shaped by electrolytic polishing, were used to measure the sputtering yield at the critical angle of incidence. The electrolytical etching solution used was 60% H_3PO_4 for copper wires and a mixture of 1 part of ethanol to 4 parts of 70% H_3PO_4 for silver wires. Some of the copper wires thus shaped were annealed at either 400 or 800 °C for 1 h at a pressure of 10^{-3}–10^{-4} Pa. The silver wires used were prepared by annealing at 700 °C for 1 h under the same vacuum conditions.

The wire samples, which were set perpendicular on the respective pure polycrystalline plates, were irradiated along their axes with a 20 keV mass-analysed-and-collimated argon beam 2 mm in diameter. The consecutive ion

*Paper presented at the International Conference on Surface Modification of Metals by Ion Beams, Kingston, Canada, July 7-11, 1986.

0025-5416/87/$3.50

irradiation and SEM observations were made repeatedly until the surface slope of the samples became stable and the surface normal made a critical incident angle $\hat{\theta}$ with respect to the ion beam direction, where the sputtering yield attained a maximum [4]. After an additional irradiation of $(2–3) \times 10^{19}$ ions cm^{-2}, the sputtering-induced erosion Δz in the ion flux direction was measured using SEM observation analysis and was obtained by averaging over the cone surface on the basis of the assumption that the axial symmetry of wire was maintained during irradiation. Thus the sputtering yield $S(\hat{\theta})$ was derived from the expression

$$S(\hat{\theta}) = \frac{\Delta z\, N}{\Delta \phi}$$

where N is the atomic density of the wire sample and $\Delta\phi$ is the additional fluence.

3. RESULTS AND DISCUSSION

Examples of the SEM micrographs used for measuring the sputtering yields of copper wires by 20 keV argon ion bombardment at the critical angle of incidence are shown in Fig. 1. Figure 1(a) and Fig. 1(b) show the tip profiles of an unannealed sample after irradiation with an ion fluence of 6.9×10^{19} ions cm^{-2} and 1.0×10^{20} ions cm^{-2} respectively. In both micrographs the incident angle at the slope of the cone surface was found to be 83°; this is considered to be the critical angle of incidence. The sputtering yield $S(\hat{\theta})$ was obtained from the micrograph analysis to be 22 ± 4 atoms ion^{-1}. The error is assumed to be 20% in the present experiment. Uncertainty due to the microscopic structures on the eroded surface was difficult to estimate even though the axial symmetry of the wire was maintained.

The method for measuring the sputtering yield was extended to the case where the incident angle did not reach the critical angle $\hat{\theta}$ in a low fluence irradiation. The results thus obtained are shown in Fig. 2 as a function of ion fluence together with the case of a high fluence irradiation. The bar shown over the ion fluence range indicates the region in which the sputtering yield was measured. It is evident from the figure that the sputtering yield attains a constant value above 2×10^{19} ions cm^{-2}.

In order to study the influence of the different intrinsic properties such as the cold-working effects or the grain size of samples

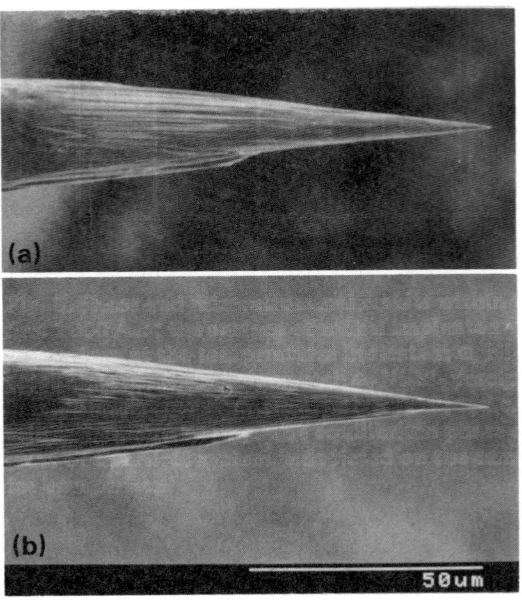

Fig. 1. An example of a pair of SEM micrographs of polycrystalline copper wires used to measure sputtering yields due to 20 keV argon ions at the critical angle of incidence: (a) after an irradiation of 6.9×10^{19} ions cm^{-2}; (b) after an irradiation of 1.0×10^{20} ions cm^{-2}.

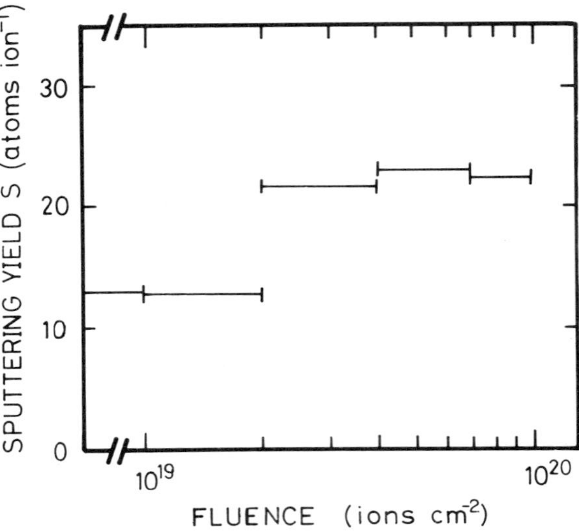

Fig. 2. Sputtering yields of a cone-shaped polycrystalline copper wire bombarded by 20 keV argon ions as a function of ion fluence.

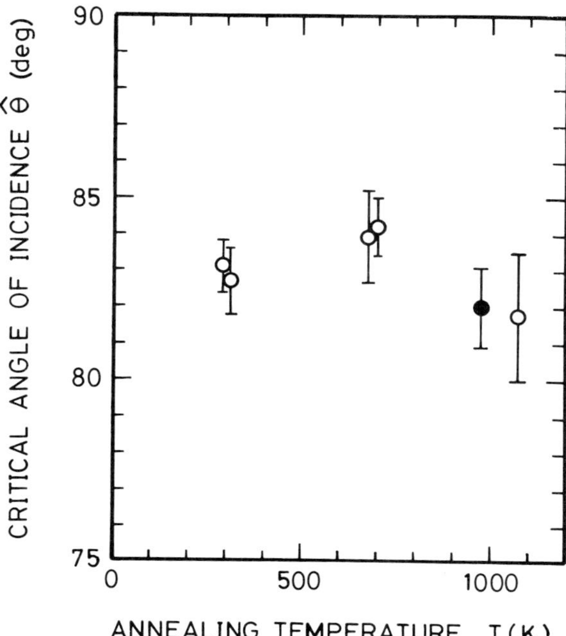

Fig. 3. The critical angle of incidence measured at various annealing temperatures. The results for copper (○) and silver (●) wires are shown. The results for an unannealed copper wire at room temperature are also indicated.

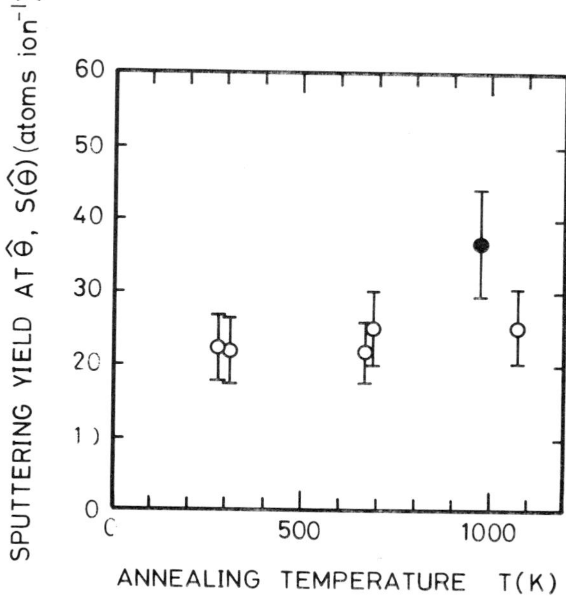

Fig. 4. Sputtering yields at the critical angle of incidence measured at various annealing temperatures. The values for copper (○) and silver (●) wires are shown. The results for an unannealed copper wire at room temperature are also indicated.

on the sputtering yield, measurements of $S(\hat{\theta})$ were performed on copper wires annealed at 400 or 800 °C and silver wires annealed at 700 °C. Above 5×10^{19} ions cm^{-2} the surface slope of a cone-shaped sample was found to be almost constant and made an angle $\hat{\theta}$ with respect to the incident ion beam direction. The critical angles measured at various temperatures are compared in Fig. 3. The results for unannealed samples at room temperature are shown. The open circles and full circles show the results for copper wires and silver wires respectively. This figure indicates that the critical angles of incidence for cone-shaped copper and silver wires are similar and equal to approximately 83°, irrespective of the different sample treatments and irrespective of the different element species.

The results of sputtering yields at the critical angle of incidence for copper wires which were unannealed, annealed at 400 °C and annealed at 800 °C and for silver wires annealed at 700 °C are compared in Fig. 4. The results for the unannealed copper wire at room temperature are also shown, as in Fig. 3. The results for copper wires and silver wires are indicated by open circles and full circles respectively. The values of $S(\hat{\theta})$ for copper wires are found to be almost constant over a wide range of annealing temperatures and are on average about 23 atoms ion^{-1}, which is in good agreement with the value of 28 atoms ion^{-1} obtained by Molchanov and Tel'kovskii [5] from 27 keV argon ion bombardment on polycrystalline copper surfaces.

For polycrystalline silver wires, however, the maximum sputtering yield $S(\hat{\theta})$ was found to be 37 atoms ion^{-1} for 20 keV argon ion bombardment in the present experiment. There are no available data for comparison with our results.

In conclusion, the method proposed here was found to be useful for measuring the sputtering yield at the critical angle of incidence. The results obtained in the present experiments show acceptable agreement with the available experimental data.

ACKNOWLEDGMENT

The authors wish to thank Mr. H. Arikawa for assisting in the experiments.

REFERENCES

1 H. H. Andersen and H. L. Bay, in R. Behrisch (ed.), *Sputtering by Particle Bombardment I, Top. Appl. Phys.*, *47* (1981) 145.
2 Y. Kanda, H. Hasuyama, K. Waki and Y. Sakamoto, *Metallography, 19* (1986) 461.
3 H. Hasuyama, Y. Kanda, T. Soeda, K. Niiya and M. Kimura, *Proc. Int. Conf. on Surface Modification of Metals by Ion Beams, Kingston, July 7–11, 1986*, in *Mater. Sci. Eng.*, *90* (1987).
4 M. J. Nobes, J. S. Colligon and G. Carter, *J. Mater. Sci.*, *4* (1969) 730.
5 V. A. Molchanov and V. G. Tel'kovskii, *Sov. Phys.-Dokl.*, *6* (1961) 137.

Materials Science and Engineering, 90 (1987) 13-19

Ion Beam Mixing of Hydrogenated Multilayer Fe–Ti and Ni–Ti Films*

J.-P. HIRVONEN†, M. A. ELVE, J. W. MAYER and H. H. JOHNSON

Department of Materials Science and Engineering, Cornell University, Bard Hall, Ithaca, NY 14853 (U.S.A.)

(Received July 10, 1986)

ABSTRACT

Multilayer Fe–Ti and Ni–Ti samples were electrolytically charged with hydrogen to a concentration corresponding to TiH_x (x = 1.5–2.3) in titanium layers. The samples were then bombarded with xenon ions (Xe^{2+}) at 600 keV to induce ion beam mixing. Rutherford backscattering and forward recoil spectroscopies were used to probe the changes in the metal and the hydrogen concentrations and profiles respectively.

A fluence of 8×10^{15} Xe^{2+} cm^{-2} caused mixing of uncharged samples of Fe–Ti and Ni–Ti. In hydrogenated samples, however, the mixing was significantly retarded in the Fe–Ti case, whereas Ni–Ti was still completely mixed. This was found to correlate with the release of hydrogen during ion bombardment. The corresponding final hydrogen concentrations were one-half of the initial concentration in Fe–Ti and one-third in Ni–Ti.

1. INTRODUCTION

Ion beam mixing is a versatile and efficient method for producing novel phases [1]. Extended solid solutions and metastable crystalline and amorphous phases have been reported in a range of systems [2–4]. In the work reported in this paper, we are concerned with the effect of hydrogen on the mixing process and the resulting structures. This question is interesting for at least two reasons. First, hydrogen is a very common contaminant, especially in metals such as titanium and zirconium because of their high affinities for hydrogen. In evaporated films, hydrogen contamination is even more pronounced because of the gettering and reaction of the residual gases with metal vapors and the consequent accumulation of hydrogen in growing films. Second, the behaviour of hydrogen in metastable crystalline and amorphous phases is not well known.

Hydrogen can of course be introduced into amorphous material and the properties of the system can be determined as has been done in earlier studies with Ti–Cu [5, 6] and Pd–Si [7]. It is also known that some crystalline structures become amorphous when hydrogen is introduced [8]. The other possible way is to try to amorphize a hydrogenated crystalline structure. An additional interesting question is whether these two methods are equivalent with respect to such characteristics as the width of the amorphous phase field and the physical properties.

If the latter way to produce hydrogenated metastable crystalline or amorphous phases is chosen, a crucial problem to solve is the effect of hydrogen on ion beam mixing, and the response of hydrogenated material to heavy ion irradiation in general. It is important to notice that irradiation effects in hydrogenated material differ from those in other materials, *e.g.* metals. Because of the extremely small atomic mass of hydrogen, it is only the metallic lattice that experiences an energy deposition through elastic nuclear collisions. This may result in a lower defect density in a cascade than that in a pure metal. In contrast, the mobility of hydrogen is very high even at room temperature, which means that the overall mobility of atoms in a cascade may be high during a thermal spike. The high mobility of hydrogen may also help to maintain thermo-

*Paper presented at the International Conference on Surface Modification of Metals by Ion Beams, Kingston, Canada, July 7–11, 1986.

†Permanent address: University of Helsinki, Department of Physics, SF-00170 Helsinki, Finland.

0025-5416/87/$3.50

dynamic equilibrium of the system. In fact, a hydrogenated material provides an extreme condition for studying irradiation effects and ion beam mixing.

In this work we report for the first time the effect of hydrogen on ion beam mixing and the response of hydrogenated material to heavy ion bombardment. We have used Ni–Ti and Fe–Ti multilayer samples. These binary samples have been shown to mix at relatively low fluences, resulting in an amorphous structure near the equiatomic composition [2, 4]. Titanium was chosen because of its high affinity for hydrogen [9] whereas the solubility of hydrogen is very low in both iron and nickel [10, 11]. Yet these samples also provide interesting differences. Nickel has an f.c.c. and iron a b.c.c. structure. Consequently, the diffusion of hydrogen in iron is faster than in nickel [12]. Although both Fe–Ti and Ni–Ti form ternary hydrides, the (Ni, Ti) hydride is said to be more stable than the (Fe, Ti) hydride [13]. In addition, at elevated temperatures, the (Ni, Ti) hydride is in equilibrium with TiH_2 and $TiNi_3$ whereas the (Fe, Ti) hydride decomposes directly to FeTi and H_2 [13]. These properties are expected to influence the ion-beam-mixing process.

2. EXPERIMENTAL METHODS AND MEASUREMENTS

Multilayer samples consisting of two titanium and two iron or nickel layers were evaporated in an oil-free system. Evaporations were performed without breaking the vacuum between the deposition of individual layers. The base pressure of the evaporation system was 2×10^{-7} Torr and it never exceeded 5×10^{-7} Torr during deposition. The nominal thickness of the titanium layers was 25 nm, and those of the nickel and iron layers were 20 nm. As substrates, SiO_2 layers of 30 mm × 15 mm area were employed. The first evaporated layer was always titanium because of its good adhesion to the substrate and, as the outermost layer, either nickel or iron was used to prevent oxidation of titanium.

Hydrogen charging was carried out in a 1 N NaOH solution using a platinum anode. The current density and the voltage across the cell were 8 mA cm^{-2} and 3.3 V respectively. Charging times from 2 to 4 h were used. After charging, the samples looked as bright as before and no bubbles were detected. The hydrogen concentration was found to be stable for at least several days after charging.

The ion bombardment was carried out using xenon ions (Xe^{2+}) at 600 keV. The range of the ions was considerably greater than the thickness of the samples [14]. Accordingly, the majority of ions penetrated into the substrate and the concentration of xenon in the samples remained very low. Fluences of 8×10^{15} and 1.6×10^{16} Xe^{2+} cm^{-2} were used for Fe–Ti, and of 4×10^{15} and 8×10^{15} Xe^{2+} cm^{-2} for Ni–Ti. In all irradiations the flux was 2×10^{13} Xe^{2+} cm^{-2} s^{-1}. For ion bombardment, one-half of each sample was shielded and used later as a control for evaluation of mixing and hydrogen concentrations. During the ion beam mixing, the samples were mounted by a heat sink paint onto an aluminum plate but no other cooling system was used.

Rutherford backscattering of α particles was used to probe metal concentration profiles, and forward recoil of protons (forward recoil spectrometry) was used to detect changes in the hydrogen concentrations. The energy of the incident helium beam was 2 MeV. For recoil measurements [15] the samples were tilted 75° and the recoil protons detected at 150° with respect to the incident beam. A Mylar film of 10.6 μm was employed as an absorber for scattered α particles. Simultaneously with the recoil measurements, backscattering spectra were also collected. These spectra were used to determine the actual charge deposition. Backscattering spectra were also acquired with the samples tilted 7°. In all cases the ion beam mixing and measurements were performed within 3 days of charging.

Both backscattering and forward recoil measurements were analyzed with the computer program RUMP [16] using the stopping powers obtained by Ziegler *et al.* [17]. For the elastic (α, p) scattering in forward recoils the differential cross-section value of 0.4×10^{-24} cm^2 sr^{-1} was used [18].

3. RESULTS

The Rutherford backscattering measurements revealed that the mixing in the Fe–Ti case was significantly retarded in hydrogenated

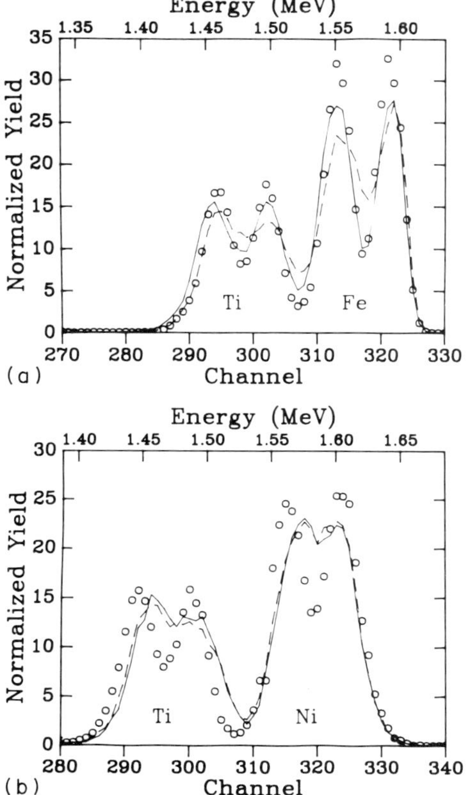

Fig. 1. Rutherford backscattering spectra from (a) Fe–Ti and (b) Ni–Ti samples, illustrating the effect of hydrogen on the mixing rate. The measurements of samples as deposited (○), samples which had been hydrogen charged and ion beam mixed (——) and samples which were uncharged and had been ion beam mixed (– –) are shown. Ion beam mixing was carried out at 600 keV 8 × 10^{15} Xe^{2+} cm^{-2}.

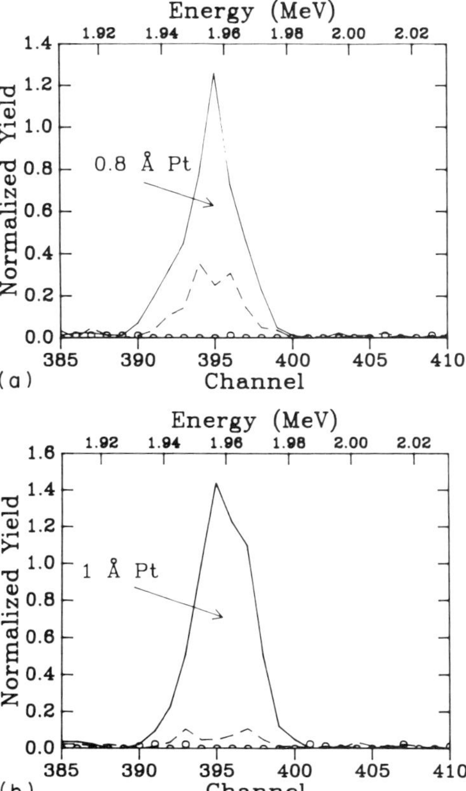

Fig. 2. Platinum contamination on (a) Fe–Ti and (b) Ni–Ti samples which had been hydrogen charged and unmixed (——) and which had been hydrogen charged and ion beam mixed (– –). For comparison, measurements from as-deposited samples (○) are also shown.

samples whereas no detectable effect was observed in the Ni–Ti case. This is illustrated in Fig. 1, where Rutherford backscattering spectra from the as-deposited samples and charged and uncharged samples bombarded to 8 × 10^{15} Xe^{2+} cm^{-2} are shown. The true compositions of the samples are 44 at.% Ti and 50 at.% Ti in the Fe–Ti and Ni–Ti cases respectively.

The contamination of the samples during the electrolytic charging was also studied and a platinum deposition approximately 1 Å thick was found on each sample, which was partly sputtered away and partly mixed during ion bombardment as shown in Fig. 2. This contamination comes from the anode. Some traces of zinc or copper contamination were also found in samples charged for longer times. This probably comes from the NaOH electrolyte. The surface contamination is believed to be responsible for the good long-term stability of hydrogen concentrations in the samples but it has no effect on the ion beam mixing.

Figure 3 and Fig. 4 present the ion beam mixing and corresponding changes in hydrogen concentrations as a function of fluence in Fe–Ti and Ni–Ti respectively. The hydrogen concentration in both samples after charging is the same. Almost all the hydrogen is in the titanium layers, as was also confirmed with bilayer samples [19]. Because of the poor depth resolution of the recoiling measurements, these hydrogen signals cannot be distinguished in the multilayer samples. By using a computer simulation [16] the hydrogen concentrations were determined as $TiH_{2.3}$ in the first and $TiH_{1.5}$ in the second titanium layer. As also can be seen from Figs. 3(b) and 4(b), the shape of the hydrogen distribution remains the same in the course of the ion

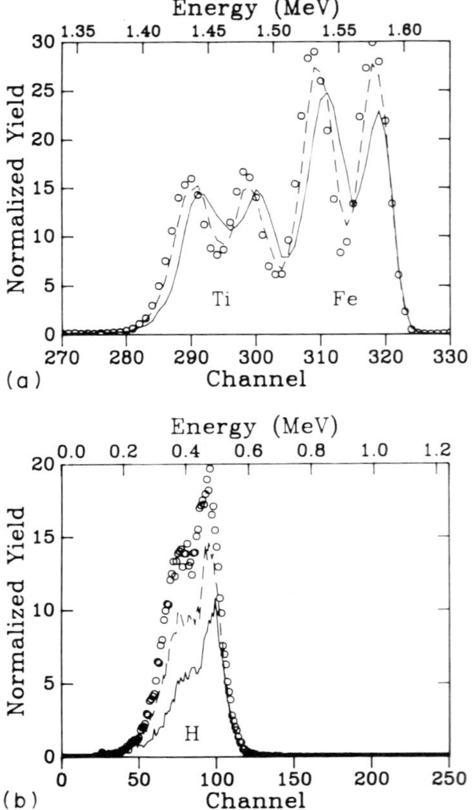

Fig. 3. (a) Rutherford backscattering spectra and (b) hydrogen recoil spectra from the hydrogenated Fe–Ti sample bombarded with various Xe^{2+} fluences at 600 keV: ○, unmixed; — —, 8×10^{15} ions cm^{-2}; ——, 1.6×10^{16} ions cm^{-2}.

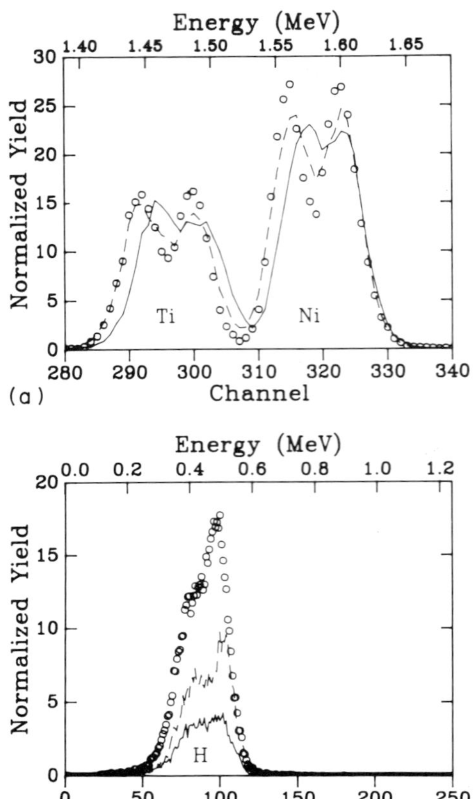

Fig. 4. (a) Rutherford backscattering spectra and (b) hydrogen recoil spectra from the hydrogenated Ni–Ti sample bombarded with various Xe^{2+} fluences at 600 keV: ○, unmixed; — —, 4×10^{15} ions cm^{-2}; ——, 8×10^{15} ions cm^{-2};

beam mixing, although the total concentration decreased with increasing fluence. This decrease in the hydrogen concentration also occurs more quickly in Ni–Ti than in Fe–Ti samples, since the xenon irradiation is more efficient at causing hydrogen outgassing of Ni–Ti. It is only after complete mixing of the metal layers that the hydrogen concentration profile becomes smooth and forms a uniform distribution throughout the sample thickness, as can be seen in the Ni–Ti case after a fluence of 8×10^{15} Xe^{2+} cm^{-2}.

In order to verify the different behaviours of hydrogen in these two types of sample, charging was carried out also using samples which were first ion beam mixed with fluences of 8×10^{15} Xe^{2+} cm^{-2}. The corresponding recoil spectra are shown in Fig. 5. When these ion-beam-mixed-and-hydrogenated samples were subsequently bombarded with a fluence of 4×10^{15} Xe^{2+} cm^{-2}, the hydrogen release was again found to be more efficient in Ni–Ti (Fig. 5(b)) than in Fe–Ti (Fig. 5(a)). In both cases the total hydrogen concentration after charging was lower than in the unmixed samples (Figs. 3 and 4), but it is not known whether this is due to the differences in the structures of the samples or to the faster diffusion of hydrogen in the ion-beam-mixed samples.

Finally the behaviour of hydrogen contamination during ion bombardment is presented. This contamination was due to the gettering of residual gases during the depositions. In both cases, nearly the same degrees of hydrogen contamination were found (Fig. 6). The ion irradiation causes changes in the Ni–Ti sample whereas no changes can be detected in the Fe–Ti sample. This is again consistent with the earlier results which showed that

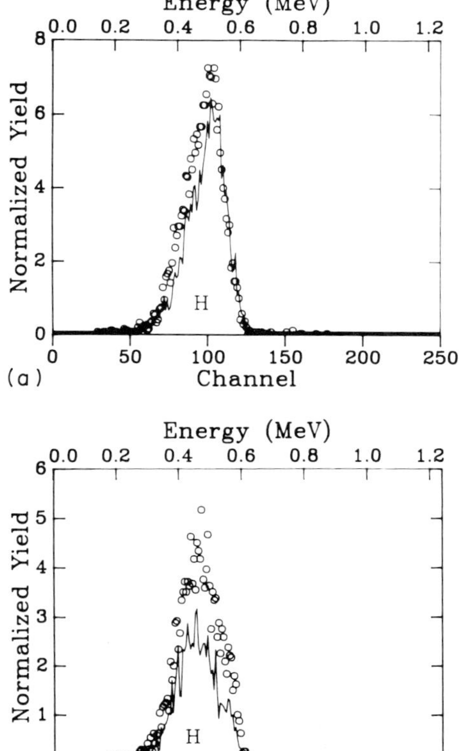

Fig. 5. Hydrogen recoil spectra (○, ——) from the (a) Fe–Ti sample and (b) the Ni–Ti sample which were first ion beam mixed and then hydrogen charged. Also shown are the spectra (+) after subsequent ion bombardment of these samples with 4×10^{15} Xe^{2+} cm^{-2}.

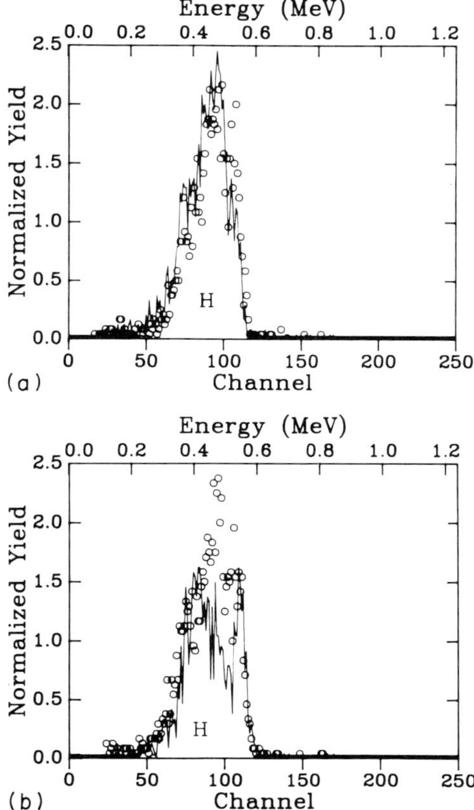

Fig. 6. Behaviour of hydrogen contamination during ion bombardment of (a) Fe–Ti and (b) Ni–Ti: ○, unmixed; ——, 8×10^{15} Xe^{2+} cm^{-2}.

the release of hydrogen is faster from Ni–Ti than from Fe–Ti.

4. DISCUSSION

We have found that the presence of hydrogen in multilayer Fe–Ti and Ni–Ti samples significantly reduces the ion beam mixing in the iron case but has almost no effect in the nickel case. Because these systems are very similar in the ballistic sense (energy deposition during ion beam mixing), there must be another reason for the difference.

According to present knowledge of irradiation effects in pure metal, a low energy cascade can be divided into three regimes [20]. During the first regime (approximately 10^{-13} s long) a high defect density is created by two-body collisions. At the end of this phase the average kinetic energy of atoms in a cascade is far higher than the average thermal energy. Because of the high concentration of vacancies and interstitials the first stage is followed by the spontaneous recombination phase. During this (about 10^{-12} s) period the majority of the defects are removed and after this phase the number of point defects corresponds to that calculated using the Kinchin and Pease model. These two ballistic regimes are followed by the thermal spike period during which point defects migrate by diffusion-like mechanisms. This phase can be considered as a collective excitation in a lattice. Molecular dynamics simulations [20] show that the mobility of vacancies during a thermal spike is higher than predicted from thermally activated diffusion, whereas interstitials migrate almost by diffusion. In less than 10^{-11} s after the arrival of a primary ion the situation in a cascade region has normalized, *i.e.* some degree of equilibrium with the surrounding material has been obtained.

It is not clear yet which of these three regimes gives the largest contribution to the ion-beam-mixing process. There is some evidence that, at least in the mixing of metal layers, most mixing occurs when the average energy of atoms is some electronvolts, *i.e.* during a thermal spike [21]. In addition, some calculations show that in some metal–metal cases the contribution from a normal vacancy mechanism is much smaller than from an interstitial-type mechanism [22]. Furthermore, at elevated temperatures, radiation-enhanced diffusion plays an important role [23].

The application of the above models to the present systems is not straightforward. The irradiation effects in hydrides are not known. Even the nature of the radiation defects in these materials is unclear. It is well established that hydrogen and vacancy-type defects may have a strong attractive interaction whereas the interaction between interstitials and hydrogen is believed to be weaker. Moreover, there is some evidence which indicates that this interaction is stronger in iron than in nickel. In fact, it is suggested that vacancy–hydrogen complexes are stable up to 720 K in iron [24] whereas, in nickel, such complexes have already dissociated above 300 K [25]. Because of high mobility of hydrogen even at room temperature, at least part of the vacancies may form hydrogen–vacancy pairs or clusters. Accordingly, the mobility of these defects may be decreased. If the binding energy between hydrogen and vacancy-type defects is stronger in iron than in nickel, then the mobility of these defects is decreased more in iron than in nickel. We can assume that in the cascade, at the TiH_x–M (M ≡ Ni or Fe) interface, hydrogen has the highest mobility. Thus, it is able to occupy vacancies in nickel or iron before considerable migration of the metal atoms can occur. However, in nickel, this occupation may be much smaller or may even be missing because of the weaker interaction. If, however, hydrogen in this way is able to lock the defect structure in the iron case, then the ion beam mixing is also decreased.

Driving forces for mixing must also be considered. It is quite generally accepted that a large negative value of a heat of mixing promotes ion beam mixing whereas systems with a positive heat of mixing are difficult or impossible to mix [25]. Both binary metal systems used here have almost the same heats of mixing, -31 kJ (g-atom)$^{-1}$ and -33 kJ (g-atom)$^{-1}$ in Fe–Ti and Ni–Ti respectively [2, 26]. The driving forces for the ternary Fe–Ti–H and Ni–Ti–H have been discussed in more detail elsewhere [19], but both systems have approximately the same negative heat of mixing. Consequently, the different ion-beam-mixing behaviours cannot be due to the different driving forces.

The release of hydrogen from the Ni–Ti sample was found to be much faster than from the Fe–Ti sample. Yet the diffusion of hydrogen in bulk iron is faster than in nickel. However, in our case the stronger interaction of hydrogen with vacancy-type defects in iron may play a role. In addition, the flow of hydrogen through nickel and iron layers may be solubility limited. According to the present measurements the release of hydrogen seems also to be a linear function of a fluence in the Fe–Ti case but non-linear in the Ni–Ti case.

An interesting question is how small a hydrogen concentration is enough to retard ion mixing in the Fe–Ti system. Because Fe–Ti and Ni–Ti are very similar from the ballistic point of view and also the driving forces are almost the same, it is attractive to think that, on the basis of our present results, hydrogen contamination is enough to cause the difference between the mixing rates of uncharged Fe–Ti and Ni–Ti.

5. CONCLUSIONS

On the basis of our measurements of ion beam mixing of hydrogenated Fe–Ti and Ni–Ti samples, we can conclude that hydrogen reduces significantly ion beam mixing in the Fe–Ti case but has almost no effect in the Ni–Ti case. Because both systems have almost equal driving forces for mixing and because also the ballistic effects are the same, it is tentatively suggested that this difference is due to the different interactions between hydrogen and vacancy-type defects in nickel and iron. The faster release of hydrogen from the Ni–Ti sample may be in part due to the different solid solubilities and in part to the stronger vacancy–hydrogen interaction in iron.

ACKNOWLEDGMENTS

This work is supported by the National Science Foundation (L. Toth). We acknowledge the National Research and Resource Facilities for Submicron Structure supported by the National Science Foundation for ion irradiation and the Materials Science Center at Cornell University for the use of their thin film facilities.

The help of Mr. L. Doolittle with the use of the computer program RUMP is gratefully acknowledged.

REFERENCES

1 J. W. Mayer and S. S. Lau, in J. M. Poate, G. Foti and D. C. Jacobson (eds.), *Proc. Conf. on Surface Modification and Alloying by Laser, Ion and Electron Beams*, in *NATO Conf. Ser., Ser. VI*, Vol. 8, Plenum, New York, 1983, p. 241.
2 J.-P. Hirvonen, M. Nastasi and J. W. Mayer, *J. Appl. Phys., 60* (1986) 980.
3 D. A. Lilienfeld, M. Nastasi, H. H. Johnson, D. G. Ast and J. W. Mayer, *J. Mater. Res., 1* (1985) 237.
4 R. S. Bhattacharya, A. K. Rai and P. P. Pronko, *Mater. Lett., 2* (1984) 483.
5 R. C. Bowman and A. J. Maeland, *Phys. Rev. B, 24* (1981) 2328.
6 R. C. Bowman, A. J. Maeland and W.-K. Rhim, *Phys. Rev. B, 26* (1982) 6362.
7 B. S. Berry and W. C. Pritchet, in G. E. Murch, H. K. Birnbaum and J. R. Cost (eds.), *Nontraditional Methods in Diffusion*, Metallurgical Society of AIME, Warrendale, PA, 1984, p. 83.
8 X. L. Yeh, K. Samwer and W. L. Johnson, *Appl. Phys. Lett., 42* (1983) 242.
9 W. M. Mueller, in W. M. Mueller, J. P. Blackledge and G. G. Libowitz (eds.), *Metal Hydrides*, Academic Press, New York, 1968, p. 336.
10 A. J. Kumnick and H. H. Johnson, *Acta Metall., 25* (1977) 891.
11 W. M. Robertson, *Z. Metallkd., 64* (1973) 436.
12 J. Völkl, in G. Bambakidis (ed.), *Metal Hydrides*, Plenum, New York, 1981, p. 105.
13 K. Yamanaka, H. Saito and M. Someno, *Nippon Kagaku Kaishi, 8* (1975) 1267.
14 J. P. Biersack and J. F. Ziegler, in H. Ryssel and H. Glawischnig (eds.), *Proc. 4th Int. Conf. on Ion Implantation: Equipment and Techniques*, in *Springer Ser. Electrophys., 10* (1982) 157.
15 L. S. Hung, E. F. Kennedy, C. L. Palmstrøm, J. O. Olowolafe and J. W. Mayer, *Appl. Phys. Lett., 47* (1985) 236.
16 L. R. Doolittle, *Nucl. Instrum. Methods B, 9* (1985) 344.
17 J. F. Ziegler, J. P. Biersack and U. Littmark, Empirical stopping powers for ions in solids, *IBM Res. Rep. RC 9250*, 1982 (IBM).
18 E. Kennedy, personal communication, 1985.
19 J.-P. Hirvonen, H. H. Johnson and J. W. Mayer, unpublished, 1986.
20 M. W. Guinan and J. H. Kinney, *J. Nucl. Mater., 103–104* (1981) 1319.
21 W. L. Johnson, Y.-T. Cheng, M. van Rossum and M.-A. Nicolet, *Nucl. Instrum. Methods B, 7–8* (1985) 657.
22 D. Peak and R. S. Averback, *Nucl. Instrum. Methods B, 7–8* (1985) 561.
23 S. Matteson, J. Roth and M.-A. Nicolet, *Radiat. Eff., 42* (1979) 217.
24 F. Wan, S. Ohnuki, H. Takahashi, T. Takeyma and R. Nagasaki, *Philos. Mag. A, 53* (1985) L21.
25 S. M. Myers, P. Nordlander, F. Besenbacher and J. K. Nørskov, *Phys. Rev. B, 33* (1986) 854.
26 M. Nastasi, J.-P. Hirvonen, M. Caro, E. Rimini and J. W. Mayer, *Appl. Phys. Lett., 50* (1987) 177.

Materials Science and Engineering, 90 (1987) 21-32

Sputtering-induced Surface Topography on F.C.C. Metals*

G. CARTER[a], I. V. KATARDJIEV[a], M. J. NOBES[a] and J. L. WHITTON[b]
[a]*Department of Electronic and Electrical Engineering, University of Salford, Salford M5 4WT (U.K.)*
[b]*Department of Physics, Stirling Hall, Queen's University, Kingston, Ontario K7L 3N6 (Canada)*
(Received July 10, 1986)

ABSTRACT

High fluence ion implantation of solids is known both to lead to new phase evolution and to modify, by sputtering, the surface topography of the solid. In the past the two processes have been separated by the use of only intermediate, but variable species, fluences in the former (e.g. by mixing of two-layer systems) or high fluence, inert gas or self-ion irradiation in the latter type of study. In the present study a variety of very high fluence (greater than 10^{19} ions cm^{-2}) implants into polycrystalline and single-crystal copper substrates have been studied to elucidate both mixing (and possible phase formation) and topographic evolution effects. The species were chosen to exert similar ballistic or collisional processes but different chemical or metallurgical effects in the substrate.

1. INTRODUCTION

In order to induce significant and useful property changes in metals by ion bombardment, it is generally necessary to incorporate substantial atomic fractions of implant species. This requires high (10^{17} ions cm^{-2}) fluences with the concomitant result of significant sputtering and erosion of substrates. At fluences greater than about 10^{17} ions cm^{-2}, it is frequently observed that a quasi-saturation or equilibrium of property change occurs because, it is suggested, of steady state competition between ion implantation and spatial redistribution of atomic species within and out from the substrate. This steady state can be only approximately valid since many studies have shown that surface structure, morphology and probably composition continue to change at ion fluences two to three orders of magnitude higher than the "property change equilibrium". It is the purpose of the present paper both to review past studies of high fluence topographic evolution of f.c.c. metals (for which most studies exist) and to outline some new data which offer further insights into the complex metallurgical and thermodynamic processes which must be occurring.

Our own [1, 2] and other work [3-5] has shown that, whilst important comparative information may be determined by studying the different morphological evolutionary patterns on different grains of polycrystalline substrates, the most enlightening results are derived from investigations with well-prepared and well-defined single-crystal substrates [3]. In the following discussion, therefore, most attention will be focused on the latter type of investigation and observations within single grains of polycrystalline materials. Particular attention will be focused on the initiation and development of the two most commonly observed features, namely the etch pit and the pyramid, on their aggregation to form dense arrays and on the dependence of the habit, size and density of such features on experimental parameters. Most technical surfaces (other than in the semiconductor industry) are polycrystalline, however; it is therefore important to know what differences occur between different grains as a result of high fluence irradiation, and it is this area of intergranular effects and dependences on experimental parameters that is explored in Section 2.

It should be clarified that little further attention will be given to phenomena which result from the accidental or deliberate provision

*Paper presented at the International Conference on Surface Modification of Metals by Ion Beams, Kingston, Canada, July 7-11, 1986.

0025-5416/87/$3.50

of a perturbational or contaminant species [3–5] at the bombarded surface or to bombardment-enhanced growth [4, 5] processes which usually require somewhat elevated substrate temperatures for their unequivocal observation. Although surface atomic redistribution, local (and extensive) feature growth and geometric modification will always be competitive with sputtering-induced erosional processes to some extent, it is only the situation in which the latter is clearly dominant which will be examined here.

Nevertheless, many technical surfaces will be contaminated either before or during irradiation and the effects on surface morphology, usually readily distinguishable from non-contaminant-induced processes, can be rather serious as depicted in Fig. 1 and Fig. 2. These figures display scanning electron micrographs of {11 3 1}-oriented copper single crystals irradiated, at 45° incidence, with 9 keV argon ions (Ar^+) and nitrogen ions (N^+) respectively [6]. The sequences (a)–(c) in these figures illustrate the effect of increasing ion fluence. What is unusual about these results is that they were obtained with crystals which were partly mechanically polished in accordance with our well-established [3] routine for preparation of flat featureless surfaces but where the later fine-polishing stages were omitted and residual debris remained on the surface. It is very evident that conical features develop, firstly in the contaminants themselves, and that this pattern is transferred to the substrate via protection of the latter from the ion flux. For non-normal incidence, these protuberances can influence large fractions of the surface, and a very rough and undulating morphology

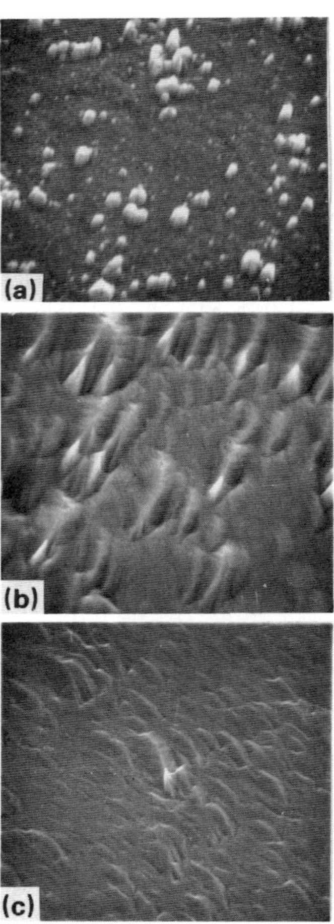

Fig. 1. The development of morphology on contaminated Cu(11 3 1) irradiated at 9 kV at $\theta = 45°$ with argon ions: (a)–(c) different regions of the irradiated zone between the edge and the centre of the zone.

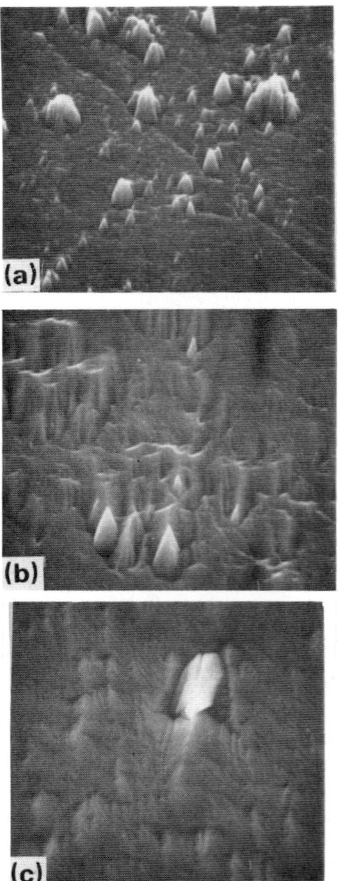

Fig. 2. The development of morphology on contaminated Cu(11 3 1) irradiated at 9 kV at $\theta = 45°$ with nitrogen ions: (a)–(c) different regions of the irradiated zone between the edge and the centre of the zone.

results. A very similar pattern of behaviour resulted not only from irradiation with argon and neon ions but also from oxygen ion (O^+), neon ion (Ne^+) and xenon ion (Xe^+) bombardment, also indicating qualitative, if not quantitative, independence of ion species. It should be remarked that the dimensions of such features are of the order of 1 μm for fluences in the 10^{19} ions cm^{-2} range suggesting that, if linear interpolation with fluence operates, features (possibly covering the whole irradiated surface) with dimensions of 0.01–0.1 μm would exist for incident fluences of 10^{17} ions cm^{-2}. This may exert a non-negligible effect on apparent changes in surface properties such as wear, friction, catalytic activity etc. Clearly, rigorous decontamination before and careful control during irradiation are required to minimize such effects.

2. INTERGRANULAR EFFECTS

Investigations with small-grained polycrystals reveal a plethora of interesting morphological features but give little useful information on the details of feature evolution. Larger-grained polycrystals yield much more insight [7] as revealed by the examples shown in Fig. 3 for polycrystalline copper substrates irradiated with normally incident 40 keV argon ions to a fluence of approximately 10^{19} ions cm^{-2}. Figure 3 shows the differing features developed in neighbouring grains and also the very different elevations of such grains. These grains were initially coplanar in the surface and it is clear that different grains sputter erode at different rates. Results such as these show that differential erosion rates exist between grains and cause grain offsets of 0.2–0.5 μm for fluences of the order of 10^{17} ions cm^{-2}. Again, such undulations across a surface might be expected to contribute to measured surface tribological changes. It should be noted that these perturbations to general surface planarity are of the order of 10% of the total sputter-eroded depth and are, therefore, not insignificant. A further informative result of Fig. 3, which is quite generally valid for all systems that we have studied, is that it is the grains which develop the greatest infrastructure which erode most rapidly. Such grains are invariably embellished with a high density of etch pits, often overlapping to form terraced to step-like patterns, whereas grains which

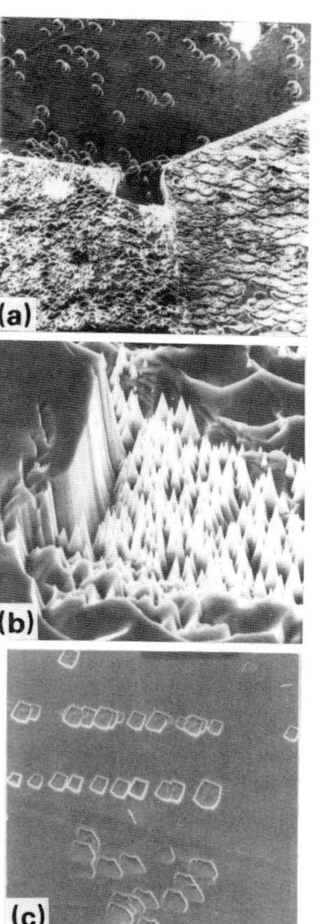

Fig. 3. (a) 40 keV argon-ion-bombarded polycrystalline copper, illustrating differential sputtering of different grains and different pit etch habits within grains; (b) 40 keV argon-ion-bombarded polycrystalline copper, illustrating the steep intergranular boundaries terraced and faceted boundaries, etch pits near boundaries, and cone structures; (c) 40 keV argon-ion-bombarded polycrystalline copper, illustrating the different habits of etch pits developed on neighbouring grains.

erode less rapidly contain low pit densities. Additionally some of, but not all, such heavily etched surfaces are decorated with cone-like structures which, on closer examination become clearly revealed as regularly faceted pyramids.

As will become evident in later discussion of single-crystal substrates, the density and habit of such features are strongly dependent on incidence conditions (ion fluence, species and energy) and substrate parameters (material and orientation) and this can be observed in our recent studies of large-grained polycrystalline copper also. In this work, polycrystals were irradiated with 35 keV krypton ions (Kr^+), iron ions (Fe^+) and nickel ions

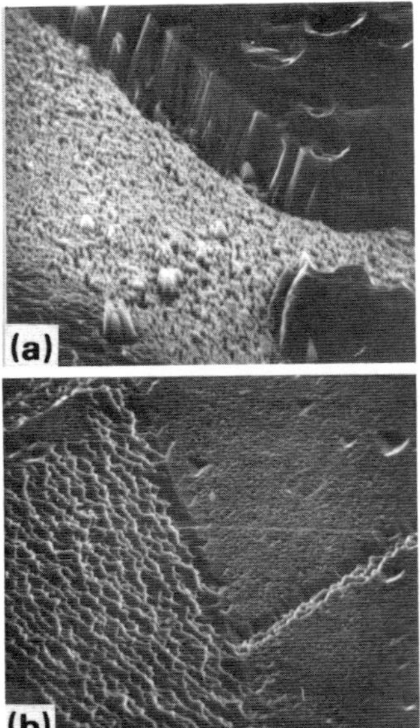

Fig. 4. The effects of bombardment with 10^{19} ions cm^{-2} on the topography evolution of large-grained polycrystalline copper: (a) krypton ion bombardment; (b) iron ion bombardment. Two neighbouring grains are shown in each case with clearly revealed differences in the etching of each grain. Krypton ion irradiation induces pyramid formation on some grains. Iron ion irradiation does not induce pyramid formation on any grains.

(Ni^+) to a fluence of approximately 10^{19} ions cm^{-2} at normal incidence with the results shown in Fig. 4(a) and Fig. 4(b) for krypton ions and iron ions respectively. These species were deliberately selected to be of mutually similar atomic mass to that of the copper substrate so that ballistic collisional processes involved in sputtering should be relatively ion species independent. The two ions were expected to behave very differently metallurgically and thermodynamically with respect to copper. Figure 4(a) shows the differential erosion pattern for krypton ions. It is similar to that observed for all inert gas irradiating species, with low and high density etch-pitted grains, the latter frequently including pyramids. Figure 4(b) for iron ions, however, whilst again showing intergranular relief and large infrastructure densities on rapidly eroding grains, indicates a complete absence of pyramidal structures. These results, which are quite typical for all iron and nickel ion irradiations, do show very high pit densities with major evidence of overlapping to produce stepped and terraced structures.

The scale of such features within grains is about 2–3 μm which, if linearly scaled with fluence would represent an undulation of 0.02 μm at a fluence of 10^{17} ions cm^{-2}. It is more probable, however, that at the lower fluences the pits would be more isolated and not overlapped so that, within even a highly eroded grain, the average perturbation to a grain surface (measured for example by the product of the fractional area occupied by pits and their mean depth) would be substantially smaller than the undulation average indicated above. More consideration of pit properties will be given in Section 3 on single-crystal investigations but it is notable from Fig. 3(c) that pit habits, as well as densities, depend on the grain (orientation) on which they are produced but perhaps less so on the species of the irradiating ion (*cf.* Figs. 4(a) and 4(b)).

As yet, we have no clear explanation as to why inert gas ions result in pyramid formation, whilst iron and nickel ions do not. Pyramids (not contaminant-related cones) always result from intra-pit and inter-pit discontinuities but, of course, not from all such discontinuities; etch pits are, undoubtedly, necessary precursors for pyramids. Since etch pits may be generated in equal profusion, and apparently similar geometry, by both inert gas ion and metallic implants, we can only speculate at this stage that the near-surface composition and defect structures induced by irradiation with different species must in some cases favour pyramidal evolution and in other cases be inimical to such feature development. Evidence in support of this proposal is found when areas of substrate are first irradiated with 10^{19} Kr^+ cm^{-2} (as in Fig. 4(a)) and then reirradiated with 10^{19} Fe^+ cm^{-2} or 10^{19} Ni^+ cm^{-2}, with the result shown, for iron ions, in Fig. 5. It is quite clear that the resulting structure is tending towards that which would be produced by iron ion irradiation alone since only a few vestigial pyramids remain within the emerging stepped and terraced surface. This quite typical result shows very convincingly that, even when conditions exist to encourage pyramid formation (*i.e.* the krypton ions had already generated them), they are obliterated as a result of the changed metallurgy of the surface. These effects will undoubtedly occur, although to a reduced extent at lower fluences, and illuminate the need

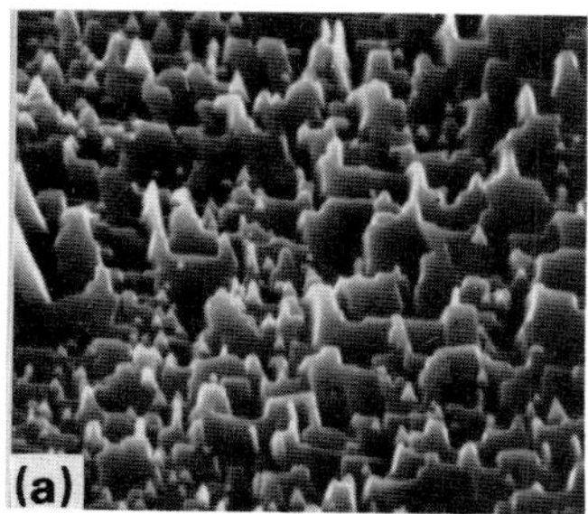

Fig. 5. The effects of sequential irradiation by krypton and iron ions (each at a fluence of 10^{19} ions cm^{-2}) on large-grained polycrystalline copper. The eradication of earlier pyramids is evident.

to employ well-defined (mass-analysed) ion beams for surface modification investigations as well as the demand for contaminant-free conditions.

3. INTRAGRANULAR EFFECTS

Since the morphological feature evolution is clearly related to crystal orientation, it is reasonable to enquire which orientations, for a given substrate material and irradiation conditions, lead to low feature densities and which orientations lead to high feature densities. Careful studies by Whitton *et al.* [8] with normally incident 40 keV argon ions onto a range of precisely oriented copper single crystals showed that large densities of features, particularly pyramids, only occurred with surface orientations in a narrow zone around the {11 3 1} plane which is, crystallographically, reminiscent of a densely packed random atomic structure to incident ions. Irradiation of low index planes such as {100}, {110} and {111} revealed very little feature evolution but, when pits did form (usually from an initial surface imperfection), they occasionally contained a few pyramids the habit of which reflected the simple symmetry of the substrate orientation. On {11 3 1} surfaces, however, pits and pyramids occur in profusion and their habits were complex and probably related to the more complex symmetry of this orientation. For pyramids, however, their facets were also disposed to include low index crystal directions [1]. Because of the plethora of features which form and are readily observed and documented with {11 3 1} surfaces the majority of our subsequent investigations have employed this crystal orientation and, in the following discussion, our findings to date on, firstly, etch pit and, secondly, pyramid formation with this orientation are summarized. The effects of varying ion flux, fluence, energy, incidence angle(s) and species are discussed for copper substrates and the influence of one species (argon ions) with a variety of different f.c.c. materials is then summarized.

3.1. *Etch pits*

Etch pits are invariably observed to increase in both density and linear dimensions with increasing ion fluence. Fairly soon after first observation, their habit becomes well defined and persists during expansion although a more detailed structure often develops within pits. When pit densities and sizes become sufficiently large, the pits overlap to form quasi-repetitive terrace structures. Examples of this latter behaviour are clear in Figs. 3(a) and 4(b). Pit expansion with fluence is clearly revealed in the sequence of micrographs shown in Fig. 6 [9] and the increase in pit dimensions with increasing fluence, determined from measurements such as Fig. 6, is found to be linear as shown in Fig. 7 [6, 9]. Figure 6 also reveals the increasing pit infrastructure and overlap with increasing fluence.

Although we have not yet made detailed measurements of pit density, before overlap, with increasing fluence this parameter appears to increase reasonably linearly with increasing fluence and, generally the pits are distributed randomly. As a result of these pit density and dimension changes with fluence, the pitted area before overlap probably increases with approximately a cube-like power law of fluence. The fluence at which overlap occurs depends on a number of ion parameters but generally lies between 10^{18} and 10^{19} ions cm^{-2}. At a fluence of 10^{17} ions cm^{-2}, therefore, the fraction of surface pitted area will be between 10^{-4}% and 10^{-1}%. Consequently, only if, for example, chemical reaction rates are strongly dependent on surface orientation will the pits moderate observed property changes substantially at a fluence of 10^{17} cm^{-2}. When pits overlap and corrugated terraces appear, their dimensions also change with fluence but their habit does not. Consequently, in this regime the surface area, although larger than the initial surface area, does not change significantly with fluence.

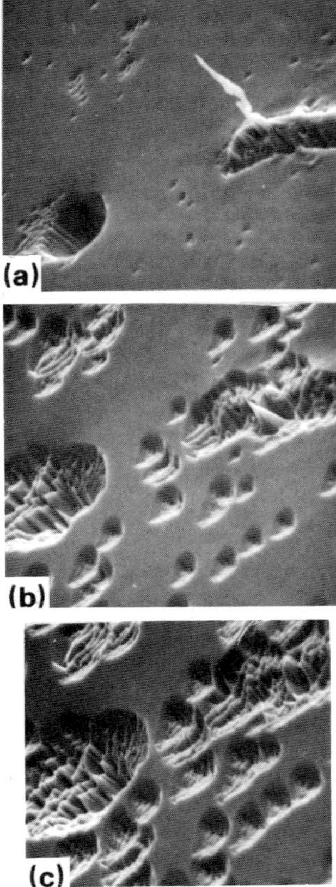

Fig. 6. The sequential development of pits on 9 kV argon-ion-bombarded copper: (a) 2×10^{19} ions cm^{-2}; (b) 6×10^{19} ions cm^{-2}; (c) 1×10^{20} ions cm^{-2}. (Magnifications, 1250X.)

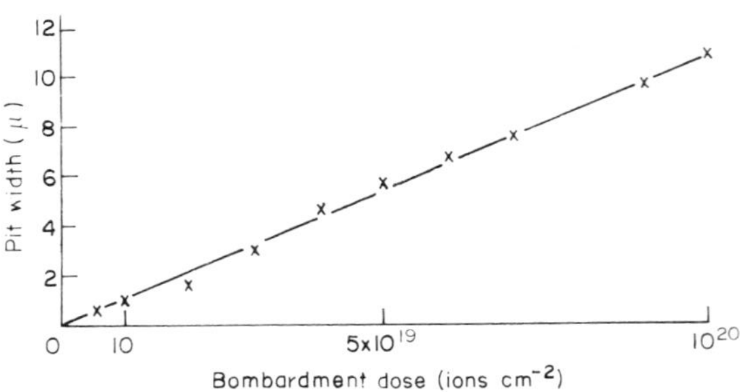

Fig. 7. Pit expansion as a function of fluence.

The pits shown in Fig. 6 were generated by 9 keV argon ion bombardment of Cu {11 3 1} at a polar angle of 45° incidence to the surface at an arbitrary azimuthal angle. After this pit formation [6, 9] the azimuthal incidence angle was changed by $\pi/2$ and irradiation continued. It was observed that the second irradiation completely eradicated the pits in a process

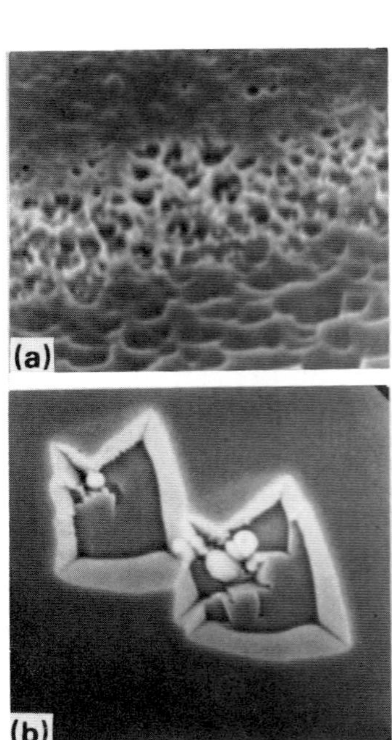

Fig. 8. (a) 40 keV neon-ion-bombarded Cu(11 3 1) in the region of the ion flux boundary; (b) 40 keV copper-ion-bombarded single-crystal Cu(11 3 1), illustrating etch pit intersection and the initiation of pyramids at such intersections and at facet jogs; (c) 40 keV krypton-ion-bombarded single-crystal Cu(11 3 1), illustrating the generation of pyramids at etch pit intersections.

where the mean surface clearly eroded faster than the pit facets. When the initial azimuthal angle of irradiation was restored, a new generation of pits uncorrelated with the first was produced. The polar angle of incidence is also observed to modify pit habit considerably. For the same irradiation conditions as above but with normal incidence, the pits produced [6, 9] are ill defined and without the clear crystalline habit shown in Fig. 6. Increasing ion energy above 20 keV does restore, however, the well-developed regular pit structure with normally incident ions onto Cu {11 3 1}. This restoration is shown in Fig. 8(a), Fig.

8(b) and Fig. 8(c) respectively for 40 keV neon, copper and krypton ion bombardment of Cu {11 3 1}. The pit habits are similar for all these species (including the copper self-ions) as they are for other inert gas ion species (argon ions and xenon ions) [2] and for iron and nickel ion irradiation, from our most recent work shown in Figs. 9(a) and 9(b). Bromine ion (Br^+) irradiation, however, produces a rather different pit structure, as illustrated in Fig. 9(c).

In contrast, pit densities do appear to depend rather strongly on projectile species and our earlier work with inert gases, nitrogen ion and copper ion irradiation indicated an increase in pit density with increasing ion mass for a given fluence (10^{19} ions cm^{-2}). Since

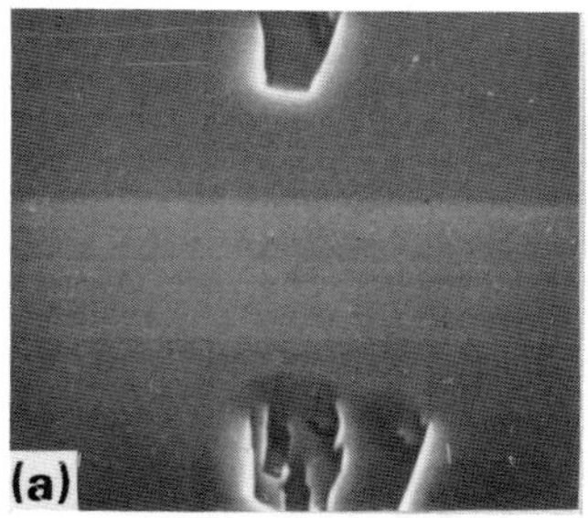

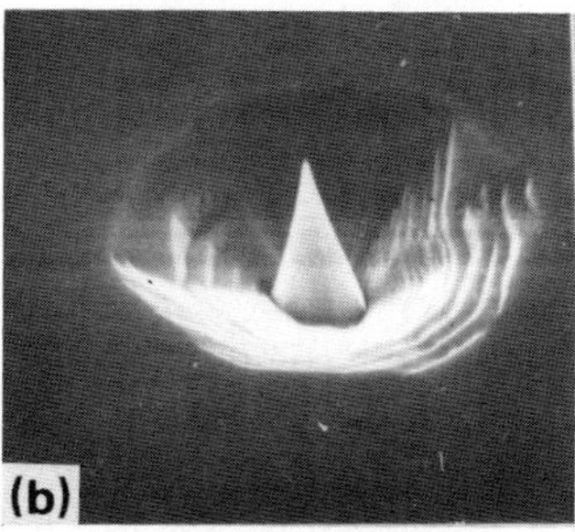

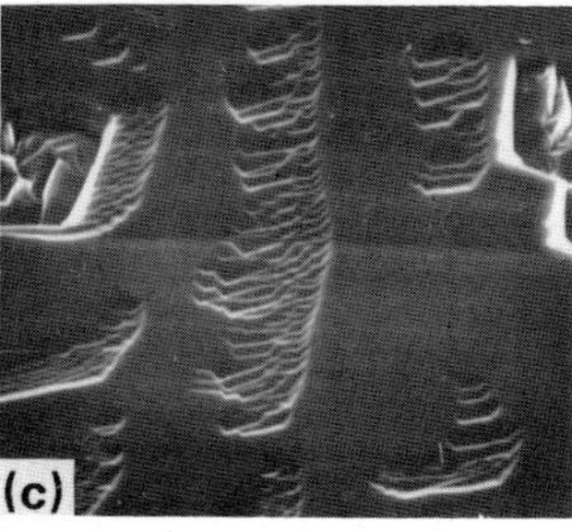

Fig. 9. (a) The effects of bombardment with 10^{19} Fe^+ cm^{-2} on the topography evolution of a single-crystal Cu(11 3 1) substrate (the etch pits are devoid of pyramids); (b) the effects of bombardment with 10^{19} Ni^+ cm^{-2} on the topography evolution of a single-crystal Cu(11 3 1) substrate (the etch pits include single pyramids); (c) bromine ion bombardment alone (no pyramids are produced).

at this energy (40 keV) the sputtering yield increases with increasing ion mass, the pit density appeared to be correlated with the sputtered depth of the substrate. The fact that this cannot be entirely true is shown by Fig. 9 for iron, nickel and bromine ions (ions of similar mass to krypton and copper ions) in which only very low pit densities were observed. However, at this stage, we begin to suspect that the {11 3 1} surface of copper, so rich a source of features for inert gas ions, may not be the most fertile orientation for other species. Figure 4(b) suggests this by indicating that, on a polycrystal of copper, many grains are densely featured at the same fluence (10^{19} ion cm^{-2}) of iron ion irradiation of Cu {11 3 1} shown in Fig. 9(a) which reveals low feature densities.

Finally, we turn our attention to irradiation of other {11 3 1}-oriented f.c.c. substrates. Figure 10(a), Fig. 10(b), Fig. 10(c) and Fig. 10(d) show that etch pits developed [10] by 25 keV argon ion bombardment of nickel, palladium, silver and gold respectively. In the first three cases the pits are similar to but not identical with those developed on copper and indeed on iridium substrates [10]. For gold, however, the pit geometry is distinctly different and both the pits and the surrounding mean surface are decorated with a low relief ripple-like faceted structure. Similar relief patterns have been observed [2] within pits on Cu {11 3 1} also but, generally, not on the surrounding unpitted surface.

In concluding this section, we would indicate that, as yet, we are not completely certain as to the origin of pits although the evidence seems to suggest a correlation with dislocation structures either present in the pristine substrate and/or introduced by ion irradiation. Collinear pit structures correlated with the drawing direction for large-grained polycrystalline substrates [3] suggest a native dislocation source, whilst the partial correlation with ion mass and energy (and thus a correlation with total sputtered depth which exposes an increasing population of bulk-included dislocation loops at the sputtered receding surface) supports such a source. The elimination of pits by modified azimuthal angle irradiation is also suggestive of such an origin since theoretical estimates [11] indicate that pits should only amplify for incident ion irradiation nearly parallel to the dislocation

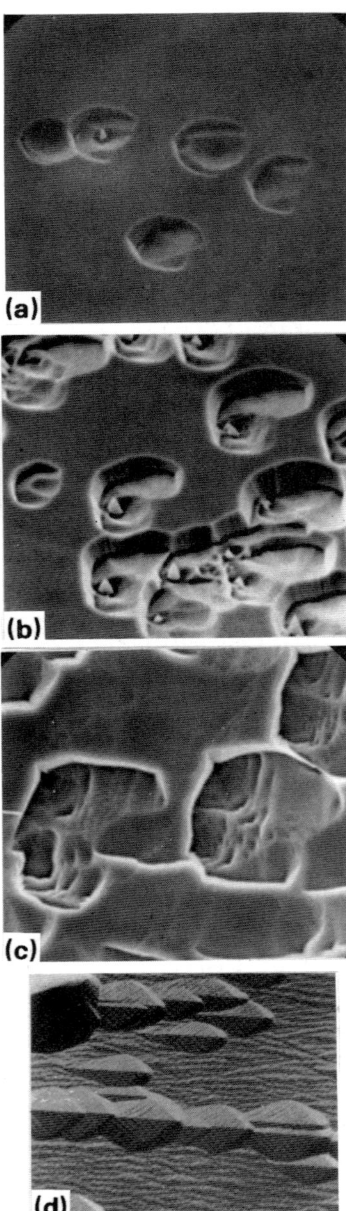

Fig. 10. Etch pit structures developed by bombardment of different (11 3 1) single-crystal f.c.c. substrates with 25 keV 10^{19} Ar^+ cm^{-2}: (a) nickel; (b) palladium; (c) silver; (d) gold.

core. The more recent studies with iron, nickel and bromine ions, however, indicate that the modified metallurgy and probable change in the rate of production of extended defects induced by irradiation must also play a significant role in etch pit initiation and evolution. The dependence of etch pit habit (and infrastructural details) on direction of ion incidence, ion energy and substrate material and the relatively small dependence on ion species for a given substrate suggest that the etch pit geometry is dictated by crystal symmetry considerations. However, the details of the sputtering process for all surface orientations, including those processes which arise from metallurgical changes induced by implantation, must also be important. It is clear that further investigations are necessary to elucidate the full details of pit initiation and evolution. It is clear that, for fluences less than or equal to about 10^{17} ions cm^{-2}, property changes will not be dramatically modified by etch pit formation unless, even for these fluences, major erosion (*i.e.* high sputtering yield) conditions obtain for a given surface orientation and material. This will, generally, not be the case for 100 keV incident ions unless the atomic mass of the incident ions is large.

One other remark is also relevant from the results shown in Fig. 8(a) for neon ions incident on Cu {11 3 1}. This figure displays the boundary region of irradiation where the incident ion flux and fluence increase from low values at the top towards high values at the bottom of the figure.

The lower part of the figure shows clearly defined etch pits, as observed at all fluences for all ion species, whereas the upper part reveals both burst and unburst blister-like features. At a periphery of a nitrogen ion irradiation a similar pattern change was observed but not with argon, krypton, copper and xenon ion irradiation. This suggests that, in copper, there may be a lateral transport of nitrogen and neon out of the main irradiation zone, which allows accumulation of gas in an only slowly sputter-eroded neighbouring zone. The accumulation may lead to bubble and blister formation of occluded species although it appears not, for copper, to be a necessary precursor for etch pit formation (*i.e.* copper ion irradiation produces etch pits). This process may be equally or even more evident for other irradiation and substrate conditions and should not be overlooked in estimating, for example, the area of surface over which a property modification occurs even when the irradiated area is well defined.

3.2. *Pyramids*

Etch pits are structures depressed relative to their immediate surroundings, *e.g.* the mean sputtered surface, whereas the pyramids to be described in the present section are elevated with respect to their immediate surroundings but not generally with respect to the mean sputtered surface. This latter behaviour allows immediate distinction to be made between pyramids and conical and whisker-like protrusions [4, 5] which are always elevated with respect to the mean sputtered surface. These arise from local contaminant protection processes which interrupt incident flux and suppress sputtering or from some gradient-driven growth process. A further distinguishing feature of pyramids is that, although they may be observed singly with a random distribution over the surface, as are cones, they are more frequently observed in dense clusters and are organized with geometric (or crystallographically related) symmetry. Moreover, although cones may but not always possess a crystallographic habit, pyramids always possess a crystallographic habit. Finally, cones usually possess a size distribution and (with respect to the mean sputtered surface) a height distribution whereas pyramids which also possess a size distribution and a height (with respect to their local base surroundings) distribution are never such that a pyramid apex is elevated above the mean sputtered surface. This latter statement indicates, in fact, just where pyramids do occur. On single crystals, they are observed within and on the boundaries of individual etch pits, at intersecting etch pit regions and at the boundary of the ion flux where the sputtered crater terminates. In polycrystals, they may also be observed on the side walls of the more elevated (after preferential grain erosion) grains. Although these are the general locations of such pyramids, they do not occur in association with every one of these features.

Examples of grain boundary and etch pit intersection sources of pyramids are already clear in Figs. 3(a) and 3(b), whilst the upper parts of Figs. 3(a) and 3(b) both indicate grains which are pitted but devoid of pyramids. As with etch pits the most detailed information on pyramids is revealed in studies with single crystals. Such studies at incident ion energies of less than 10 keV, as indicated in Fig. 6, displayed etch pits but pyramids were not evolved for any incident angular conditions. For energies greater than 20 keV and at least up to about 45 keV, pyramids are developed quite routinely on Cu {11 3 1} under normal incidence conditions. The variation in substrate temperature [2] from −196 to +500 °C appears not to influence the propensity for either pyramid formation or pyramid habit.

A series of studies [12], in which ion mass was varied, revealed interesting results, examples of which are shown in Fig. 11, Fig. 12, Fig. 13 and Fig. 14 respectively for argon, krypton, copper and xenon ion irradiations to a fluence of 10^{19} ions cm^{-2}. Pyramid populations are clear in every case and were also observed, but in lower density, with nitrogen

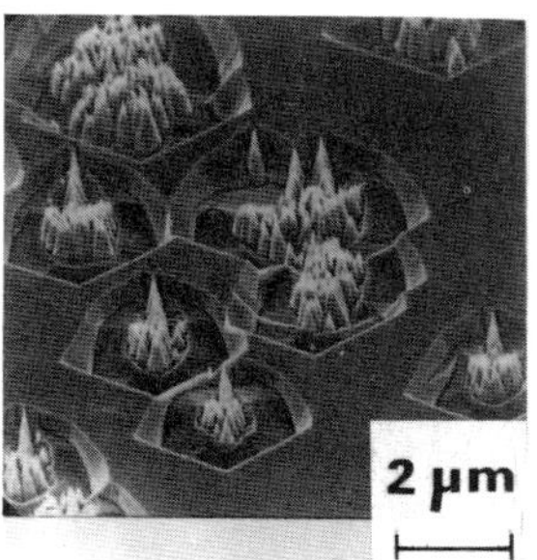

Fig. 11. 40 keV argon-ion-bombarded single-crystal Cu(11 3 1), illustrating the initiation of pyramids at etch pit corners and jogs.

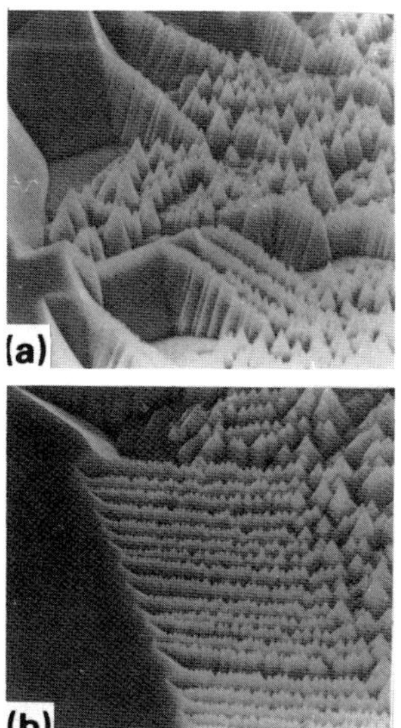

Fig. 12. (a), (b) Two examples of 40 keV krypton-ion-bombarded single-crystal Cu(11 3 1), illustrating collinear pyramid geometries and their emergence, either in isolation from etch pit jogs or repetitively from a sequence of jogs.

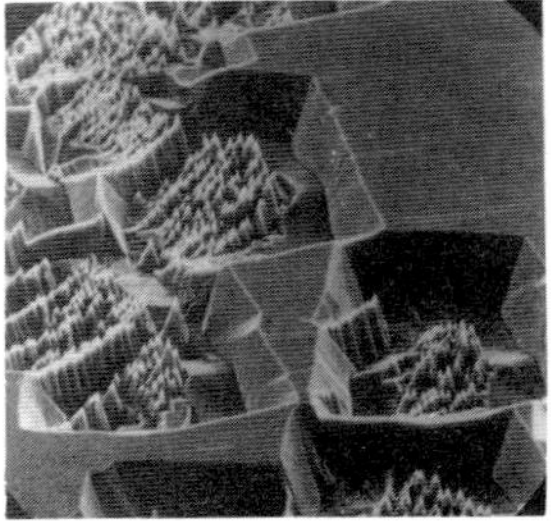

Fig. 13. 40 keV copper-ion-bombarded single-crystal Cu(11 3 1), illustrating the high density of well-defined geometric arrays of pyramids in etch pits.

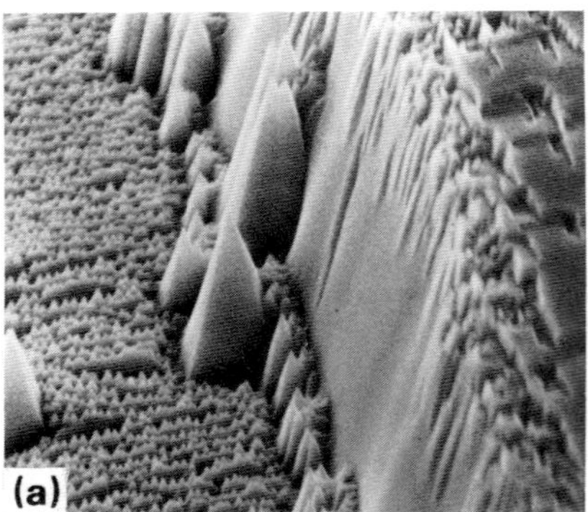

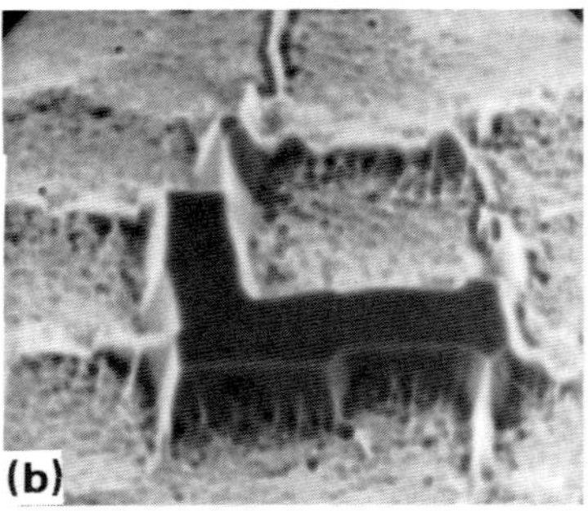

Fig. 14. (a) 40 keV xenon-ion-bombarded single-crystal Cu(11 3 1), illustrating pyramid generation where there is a steep gradient in the ion flux and a steep boundary is formed (the etch pits near the boundary in the low flux density region are notable); (b) 40 keV xenon-ion-bombarded single-crystal copper, illustrating plateau formation by neighbouring enhanced erosion of a heavily structured surface.

and neon ion bombardment. Many of the aspects of pyramids described earlier are well displayed in these figures. For example, Fig. 11 for argon ions, Fig. 12 for krypton ions and Fig. 13 for copper ions show very clearly the regular array structures of pyramid populations, the association of pyramid rows with jog-like discontinuities on etch pit side facets and the initiation of pyramids at etch pit intersections. Figure 14 for xenon ions shows the extremely high density of pyramids which can result when quite extensive erosion occurs as with this heavy ion species of high sputtering yield. In this case, almost all evidence of precursor pits has been lost, except where the ion flux was very low at the beam boundary as in Fig. 14(a), but even in this densely pyramidal situation there are still zones of the crystal which remain unpitted as in Fig. 14(b). However, both of these figures do reveal, that, even though the pyramid apices do initially coincide with the mean surface, they erode and become smaller. This feature, together with all other parameters associated with pyramids, indicates that pyramids are associated with the crystallography of the substrate and are the result of local differential erosion phenomena; they are not contaminant- or growth-related features.

The ability to generate pyramids with all inert gas ion species and with the copper self-ion is a clear indication that these features do not arise as a result of bulk-included contaminants since the inert gases, at the high fluences employed, would be expected to precipitate into a well-defined phase of bubbles within an otherwise well-ordered copper substrate (which will, of course, include point and extended defects), whilst the copper ion irradiation would be expected to generate the defect population only. It may be suggested, therefore, that the initiation and evolution of pyramids are associated with an organized defect population and/or organized gas bubble population but that, in order to form at all, the bombardment must generate an appropriate etch pit morphology. It appears that, if the crystallography of the system and the interdependent details of the sputtering yield on orientation give rise to pits (and inter-pit regions) with both elevated facets and jog-like discontinuities, then pyramid formation is favoured. Unfortunately, the present data do not allow any more detailed understanding of the processes leading to pyramid formation than the qualitative statements above.

Recent studies have shown that substrate composition may also be an important factor in pyramid generation. Reference has already been made to studies of iron, nickel and bromine ion irradiation of Cu {11 3 1} and polycrystalline copper where the latter studies, depicted in Fig. 4(b) for example, showed no pyramid evolution on any copper grain for these implants. Moreover, irradiation of Cu {11 3 1} by these species showed virtually no evidence of pyramid formation in the relatively low density pits generated as shown in Fig. 9. It might be remarked here that the nickel ion result is open to some doubt since, in one series of studies, nickel ion bombardment generated pits with a solitary included pyramid in each pit, whilst a second series revealed pits devoid of pyramids. In general, therefore, it appears that, at least for the ion fluence employed, incident species which modify the metallurgy and chemistry of the substrate can inhibit, at least for Cu {11 3 1}, the generation of both pits and pyramids.

A further example of the importance of this modification was shown in Fig. 5 where, for polycrystalline copper, iron ion irradiation eliminated krypton-ion-induced pyramids. This has also been confirmed for {11 3 1}-oriented single-crystal copper, as shown in Fig. 15, where a post-irradiation by 35 keV bromine ions onto a krypton-ion-irradiated zone again results in pyramid removal. In contrast, when iron or nickel ions were first implanted, subsequent krypton ion irradiation appears to generate pyramids within the infrastructure of pits developed by the preceding irradiation. All these results indicate that the crystallography and metallurgy of the system formed by irradiation play important roles in the development of both etch pits and pyramids and that, as yet, no simple rules of behaviour can be formulated for all ion–substrate combinations. For example, although the Cu {11 3 1} surface is beneficial for feature elaboration by many ion species, other surfaces seem to be favoured for different species. It should also be noted here that, as with etch pit evolution, studies of pyramid generation on different {11 3 1}-oriented f.c.c. materials with 25 keV argon ion irradiation have been conducted [10]. Again, pyramids were found to be a quite general and reproducible feature and arose from the same type of pit and inter-pit sources as with copper. There did appear to be a difference in pyramid density with material type, with iridium and gold revealing lower densities than copper, silver and palladium. This, again, seems to be associated with an overall lower sputtering rate for the former substrate and a less elaborate etch pit system. As for etch pits the detailed mechanisms of pyramid formation are unclear and substantial further work is required.

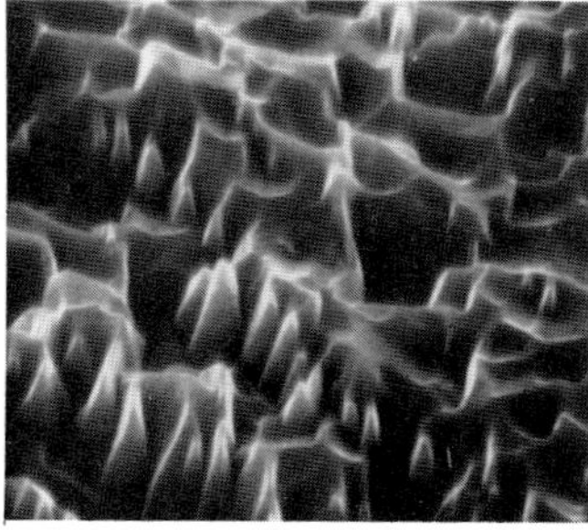

Fig. 15. Krypton ion bombardment followed by bromine ion bombardment. The eradication of pyramids is evident on this bombarded Cu(11 3 1) surface.

Nevertheless, estimates can be made from the results of Figs. 11–15 of the influence of pyramidal structure on other measured property changes of implanted substrates. The lighter ion bombardments (*cf.* Fig. 11) probably represent the effects of heavier ions but at a lower fluence of these higher sputtering yield species. Consequently, in the 10^{17} ions cm^{-2} fluence regime, etch pit densities will probably be low, their areas will be small and their included pyramid densities and sizes will also be small. The total surface area, important in chemical reaction processes, will therefore not differ significantly from the near-plane surface area. The ratio of areas is, in fact, $1 + \pi r p_d (l - r)$ where p_d is the areal density of pyramids of effective base radius r and slant height l. Again, only if reaction rates are strongly crystal orientation dependent and exhibit very different values on the pyramid facets and apices than on the mean surface, will substantial influence on the reaction rates occur.

Similar statements will obtain with respect to other surface tribological effects such as friction, wear and hardness since, although contact areas and angles to probes will be very different for pyramid-covered zones, these will generally represent a relatively small fraction of the total tested surface area.

4. CONCLUSIONS

Although, in the preceding discussion, it has been shown that major morphological changes occur during implantation and attendant sputter erosion of substrates, the mechanisms of formation and evolution of the two major features, namely etch pits and pyramids, are still not clear. Their appearance and density depend on the ion species and energy, substrate material and orientation and, it is becoming evident, also on the metallurgical changes induced as well as the crystallography of the substrate. However, the major influence is the ion fluence and it has been shown that, for fluences of interest in surface property modification applications (10^{17} ions cm^{-2}), the morphology is only developed on a minor scale and should not substantially moderate measured property changes.

REFERENCES

1 J. L. Whitton and G. Carter, in P. Varga, G. Betz and F. P. Vienbock (eds.), *Proc. Symp. on Sputtering, Vienna, 1980*, Institut für Allgemeine Physik, Technische Universität Wien, Vienna, 1980, p. 552.

2 G. Carter, M. J. Nobes and J. L. Whitton, *Appl. Phys. A, 38* (1985) 77.

3 G. Carter, B. Navinsek and J. L. Whitton, in R. Behrisch (ed.), *Sputtering by Particle Bombardment II, Top. Appl. Phys., 52* (1983) 231–269.

4 G. K. Wehner, *J. Vac. Sci. Technol. A, 3* (1985) 1821.

5 R. S. Rossnagel, in G. Kiriakidis, J. L. Whitton and G. Carter (eds.), *Proc. NATO Advanced Study Institute on Erosion and Growth of Solids Stimulated by Atom and Ion Beams, Crete, 1985,* in *NATO Adv. Study Inst. Ser., Ser. E, 12* (1986) 181–199.

6 G. W. Lewis, G. Kiriakidis, G. Carter and M. J. Nobes, *Surf. Interface Anal., 4* (1982) 141.

7 J. L. Whitton, G. Carter, M. J. Nobes and J. S. Williams, *Radiat. Eff., 32* (1977) 129.

8 J. L. Whitton, L. Tanovic and J. S. Williams, *Appl. Surf. Sci., 1* (1978) 408.

9 G. Carter, M. J. Nobes and G. W. Lewis, *Nucl. Instrum. Methods, 194* (1982) 509.

10 J. L. Whitton, G. Kiriakidis, G. Carter, G. W. Lewis and M. J. Nobes, *Nucl. Instrum. Methods, 230* (1984) 640.

11 G. Carter and M. J. Nobes, *Nucl. Instrum. Methods, 230* (1984) 635.

12 G. Carter, M. J. Nobes, G. W. Lewis, J. L. Whitton and G. Kiriakidis, *Vacuum, 34* (1984) 167.

Materials Science and Engineering, 90 (1987) 33-36

Ion-beam-induced Morphological Evolution of Cone-shaped Thin Wires*

H. HASUYAMA, Y. KANDA, T. SOEDA, K. NIIYA and M. KIMURA

Department of Energy Conversion Engineering, Kyushu University, Kasuga, Fukuoka 816 (Japan)

(Received July 10, 1986)

ABSTRACT

The morphological evolution of cone-shaped polycrystalline copper wires 30-50 μm in diameter has been studied as a function of consecutive irradiation of 20 keV argon ions from 1×10^{17} *to* 1×10^{20} *ions* cm^{-2}*. In order to study the dependence of the ion-beam-induced development of the cone-shaped specimens on initial tip profiles, two types of tip were used with initial apex angles of 10° and 50°. The influence of microstructure on the tip evolution was also investigated through the use of different pre-irradiation annealing treatments. It is found, by scanning electron microscopy, that the cones develop with similar features, having an apex angle of 15° after an ion irradiation fluence greater than* 5×10^{19} Ar^+ cm^{-2}*, irrespective of the differences in the initial tip profile of copper wires and also irrespective of the specimen treatment before the irradiation.*

1. INTRODUCTION

Although the appearance of characteristic features such as cones or pyramids on a solid surface due to ion sputtering during ion bombardment has been observed by many investigators, the mechanisms of their generation and process of evolution remain at present controversial [1]. Systematic studies on the evolution of cones are complicated by the effect of sputtered atoms or recoiled ions and reflected ions from neighbouring cones [2, 3].

In previous work [4] a thin copper wire approximately 50 μm in diameter was alternately irradiated with argon ions and observed using scanning electron microscopy (SEM) in order to study the sputtering-induced evolution of a single cone. It was found that the round tip of the wire developed to a cone after a high irradiation. The experimental conditions used also ensured that the evolution of the cone during irradiation was never affected by reflected ions from neighbouring cones. The previous study has been extended in the present work by measuring quantitatively the apex angle as a function of ion fluence. Once a conical shape is produced, the development of the cone-shaped features is considered to be determined by the angular dependence of the sputtering yield, whatever mechanisms are attributed to their origins. Carter and coworkers [2, 5] have shown that the cone surface normal makes a critical angle $\hat{\theta}$, at which the sputtering yield reaches a maximum, with respect to the ion beam direction at the equilibrium situation. Thus the apex angle α of the cone is related to $\hat{\theta}$ by $\alpha = \pi - 2\hat{\theta}$. Since Wilson and Kidd [6] first measured the apex angles of cones developed on polycrystalline gold surface after ion irradiation, there have been a few measurements [7, 8] and theoretical calculations [9-11] of cone apex angles.

In the present experiment, we prepared single cone-shaped polycrystalline copper wires 30-50 μm in diameter with different apex angles to study the influence of the initial shape on the evolution of the tip of samples after a consecutive irradiation with 20 keV argon ions. The effect of the specimen treatments before irradiation was also examined by annealing the specimens at different temperatures. The evolution of cones was observed using SEM after every irradiation of ion fluence from 1×10^{17} to 1×10^{20} ions cm^{-2}.

*Paper presented at the International Conference on Surface Modification of Metals by Ion Beams, Kingston, Canada, July 7-11, 1986.

0025-5416/87/$3.50

2. EXPERIMENTAL DETAILS

The cone-shaped thin wires were formed out of the as-drawn polycrystalline copper wires (chemical purity, 99.9%) with an original diameter of 100 μm. The tip of the sample, which had been fixed perpendicularly through a hole at the centre of a copper plate, was shaped by electrolytically etching in a solution of 60% H_3PO_4. The final feature of the tip can be shaped into a cone with a diameter of 30–50 μm by drawing the wire slightly along its axis and 50 μm from the surface of the electrolytic solution.

In order to study the influence of the initial shape of the tip on its morphological change during ion beam irradiation, two types of tip were prepared with apex angles of about 10° and 50°. To investigate the influence of the specimen microstructure on the change in sample profile during irradiation, some specimens whose tips were conically shaped were annealed at either 400 or 800 °C for 1 h at 10^{-3}–10^{-4} Pa. The grain sizes obtained were 20 μm and 50 μm at annealing temperatures of 400 °C and 800 °C respectively. The samples were irradiated with mass-analysed 20 keV argon ions along their axes in the target chamber. A typical ion beam current density was about 100 μA cm^{-2} with a beam spot 2 mm in diameter. The ion beam irradiation and the SEM observation were performed alternately for the same wire. The change in tip profiles in the fluence range from 1×10^{17} to 1×10^{20} ions cm^{-2} was observed using the SEM micrograph.

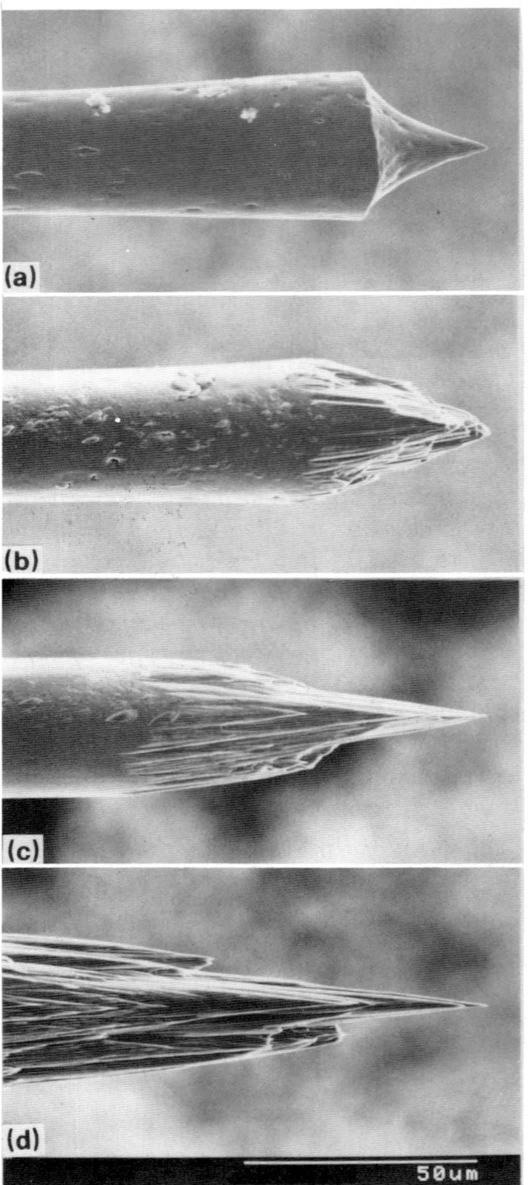

Fig. 1. A series of SEM micrographs of a thin copper wire with a conical tip with an apex angle of 50° irradiated by 20 keV argon ions for various ion fluences: (a) 0 ions cm^{-2} (unirradiated); (b) 1.5×10^{19} ions cm^{-2}; (c) 4.0×10^{19} ions cm^{-2}; (d) 1.0×10^{20} ions cm^{-2}.

3. RESULTS AND DISCUSSION

Examples of SEM micrographs showing the change in the tip profile for the copper wire by successive irradiation with 20 keV argon beams are given in Fig. 1. Figure 1(a) shows the original profile of the sample before irradiation; the slope at the tip of the sample is concave with an initial apex angle of 50° measured 10 μm from the tip. No remarkable change was observed on the surface of the tip of samples on irradiation up to an ion fluence of the order of 10^{17} ions cm^{-2}. After an increase in the ion fluence to higher than 3×10^{18} ions cm^{-2}, the concave slope of sample surface began to change gradually to be round and then became convex at 1.5×10^{19} ions cm^{-2}, as shown in Fig. 1(b). When the ion fluence was increased to 4×10^{19} ions cm^{-2}, the round tip of the convex cone changed to a straight edge, as the tip became sharp, like the tip of a spear (Fig. 1(c)). After irradiation to a fluence of 1×10^{20} ions cm^{-2}, the whole tip developed to a large cone, as shown in

Fig. 1(d). Although the surface was observed to be covered with many small cones which had a similar apex angle to that of a large cone, which was of interest from the viewpoint of its evolution, the data obtained so far were too few to determine how the small cones would develop. At present, our interest in the evolution of the large cone centred on measuring the apex angle as a function of fluence, as a typical example of a single cone. The micrographs show that the apex angle α of the tip of the whole sample is 15°, which corresponds to an incident angle of 83°.

In order to study the dependence of the development of the cone shape on the initial apex angle, the sample with an apex angle of 10° was also consecutively irradiated by 20 keV argon ions. The variations in the apex angle and the incident angle for both thin copper wires with a conical tip are shown in Fig. 2 as a function of ion fluence. The results measured at the different positions from the tip of the samples with initial apex angles of 50° and 10° are indicated by full symbols and open symbols respectively. This figure shows that for the 50° sample the variation in the angles begins to change at an ion fluence of about 1×10^{19} ions cm^{-2}, which corresponds to the change in surface slope of the samples from concave to convex. It is found for both samples that above a fluence of 5×10^{19} ions cm^{-2} the incident angle reaches a constant value of 83° with an apex angle of about 15°, irrespective of the different initial shapes of the samples.

To investigate the influence of the specimen treatments, the ion beam irradiations and the SEM observations were repeated for the samples annealed at 400 and 800 °C using the same procedure as for unannealed samples. It was expected that the elimination of cold-working effects could be achieved by 400 °C annealing (the temperature of 400 °C is much higher than the temperature of recrystallization of copper (about 250 °C)) and that enlargement of the grain size up to 50 μm would be attained by the 800 °C anneal. The latter might change the processes of cone evolution of the copper wire and microscopic structures on the cone surface since the grain size would be similar to the wire diameter. An example of the variations in apex angle and incident angle, which were measured at different positions from the top, for the sample annealed

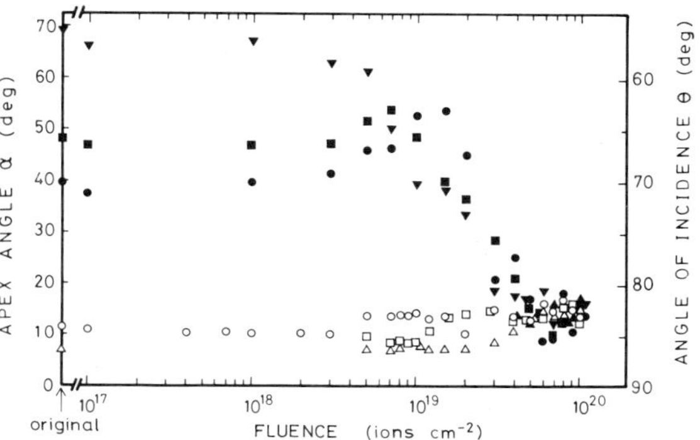

Fig. 2. Variations in the apex angle and incidence angle for a thin copper wire with a conical tip as a function of 20 keV argon ion fluence, showing the results measured at different positions (●, 2 μm; ■, 10 μm; ▼, 20 μm; ▲, 40 μm) from the top for an initial apex angle of 50° and the results measured at different positions (○, 2 μm; □, 10 μm; △, 40 μm) from the top for an initial apex angle of 10°.

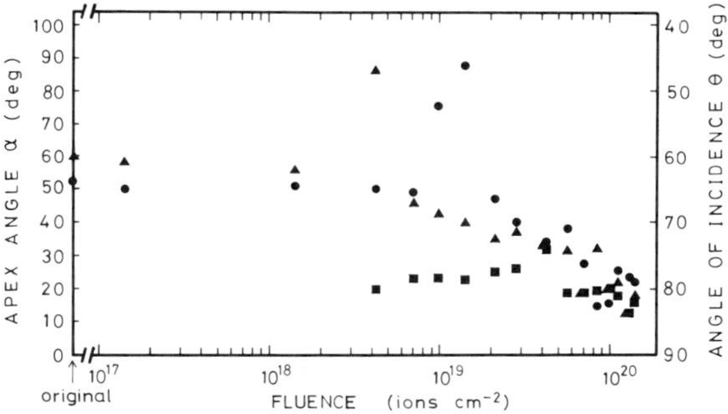

Fig. 3. Variations in the apex angle and incidence angle measured at different positions (●, 10 μm; ▲, 20 μm; ■, 40 μm) from the tip for a thin copper wire, which had been annealed at 800 °C for 1 h, as a function of 20 keV argon ion fluence.

at 800 °C is shown in Fig. 3 as a function of ion fluence. This figure shows that the apex angle converges to about 20° for an ion fluence in excess of 5×10^{19} ions cm^{-2}, which is equivalent to an incident angle of 80°. For copper wires annealed at 400 °C, the apex angle was found to show a similar variation with ion fluence to that of the copper wires annealed at 800 °C and became constant at almost the same value of about 15° after a high irradiation. The results of the ion-induced variation in apex angle obtained for thin copper wires annealed at both 400 and 800 °C were the same as those obtained for unannealed samples. Therefore, it is concluded that the high-irradiation-induced cone evolution leads to the equilibrium feature of a constant apex angle and is independent of cold-working effects and the grain size of the copper wires.

In summary, the present results reveal that the apex angle of cone-shaped thin copper wire samples attained a constant value after a high fluence irradiation of 5×10^{19} ions cm^{-2} with 20 keV argon ions, irrespective of the different initial features of the specimens and also of the specimen treatments before irradiation even though the intrinsic properties and grain size differ considerably in various specimens. It is also confirmed that, irrespective of whether $\theta > \hat{\theta}$ or $\theta < \hat{\theta}$, the cone-shaped tip of the specimens is eroded at the position where the incident angle θ is near the critical angle $\hat{\theta}$ and finally develops a stable feature with a slope of incident angle $\hat{\theta}$.

REFERENCES

1 G. Carter, B. Navinšek and J. L. Whitton, in R. Behrisch (ed.), *Sputtering by Particle Bombardment II, Top. Appl. Phys., 52* (1983) 231.

2 G. Carter, J. S. Colligon and M. J. Nobes, *J. Mater. Sci., 6* (1971) 115.

3 O. Auciello, R. Kelly and R. Iricibar, *Radiat. Eff. Lett., 43* (1979) 37.

4 Y. Kanda, H. Hasuyama, K. Waki and Y. Sakamoto, *Metallography, 19* (1986) 461.

5 M. J. Nobes, J. S. Colligon and G. Carter, *J. Mater. Sci., 4* (1969) 730.

6 I. H. Wilson and M. W. Kidd, *J. Mater. Sci., 6* (1971) 1362.

7 B. B. Meckel, T. Nenadovic, B. Perovic and A. Vlahov, *J. Mater. Sci., 10* (1975) 1188.

8 D. Ghose, D. Basu and S. B. Karmohapatro, *Nucl. Instrum. Methods B, 1* (1984) 26.

9 A. D. G. Stewart and M. W. Thompson, *J. Mater. Sci., 4* (1969) 56.

10 M. J. Witcomb, *J. Mater. Sci., 9* (1974) 1227.

11 D. Ghose, *Jpn. J. Appl. Phys., 18* (1979) 1847.

Materials Science and Engineering, *90* (1987) 37-43

Surface Roughening During Elevated Temperature Implantation of Nickel*

M. AHMED, K. RUFFING and D. I. POTTER

Metallurgy Department, and Institute of Materials Science, University of Connecticut, Storrs, CT 06268 (U.S.A.)

(Received July 10, 1986)

ABSTRACT

Ion implantation at fluences near 10^{17} ions cm^{-2} causes roughening and loss of reflectivity on electropolished nickel surfaces. The roughening is strongly temperature sensitive, not occurring at fluences well above 10^{17} ions cm^{-2} at $T < 300$ °C or $T \gtrsim 700$ °C, and reaching a maximum near 600 °C. The roughening is caused by pits that form at fluences near 3×10^{17} ions cm^{-2}, with pit overlap taking place by two to three times this fluence. Internal voids are observed in the same temperature interval as the pits but at lower fluences, and the voids act as nuclei for the pits.

1. INTRODUCTION

Ion implantation at elevated temperatures is being developed in order to provide deeper penetration of implanted solute, via diffusion, and in order to produce phases stable at higher temperatures but metastable during room temperature implantation [1]. As reported here, roughening of the surface accompanies the implantation at temperatures from about 300 to about 700 °C. The roughening occurs at fluences one to two orders of magnitude less than that required to cause the same amount of roughening at room temperature, as judged by our own work as well as that of others [2]. These two observations, the strong temperature dependence of roughening and its occurrence at such low fluences, led us to suspect that its origins were different from that of the roughening observed at higher fluences and temperatures less than 300 °C. The nature and causes of the phenomena, which originate from void formation beneath the surface, are explored in this paper.

2. EXPERIMENTAL PROCEDURES

Nickel discs of 3 mm diameter were punched from a sheet of 99.99% purity nickel 0.25 mm thick. The discs were encapsulated in a quartz tube at a pressure of about 1×10^{-5} Pa and subsequently annealed at 850 °C for 6 h. The surfaces of the discs were then electropolished with 20% perchloric acid in ethanol.

The specimens were implanted with 180 keV aluminum (Al^{+}) or nickel (Ni^{+}) ions, at fluxes near 10^{14} ions cm^{-2} s^{-1} and at pressures near 10^{-6} Pa. Electron emission furnaces provided specimen heating at temperatures fixed to within ±2 °C. The temperatures were monitored using thermocouples attached to selected specimens. The temperature of each specimen was measured with an IR pyrometer. Microstructures following implantation were observed with a scanning transmission electron microscope operated, as appropriate, in the secondary electron scanning mode (scanning electron microscopy (SEM)) for surface imaging or in the transmission mode (transmission electron microscopy (TEM)) for imaging internal structure. Images, unless otherwise noted, were recorded with the electron beam normal to the sample surface.

3. RESULTS

3.1. Surface roughening and reflectivity losses

The progression of surface topography was examined as a function of fluence, from 10^{17} to 2.4×10^{18} ions cm^{-2}, and at fixed temperatures between 25 and 750 °C. The results of this examination are presented below.

*Paper presented at the International Conference on Surface Modification of Metals by Ion Beams, Kingston, Canada, July 7-11, 1986.

0025-5416/87/$3.50

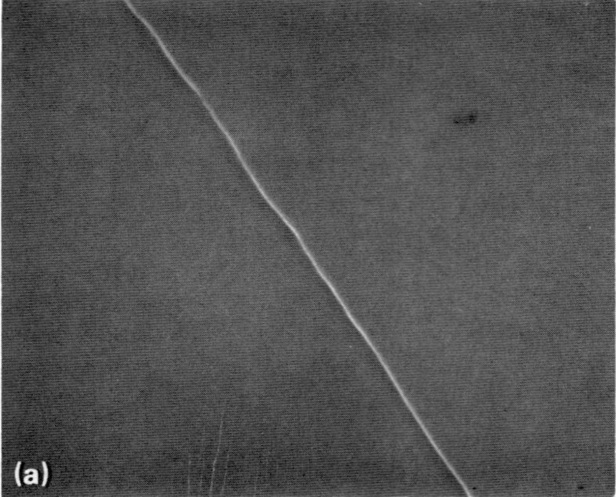

(b)

Fig. 1. SEM images after implanting with 1.8×10^{18} Al^{+} cm^{-2}: (a) 25 °C implantation; (b) 300 °C implantation.

The SEM micrographs presented in Fig. 1 show the surfaces resulting from 25 and 300 °C implantation with 1.8×10^{18} Al^{+} cm^{-2}. These surfaces are typical of those found for implantation temperatures up to 300 °C and fluences to 2.4×10^{18} ions cm^{-2}, the highest fluence investigated. Generally, the surfaces retained most of their reflectivity, as assessed visually. Grain boundaries, not seen in the original electropolished surfaces, are evident after 25 °C implantation. A small amount of surface roughening develops within the grains (Fig. 1(b)) as the implantation temperature is increased to 300 °C.

The reflectivity of the surfaces progressively decreased with increasing fluence at implantation temperatures above 500 °C but less than 700 °C. This was the case for implantation with either nickel or aluminum ions, although the rate of reflectivity loss was greater for nickel ions than for aluminum ions. The optical micrographs in Fig. 2 demonstrate this reflectivity loss. The unimplanted surface is featureless (Fig. 2(a)), while grain boundaries can be seen after implanting with 1×10^{17} Ni^{+} cm^{-2} (Fig. 2(b)). Intragranular reflectivity losses were noted by 3×10^{17} Ni^{+} cm^{-2} (Fig. 2(c)) with increasing intergranular contrast exhibited with increasing fluence (Fig. 2(d)). Microscopic features to be explored shortly cause the contrast. Roughly speaking, three levels of contrast, labeled a, b and c in Fig. 2, can be noted. The examination of the specimens using selected area electron channeling patterns revealed that the degree of roughness was related to grain orientation. The grains oriented with the surface normal parallel to ⟨100⟩ depicted minimum surface roughness. The maximum loss of reflectivity was observed in grains surfaces normal to ⟨111⟩ orientation. Different sputtering rates in these orientations [3] may account for these contrast groupings. Grain relief of about 1000 Å was measured with an interference microscope from adjacent type a and c grains in Fig. 2(d). It should also be noted, in Fig. 2(d), that the portion of the specimen masked from the ion beam exhibits no contrast, showing that surface roughening does not result from time at temperature alone.

Similar reflectivity losses were observed for aluminum-implanted nickel surfaces. Contrast groupings, as above, were recognized at 6×10^{17} Al^{+} cm^{-2}. Fluences three times this left grains of quite uniform contrast, because of the great reflectivity losses of all grains. The fluences for equal reflectivity losses, judged visually, for aluminum-implanted nickel were roughly three times as great as the corresponding nickel fluences. This is reasonable, to the extent that sputtering causes these developments, in that the calculated sputtering yield of 180 keV nickel ions on nickel is about three times that for aluminum ions on nickel [4]. This agrees with sputtering yields measured in our work using interference microscopy and partially masked specimens.

3.2. *Intragranular pitting*

The causes of the contrast noted above were investigated further. Higher magnifica-

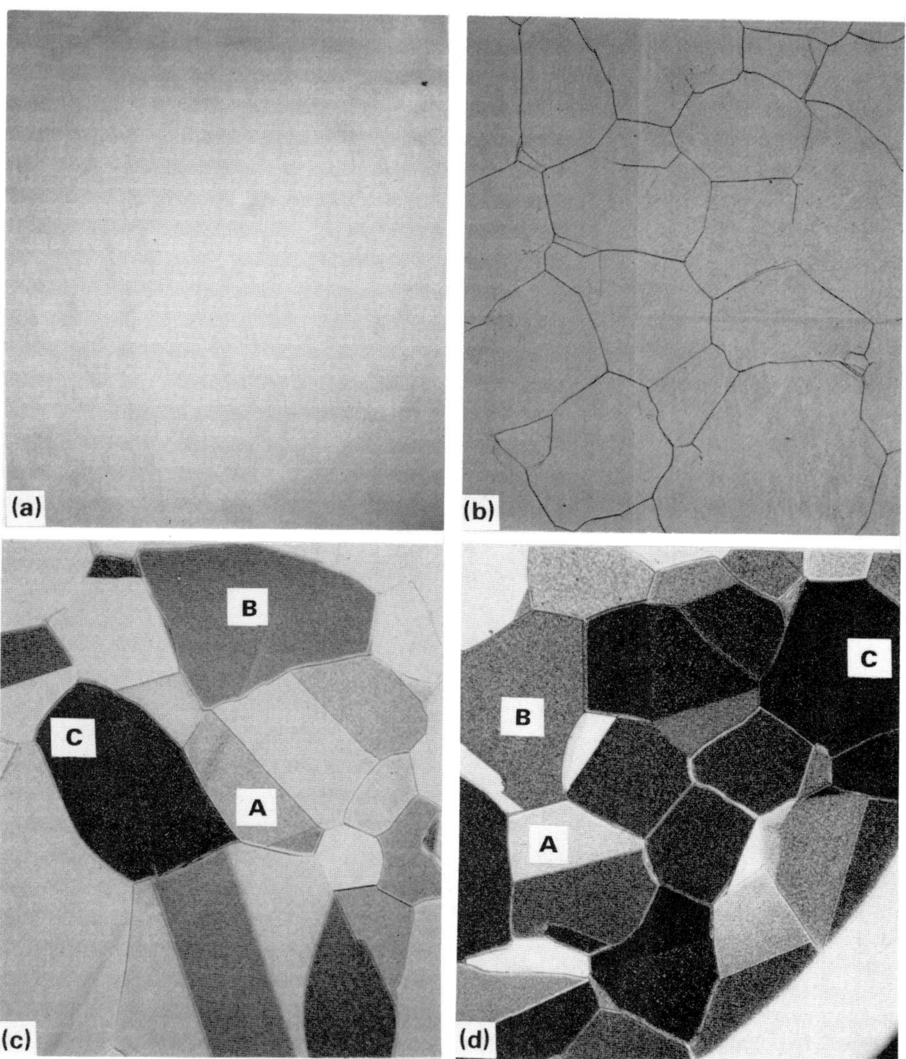

Fig. 2. Optical micrographs showing (a) the starting surface and (b)–(d) the surfaces after implanting at 650 °C with nickel ions to fluences of (b) 1×10^{17} ions cm^{-2}, (c) 3×10^{17} ions cm^{-2} and (d) 6×10^{17} ions cm^{-2}. The arc-shaped shadow of the sample holder (lower right-hand corner of (d)) should be noted. (Magnifications, 100×.)

tion SEM images revealed that the reflectivity loss occurs as a result of the formation of microcavities or pits. These pits are faceted and oriented in accord with the orientation of the grains in which they form, sharing this and some other features in common with pits observed in sputtered copper [1]. However, it should be noted that the pits described here are evident at much lower fluences than those observed in copper, *i.e.* about 10^{17} ions cm^{-2}, compared with fluences ten to 100 times higher for copper.

Figure 3 shows the development of these pits at 650 °C during nickel ion implantation of nickel. Well-defined faceted pits form by 3×10^{17} ions cm^{-2} (Fig. 3(a)). The areal density and shape of these pits varied considerably from grain to grain, a density of about 10^7 pits cm^{-2} being typical of a grain showing maximum contrast. For nickel-implanted nickel, this density increased by a factor of 5 at 6×10^{17} ions cm^{-2} (Fig. 3(b)). Pit overlap at higher fluences caused the topography seen in Fig. 3(c).

The pit structures observed at temperatures near 500 °C were similar to those at 650 °C, except that the pits were smaller and their densities were higher for the same fluences.

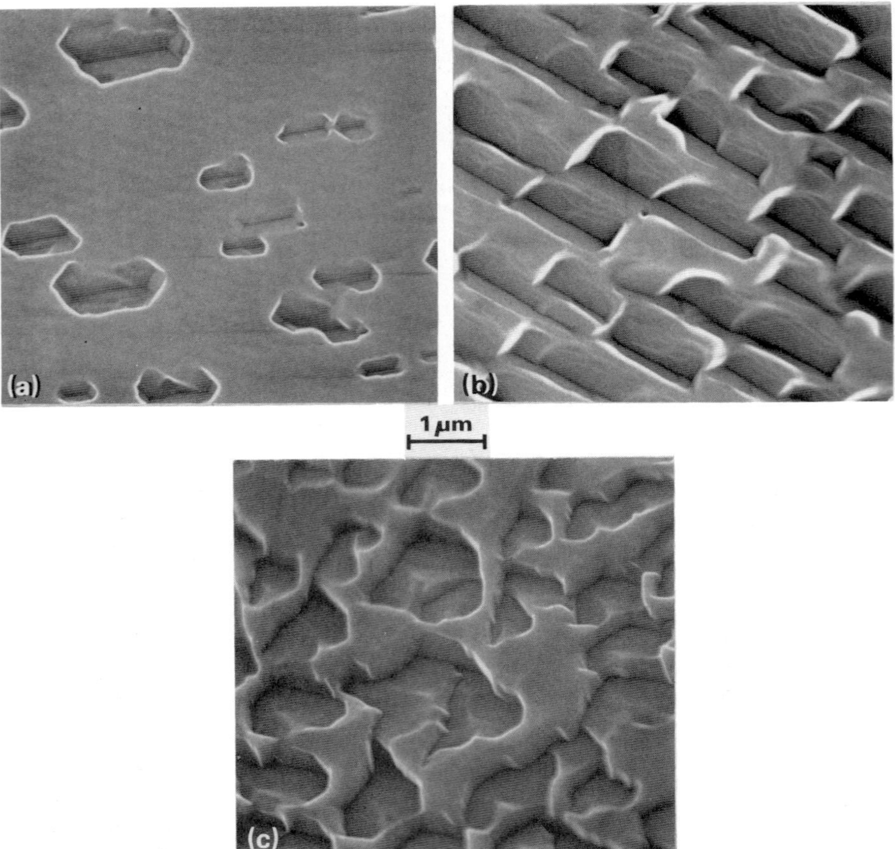

Fig. 3. SEM images showing pit development during 650 °C implantation with nickel (Ni^+) ions: (a) 3×10^{17} ions cm^{-2}; (b) 6×10^{17} ions cm^{-2}; (c) 1.8×10^{18} ions cm^{-2}.

A marked reduction in surface roughness, *i.e.* retention of surface reflectivity, occurred for specimens implanted at 750 °C (Fig. 4). The optical micrograph in Fig. 4(a), following implantation with 3×10^{17} Ni^+ cm^{-2}, should be compared with Fig. 2(c) which was recorded after the same fluence at 650 °C. It is clear from the near absence of grain-to-grain contrast that surface roughening is greatly reduced at the higher temperature. The SEM image in Fig. 4(b) shows that the intragranular pits are essentially gone but that grain boundary etching remains severe.

The pit structures and their temperature dependence for aluminum-implanted nickel were similar to those reported above for nickel ion implantation, except for three features. One, already noted, was that comparable structures developed at earlier fluences in the latter case than in the former. Secondly, the pits in aluminum-implanted nickel were about an order of magnitude smaller than those in nickel-implanted nickel, *i.e.* about 0.2 μm *vs.* about 2 μm. A third feature was that the areal density of pits in aluminum-implanted nickel specimens rose quickly to a fixed value, *e.g.* about 5×10^8 pits cm^{-2} at 6×10^{17} Al^+ cm^{-2} and 550 °C, and then remained constant. Pit overlap, governed by the growth of pits, began to occur near 1.8×10^{18} ions cm^{-2} at this temperature.

3.3. *Voids in the implanted layers*

The microstructural developments within the implanted layer were investigated with TEM. The results show that dislocations and voids form under certain implantation conditions, governed mainly by the implantation temperature. These developments were anticipated to some extent, based on previous research by many investigators exploring void

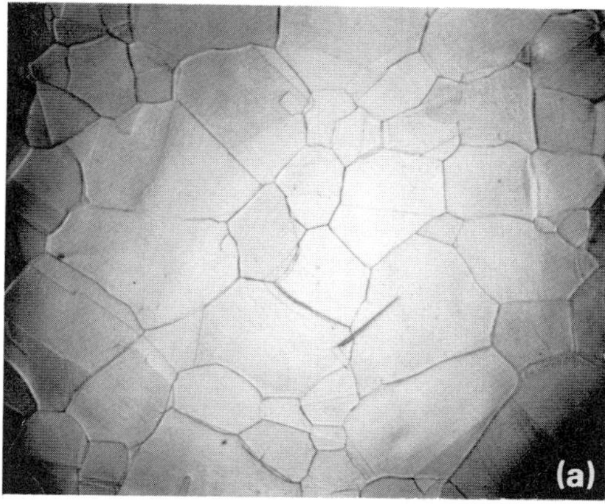

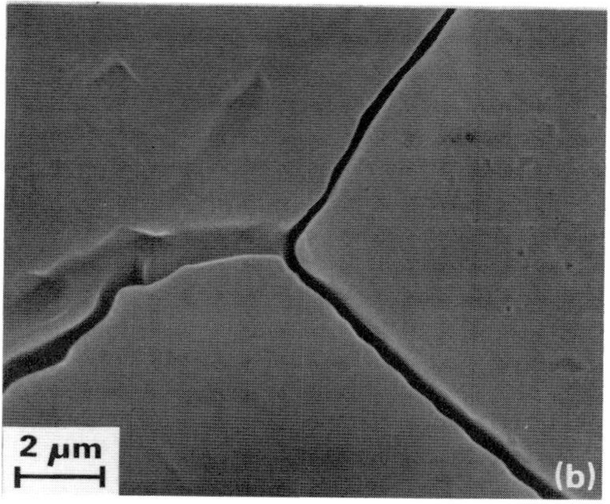

Fig. 4. Surface structure after implanting 3×10^{17} Ni^+ cm^{-2} at 750 °C: (a) optical micrograph; (b) SEM image. (Magnification: (a) 50×.)

swelling in nickel [5, 6] and other nuclear reactor materials [7].

The microstructure after implanting with nickel ions at 550 °C is shown in Fig. 5(a). Uniformly distributed voids about 150 Å in size and with a density of about 10^{15} voids cm^{-3} can be noted. Small dislocation loops are also evident, with a size of about 100 Å and a density of about 10^{16} loops cm^{-3}. The void size increases to about 500 Å and the density falls to about 10^{14} cm^{-3} for nickel implanted at 650 °C (Fig. 5(b)) while line dislocations replace the dislocation loops. Voids were not observed following 750 °C implantation (Fig. 5(c)) but a very low density of line dislocations can be noted. Increasing the fluence at a particular temperature resulted in an increase in void density (compare Fig. 5(d) with Fig. 5(a)).

A similar temperature dependence for void size and density occurred for aluminum-implanted nickel at fluences less than or near 6×10^{17} ions cm^{-2}. Voids were not observed at fluences beyond about 1.2×10^{18} ions cm^{-2}, where the β' (NiAl) phase forms [1, 8], consistent with previous work [9] which also showed that voids do not form in β'.

4. SUMMARY AND DISCUSSION

Surface roughening occurs during elevated temperature implantation of nickel with aluminum (Al^+) and nickel (Ni^+) ions. The roughening and associated reflectivity loss occur at fluences near 10^{17} ions cm^{-2}, one to two orders of magnitude less than that causing comparable roughening at room temperature. The roughening rate is strongly temperature sensitive, being near zero at temperatures below 300 °C and above about 700 °C. It reaches a maximum near approximately 600 °C. The roughening rate also depends on the ion implanted and is about three times faster for nickel than for aluminum ions.

Pits in the surface cause the roughening and reflectivity loss. These pits are about 0.2 μm in size for aluminum ions and about 2.0 μm in size for nickel ions and they are faceted. In the most roughened grains, the areal densities of pits are near 10^7 cm^{-2} and 5×10^7 cm^{-2} after implanting 3×10^{17} Ni^+ cm^{-2} and 6×10^{17} Ni^+ cm^{-2} respectively at 650 °C. The pit densities increase with decreasing implantation temperatures.

Voids contained beneath the surface and within the implanted layer were observed over the same temperature interval in which the pits were observed. At 650 °C, for example, the void size was about 500 Å and their number density was about 10^{14} cm^{-3} after implanting 1×10^{17} Ni^+ cm^{-2}. Like pits, the voids were not observed at temperatures below 300 °C or above about 700 °C, and the void volume fraction passes through a maximum with temperature near 600 °C [5, 6]. The voids are faceted, as were the pits.

Several observations suggest a causal relationship between voids and the pits present here.

(1) Pits are observed only when voids are observed.

(2) The presence of pits and voids both show the same strong temperature dependence.

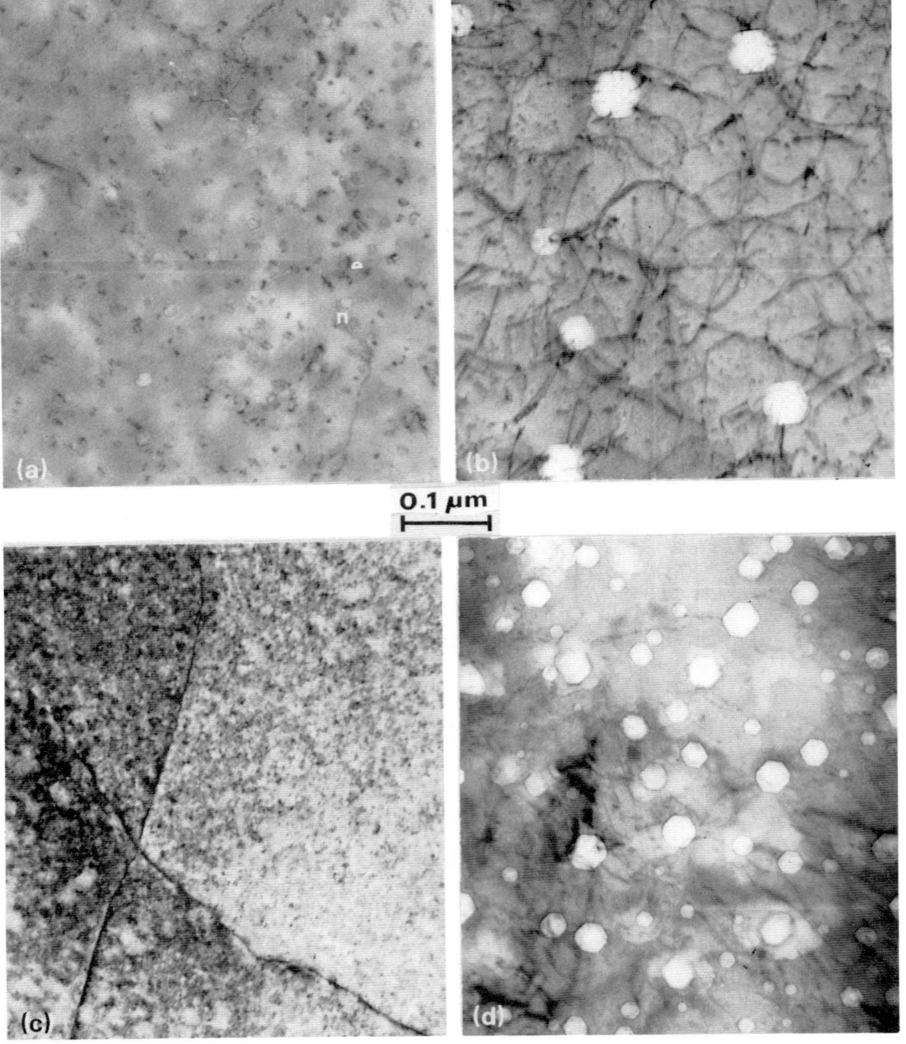

Fig. 5. TEM images of the implanted layer: (a) 1×10^{17} Ni^+ cm^{-2}, 550 °C; (b) 1×10^{17} Ni^+ cm^{-2}, 650 °C; (c) 1×10^{17} Ni^+ cm^{-2}, 750 °C; (d) 3×10^{17} Ni^+ cm^{-2}, 550 °C.

(3) The pit sizes increase and areal densities decrease with increasing implantation temperature; void sizes and densities respond similarly.

We also note that an incubation period precedes pit formation. For example, no pits are seen at 1×10^{17} Ni^+ cm^{-2} at 650 °C, but well-developed pits are noted at 3×10^{17} Ni^+ cm^{-2}. This is consistent with a void-denuded layer existing, initially, next to the surface, a phenomenon substantiated by high voltage electron microscopy experiments [10]. The incubation period corresponds to the time for removal of this layer by sputtering, and the arrival of voids at the free surface. Using our measured sputtering yield of about 3 for 180 kV nickel ions on nickel, we calculate a denuded layer of about 400 Å, quite reasonable for the high damage rates accompanying the relatively high fluxes in our work.

Lastly, we note that not all voids can result in stable pits. A large fraction must not grow as implantation proceeds. This is so since the void density is, for example, about 10^{14} cm^{-3} or more for nickel ion implantation at 650 °C. The pit density increases by 4×10^7 cm^{-2} between fluences of 3×10^{17} and 6×10^{17} Ni^+ cm^{-2}. The areal density of voids which surface during this fluence interval is about 5×10^8

cm^{-2}. Only about one void in ten nucleates a pit that grows with further implantation. Based on similar reasoning, the pit density is expected to increase by a factor of 5 in this fluence interval, consistent with observations in Fig. 3.

ACKNOWLEDGMENTS

We thank the following people for contributing to this research: H. Hayden and C. Koch for assisting in the implantations; L. McCurdy for TEM assistance; L. Witherell for typing.

This material is based on work supported by the National Science Foundation under Grant DMR8507641. The transmission electron microscope was purchased with National Science Foundation support under Grant DMR8207266 and with State of Connecticut funding.

REFERENCES

1 M. Ahmed and D. I. Potter, *Acta Metall.*, to be published.

2 G. Carter, M. J. Nobes and J. L. Whitton, *Appl. Phys. A*, *38* (1985) 77.

3 J. J. Ph. Elich, H. E. Roosendaal and D. Onderdelinders, *Radiat. Eff.*, *14* (1972) 93.

4 P. Sigmund, *Phys. Rev.*, *184* (1969) 383.

5 T. D. Ryan, *Ph.D. Thesis*, University of Michigan, Ann Arbor, MI, 1975.

6 J. A. Sprague, J. E. Westmoreland, F. A. Smidt, Jr., and P. R. Malmberg, in *Properties of Reactor Structural Alloys After Neutron or Particle Irradiation, ASTM Spec. Tech. Publ. 570*, 1975, p. 505.

7 M. L. Bleiberg and J. W. Bennett (eds.), *Radiation Effects in Breeder Reactor Structural Materials*, AIME, New York, 1977.

8 D. I. Potter, M. Ahmed and S. Lamond, in G. K. Hubler, C. W. White, O. K. Holland and C. R. Clayton (eds.), *Ion Implantation and Ion Beam Processing of Materials, Materials Research Society Symp. Proc.*, Vol. 27, Elsevier, New York, 1984, p. 117.

9 H. C. Liu, C. Kinoshita and T. E. Mitchell, in J. R. Holland, L. K. Mansur and D. I. Potter (eds.), *Phase Stability During Irradiation*, AIME, New York, 1981, p. 343.

10 D. I. R. Norris, *Radiat. Eff.*, *15* (1972) 1.

Materials Science and Engineering, 90 (1987) 45–49

Scanning Electron Microscopy Observation of High Energy α-bombarded Molybdenum Surfaces*

S. V. NAIDU† and P. SEN

Saha Institute of Nuclear Physics, Calcutta 700009 (India)

(Received July 10, 1986)

ABSTRACT

The surface morphologies of two sets of high energy α-bombarded molybdenum specimens were studied using scanning electron microscopy. The irradiation was performed with 33.5 and 47 MeV α particles to a total dose of about $(1–2) \times 10^{17}$ α cm^{-2}. The micrographs of the bombarded surfaces show interesting features. Circular pits of diameter 5–12 μm, small pinhole pits and few elongated blister chains are found. The irregular shape of the pits on the second set of specimens reveal that the reference surface conditions have an important bearing in determining the shape of the pits. The results indicate that, although the lateral stress in the low energy ion implantation plays only a small role in the surface erosion, at high energies the compressive stress created by the lattice defects has an important role as the cascade displacement damage and the thermal spike phenomena are more probable in the near-surface region. These observations also support the idea that the initiation of the fracture plane need not always be around the depth of the peak helium concentration; rather it may develop randomly along the depth distribution of implantation.

1. INTRODUCTION

The phenomena of surface damage and erosion have been the subject of extensive studies in recent years because of their technological importance in thermonuclear fusion devices and reactors [1–3]. The surface erosion takes place by various processes: sputtering, blistering and flaking. At higher ion energies the last two processes are the most effective and they are considered to result from one and the same phenomenon [4, 5]. Several models have been proposed to explain the formation of surface blistering on helium-bombarded metal surfaces. These models are based on (1) gas bubble coalescence and the build-up of internal gas pressure, (2) the percolation of helium into the lattice and (3) the build-up of stresses in the implanted layer, which has been described well in ref. 6. In the gas-driven model the blister lid thickness is expected to be equal to the depth of the helium peak. Objections were raised against this model as the discovery of blister lid thickness exceeding the depth of the helium peak [7–9] and the lateral stress model was developed. This model suggests that the key factor is the relief of the compressive stress system set up as a result of the irradiation damage layer. Most of these studies have been carried out in the low and medium ion energy region but, in the megaelectronvolt region, relatively few experiments have been performed. As we go from the low ion energy to the high ion energy region the blistering process becomes more complex with increasing probability of flaking and exfoliation [3, 10, 11]. The results obtained in the high ion energy region are quite interesting [3, 12, 13] as the blister skin thickness was found to be considerably smaller than the projected range of the implanted ion. The study [12] on 30 MeV α-bombarded tantalum foils showed that blisters can develop on foils which are very thin, about one-tenth of the projected range of the ion. In this paper we present

---

*Paper presented at the International Conference on Surface Modification of Metals by Ion Beams, Kingston, Canada, July 7–11, 1986.

†Present address: Center for Positron Studies, Physics Department, The University of Texas at Arlington, Arlington, TX, U.S.A.

0025-5416/87/$3.50

TABLE 1

Summary of the α irradiation conditions for specimens 1 and 2

Specimen	*α particle energy* (MeV)	*Average current* (μA)	*Total dose* (α cm^{-2})	*Temperature of the specimens during irradiation* (°C)	*Mean ion depth* (μm)	*Longitudinal straggling* (μm)	*Transverse straggling* (μm)
1	33.5	≤ 2	1.25×10^{17}	<150	175	3	7.5
2	47	≤ 3	2.1×10^{17}	<150	310	5	10

scanning electron microscopy (SEM) observations of high energy (33.5 and 47 MeV) α-bombarded molybdenum surfaces which may help us to understand further the blistering phenomenon.

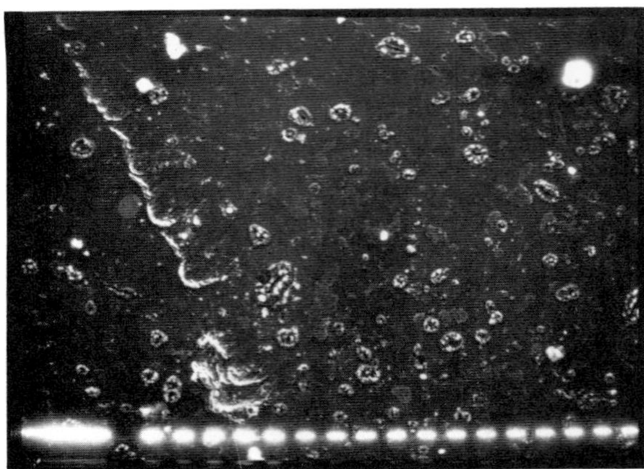

Fig. 1. Scanning electron micrograph of 33.5 MeV α-bombarded molybdenum (specimen 1). The white dashes represent 10 μm.

2. EXPERIMENTAL DETAILS

Very pure molybdenum specimens about 0.28 mm thick were well polished and annealed at 1200 °C for several hours. All these specimens were characterized in the analytical laboratory and the total impurity content was estimated to be less than 30–40 ppm. Two sets of specimens were subjected to α irradiation at normal incidence by 33.5 and 47 MeV α particles at the Variable Energy Cyclotron Centre, Calcutta. Prior to the bombardment the α beam was passed through thin tantalum degrader foils. The specimens were in good thermal contact with a water-cooled copper target flange. The temperature of the specimens during the irradiation was monitored with a temperature sensor and kept below 150 °C, the temperature at which monovacancies migrate in molybdenum [14]. The beam current was about 2–3 μA and the total dose was $1\text{–}2 \times 10^{17}$ α cm^{-2}. The irradiation conditions for the two sets of specimens are summarized in Table 1. After irradiation the specimen surfaces were examined using the Philips PSEM 500 scanning electron microscope at the Bose Institute, Calcutta.

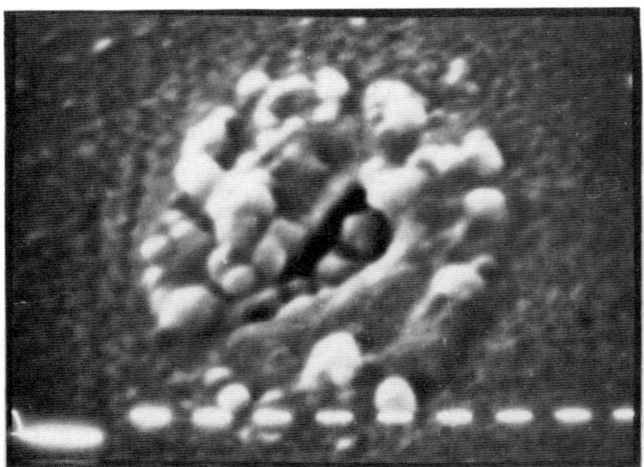

Fig. 2. An enlarged view of one of the circular pits shown in Fig. 1. The white dashes represent 1 μm.

3. RESULTS

The scanning electron micrographs are shown in Figs. 1 and 2 for specimen 1. Circular pits with diameters of 5–12 μm were observed to occur in a random fashion at concentrations of about 8×10^6 pits cm^{-2}. At several places in the irradiated area, chains of elongated blisters are seen. One of the chains about 200 μm in length is shown in Fig. 1. The pits seem to be due to a process close to that of exfoliation of small blisters. The exfoliated part of the material can be seen around the pit as white spots. Apart from the larger circular pits of 5–12 μm size, several small and irregularly shaped pits appear as black spots. Such small black spot pits are usually referred to as pinholes [15]. An enlarged view of the circular pits is shown in Fig. 2 at 30° tilt from the normal to the surface. The exfoliation around the pit is evident. The pit is deep and the pit wall has several layers of thickness about 1–2 μm. About four such layers can be seen in Fig. 2.

The scanning electron micrographs of specimen 2 are shown in Figs. 3–5. This polishing of the specimen surface prior to α bombardment was relatively poor compared with that for specimen 1. The three scratches on the surface shown in Fig. 3 are due to mishandling of the specimen during mounting it on the irradiation flange. Contrary to specimen 1, specimen 2 shows pits of irregular shapes and sizes. Pits of large size (about 10–15 μm) and very small size (even less than 1 μm) are seen. One of the large pits is shown in Fig. 4. The pit formation process seems to be similar to that for specimen 1 and is due to plastic deformation caused by a high internal pressure and a local temperature rise which together resulted in exfoliation. Unlike the walls of the pits on specimen 1, the walls of these pits do not show a multilayered structure and the pits have no particular shape. A small (about 60 μm wide and 140 μm long) portion of the irradiated area shows the development of the sponge-like structure. The center region of the sponge-like structure spot is shown in Fig. 5. This region might be exposed to a high α dose that could have been due to the presence of a hot spot in the α beam, a typical cyclotron beam characteristic.

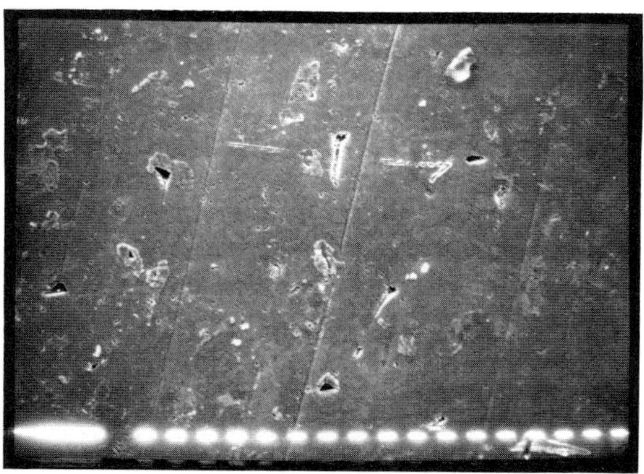

Fig. 3. Scanning electron micrograph of 47 MeV α-bombarded molybdenum (specimen 2). The white dashes represent 10 μm.

4. DISCUSSION

The above observations show some interesting features. It is hard to explain the mechanism involved in forming such structures. The sputtering phenomena cannot explain these observations, *i.e.* mainly the exfoliations around the pits, which indicate some type of internal pressure build-up near the surface, finally resulting in bursts. Also at high ion energies the experimental conditions are far from those of sputtering investigations [16].

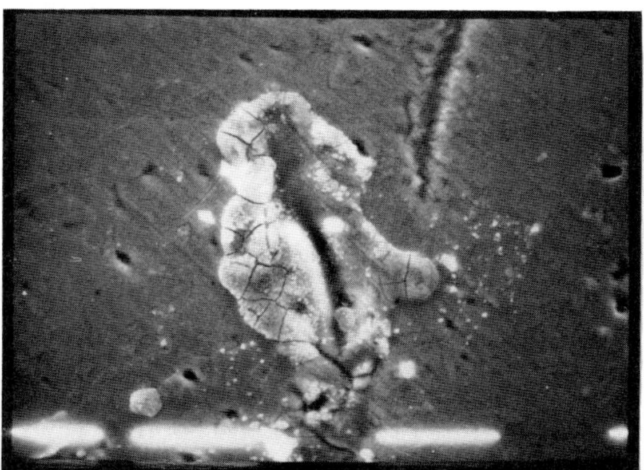

Fig. 4. Enlarged view of one of the pits shown in Fig. 3. The white dashes represent 10 μm.

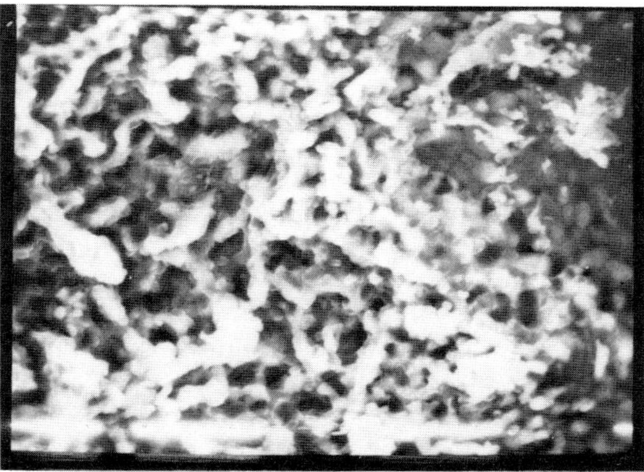

Fig. 5. Central region of the sponge-like structure on a small area of specimen 2. The white dashes represent 10 μm.

The helium gas-driven model is not appropriate for these observations because, under these irradiation conditions, helium bubble formation would not be expected near the surface regions. The mean ion depths, longitudinal and transverse straggling ranges [17] at the energies bombarded are given in Table 1. The mean ranges of the α particles are about 175 and 310 μm, and the straggling ranges are between 3 and 10 μm. This shows that the α particles are stopped within a narrow width at the end of their range (or are transmitted out of the specimen), leaving behind a high concentration of lattice defects throughout their path. So, near to the surface regions, we expect a high concentration of lattice defects and defect clustering with slight trapping of helium. These interpretations are supported by our positron annihilation studies [14] on 30 MeV α-irradiated molybdenum to a total dose of about 1.2×10^{17} $\alpha\ cm^{-2}$, irradiated under similar conditions. The positron studies indicate the presence of vacancies, dislocation–vacancy loops and microvoids at high concentrations in as-irradiated specimens and no evidence of helium bubbles except vacancy-rich helium–vacancy complexes. This indicates that the deformation of the irradiated surface might be the result of the compressive stress created by radiation-induced lattice defects, defect clusters and a small fraction of implanted helium that is not necessarily connected with the helium bubble pressure. These interpretations are consistent with that of bright field and dark field electron microscope images of the silicon layer implanted with iron ions [18], where anomalous "bulges" are seen because of the bowing of certain surface regions to accommodate the volume expansion accompanying the compressive stress of lattice damage and inclusion of additional atoms. The surface-roughening studies on ion-implanted nickel [19] have also shown good correlation between surface pits and internal voids and conclude that the voids act as nuclei for the pits. Unlike the situation in low energy ion implantations, the high energy ion bombardments create a high density of lattice defects, and the high probability of the displacement cascade damage process and the local heating thermal spikes within the collision cascades result in larger defect clusters [14]. Hence in the high energy ion implantation the surface morphology might be dominated by the defect structures and quite different from that in the low energy ion implantation.

The multilayered structure around the circular pits as shown in Fig. 2 may occur as a result of the same terrace structure observed in helium-implanted Inconel 625 [2] where up to 39 exfoliations were observed at high doses of about 3×10^{19} ions cm^{-2}. Similar observations [20] on helium- and heavy-ion-implanted muscovite mica combining annealing and etching, although indirect, have given rise to the interpretation that the multilayered structures are made up of point defects and defect clusters.

The regular circular shape of the pits obtained on specimen 1 is quite interesting. Circularly shaped pits are usually assumed to be due to the gas-driven mechanism. However, recent theoretical studies [21] have shown that, because of the strain created by lattice defects such as dislocations, pits of well-defined shape with a finite depth can form. The pit shape is determined by the surface energy, the strain energy of the dislocation, the dislocation density and the dislocation core size. Here the question arises whether the extended blister chain seen in Fig. 1 is due to the dislocation lines parallel to the surface. Contrary to specimen 1, the pits formed on specimen 2 have no regular shape. It should be noted that the surface polishing of specimen 2 was comparatively poor. This indicates that the reference surface conditions have an important bearing in determining the shape of the pits formed during ion implantation.

The sponge-like configuration seen in Fig. 5 is expected to be a result of the hot spot of the α beam. Such structures are usually formed at high temperatures [15] where vacancies are mobile and the vacancy concentration is high. At the hot spot of the α beam, both these conditions are favorable because of the high density of α particles. Our previous α irradiation studies (under similar conditions to those in the present investigation) with a thermo-sensor system indicate that the temperature of the molybdenum specimen in the hot-spot region could be as high as about 400 °C whereas the rest of the specimen temperature is maintained below 150 °C. The local temperature rise of a small surface area during ion implantation is possible. According to the SEM *in situ* observation of exfoliation on stainless steel and Inconel produced by 4 MeV helium ion bombardment [3], right after exfoliation the middle part of the skin appeared to be red hot. The estimated temperature of the exfoliated skin is about 800 °C, whereas the temperature rise of the bulk is only about 250 °C.

In conclusion, these observations (1) support the idea that the initiation of the fracture plane need not always be around the depth of the peak helium concentration [12] (rather it may develop randomly along the depth distribution of implantation), (2) indicate that, although the lateral stress in the implanted surface plays a small role in blister formation at low ion energies [6], at high energies the compressive stress and the elastic instabilities created by the lattice defects is important as the cascade displacement damage and the thermal spikes in the depleted zones are more probable in the near-surface regions and (3) reveal that the reference surface conditions have an important bearing in determining the shape of the pits formed during ion implantation.

ACKNOWLEDGMENTS

We thank Professor S. C. Sharma for useful discussions, Dr. B. Viswanathan for supplying the reference specimens and A. Sen Gupta and R. K. Bhandari for their help during the irradiation experiments. The crew of the Variable Energy Cyclotron Centre, Calcutta, is also acknowledged for the α irradiation experiments.

REFERENCES

1 J. Gittus, *Irradiation Effects in Crystalline Solids*, Applied Science, London, 1978, p. 484.
2 J. L. Whitton, C. H. Ming, U. Littmark and B. Emmoth, *Nucl. Instrum. Methods, 182-183* (1981) 291.
3 F. Paszti, G. Mezey, L. Pogany, M. Fried, A. Manuaba, E. Kotai, T. Lohner and L. Pocs, *Nucl. Instrum. Methods, 209-210* (1983) 1001.
4 V. M. Gusev, M. I. Guseva, Yu. V. Martinenko, A. N. Mansurova, V. N. Morozov and O. I. Chelnokov, *J. Nucl. Mater., 85-86* (1979) 1101; *Radiat. Eff., 40* (1979) 37.
5 B. Terreault, in H. Vensickel, R. Behrisch, B. M. U. Scherzer and F. Wagner (eds.), *Proc. 4th Int. Conf. on Plasma-Surface Interactions in Controlled Fusion Devices, Garmisch-Partenkirchen, 1980*, in *J. Nucl. Mater., 93-94* (1980) 707.
6 M. Kaminsky and S. K. Das, in R. K. Janev (ev.), *The Physics of Ionized Gases*, Symp. on the Physics of Ionized Gases, Dubrovnik, August 28-September 2, 1978, Institute of Physics, Beograd, 1978.
7 R. Behrisch, J. Bottiger, W. Eckstein, U. Littmark, J. Roth and B. M. U. Scherzer, *Appl. Phys. Lett., 27* (1975) 199.
8 M. Risch, J. Roth and B. M. U. Scherzer, *Proc. Int. Symp. on Plasma-Wall Interactions*, Pergamon, Oxford, 1977, p. 391.
9 E. P. EerNiss and S. T. Picraux, *J. Appl. Phys., 48* (1977) 9.
10 J. Roth, in G. Carter, J. S. Colligon and W. A. Grant (eds.), *Proc. Int. Conf. on Applications of Ion Beams to Materials, Warwick, 1975*, in *Inst. Phys. Conf. Ser., 28* (1976) 280.
11 A. Manuaba, F. Paszti, L. Pogany, M. Fried, E. Kotal, G. Mezey, T. Lohner, I. Lovas, L. Poes and J. Gyulia, *Nucl. Instrum. Methods, 199* (1982) 409.
12 D. Ghose, D. Basu and S. B. Karmohapatro, *J. Nucl. Mater., 125* (1984) 342.
13 S. V. Naidu and P. Sen, *Proc. Conf. on the Application of Accelerators in Research and Industry, Denton, TX, November 1986*, in *Nucl. Instrum. Methods*, to be published.
14 S. V. Naidu, A. Sengupta, R. Roy, P. Sen and R. K. Bhandari, *Philos. Mag. A, 52* (1985) 255.
15 J. H. Evans, R. Williamson and D. S. Whitmell, in F. A. Garner and J. S. Perrin (eds.), *Proc. 12th Int. Symp. on the Effects of Radiation on Materials*, in *ASTM Spec. Tech. Publ. 870*, 1985, p. 1225.
16 W. G. Johnston, J. H. Rosolowski, A. M. Turkalo and T. Lauritzen, in R. S. Nelson (ed.), *Physics of Irradiation Produced Voids, AERE Rep. R7934*, 1975, p. 101 (Atomic Energy Research Establishment).
17 U. Littmark and J. F. Ziegler, *Range Distributions for Energetic Ions in all Elements*, Vol. 6, Pergamon, Oxford, 1977.
18 L. E. Thailamani and M. C. Joshi, *Nucl. Instrum. Methods, 191* (1981) 87.
19 M. Ahmed, K. Ruffing and D. I. Potter, *Proc. Int. Conf. on Surface Modification of Metals by Ion Beams, Kingston, July 7-11, 1986*, in *Mater. Sci. Eng., 90* (1987).
20 J. C. Dran, Y. Langevin, J. C. Petit, J. Chaumont and B. Vassent, *Nucl. Instrum. Methods B, 1* (1984) 402.
21 D. J. Srolovitz and S. A. Safran, *Philos. Mag. A, 52* (1985) 793.

Materials Science and Engineering, *90* (1987) 51-54

Surface Texture Induced by Heavy Ion Bombardment: Ion Reflection Analysis*

J. BUDZIOCH and W. SOSZKA

Institute of Physics, Jagiellonian University, Krakow (Poland)

(Received July 10, 1986)

ABSTRACT

The structure of the surface of a polycrystalline platinum target bombarded with krypton ions was tested using the effect of ion reflection by atomic pairs. It was shown that the facets for which the local incidence angle is smaller than the incidence angle at which the knocking-out of one atom from the atomic pair is possible will not be etched. After a long ion irradiation time, only crystallites which have a prism shape with an eroded top will protrude.

1. INTRODUCTION

The problem of the ion etching of polycrystalline metal targets seems to be more complicated than that of crystalline samples. At the beginning of ion bombardment of a polycrystalline target the areas between grains are etched and then the planes of crystallites with a large sputtering coefficient (usually with a high index). These facets become smooth relative to the initial surface of the target. After ion irradiation for some time, if the incidence angle is not changed, the ion beam will be blocked by protruding grains so crystal planes with a high index are hardly "visible" to the ion beam. The crystal structure of the protruding crystallites will be dominant and the target surface can be treated as a textured surface. However, the fact that the surface irregularity has a dynamic character should be taken into account.

Thus the dynamic texture of a metallic target should be observed by the use of some effect connected with ion bombardment and sensitive to the crystal structure of the target. In the work reported in this paper the ion-scattering effect due to atomic pairs [1, 2] was chosen for this purpose, and a platinum surface under krypton ion (Kr^+) bombardment was examined.

2. EXPERIMENTAL DETAILS

The polycrystalline platinum target (purity, 99.99%) was mounted in the head of a goniometer (the geometrical arrangement of the experiment is presented in Fig. 1) placed in a vacuum chamber with a residual gas pressure of about 6×10^{-10} Torr. A monoenergetic ion beam formed by a tube collimator with an aperture of 1 mm and an angular divergence of about 0.3° hits the target surface with an incidence angle Ψ of between 5° and 65°. The primary energy of the krypton ions was 8 keV. The ion current was measured by a movable Faraday cup and was about 10 nA. The rotatable ion detector system consisted of a cylindrical energy analyser with an energy resolution of better than 1% and a channeltron multiplier.

Before the target was mounted in the vacuum chamber, its surface was mechanically polished. Inside the chamber the sample was cleaned by argon sputtering and annealing at a temperature above 1000 K.

3. RESULTS AND DISCUSSION

The energy spectra of 8 keV krypton ions (Kr^+) backscattered by the polycrystalline platinum target are presented in Fig. 1. The curve consists of the first maximum indicated by I and the second maxima indicated by II_A and II_B. It can be seen that the second maxima for the new spot are small and they rapidly increase with increasing ion bombardment time.

---

*Paper presented at the International Conference on Surface Modification of Metals by Ion Beams, Kingston, Canada, July 7-11, 1986.

0025-5416/87/$3.50

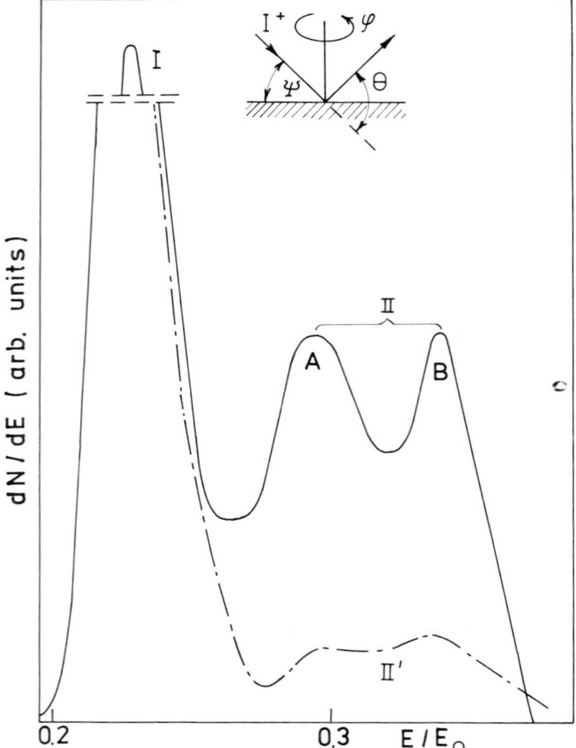

Fig. 1. Energy distributions of krypton ions (Kr^+) backscattered by platinum target bombardment with 8 keV ions to fluences of 5×10^{14} and 5×10^{17} ions cm^{-2} ($\theta_0 = 126°$; $\Psi = 40°$): —·—, new spot; ——, after ion etching.

According to ref. 1, the first maximum is considered to be a result of single scattering of ions and the second maximum, usually attributed to the crystalline target, is connected with scattering by atomic chains (Fig. 2). For heavy colliding atoms and large scattering angles the scattering by an atomic row can be treated as double scattering because the contribution from further atoms is negligible (Fig. 2).

Under the assumption that peak II_A (Fig. 1) corresponds to double scattering in the ⟨110⟩ direction of a platinum crystallite the incidence angle Ψ can be estimated and compared with the experimental value.

The energy position E/E_0 of each second peak is given by the following kinematic formula:

$$\frac{E}{E_0} = K(\theta_1)\,K(\theta_2)$$

where θ_1 and θ_2 are the scattering angles for the first and second collisions in the laboratory

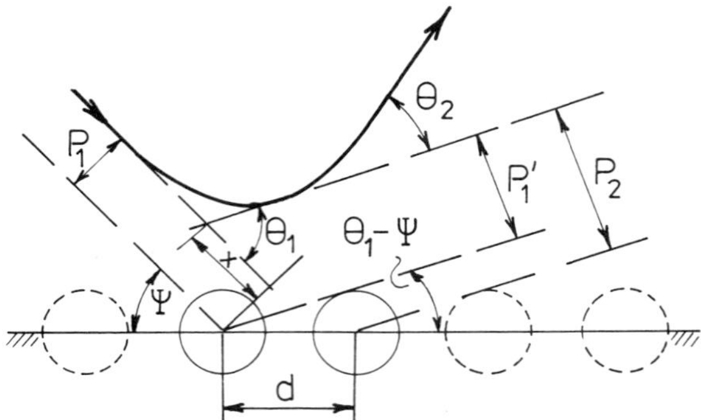

Fig. 2. Illustration of scattering by an atomic pair.

system. The function $K(\theta)$ can be expressed by

$$K(\theta) = \left(\frac{m_1}{m_1 + m_2}\right)^2 \left[\cos\theta + \left\{\left(\frac{m_2}{m_1}\right)^2 - \sin^2\theta\right\}^{1/2}\right]^2 \tag{1}$$

where m_1 is the mass of the projectile and m_2 is the mass of the target atom. The scattering angles θ_1 and θ_2 satisfy the condition

$$\tan\theta = \frac{\sin\chi}{m_1/m_2 + \cos\chi} \tag{2}$$

where χ is the scattering angle in the centre-of-mass system:

$$\chi = \pi - 2\int_{r_m}^{\infty} \frac{p\,dr}{r^2\{1 - V(r)/E_r - p^2/r^2\}^{1/2}} \tag{3}$$

where p is the impact parameter, $V(r)$ is the interaction potential at the internuclear distance r, r_m is the distance of closest approach and $E_r = m_2/(m_1 + m_2)E_0$. The correlated impact parameter p_2 for the second scattering depends on the scattering angle θ_1, the incidence angle Ψ and the distance d between atoms; it is given by the following expression [3]:

$$p_2 = p_1\cos\theta_1 + x\sin\theta_1 + d\sin(\theta_1 - \Psi) \tag{4}$$

where (Fig. 2)

$$x = \frac{2m_1}{m_1 + m_2} \times \left(r_m - \int_{r_m}^{\infty}\left[\left\{1 - \frac{V(r)}{E_r} - \frac{p^2}{r^2}\right\}^{-1/2} - 1\right]dr\right) + \frac{p_2(m_2 - m_1)}{m_1 + m_2}\tan\left(\frac{\chi}{2}\right) \tag{5}$$

The calculations were done under the assumption that the bombardment of platinum with krypton ions can be described by the Thomas–Fermi–Firsov potential

$$V(r) = \frac{A}{r^2}$$

where [4]

$$A = \frac{3.05 \times 10^{-20} Z_1 Z_2}{(Z_1^{1/2} + Z_2^{1/2})^{2/3}} \text{ eV m}^2$$

and Z_1 and Z_2 are the atomic numbers of the colliding particles. The results of the calculations are presented in Fig. 3. The experimental value of the energy of peak II_A corresponds to $\Psi_1 = 33.6°$ compared with the experimental value of the incidence angle Ψ_0 of 40°. The difference between the calculated value of the incidence angle and its experimental value can be explained by assuming that the surface of most crystallites is not the same as the initial surface of the target.

The following model of a platinum surface bombarded with krypton ions is proposed. At the beginning the area between grains, and the planes of grains with a high index, will be etched [5]. After ion bombardment for some time, those crystallites for which the {100} planes are almost parallel to the intial surface of the sample remain protruding. So for each grain the local incidence angle should be taken. Any azimuthal angle between the ⟨110⟩ direction of those planes and the scattering plane is possible. If the ion irradiation is continued and the incidence angle is not changed, then the etching of the protruding grains begins, but the etching rate depends on the azimuthal orientation and local incidence angle. For some incidence angles, penetration of atomic pairs by an ion is possible [2] and one atom from the atomic pair can be knocked out. For example, when the ⟨110⟩ surface atomic row is parallel to the scattering plane, the knocking-out of atoms from this row occurs at an incidence angle Ψ_p of about 34° or more. This mechanism seems to be a very effective component of the etching process.

If $\Psi < \Psi_p$, the double-scattering effect reduces ion penetration and it occurs over the maximum tilt angle α range for the most close-packed axes, which for platinum are ⟨110⟩. It is reasonable to assume that after a long ion irradiation time at $\Psi_0 = 40°$, for most of the protruding grains, α will be approximately $\Psi_0 - \Psi_p \approx 6°$, which corresponds to the experimental value of α_1 for peak II_A.

The hypothetical shape of the prism-like protruding part of the grains is illustrated in Fig. 4. It is assumed that furrows perpendicular to the scattering plane can be created on the eroded top surface of the grains. A series of {100} terraces will result from the greater probability of displacing an atom from the unprotected leading edge of each step. This effect will increase as α approaches Ψ_0, which may explain the optimum α value of 6° [6]. Because of this, the grains for which the ⟨100⟩surface atomic rows are parallel to the scattering plane ($\Psi_p \approx 25°$ and $\alpha \approx 15°$) will be etched very quickly. The structure can be confirmed directly from the energy position of the sharp peak II_B, which corresponds to two uncorrelated scatterings (*i.e.* from some distance apart) with the intermediate trajectory parallel to the initial surface $K(\theta_1)$ $K(\theta_2) = 0.346$. This would occur for atoms at successive step edges and a purely geomet-

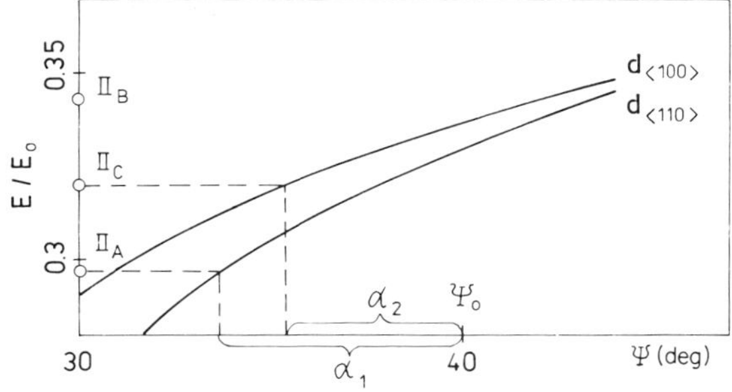

Fig. 3. Dependence of the energy of doubly scattered krypton ions (Kr^+) (on a platinum target) on the incidence angle Ψ for $d_{\langle 110 \rangle} = 2.77$ Å and $d_{\langle 100 \rangle} = 3.92$ Å.

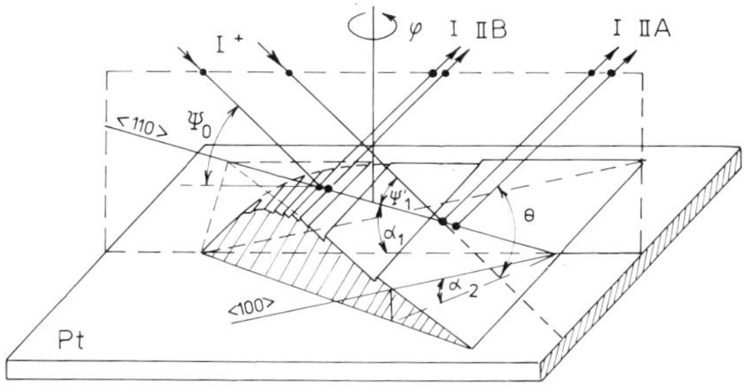

Fig. 4. Shape of the prism-like protruding part of one grain with an eroded top.

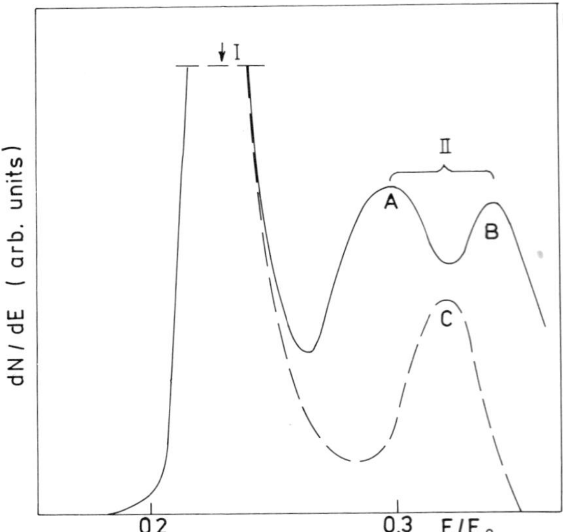

Fig. 5. Energy distributions of krypton ions (Kr^+) scattered by a platinum target for two values of the azimuthal angle ($\theta_0 = 126°$; $\Psi_0 = 40°$): ——, $\varphi = 0°$; — —, $\varphi = 45°$.

rical estimate of the furrow width gives $d_f = d_{\langle 100 \rangle}/(2 \sin \alpha) = 19$ Å [6].

The proposed model of a bombarded platinum surface was tested by azimuthal rotation of the target. In Fig. 5 a comparison of the ion energy spectra for $\varphi = 0°$ and $\varphi = 45°$ is shown. The characteristic feature of the energy spectra at $\varphi = 45°$ is the occurrence of only one peak connected with double scattering. At $\varphi = 45°$ the ⟨100⟩ surface atomic rows are parallel to the scattering plane and $\alpha_1 \to \alpha_2$ where $\alpha_2 = \arcsin(\sin \alpha_1 \cos \varphi) \approx 4.6°$. In this case the calculated energy E/E_0 after scattering by an atomic pair with $d = 3.92$ Å is $E/E_0 = 0.32$, consistent with the experimental position of peak II_C in Fig. 5. At $\varphi = 45°$ the distance d in the region of the furrowed surface of the grain is large and the probability of realization of two scatterings which give $\theta_0 = \theta_1 + \theta_2$ seems to be small. No peak originating from the furrowed area is observed in the experiment (Fig. 5).

ACKNOWLEDGMENT

This work was supported by Research Project CPB 01.06.08.02.

REFERENCES

1 E. S. Mashkova and V. A. Molchanov, *Medium-energy Ion Reflection from Solids*, North-Holland, Amsterdam, 1985.
2 E. S. Parilis, *Proc. 7th Int. Conf. on Phenomena in Ionized Gases, Belgrade, 1965.*
3 C. Lehmann, *Interaction of Radiation with Solids*, North-Holland, Amsterdam, 1977.
4 O. B. Firsov, *Zh. Eksp. Teor. Fiz.*, *36* (1959) 1617.
5 P. D. Towsend, J. C. Kelly and N. E. W. Hartley, *Ion Implantation, Sputtering and their Applications*, Academic Press, London, 1976.
6 P. D. Townsend, J. C. Kelly and N. E. W. Hartley, personal communication 1986.

Materials Science and Engineering, *90* (1987) 55–68

Ion-beam-induced Amorphization*

P. M. OSSI

Istituto di Ingegneria Nucleare, Centro di Studi Nucleari E. Fermi, Politecnico di Milano, Via Ponzio 34/3, 20133 Milan (Italy)

(Received July 10, 1986)

ABSTRACT

Ion beams are capable of inducing the formation of metastable amorphous phases in metallic systems, depending on the coupling between irradiation conditions, namely temperature, implanted species, energy and fluence, and target characteristics such as constituents, stoichiometry and structure. Experimental results obtained by the implantation technique are compared with those obtained by ion mixing, taking into account a number of systems extracted from the various classes of studied materials.

Material parameters mainly of an atomistic, thermodynamic and electronic nature have been used to devise several empirical criteria and rules to explain and possibly to predict the ease of glass formation; they are examined both to find unifying concepts and to correlate these with the features of leading viewpoints on ion-beam-induced amorphization processes.

An atomistic model of glass formation under irradiation, based on spike evolution and on preferential sputtering, is presented and discussed with reference to some experimental results.

1. INTRODUCTION

Amorphous materials are characterized by the high number of configurational degrees of freedom which in principle may be attained whenever the restrictions imposed by crystal symmetry are removed. In non-periodic systems, atoms can be arranged in spatial configurations with the only bounds imposed by electrical and chemical local fields experienced by single atoms in their environment. For metals and alloys, these result in short-range order (SRO) similar to that of the liquid state.

The space-varying set of boundary conditions allows for the many achievable local equilibrium conditions. These in turn are strongly influenced by the dynamic characteristics typical of various useful processes used to prepare amorphous solids, ranging from sputtering to chemical and physical vapour deposition, electrochemical deposition, ion implantation and ion mixing.

Metal glasses were first prepared by high fluence ion implantation in 1973 [1] while, in 1981 [2], intermetallic compounds were amorphized by ion mixing. In the last decade, new metastable alloys were prepared, and improved surface properties of materials obtained by ion beams were tested. Following high rate data accumulation, it was difficult to discriminate between complex ion–solid interaction phenomena in order to attribute material alterations to single well-defined mechanisms.

A typical experiment consists in bombarding a film with energetic ions which penetrate into it undergoing a number of energy loss collisions, both of elastic displacement and of inelastic ionizing nature, until they come to rest as implanted impurities at a depth of some hundreds of atom layers. After an elastic collision a recoil target atom collides with other atoms; many similar higher generation events give rise to cascade formation.

The total penetration depth depends on ion energy and nature, sample constituents, crystalline structure and orientation. The final distribution of implanted atoms corresponds to a gaussian, whose maximum coincides with the most probable ion range R_p. The target properties are affected in a complicated way by collision-induced elementary

---

*Paper presented at the International Conference on Surface Modification of Metals by Ion Beams, Kingston, Canada, July 7–11, 1986.

0025-5416/87/$3.50

processes, including defect production and interaction (for a recent review of the subject see ref. 3).

Depending on the experimental conditions, we can distinguish between ion implantation and ion mixing. In the former a mass-selected beam of chemically active species B impinges onto a crystalline elemental A film, so inducing the formation of A–B alloy of varying composition with depth, rarely exceeding 20 at.% B; otherwise, an A–B alloy sample of given stoichiometry is irradiated to modify the condition of near-surface layers. Ion energies are usually in the range 50–500 keV. The implanted (high) fluence, of the order of 10^{18} ions cm^{-2}, is calculated by integration of the ion current. Good precision in mass selection and vacuum conditions of usually better than 10^{-4} Pa result in a very clean process. Ion implantation is the only technique by which it is possible to correlate ion-induced processes (specifically amorphization) with the concentration of one of the system constituents.

Typical ion-mixing samples consist of alternate thin layers of two different elements evaporated or sputter deposited on a substrate, up to a total thickness of 100–200 nm. The alloy is formed by irradiation with inert gas species (usually Ar^+, Kr^+ or Xe^+) in the energy range 100–500 keV, at room temperature or at a low temperature, at comparatively low fluences (10^{13}–10^{16} ions cm^{-2}), the total retained amount not exceeding 0.5 at.%. The presence of interfaces between alternate layers in the sample is a favourable requisite for mixing; the upper limit to the thickness of the mixed region is given by R_p for the chosen ion species [4]. Ion–target interaction encompasses prompt events which last about one lattice vibration (*i.e.* 10^{-13}–10^{-12} s following the initial collision, typical ion-stopping time) and which are unaffected by thermal relaxation (athermal processes), as well as delayed processes (up to 10^{-10} s) which consist of rapid cooling and which are important for ion implantation and even more so for ion mixing. Such time scales permit the production of metastable phases unattainable by conventional thermal alloying processes (and often also by fast quenching techniques).

A unique characteristic of ion mixing is the availability of alloy compositions in practice over the complete binary composition range, the stoichiometry being determined by the thickness of the alternate layers and consequently constant over the entire mixed depth. Moreover, ion mixing overcomes the compositional limits set by sputtering effects, a major drawback in ion implantation experiments, and also reduces processing time.

In this review, experimental results on the formation of ion-implantation- and ion-mixing-induced amorphous metallic systems will first be examined, taking into account empirical rules, criteria and models proposed to explain glass formation under irradiation. A new model is then outlined, which is based on the features of a highly non-equilibrium state created by irradiation within the target.

2. AMORPHOUS METALLIC SYSTEMS BY ION IMPLANTATION

Without presenting a detailed list of results, we try to identify some systematic trends; the experimental conditions pertaining to a broad selection of amorphized metallic systems are summarized in Table 1, and we confine ourselves to discussing some cases in which amorphization is induced both by ion implantation and by ion mixing, picking out specific examples from the various classes of materials studied.

Metalloid elements m from the second and third row in the periodic table were implanted into pure metals M, mainly at room temperature [1, 5-9]. The choice is justified by the high (up to 20 at.%) concentrations achievable with such projectiles, which can lead to second-phase formation because of elevated packing efficiency. As prototype systems we consider the implantation of boron ions (B^+) in nickel [5] and boron ions (B^+) in iron [5, 8]. Low energies (30–50 keV) induce the formation of glassy alloys with stoichiometry near $M_{75}m_{25}$. Characterization of the phases obtained is based on transmission electron microscopy (TEM) observations, possibly complemented by measurements of electrical resistivity *vs.* temperature. Two conclusions result: first, some type of chemical interaction is needed to stabilize the disorder introduced by the ions in the target matrix, an idea supported by the impossibility of obtaining amorphous phases by pure metal self-irradiations, even at very high fluences; sec-

TABLE 1

Collection of experimental results for metallic systems amorphized by ion implantation

Projectile	*Energy* (keV)	*Target*	*Threshold fluence* ($\times 10^{16}$ ions cm^{-2})	*Damage* (dpa)	*Temperature* (K)	*Alloy Composition* (at.%)	*Reference*
As^+	40	Al	6	—	300	—	[1]
B^+	40	Fe; Co; Ni	3	—	300	$Fe_{75}B_{25}$; $Co_{75}B_{25}$; $Ni_{75}B_{25}$	[5]
P^+	40	Fe; Co; Ni	3	—	300	$Fe_{75}P_{25}$; $Co_{75}P_{25}$; $Ni_{75}P_{25}$	[5]
Sb^+	40	Ni	1–10	—	300	—	[6]
As^+	40	Ni	8	—	300	—	[6]
Bi^+	40	Ni	1–10	—	300	—	[6]
Sb^+	80	Co	0.1–10	—	300	$Co_{95}Sb_5$	[7]
Te^+	80	Co	0.1–10	—	300	$Co_{95}Te_5$	[7]
B^+	30; 50	Fe	3–5	—	300	$Fe_{75}B_{25}$; $Fe_{67}B_{33}$	[8]
P^+	30; 50	Fe	5	—	300	$Fe_{75}P_{25}$	[8]
Si^+	30; 50	Fe	1	—	300	—	[8]
P^+	100	Ni	≈10; ≈18	—	90; 300	$Ni_{85}P_{15}$; $Ni_{81}P_{19}$	[9]
Dy^+	20	Ni	1	—	300	$Ni_{90}Dy_{10}$	[10]
Ta^+	150	Cu	≈5	—	300	$Cu_{90}Ta_{10}$	[11]
Au^+	1400	Pt	≈10	—	300	$Pt_{85}Au_{15}$	[12]
W^+	50	Cu	5	—	40	$Cu_{90}W_{10}$	[13]
Ar^+	500–2000	$Zr_{75}Al_{25}$	0.05; 0.1	—	30; 285–693	[a]	[14]
Ar^+	4000	$Fe_{75}B_{25}$	—	0.15	300	[a]	[15]
Ni^+	3000	$Nb_{40}Ni_{60}$	[b]	—	300; 900	[a]	[16]
Ni^+	3000	$Nb_{40}Ni_{60}$	—	1–2	300; 900	$Nb_{40}Ni_{60}$; $Nb_{50}Ni_{50}$	[16]
Ni^{2+}	5000	$Mo_{50}Ni_{50}$	—	20	570	[a]	[17]
Ni^+	2500	$Ni_{33}Ti_{67}$; $Ni_{50}Ti_{50}$	—	0.5	300	[a]	[18]
Ni^+	2500	$Ni_{25}Al_{75}$	—	⩽10	300	[a]	[18]
Ni^+	2500	$Ti_{33}Fe_{67}$; $Ti_{50}Fe_{50}$	—	1	300	[a]	[18]
Ni^+	2500	$Mo_{50}Ni_{50}$	—	<10	300	[a]	[18]
Ni^+	2500	$Re_{75}Ta_{25}$	—	<10	300	[a]	[18]
Xe^+	300	$Ni_{25}Al_{75}$	—	0.4	≈100; 300	[a]	[19]
Ne^+	60	$Ni_{25}Al_{75}$	—	0.4	≈100; 300	[a]	[19]
e^-	1000	$Ni_{25}Al_{75}$	—	0.4	≈100; 300	[a]	[19]
Ni^+	50–250	$Ni_{50}Ti_{50}$	0.007	—	300	[a]	[20]
Si^+	50–250	$Ni_{50}Ti_{50}$	1	—	300	[a]	[20]
Ni^+	390	$Ni_{50}Ti_{50}$	0.0015	—	300	[a]	[21]
He^+	100	$Te_{90-10}Au_{10-90}$	≈0.05	—	250	[a]	[22]
Ar^+	350	$Au_{10-80}Bi_{90-20}$	—	—	<10	[a]	[23]
V^+	⩽400	Ti	—	—	<5	⩽10 V	[24]
Cr^+	⩽400	Ti	—	—	<5	⩽10 Cr	[24]
Fe^+	⩽400	Ti	—	—	<5	⩽10 Fe	[24]
Cu^+	⩽400	Ti	—	—	<5	⩽15 Cu	[24]
V^+	⩽400	Zr	—	—	<5	⩽10 V	[24]
Cr^+	⩽400	Zr	—	—	<5	⩽10 Cr	[24]
Mn^+	⩽400	Zr	—	—	<5	⩽10 Mn	[24]
Fe^+	⩽400	Zr	—	—	<5	⩽10 Fe	[24]
Ni^+	⩽400	Zr	—	—	<5	⩽10 Ni	[24]
Cu^+	⩽400	Zr	—	—	<5	⩽15 Cu	[24]
Cr^+	⩽400	Hf	—	—	<5	⩽10 Cr	[24]
Fe^+	⩽400	Hf	—	—	<5	⩽10 Fe	[24]
Cu^+	⩽400	Hf	—	—	<5	⩽15 Cu	[24]
B^+	20–160	Nb	—	—	77	$Nb_{92}B_8$	[25]
B^+	20–160	Mo	—	—	77	$Mo_{91}B_9$	[26]
Ar^+	275	α-Ga	0.02	—	<10	[a]	[27]
Ne^+	250	α-Ga	2	—	<10	[a]	[27]
Ar^+	250	$Al_{50}Au_{50}$	0.025	—	<10; 80	[a]	[28]
He^+	200	$Al_{50}Au_{50}$	2	—	<10	[a]	[28]

[a] Initial sample composition is unaltered after irradiation.
[b] Initially, amorphous samples do not recrystallize under ion bombardment.

ond, comparable values of threshold fluences, beyond which the increasingly disordered structure converts to the amorphous state, are found. A careful study on boron-implanted niobium [25] and molybdenum [26] shows that, after a first stage of lattice homogeneous deformation due to interstitial metalloid incorporation (up to 5 at.% B), transition to an amorphous phase occurs over a narrow boron concentration, around 9 at.%. Systematization of the results for M–m alloys [8] hinges on the Hägg size rule, similar to the Hume–Rothery rules, to decide whether a glass should form or a crystalline structure should be retained after ion implantation. Critical lower and upper values of the ratio R of the metalloid to metal atom radii are found, which define a glass formation region ($0.59 \leqslant R \leqslant 0.88$); the upper limit, however, is experimentally contradicted [6].

The amorphization process in nickel single crystals implanted with 100 keV phosphorus ions (P^+) was carefully analysed [9]. TEM observations show that, with increasing ion fluence, dislocation production increases until a dense network forms throughout the whole implanted layer. Such a structure is progressively removed with glass formation, which is completed at 15–20 at.% P concentration, according to X-ray diffraction. *In situ* Rutherford backscattering and channelling experiments at room temperature and at 90 K show no phosphorus mobility in nickel at low temperatures, short-range motions over a few interatomic distances occurring in implantation at room temperature. The level of induced disorder, up to amorphization, is analysed in terms of amorphous cluster formation and accumulation; at 90 K the cluster radius is about 1 nm (four interatomic distances), a typical value for an amorphous structure correlation length [29]. The fraction of glassy material present within layers implanted at low temperatures is 0.5 for $Ni_{88}P_{12}$ and 1 for $Ni_{85}P_{15}$. At room temperature the phosphorus concentration within a cluster corresponds to the eutectic (19 at.% P), with an amorphous fraction of 1. Chemical SRO due to solute atoms is thought to be an essential factor in stabilizing lattice disorder, the role of damage cascades being to provide a spectrum of geometrical and compositional fluctuations, among which elementary atomic assemblies are allowed to pick up a compatible equilibrium configuration. Similar observation and characterization of amorphous clusters have been obtained recently for aluminium implanted with nickel ions (Ni^+) [30].

Progressive structure disordering, coupled to extremely severe stress levels (about 10^{10} Pa), induces, as a result of damage accumulation, a free energy increase up to destabilization of the defective crystal lattice towards a completely disordered state. Such a transformation path has been followed in a highly homogeneous crystalline Fe_3B alloy bombarded with argon ions (Ar^+) at room temperature [15]; amorphization is induced at a level of 0.15 dpa, the lowest level reported until now. Also earlier metal implantation into metal [10–13], characterized by a very low solubility or immiscibility of the element couples, reveals the existence of a threshold in the implanted fluence near 10 at.%, separating metastable substitutional solid solutions from glassy alloys.

Contrasting results are obtained on the stability of the amorphous phase: amorphous $Nb_{40}Ni_{60}$ (a-$Nb_{40}Ni_{60}$) ribbons (recrystallization temperature $T_r \approx 910$ K) implanted at room temperature and at 900 K [16] remain non-crystalline; in contrast, nearly equiatomic crystalline $Mo_{\approx 50}Ni_{\approx 50}$ (c-$Mo_{\approx 50}Ni_{\approx 50}$) and a-$Mo_{\approx 50}Ni_{\approx 50}$ ($T_r > 1000$ K) becomes or remains amorphous at 570 K, crystalline structures being formed at 875 K [17].

The relation between the ease of glass formation and binary phase diagram features has attracted attention since a systematic study [18] of intermetallic compounds irradiated at room temperature with high energy nickel ions. $Ni_{50}Ti_{50}$, $Ni_{33}Ti_{67}$ and $Ni_{25}Al_{75}$ are amorphized, while both $Ni_{50}Al_{50}$ and $Ni_{75}Al_{25}$ remain crystalline. The irradiation temperature was kept below $0.3T_g/T_m$ (where T_m is the melting point) in order to fulfil a temperature criterion [31] originally found for amorphization in non-metallic systems. While the companion ionicity criterion [31] fails (an ionicity of 0.47 or less should imply glass formation) as all considered intermetallics (both crystalline and amorphous under bombardment) exhibit ionicities of 0.1 or less, the temperature criterion is met; such a factor expresses the overall thermal stability and is correlated to the observation that compounds with limited solubility tend to become amorphous during irradiation. Published data also support both the idea that a critical density of defects (and thus a critical energy density) must be reached to trigger a crystalline-to-glass transition and also the observation that an exiguous compositional range results in the retaining of basic compositional units and therefore in a high degree of SRO within an amorphous structure.

The effect of different phases (and in particular of a martensitic phase) on glass formation was studied in Ni–Ti [20], and the data indicated a transition path from martensite through austenite to an amorphous phase. Later experiments reveal direct amorphization of the initially present martensite, however; a study of amorphous volume increase *vs.* increasing ion fluence shows that expansion of the glassy zone is correlated not with the enlargement of the already present glassy nuclei but with an increase in their number. Computer simulation supports the attribution of the effect to a cascade overlap mechanism. The irradiation-induced amorphous microstructure is similar to that of vapour-deposited samples of the same composition [32]; this holds for first-neighbour coordination numbers (about 12) and for Ni—Ti, Ni—Ni and Ti—Ti bond lengths.

A striking demonstration of the dependence of the ease of glass formation on sample structure is given by the data on gallium [27]; irradiation of β-Ga with argon ions does not produce an amorphous state, while α-Ga transforms under identical conditions. β-Ga SRO is similar to liquid gallium SRO, thus resulting in lowered stability of the amorphous state with respect to nucleation of crystalline regions (triggered by the surrounding lattice) inside the glass structure. The retaining of amorphization in α-Ga is due to the difference between its SRO and that of liquid metal.

The same experiment throws light on the role of spikes in the glass-forming processes. The amorphization of α-Ga by low temperature (less than 10 K) argon ions(Ar^+) and neon ions(Ne^+) but not by helium ions(He^+) [27] shows that the same total energy deposition by atomic collisions does not produce identical results if massive or light projectiles are used. The inability of helium ions to induce amorphization is attributed to a cascade density which is too low in contrast with the high density of atomic displacements produced by argon and neon ions. Glass nucleation within dense ion cascades can be attributed to the existence and fast quenching of spikes.

After the first prompt event regime the spike volume, which encompasses some 10^3 atoms, is assumed to be a state of collective atomic excitation. The energy of single atoms (0.1–2 eV) is high enough for a strongly non-equilibrium hot state to persist but insufficient to cause further recoil collisions. Thus, collective energy sharing among neighbouring atoms occurs, lasting about 10^{-12}–10^{-10} s, followed by relaxation to the surrounding lattice via the usual thermal conduction processes. Typical spike temperatures are estimated to be of the order of 10^4 K, resulting in cooling rates near 10^{14} K s^{-1}. The resulting kinetic constraints are such that, if radiation-enhanced diffusion is inhibited in low temperature experiments, amorphization is strongly favoured. A further hypothesis concerning particle dynamics within a spike is suggested by considering the early stage of a heavy ion path through a solid [33–35]. At the high rate of energy loss per atomic layer traversed (about 10^2 eV), target atoms move with an elevated speed, the energy transport process being in the domain of shock wave propagation phenomena. The spike core is thought to act as a source of thin shock fronts continuously evolving, with rapid energy dissipation to the surrounding (isotropic) medium, until these decay into standard sound waves.

The interplay of temperature and projectile features determine the relative importance of a disorder-based or a spike-based model of glass formation; implantation of Al–Au [28] at low temperatures (less than 10 K) with argon or helium ions induces amorphization; at 80 K, even with a strongly increasing helium ion fluence, the alloy remains crystalline, while it is transformed by argon ions under identical conditions. It seems therefore that, provided that the temperature is low enough to freeze defects, spikes are unnecessary to produce amorphization, their role being essential at higher temperatures where defect mobility plays a role. The above results are confirmed for c-$Ni_{25}Al_{75}$ [19] irradiated at 100 K and at room temperature with xenon ions, neon ions on electrons; at low temperatures, 0.40 dpa damage for all projectiles induces amorphization while, at room temperature, xenon ions cause complete transformation, neon ions only partially amorphize the alloy and electrons are unable to produce even disordering of the c-$Ni_{25}Al_{75}$.

3. ION-MIXING-INDUCED AMORPHIZATION

Although ion mixing is quite a recent technique, it has been extensively applied to a large number of systems; quantitative experiments up to 1985 were reviewed together with the basic mechanisms in ref. 36. Table 2 presents a selection of experimental conditions and results. Incoming ions are responsible for many random walk relocations involving a relatively low energy; this displacement phase is followed by the relaxation of the spikes. To achieve amorphization, the irradiation temperature must be sufficiently low to suppress radiation-enhanced diffusion.

Among M–m systems, $Fe_{70}B_{30}$ and $Fe_{60}B_{40}$ [40] are amorphous at 77 K, while ambiguities are present for samples treated at room tem-

TABLE 2

Collection of experimental results for metallic alloys amorphized by ion mixing

Projectile	*Energy* (keV)	*Target*	*Threshold fluence* ($\times 10^{15}$ ions cm^{-2})	*Damage* (dpa)	*Temperature* (K)	*Alloy composition* (at.%)	*Reference*
Xe^+	100–300	Au–Co	⩽2	—	77	$Au_{75}Co_{25}$	[2]
Xe^+	100–300	Au–V	⩽2	—	77	$Au_{40}V_{60}$	[2]
Xe^+	250–500	Fe–W	5	—	300	$Fe_{70}W_{30}$	[37]
Xe^+	300	Al–Nb	4	—	300	$Al_{55}Nb_{45}$	[38]
Xe^+	300	Ni–Mo	7; 3; 17	—	300	$Ni_{65}Mo_{35}$; $Ni_{50}Mo_{50}$; $Ni_{35}Mo_{65}$	[38]
Xe^+	300	Ni–Nb	5	—	300	$Ni_{55}Nb_{45}$; $Ni_{35}Nb_{65}$	[38]
Xe^+	300	Mo–Ru	10	—	300	$Mo_{55}Ru_{45}$	[38]
Xe^+	300	Mo–Co	7; 10	—	300	$Mo_{35}Co_{65}$; $Mo_{65}Co_{35}$	[38]
Xe^+	300	Ti–Au	5; 4	—	300	$Ti_{65}Au_{35}$; $Ti_{40}Au_{60}$	[38]
Xe^+	300	Ti–Ni	5	—	300	$Ti_{50}Ni_{50}$	[38]
Xe^+	300	Er–Ni	5	—	300	$Er_{65}Ni_{35}$; $Er_{40}Ni_{60}$	[38]
Xe^+	300	Ni–Hf	1	—	300	$Ni_{80}Hf_{20}$; $Ni_{55}Hf_{45}$; $Ni_{20}Hf_{80}$	[39]
Xe^+	300	Fe–Si	5	—	77	$Fe_{77-30}Si_{23-70}$	[40]
Xe^+	300	Fe–B	5	—	77	$Fe_{70}B_{30}$; $Fe_{60}B_{40}$	[40]
Xe^+	300	Co–Nb	3	—	300	$Co_{90}Nb_{10}$; $Co_{86}Nb_{14}$ $Co_{75}Nb_{25}$	[40]
Kr^+	180	Ni–B	20	—	170	$Ni_{80}B_{20}$	[41]
Xe^+	100	Ni–Sn	5	—	300	$Ni_{75}Sn_{25}$	[42]
Xe^+	500	Pt–Al	0.2	—	300	$Pt_{33}Al_{67}$; $Pt_{67}Al_{33}$; $Pt_{75}Al_{25}$	[43]
Xe^+	500	Pd–Al	0.2	—	300	$Pd_{25}Al_{75}$; $Pd_{40}Al_{60}$; $Pd_{67}Al_{33}$	[43]
Xe^+	500	Ni–Al	0.2	—	300	$Ni_{25}Al_{75}$; $Ni_{40}Al_{60}$; $Ni_{75}Al_{25}$	[43]
Xe^+	340	Ni–Al	11	—	77	$Ni_{50}Al_{50}$	[44]
Kr^+	500	Ni–Al	0.5	—	80; 300	Various Al rich	[45]
Ar^+	150	Ti–Ni	1	—	300	$Ti_{50}Ni_{50}$	[46]
Au^+	1000	Ti–Ni	10	—	300	$Ti_{50}Ni_{50}$	[47]
Kr^+	350	Cu–Bi	0.05	—	10	Various	[48]
Kr^+	380	Au–Bi	<10	—	10	$Au_{10-90}Bi_{90-10}$	[48]
Xe^+	320	Co–Al	5	—	77	$Co_{50}Al_{50}$	[49]
Au^+	1000	Mo–Ni	8	—	300	$Mo_{50}Ni_{50}$	[50]
Xe^+	500	Fe–Zr	11	—	40	$Fe_{89}Zr_{11}$; $Fe_{87}Zr_{13}$; $Fe_{85}Zr_{15}$	[51]
Kr^+	500	Fe–Zr	20	—	40; 300; 320; 350	$Fe_{83}Zr_{17}$	[51]
Ar^+	500	Fe–Zr	70	—	40	$Fe_{85}Zr_{15}$	[51]
Xe^+	300	Zr–Ru	7; 20	—	300	$Zr_{75}Ru_{25}$; $Zr_{25}Ru_{75}$	[52]
Kr^+	750	Cu–Nb	21	—	6	$Cu_{50}Nb_{50}$	[53]
Kr^+	1000	Cu–Bi	5.7	—	6	$Cu_{50}Bi_{50}$	[53]
Xe^+	100–300	Al–Fe	10–50	—	300	$Al_{90-20}Fe_{10-80}$	[54]
Ar^+	100–300	Al–Fe	10–50	—	300	$Al_{90-20}Fe_{10-80}$	[54]
Xe^+	400	Al–Mn	10	—	300–500	$Al_{84}Mn_{16}$	[55]

perature; $Fe_{80}B_{20}$ and $Ni_{80}B_{20}$ [41] at room temperature are not amorphized, while a-$Ni_{80}B_{20}$ is obtained at 170 K. The qualitative difference between Fe–B and Ni–B alloys is attributed to better boron mixing with nickel than with iron (as shown by Auger electron spectroscopy) and to an Fe–B reaction resulting in chemical bonding. Thus a lower irradiation temperature is needed to induce the amorphization of $Fe_{80}B_{20}$ than that needed to induce the amorphization of $Ni_{80}B_{20}$.

Annealing studies show that mixed a-$Mo_{\approx 50}Ni_{\approx 50}$ [50] is stable up to about 873 K. The value of T_g may help to explain the impossibility of amorphizing the same alloy under direct implantation temperatures close to T_g [17]. Mixed a-Mo–Ni alloys exhibit better corrosion resistance than crystalline samples. The improvement in surface-sensitive properties of metallic systems, such as wear and corrosion resistance, has been verified in various studies; we mention only recent results on the Ni_{75}-Sn_{25} new phase [56] obtained so far only by ion-mixing [42].

Ni–Al alloys of various compositions at different temperatures and ion energies have been mixed [43, 44]. Amorphous alloys result whenever the composition does not allow formation of intermetallics with very simple crystal structures. Ion mixing, as well as the implantation of aluminium ions (Al^+) into nickel, the implantation of nickel ions (Ni^+) into aluminium and the irradiation of Al–Ni, results in amorphous structures or simple c-$Al_{50}Ni_{50}$ or a mixture of the two. Complex $Al_{75}Ni_{25}$, $Al_{60}Ni_{40}$ and $Ni_{75}Al_{25}$ are directly amorphized or decompose into mixtures of amorphous material and c-$Al_{50}Ni_{50}$. Also metastable crystalline phases (mainly in nickel-rich alloys and amorphous aluminium-rich alloys) have been observed which are not present in the equilibrium Al–Ni phase diagram [45]. A recent comparison between low temperature mixed equiatomic $Ni_{50}Al_{50}$ and $Fe_{50}Al_{50}$ [57] attributes amorphization of the former to the presence of very strong chemical order, the high chemical disordering energy leading to instability of the already induced disordered crystalline phase. Such a driving force is exiguous in $Fe_{50}Al_{50}$ which transforms under irradiation to a disordered but crystalline phase. The difficulty of obtaining a-$Fe_{50}Al_{50}$ alloys is confirmed by a rather critical dependence on irradiation dose and on stoichiometry [54], leading to an a-$Fe_{50}Al_{50}$ phase and to various metastable crystalline alloys, c-$Fe_{75}Al_{25}$, c-$Fe_{50}Al_{50}$ and c-$Fe_{25}Al_{75}$.

This set of experiments is in agreement with the glass formation criterion requiring that a large number of phases present in a binary phase diagram favours amorphization, and with another predictive rule by which, if the nucleation time of a given crystalline phase is shorter than the required cooling time, then that phase will form. The nucleation time strongly depends on the crystalline structure as it requires atomic mobility over a sufficient time to realize the regular spatial atomic arrangement of the crystalline unit cell.

Such concepts are part of a family of related ideas the ease of glass formation connecting with atomic properties of alloy constituents (such as the atomic size differences, the electronegativities and the elemental crystal structures) as well as with bulk alloy properties (such as the heat of formation, the Fermi level state densities, the number of structures of phases present in phase diagram, the relative constituent solubility, and the presence and depth of eutectics). A specific rule for glass formation in mixed binary metal systems is the "structural rule" [58]: it states that a necessary condition for a system to be amorphized is that elements entering the alloy have different crystal structures. The rule is strongly reminiscent of the eutectic criterion; both are based on the idea that crystallization of an A–B alloy is frustrated if A-rich regions, with solid A SRO may coexist with regions enriched in B, of nearly $A_{50}B_{50}$ composition. This happens near a eutectic, where solid stability is not significantly higher than liquid stability. The absence of intermediate equilibrium compounds thus appears to be a necessary condition. The structural rule encounters difficulties with some systems (*e.g.* Fe–Au) which probably require extremely low temperatures to achieve amorphization. However, the mixing of Al–Ni (both aluminium and nickel are f.c.c.) at 77 K [44] shows direct impact amorphization whilst the crystalline structure is retained at room temperature.

A predictive rule, contrasting with structural difference criterion, correlates the ratio of the atomic radii of the constituents with the heat ΔH of formation of a given alloy [59] calculated by Miedema's approach. When the data for binary alloys amorphized by ion mixing are arranged in a two-coordinate map, an empirical straight line separating the amorphous and crystalline regions can be drawn. Equilibrium ordered compounds are needed to favour the occurrence of an amorphous phase. Their presence is indicated by negative values of the heat of compound formation, which is observed to be associated with SRO development, usually excluding the possibility of solute–solute atoms as nearest neighbours. However, systematic experiments demonstrate [60] the possibility of inducing partial amorphization of systems with positive ΔH (Ag–Cu, Ag–Ni, Ag–Cr, Co–Cu, Cu–Fe, Fe–Mo and Fe–Nd), resulting in amorphous phases interspersed within high fluence mixed targets. Stoichiometry determination of the glassy zones invariably indicates narrow composition ranges, probably the origin of the difficulty in obtaining amorphous alloys in such systems. Also large solid solubilities, even in the presence of negative ΔH, show reduced glass formation (Mo–Ru). The requirement of a very limited solid solubility, already formulated for ion implantation experiments [18], appears to be crucial to ensure the ease of glass formation.

$Ni_{50}Ti_{50}$ equiatomic alloys have been prepared at room temperature [46] and the experiments identify a critical fluence for the onset of amorphization. The process is observed to coincide with mixing at the interface, which extends without prior formation of any dislocation structure and is attributed to the cooperative effect of atomic mixing, favoured by radiation-induced lattice destabilization. Again, significantly improved corrosion resistance of a-$Ni_{50}Ti_{50}$ was observed on quite thick samples [47].

Metastable superconducting alloys have been formed by ion mixing (and also by ion implantation); as an example, Au–Bi [48] in the concentration range 30 at.% $\leqslant C_{\text{Bi}} \leqslant$ 85 at.% is considered. Amorphization is induced at low temperatures (less than 6 K). The values of the superconducting transition temperature T_c of mixed alloys are nearly constant and independent of concentration, as the interface always presents an amorphous alloy sandwiched by pure gold and bismuth; T_c values are higher than the maximum values for implanted alloys [23]. A study on the rather similar Cu–Bi, Cu–Nb and Cu–Mo systems [53] shows that Cu–Bi irradiated at 10 K is amorphous, superconductive and very stable, complete recrystallization occurring at room temperature. Mixing is efficient at low temperatures, while it is in practice absent at room temperature. The same behaviour is displayed by Cu–Nb, both systems having a low solubility in the solid state, while they are miscible as liquids. For comparison, Mo–Cu, immiscible both in the solid and in the liquid state, can be mixed neither at low temperatures nor at room temperature. Chemical forces are thus confirmed as essential in the mixing process, together with atomic defect migration after their production in the first stage of the dynamic cascade.

4. A MODEL FOR ION-BEAM-INDUCED AMORPHIZATION

An atomistic model has been formulated to explain glass transitions under irradiation in a variety of metallic binary systems [61].

We take into account the dynamic character of the multiple ion–atom collision processes, considering energy spike evolution as well as strong surface stoichiometry alterations induced by high ion fluences [62, 63] (preferential sputtering). Bombardment leads to preferential segregation even at room temperature, the segregated species being that which is preferentially lost by sputtering. A connection exists with the processes of defect production and migration, giving rise to radiation-enhanced diffusion and to radiation-induced segregation [64]. A temperature interval, lying between $0.3T_m$ and $0.6T_m$, is found in which radiation-induced segregation is highly effective in producing enrichment of undersized solute atoms and depletion of oversized atoms near the sample surface.

For simplicity we consider, in the ion-mixing scheme, an A–B multilayered film of given initial composition (and assume that B is the solute); bombardment by inert gas ions is supposed to mix the layers efficiently. The spike concept is adopted and we assume that the spike volume is enclosed by surfaces marking a discontinuity in temperature and physical state with the surrounding lattice. The late stage of the spike life is studied, when the temperature is still sufficiently high to allow local radiation-induced segregation within the spike volume. We focus on the composition evolution of the interface region, being enriched in one component, as a consequence of segregation.

Also ion implantation may be considered. The essential feature of preferential segregation at some surface of one component is preserved even in low temperature experiments; indeed, evidence of segregation of the implanted solute into pure metal targets was found [65] at liquid nitrogen temperature, irrespective of the size misfit rule.

The choice of the segregating element is based on experimental evidence from radiation-induced segregation and preferential sputtering as well as thermally induced segregation. Agreement is generally found for a given alloy among various data; the segregating element is the same under different experimental conditions. The model assumes that the element experimentally found to segregate is that enriched at a spike surface; for example, let us suppose that B segregates.

By the valence electron localization degree model [66], as a consequence of stoichiometry alteration, the atom–atom interaction at the surface is represented by an electronic elementary charge transfer process between two atoms, one atom of the surface-enriched species B and one atom of A type. Such interaction is governed by the condition that the segregating element behaves in an electron acceptor way, thus giving

$$\begin{aligned} &A(-1e^-) \rightarrow C_{eff} \\ &B(+1e^-) \rightarrow D_{eff} \end{aligned} \qquad (1)$$

the C–D effective atom couple being considered as a microalloy cluster. For such microalloys, segregation of the effective alloy component corresponding to the segregating component in the original alloy is assumed (in the example, D (corresponding to B) is assumed to segregate with respect to C).

The electronic energy variation ΔE associated with the A–B → C–D transition is calculated [67].

The segregation behaviour of the elements in the original A–B and effective C–D alloys is tested [68] using the Miedema φ^* and $n_{WS}{}^{1/3}$ parameters. The quantity $\Delta\varphi^*$ is strictly connected to the difference between the surface energy γ_A of element A and the surface energy γ_B of element B:

$$\begin{aligned} \Delta\varphi^* &= \varphi_B{}^* - \varphi_A{}^* \\ &= a(\gamma_B - \gamma_A) + b\ \Delta n_{WS}{}^{1/3} \end{aligned} \qquad (2)$$

with

$$b = SN$$

and [69]

$$N = \frac{\bar{\varphi} - 0.6}{\bar{n}_{WS}{}^{1/3}}$$

the bars indicating average values of the corresponding quantities for element A and B.

The variation in segregation ability between the effective and original segregants is

$$\begin{aligned} \Delta(\Delta\varphi^*) &= \Delta\varphi_{eff}{}^* - \Delta\varphi_{orig}{}^* \\ &= (\gamma_D - \gamma_C) - (\gamma_B - \gamma_A) \end{aligned} \qquad (3)$$

and

$$\Delta(\Delta n_{WS}{}^{1/3}) = \Delta n_{WS}{}^{1/3}{}_{eff} - \Delta n_{WS}{}^{1/3}{}_{orig} \qquad (4)$$

Values for the φ^* and $n_{WS}{}^{1/3}$ parameters are taken from ref. 70.

We report in Tables 3 and 4 details of the calculations for glass-forming systems as well as for crystalline systems; the choice of the alloys is determined by the availability of surface segregation data. In Table 3, electronic energy variations ΔE at the original → effective alloy transitions are always positive, while they are invariably negative for crystalline systems (see Table 4).

In the tables a comparison with the eutectic criterion is also reported. This criterion reflects the thermodynamic and chemicophysical interactions influencing glass formation. In the *Eutectic* the word Yes is used for glassy systems in which localized stoichiometry changes are directed towards the bottom of a eutectic in the pertinent phase diagram region; otherwise, the word No is stated. For crystalline alloys, the word Yes corresponds to the absence of a eutectic in the region of interest or to departure from the eutectic bottom, while the word No stands for the other possibilities.

In Figs. 1 and 2, $\Delta(\Delta\varphi^*)$ is plotted *vs.* $\Delta(\Delta n_{WS}{}^{1/3})$ for both categories of materials; in the former case the points lie on the right-hand side of the line [68]

$$\Delta(\Delta\varphi^*) = \frac{5}{2} N \Delta(\Delta n_{WS}{}^{1/3}) \qquad (5)$$

indicating a reduced segregation ability in the effective alloys compared with the original

TABLE 3

Collection of experimental data and calculation details for metal glasses (all alloys except Pt–Au were obtained by ion mixing; Pt–Au was produced by gold implantation into platinum)

Original alloy	*Segregating element*	*Eutectic*[a]	*Interaction scheme*	*Effective alloy*	*Segre-gating element*	*Eutectic*[a]	ΔE (eV)
$Ni_{75}Al_{25}$ [43]	Ni [63, 72]	Yes	$Ni(+1e^-)$-$Al(-1e^-)$	Cu–Mg	Cu	No	+0.12
$Al_{75}Ni_{25}$ [43]	Al [73]	No	$Al(+1e^-)$-$Ni(-1e^-)$	Si–Co	Si	Yes	+0.26
$Pt_{85}Au_{15}$ [12]	Au [63, 74]	No	$Pt(-1e^-)$-$Au(+1e^-)$	Ir–Hg	Hg	?	+0.25
$Ni_{50}Mo_{50}$; $Ni_{65}Mo_{35}$ [52]	Ni [72]	Yes	$Ni(+1e^-)$-$Mo(-1e^-)$	Cu–Nb	Cu [75]	No	+0.15
$Ni_{65}Ti_{35}$ [46]	Ni [72]	Yes	$Ni(+1e^-)$-$Ti(-1e^-)$	Cu–Sc	Cu [75]	Yes	+0.30
$Bi_{>50}Cu_{<50}$ [48]	Bi [76]	Yes	$Bi(+1e^-)$-$Cu(-1e^-)$	Po–Ni[b]	Po[b]	No	+0.26
$Fe_{70}Si_{30}$; $Fe_{60}Si_{40}$ [40]	Si [77]	No	$Fe(-1e^-)$-$Si(+1e^-)$	Mn–P	P	Yes	+0.43
$Al_{75}Pd_{25}$[c] [43]	Al [73]	Yes	$Al(+1e^-)$-$Pd(-1e^-)$	Si–Rh	Si	?	+0.29
$Pd_{67}Al_{33}$[c] [43]	Al [62]	Yes	$Pd(-1e^-)$-$Al(+1e^-)$	Rh–Si	Si	?	+0.29
$Fe_{72}Zr_{28}$ [51]	Zr [78]	No	$Fe(-1e^-)$-$Zr(+1e^-)$	Mn–Nb	Nb [75]	?	+0.15
$Al_{67}Au_{33}$ [28]	Al [62]	Yes	$Al(+1e^-)$-$Au(-1e^-)$	Si–Pt	Si	Yes	+0.23
$Ni_{75}Sn_{25}$ [42]	Sn [79]	Yes	$Ni(-1e^-)$-$Sn(+1e^-)$	Co–Sb	Sb	No	+0.27
$Ni_{80}B_{20}$ [41]	B [41]	No	$Ni(-1e^-)$-$B(+1e^-)$	Co–C	C	No	+0.49

?, phase diagram not available.
[a]Phase diagram data from ref. 71.
[b]No segregation analysis for the Bi–Cu → Po–Ni transition, as the φ^* and $n_{WS}{}^{1/3}$ values are not available for polonium.
[c]The calculated ΔE for $Al_{75}Pd_{25}$ is equal to that for $Pd_{67}Al_{33}$, as the microalloys obtained coincide in the two cases because of aluminium segregation when both the solvent and the solute are in the original alloys.

TABLE 4

Collection of experimental data and calculation details for crystalline metallic systems formed via ion mixing

Original alloy	*Segregating element*	*Eutectic*[a]	*Interaction scheme*	*Effective alloy*	*Segre-gating element*	*Eutectic*[a]	ΔE (eV)
$Ag_{80}Cu_{20}$; $Ag_{60}Cu_{40}$ [80]	Ag [81]	Yes	$Ag(+1e^-)$-$Cu(-1e^-)$	Cd–Ni	Cd	No	−0.44
$Cu_{83}Ag_{17}$; $Cu_{80}Ag_{20}$; $Cu_{60}Ag_{40}$ [80]	Cu [72]	Yes	$Cu(+1e^-)$-$Ag(-1e^-)$	Zn–Pd	Zn	Yes	−0.80
$Al_{>50}Cu_{<50}$ [82]	Al [62]	Yes	$Al(+1e^-)$-$Cu(-1e^-)$	Si–Ni	Si	No	−0.26
$Ag_{80}Ni_{20}$; $Ag_{60}Ni_{40}$[b] [83]	Ag [84]	Yes	$Ag(+1e^-)$-$Ni(-1e^-)$	Cd–Co	Cd	?	−0.13
$Ni_{80}Ag_{20}$; $Ni_{60}Ag_{40}$[b] [83]	Ag [84]	Yes	$Ni(-1e^-)$-$Ag(+1e^-)$	Co–Cd	Cd	?	−0.13
$Pt_{60}Fe_{40}$ [85]	Fe [86]	Yes	$Pt(-1e^-)$-$Fe(+1e^-)$	Ir–Co	Co [75]	Yes	−0.10
$Au_{75}Ni_{25}$ [87]	Au [62]	Yes	$Au(+1e^-)$-$Ni(-1e^-)$	Hg–Co	Hg	?	−0.17
$Ni_{75}Au_{25}$ [87]	Ni [72]	Yes	$Ni(+1e^-)$-$Au(-1e^-)$	Cu–Pt	Cu [75, 88]	?	−0.31

?, phase diagram not available
[a]Phase diagram data from ref. 71.
[b]The cases of $Ag_{80}Ni_{20}$ and $Ag_{60}Ni_{40}$ and of $Ni_{80}Ag_{20}$ and $Ni_{60}Ag_{40}$ are analogous to those of $Al_{75}Pd_{25}$ and $Pd_{67}Al_{33}$ in Table 3.

alloys; moreover, solute segregation results in positive $\Delta(\Delta\varphi^*)$ and $\Delta(\Delta n_{WS}{}^{1/3})$, parameters which are negative for solvent segregation.

For crystalline systems (Fig. 2) the opposite holds; points lie on the left-hand side of the line given by eqn. (5), with enhanced segregation ability in effective alloys. $\Delta(\Delta\varphi^*)$ and $\Delta(\Delta n_{WS}{}^{1/3})$ are both negative for solute segregation and positive for solvent segregation.

Figures 3 and 4 display plots of $\Delta(\Delta\varphi^*)$ *vs.* ΔE for both categories of materials. A linear relation holds; surface energy variations

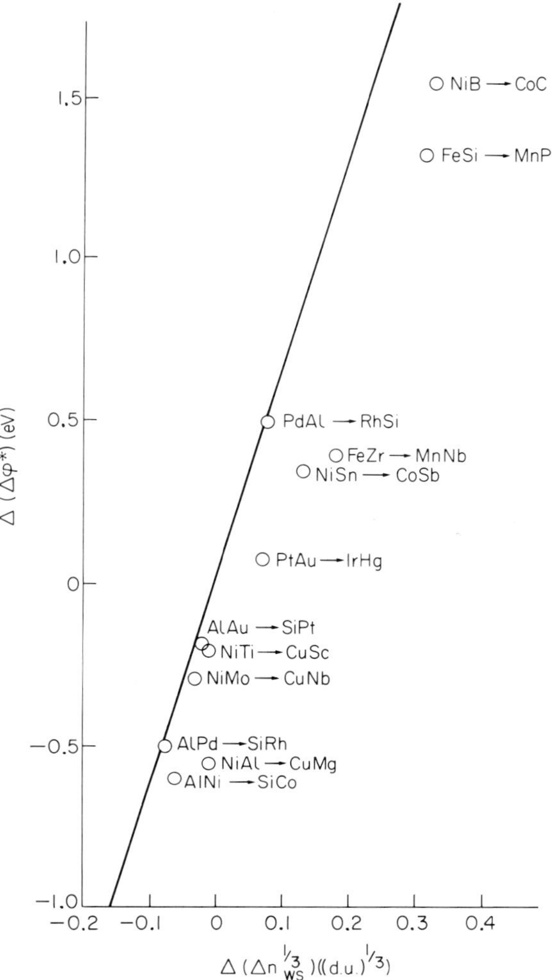

Fig. 1. Calculated values of $\Delta(\Delta\varphi^*)$ *vs.* $\Delta(\Delta n_{ws}{}^{1/3})$ for original → effective alloy transitions in metal glasses (d.u.,):

$\Delta(\Delta\varphi^*) = +2.60 \times \frac{5}{2}\Delta(\Delta n_{WS}{}^{1/3})$

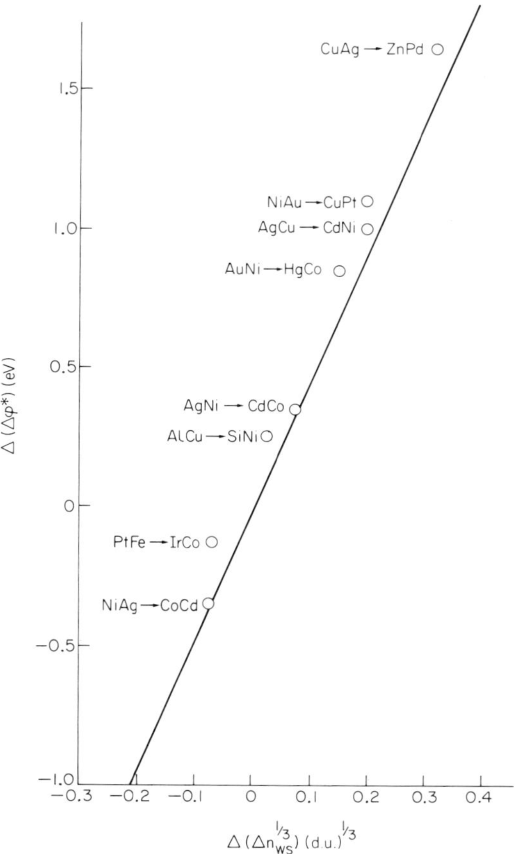

Fig. 2. Calculated values of $\Delta(\Delta\varphi^*)$ *vs.* $\Delta(\Delta n_{WS}{}^{1/3})$ for original → effective alloy transitions in crystalline systems:

$\Delta(\Delta\varphi^*) = +1.91 \times \frac{5}{2}\Delta(\Delta n_{WS}{}^{1/3})$

are connected to electronic energy changes due to charge transfer processes. $\Delta(\Delta\varphi^*)$ expresses the variation in interaction strength between effective alloy atoms with respect to that between original alloy atoms, while ΔE shows the energy contribution needed to create an effective atom couple. We observe a striking regularity of the observed behaviour of the different classes of systems which may be explained in terms of current concepts on glass formation and stability.

A positive ΔE contribution in amorphous alloys indicates a reduced system stability concerning both chemical and structural atomic arrangements; a negative surface energy contribution $\Delta(\Delta\varphi^*)$ obtained for solvent segregation implies a reduction in the surface energy component of effective microalloys with respect to original alloys and thus the atomic surface mobility increases. There is tendency towards local lowering of atomic coordination numbers, typically towards those of a liquid, with many equivalent configurations. Fast quenching of the highly uncorrelated atomic motions favours glass formation; the metastable microalloys are considered to be amorphization nuclei.

A positive $\Delta(\Delta\varphi^*)$ value for solute segregation indicates a surface energy increase and thus a lowered atomic mobility; this should oppose glass formation but the exiguous number of available solute atoms over the typical lifetime of a spike renders irrelevant the probability of spatial correlation among effective

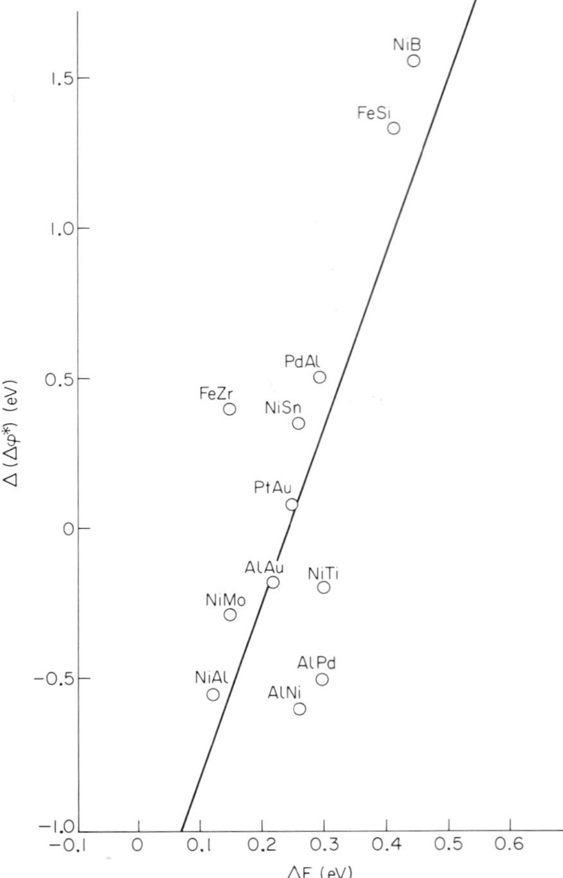

Fig. 3. Calculated values of $\Delta(\Delta\varphi)$ *vs.* ΔE for original → effective alloy transitions in metal glasses:

$\Delta(\Delta\varphi^*) = -1.37 + 5.70\ \Delta E$

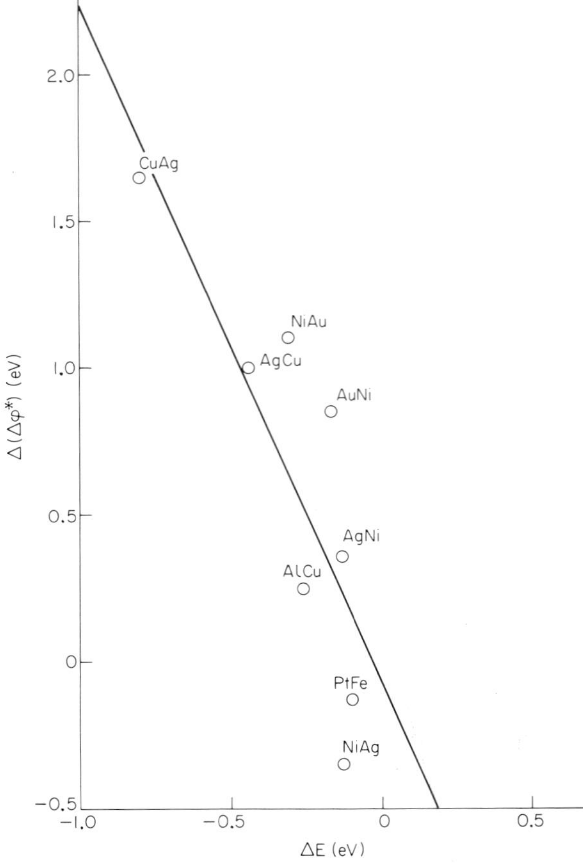

Fig. 4. Calculated values of $\Delta(\Delta\varphi^*)$ *vs.* ΔE for original → effective alloy transitions in crystalline systems:

$\Delta(\Delta\varphi^*) = -0.08 - 2.32\ \Delta E$

pairs which could act as recrystallization nuclei.

A negative electronic energy variation ΔE shows that, under bombardment, atomic rearrangements of the alloys via charge transfer are directed towards the enhanced stability typical of the retaining of crystalline structure.

An increasing surface energy when the solvent segregates indicates stronger interatomic correlation to achieve a static configuration; because of the rather high density of effective atom pairs present, the mechanism may contribute significantly to induce crystallization. On the contrary, a negative $\Delta(\Delta\varphi^*)$ term which occurs when solute segregates implies, as already explained, an enhanced atomic mobility opposing crystallization; the low density of amorphous phase nuclei is then thought to be ineffective against overall system stabilization.

ACKNOWLEDGMENTS

Useful discussions with Professor E. Lüscher, Technischen Universität München, on modelling amorphization processes and amorphous solid structures are gratefully acknowledged.

The author is indebted to Professor L. Guzman, Istituto per la Ricerca Scientifica e Tecnologica, Trento, for stimulating discussions on ion implantation- and ion-mixing-induced material modifications and to Dr. A. Cavalleri and Dr. Ing. F. Giacomozzi for valuable comments.

This work was supported financially by the Ministero Pubblica Istruzione.

REFERENCES

1 P. V. Pavlov, E. I. Zorin, D. I. Tetelbaum, V. P. Lesnikov, G. M. Ryzhkov and A. V. Pavlov, *Phys.*

Status Solidi A, 19 (1973) 373.

2 B. Y. Tsaur, S. S. Lau, L. S. Hung and J. W. Mayer, *Nucl. Instrum. Methods, 182–183* (1981) 67.

3 B. R. Appleton, in J. S. Williams and J. M. Poate (eds.), *Ion Implantation and Beam Processing*, Academic Press, New York, 1984, p. 189.

4 J. W. Mayer, B. Y. Tsaur, S. S. Lau and L. S. Hung, *Nucl. Instrum. Methods, 182–183* (1983) 1.

5 A. Ali, W. A. Grant and P. J. Grundy, *Philos. Mag. B, 37* (1978) 353.

6 W. A. Grant, *Nucl. Instrum. Methods, 182–183* (1981) 809.

7 W. Z. Li, Z. Al-Tamini and W. G. Grant, *5th Int. Conf. on Ion Beam Modification of Materials, Catania, June 9–13, 1986, Nucl. Instrum. Methods B*, (1987) 566.

8 B. Rauschenbach and K. Hohmuth, *Phys. Status Solidi A, 72* (1982) 667.

9 C. Cohen, A. Benyagoub, H. Bernas, J. Chaumont, L. Thomè, M. Berti and A. V. Drigo, *Phys. Rev. B, 31* (1985) 5.

10 R. Andrew, W. A. Grant, P. J. Grundy, J. S. Williams and L. T. Chadderton, *Nature (London), 262* (1976) 380.

11 A. G. Cullis, J. A. Borders, J. K. Hirvonen and J. M. Poate, *Philos. Mag. B, 37* (1978) 615.

12 S. P. Singhal, H. Herman and J. K. Hirvonen, *Appl. Phys. Lett., 32* (1978) 25.

13 A. G. Cullis, J. M. Poate and J. A. Borders, *Appl. Phys. Lett., 28* (1976) 316.

14 L. M. Howe and M. H. Rainville, *J. Nucl. Mater., 68* (1977) 215.

15 A. E. Berkowitz, W. G. Johnston, A. Mogro-Campero, J. L. Walter and H. Bakhru, in S. T. Picraux and W. J. Choyke (eds.), *Metastable Materials Formation by Ion Implantation, Materials Research Society Symp. Proc.*, Vol. 7, Elsevier, New York, 1982, p. 195.

16 M. D. Rechtin, J. Vander Sande and P. M. Baldo, *Scr. Metall., 12* (1978) 639.

17 J. L. Brimhall, L. A. Charlot and R. Wang, *Scr. Metall., 13* (1979) 217.

18 J. L. Brimhall, H. E. Kissinger and L. A. Charlot, *Radiat. Eff., 77* (1983) 273.
J. L. Brimhall and E. P. Simonen, *Nucl. Instrum. Methods B, 16* (1986) 187.

19 M. Nastasi, J. M. Williams, E. A. Kenik and J. W. Mayer, *5th Int. Conf. on Ion Beam Modification of Materials, Catania, June 9–13, 1986, Nucl. Instrum. Methods B*, (1987) 543.

20 P. Moine, J. P. Eymery, R. J. Gaboriaud and J. Delafond, *Nucl. Instrum. Methods, 209–210* (1983) 267.

21 P. Moine, J. P. Riviere, M. O. Ruault, J. Chaumont, A. Pelton and R. Sinclair, *Nucl. Instrum. Methods B, 7–8* (1985) 20.

22 J. D. Meyer and B. Stritzker, *Z. Phys. B, 36* (1979) 47.

23 A. Wolthuis and B. Stritzker, *J. Phys. (Paris), Colloq. C5, 44* (1983) 489.

24 J. D. Meyer and B. Stritzker, *Z. Phys. B, 54* (1983) 25.

25 G. Linker, *Mater. Sci. Eng., 69* (1985) 105.

26 G. Linker, *5th Int. Conf. on Ion Beam Modification Materials, Catania, June 9–13, 1986, Nucl. Instrum. Methods B*, (1987) 526.

27 M. Holz, P. Ziemann and W. Buckel, *Phys. Rev. Lett., 51* (1983) 1584.

28 A. Schmid and P. Ziemann, *Nucl. Instrum. Methods B, 7–8* (1985) 581.

29 J. Wong and H. H. Liebermann, *Phys. Rev. B, 29* (1984) 651.

30 A. V. Drigo, M. Berti, L. Thomè, J. Chaumont, H. Bernas, A. Benyagoub, C. Cohen, J. C. Pivin and F. Pons, *5th Int. Conf. on Ion Beam Modification of Materials, Catania, June 9–13, 1986, Nucl. Instrum. Methods B*, (1987) 533.

31 H. M. Naguib and R. Kelly, *Radiat. Eff., 25* (1977) 1.

32 D. Bodin, P. Moine and J. P. Eymery, *5th Int. Conf. on Ion Beam Modification of Materials, Catania, June 9–13, 1986.*

33 G. Carter, *Radiat. Eff. Lett., 43* (1979) 193.

34 G. Carter, *Radiat. Eff. Lett., 50* (1980) 105.

35 B. Rauschenbach and K. Hohmuth, *Phys. Status Solidi A, 75* (1983) 159.

36 B. M. Paine and R. S. Averback, *Nucl. Instrum. Methods B, 7–8* (1985) 666.

37 G. Göltz, R. Fernandez, M. A. Nicolet and D. J. K. Sadana, in S. T. Picraux and W. J. Choyke (eds.), *Metastable Materials Formation by Ion Implantation, Materials Research Society Symp. Proc.*, Vol. 7, Elsevier, New York, 1982, p. 227.

38 B. X. Liu, W. L. Johnson, M. A. Nicolet and S. S. Lau, *Nucl. Instrum. Methods, 209–210* (1983) 229.

39 M. Van Rossum, V. Shreter, W. L. Johnson and M. A. Nicolet, in G. K. Hubler, O. W. Holland, C. R. Clayton and C. W. White (eds.), *Ion Implantation and Ion Beam Processing of Materials, Materials Research Society Symp. Proc.*, Vol. 27, Elsevier, New York, 1984, p. 127.

40 M. Van Rossum, M. A. Nicolet and C. H. Wilts, *J. Appl. Phys., 56* (1984) 1032.

41 M. Elena, L. Guzman, G. Giunta, F. Marchetti, P. M. Ossi, G. Riontino, and V. Zanini, in E. Lüscher and G. Fritsch (eds.), *Proc. Conf. on Amorphous and Liquid Materials*, in *NATO Adv. Study Inst. Ser.*, to be published.

42 L. Calliari, L. M. Gratton, L. Guzman, G. Principi and C. Tosello, in G. K. Hubler, O. W. Holland, C. R. Clayton and C. W. White (eds.), *Ion Implantation and Ion Beam Processing of Materials, Materials Research Society Symp. Proc.*, Vol. 27, Elsevier, New York, 1984, p. 85.

43 L. S. Hung, M. Nastasi, J. Gyulai and J. W. Mayer, *Appl. Phys. Lett., 42* (1983) 672.

44 C. Jaouen, J. P. Riviere, R. J. Gaboriaud and J. Delafond, in M. Von Allmen (ed.), *Amorphous Metals and Nonequilibrium Processing*, Les Editions de Physique, Paris, 1984, p. 117.

45 J. Eridon, L. Rehn and G. Was, *5th Int. Conf. on Ion Beam Modification of Materials, Catania, June 9–13, 1986, Nucl. Instrum. Methods B*, (1987) 626.

46 K. Saito and M. Iwaki, *J. Appl. Phys., 55* (1984) 4447.

47 R. S. Bhattacharya, A. K. Rai and P. P. Pronko, *Mater. Lett., 2* (1984) 483.

48 B. Stritzker, in M. Von Allmen (ed.), *Amorphous Metals and Nonequilibrium Processing*, Les Editions de Physique, Paris, 1984, p. 141.
49 C. Jaouen, J. P. Riviere, A. Bellara and J. Delafond, *Nucl. Instrum. Methods B, 7–8* (1985) 591.
50 A. K. Rai, R. S. Bhattacharya, A. W. McCornik, P. P. Pronko and M. Kholaib, *Appl. Surf. Sci., 21* (1985) 95.
51 J. Bøttiger, N. J. Mikkelsen, S. K. Nielsen, G. Weyer and K. Pampus, *J. Non-Cryst. Solids, 76* (1985) 303.
52 B. X. Liu, *Nucl. Instrum. Methods B, 7–8* (1985) 547.
53 R. S. Averback, D. Peak and J. L. Thompson, *Appl. Phys. A, 39* (1986) 59.
54 B. Rauschenbach and K. Hohmuth, *5th Int. Conf. on Ion Beam Modification of Materials, Catania, June 9–13, 1986.*
55 D. M. Follstaedt and J. A. Knapp, *5th Int. Conf. on Ion Beam Modification of Materials, Catania, June 9–13, 1986, Nucl. Instrum. Methods B,* (1987) 611.
56 L. Fedrizzi, L. Guzman, A. Molinari, S. Girardi and P. L. Bonora, *Nucl. Instrum. Methods B, 7–8* (1985) 711.
57 C. Jaouen, J. P. Riviere and J. Delafond, *5th Int. Conf. on Ion Beam Modification of Materials, Catania, June 9–13, 1986.*
58 B. X. Liu, W. L. Johnson and M. A. Nicolet, *Appl. Phys. Lett., 42* (1983) 45.
59 J. A. Alonso and S. Simozar, *Solid State Commun., 48* (1983) 765.
60 B. X. Liu, E. Ma, L. J. Huang and J. Li, *5th Int. Conf. on Ion Beam Modification of Materials, Catania, June 9–13, 1986.*
61 P. M. Ossi, *I Congr. Naz. di Fisica della Materia, Genova, June 24–27, 1986.*
62 R. Kelly, in R. Wanselow and R. Howe (eds.), *Chemistry and Physics of Solid Surfaces V*, Springer, Berlin, 1984, p. 195.
63 R. Kelly, *Surf. Sci., 100* (1980) 85.
64 L. E. Rehn, R. S. Averback, P. R. Okamoto, *Mater. Sci. Eng., 69* (1985) 1.
65 L. Kornblit, A. R. Zamorrodian, S. Tougaard and A. Ignatiev, *Radiat. Eff., 91* (1985) 97.
66 P. M. Ossi, *Mater. Sci. Eng., 77* (1985) L5.
67 F. Herman and S. Skillman, *Atomic Structure Calculations*, Prentice Hall, Englewood Cliffs, NJ, 1963.
68 J. C. Hamilton, *Phys. Rev. Lett., 42* (1979) 989.
69 A. R. Miedema, *Z. Metallkd., 69* (1978) 287.
70 A. K. Niessen, F. R. de Boer, R. Boom, P. F. de Châtel, W. C. M. Mattens and A. R. Miedema, *Calphad, 7* (1983) 51.
71 W. G. Moffatt, *The Handbook of Binary Phase Diagrams*, Vols. I–III, General Electric Co., Schenectady, NY, 1978.
72 L. E. Rehn, in S. T. Picraux and W. J. Choyke (eds.), *Metastable Materials Formation by Ion Implantation, Materials Research Society Symp. Proc.*, Vol. 7, Elsevier, New York, 1982, p. 17.
73 L. Z. Mezey and J. Giber, *Surf. Sci., 162* (1985) 514.
74 T. T. Tsong, Ng. S. Yee and S. B. McLane, Jr., *J. Chem. Phys., 73* (1980) 1464.
75 J. R. Chelikowsky, *Surf. Sci., 139* (1984) L197.
76 V. Vitek and J. G. Wang, *Surf. Sci., 144* (1984) 110.
77 J. Du Plessis and P. E. Viljoen, *Surf. Sci., 131* (1983) 321.
78 R. S. Polizzotti and J. J. Burton, *J. Vac. Sci. Technol., 14* (1977) 347.
79 L. Guzman, personal communication, 1986.
80 B. Y. Tsaur, S. S. Lau and J. W. Mayer, *Appl. Phys. Lett., 36* (1980) 823.
81 P. Braun and W. Farber, *Surf. Sci., 47* (1975) 57.
82 F. Besenbacher, J. Bøttiger, S. K. Nielsen and H. J. Whitlow, *Appl. Phys. A, 29* (1982) 141.
83 B. Y. Tsaur and J. W. Mayer, *Appl. Phys. Lett., 37* (1980) 389.
84 J. Fine, T. D. Andreadis and F. Davarya, *Nucl. Instrum. Methods 209–210* (1983) 521.
85 G. Battaglin, A.. Carnera, V. N. Kulkarni, S. Lo Russo, P. Mazzoldi and G. Celotti, *Thin Solid Films, 131* (1985) 69.
86 J. J. Burton and R. S. Polizzotti, *Surf. Sci., 66* (1977) 1.
87 B. Y. Tsaur, S. S. Lau, L. S. Hung and J. W. Mayer, *Nucl. Instrum. Methods, 182–183* (1981) 67.
88 H. H. Andersen, V. Chernysh, B. Stenum, T. Sørensen and H. J. Whitlow, *Surf. Sci., 123* (1982) 39.

Amorphization of Intermetallic Compounds by Ion Bombardment*

D. F. PEDRAZA

Metals and Ceramics Division, Oak Ridge National Laboratory, Oak Ridge, TN 37831 (U.S.A.)

(Received July 10, 1986)

ABSTRACT

A theory of amorphization of ordered intermetallic alloys subject to electron and ion bombardment has been previously developed. It was proposed that a basic mechanism of amorphization is the build-up to a critical level of a complex defect consisting of a vacancy–interstitial associated pair. When that level is reached in a certain region, the crystalline lattice is destabilized and that region becomes amorphous. Complex clusters were proposed for modeling the more localized damage occurring under ion bombardment. The possibility of a mechanism of direct amorphization in the displacement cascade, in addition to that of complex accumulation, was also suggested. In the present paper, further numerical calculations of amorphization kinetics are conducted for the equiatomic compound TiNi. The results of using different cluster sizes, including the possibility of direct amorphization in the cascade, are compared. The dependences on irradiation temperature, dose rate, free-defect generation efficiency and existing point defect sink microstructure are analyzed. The important role of interstitial concentration evolution as a rate-controlling factor of the amorphization process is demonstrated, under the accompanying hypothesis of vacancy immobility in the temperature range under consideration. It is predicted that amorphization is inhibited above a cut-off temperature and below a cut-off dose rate.

1. INTRODUCTION

Several conditions may influence the response of a crystalline solid to displacement damage caused by bombardment with energetic particles. Those conditions are the alloy chemistry and initial microstructure of the material as well as the irradiation temperature, the particle species, flux and energy and the fluence. The result of irradiation can be a highly defective structure, particularly in those regimes where Frenkel pair annihilation, both direct and at sinks, is very limited.

Amorphization of a crystalline solid can be viewed as the most dramatic case of phase instability in as much as the crystalline structure itself is upset. The amorphous state, having a higher energy than the undamaged crystal, may be produced by irradiation if the increasingly defective crystal becomes unstable. The question that ensues is what type of defect causes the instability. A detailed description of such a defect may also allow us to understand the mechanistic aspects of the transition.

The case of ordered intermetallic alloys is very peculiar. Electron irradiation typically seems to produce long-range (chemical) disorder and thus a higher energy state than that of an ordered crystal which under thermal conditions is stable up to the melting point. However, disordering alone is apparently not a cause of amorphization, as can be inferred from experimental observation. For instance, it is known that Zr_3Al is easily rendered amorphous in a wide temperature range when bombarded with ions [1, 2]. However, electron-irradiated Zr_3Al, studied at 130 K by Carpenter and Schulson [3], becomes disordered after a low dose but does not amorphize after a fairly high dose. Chemical disorder is not sufficient to cause lattice instability; the latter, however, can arise if topological disorder is also produced.

Although the crystalline-to-amorphous transition is a very different phenomenon from a first-order phase transformation, it

*Paper presented at the International Conference on Surface Modification of Metals by Ion Beams, Kingston, Canada, July 7–11, 1986.

0025-5416/87/$3.50

bears some similarity to the latter's nucleation stage in the sense that amorphization appears to be a very localized effect. We have thus adopted the criterion that, when the defect build-up in a small region consisting of a few hundred atoms attains a certain level, that region becomes unstable and turns amorphous.

We shall here consider self-ion implantation where, in many cases, amorphization occurs after a fairly low ion fluence. The possible effects of compositional changes can be considered to be negligible, and we may focus our attention on the effects of displacement damage alone. In previous work [4, 5] the nature of the defect that causes the lattice instability during electron or ion bombardment was suggested. We shall briefly review the theory in Sections 2 and 3 and present the mathematical modeling in Section 4, especially as it applies to ion bombardment. Next, a study of amorphization kinetics as a function of irradiation temperature and existing point defect sink microstructure is presented for various cases of cascade damage. Results are given for the ion-induced amorphization of TiNi.

2. COMPLEX DEFECTS

Amorphization can be induced by electron or by ion irradiation. When energetic electrons are used, there is no displacement cascade. Frenkel pairs are uniformly generated in the electron-irradiated material. Hence, if a mechanism for the transition exists that is common to both cases, it must be related to point defect behavior. This does not preclude, for ions, additional or separate mechanisms related to cascade production.

Two experimental observations can be singled out in metallic systems that become amorphous under irradiation. First, no aggregation of like point defects appears to occur which might evolve into a void or a loop microstructure. Second, there is a temperature range, usually narrow for electron irradiation, through which the fluence necessary for promoting a complete crystalline-to-amorphous transition increases quite abruptly. A cut-off temperature for amorphization can, in this case, be defined within that range. This temperature dependence can be related to point defect mobility, indicating that, in the range where amorphization occurs without any appreciable hindrance, at least one point defect class has to remain essentially immobile.

For electron irradiation, we have therefore concluded that the retention in the crystalline lattice of some sort of defect ought to be the mechanism responsible for amorphization and assumed that the local build-up should be of the order of 2% [4, 5]. This defect level can be estimated from measurements of crystallization enthalpies of amorphous metallic systems. First, the implications of an accumulation to that level of free point defects, interstitials or vacancies or both, was analyzed [4]. It can easily be calculated that, if 1%–2% is attained, point defects must resist both recombination at very short separation distances (about $2b$ or less, where b is the nearest-neighbor interatomic distance) as well as like-defect aggregation.

We next reasoned that, in order for the defect and antidefect to coexist within such short distances, they should be able to produce a more stable relaxed defect configuration involving both types. We thus proposed, as the simplest configuration of this kind, a vacancy–interstitial complex bound within a short interaction distance, as shown in Figs. 1(a) and 1(b). For the case in Fig. 1(a), we assume the interstitial to be a single atom occupying an octahedral interstice. For the case shown in Fig. 1(b) the interstitial is in a dumbbell configuration. Very similar numerical results are obtained [4] using either assumption with nearly the same values of the physical parameters involved in the modeling.

The production of a complex as just described amounts to a sort of incomplete recombination. This implies that the complex is a different entity from (free) point defects. The trapping effect, moreover, relaxes the requirement that both (free) point defect types should be immobile [4] because a significant complex build-up can be obtained if only one type, *e.g.* the vacancy, is assumed to be immobile at a given irradiation temperature. When complex generation is taken into account, the calculated concentrations of free vacancies and interstitials have similar magnitudes to those that correspond to the case where both defects can annihilate at a significant rate at any existing sink.

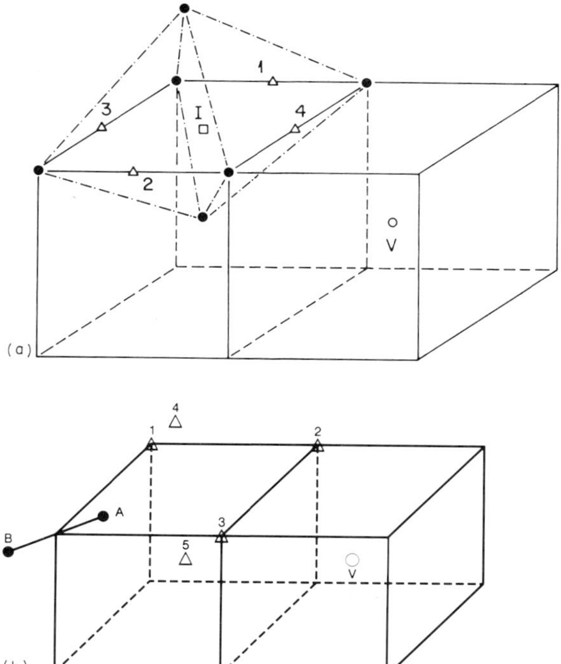

Fig. 1. Vacancy–interstitial complex model for unrelaxed configurations: (a) non-split interstitial (out of the six atoms surrounding the interstitial at I, at least five are of the same species (different from the interstitial)); (b) dumbbell interstitial (the atoms at A and B are different and, out of the atoms at sites 1, 2, 3, 4 and 5, at least four are of the same species as that at A).

The physical reasons for the existence of complexes can be found in the nature of the systems under study and in experimental observations. Under electron irradiation, there is a transition sequence that ends in amorphization. For NiTi, the well-known shape memory alloy, if the initial structure is martensitic, it will first transform to ordered B2, next it will disorder and finally the crystalline-to-amorphous transition will take place [6–9]. Under ion irradiation, such a sequence is not always observed, *e.g.* martensite appears to be retained in the remaining crystal while amorphization proceeds [9, 10]. The production of the collision cascade can be thought of as inducing, together with a localized topological disorder, a chemical disorder. If the structure is being disordered, a self-interstitial atom may have a higher probability of remaining in a given site if it has the appropriate neighbors, *i.e.* if it can establish a sort of short-range order that will resemble its chemical environment in the more stable (ordered) structure. However, being still in an interstice, it introduces stresses. These can be relieved by a nearby vacancy which thus allows for partial volume relaxation.

We thus postulate that the factors that lead to complex formation are the tendency to ordering assisted by a partial volume relaxation. Clearly, the way that this ordering is achieved introduces another element: topological disorder. So, we may class the complex as having the property of creating a center of short-range order and a focus of topological disorder under irradiation conditions. Another property of the complex is that it is immobile because it involves not only the defect–antidefect pair but also their atomic environment. However, the identity of both point defects is maintained as long as the crystalline structure prevails.

3. CASCADE EFFECTS

As expressed by Doran and Schiffgens [11], the processes taking place after a primary knock-on atom is generated in a crystalline solid can be described as occurring in three phases. The first extends from the production of the primary knock-on atom until no more atoms can be displaced from their lattice sites and lasts for about 10^{-14}–10^{-13} s. During the second phase, subthreshold events take place, which produce a highly excited region. When this phase concludes, stable vacancies, interstitials and clusters are left, and an excess energy has been dissipated to the surrounding crystal. The duration of this phase is of the order of 10^{-12}–10^{-11} s. The final phase involves in-cascade recombination, clustering and point defect escape from the region.

If no disruption of that sequence of events occurs, cascade overlap tends to reduce the free-energy increase per cascade because the displacement efficiency decreases with successive overlaps [12]. A free-energy decrease can also be accomplished by redistributing defects or, for instance, by favoring the precipitation of vacancy loops. However, as we have already mentioned, loop formation does not occur in systems that will become amorphous. One possibility is that the transition takes place during the second phase of cascade evolution. Otherwise, it may be concluded that some process that allows for point defect accumulation must be involved in the amorphization phenomenon, as for electrons.

If there is a defect build-up and cascade overlap is required to induce amorphization, it is because other types of defects, more complex than free point defects, have to be formed during the last two stages of cascade evolution. A more complex defect type that remains stable must be generated, and its existence may account for the fact that, in certain alloys, amorphization can occur while, in others, it does not. Furthermore, two experimental facts indicate that defect formation mainly occurs in the cascade region. First, amorphization seems to be initiated at lower displacement doses under ion irradiation than under electron irradiation [9]. In many instances, even though it occurs at a fairly low dose under ion bombardment at room temperature, it is not observed under electron irradiation. Second, at variance with the electron case, amorphization can progress while the remaining crystal keeps its long-range order, *e.g.* in TiNi bombarded with nickel (Ni^+) ions [9, 10]. This means that the cascade efficiency for producing free point defects is probably low.

Using a rate theory approach, it can easily be shown that a process of direct amorphization during cascade production leads to much faster kinetics than experimentally observed in a number of cases [9]. We thus propose that there is complex generation in the third phase of cascade evolution at the expense of direct recombination. Owing to cascade geometry, a more complex defect configuration than the simple entity suggested for electrons might form. In order to conduct a quantitative analysis, we shall assume that more complex defect configurations can be modeled as small clusters of the simpler entities described earlier.

4. MATHEMATICAL MODELING

We have previously modeled the formation of more complex configurations, which may be produced in the collision cascade, as clusters of the single complex [5]. Since complexes are immobile and the collision cascades occur at random, the complex distribution is random as well. We have also considered the temperature range where vacancy mobility (for annihilating at point defect sinks at a meaningful rate) can be neglected, in which case we may assume that amorphized regions do not shrink. Following our previous treatment [4, 5], let us divide the material into elemental volumes v_0 that contain n_0 atoms and prescribe that, when in any such volume the complex concentration attains a critical value c_t^0, that volume turns amorphous. Let ζ_l be the volume fraction of elemental volumes v_0 that contain l complexes and c_t their average rate of increase, where c_t is the average complex concentration per atom. The rate of change of ζ_l is therefore given by

$$\frac{d\zeta_l}{dt} = -n_0 \sum_{r=1}^{s} \dot{c}_t(r)\zeta_l + n_0 \sum_{r=1}^{l \text{ or } s}{}' \dot{c}_t(r)\zeta_{l-r} - K\epsilon_a{}'\zeta_l \qquad (1)$$

where s denotes the number of complexes in the largest cluster that may form in the cascade, Σ' indicates that the sum is to the lower index l or s, and $l = 1$ to m. The critical number of complexes is $m + 1 = c_t^0 n_0$. For the fraction containing no complexes, the rate of change is given by

$$\frac{d\zeta_0}{dt} = -n_0 \sum_{r=1}^{s} \dot{c}_t(r)\zeta_0 - K\epsilon_a{}'\zeta_0 \qquad (2)$$

The first terms in eqns. (1) and (2) account for the rate of decrease in ζ_l due to additional complex formation in the regions already containing l complexes. The last terms yield the rate of direct amorphization in the displacement cascade, which is assumed to be proportional to the corresponding volume fraction ζ_l. The proportionality constant has been factorized into the displacement rate K (displacements per atom (dpa) per second) and a dimensionless constant $\epsilon_a{}'$. The second term in eqn. (1) yields the rate of increase in ζ_l due to complex formation, totaling l, in regions which contain less than l complexes. The integral of the system of eqns. (1) and (2) is given by the expression

$$\zeta_l = \exp(-K\epsilon_a{}'t) \exp\left\{-n_0 \sum_{r=1}^{s} c_t(r)\right\} \times \sum_{m_l} \prod_{r=1}^{s} \frac{\{n_0 c_t(r)\}^{j_r(m_l)}}{\{j_r(m_l)\}!} \qquad (3)$$

for $l = 0$ to m, where

$$\sum_{r=1}^{s} r j_r(m_l) = l \tag{4}$$

Owing to the conservation of the total number of atoms, the amorphous fraction is given by

$$\zeta_a = 1 - \sum_{l=0}^{m} \zeta_l \tag{5}$$

A rate theory approach is used for calculating the build-up of complexes and of complex clusters [5]. Because the vacancy and the interstitial that form the complex maintain their identity as lattice defects (as long as the crystalline structure persists), complexes can be destroyed. Two processes can be considered, *i.e.* thermal decomposition which consists of a random detrapping jump of the interstitial, and recombination involving the vacancy member with another interstitial. For $l = 2$ to $s-1$,

$$\frac{dc_t(l)}{dt} = K\epsilon_l + \frac{c_t(l+1)}{\tau_{l+1}} - \frac{c_t(l)}{\tau_l} - \frac{1}{\tau_{al}} \tag{6}$$

where ϵ_l is the efficiency of complex production in the cascades relative to the displacement rate K and

$$\tau_l^{-1} = \beta_l \exp\left(-\frac{E_{Bl}}{kT}\right) + \gamma_l c_i \tag{7}$$

is the rate of destruction of one complex in the cluster of size l. c_i is the (free) interstitial concentration, E_{Bl} is the binding energy of a complex in a cluster of size l, β_l is the rate coefficient for thermal decomposition and γ_l is the rate coefficient for indirect recombination. The last term in eqn. (6) accounts for cluster removal from the distribution owing to amorphization and is given by (see Appendix A)

$$\tau_{al}^{-1} = \frac{c_t(l)}{\sum_{j=1}^{s} j c_t(j)} \times \left\{\frac{m+1}{n_0}\frac{d\zeta_a}{dt} - \frac{K\epsilon_a'}{n_0}\sum_{i=0}^{m}(m+1-i)\zeta_i\right\} \tag{8}$$

For the cluster of largest size s,

$$\frac{dc_t(s)}{dt} = K\epsilon_s - \frac{c_t(s)}{\tau_s} - \frac{1}{\tau_{as}} \tag{9}$$

and, for single complexes,

$$\frac{c_t(1)}{dt} = K\epsilon_1 + \alpha c_i c_v + \frac{c_t(2)}{\tau_2} - \frac{c_t(1)}{\tau_1} - \frac{1}{\tau_{a1}} \tag{10}$$

where c_v is the (free-)vacancy concentration, and the rate coefficients τ_1, τ_2, τ_s, τ_{al} and τ_{as} are given by similar expressions to eqns. (7) and (8). The second term in eqn. (10) accounts for the probability of complex formation by random encounter of an interstitial with a vacancy. The rate coefficient α depends on the degree s of long-range order [4]. It is maximum for $s = 0$ and zero for $s = 1$.

The rate equations for vacancies and interstitials are respectively

$$\frac{dc_v}{dt} = G - (R+\alpha)c_i c_v + \sum_{l=1}^{s} \beta_l' c_t(l) \exp\left(-\frac{E_{Bl}}{kT}\right) \tag{11}$$

and

$$\frac{dc_i}{dt} = G - (R+\alpha)c_i c_v - k_{is}^2 D_i c_i + \sum_{l=1}^{s} \beta_l' c_t(l) \exp\left(-\frac{E_{Bl}}{kT}\right) \tag{12}$$

R is the rate coefficient for direct recombination of vacancies and interstitials, $k_{is}^2 D_i$ is the overall loss rate of interstitials to extended sinks (*e.g.* dislocations and grain boundaries) and β_l'/β_l is the fraction of thermal decomposition events that do not lead to direct recombination of the pair. The generation rate of free point defects is given by

$$G = K\left(1 - \sum_{r=1}^{s} r\epsilon_r - \eta - \frac{m+1}{n_0}\epsilon_a'\right) \tag{13}$$

In eqn. (13), ηK is the in-cascade recombination rate of Frenkel pairs and the last term accounts for a reduction of free-point-defect production due to direct amorphization. We have assumed that direct amorphization in the cascade is equivalent to generating the critical number of complexes in a region containing n_0 atoms.

5. AMORPHIZATION KINETICS OF NiTi

We have performed some model calculations for the equiatomic NiTi compound. The crystalline structure of the ordered intermetallic is B2 up to the melting point, although it may transform martensitically into a monoclinic distortion of the orthorhombic B19 structure below room temperature [13]. When the B2 compound disorders under the effect of particle bombardment, the unit cell becomes b.c.c.

We have previously calculated the rate coefficients that appear in eqns. (6)–(12) for the cases where the interstitial is assumed to be in a non-split and in a dumbbell configuration [4]. These rate coefficients are functions of geometric factors that can be calculated for any given specific crystalline structure, of the degree of long-range order and of various physical properties of the system under consideration. They are given in Table 1.

The important physical parameters that appear in the rate equations are (1) the interstitial diffusivity D_i, (2) the complex binding energy E_{Bl}, (3) the barrier ΔE_l for indirect recombination and (4) the various coefficients that account for point defect partitioning among the different defects that originate following a cascade collapse.

The values of the diffusion coefficient, the complex binding energy and the barrier for indirect recombination, and their relative magnitudes, determine the cut-off temperature and the steepness of the dependence of the required dose for complete amorphization on temperature. The latter dependence is very sensitive to the magnitude of ΔE. The values of those three parameters were adjusted using available data of the dose required for complete amorphization *vs.* temperature in electron-irradiated NiTi [4]. These values are given in Table 2, together with the geometric factors calculated for the case where the interstitial is assumed to be in a dumbbell configuration. Figure 2, taken from ref. 5, shows the excellent agreement of theory and experiment for that set of parameters. We have adopted these same values in the present work and analyzed the amorphization kinetics as a function of dose, dose rate, temperature and microstructure, for ion bombardment.

The additional effect of displacement cascades can clearly be of considerable weight, as demonstrated by the fact that some systems resist amorphization under electrons whereas they become amorphous after relatively low ion fluences. However, the measured amorphization kinetics of NiTi at about 300 K cannot be accounted for by in-cascade direct amorphization alone [9,10].

TABLE 1

Rate coefficients describing the elementary processes undertaken by vacancies and self-interstitials (the dumbbell configuration)

Process	*Rate coefficient*
Complex formation (one possible configuration)	$\alpha_{Ni} = 10.40 \dfrac{D_i}{b^2}$
Direct recombination (second-nearest neighbors)	$R = 26.5 \dfrac{D_i}{b^2}$
Indirect recombination	$\gamma = 26.5\phi \exp\left(-\dfrac{\Delta E}{kT}\right)\dfrac{D_i}{b^2}$
Thermal decomposition	$\beta = 3 \dfrac{D_i}{b^2}$
Thermal decomposition not leading to direct recombination	$\beta' = 2.5 \dfrac{D_i}{b^2}$

TABLE 2

Physical parameters used for calculating the amorphization kinetics of ion-bombarded NiTi

$D_i = 10^{-6} \exp\left(-\dfrac{0.29\ \text{eV}}{kT}\right)\ \text{m s}^{-2}$
$K = 10^{-3}$ dpa s^{-1}
$k_{is} = 3.14 \times 10^6\ \text{m}^{-1}$
$c_t{}^c = 2\%$
$n_0 = 400$ atoms
$\eta = 0.20$
$E_{B1} = 1.0$ eV
$E_{B2} = E_{B3} = 1.5$ eV
$\Delta E_1 = 0.15$ eV
$\Delta E_2 = \Delta E_3 = 0.23$ eV
$\phi = 0.53$
$\eta = 0.20$
$\gamma_1 = \gamma_2 = \gamma_3$
$\beta_1 = \beta_2 = \beta_3$
For $s = 1$, $\epsilon_1 = 0.60$
For $s = 2$, $\epsilon_1 = 0.10$, $\epsilon_2 = 0.25$
For $s = 3$, $\epsilon_1 = \epsilon_2 = 0.05$, $\epsilon_3 = 0.15$
For $s = 4$, $\epsilon_a{}' = 5$, $\epsilon_1 = \epsilon_3 = 0.10$, $\epsilon_2 = 0.05$

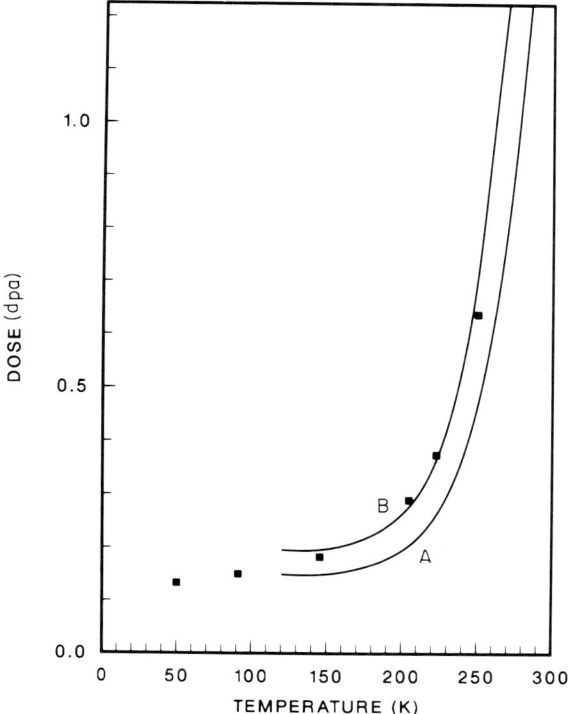

Fig. 2. Dose required to attain a given amorphous fraction under electron irradiation, as a function of temperature (dpa,): curve A, $\zeta_a = 80\%$; curve B, $\zeta_a = 97\%$; ■, experimental results of Mori and Fujita [6]. (From ref. 5.)

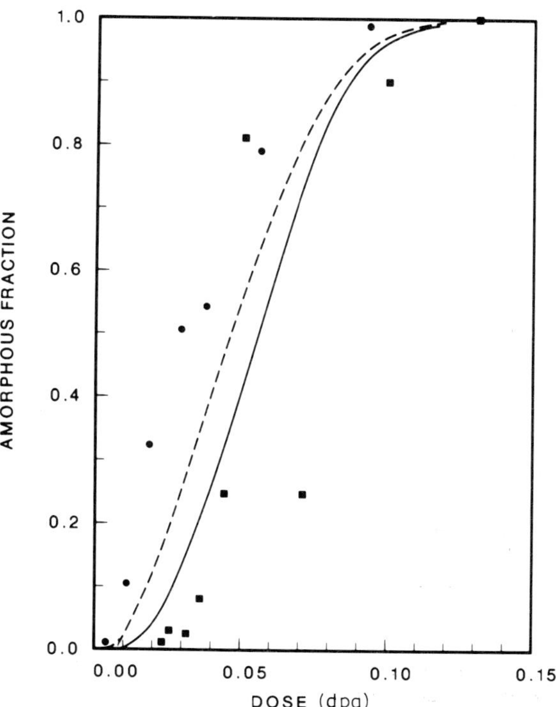

Fig. 3. Amorphous fraction *vs.* ion dose at 300 K: ——, simple complex configuration ($s = 1$); - - -, two complex clusters and single complexes ($s = 2$); ■, experimental data of Brimhall *et al.* [10]; ●, experimental data of Moine *et al.* [9]. (From ref. 5.)

We shall analyze, in the following, amorphization kinetics under various models of complex formation in the cascade. These are as follows: (A) single complexes, as in electron bombardment but at an enhanced rate; (B) clusters of one and two complexes; (C) same as in (B) plus clusters of three complexes; (D) same as in (C) plus direct amorphization in a cascade event. In addition to the physical parameters discussed above, the value of the efficiency coefficient ϵ related to complex production in the cascade becomes very important. Calculations were carried out previously [5] for NiTi, assuming single and double complex clusters and no direct amorphization in order to adjust the value of ϵ by comparison with available kinetic data at 300 K [9, 10]. Figure 3, reproduced from ref. 5, shows the very good agreement obtained for a total value of $\epsilon = 0.60$, and for the given partitioning of $\epsilon_1(\mathrm{B})$ and $\epsilon_2(\mathrm{B})$. In order to compare the different hypotheses, we have assumed the same total complex production coefficient ϵ, *i.e.*

$$\begin{aligned}\epsilon &= \epsilon_1(\mathrm{A}) \\ &= \epsilon_1(\mathrm{B}) + 2\epsilon_2(\mathrm{B}) \\ &= \epsilon_1(\mathrm{C}) + 2\epsilon_2(\mathrm{C}) + 3\epsilon_3(\mathrm{C}) \end{aligned} \qquad (14)$$

for assumptions (A)–(C), and

$$\epsilon = \epsilon_1(\mathrm{D}) + 2\epsilon_2(\mathrm{D}) + 3\epsilon_3(\mathrm{D}) + \frac{m+1}{n_0}\,\epsilon' \qquad (15)$$

Moreover, for hypotheses (C) and (D), the values of the various coefficients appreciably favoring the production of the largest cluster were chosen, in order to emphasize possible differences in the relative kinetics. The values of all the parameters used here are listed in Table 2.

The dose required to induce complete amorphization ($\zeta_a = 99\%$) is plotted in Fig. 4 as a function of temperature for the various hypotheses just described. Except for case (D), it can be seen that there is a well-defined cut-off temperature, near 400 K, above which complete amorphization cannot be expected. Such cut-off will also be obtained if higher

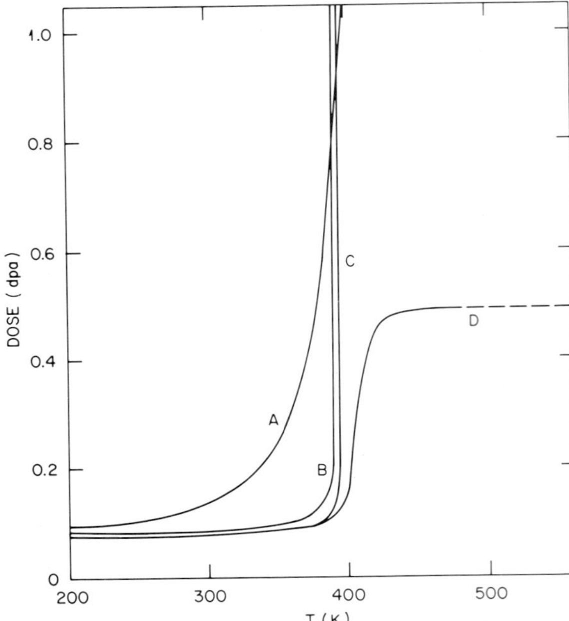

Fig. 4. Dose required to attain complete amorphization as a function of bombarding temperature, for various hypotheses regarding the maximum cluster size: curve A, hypothesis (A), $s = 1$; curve B, hypothesis (B), $s = 2$; curve C, hypothesis (C), $s = 3$; curve D, hypothesis (D), same as hypothesis (C) plus direct in-cascade amorphization.

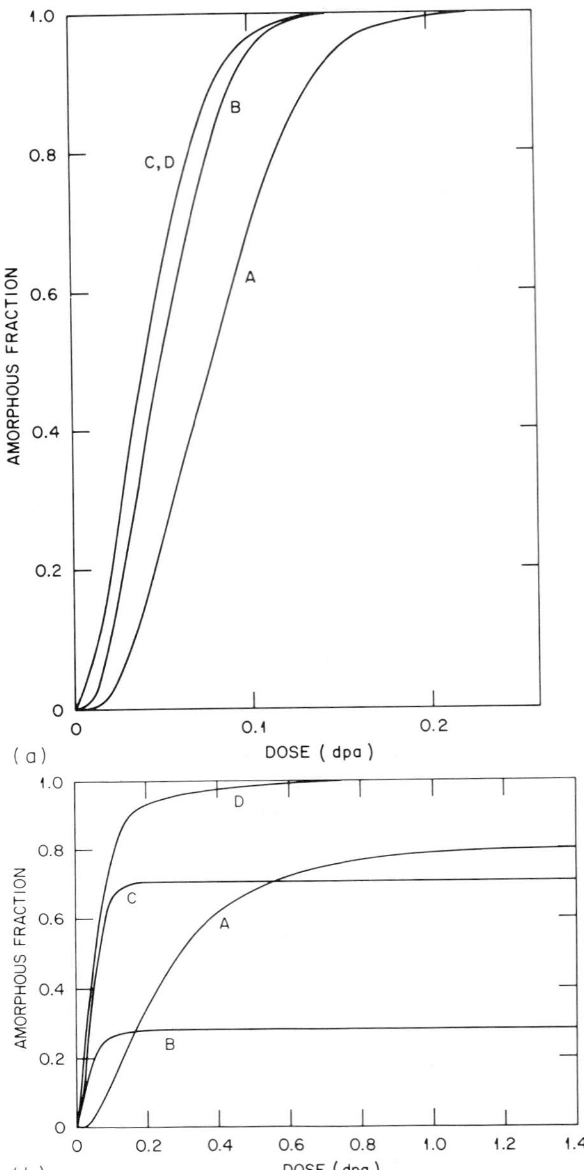

Fig. 5. Amorphous fraction *vs.* ion dose at (a) 330 K and (b) 400 K, for various hypotheses regarding the maximum cluster size: curves A, hypothesis (A), $s = 1$; curves B, hypothesis (B), $s = 2$; curves C, hypothesis (C), $s = 3$; curves D, hypothesis (D), same as hypothesis (C) plus direct in-cascade amorphization.

clusters (up to $s = m$) are assumed for modeling cascade-formed defects. The closeness of the curves for cases (B) and (C), furthermore, shows that higher cluster sizes will not introduce a substantial difference in the temperature dependence of the dose for complete amorphization. The cut-off temperature occurs because of the increasing importance of the processes that produce complex destruction, preventing the complex population from building up to the critical value in the entire specimen.

The results are different for case (D), but only near the cut-off temperature where direct amorphization in the cascade starts to contribute. In fact, at temperatures above about 420 K, it becomes the only contributing process. In this case, however, a cut-off temperature must also exist, but at the value where vacancy mobility allows for recrystallization of amorphized regions, a process which we have not considered here.

The amorphization kinetics that are obtained under the various hypotheses are illustrated in Fig. 5 at two different temperatures. At 300 K (Fig. 5(a)), it can be seen that the kinetics are faster when higher complex cluster sizes are assumed although, as pointed out above, the dose at which complete amorphization is attained is almost independent of cluster size for $s > 1$. At 400 K, as shown in Fig. 5(b), hypotheses (B) and (C) lead to a saturation in the amorphous fraction at fairly low doses. The saturation occurs because a steady state has been attained in the complex population such that the complex creation rate equals that of complex destruction. For case (A), however, the kinetics yield a persistent increase in the amorphous fraction with increasing dose, leading to a higher value of the amorphous fraction after a certain dose. In fact, the ζ_a value of 81% shown in the figure is the saturation level, which drops to 33% at 410 K and to 5% at 420 K.

The reason why ζ_a increases for case (A), at temperatures close to the cut-off, above the saturation values of cases (B) and (C) is as follows. In cases (B) and (C), larger fractions of regions containing a larger number l of complexes are more rapidly generated at a lower dose. Thus, although the rate of total complex generation and the rate of free vacancies and interstitials are assumed to be the same, the complex build-up to attain locally the critical value is initially faster for the cases where clusters form. This yields more rapid amorphization kinetics at a low dose, but it also allows for a higher interstitial build-up. As the dose increases, this situation persists, *i.e.* the interstitial population is always smaller in case (A) than in the other cases. Since the attainment of saturation depends on having the same rates of complex generation and destruction, it is clear that this will be obtained at a higher value of c_t, and hence of the amorphous fraction, for case (A).

Figure 6 illustrates the effect of the magnitude of the sink strength for interstitial absorption in the specimen, attributed in this example to a different dislocation concentration. The effect is the same for any of the hypotheses being analyzed, provided that direct amorphization in the cascade region is not the rate-controlling mechanism of amorphization. The acceleration of the amorphization process, and even its completion as in the example shown in Fig. 6, can again be understood in terms of the evolution of the free-interstitial concentration as the complex concentration keeps increasing. The presence of more interstitial sinks decreases the interstitial population and thus the rate of destruction of complexes by indirect recombination, allowing for continuing complex build-up.

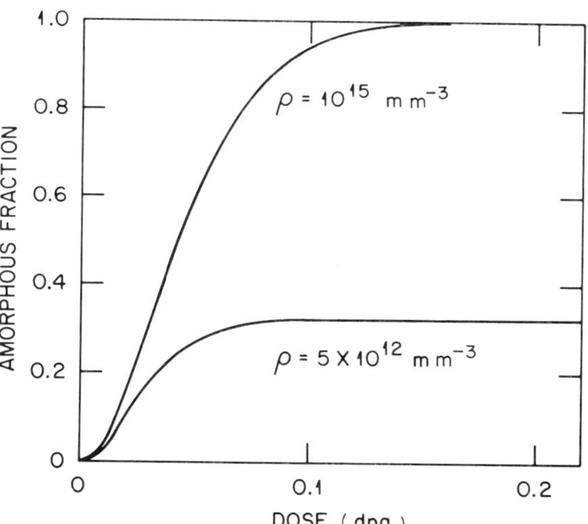

Fig. 6. Amorphization kinetics as a function of dislocation density ρ ($T = 410$ K; $s = 3$).

The magnitude of the interstitial concentration is seen, from the above analyses, to play a very important role in determining the amorphization kinetics. Another factor that determines concentration is clearly the fraction of free point defects that escape the cascade region, which is quantified in our equations by the value of G in eqns. (11) and (12). Figure 7 illustrates the kinetics at 380 K for two cases where the magnitudes of all the terms in eqn. (13) were kept constant, except for η which was given the values 0.3 and 0.1, to yield respectively $G = 0.1K$ and $G = 0.3K$. It can be seen, as expected, that the kinetics are faster in the former than in the latter case. The effect is less pronounced at lower temperatures where the interstitial mobility is less, and it grows, as temperature is increased, up to a maximum value.

The effect of the dose rate on the kinetics is shown in Fig. 8. Above a certain dose rate and up to a certain temperature, the kinetics are dose rate independent. When the displacement rate is very low compared with the rate of complex destruction by thermal decomposition and/or the rate of indirect recombination, no substantial complex build-up can take place, and amorphization is greatly reduced or cancelled. At low enough temperatures, where the processes of complex destruction are, in turn, insignificant, the amorphization kinetics are again dose rate independent.

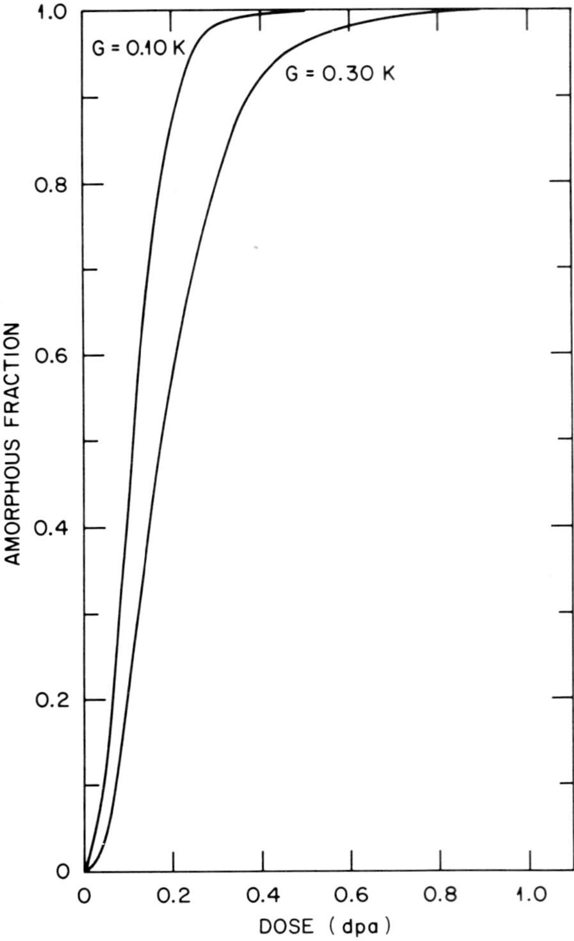

Fig. 7. Dependence of amorphization kinetics on free-point-defect generation rate G ($T = 380$ K; $s = 1$).

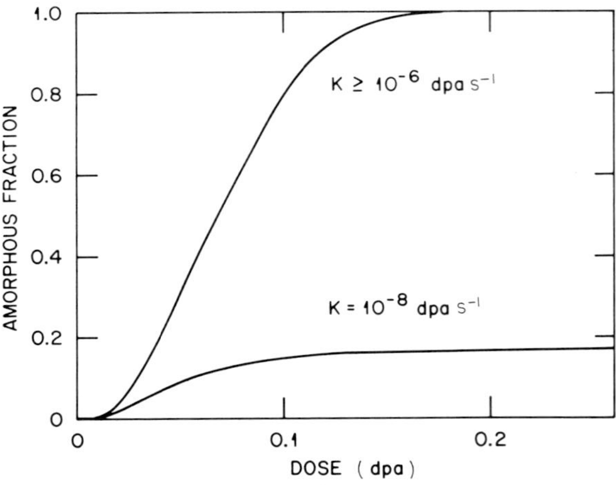

Fig. 8. Amorphization kinetics as a function of dose rate ($T = 300$ K; $s = 1$).

6. SUMMARY AND CONCLUSIONS

Further analyses of the amorphization kinetics under ion bombardment have been presented using a previously developed theory [5]. This theory proposes that irradiation induces a defective lattice which is turned locally unstable when the defect build-up attains a critical value. Experimental observations of defect evolution prior to amorphization and its differences with systems that do not undergo the transition were used as a basis of the theory [4,5]. The defect responsible for the crystalline instability leading to amorphization was suggested to be an interstitial–vacancy pair. A decrease in local free energy can occur when the self-interstitial becomes trapped in a site so that, with its atomic environment, it gives rise to a short-range order. The trapping is made possible by the presence of a nearby vacancy that allows for partial volume relaxation.

It has been suggested that direct accumulation of vacancies plus interstitials to the critical level required to induce amorphization can be promoted by low point defect mobilities [14]. However, as discussed in ref. 4, the recombination volume should be quite small, in which case the resistance to like-point-defect aggregation must be accounted for. Thus a complex defect including both point defect species should also form for low defect mobilities. Rate equations including the pertinent non-diffusional processes, different from eqns. (6)–(12), must be formulated in this instance.

The proposed defect build-up is thus an alternative manner of point defect behavior to that in systems where they either recombine or are absorbed at sinks. It furthermore gives a path for the transition to occur, as topological disorder is created by an atom that is stabilized in an interstitial configuration together with lattice distortions in a certain region around it that must accompany the trapping.

The very localized defect production during ion bombardment leads us to suggest that a more complex defect might be induced in the displacement cascade, in addition to possible direct amorphization. We have assumed that the latter is proportional to the total displacement damage and modeled more complex defects as clusters containing multiple elemental complexes consisting of a vacancy–interstitial pair. Previous calculations [5] for NiTi allowed us to adjust model parameters with a very good fitting to available experimental data of amorphization kinetics under nickel (Ni^+) ion bombardment at room temperature [9, 10].

We have here studied the temperature dependence of the required dose to induce complete amorphization by extending our calculations to analyze the importance of cluster size in describing the amorphization kinetics and the additional effects due to direct amorphization in the cascade. We have shown that, similar to electrons, there exists a cut-off temperature above which amorphization becomes rapidly inhibited. This result is a consequence of a higher indirect recombination rate, due to the higher interstitial mobility, and of increased complex thermal decomposition.

For the case where direct amorphization in the cascade was also considered, no cut-off resulted from the calculations. Instead, at higher temperatures, a temperature-independent dose was obtained for inducing complete amorphization. This result was explained by the direct process, since defect build-up becomes irrelevant at higher temperatures. However, although systematic studies of ion bombardment of NiTi at different temperatures have not been reported, some experiments by Brimhall *et al.*]10] indicate a very pronounced increase in amorphization dose with temperature at 450 K. Therefore, another process not included in the present theoretical model must gain importance as the temperature is further increased. We suggest that, as vacancies become mobile, recrystallization of the athermally amorphized regions takes place, and a cut-off temperature should be obtained as a consequence, as well.

It was shown that different hypotheses regarding cluster size do not influence the total time required to attain complete amorphization, when the maximum cluster size is larger than unity. However, the dose dependence of the amorphous fraction exhibits a strong dependence on maximum cluster size in the vicinity of the cut-off temperature when only partial amorphization can be obtained.

The influence of the microstructure appears in this approach as a kinetic effect. The microstructure regulates the interstitial concentration, which in turn controls the kinetics. It is thus a logical result of the theory that a higher dislocation concentration should lead to faster amorphization kinetics. However, no substantial reduction in the dislocation concentration has been included in the present approach. Such a reduction may occur if amorphization progresses preferentially in regions of high sink density.

Finally, we have analyzed the dose rate dependence of the kinetics and obtained what we may term a dose rate cut-off. Such a cut-off is related again to the interstitial diffusion rate relative to the displacement rate. When the former is fast enough to destroy complexes at the same rate as they are created before any meaningful complex accumulation can take place, no amorphization will occur.

ACKNOWLEDGMENTS

The research reported in this paper was sponsored by the Division of Materials Sciences, U.S. Department of Energy, under Contract DE-AC05-840R21400 with Martin Marietta Energy Systems Inc.

REFERENCES

1 L. M. Howe and M. H. Rainville, *J. Nucl. Mater.*, *68* (1977) 215.
2 L. M. Howe and M. H. Rainville, *Radiat. Eff.*, *48* (1980) 151.
3 G. J. C. Carpenter and E. M. Schulson, *J. Nucl. Mater.*, *73* (1978) 180.
4 D. F. Pedraza, *J. Mater. Res.*, *1* (1986) 425.
5 D. F. Pedraza and L. K. Mansur, *Nucl. Instrum. Methods B*, *16* (1986) 203.
6 H. Mori and H. Fujita, *Jpn. J. Appl. Phys.*, *21* (1982) L494.
7 H. Mori, H. Fujita and M. Fujita, *Jpn. J. Appl. Phys.*, *22* (1983) L94.
8 A. R. Pelton, *Proc. 7th Int. Conf. on High Voltage Electron Microscopy*, in *Rep. LBL-16031, UC-25, CONF-830819*, 1983, p. 245 (University of California, Berkeley, CA).
9 P. Moine, J. P. Rivière, M. O. Rouault, J. Chaumont, A. Pelton and R. Sinclair, *Nucl. Instrum. Methods B*, *7–8* (1985) 20.
10 J. L. Brimhall, H. E. Kissinger and A. R. Pelton, in G. K. Hubler, C. W. White, O. W. Holland and C. R. Clayton (eds.), *Ion Implantation and Ion Beam Processing of Materials, Materials Research Society Symp. Proc.*, Vol. 27, Elsevier, New York, 1984, p. 163.

11 D. G. Doran and J. O. Schiffgens, *Proc. Workshop on Correlation of Neutron and Charged Particle Damage*, in *Rep. ORNL CONF-760673*, 1976, p. 3 (Oak Ridge National Laboratory, Oak Ridge, TN).

12 J. R. Beeler, Jr., in J. W. Corbett and L. C. Ianiello (eds.), *Radiation Induced Voids in Metals*, U.S. Atomic Energy Commission, Office of Information Services, Washington, DC, 1971, p. 684.

13 G. D. Sandrock, A. J. Perkins and R. F. Hehemann, *Metall. Trans.*, *2* (1971) 2769.

14 J. L. Brimhall and E. P. Simonen, *Nucl. Instrum. Methods B*, *16* (1986) 187.

APPENDIX A

A.1. Adjustment of the complex concentration due to the increase in amorphous fraction

Amorphization may occur by two mechanisms, *i.e.* directly in the displacement cascade and by complex accumulation to the critical level. For simplicity, we shall here only consider single complexes. The extension to multicomplex clusters is trivial. Let $\Delta\zeta_1$ be the volume fraction that becomes amorphous by the first process, and $\Delta\zeta_2$ the volume fraction due to the second process, in the time interval Δt. Let $c_t(t)$ be the complex concentration at time t, and $c_t(t+\Delta t)$ that after the time increment Δt. The difference between the two complex concentrations due to amorphization alone is given by

$$-\tau_a^{-1}\,\Delta t = c(t+\Delta t) - c(t) = -\frac{m+1}{n_0}(\Delta\zeta - \Delta\zeta_2) - \sum_{i=0}^{m}\frac{i}{n_0}\zeta_i K\epsilon_a'\,\Delta t \qquad \text{(A1)}$$

where

$$\Delta\zeta_2 = \sum_{i=0}^{m}\zeta_i K\epsilon_a'\,\Delta t \qquad \text{(A2)}$$

and

$$\Delta\zeta = \Delta\zeta_1 + \Delta\zeta_2 \qquad \text{(A3)}$$

is the total volume fraction that has become amorphous in Δt. Substituting eqn. (A2) for $\Delta\zeta_2$ and letting $\Delta\zeta \to 0$, while $\Delta t \to 0$, we obtain, after some algebra,

$$\tau_a^{-1} = \frac{m+1}{n_0}\frac{\mathrm{d}\zeta}{\mathrm{d}t} - \frac{K\epsilon_a}{n_0}\sum_{i=0}^{m}(m+1-i)\zeta_i \qquad \text{(A4)}$$

If there are clusters of size l ($l = 1, s$), eqn. (A4) for each cluster size has to be multiplied by the factor

$$\frac{c_t(l)}{\sum_{j=1}^{s} j c_t(j)} \qquad \text{(A5)}$$

Amorphous Phase Formation in Aluminum-ion-implanted Refractory Metals*

M. SAQIB and D. I. POTTER

Metallurgy Department, and Institute of Materials Science, University of Connecticut, Storrs, CT 06268 (U.S.A.)

(Received July 10, 1986)

ABSTRACT

Implantation with aluminum ions (Al^+) was investigated for producing amorphous, microcrystalline and crystalline phases in the refractory metals tantalum, niobium and vanadium. Surface alloys with a substantial aluminum content resulted from the implantation, reaching more than 70 at.% Al in tantalum. Surface roughening occurred at the higher fluences, a factor important in interpreting microstructures observed by transmission electron microscopy. B.c.c. solid solutions persisted to fluences greater than 1.2×10^{18} Al^+ cm^{-2} (near about 60 at.% Al for tantalum) for all three refractory metals. At higher fluences the implanted layers transformed to microcrystalline and then amorphous phases (tantalum and niobium) or to another crystalline phase (vanadium).

Microstructural observations show that dislocations are present at fluences of 6×10^{17} Al^+ cm^{-2} and persist to at least twice this fluence. The microcrystalline phase nucleates beneath the surface and first appears at the surface in sputtered depressions. Further implantation leaves islands of the b.c.c. phase surrounded by the microcrystalline phase. The results are discussed in terms of the calculated free energies for the alloy systems and in terms of the phase instabilities caused by the radiation damage accompanying implantation.

1. INTRODUCTION

Aluminide layers on refractory metals are being investigated as barriers to oxidation and corrosion, extending the work by Beaver *et al.* [1]. In particular, amorphous layers produced by aluminum ion (Al^+) implantation will be investigated and their diffusive transport compared with crystalline layers of the same chemical composition. Here we report the formation of microcrystalline and amorphous layers on tantalum and niobium metal, and a crystalline transformation in vanadium. The presence of these phases is discussed in terms of their free energies and in terms of the radiation damage accompanying implantation.

2. EXPERIMENTAL PROCEDURES

Electropolished specimens of tantalum, niobium and vanadium metal were implanted with 180 keV aluminum ions at fluxes near 10^{14} ions cm^{-2} s^{-1} and in residual gas pressures near 10^{-6} Pa. The specimen temperatures were monitored during implantation with an IR pyrometer and did not exceed 50 °C. Subsequent examination proceeded using transmission electron microscopy (TEM) and electron diffraction to characterize the microstructures and phases in the implanted layer extending from the surface to a depth of about 1000 Å. Rutherford backscattering spectroscopy with 1.4 MeV helium ions (He^+) and Auger depth profiling yielded composition profiles after implantation. The average compositions in TEM foils were measured with energy-dispersive X-ray analysis, with precipitated $TaAl_3$ as a reference to fix the proportionality constant relating concentrations with X-ray intensities.

3. RESULTS

The results are presented in the following sections. First, in Section 3.1, we describe the

*Paper presented at the International Conference on Surface Modification of Metals by Ion Beams, Kingston, Canada, July 7-11, 1986.

0025-5416/87/$3.50

roughening of the surface due to implantation, a process leading to surface depressions about 500 Å in scale. An appreciation of this roughening is needed to understand the morphologies of the phases examined in Section 3.2. The compositions at which these phase transformations occur are described in Section 3.3. Annealing of selected implanted specimens is explored in Section 3.4. These last results bear directly on the interpretation of phase stability in these metals, a topic pursued further in Section 4.

3.1. Surface roughening from implantation

The implanted surfaces remained shiny to fluences of about 1.6×10^{18} Al^+ cm^{-2}. Thereafter, their surface reflectivity decreased as the fluence was increased. The reflectivity losses are caused by surface roughening. The slight extent of this roughening compared for example with that for nickel implanted at elevated temperatures [2] precluded its observation with scanning electron microscopy. It was observable in thin foils using TEM (Fig. 1). For reference, Fig. 1(a) shows an

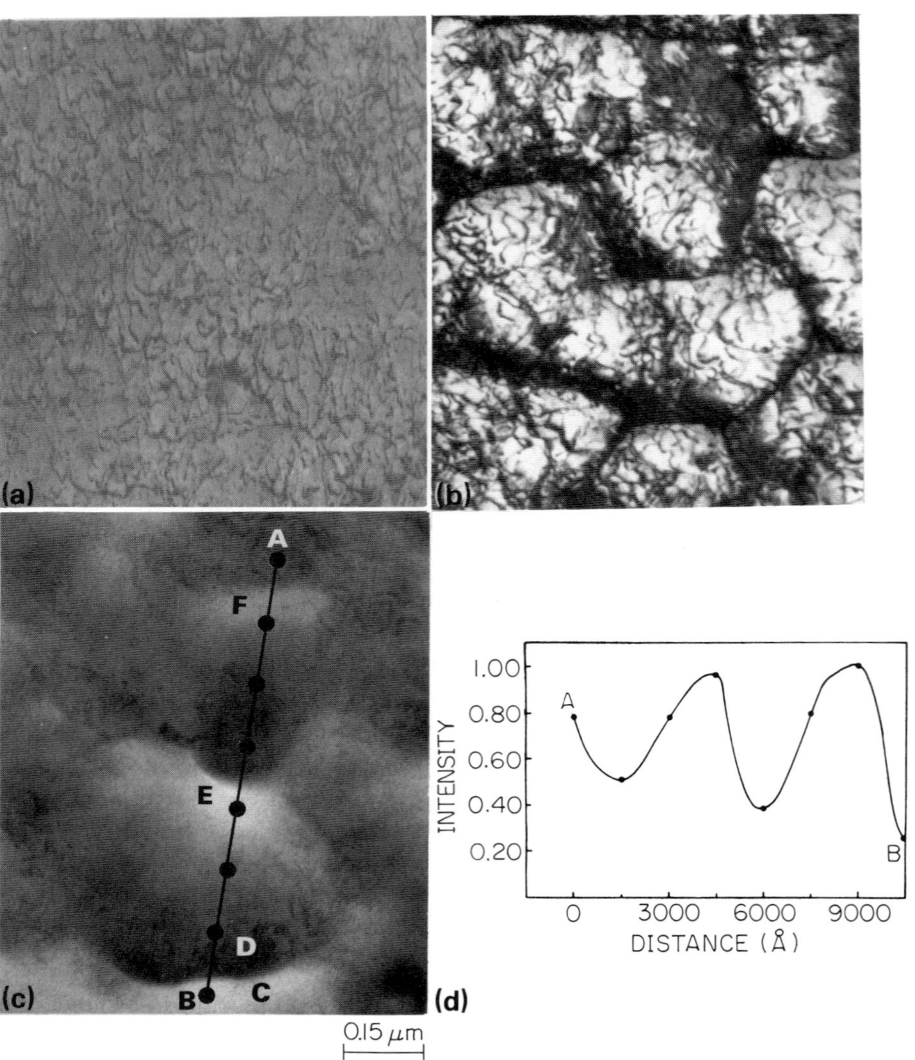

Fig. 1. Development of surface roughening on tantalum: (a) dislocations in foil of uniform thickness, 6×10^{17} Al^+ cm^{-2}; (b) cell structure imaged with diffraction contrast, 1.2×10^{18} Al^+ cm^{-2}; (c) cell structure imaged with absorption contrast, 1.5×10^{18} Al^+ cm^{-2}; (d) X-ray intensity along the line indicated in (c).

image obtained after implanting the bulk specimen to 6×10^{17} Al^+ cm^{-2} and then thinning the specimen from the back to the implanted surface. Aside from the presence of dislocations in the microstructure, we point out that the field of view is of uniform contrast. This uniform contrast, together with the absence of bend and thickness contours [3], shows that the foil is of constant thickness. In turn, then the implanted surface remains flat at least to this fluence.

The remainder of Fig. 1 shows that some roughening of the surface occurs at higher fluences. This was first noted when the TEM diffraction conditions were optimum for observing images of dislocations (Fig. 1(b)). A cellular structure is evident here and is composed of darker, strongly diffracting regions and interspersed lighter regions diffracting less strongly. Dislocations are easy to see in the regions of strong contrast but difficult to see in the lighter regions. We initially interpreted images such as Fig. 1(b) as cellular structures caused by dislocation-free volumes alternating with volumes very high in dislocation density [4]. Two further observations have led us to reject this interpretation: (1) the light and dark regions, and with them the regions of alternating dislocation density, interchange in contrast as the foil is tilted slightly near the two-beam condition, *i.e.* as the diffraction error s is varied by about 10^{-3} $Å^{-1}$ near the $s > 0$, two-beam, $\langle 110 \rangle$ $\mathbf{g}$; (2) a cell-liked structure is observed under absorption contrast conditions (Fig. 1(c)). The images of dislocations depended on the local diffraction conditions and, by carefully controlling these conditions, observation (1), we found that the location-to-location dislocation density did not change radically, *i.e.* it varied by less than 25%. The cell-like structure in Fig. 1 is not caused by dislocation density variations but results from thickness variations across the field of view (Fig. 1(c)). The lighter contrast near the cell boundaries in Fig. 1(c) indicates that these regions are more transparent to electrons and are probably thinner than the cell interiors. This was confimed by examining stereo images and by the results that follow.

The relative foil thicknesses in cell interiors and near the cell boundaries were investigated using a 120 keV electron probe 100 Å in diameter from an analytical transmission electron microscope. The intensities of X-rays generated from a thin foil examined with such a probe are directly proportional to the foil thickness [5]. The relative X-ray intensities measured along the line indicated in Fig. 1(c) are plotted in Fig. 1(d). The boundary thickness is less than the cell interior thickness by a factor of roughly 2. Carbon spot marking [6] showed that the foil thickness at the boundary C (Fig. 1) was about 500 Å, while the cell interior D was about 1170 Å thick. Thus the cells represent surface undulations of about 500 Å, measured from the tops of relatively flat-topped "hills" to the bottoms of the intercellular "valleys". Sharp thickness transitions cause well-defined boundaries while more gentle changes give diffuse boundaries F (Fig. 1(d)).

3.2. Phase transformations induced by aluminum ion implantation

The microstructures and phases resulting from implantation will be summarized in this section. The b.c.c. solid solution was retained to high fluences and aluminum concentrations for all three metals investigated. With higher fluences, the b.c.c. phase of tantalum and niobium transformed to a microcrystalline phase that subsequently transformed to an amorphous phase. The vanadium b.c.c. phase transformed to another crystalline phase with appreciable crystallite size.

Electron diffraction patterns from tantalum implanted with aluminum up to fluences of 1.4×10^{18} ions cm^{-2} (Fig. 2(a)) contained beams diffracted by a b.c.c. structure and no other beams. Thus, implantation extends the solubility of aluminum in tantalum, and in niobium since it behaved similarly, well beyond their equilibrium compositions of about 1 at.% and about 10 at.% for tantalum [7] and niobium [8] respectively. Diffuse rings were increasingly apparent in the patterns at fluences beyond 1.6×10^{18} Al^+ cm^{-2}. These diffuse rings (Figs. 2(b)-2(d)) were the only features observed in diffraction patterns recorded beyond 1.8×10^{18} Al^+ cm^{-2}. Dark field imaging revealed a structure composed of microcrystals about 10 Å in size at 1.8×10^{18} Al^+ cm^{-2} (Fig. 2(c)). These particles were increasingly difficult to image as the fluence increased and, by 2.4×10^{18} Al^+ cm^{-2} (Fig. 2(d)), they were no longer resolvable. Thus, with increasing fluence the b.c.c. phase transforms to a microcrystalline phase, and then the latter gives way to an amorphous phase. The fluence intervals over which these transformations occurred were the same for tantalum and niobium.

The microstructural changes occurring near the surface during the b.c.c. → microcrystalline transformation were investigated and are summarized in Fig. 3. Three images of the same foil area are shown in Figs. 3(a), 3(b) and 3(c). The images record a stage in the transformation where the near-surface region is still crystalline b.c.c. phase. The cell structure of Fig. 1 can be descerned in Fig. 3 and we note that the microcrystalline phase is present as wedge-shaped regions between the cells. The two phases are easily recognized in the bright field image (Fig. 3(a)) by noting that the b.c.c. phase is bright in the dark field image from an electron beam diffracted by the b.c.c. phase (Fig. 3(b)). The microcrystalline phase is bright in the dark field image from a segment of the diffuse ring (Fig. 3(c)). These observations are consistent with the following two interpretations: (1) the transformation nucleates heterogeneously in intercellular regions, at least some of which are adjacent to the implanted surface, or (2) a nearly planar interface parallel to the implanted surface separates the b.c.c. phase on the surface side from a slab of microcrystalline phase on the substrate side of the interface. In case (1) the wedge-shaped microcrystalline regions would either extend completely through the approximately 1000 Å TEM foil or have the b.c.c. phase underlying them if they exist only very near to the surface. The total lack of contrast in the wedges in Fig. 3(b) shows that the wedges extend through the foil. The viability of case (1) now rests on the fact that the b.c.c. phase also passes completely through the TEM foil. This is not the case, however, as contrast remains in the b.c.c. regions (Fig. 3(c)) when images from the diffuse ring are recorded. These observations are consistent with the second interpretation where the microcrystalline phase lies beneath the b.c.c. phase. The case is made even more clear by the images in

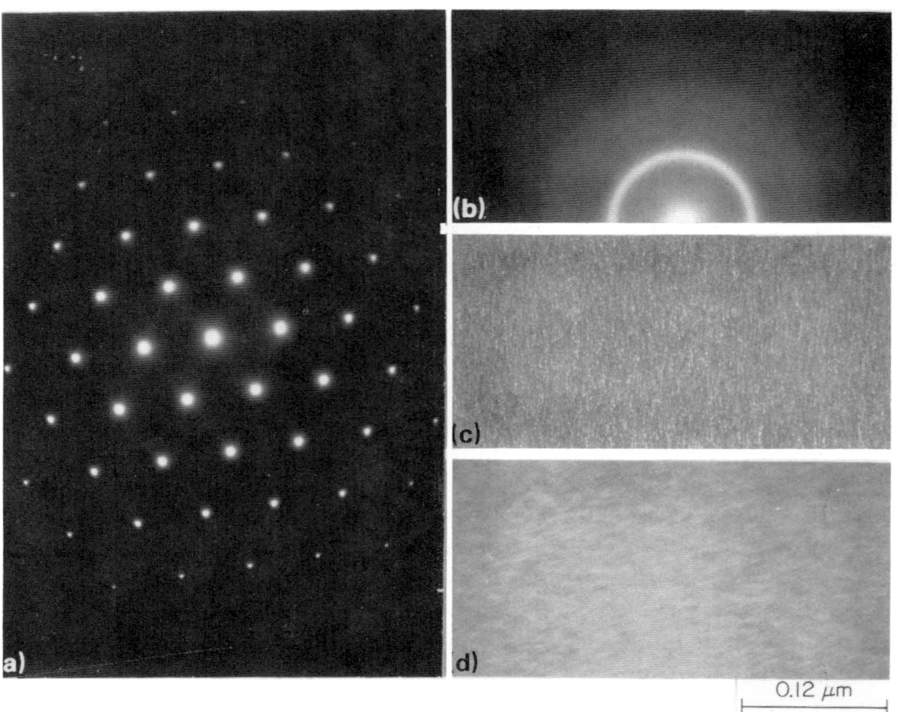

Fig. 2. Transition from b.c.c. to amorphous phase: (a) b.c.c. (111) diffraction pattern, 1.2×10^{18} Al^+ cm^{-2} into tantalum; (b) aluminum-implanted niobium, diffraction pattern at fluences beyond 1.8×10^{18} Al^+ cm^{-2}; (c) microcrystalline aluminum phase at 1.8×10^{18} Al^+ cm^{-2}; (d) amorphous aluminum phase at 2.4×10^{18} Al^+ cm^{-2}. The dark field images in (b)–(d) were obtained using the most intense diffuse ring.

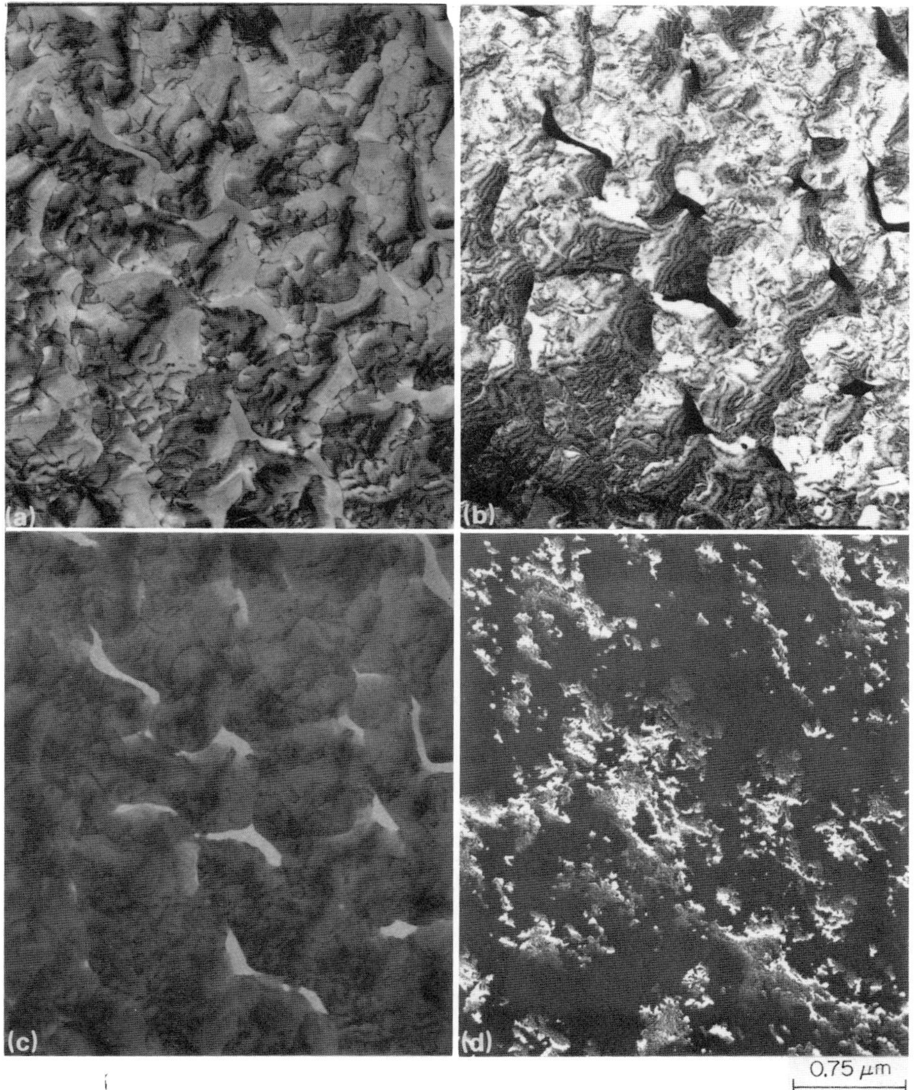

Fig. 3. Presence of microcrystalline phase at b.c.c. cell boundaries: (a) bright field image after implantation of 1.65×10^{18} Al^{+} cm^{-2} into tantalum; (b) crystalline dark field image after implantation of 1.65×10^{18} Al^{+} cm^{-2} into tantalum; (c) microcrystalline dark field image (as in Fig. 2(b)) after implantation of 1.65×10^{18} Al^{+} cm^{-2}; into tantalum (d) crystalline dark field image after implantation of 1.75×10^{18} Al^{+} cm^{-2} into tantalum.

Fig. 3(d), recorded after a fluence of 1.75×10^{18} Al^{+} cm^{-2}. The b.c.c. phase is even thinner at this fluence, and more of the microcrystalline phase (on whose surface the b.c.c. phase rests) is evident.

Unlike aluminum-implanted tantalum and niobium, the b.c.c. phase of vanadium did not transform to a microcrystalline or amorphous phase. The b.c.c. phase was retained to fluences of about 1.8×10^{18} Al^{+} cm^{-2}, beyond which another crystalline phase was observed. Samples implanted to 2.4×10^{18} Al^{+} cm^{-2} contained only the new phase. The crystal structure of this phase is not known at present but many of its diffracted electron beams appear at angles close to those from the b.c.c. phase (Fig. 4(b)). The array of weaker beams at one-half of the spacing of the stronger beams indicates that the aluminum and vanadium atoms are ordered in the crystal structure. The morphology of the phase is shown in the micrograph in Fig. 4(a). Several orientations

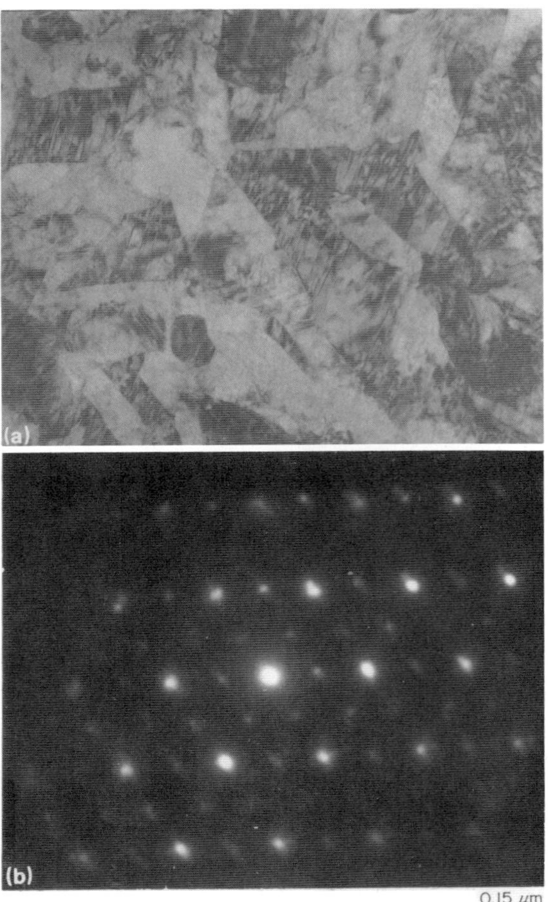

Fig. 4. Vanadium after implantation with 2.4×10^{18} Al^+ cm^{-2}: (a) bright field image of twinned crystals; (b) electron diffraction pattern from new phase, near the ⟨111⟩ direction of original b.c.c. phase.

of the phase, which exhibits twinning, are present. Only a single twin variant operates in a given orientation.

3.3. Compositions in the implanted layers

The aluminum concentration *vs.* depth from the implanted surface was measured with Rutherford backscattering spectroscopy (RBS). Composition variations were also detectable using Auger electron spectroscopy (AES) and sputter etching. Actual compositions were not determined by AES because of the lack of refractory metal–aluminum standards. Energy-dispersive X-ray analysis (EDXA) from the TEM foils under examination provided the aluminum concentration averaged through the foil thickness.

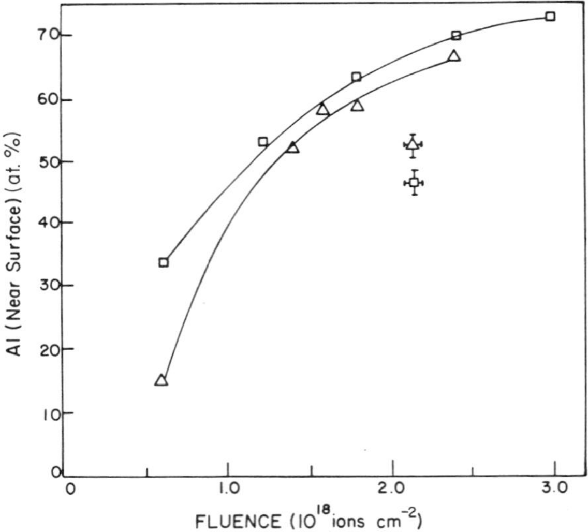

Fig. 5. Aluminum concentration in tantalum TEM foils *vs.* fluence, measured with energy-dispersive spectroscopy (△) and by averaging RBS contents (□) within 700 Å of the surface.

The aluminum content measured as a function of fluence using EDXA is shown in Fig. 5. Also shown are the results of averaging the compositions measured by RBS in the 700 Å of material adjacent to the surface. 700 Å was a typical TEM foil thickness. Except at the lowest fluence, the two techniques agree within the limits of experimental error. The aluminum content increases rapidly with increasing fluence, from about 20 at.% at 6×10^{17} Al^+ cm^{-2} to about 50 at.% by 1.2×10^{18} Al^+ cm^{-2}. Thereafter, it increases more slowly, reaching only about 60 at.% by 1.8×10^{18} Al^+ cm^{-2} and finally leveling off near 70 at.% beyond 2.4×10^{18} Al^+ cm^{-2}. Both RBS and AES showed some aluminum variation through the typical TEM foil thickness [9] (Table 1). At 6×10^{17} Al^+ cm^{-2}, the content was less at the surface and more at 1000 Å, relative to the average, by −10 at.% and +10 at.% respectively. By 1.8×10^{18} Al^+ cm^{-2}, this variation was only ±4 at.%, and at fluences beyond 2.4×10^{18} Al^+ cm^{-2} it was even less than this.

3.4. Annealing of the amorphous Ta–Al phase

Foils of tantalum implanted with 2.4×10^{18} Al^+ cm^{-2} were annealed at temperatures between 600 and 800 °C. Dendrites formed within 30 min at 600 °C (Fig. 6). Analysis of single-crystal patterns from these dendrites, such as the inset on the left in Fig. 6, showed that the dendrites were composed of the $D0_{22}$ $TaAl_3$ phase found on the equilibrium phase diagram. The other inset pattern on the right in Fig. 6 is from the matrix. At least five diffuse rings are visible in the original print of this pattern. These same rings could be found, but were not nearly as distinct, in patterns from the as-implanted microcrystalline and amorphous Ta–Al and Nb–Al phases. These same five rings are shared by polycrystalline b.c.c. tantalum. However, one of the b.c.c. rings is missing from the Ta–Al patterns, namely that with the indices 200. Thus the microcrystalline phase and amorphous phases probably originate from implantation destabilizing $TaAl_3$ rather than from the b.c.c. solid solution, a point developed further in Section 4.

TABLE 1

Aluminum content at selected fluences for aluminum-implanted tantalum

Fluence (Al^+ cm^{-2})	*Surface Al concentration* (at.%)	*Al concentration at 1000 Å* (at.%)
6×10^{17}	30	37
1.2×10^{18}	48	55
1.8×10^{18}	60	65
2.4×10^{18}	67	70
3.0×10^{18}	70	72

4. DISCUSSION

The transformation to microcrystalline and amorphous phases was observed in both tantalum and niobium as a result of aluminum ion implantation. This will now be discussed in terms of the free energies of the various phases involved, paralleling the discussion by other researchers for other alloy systems [10, 11]. The free-energy diagram for the Nb–Al system, the only system of the three studied here for which thermodynamic parameters were complete, was calculated from data provided by Kaufman and Nesor [12]. This is presented in Fig. 7. The corresponding free-energy diagram for the Ta–Al system would be quite similar, since the phase diagrams of Nb–Al and Ta–Al share many features in common [13, 14].

Under equilibrium conditions, the tangent rule applied to Fig. 7 shows that the following

Fig. 6. Dendrites of $TaAl_3$ formed during 30 min at 600 °C, after implantation with 2.4×10^{18} Al^+ cm^{-2} into tantalum at 25 °C. The inset diffraction patterns are from crystalline $TaAl_3$ (left) and from the microcrystalline matrix (right).

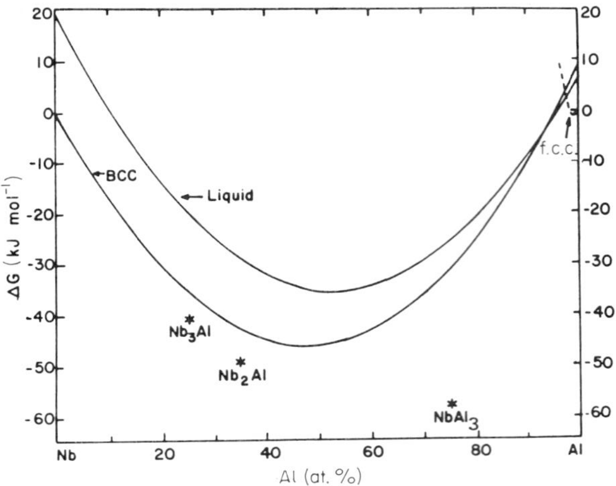

Fig. 7. Free energies of Nb–Al solutions and compounds at 25 °C, plotted *vs.* aluminum concentration.

phases are stable: b.c.c. niobium solid solution, less than 10 at.% Al; b.c.c. solid solution $+Nb_3Al$, about 10–25 at.% Al; $Nb_3Al + Nb_2Al$, 25–33 at.% Al; $Nb_2Al + Nb_2Al_3$, 33–60 at.% Al; $NbAl_3$ + f.c.c. solid solution 60–99 + at.% Al; f.c.c. solid solution, greater than 99 + at.% Al. Brimhall *et al.* [11] have suggested that the free energies of compounds with narrow homogeneity ranges will increase rapidly during irradiation. This is due to microscopic spatial variations in chemical composition caused by the irradiation and the marked sensitivity of the compound's free energy to these chemical changes. The close proximities of the free energies of Nb_3Al and Nb_2Al to the b.c.c. free-energy curve, coupled with this irradiation effect, foretell that these compounds will be unstable during implantation and will be replaced by the b.c.c. solid solution. Our observations of the extensions of the b.c.c. solubility range to aluminum concentrations much greater than found under equilibrium, from less than 1 at.% to about 50 at.% for tantalum, is consistent with this interpretation.

The remaining intermetallic compound, $NbAl_3$, appears to be quite stable under equilibrium conditions relative to the b.c.c. solid solution (Fig. 7). Its free energy lies well below that of the b.c.c. phase, unlike the case above. The homogeneity ranges of $NbAl_3$ and $TaAl_3$ are extremely limited [13, 14], *i.e.* they are line compounds. In accord with Brimhall *et al.*, we identify them as highly

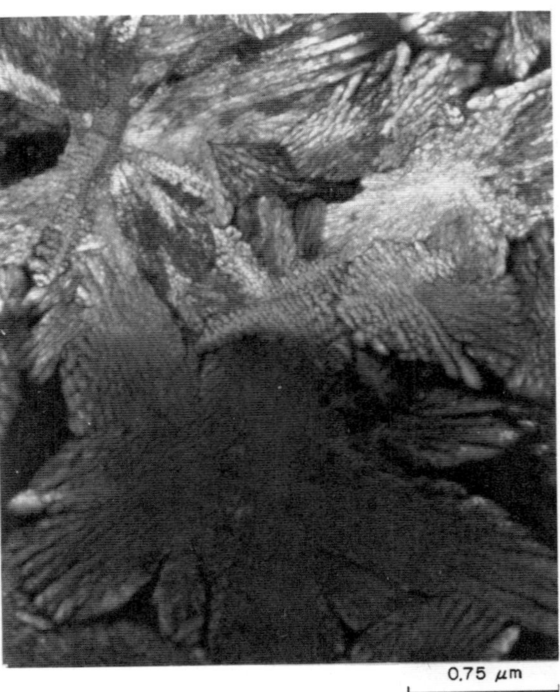

Fig. 8. Specimen treated as in Fig. 6, partially masked from the ion beam and then implanted at 25 °C with 1×10^{15} Al^+ cm^{-2}. The upper portion was shielded from the beam, and the lower portion was not.

unstable to irradiation. This is confirmed by experiment (Fig. 8). Dendrites of crystalline $TaAl_3$ were formed by annealing an implanted foil, as in Fig. 6. The foil was then partially masked and implanted (irradiated) further with aluminum ions. A fluence as small as 1×10^{15} Al^+ cm^{-2} causes a homogeneous crystalline-to-amorphous transformation, as can be noted by the reduced diffraction contrast of the newly implanted portions of $TaAl_3$ dendrites in Fig. 8. By 10^{17} Al^+ cm^{-2} the dendrites were indistinguishable from the amorphous matrix, *i.e.* the transformation was complete before this fluence.

Strict adherence to the free-energy diagram (Fig. 7) leads us to expect that the b.c.c. phase would replace the $NbAl_3$ phase in implanted specimens, since the b.c.c. free energy is less than the liquid phase at 75 at.% Al. Instead, the amorphous (liquid) phase is observed in our work. We explain this by noting that the free-energy calculation for the liquid phase takes no account of clustering or ordering which may occur at lower temperatures. Indeed here we note the strong tendency to order, exhibited by the microcrystalline phase field. Patterns from the microcrystalline phase and the amorphous phase both show strong associations with the $NbAl_3$ and $TaAl_3$ structures. Such ordering would reduce the free-energy curve for the liquid, were it included in the calculation. The b.c.c. and liquid curves (Fig. 7) are already close to one another at 75 at.% Al, and this would cause the liquid phase to have a lower free energy, consistent with the experimentally observed presence of the amorphous phase. Finally, we note, on the basis of Fig. 7, that the amorphous phase forms at very low fluences and tantalum contents (well below those required to produce $TaAl_3$) when tantalum ions are implanted into aluminum.

In closing, we note again that the V–Al system did not exhibit amorphous phase formation but instead underwent a crystalline-to-crystalline transformation. Lacking both a complete characterization of the structure of the new phase and the free-energy data for the V–Al system, we postpone discussion of this system to a later date.

5. SUMMARY

The solubility of aluminum in the solid solutions of the refractory metals tantalum and niobium is greatly extended as a result of ion implantation, from less than 1 at.% under equilibrium to about 50 at.% during implantation of tantalum. The b.c.c. phase transforms to a microcrystalline phase which, in turn, transforms to amorphous phase with increasing aluminum ion fluence. The transformation from b.c.c. phase begins near 1.6×10^{18} Al^+ cm^{-2} (about 60 at.% Al in tantalum) and ends near 1.8×10^{18} Al^+ cm^{-2}. The microcrystalline and amorphous phase diffraction patterns reflect the retention of structure from the $D0_{22}$-type MAl_3 phases. Crystalline $TaAl_3$ transformed to amorphous phase during low fluence (10^{15}–10^{17} Al^+ cm^{-2}) implantation. The free-energy diagram of Nb–Al was consistent with these experimental observations.

Vanadium implanted with aluminum ions did not become amorphous but instead exhibited a transformation from the b.c.c. structure to another crystalline structure.

ACKNOWLEDGMENTS

We thank the following people for contributing to this research: H. Hayden and C. Koch for assisting in the ion implantations; L. McCurdy for TEM assistance; Lane Witherell for typing.

This material is based on work supported by the National Science Foundation under Grant DMR8507641. The transmission electron microscope was purchased with support from the National Science Foundation under Grant DMR8207266 and the State of Connecticut.

REFERENCES

1 W. W. Beaver, A. J. Stonehouse and R. M. Payne, *Proc. Plansee Semin., 1965*, Metallwerk Plansee, Reutte, 1965, p. 682.

2 M. Ahmed, K. Ruffing and D. I. Potter, *Proc. Int. Conf. on Surface Modification of Metals by Ion Beams, Kingston, July 7–11, 1986*, in *Mater. Sci. Eng., 90* (1987).

3 P. B. Hirsch, A. Howie, R. B. Nicholson, D. W. Pashley and M. J. Whelan, *Electron Microscopy of Thin Crystals*, 1st edn, Butterworths, London, 1965, p. 159.

4 M. Saqib and D. I. Potter, *Proc. Int. Conf. on Surface Modification of Metals by Ion Beams, Kingston, July 7–11, 1986*, in *Mater. Sci. Eng., 90* (1987).

5 J. I. Goldstein, in J. J. Hren, J. I. Goldstein and D. C. Joy (eds.), *Introduction to Analytical Electron Microscopy*, Plenum, New York, 1979, p. 83.

6 J. J. Hren, in J. J. Hren, J. I. Goldstein and D. C. Joy (eds.) *Introduction to Analytical Electron Microscopy*, Plenum, New York, 1979, p. 481.

7 R. P. Elliot (ed.), *Constitution of Binary Alloys, First Supplement*, McGraw-Hill, New York, 1965, p. 56.

8 C. E. Lundin and A. S. Yamamoto, *Trans. Metall. Soc. AIME, 236* (1966) 863.

9 M. Saqib, *M.S. Thesis*, Metallurgy Department, University of Connecticut, Storrs, CT, 1984.

10 N. Saunders and A. P. Miodownik, *Ber. Bunsenges. Phys. Chem., 87* (1983) 830.

11 J. L. Brimhall, H. E. Kissinger and L. A. Charlot, *Radiat. Eff., 77* (1983) 237.

12 L. Kaufman and H. Nesor, *Calphad, 2* (1978) 325.

13 *Bull. Alloy Phase Diagrams, 2* (1981) 75.

14 J. C. Schuster, *Z. Metallkd., 76* (1985) 724.

Materials Science and Engineering, 90 (1987) 91-97

Structural Changes in a Cobalt-based Alloy after High Fluence Ion Implantation*

S. A. DILLICH and R. R. BIEDERMAN

Worcester Polytechnic Institute, Worcester, MA 01609 (U.S.A.)

(Received July 10, 1986)

ABSTRACT

Structural changes in a cobalt-based alloy (Co-31Cr-12.5W-2.2C where the composition is in approximate weight per cent; Stoody 3) as a result of high fluence nitrogen or titanium ion implantations were investigated via transmission electron microscopy and selected area diffraction examinations of unimplanted and implanted foils. The alloy microstructure was found to consist of several morphologies of single-crystal carbides (Cr-Co-W solutions) in a cobalt-rich f.c.c. matrix phase of high planar defect density. Titanium implantation (5×10^{17} Ti^+ cm^{-2} at 190 keV) produced a surface layer with an amorphous matrix phase and recrystallized carbides, while nitrogen implantation (4×10^{17} N^+ cm^{-2} at 50 keV) greatly increased the planar fault density in the matrix phase. Previously reported effects of titanium and nitrogen implantations on the tribological behavior of the alloy are discussed in terms of the results of these investigations.

1. INTRODUCTION

Cobalt-based alloys are among the most wear-resistant materials commercially available. These alloys consist of hard chromium and tungsten carbides (most often M_7C_3 and M_6C) dispersed in softer, more ductile cobalt-rich solid solutions. Although Co-Cr-W solid solutions normally have h.c.p. structures at low temperatures, for most applications the alloys are designed to have the more ductile metastable f.c.c. phase. The superior abrasion and erosion resistance of these alloys is attributed to their low stacking fault energies which allow low temperature localized martensitic f.c.c.-to-h.c.p. transformations to occur with strain at an abraded or eroded surface [1-3]. Thus, it is believed, the ductile bulk alloys are protected by surfaces with high strain-hardening characteristics.

Ion implantations of a cobalt-based superalloy (Co-31Cr-12.5W-2.2C where the composition is in approximate weight per cent; commercial designation, Stoody 3) with titanium and nitrogen have been found to produce significant changes in the tribological behavior of the alloy, as observed during abrasive wear, dry sliding friction and cavitation erosion tests [4-6]. The abrasion resistances of nitrogen- and titanium-implanted samples relative to that of the unimplanted alloy can be seen in Fig. 1. Titanium implantation of the alloy created abrasion-resistant surfaces which have also exhibited improved cavitation erosion resistance [6] and a 50%-70% reduction in dry sliding friction [4, 5]. Scanning Auger microscope analysis of titanium-implanted surfaces revealed that a vacuum-carburized titanium layer was produced on the alloy in both matrix and carbide phases during implantation [4]. Similar vacuum-carburized (Fe-Ti-C) surfaces found on titanium-implanted steels have been shown to be amorphous [7-9]. However, the existence of disordered phases in titanium-implanted Stoody 3 surfaces remained to be determined.

In contrast, the abrasive wear resistance of the Stoody 3 alloy dropped to about one-half of the bulk value after nitrogen implantation (Fig. 1). Although friction and erosion of the alloy remained high after nitrogen implantation, changes in the wear mode during dry sliding [5] and cavitation erosion tests [10] were observed. Nitrogen stabilization of the metastable f.c.c. matrix phase, with a resultant decrease in surface work hardening during sliding and abrasion, was offered as a possible

*Paper presented at the International Conference on Surface Modification of Metals by Ion Beams, Kingston, Canada, July 7-11, 1986.

0025-5416/87/$3.50

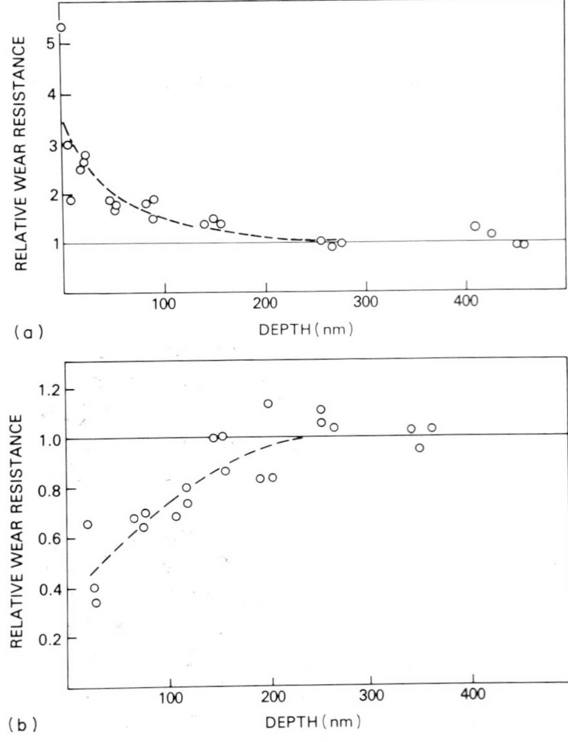

Fig. 1. Relative wear resistances of (a) titanium-im-implanted (5×10^{17} Ti^+ cm^{-2}; 190 keV) Stoody 3 samples and (b) nitrogen-implanted (4×10^{17} N^+ cm^{-2}; 50 keV) Stoody 3 samples, abraded with 1-5 μm diamond *vs.* depth [5]: ——, unimplanted Stoody 3 samples.

but unproven explanation of the observed behavior [5]. Since nitrogen has little or no solid solubility in cobalt, its effects as a phase stabilizer have not been documented. However, nitrogen stabilization of the f.c.c. phase in the binder alloy of a nitrogen-implanted cobalt-cemented tungsten carbide has been reported [11].

In this work, microstructural changes responsible for the observed tribological behavior of the titanium- and nitrogen-implanted alloy surfaces were investigated by transmission electron microscopy (TEM) and selected area diffraction (SAD) analyses of titanium-implanted, nitrogen-implanted and unimplanted alloy foils.

2. SAMPLE PREPARATION

TEM foils (3 mm in diameter) were punched from hand-ground disks (12.7 mm in diameter and 0.1 mm thick) and then polished on both sides to a 1 μm or less finish. Implant conditions for the foils were chosen to correspond to those used in the friction, wear and erosion studies discussed above, *i.e.* 5×10^{17} Ti^+ cm^{-2} at 190 keV or 4×10^{17} N^+ cm^{-2} at 50 keV. Implantations were performed in a modified model 200-20A2F Varion-Extrion ion implanter with a hot-cathode arc discharge-type ion source. The samples were heat sunk onto a water-cooled holder during implantation to limit the temperature rise at the surfaces to less than 50 °C.

The foils were thinned by one-sided electropolishing at −50 °C, in a solution of 15% HNO in methanol. Plastic "dummy" foils protected the surface of interest (*i.e.* the implanted surface or for the unimplanted foils the as-polished surface) from the electrolyte during thinning.

3. RESULTS

Thinned regions on the foils were examined in a 100 kV JEOL 100C scanning transmission electron microscope. Transmission of electrons at this energy is limited to regions with a maximum thickness of about 150 nm, *i.e.* corresponding to the range of ion penetration on the implanted surfaces [5].

Microstructural features of the unimplanted material were first examined to establish a reference for the implanted surfaces. A scanning electron microscopy (SEM) micrograph of an unimplanted foil is shown in Fig. 2. Several distinct carbide morphologies can be distinguished in the alloy: large dark lath- or block-shaped carbides, a light "script" phase, and smaller light and dark elliptically shaped carbides dispersed between the larger carbides. Energy-dispersive X-ray analysis revealed the metal components of the carbides to be Cr–Co–W solutions, with the lighter carbides having higher tungsten concentrations than the darker carbides. Exact stoichiometries of these carbides have not yet been determined.

Bright field TEM micrographs and SAD patterns from unimplanted foils are shown in Figs. 3-5. The matrix phase of the alloy is characterized by networks of planar defects (predominantly stacking faults) running through the grains and extending occasionally from one matrix grain to another (Figs. 3 and

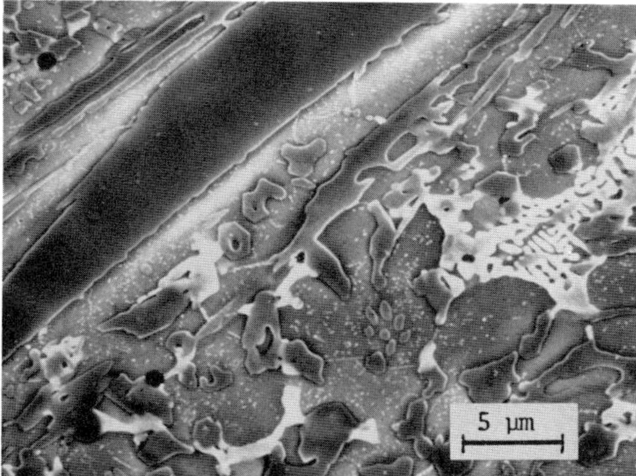

Fig. 2. SEM micrograph of an unimplanted foil. Several distinct carbide morphologies can be distinguished in the alloy.

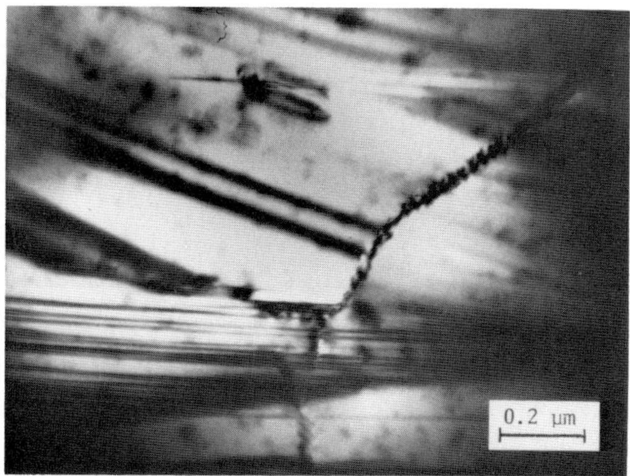

Fig. 3. Bright field micrograph of an unimplanted foil. The matrix phase was characterized by networks of planar defects running through the grains.

4(a)). Dislocation tangles could be found, infrequently, in combination with the stacking faults or concentrated at grain boundaries.

Single-crystal diffraction patterns from the cobalt-rich matrix phase indicated only the presence of the metastable f.c.c. phase (*e.g.* Fig. 4(b)), suggesting that cold working during the polishing process was not sufficiently severe to produce the martensitic f.c.c.-to-h.c.p. transformation at the surface. The large carbides appeared to be semiopaque and virtually defect free (Fig. 5(a)). SAD patterns from these carbides (Fig. 5(b)) were typical of those produced by single-crystal complex carbides.

Fig. 4. (a) Bright field micrograph of the matrix phase in an unimplanted foil; (b) the corresponding SAD pattern (110 zone; f.c.c.).

In order to investigate the effects of cold-work deformation on the alloy microstructure, observations were also made on an unpolished (600 grit SiC paper finish) foil (Fig. 6(a)). A much denser dislocation substructure was observed in this foil than in the fine-polished samples (compare Figs. 4(a) and 6(a)). Deformation of the sample surface produced streaking and ring segments on the SAD patterns (Fig. 6(b)). Again, the cubic phase was predominant in all patterns from the matrix phase; however, the presence of extra spots and streaks superimposed on the basic f.c.c. pattern in Fig. 6 suggests the possibility that the f.c.c. matrix contained some regions of transformed h.c.p. material.

The effects of titanium implantation on the alloy microstructure are shown in Figs. 7–9. Implantation produced an amorphous matrix phase, as evidenced by the diffuse rings in the SAD patterns from this phase, and by the

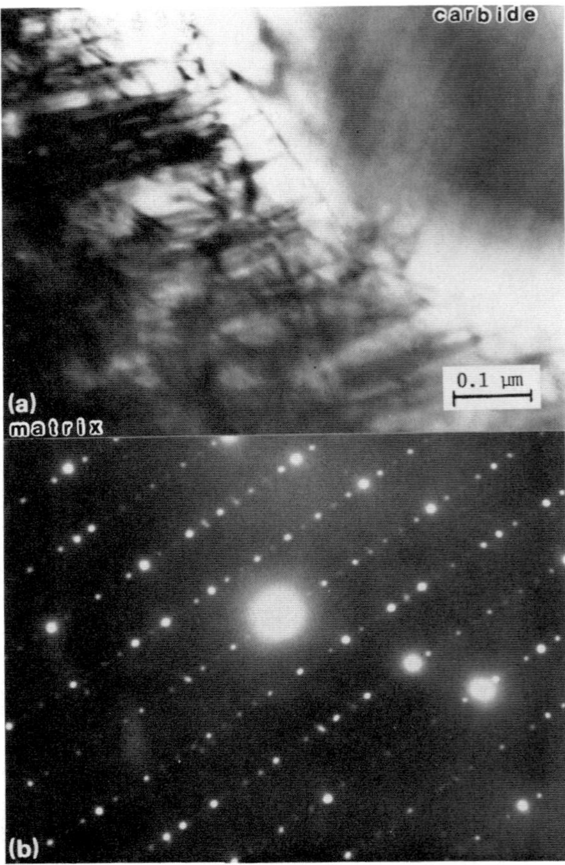

Fig. 5. (a) Bright field micrograph of a carbide–matrix interface; (b) an SAD pattern from the carbide.

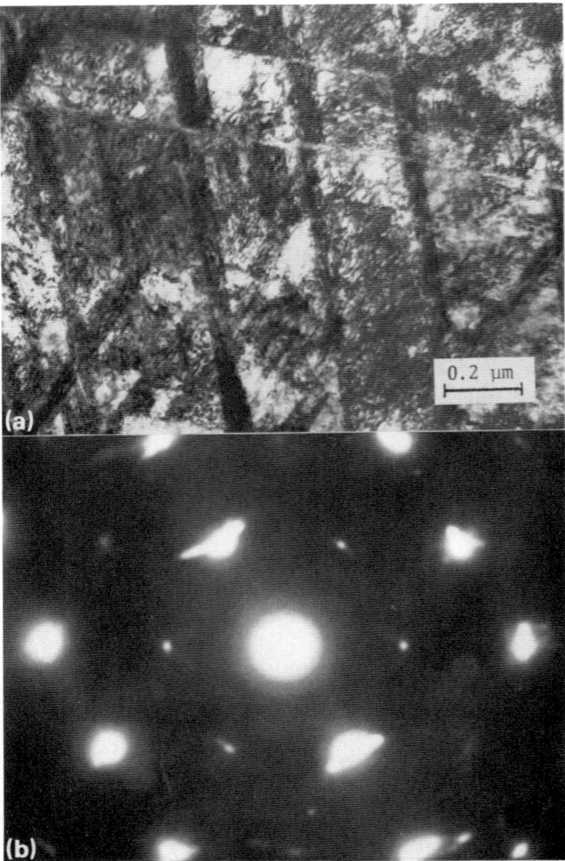

Fig. 6. (a) Bright field micrograph of the matrix phase in an unpolished foil; (b) the corresponding SAD pattern (211 zone; f.c.c.).

absence of defects, *i.e.* dislocations and planar faults (Fig. 7). Similarly, the carbides were transformed from monolithic single crystals to fine-grained polycrystals (Figs. 8 and 9), with diffraction spots forming ring patterns (Fig. 8). Partial amorphization of the carbide surfaces is suggested by the diffuse nature of the rings.

The carbides were outlined by borders of light contrast matrix material which, on close inspection, appeared to be carbide depleted. Energy-dispersive X-ray analysis of the border region shown in Fig. 9 indicated higher titanium concentrations in the carbides and bordering matrix than in the neighboring darker contrast matrix material. Preferential carburization (*i.e.* the formation of a titanium-plus-carbon-enriched surface layer [5]) at or near the carbides during implantation is a possible explanation of this result.

Micrographs from a nitrogen-implanted foil can be seen in Figs. 10 and 11. Single-crystal SAD patterns were obtained from both matrix and carbide phases on the foil. As was the case for the unimplanted foils, the matrix phase was found to be cubic (Fig. 10). Although nitrogen implantation did not produce changes in the crystallinity of the surface, a striking increase in the fault density was observed (compare Figs. 10 and 11 with Figs. 3 and 4).

4. DISCUSSION

In a previous work, it was found that titanium implantation modifies the chemistry of the Stoody 3 surface via vacuum carburization, *i.e.* the introduction of excess carbon atoms into the surface during implantation [4]. The TEM investigations described above revealed that microstructural changes (most notably a non-crystalline matrix phase) are

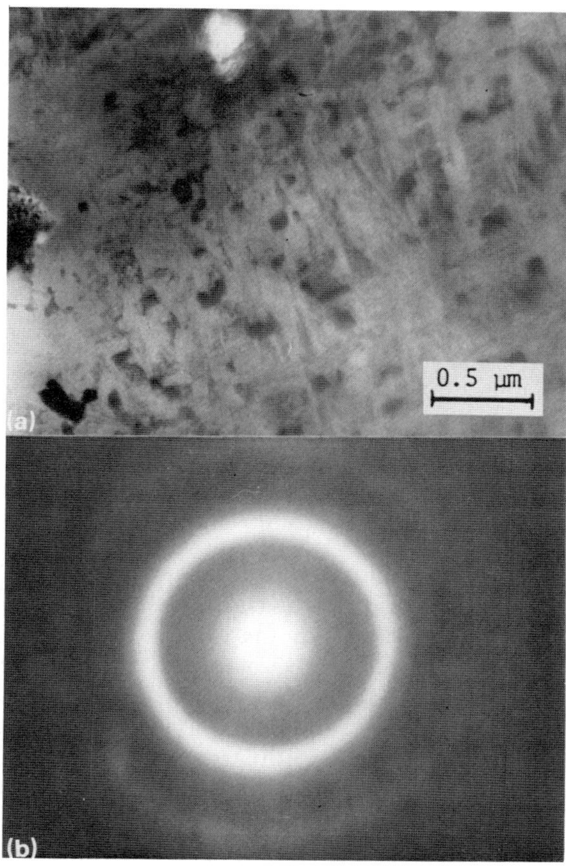

Fig. 7. (a) Bright field micrograph of the matrix phase in a titanium-implanted foil; (b) the corresponding SAD pattern.

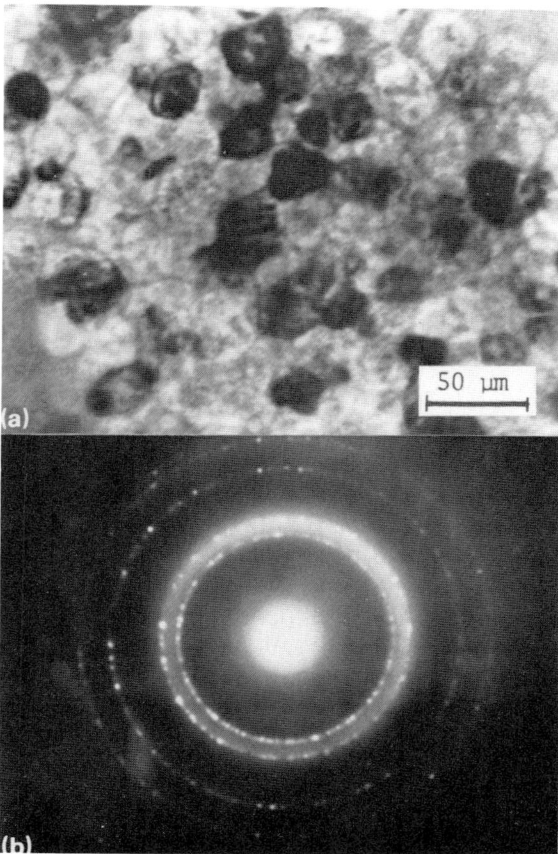

Fig. 8. (a) Bright field micrograph of a carbide in the titanium-implanted foil; (b) the corresponding SAD pattern. The rings in the pattern indicated a polycrystalline phase. Partial amorphization of the carbide surface is suggested by the diffuse nature of the rings.

Fig. 9. Bright field micrograph of the titanium-implanted foil and energy-dispersive X-ray spectra from (a) the matrix, (b) the matrix material bordering the carbide and (c) the carbide. The matrix appeared to have a lower surface concentration of titanium than the carbide or bordering matrix material does.

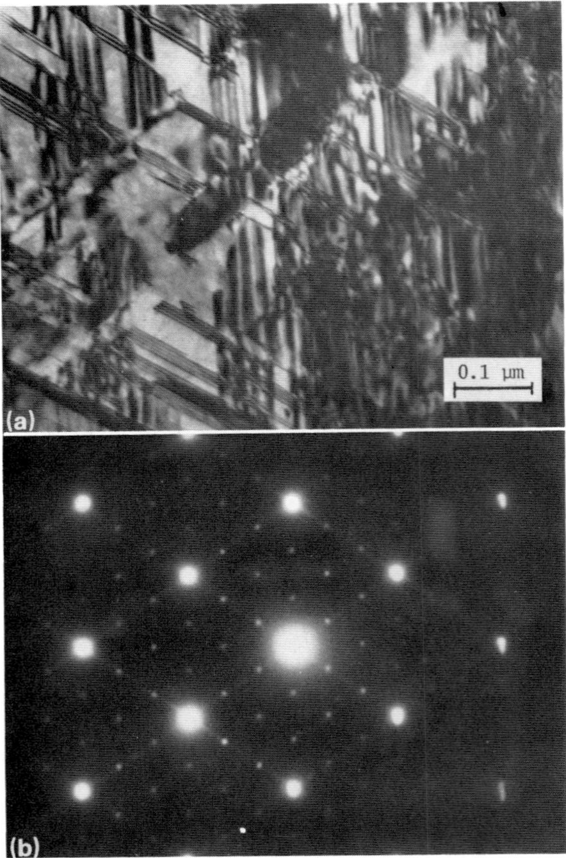

Fig. 10. (a) Bright field micrograph of the matrix phase in a nitrogen-implanted foil; (b) the corresponding SAD pattern (110 zone; f.c.c.). The extra spots grouped in clusters around the basic f.c.c. cobalt spots are believed to originate from carbides dispersed in the matrix.

Fig. 11. Bright field micrograph of matrix regions in a nitrogen-implanted foil. Very high stacking fault densities were seen in this phase.

also produced, the result presumably of ion bombardment damage to the original crystalline lattice. This amorphous carburized layer provides a low friction wear-resistant alloy surface, as shown by dry sliding friction, abrasion and cavitation erosion tests.

SEM examinations of cavitation-eroded Stoody 3 surfaces have shown that the titanium implant layer inhibits debonding at carbide–matrix interfaces and contributes additional toughness to the cobalt-based matrix, thereby delaying the loss of matrix phase material from the surface [6]. A corresponding increase in wear resistance of the carbides was not indicated, a consequence perhaps of recrystallization softening during implantation. This is consistent with results of other studies in which high fluence ion implantation has been found to cause softening [11, 12] and loss of wear resistance [13] in ceramic and carbide materials.

The superior wear resistance of the titanium-implanted Stoody alloy surfaces can, it appears, be attributed primarily to the matrix phase implant layer. However, the enhanced cohesion at carbide–matrix interfaces observed on eroded titanium-implanted surfaces suggests that these regions also play an important role in determining alloy erosion and abrasion resistance. Further study is needed to characterize fully the chemistry and structure of the interface regions and their influence on the early stages of wear.

No attempt was made in this study to determine the minimum titanium ion fluence or energy necessary to drive the surface amorphous or to correlate improved tribological behavior with a critical surface concentration of carbon, nor has the role of carbon in the stabilization of the disordered phase been identified. These topics remain to be addressed in more extensive studies.

After nitrogen implantation the dominant microstructural change observed in the alloy was the extremely high stacking fault density in the matrix phase. The origin of this particular feature, *i.e.* whether it is due to implantation-induced damage or to the presence of nitrogen in the alloy lattice, or to both, is unclear. Also a complete explanation of the effects of nitrogen on the alloy tribology is not yet available. Since the metastable f.c.c. phase was found in both unimplanted and nitrogen-implanted alloy foils, the possible role of nitrogen as an f.c.c. phase stabilizer cannot be commented on. It is possible that the presence of a greatly increased fault density in the implant layer limited the ability of the alloy to respond plastically under stress and that this effect, in and by itself, produced the poor abrasion resistance exhibited by the nitrogen-implanted Stoody 3 surfaces.

5. SUMMARY

TEM studies made on unimplanted and implanted foils revealed the microstructural changes resulting from titanium and nitrogen implantations. Amorphization of the cobalt-based matrix phase of the alloy after titanium implantation produces a low friction wear-resistant surface. Although nitrogen implantation does not alter the crystallinity of the alloy, a greatly increased stacking fault density is produced in the implant layer which results in a high friction, low wear resistance surface.

ACKNOWLEDGMENTS

We thank the Surface Modification and Materials Analysis Group, Naval Research Laboratory, Washington, DC, for their cooperation with implantations and James Steele (University of Connecticut) for the use of his diffraction pattern analysis software.

This work was supported through the materials division of the Office of Naval Research.

REFERENCES

1 K. C. Antony, *J. Met., 25* (February 1973) 52.
2 K. J. Bhansali and A. E. Miller, *Wear, 75* (1982) 241.
3 C. J. Heathcock, A. Ball and B. E. Protheroe, *Wear, 74* (1981-1982) 254.
4 S. A. Dillich and I. L. Singer, *Thin Solid Films, 73* (1981) 219.
5 S. A. Dillich, R. N. Bolster and I. L. Singer, in G. K. Hubler, O. W. Holland, C. R. Clayton and C. W. White (eds.), *Ion Implantation and Ion Beam Processing of Materials, Materials Research Society Symp. Proc., Boston, MA, November 14-17, 1983*, Vol. 27, Elsevier, New York, 1984, p. 637.
6 N. V. H. Gately and S. A. Dillich, *Proc. Int. Conf. on Surface Modification of Metals by Ion Beams, Kingston, July 7-11, 1986*, in *Mater. Sci. Eng., 90* (1987).
7 I. L. Singer, C. A. Carosella and J. R. Reed, *Nucl. Instrum. Methods, 182-183* (1981) 923.
8 D. M. Follstaedt, F. G. Yost and L. E. Pope, in G. K. Hubler, O. W. Holland, C. R. Clayton and C. W. White (eds.), *Ion Implantation and Ion Beam Processing of Materials, Materials Research Society Symp. Proc., Boston, MA, November 14-17, 1983*, Vol. 27, Elsevier, New York, 1984, p. 661.
9 J. A. Knapp, D. M. Follstaedt and B. L. Doyle, *Nucl. Instrum. Methods B, 7-8* (1985) 38.
10 N. V. H. Gately and S. A. Dillich, unpublished work, 1987.
11 G. Dearnaley, B. James, D. J. Mazey and F. J. Minter, *Proc. Plansee Conf., 1985*, Metallwerk Plansee, Reutte, in the press.
12 P. J. Burnett and T. F. Page, in E. A. Almond, C. A. Brookes and R. Warren (eds.), *Proc. 2nd Int. Conf. on the Science of Hard Materials, Rhodes, September 23-28, 1984*, in *Inst. Phys. Conf. Ser., 75* (1986) 789.
13 S. A. Dillich and I. L. Singer, *Proc. Int. Conf. on Surface Modifications and Coatings*, 1985, American Society for Metals, Metals Park, OH, in the press.

Materials Science and Engineering, *90* (1987) 99-109

Nitrogen Implantation into Steels*

N. MONCOFFRE

Institut de Physique Nucléaire et de Physique des Particules, Université Claude Bernard Lyon-I, 43 boulevard du 11 Novembre 1918, 69622 Villeurbanne Cédex (France)

(Received July 10, 1986)

ABSTRACT

Nitrogen implantation into steels is certainly the most well-known case of the improvement obtained in wear resistance using implantation. To a large extent, tribological results depend on the steel composition but also on the implantation conditions (fluence, temperature and vacuum quality). Both aspects will be discussed. In the evolution of the implanted layer, the combined influence of the newly created phases, the residual stresses and the nitride grain size must be of primary importance even if they are not always fully controlled. These are studied as a result of the thorough characterization of the implanted region which has been performed by numerous researchers and will be reviewed here. The study consists first in plotting the distribution of the implanted element with a good depth resolution (Auger electron spectroscopy, secondary ion mass spectrometry and nuclear reaction analysis), secondly in measuring the grain sizes (transmission electron microscopy) and thirdly in identifying the compounds formed during implantation (conversion electron Mössbauer spectroscopy, grazing-angle X-ray diffraction, electron spectroscopy for chemical analysis, transmission electron microscopy etc.). With respect to the characterization of the implanted region, recent results will be reported and in particular those concerning the preferential orientation of the nitride and carbonitride phases under irradiation.

*Paper presented at the International Conference on Surface Modification of Metals by Ion Beams, Kingston, Canada, July 7-11, 1986.

1. INTRODUCTION

Ion implantation is a surface treatment technique for metals that has been particularly developed over the last decade. The earliest work on the subject was by Hartley and co-workers [1-3] who investigated the wear resistance and other mechanical properties of steels using high fluence nitrogen implantation. Since that time, it has been demonstrated that the fatigue, friction and corrosion of different metals and alloys are surface properties that can be significantly modified by introducing appropriate ions via ion implantation. The aim of the research performed in many laboratories is increasingly a better understanding of the phenomena, allowing improvement in the performance of the treated materials. This work implies necessarily excellent characterization of the implanted layer, correlated when possible with tribological tests before and after implantation. However, in order to be precise about the purposes of the research in this field, we shall divide it into two important and complementary areas: study of the physical and chemical mechanisms of modifications of the implanted layer; understanding of the mechanisms and processes underlying the origin of the improvement in mechanical properties. Our work has essentially focused on the first and, in this paper, we shall examine only nitrogen implantation into steels in order to increase wear resistance. For nitrogen implantation, it is already known that the wear rate of steels is often reduced, thereby extending the lifetime of some tools [4]. These improvements mainly depend on the steel composition, on the implantation fluences and temperatures and also on the tribological tests. We shall consider the following parameters in succession in more detail:

0025-5416/87/$3.50

the nitrogen profile evolution, the nitrogen chemical state and microstructure evolution *vs.* fluence and temperature, the two last parameters being of primary importance.

2. CHARACTERIZATION OF THE IMPLANTED SURFACE

2.1. Profiling of the implanted element

The theoretical distributions of implanted ions into a material can be obtained for example from the tables of Winterbon [5]. As a first approximation, these distributions can be considered to have an almost gaussian shape. They are thus characterized by their R_p (mean projected range) and ΔR_p (standard deviation or projected range straggling) values. It is possible, however, to refine these distributions, taking into account other shape parameters such as the skewness γ_1 and kurtosis γ_2. For a 40 keV nitrogen implantation the theoretical curve thus obtained fits the experimental spectra very well when the implantation fluence is low ($\phi < 10^{16}$ ions cm^{-2}) and the temperature is about 20 °C. For higher fluences, progressive erosion of the material is quite significant and, in addition to the above four parameters, the sputtering yield S has to be calculated. Figure 1 compares theory and experience for a 10^{17} N^+ cm^{-2} implantation at room temperature into AISI 1006 steel. Good agreement between the measured and predicted profiles is observed under these conditions.

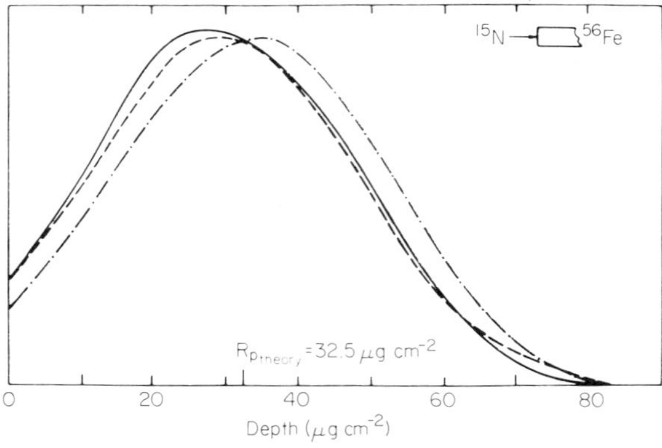

Fig. 1. Distribution profiles of 40 keV nitrogen ions ($^{15}N^+$) implanted in iron: - - -, experimental ($\phi = 10^{17}$ N^+ cm^{-2}); ——, theoretical (considering the first four moments); —·—, theoretical (considering the sputtering yield $S = 0.79$ and the first four moments).

The experimental techniques more commonly used to obtain nitrogen profiles are Auger electron spectroscopy [6–8] and the nuclear reactions $^{14}N(\alpha, \gamma)$ [6, 9, 10] and $^{14}N(d, p)$ [11]; $^{14}N(d, \alpha)$ [12] can also be used to quantify the Auger technique [6]. Rutherford backscattering [13, 14], secondary ion mass spectrometry (SIMS) [15] and X-ray photoelectron spectroscopy [15] are also often mentioned in the literature. It seems important, however, to emphasize that nuclear reactions present some very interesting features and advantages; they are very often nondestructive methods and they allow good depth resolutions to be obtained. In particular, the resonant reaction $^{15}N(p, \alpha\gamma)^{12}C$ that we have often used [16] gives a depth resolution of better than 5 nm.

Irrespective of the technique, a saturation fluence is always obtained when the implantation fluence increases at a given energy. It has been shown for example [16] that, for AISI 1006 steel implanted with 40 keV nitrogen ions, this fluence is 2×10^{17} ions cm^{-2} corresponding to about 30 at.% N (Fig. 2). These results are in agreement with Singer's [6] data. This saturation phenomenon occurs before the predicted theoretical value, taking into account the sputtering yield. To explain this fact, Barnavon *et al.* [16] proposed two models; one of these is based on radiolytic equilibrium, *i.e.* nitride formation and destruction. The chemical equilibrium

$$Fe_xN \underset{2}{\overset{1}{\rightleftharpoons}} x\,Fe + N$$

under ion beam irradiation is probably displaced in the direction 1, thus liberating free

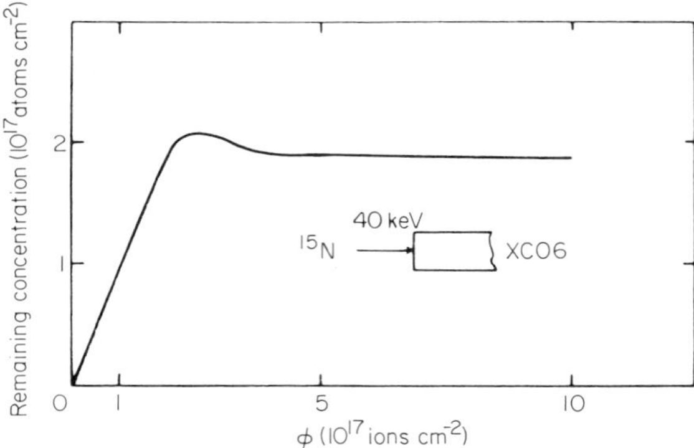

Fig. 2. Remaining nitrogen concentration *vs.* the implanted fluence in AISI 1006 steel.

nitrogen. This model is based on nitrogen diffusion linked to chemical reactions under ion beam irradiation.

Let us examine the effect of temperature, a very important implantation parameter. It has been shown, in particular by Rauschenbach [8] and by Moncoffre *et al.* [17] that, when the temperature during implantation is higher than 20 °C, the nitrogen profiles change rather rapidly. This evolution depends on the steel composition. In AISI 1006 steel, for implantation temperatures in the range 50–200 °C and fluence of 10^{17} N^+ cm^{-2}, the nitrogen has been proved to migrate progressively towards the surface, thus forming a superficial peak so that at 200 °C the whole nitrogen is redistributed in the first 60 nm [17] (Fig. 3). Up to 200 °C, the total remaining nitrogen is kept constant but, above this temperature, the nitrogen concentration decreases corresponding to out-diffusion. This means that above

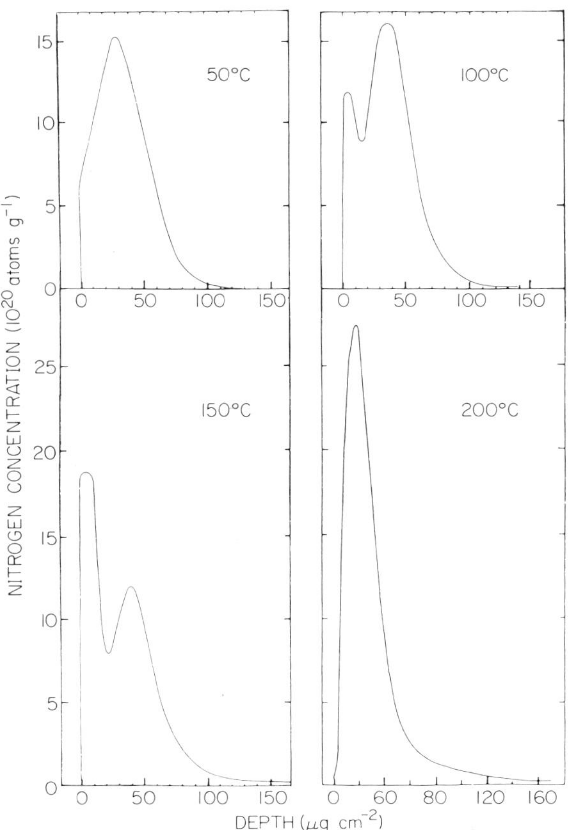

Fig. 3. Nitrogen distribution profiles in AISI 1006 steel implanted at various temperatures as indicated and at 10^{17} N^+ cm^{-2} (10 μg cm^{-2} roughly correspond to 13 nm).

50 °C the experimental spectra no longer fit the standard calculations. We shall discuss later the chemical phase change associated with this profile evolution.

In contrast, the temperature effects after implantation are completely different. Annealings of implanted samples in air or under vacuum will modify both the distribution profiles of the remaining nitrogen concentration and the chemical phases. Using conversion electron Mössbauer spectroscopy (CEMS), Longworth and Hartley [18] have shown that the annealing under vacuum of samples implanted with nitrogen at 20 °C drastically changes the phase nature of the compounds and enhances nitrogen diffusion towards the bulk.

Using the same technique, dos Santos *et al.* [19] demonstrate that, in AISI 1020 steel implanted with a high fluence of nitrogen and then annealed under vacuum, carbonitride phases completely dissolve above 450 °C and nitrogen migrates deeply into the bulk. Rauschenbach *et al.* [20] have also studied, by electron microscopy, nitride formation resulting from implantations and post-thermal treatments under vacuum. They deduced a phase diagram depending on fluence and temperature. Barnavon *et al.* [21] have performed a study whose aim was to compare the effects of annealing in air and under vacuum. They noticed in both cases a rapid nitrogen distribution narrowing. However, nitrogen removal is much more rapid under vacuum and at 175 °C a progressive decrease in the maximum nitrogen concentration has already been observed; in contrast, in air, no modification appears up to 250 °C. We shall discuss later a hypothesis that could explain the above behaviour as a function of temperature.

The chemical composition of the steels can also have a noticeable influence on the profile shapes. Singer [6, 22] suggested that the chromium content of the steel may play a role in the more or less gaussian shapes of the nitrogen distributions. In high chromium alloyed steels such as type 304 steel, he noticed an almost gaussian form of the profiles, suggesting that chromium which he called a strong nitride former "held" the nitrogen. In the opposite case, *i.e.* low chromium alloyed steels, a bulge appeared at the distribution surfaces. These observations seem to confirm the hypothesis of radiolytic equilibrium or at least nitrogen migration. The influence of alloying elements has also been mentioned by other researchers [23].

2.2. *Chemical states*

The formation of new phases at the surface of metals and alloys by ion implantation is probably one of the main reasons for the wear resistance improvements. The experimental techniques often used to characterize the implanted layers are CEMS [18, 19, 24-26], electron microscopy [27-29] and also more recently grazing-angle X-ray diffraction [30].

The phases almost always identified for 40 keV nitrogen ion (N^+) implantation in low alloyed steels, can be summarized as follows.

(i) For rather low fluences ($\phi < 10^{17}$ N^+ cm^{-2}) the detected phases are mainly nitrogen solid solutions (γ-austenite and α-martensite), nitrogen-poor nitrides (α''-$Fe_{16}N_2$) and small amounts of ϵ-nitrides and carbonitrides.

(ii) For fluences higher than 10^{17} N^+ cm^{-2} the hexagonal ϵ-nitride and carbonitrides are observed (ϵ-Fe_{2+x}(C, N), x varying from 0 to 1). The steel composition is of course of primary importance in the nature of the phases obtained but it is not the purpose here to review completely all the possible compounds that can be formed by implantation.

2.3. *Microstructures*

Grazing-angle X-ray diffraction is a technique of great interest for the study of implanted layers and can be used in a complementary way to CEMS to determine the phase composition after implantation. In particular, its advantage is depth sensitivity. By varying the X-ray beam incident angle ($0.3° < \alpha_i < 1°$), it is possible to follow the phase evolution for depths from 5 to 100 nm, which is the maximum range of 40 keV implanted nitrogen ions. Using this method, Moncoffre *et al.* [30] have investigated the phase evolution due to nitrogen implantation as a function of fluence (10^{17}-4×10^{17} N^+ cm^{-2}) and temperature (20-200 °C). This study gives evidence that, at 20 °C and 10^{17} N^+ cm^{-2}, N-martensite is preferentially located at the surface. Moreover, up to 100 °C, no nitride phases are detected (Fig. 4). At a fluence of 2×10^{17} N^+ cm^{-2} and even at 20 °C, carbonitrides ϵ-Fe_{2+x}(C, N) are detected (Fig. 5). Their average compositions can be obtained using empirical relations established by Firrao

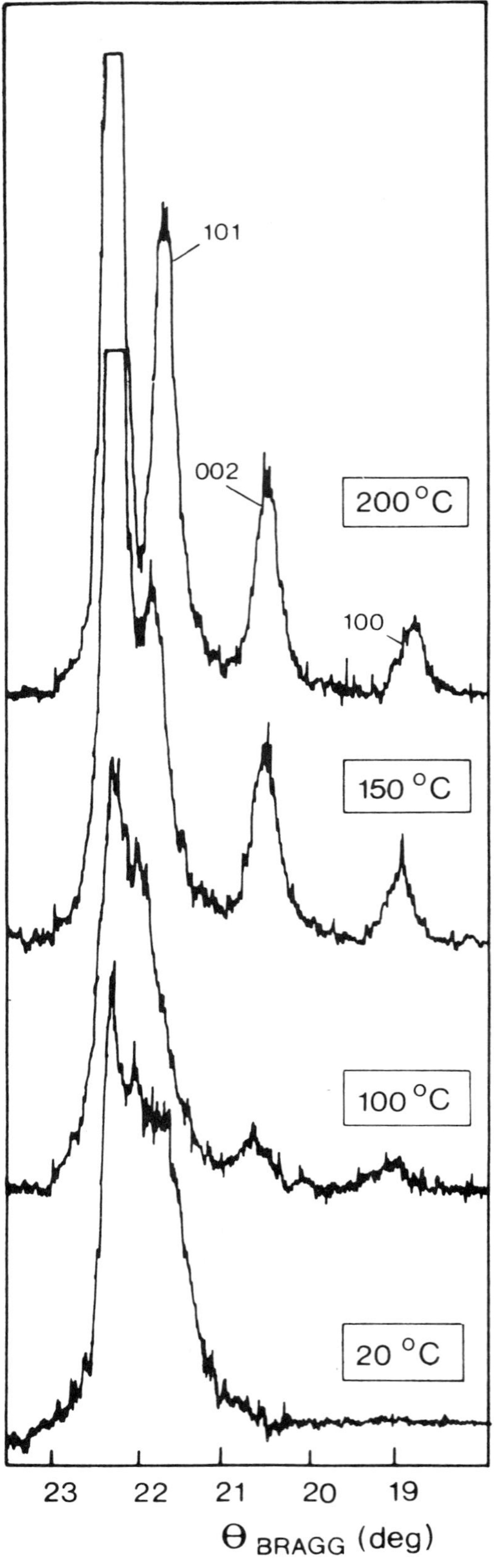

Fig. 4. X-ray diffraction spectra of 40 keV AISI 1006 implanted steel ($\alpha = 0.65°$; $\phi = 10^{17}$ N^+ cm^{-2}).

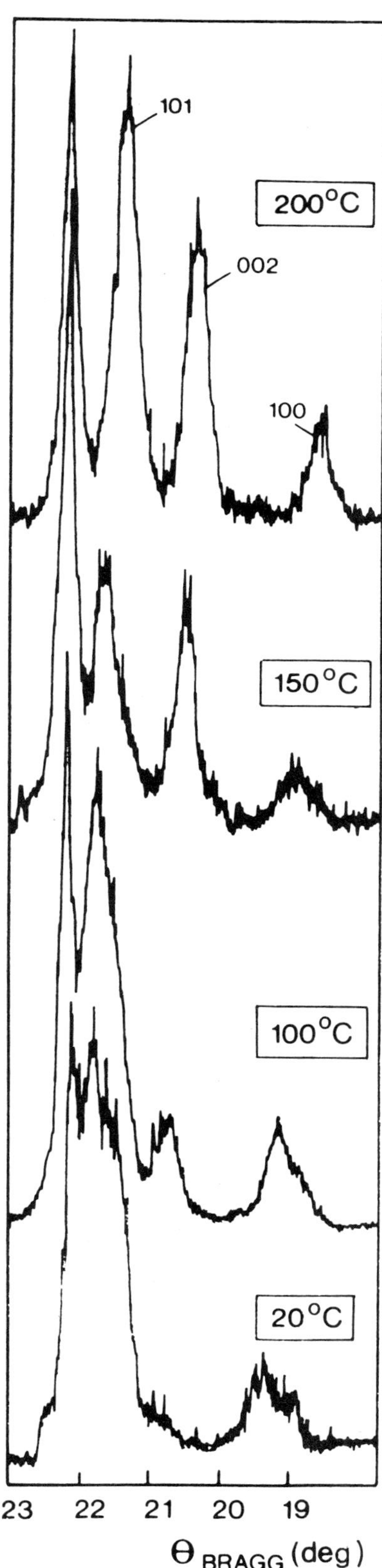

Fig. 5. X-ray diffraction spectra of 40 keV AISI 1006 implanted steel ($\alpha = 0.5°$; $\phi = 2 \times 10^{17}$ N^+ cm^{-2}).

et al. [31]. Nitrogen enrichment is observed at depths at least equal to R_p. At 4×10^{17} N^+ cm^{-2}, for a 1° incident angle, the calculated composition corresponds to the ϵ-$Fe_2(C,N)$ carbonitride. A phase analysis *vs.* implantation temperature shows nitrogen enrichment of nitride at high temperatures to the detriment of N-martensite or α''-$Fe_{16}N_2$ which both progressively disappear. These results are in good agreement with CEMS data [17, 26] and in particular with the phase sequence proposed by Carbucicchio *et al.* [32] who used depth-selective CEMS. Moreover, it is observed in the diffraction spectra that the diffraction peaks become narrower as the temperature increases, perhaps because the crystallite size increases but mainly because of a narrowing of the nitride composition range. However, these peaks remain relatively wide, preventing the distinction between nitride and carbonitride phases. When the CEMS results and surface carbon analysis (performed using 5.7 MeV α particle Rutherford backscattering) are taken into account, it seems obvious that the present phases are essentially carbonitrides.

Glancing X-ray diffraction measurements have not identified nitride phases at 10^{17} N^+ cm^{-2} and 20 °C. Under the same implantation conditions, nitride phases have been observed by CEMS. At such low fluences, nitrides are probably present as very fine and dispersed microcrystallites but not yet detectable by diffraction, a technique less sensitive to local order than is CEMS. As soon as the temperature reaches 100 °C, these crystallites grow and become detectable with the X-ray diffraction technique. For TEM studies, we can assume that the sample thinness necessary for this analysis probably favours a temperature increase and so an increase in the crystallite size. It could explain why Fayeulle *et al.* [28] observe small amounts of nitrides in Fe–Cr–Ni alloys implanted at 10^{17} N^+ cm^{-2} and 20 °C, using TEM.

The texture of the implanted layer was studied in the AISI 1006 steel using glancing X-ray diffraction. A preferential orientation of the nitrides appears in the diffraction spectra (and in particular is demonstrated better at high fluences, as shown in Fig. 6), when considering the abnormal intensity ratio of the 100, 002 or 101 reflections. In fact, at 20 °C and irrespective of the fluence, the 002

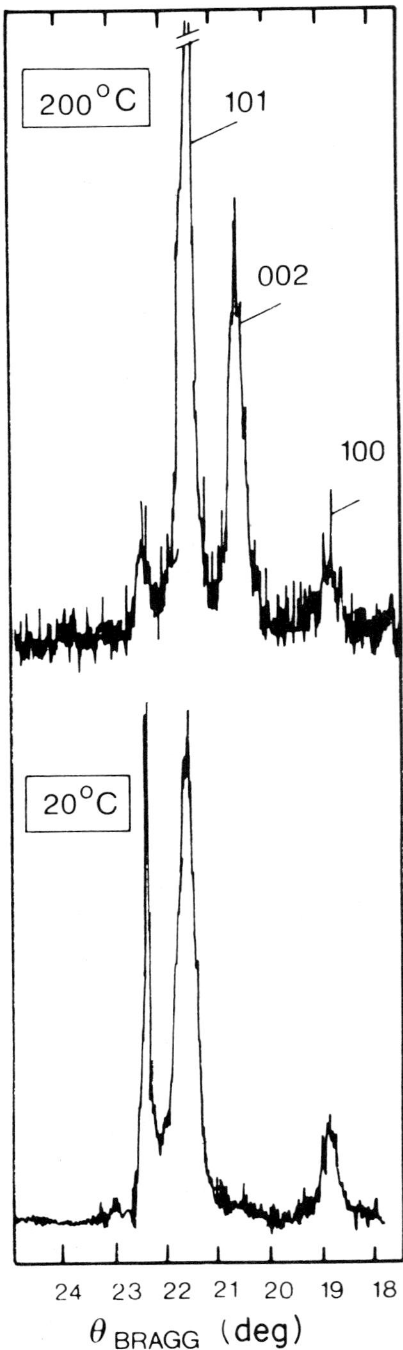

Fig. 6. X-ray diffraction spectra of 40 keV AISI 1006 implanted steel ($\phi = 4 \times 10^{17}$ N^+ cm^{-2}).

reflection is never observed when the diffracted beam is detected in the incident plane. Its intensity is a maximum in the surface plane. At 100 °C or above, the 002 reflection appears in the incident plane, becoming abnormally intense for an implantation temperature of 200 °C.

From these observations, we conclude that at 20 °C the *c* axis of the nitride crystallites is preferentially oriented in the surface plane, this orientation changing with the temperature but being fluence independent. It was observed that the beam direction is not the cause of the texture as was suggested by Van Wyk [33, 34] for copper crystallites. Texture of the implanted layers has already been mentioned by Hutchings [35] who used TEM for nitrogen-implanted chromium plates and by Fayeulle and Tréheux [36] for type 304 stainless steel.

The hypothesis that we propose to explain this phenomenon, but which would of course imply further investigations, is linked to the strong strains resulting from the implantation. Possible techniques to analyse these strains are grazing-angle diffraction and also the nuclear methods using channelling geometry.

2.4. The role of surface carbon

Mössbauer spectroscopy almost always detects ε-carbonitrides in nitrogen-implanted steels, giving rise to the question of the origin of the carbon. We performed a detailed study of the surface carbon contamination of nitrogen-implanted samples as a function of implantation temperature. We used 5.7 MeV α particle backscattering on a 60°-tilted sample of AISI 1006 steel so that the depth resolution thus obtained is about 7 nm. We have shown [37] that for a 10^{17} N^+ cm^{-2} implantation at 150 °C the surface carbon content increases by a factor of 3 over that for a 20 °C implantation. The carbon contamination reaches 1.2×10^{16} atoms cm^{-2} at 150 °C.

Singer [6] suggests that this carbon increase could be due to carbon migration from the bulk to the surface under an irradiation effect and to the great number of dislocations near the surface. Confirmation of this migration is given by SIMS experiments [38], displaying a carbon depression inside the implanted layer. However, experiments performed on pure iron [37] also indicate the presence of carbon at the surface (although in a smaller quantity). Thus, both origins for carbon must be taken into account.

The mechanism proposed to explain the growth of contamination carbon under beam consists of three steps, the main step being radiolysis. This can be considered as destructions due to atomic collisions and radiolytic evolutions under important X-ray and UV

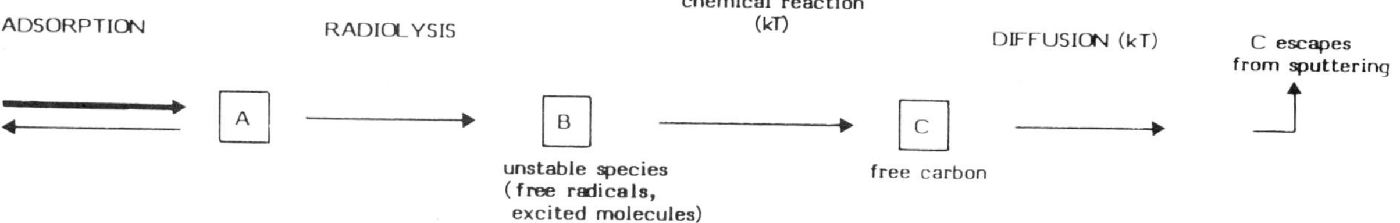

Fig. 7. Schematic representation of the carbon growth mechanism beam under ion beam irradiation.

photon flows. It leads to the transformation of stable molecules (molecules A) to other unstable molecules (molecules B). B can react chemically to give rise to stable compounds (C), the reaction kinetics being temperature dependent.

In this model, species A are carbonaceous coming from the residual vacuum of the implantation chamber (in spite of the nitrogen cold traps, they can be from the diffusion pumping, the graphite inner sides of the ion source etc.). They lead to stable phases that are, in this case, free carbon. This carbon diffusion into the bulk thus escapes sputtering (Fig. 7).

The step B→C and the diffusion process are thermally activated and could accelerate growth of carbon under ion beam irradiation with increasing temperature. This carbon contamination growth has to be correlated with the nuclear reaction profiles and with a CEMS study of the implanted region. It is shown [26] that the superficial peak mainly corresponds to the carbonitrides ϵ-$Fe_{2+x}(C,N)$. In these chemical phases, carbon forms very stable compounds with nitrogen, preventing nitrogen out-diffusion from the implanted zone. These new phases could play a role in the hardening of the implanted surface, as has already been suggested by dos Santos *et al.* [19].

3. TRIBOLOGICAL PROPERTY IMPROVEMENT

Because of the complexity of the problem [39], complete comprehension of the hardening mechanisms of implanted systems has not yet been attained. As well as the various implantation parameters (energy, fluence, temperature, vacuum quality etc.), the type of tribological test (lubricant, load, relative sliding velocity etc.) and the nature of the steels (chemical composition, surface state and

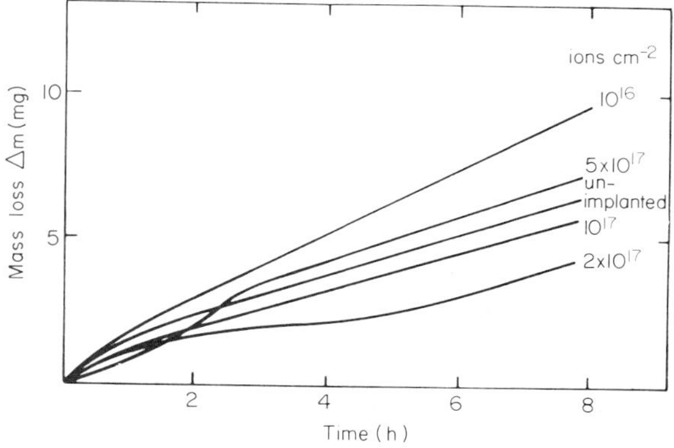

Fig. 8. Improvement in wear resistance in a high speed steel nitrogen implanted at various fluences and 20 °C [43].

thermal treatments) have to be taken into consideration. In order first to restrict the number of variables, let us consider an abrasive wear test. In this case, the improvement in wear resistance is essentially dependent on fluence, temperature and steel composition.

3.1. *Fluence*

Several researchers have already observed the existence of an optimal implantation fluence for a given material [12, 40–42]. For steels nitrogen implanted at 40 keV, this fluence is of the order of 2×10^{17} $N^+ cm^{-2}$. Figure 8 is taken from a study by Fayeulle [43]. It displays the wear resistance in a high speed steel (18 wt.%W–4 wt.%Cr–4 wt.%Co–1 wt.%V), implanted at 20 °C under various fluences. The results show clearly that the implantation process can also have a negative effect if the fluence is not adequately chosen as in the present fluences of 10^{16} and 5×10^{17} $N^+ cm^{-2}$. It has been said previously that in the two steels AISI 1006 and AISI 52100 a fluence of 2×10^{17} $N^+ cm^{-2}$ (*i.e.* a low influence) corresponds to the formation of well-defined crystallites. So at higher fluences the degradation noticed could be due to an increase in grain size or to dislocation enhancement. The ideal fluence is also evident when considering Fig. 9 which represents the wear tracks of type 304 steel implanted at three different fluences [36].

3.2. Temperature

We have already mentioned the great influence of implantation temperature on the nitrogen distribution profiles. Figure 10 illustrates the evolution of the wear rate for AISI 52100 steel implanted with 2×10^{17} N^{+} cm^{-2} *vs.* implantation temperature [44]. It appears that the best results are obtained at 150 °C. First, at this temperature, Rutherford backscattering data display a higher carbon concentration. Secondly, CEMS detects a larger carbonitride concentration. As suggested by dos Santos *et al.* [19], these phases could be, in part, a possible way of improving the wear properties. Moreover, glancing X-ray diffraction indicates that at this temperature the *c* axis of the nitride crystallites is preferentially oriented perpendicular to the surface. It will be very interesting and fundamental to answer the question whether this texture plays an important role in the tribological behaviour of the outer implanted layers.

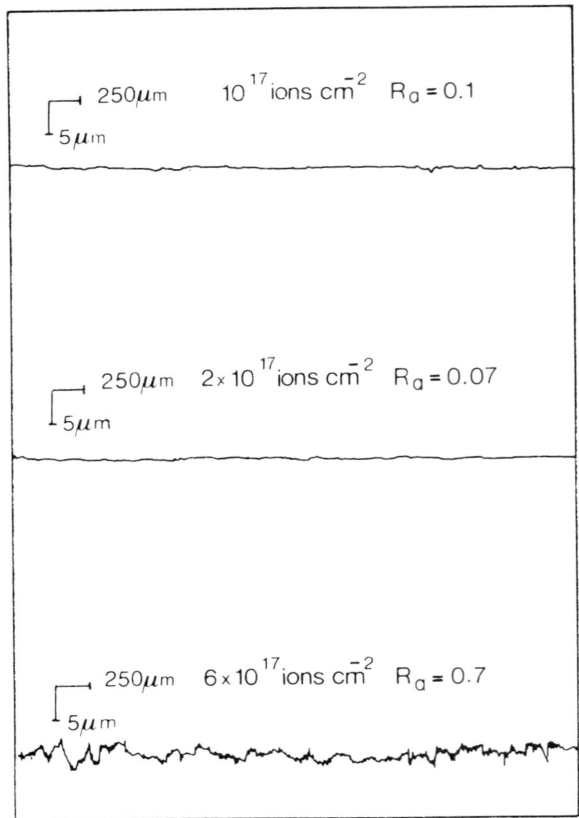

Fig. 9. Profiles of wear tracks in implanted type 304 stainless steel [36].

3.3. Steel composition

Several papers have shown the influence of alloying elements. In particular the role of chromium has often been studied [6, 45]. Principi *et al.* [45] have performed a detailed CEMS study of the chemical phases formed by nitrogen implantation of various alloyed steels (100C6, X30C13 and X10C17). For a given fluence, they call ρ the ratio

$$\rho = \frac{\Sigma(\text{iron nitrides and carbonitrides})}{\text{iron matrix}}$$

This ratio is shown to increase with increasing chromium content of the matrix. We also check this effect by implanting steels of different compositions [26] and characterizing them by nuclear reactions. In high chromium steels the profiles usually display tails, giving evidence of the formation of chromium nitrides and of the slowing-down of the inward diffusion of nitrogen. Moreover, it seems that nitrogen migration is closely linked to the nature of the phase formed by implantation. This nitrogen migration during wear has often

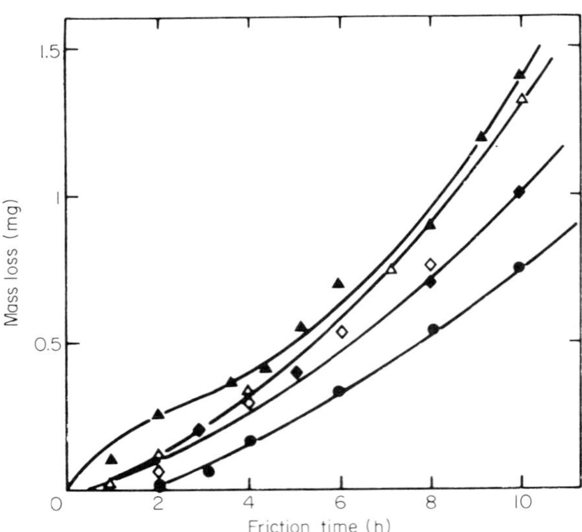

Fig. 10. Mass loss after wear test (AISI 52100 on AISI 52100) of implanted samples ($\phi = 2 \times 10^{17}$ N^{+} cm^{-2}): ▲, unimplanted; △, 20 °C; ◆, 100 °C; ●, 150 °C; ◇, 200 °C [44].

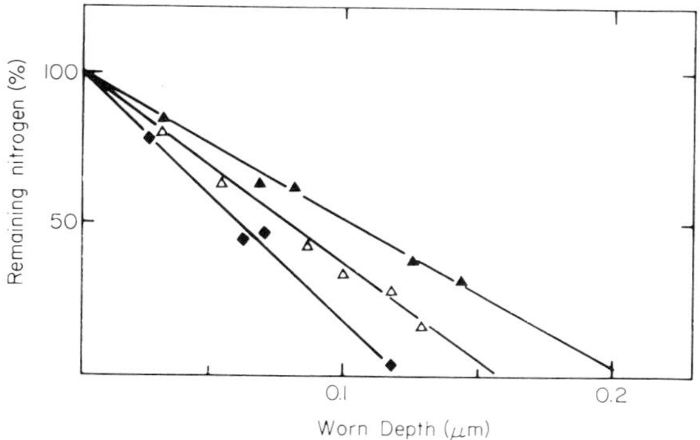

Fig. 11. Remaining nitrogen concentration during abrasive wear tests in various alloyed steels implanted under similar conditions ($\phi = 10^{17}$ N^+ cm^{-2}; $E = 40$ keV; $T = 20$ °C) [43]: ▲, 42CD4; △, 100C6; ◆, Z200C13.

been discussed [12, 15]. Friction could enhance phase destabilization; the free nitrogen thus liberated could migrate deeper in the sample and there form new nitrides. Marest *et al.* [46] have performed a study of nitrogen migration in the abrasive wear of AISI 4142 steel, implanted at 20 °C and 2×10^{17} N^+ cm^{-2} at 40 keV. By CEMS, they showed that, after the removal of 180 nm, iron nitrides were still detected (40 keV implanted nitrogen is concentrated in the first 100 nm). However, this migration is much smaller than that proposed in other work [12, 15]. Figure 11 [43] represents the remaining nitrogen concentration *vs.* worn depth in some steels. The higher chromium content steels show less nitrogen migration. This conclusion is in agreement with the idea that stable compounds prevent nitrogen migration.

4. DISCUSSION

Perhaps it is very ambitious and untimely to desire perfect comprehension of the whole wear mechanisms of ion implantation, as the fundamental basis still remains rather vague. However, it is possible to sum up some very important results by considering a few recent publications [39, 47, 48] as well as our own work. Schematically, it can be said that nitrogen implantation strengthens a ferritic steel phase but only to a small degree a martensitic steel phase. Thus the improvement in the wear resistance of the steel will depend on its carbon content and on its previous thermal treatments. Nitrogen implantation will be more effective on low carbon steels or on high carbon steels tempered to a low hardness. On the contrary, it will have less effect on high carbon steels in the quenched state [39].

However, the friction coefficient is almost never changed by nitrogen implantation (as it is by titanium implantation). The improvement in wear properties is linked to surface strengthening and more precisely to the formation of microstructures very resistant to the run-in period.

Hubler [47] proposed six possible mechanisms to explain the material surface strengthening, each one restricting the dislocation motion. The most important of these are as follows.

(1) *Surface compressive stress strengthening.* The fact that the strain has an effect on the wear resistance is well known in spite of the difficulties in measuring it. A correlation between strain and hardness has not yet been clearly established [49]. Concerning our study, we have suggested that the strains were responsible for the preferential orientation of ϵ-carbonitride, which means texture formation of the implanted zone. It will be very interesting to pursue these experiments in order to attain a more complete interpretation of this texture phenomenon.

(2) *Grain size strengthening.* The strength of a material increases as the grain size decreases because grain boundaries are obstacles to dislocation motion. Fayeulle [50] has shown for example that the very small nitride crystallite size corresponding to maximum implantation effectiveness was of the order of 3–5 nm in an Fe–Cr alloy. These dimensions have to be compared with the value of 1–100 μm obtained when using the classical nitration process. This could perhaps explain the possible contribution of nitrogen implantation in steels which had already been nitrided [51]. Using electron microscopy, dos Santos *et al.* [19] have also demonstrated that in AISI 1020 steel there was evidence for an optimal grain nitride size that ranged between 2 and 10 nm. This optimal grain size could be a factor behind the origin of the maximal resistance improvement often noted as a function of fluence. In fact, it is not easy to explain it when considering only phase chemical compositions or remaining nitrogen concentrations that become constant rather rapidly.

(3) *Interstitial solid solution strengthening.* Martensitic transformation is an example of this.

(4) *Second-phase strengthening.* We have developed this point previously in the main part of this paper.

However, it remains that, in many cases, implantation concerns the first 100 nm from the surface and in fact noticeable migration has never been really proved. So this treatment could be an "initiator" [43] of the wear mechanisms, allowing good and effective strain hardening of the surface in order to reduce significantly the wear rate in the initial period of wear. This initiation period is undoubtedly the most decisive period of the test. Weak wear conditions are thus often required to start the friction process and to avoid too severe phenomena such as seizing.

5. CONCLUSION

Phase identification has been probably one of the most studied aspects in nitrogen implantation research and it can be considered that the new phases formed by implantation and depending on fluence, temperature and steel composition are now quite well known. This appears to be necessary in order to predict the best experimental implantation conditions that can allow maximal wear resistance improvement.

However, the localization of these compounds inside the implanted layer remains rather vague and could still be improved, in particular using such methods as depth-selective CEMS, grazing-angle X-ray diffraction or TEM.

An effort could be made to study the role of the compressive strains, however difficult these are to measure. The most important step now is the correlation between the various improvement factors which correspond to comprehension of the purely mechanical mechanisms of the implantation process. Perhaps there is not only a cumulative action of all the factors (new phases, grain sizes, compressive strains etc.) but also a synergistic effect of these numerous contributions.

ACKNOWLEDGMENTS

The author would like to thank J. Tousset, G. Marest, S. Fayeulle and F. Zawislak for very interesting and helpful discussions.

This work is supported in part by a Ministère de la Recherche et de l'Industrie contract in collaboration with UNIREC (USINOR).

REFERENCES

1 G. Dearnaley and N. E. W. Hartley, *Proc. 4th Conf. on the Scientific and Industrial Applications of Small Accelerators, Denton, TX, 1975*, IEEE, New York, 1976, p. 20.
2 N. E. W. Hartley and R. E. J. Watkins, in G. Dearnaley (ed.), *Ion Implantation for Improved Resistance to Wear and Corrosion, Mater. Eng. Appl., 1* (1978) 28.
3 G. Dearnaley and N. E. W. Hartley, *Thin Solid Films, 54* (1978) 215.
4 J. K. Hirvonen, in G. K. Hubler, O. W. Holland, C. R. Clayton and C. W. White (eds.), *Ion Implantation and Ion Beam Processing of Materials, Materials Research Society Symp. Proc., Boston, MA, November 14–17, 1983*, Vol. 27, Elsevier, New York, 1984, p. 621.
5 K. B. Winterbon, *Ion Implantation Range and Energy Deposition Distributions*, Vol. 2, *Low Incident Ion Energies*, IFI Plenum, New York, 1975.
6 I. L. Singer, *Vacuum, 34* (10–11) (1984) 853.
7 B. Rauschenbach, *Phys. Status Solidi A, 85* (1984) 473.
8 B. Rauschenbach, *Nucl. Instrum. Methods B, 15* (1986) 756.
9 P. D. Goode and I. J. R. Baumvol, *Nucl. Instrum. Methods, 189* (1981) 161.
10 W. M. Bone, R. J. Colton, I. L. Singer and C. R. Gossett, *J. Vac. Sci. Technol. A, 2* (2) (1984) 788.
11 E. Ramous, G. Principi, L. Giordano, S. Lo Russo and C. Tosello, *Thin Solid Films, 102* (1983) 97.
12 S. Lo Russo, P. Mazzoldi, I. Scotoni, C. Tosello and S. Tosto, *Appl. Phys. Lett., 34* (10) (1979) 627.
13 B. L. Doyle, D. M. Follstaedt, S. T. Picraux, F. G. Yost, L. E. Pope and J. A. Knapp, *Nucl. Instrum. Methods B, 7–8* (1985) 166.
14 J. T. A. Pollock, M. J. Kenny and P. J. K. Paterson, in G. K. Hubler, O. W. Holland, C. R. Clayton and C. W. White (eds.), *Ion Implantation and Ion Beam Processing of Materials, Materials Research Society Symp. Proc., Boston, MA, November 14–17, 1983*, Vol. 27, Elsevier, New York, 1984, p. 691.
15 H. G. Feller, R. Klinger and W. Benecke, in G. K. Wolf, W. A. Grant and R. P. M. Procter (eds.), *Proc. Int. Conf. on Surface Modification of Metals by Ion Beams, Heidelberg, September 17–21, 1984*, in *Mater. Sci. Eng., 69* (1985) 173.
16 T. Barnavon, J. Tousset, S. Fayeulle, P. Guiraldenq, D. Tréheux and M. Robelet, *Radiat. Eff., 77* (1983) 249.
17 N. Moncoffre, G. Hollinger, H. Jaffrezic, G. Marest and J. Tousset, *Nucl. Instrum. Methods B, 7–8* (1985) 177.

18 G. Longworth and N. E. W. Hartley, *Thin Solid Films, 48* (1978) 95.
19 C. A. dos Santos, M. Behar and I. J. R. Baumvol, *J. Appl. Phys. D, 17* (1984) 551.
20 B. Rauschenbach, A. Kolitsch and K. Hohmuth, *Phys. Status Solidi A, 80* (1983) 471.
21 Th. Barnavon, H. Jaffrezic, G. Marest, N. Moncoffre, J. Tousset and S. Fayeulle, in G. K. Wolf, W. A. Grant and R. P. M. Procter (eds.), *Proc. Int. Conf. on Surface Modification of Metals by Ion Beams, Heidelberg, September 17–21, 1984*, in *Mater. Sci. Eng., 69* (1985) 531.
22 I. L. Singer, *Appl. Surf. Sci., 18* (1984) 28–62.
23 H. Dimigen, K. Kobs, R. Leutenecker, H. Ryssel and P. Eichinger, in G. K. Wolf, W. A. Grant and R. P. M. Procter (eds.), *Proc. Int. Conf. on Surface Modification of Metals by Ion Beams, Heidelberg, September 17–21, 1984*, in *Mater. Sci. Eng., 69* (1985) 181.
24 D. Firrao, M. Rosso, G. Principi and R. Frattini, *J. Mater. Sci., 17* (1982) 1773.
25 R. Frattini, G. Principi, S. Lo Russo, B. Tiveron and C. Tosello, *J. Mater. Sci., 17* (1983) 1683.
26 N. Moncoffre, G. Marest, S. Hiadsi and J. Tousset, *Nucl. Instrum. Methods B, 15* (1986) 620.
27 B. Rauschenbach and A. Kolitsch, *Phys. Status Solidi A, 80* (1983) 211.
28 S. Fayeulle, D. Tréheux and C. Esnouf, *Appl. Surf. Sci., 25* (3) (1986) 288.
29 J. L. Whitton, G. T. Ewan, M. M. Ferguson, T. Laursen, I. V. Mitchell, H. H. Plattner, M. L. Swanson, A. V. Drigo, G. Celotti and W. A. Grant, in G. K. Wolf, W. A. Grant and R. P. M. Procter (eds.), *Proc. Int. Conf. on Surface Modification of Metals by Ion Beams, Heidelberg, September 17–21, 1984*, in *Mater. Sci. Eng., 69* (1985) 111.
30 N. Moncoffre, M. Brunel, P. Deydier and J. Tousset, *Surf. Interface Anal.*, to be published.
31 D. Firrao, B. Debenedetti and M. Rosso, *Metall. Ital., 71* (1979) 373.
32 M. Carbucicchio, L. Bardani and S. Tosto, *J. Appl. Phys., 52* (7) (1981) 4589.
33 G. N. Van Wyk, *Rad. Eff. Lett., 57* (1980) 45.
34 G. N. Van Wyk, *Rad. Eff. Lett., 57* (1981) 187.
35 R. Hutchings, in G. K. Wolf, W. A. Grant and R. P. M. Procter (eds.), *Proc. Int. Conf. on Surface Modifications of Metals by Ion Beams, Heidelberg, September 17–21, 1984*, in *Mater. Sci. Eng., 69* (1985) 129.
36 S. Fayeulle and D. Tréheux, *Proc. Int. Conf. on Ion Beam Modification of Materials, Catania, June 1986, Nucl. Instrum. Methods*, to be published.
37 N. Moncoffre, *Diplôme de Doctorat*, Institut de Physique Nucléaire Lyon-1, 1986.
38 J. Brissot, Science et surface, personal communication, 1986.
39 G. K. Hubler and F. A. Smidt, *Nucl. Instrum. Methods B, 7–8* (1985) 151.
40 T. Varjoranta, J. Hirvonen and A. Anttila, *Thin Solid Films, 75* (1981) 241.
41 R. Hutchings and W. C. Oliver, *Wear, 92* (1983) 143.
42 P. D. Goode, A. T. Peacock and J. Asher, *Nucl. Instrum. Methods, 209–210* (1983) 925.
43 S. Fayeulle, *Wear, 107* (1) (1986) 61.
44 S. Fayeulle, unpublished results, 1986.
45 G. Principi, S. Lo Russo and C. Tosello, *J. Phys. F*, to be published.
46 G. Marest, N. Moncoffre and S. Fayeulle, *Appl. Surf. Sci., 20* (1985) 205.
47 G. K. Hubler, in S. T. Picraux and W. J. Choyke (eds.), *Metastable Materials Formation by Ion Implantation, Materials Research Society Symp. Proc.*, Vol. 7, Elsevier, New York, 1982, p. 341.
48 G. Dearnaley, *Nucl. Instrum. Methods B, 7–8* (1985) 158.
49 J. Y. Robic, J. Piaquet and J. P. Gaillard, *Nucl. Instrum. Methods, 209–210* (1983) 919.
50 S. Fayeulle, *Thèse Docteur Ingénieur*, Ecole Centrale de Lyon, 1984.
51 C. A. dos Santos, I. J. R. Baumvol, in H. Ryssel and H. Glawischnig (eds.), *Proc. 4th Int. Conf. on Ion Implantation: Equipment and Techniques, Berchtesgaden, September 13–17, 1982*, Springer, New York, 1983, p. 347.

Migration Studies of Nitrogen-implanted-iron by Improved Depth Profiling*

F. BODART, G. TERWAGNE and M. PIETTE

Laboratoire d'Analyse par Réactions Nucléaires (LARN), Facultés Universitaires de Namur, 22 rue Muzet, B-5000 Namur (Belgium)

(Received July 10, 1986)

ABSTRACT

It has been established that the implantation of nitrogen ions into steel can improve the tribological properties of the metal. The mechanism responsible for these improvements is not well understood. In many cases, properties such as wear, corrosion and friction were improved to a depth greater than the range of the implanted ions.

In an attempt to elucidate the mechanism of the nitrogen ion diffusion under thermal treatment, pure iron foils were implanted at 50 keV with fluences ranging between 10^{16} *and* 2×10^{17} $N^+ cm^{-2}$ *with enriched nitrogen (70%* ^{14}N *plus 30%* ^{15}N*). The depth profiling of* ^{15}N *and* ^{14}N *is performed on the same sample using the* $^{15}N(p, \alpha\gamma)$ *reaction at* $E_p = 429$ *keV and the* $^{14}N(\alpha, \gamma)$ *reaction at* $E_\alpha = 1531$ *keV.*

Initial results indicate a well-resolved two-component distribution in the species implanted. Profiling ^{15}N *using the 429 keV resonance does not resolve this structure because of the poor depth resolution, but with the* $^{14}N(\alpha, \gamma)$ *reaction it is possible to observe the migration kinetics of each component vs. the temperature of implantation and vs. the fluence of the nitrogen ions implanted.*

1. INTRODUCTION

Ion implantation allows the introduction of a controlled amount of one atomic species into the surface layers of a substrate. When the ions penetrate the target material, they lose energy (mainly in elastic collisions) before coming to rest. The statistical nature of the collision process leads to a distribution of ion ranges within the target, which is roughly gaussian in shape for polycrystalline materials [1].

Implanted nitrogen profiles have previously been measured using secondary ion mass spectrometry [2], characteristic X-ray detection [3] and several nuclear scattering techniques including $^{14}N(d, \alpha)$ [4], $^{15}N(p, \alpha\gamma)$ [5], $^{14}N(\alpha, \gamma)$ [6], $^{14}N(\alpha, p)$ and helium Rutherford backscattering [7].

It has been established that nitrogen implantation into steel resulted in a modification of the tribological properties [8-10]. A particularly beneficial effect of nitrogen implantation in some types of steel is wear reduction. Tribological tests and industrial applications are realized on samples implanted with ^{14}N [11], but the depth profiling of nitrogen into steel is usually carried out using the well-known resonant $^{15}N(p, \alpha\gamma)^{12}C$ reaction at $E_p = 429$ keV when implantations are made with enriched ^{15}N.

However, high resolution depth profiles of ^{14}N are also measurable with an α capture resonant reaction. In this paper, we compare the profiles obtained on the same sample using these two nuclear reactions.

A fortuitous effect observed in some laboratories is a persistence in the benefit of the nitrogen implantation well beyond the point of removal of the implanted layer and it is of fundamental interest to study the interaction between implants and bombardment-induced defects, the compositional redistribution *vs.* its dependence on implantation temperature and the local phase formation induced by nitrogen implantation.

2. EXPERIMENTAL PROCEDURE

2.1. Sample preparation

Disc samples of polycrystalline pure iron (diameter, 12 mm; thickness, 0.5 mm) were

*Paper presented at the International Conference on Surface Modification of Metals by Ion Beams, Kingston, Canada, July 7-11, 1986.

0025-5416/87/$3.50

mechanically polished with diamond paste (15–1 μm) and then with alumina powder (1–0.05 μm). They were then implanted with 50 keV enriched nitrogen ions (N^+) (70% ^{14}N plus 30% ^{15}N) using a SAMES accelerator; the beams of ^{14}N ions and ^{15}N ions were not separated and the foils were implanted with an homogeneous fluence of 10^{16}–2×10^{17} N^+ cm^{-2}. Implantations were performed at temperatures of 25, 60, 80, 100, 125 and 140 °C. The sample temperature was kept at 25 °C with a water-cooling device which supported the sample. For heating the target, an air-heating system was coupled to the target holder which acted as a sink.

2.2. Depth profile measurements

Nitrogen distributions were measured by using two narrow resonance reactions: $^{14}N(\alpha, \gamma)^{18}F$ at $E_\alpha = 1531$ keV and $^{15}N(p, \alpha\gamma)^{12}C$ at $E_p = 429$ keV. Nitrogen depth profiles were obtained with an automatic energy scan installed on a 2.5 MV Van de Graaff accelerator [12,13]. As shown in Fig. 1, the ion beam was deflected before the analysis slits by modifying the electric potential on the deflectors. To obtain an excitation curve, the energy of the beam was changed automatically by small energy steps over the whole energy range. Typical ion current densities of about 10–15 μA cm^{-2} were used during implantation.

2.3. ^{14}N determination

The $^{14}N(\alpha, \gamma)^{18}F$ reaction shows a narrow resonance at $E_\alpha = 1531$ keV as studied by Gosset [6]. The nearest resonance appears at 1618 keV, thus providing a clear region of bombarding energy, of 87 keV, for profiling. The natural width of this resonance (0.6 keV) and the beam energy spread given by the accelerator (1.5 keV) produce a depth resolution of 2.8 nm at the surface of the sample. The beam energy spreading in the iron foil is principally due to the straggling effect and affects the profile depth resolution. In this reaction the radiation of γ rays in the range 3.0–5.3 MeV was detected in a 4 in × 4 in pit NaI detector. The low cross-section from the α capture reaction requires the use of a high beam current (500 nA) and a long exposure (2 h) to obtain adequate statistics for the yield of γ rays as a function of the bombarding energy. The dissipation of the heat generated in stopping megaelectronvolt α particles is taken care of by placing a cold trap near the sample. For long exposures the accumulation of α particles (5×10^{17} He^+ cm^{-2}) in the iron foil leads to the formation of bubbles which are

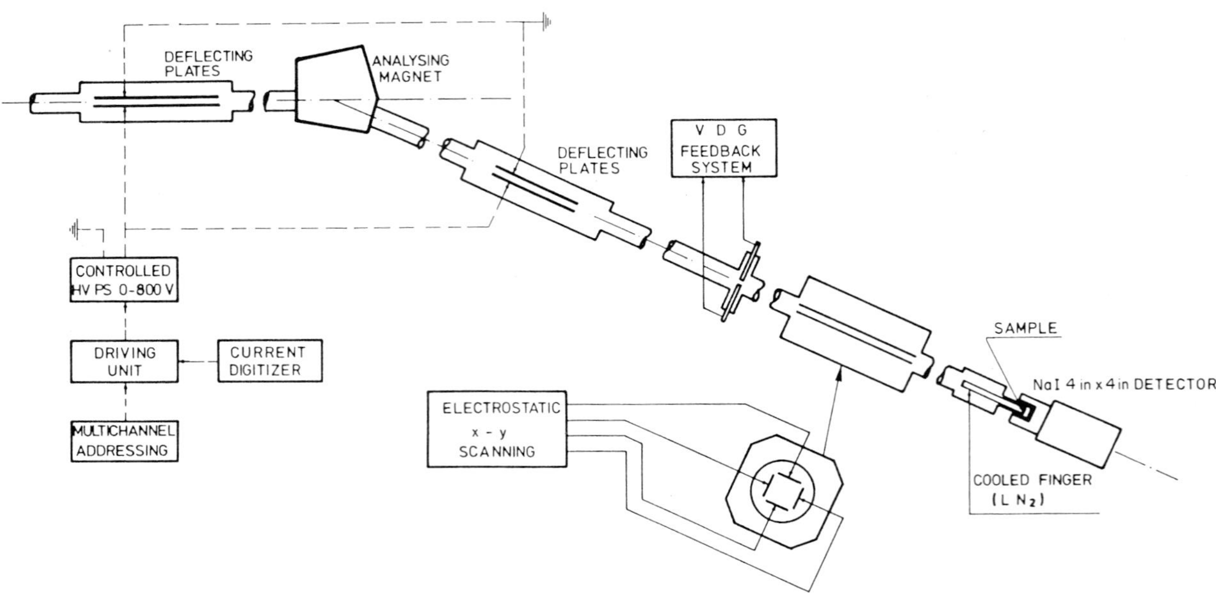

Fig. 1. Diagram of the experimental device showing the automatic change in energy with the deflecting plates and the beam electrostatic x–y scanning on the sample: VDG, Van de Graaff; LN_2, liquid nitrogen.

unacceptable. To avoid this, a scanning electrostatic deflector system is used to sweep the α particle spot on the sample over an area of 25 mm^2 (Fig. 1).

In Fig. 2 we have reproduced the measured ^{14}N depth profile of an iron foil implanted with 50 keV enriched ^{15}N at a fluence of 10^{17} ions cm^{-2}. The background measured on a unimplanted iron foil was subtracted.

2.4. ^{15}N determination

^{15}N was analysed by using the 429 keV resonance from the ^{15}N(p, $\alpha\gamma$)^{12}C reaction. γ ray radiation was detected in the geometry used for ^{14}N determination. This very intense resonance is isolated and there is no background under the γ ray peak. A natural width of 120 eV from this resonance, measured by Maurel and Amsel [14], and the normal energy spread of the beam produce a depth resolution of 3 nm at the surface of iron. Unfortunately, the beam energy spreading due to straggling completely destroys the depth resolution of this resonant reaction. Figure 3 illustrates the comparison between the two resonant reactions; the depth resolution for the ^{15}N(p, $\alpha\gamma$)^{12}C reaction is twice that of the ^{14}N(α, γ)^{18}F reaction. In Fig. 2, the comparison of the ^{14}N and ^{15}N depth profiles for the same sample proves this effect; the experimental points for the ^{14}N(α, γ)^{18}F reaction clearly show a surface component while ^{15}N measurements show only a shoulder in the front of the depth profile.

2.5. X-ray photoelectron spectra

In order to determine the chemical state of nitrogen implanted in iron, X-ray photoelectron spectroscopy measurements were performed on a sample implanted at 140 °C with a fluence of 5×10^{16} N$^+$ cm^{-2}, using a Hewlett-Packard spectrometer equipped with monochromatized Al Kα (1486.6 eV) radiation. Depth-profiling experiments were carried out using an argon ion (Ar$^+$) gun and spectra were recorded after the sample had been etched to 3, 6 and 30 nm.

3. RESULTS AND DISCUSSION

3.1. Deconvolution technique

The observed excitation curves $Y(E_0)$ are in fact the convolution of the true concentra-

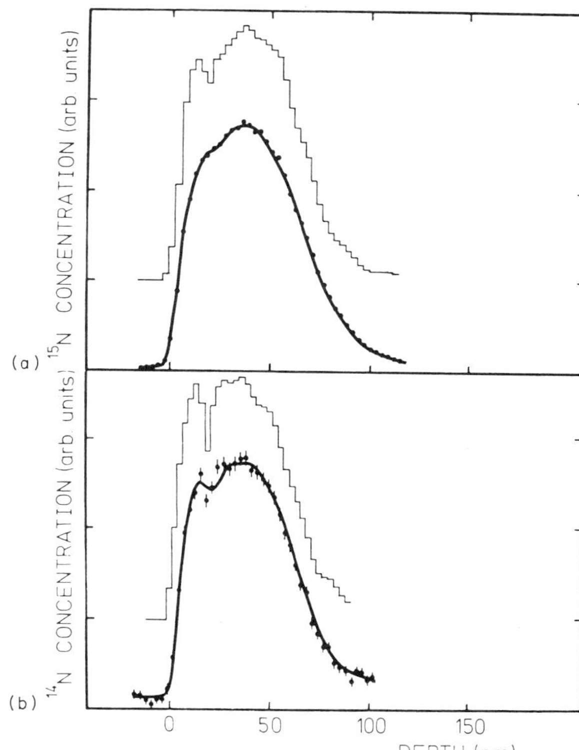

Fig. 2. Measurements of nitrogen depth profiles on an iron foil implanted at 10^{17} N$^+$ cm^{-2} with 50 keV enriched nitrogen (70% ^{14}N plus 30% ^{15}N): ●, experimental data; ——, deconvoluted profile. The characteristics of the resonant nuclear reactions are (a) ^{15}N(p, $\alpha\gamma$)^{12}C (E_p = 429 keV; Γ_R = 120 eV; E_γ = 3-5 MeV) and (b) ^{14}N(α, γ)^{18}F (E_α = 1531 keV; Γ_R = 0.6 keV; E_γ = 3-5.3 MeV) and the deconvoluted profiles are represented by histogram curves.

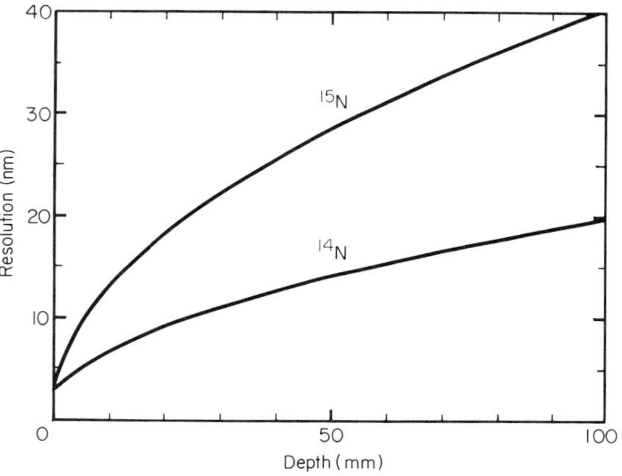

Fig. 3. Comparison of the depth resolution for nitrogen profiling into iron. The two resonant nuclear reactions used are ^{14}N(α, γ)^{18}F at E_α = 1551 keV and ^{15}N(p, $\alpha\gamma$)^{12}C at E_p = 429 keV.

tion $C(x)$ and the beam energy distribution at the moment that the nuclear reaction takes place. This distribution $F(E_0, x)$ is a function of the depth where the reaction occurs and the experimental data are described by the formula

$$Y(E_0) = k \int_{x=0}^{R} C(x)F(E_0,x)\,\mathrm{d}x$$

where k is a constant for given detection conditions. The function $F(E_0,x)$ is in fact the "resolution" of the nuclear resonance considered as a probe for measuring the concentration depth (Fig. 3). This resolution function is a convolution between beam energy spread as produced by the accelerator, the width of the nuclear cross-section and the straggling function.

The "deconvolution" procedure has been described by Deconninck and Van Oystaeyen [13] in a previous paper. As a first approximation the stopping power used for this calculation is constant and corresponds to a mixture of 80% iron and 20% nitrogen [15].

The ^{14}N and ^{15}N deconvoluted profiles for an iron foil implanted with 50 keV enriched nitrogen are presented in Fig. 2.

3.2. ^{14}N and ^{15}N profiles

Let us examine the nitrogen profiles obtained from the iron implanted at room temperature (25 °C) and at different fluences (Fig. 4). At a fluence of 10^{17} ions cm^{-2}, the

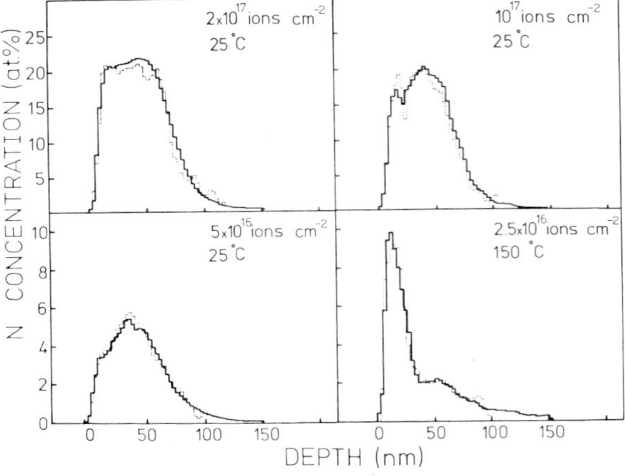

Fig. 4. Comparison between ^{14}N (· · ·) and ^{15}N (——) depth profiles in iron at various fluences and various implantation temperatures.

^{14}N concentration clearly shows that the depth profile is composed of a narrow surface component and a deep component. For the deep contribution the experimental projected range (R_p = 40 nm) and the standard deviation (ΔR_p = 16 nm) are in agreement with the theoretical values calculated by Winterbon [16]. At lower fluences a bulge is observed in the near-surface region, and at higher fluences the surface component seems to be saturated. ^{15}N profiles confirm this tendency and there are no noticeable differences between the ^{14}N and the ^{15}N depth profiles [17].

The distribution of nitrogen in iron at a higher implantation temperature (150 °C) is completely different; at this temperature, complete migration of nitrogen towards the surface takes place; this result confirms those obtained on other steels [18, 19]. ^{14}N and ^{15}N depth profiles seem to be the same. For ^{14}N determination the measurements are long because of the low $^{14}\mathrm{N}(\alpha,\gamma)^{18}\mathrm{F}$ reaction cross-section. It is possible that long α irradiation changes the shape of the depth profile. An iron foil was implanted with enriched nitrogen at a fluence of 10^{17} ions cm^{-2} and was irradiated for a long time with 1.53 MeV α particles (5×10^{17} He$^+$ cm^{-2}) on a well-defined surface area of the sample. The ^{15}N depth profiles measured before and after α irradiation show exactly the same features (Fig. 5). Thus the ^{14}N depth profile measurement does not change the nitrogen distribution. After thermal annealing of this sample, a marked difference arose in the ^{15}N depth profile measurements made in and outside the surface area irradiated with α particles (Fig. 5). The nitrogen diffusion after thermal annealing is enhanced by ^{14}N depth profile measurements but there is no nitrogen diffusion during the α irradiation.

3.3. Implantation temperature dependence

Six samples were implanted at a fluence of about 5×10^{16} N$^+$ cm^{-2} and at temperatures of 25, 60, 80, 100, 125 and 140 °C. In Fig. 6 the profiles are presented together with a fit by a two-component distribution. A summary of the parameters obtained by this calculation is given in Fig. 7. It is worth noting that the area under each nitrogen distribution is nearly constant; this means that the remaining amount of nitrogen does not decrease drastically with the implantation temperature (at

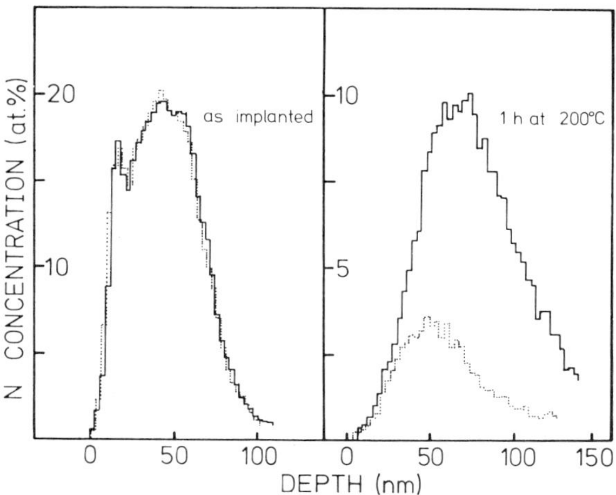

Fig. 5. Influence of α irradiation (10^{17} He^+ cm^{-2}) on nitrogen diffusion: ——, before α irradiation; · · ·, after α irradiation. On the left-hand side the ^{15}N depth profile was measured before and after α irradiation, and on the right-hand side the same measurements were performed on the sample which had been annealed for 1 h at 200 °C in an atmosphere of nitrogen.

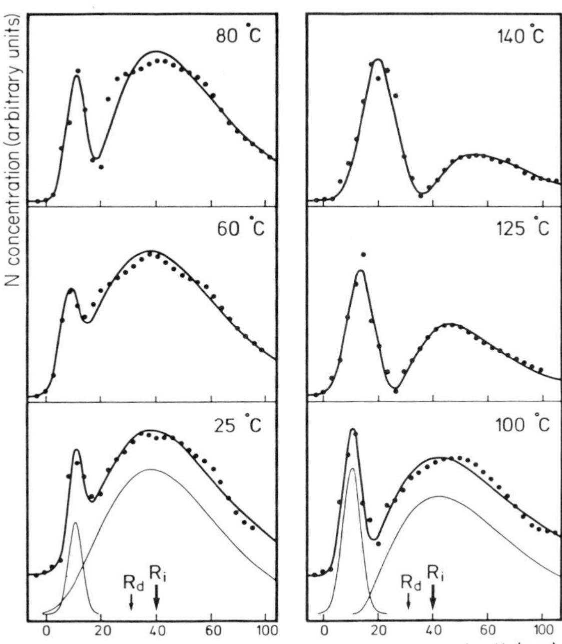

Fig. 6. Profiles obtained after deconvolution of the experimental data for six iron samples implanted at a fluence of about 5×10^{16} $^{15}N^+$ cm^{-2} and at temperatures of 25, 60, 80, 100, 125 and 140 °C. The fit is obtained by the convolution of the two distributions as shown in the lower part of the figure. (R_i and R_d indicate the projected ranges for ion implantation and defects respectively.)

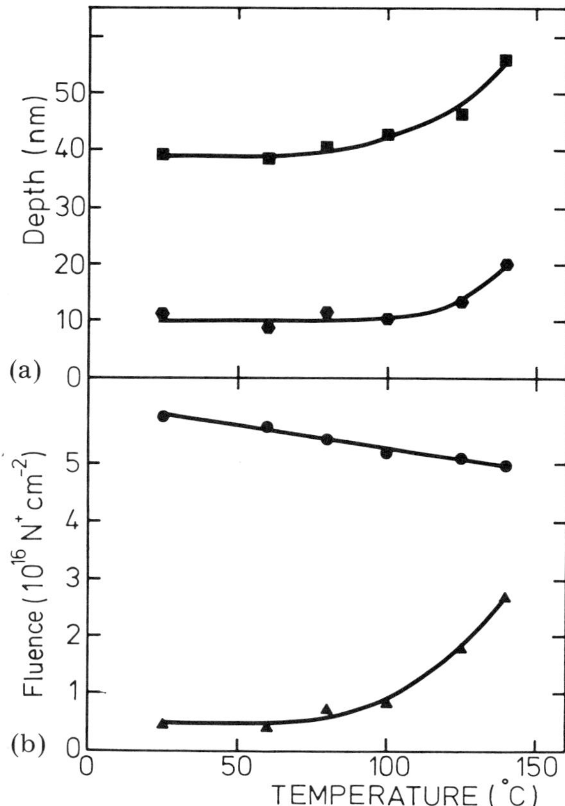

Fig. 7. (a) Depth (*i.e.* the positions of the two distributions used for the fit in Fig. 6) *vs.* temperature (■, volume peak; ●, surface peak); (b) fluence *vs.* temperature (●, total; ▲, surface peak).

least until 140 °C) as already mentioned in the literature [20]. A surface peak appears at room temperature and becomes more important when the implantation temperature increases; this means that nitrogen diffuses from the bulk to the near-surface peak but does not diffuse out of the sample. The centre of these distributions seems to be stable and a slow displacement can be attributed to a thin oxide layer formation on the sample above 100 °C.

3.4. *Fluence dependence*

Characteristics of five samples implanted with ^{15}N ions at 100 °C with fluences of 0.9×10^{16}, 1.25×10^{16}, 2.5×10^{16}, 5×10^{16} and 10×10^{16} ions cm^{-2} are presented in Fig. 8. The positions of the near-surface peak and volume peak are also nearly constant. The total amount of nitrogen increases linearly with increasing fluence, which means that the saturated configuration has not yet been

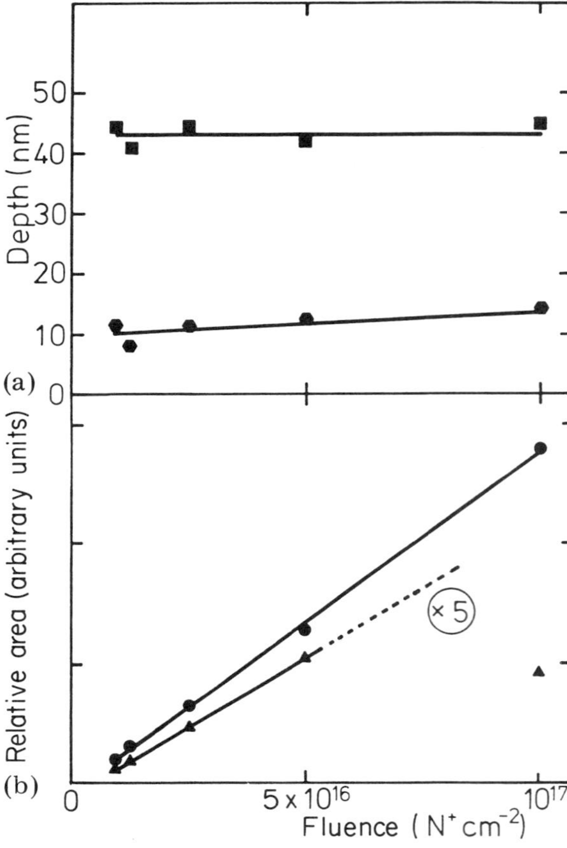

Fig. 8. (a) The variation in the positions of the near-surface (●) and volume (■) peak with fluence; (b) the variation in the total fluence (●) and the area under the near-surface distribution (▲) is shown for five iron samples implanted at 100 °C with various fluences.

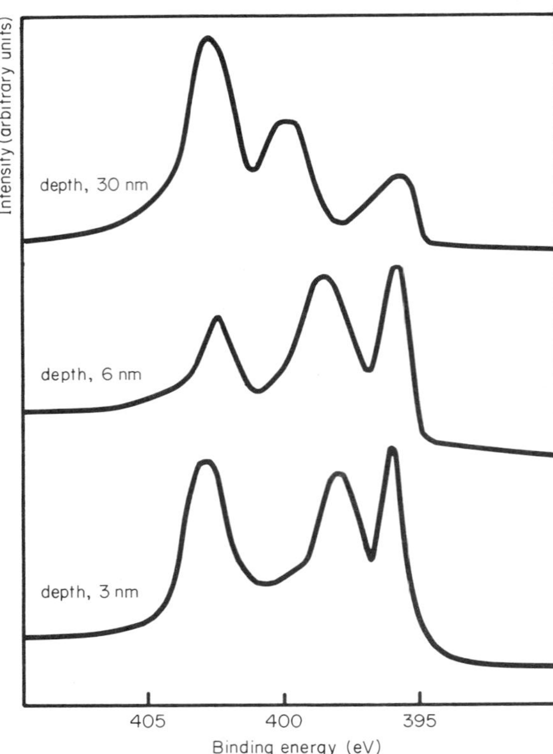

Fig. 9. X-ray photoelectron spectra of the N 1s level of an iron sample implanted with 5×10^{16} $^{15}N^{+}$ cm^{-2} at 140 °C, which had been etched to 3, 6 and 30 nm to reveal the implanted region.

reached. The area under the near-surface peak increases slowly, except for the 10^{17} N^{+} cm^{-2} implantation. More complex deconvolution must be adopted for this situation according to the variation in the defect migration probably due to nitrogen concentration and the possibility of local phase transformations during implantation at high fluences.

3.5. Nitrogen chemical bonding

Figure 9 shows the X-ray photoelectron spectra for N 1s recorded at three different depths (3, 6 and 30 nm). From the literature [20, 21], it seems that the peak at 402.5 eV should be characteristic of nitrogen bound to oxygen. Two other peaks (398 and 396 keV) indicate the formation of nitrides during implantation near the surface. The upper spectra reveal a nitrogen signal at 400 eV corresponding to free nitrogen gas, whereas the intensity of the signal corresponding to nitride is reduced.

4. CONCLUSIONS

Very high resolution profiling of ^{14}N is measurable by using the $^{14}N(\alpha,\gamma)^{18}F$ resonant reaction at $E_\alpha = 1531$ keV, but the low cross-section of this capture reaction limits the sensibility to 10^{16} N^{+} cm^{-2}; the presence of another resonant reaction at $E_\alpha = 1618$ keV restricts the measurable depth to 110 nm but is sufficient for measuring the iron foil implanted with 50 keV nitrogen ions. The ^{14}N profiling does not change the nitrogen distribution during the measurements of the depth concentration but modifies the behaviour of the region irradiated by α particles for post-thermal annealing.

The most important fact displayed in the nitrogen profile spectra for implantation fluences below 10^{17} N^{+} cm^{-2} is the existence of a near-surface peak whose intensity increases with increasing temperature. This surface accumulation could be explained by the nitrogen mobility through the structural disorder due to irradiation. During implantation, under the influence of damage rate gradients, point defects flow out of the peak damage region towards the surface and into the bulk region. This diffusion is enhanced by increasing the temperature. However, nitrogen could accumulate behind the surface as a result of the formation of stable phases. During implantation at high fluences, some local phase transformations could occur and give rise to redistribution of implanted solutes and variations in defect migration.

ACKNOWLEDGMENT

The authors wish to thank J. J. Pireaux who performed the X-ray photoelectron spectroscopy analysis at the Electron Spectroscopy for Chemical Analysis Laboratory in Namur.

REFERENCES

1 J. Lindhard, M. Scharff and H. E. Schiott, *Mat. Fys. Medd. K. Dan. Vidensk. Selsk., 33* (14) (1963).
2 F. Z. Cui, H. D. Li and X. Z. Zhang, *Nucl. Instrum. Methods, 209-210* (1983) 881.
3 R. Hutchings and W. C. Oliver, *Wear, 92* (1983) 143.
4 S. Lo Russo, P. Mazzoldi, I. Scontoni, C. Tosello and S. Tosto, *Appl. Phys. Lett., 34* (1979) 627.
5 S. Fayeulle, D. Treheux, P. Guiraldeno, T. Barnavon, J. Tousset and M. Robelet, *Scr. Metall., 17* (1983) 459.
6 C. R. Gosset, *Nucl. Instrum. Methods B, 10-11* (1985) 722.
7 B. L. Doyle, D. M. Follstaedt, S. T. Picraux, F. G. Yost, L. E. Pope and J. A. Knapp, *Nucl. Instrum. Methods B, 7-8* (1985) 166.
8 N. E. W. Hartley, in J. K. Hirvonen (ed.), *Ion Implantation, Treatise Mater. Sci. Technol., 18* (1980) 321.
9 J. K. Hirvonen (ed.), *Ion Implantation, Treatise Mater. Sci. Technol., 18* (1980).
10 G. Dearnaley, in G. K. Wolf, W. A. Grant and R. P. M. Procter (eds.), *Proc. Int. Conf. on Surface Modification of Metals by Ion Beams, Heidelberg, September 17-21, 1984*, in *Mater. Sci. Eng., 69* (1985) 139.
11 B. L. Doyle, D. M. Follstaedt, S. T. Picraux, F. G. Yost, L. E. Pope and J. A. Knapp, *Nucl. Instrum. Methods B, 7-8* (1985) 166.
12 G. Amsel, E. D'Artemare and E. Girard, *Nucl. Instrum. Methods, 205* (1983) 5.
13 G. Deconninck and B. Van Oystaeyen, *Nucl. Instrum. Methods, 218* (1983) 165.
14 B. Maurel and G. Amsel, *Nucl. Instrum. Methods, 218* (1983) 159.
15 J. F. Ziegler and H. H. Andersen, *The Stopping Power and Ranges of Ions in Matter*, Vols. 3, 4, Pergamon, Oxford, 1977.
16 K. B. Winterbon, *Ion Implantation Range and Energy Deposition Distribution*, Vol. 2, *Low Incident Ion Energies*, IFI-Plenum, New York, 1975.
17 G. Terwagne, M. Piette and F. Bodart, *Proc. 5th Int. Conf. on Ion Beam Modification of Materials, Catania, June 9-13, 1986*, in *Nucl. Instrum. Methods*, to be published.
18 N. Moncoffre, G. Marest, S. Hiadsi and J. Tousset, *Nucl. Instrum. Methods B, 15* (1986) 620.
19 B. Rauschenbach, *Nucl. Instrum. Methods B, 15* (1986) 756.
20 N. Moncoffre, G. Hollinger, H. Jaffrezic, G. Marest and J. Tousset, *Nucl. Instrum. Methods B, 7-8* (1985) 177.
21 D. C. Kothari, M. R. Nair, A. A. Rangwala and K. B. Lal, *Nucl. Instrum. Methods B, 7-8* (1985) 235.

Quantitative Analysis of Boron in Steel by Secondary Ion Mass Spectrometry using Ion-implanted Iron as Standard*

S. HASHIMOTO[a], S. DOI[a], K. TAKAHASHI[a], M. TERASAKA[a] and M. IWAKI[b]

[a]Technical Research Center, Nippon Kokan K.K., Kawasaki, Kanagawa 210 (Japan)
[b]RIKEN (Institute of Physical and Chemical Research), Wako, Saitama 351-01 (Japan)

(Received July 10, 1986)

ABSTRACT

The trace boron in steels was quantitatively analysed by secondary ion mass spectrometry using ion-implanted iron as standards. The calibration curve obtained ranges from the per cent to the parts per billion level. The detection limit for boron using BO_2^- secondary ions was found to be 50 wt.ppb.

The grain boundary segregation and precipitation of boron in steel containing a boron concentration of 10 wt.ppm were observed as BO_2^- secondary ion images. The quantitative analysis of the boron concentration segregated to the grain boundaries was performed by application of the above calibration curve.

1. INTRODUCTION

Recently the control of trace elements has become important in the research and development of steels. Such trace elements not only exist as a solid solution but also segregate to the grain boundary or surface and moreover precipitate in the matrix.

Boron is one such trace element in steels. In the 60 kgf mm^{-2} high tensile steel, about 10 wt.ppm B are added in order to obtain hardenability. Quenching from austenite causes hardening to occur, but cooling in air reduces the hardness. It has been reported that segregation of boron at the prior austenite grain boundary gives rise to a higher hardenability than precipitation does [1]. The precipitation or segregation of boron in the concentration range 10–200 wt.ppm was studied using secondary ion mass spectrometry (SIMS) [2–4]. However, these studies were not performed on low boron steels and no quantitative analysis was carried out using SIMS.

SIMS is a surface analysis technique with high sensitivity for many elements. SIMS is capable of microanalysis so that it is useful for studying the behaviour of trace elements.

In general for SIMS techniques, the quantitative analysis based on theoretical correction is very difficult because the secondary ion yield (the ratio of the number of secondary ions to the number of primary ions) is easily changed and is markedly dependent on many factors, *e.g.* the chemical state [5, 6], alloying [7] and crystal orientation [8]. So quantitative analysis is mainly achieved by calibration methods using standard samples [9, 10]. However, quantitative analysis of the trace elements in steel by SIMS is not common because there are very few homogeneous standard samples with a highly controlled concentration.

In ion implantation, high energy ions of about 100 keV are implanted into the substrate while being rastered, after specific ions have been selected by the mass analyser. Therefore, this technique has some advantages, as follows.

(1) Any elements are able to be implanted.

(2) The distribution of implanted elements is homogeneous over a wide area.

(3) The purity of implanted elements is very high.

(4) The concentration is well controlled even in the low range.

These conditions are required to make standard samples for the calibration curve using SIMS.

So, using iron samples implanted with boron ions as standards, the calibration curve of

*Paper presented at the International Conference on Surface Modification of Metals by Ion Beams, Kingston, Canada, July 7–11, 1986.

0025-5416/87/$3.50

boron was obtained, and the boron distributions in 40, 25 and 10 wt.ppm steels were observed as secondary ion images. Moreover, the boron concentration in steels was quantitatively estimated using the above calibration curve.

2. QUANTITATIVE ANALYSIS OF BORON IMPLANTED INTO STEEL

Generally, the intensity I_s of the secondary ions is given by the following equation:

$$I_s = kYCI_p \tag{1}$$

where I_p, Y, C and k are the intensity of the primary ions, the secondary ion yield, the concentration of the element and a constant related to the apparatus respectively.

The distribution $I(z)$ of the ion intensity for elements implanted into the substrate was a gaussian shape, where $I(z)$ is the ion intensity measured at the depth z. If the secondary ion yield does not change in the range of the concentration, $I(z)$ is directly proportional to the atomic number $n(z)$:

$$I(z) = Kn(z) \tag{2}$$

where K is a constant. For ion-implanted samples the dose D is equal to the total number of implanted ions within the maximum projected range R_m:

$$D = \int_0^{R_m} n(z)\,\mathrm{d}z \tag{3}$$

The number of elements at z is obtained from eqns. (2) and (3) and the total intensity I_{tot} of secondary ions measured within R_m:

$$n(z) = \frac{DI(z)}{I_{tot}} \tag{4}$$

Quantitative analysis of boron implanted into steel was carried out using eqn. (4).

3. EXPERIMENT

3.1. The calibration curve obtained from the ion-implanted sample

The ion-implanted technique was adopted to make standard samples of the Fe–B binary system. Pure iron (Johnson Matthey Chemicals Limited; purity, 99.99%) for the substrate was polished using diamond and alumina pastes.

Boron was implanted into iron at an accelerating voltage of 150 kV in the dose range 1×10^{13}–5×10^{16} ions cm^{-2}.

Using SIMS (an IMS-3F CAMECA instrument) the depth profiles of implanted boron were obtained. The experimental conditions are shown in Table 1. The primary ions were oxygen ions (O_2^+) and the secondary ions of BO_2^- were measured. The measured ion intensity of BO_2^- is normalized by that of the oxygen ions (O^-) to avoid variation in the experimental conditions.

The calibration curve of boron was obtained from the depth profile, the sputtered depth measured with a surface roughness meter and eqn. (4).

3.2. Analysis of boron in steel

The chemical compositions of the steels are shown in Table 2. Each sample was treated with two types of heat treatment, as shown in Fig. 1. The Vickers hardness of each sample was measured using a Vickers-hardness-testing machine.

The images of the distributions of boron and iron were obtained using the conditions given in Table 1. The diameter of the observed area was 150 μm, and the BO_2^- ion and iron ion (Fe^+) images were used to obtain the distributions of boron and iron respectively.

A line analysis was performed on 25 wt.ppm B steel which had been subjected to heat treatments (a) and (b). The conditions are as shown in Table 1 except for the analysed area (8 μm). The sample was moved at intervals of 5 μm using a stepping motor.

TABLE 1

Experimental conditions

Analyser	IMS-3F (CAMECA)
Primary ion parameters	
Primary ions	O_2^+
Incident angle	60°
Accelerating voltage	15 kV
Ion current	2 μA
Rastering area	250 μm × 250 μm
Secondary ion parameters	
Secondary ions	$^{43}BO_2^-$ (normalized by $^{16}O^-$)
Diameter of analysed area	60 μm

TABLE 2

Chemical compositions of the steels

Steel	*Amount* (wt.%) *of the following elements*										
	C	*Si*	*Mn*	*P*	*S*	*Cr*	*Mo*	*V*	*B*	*Al (soluble)*	*TN*
10 wt.ppm B steel	0.10	0.28	0.95	0.004	0.004	0.68	0.35	0.042	0.0010	0.045	0.0047
25 wt.ppm B steel	0.10	0.29	0.94	0.004	0.004	0.67	0.37	0.042	0.0025	0.052	0.0029
40 wt.ppm B steel	0.10	0.27	0.88	0.004	0.004	0.64	0.33	0.040	0.0042	0.052	0.0039

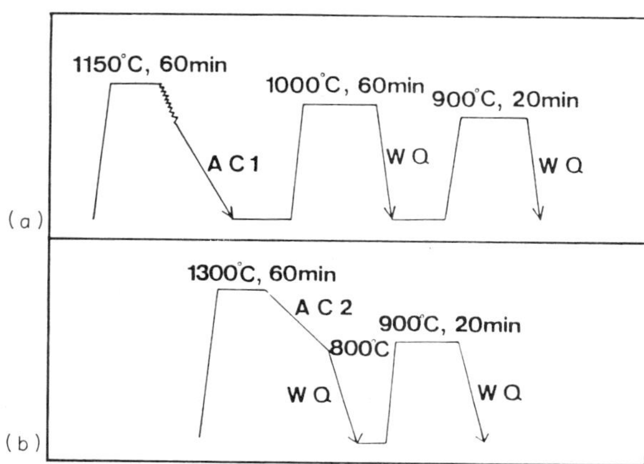

Fig. 1. The two types of heat treatment of boron steels: WQ, water quenching; AC, air cooling.

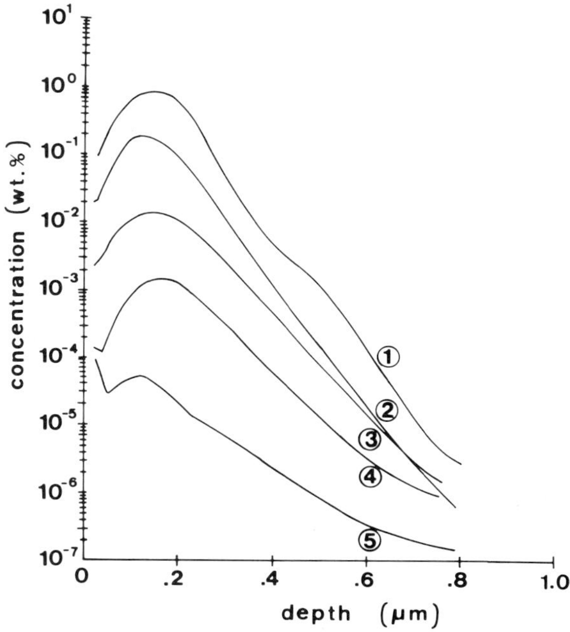

Fig. 2. The depth profile of boron ions implanted into iron (primary ions, oxygen (O_2^+) (15 kV); secondary ions, BO_2^-) at various doses: curve 1, 5×10^{16} ions cm^{-2}; curve 2, 1×10^{16} ions cm^{-2}; curve 3, 1×10^{15} ions cm^{-2}; curve 4, 1×10^{14} ions cm^{-2}; curve 5, 1×10^{13} ions cm^{-2}.

4. RESULTS AND DISCUSSION

4.1. *Ion-implanted samples*

The depth profiles of boron implanted into iron at a dose of 1×10^{13}–5×10^{16} ions cm^{-2} are shown in Fig. 2, where the weight concentration converted from BO_2^- ion intensity based on eqn. (4) is shown as a function of the depth. The distribution curves of boron are approximately gaussian and the peak is 0.18 μm for all samples. The distribution of oxygen in the samples was measured using caesium primary ions (Cs^+) to check the uniformity. It was found that oxygen is homogeneously distributed in the samples.

The same samples were analysed using secondary boron ions (B^+) to compare with the above results. The SIMS intensity of these ions (B^+) was a hundred times less than that of BO_2^- for the same sample. The distribution curves are displayed in Fig. 3, which is very similar to Fig. 2. Therefore, it is considered that BO_2^- ions are produced by a combination of the atoms in the samples and the oxygen primary ions, which has been reported for many elements [8, 11].

The calibration curve against the content in weight per cent obtained from the depth profile of boron is shown in Fig. 4. It was found that the boron concentration is proportional to BO_2^- ion intensity and that the slope is unity. Therefore the boron concentration C_B can be calculated from

$$C_B = 0.38 \frac{I_{BO_2^-}}{I_{O^-}} \tag{5}$$

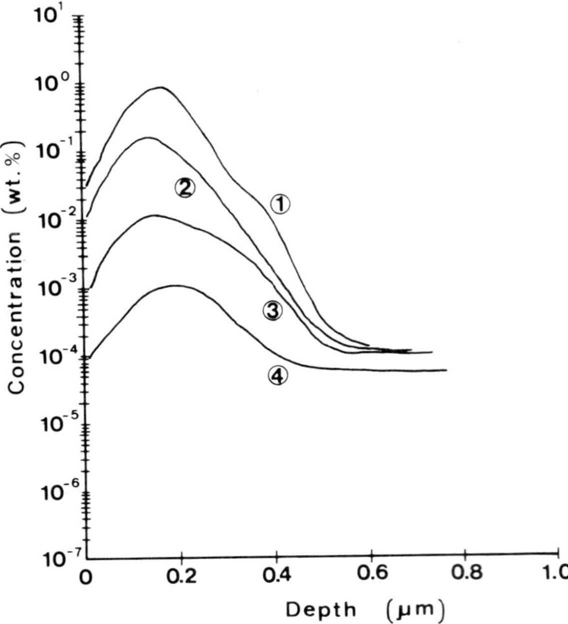

Fig. 3. The depth profile of boron ions into iron (primary ions, oxygen (O_2^+) (12.5 kV); secondary ions, boron (B^+)) at various doses: curve 1, 5×10^{16} ions cm^{-2}; curve 2, 1×10^{16} ions cm^{-2}; curve 3, 1×10^{15} ions cm^{-2}; curve 4, 1×10^{14} ions cm^{-2}.

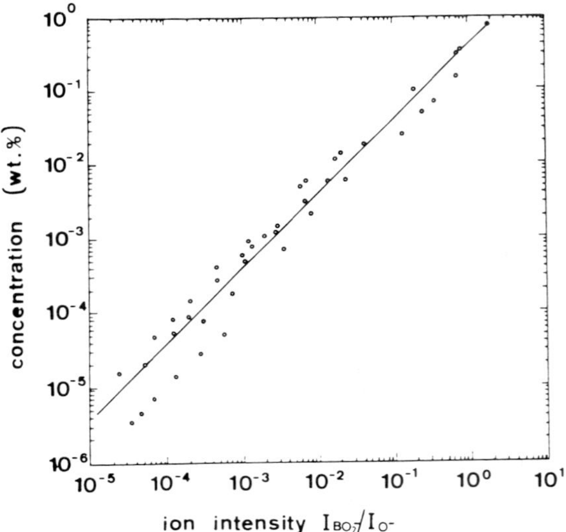

Fig. 4. The calibration curve of boron (primary ions, oxygen (O_2^+) (15 kV); secondary ions, BO_2^-).

The variation in the calibration curves obtained is within ±200%. The errors could be caused by various fluctuations in the ion implantation (the variation in the dose and heterogeneity of the implanted ion) and in the SIMS measurement (the variation in the primary ion currents and the electro-optical conditions for secondary ion detection).

The calibration curve obtained ranges from per cent to parts per billion and it will be possible to analyse the trace boron using this curve. The detection limit is estimated to be 50 wt.ppb.

4.2. Boron in steel

The secondary ion images of BO_2^- and Fe^{2+} for 10 wt.ppm B steel, 25 wt.ppm B steel and 42 wt.ppm B steel subjected to heat treatments (a) and (b) are shown in Fig. 5, Fig. 6 and Fig. 7 respectively, and the microstructures of the samples are also displayed. For each boron steel subjected to heat treatment (a), network contrast is observed. The size of the network is similar to the grain size in the microstructures. These results suggest that the network corresponds to boron segregated to the grain boundaries. No strong contrast for the iron ion image was observed in the photographs.

In contrast, no network was observed but some spots with a high intensity were found for the samples subjected to heat treatment (b). These spots were identified as precipitates of boron nitride using analytical electron microscopy.

It was found that the hardnesses of samples subjected to heat treatment (a) are higher than those for samples subjected to heat treatment (b), as shown in Table 3. It is well known [1] that such treatment causes hardness in steels, which has been explained as due to the segregation of boron to the grain boundaries although there is little direct evidence of this [12]. Our SIMS images are direct evidence of the segregation of boron and they reveal that the hardness of the steel is due to the above mechanism.

A line analysis of 25 wt.ppm B steel was carried out to estimate the boron concentration segregated to the grain boundaries. The secondary ion intensity of BO_2^- was converted to the weight concentration using the above calibration curve. The results are shown in Fig. 8. Peaks of 45 wt.ppm (arrows) are observed for samples subjected to heat treatment (a). The real concentration would be higher than this value because the diameter of the analysed area (8 μm) is wider than the segre-

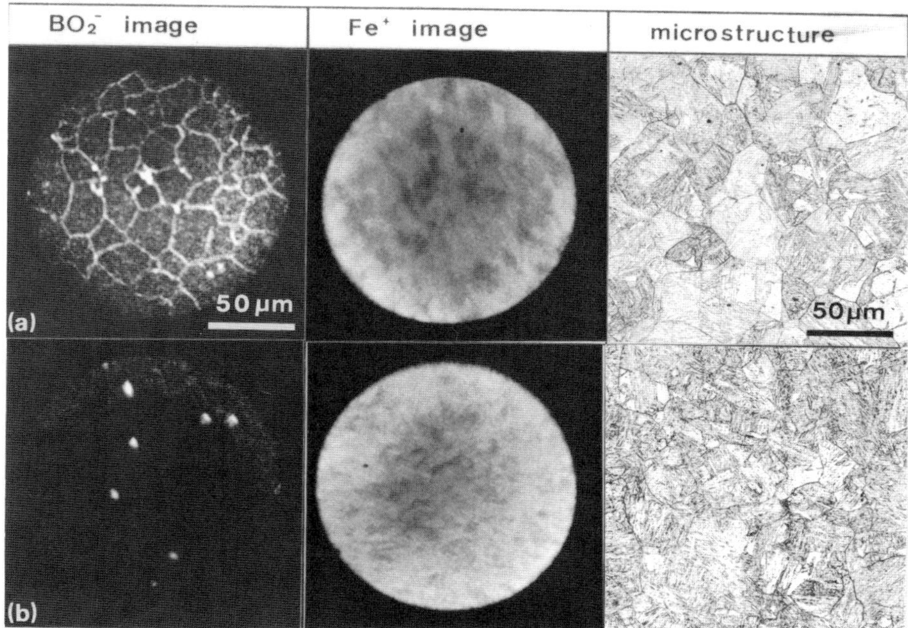

Fig. 5. BO_2^- ion and iron ion (Fe^+) images and related microstructures of the 10 wt.ppm B steel: (a) sample with heat treatment (a); (b) sample with heat treatment (b).

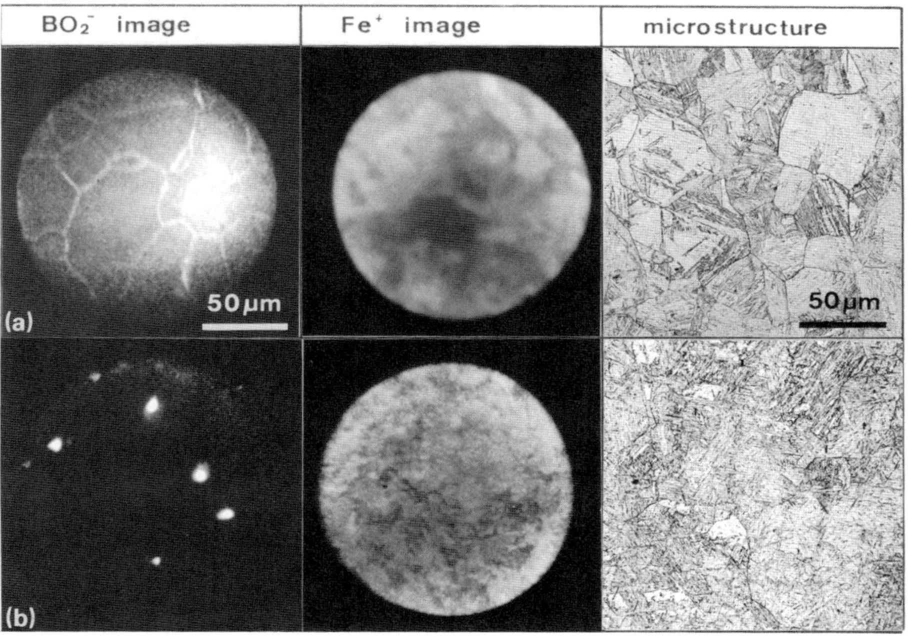

Fig. 6. BO_2^- ion and iron ion (Fe^+) images and related microstructures of the 25 wt.ppm B steel: (a) sample with heat treatment (a); (b) sample with heat treatment (b).

gated region of boron. The distance between the peaks is about 40 μm corresponding to the grain size. The concentration of the base level for the sample subjected to heat treatments (a) and (b) is about 20 wt.ppm B, which is quite reasonable because there is 25 wt.ppm B in this sample.

A strong peak corresponding to a precipitate is obtained for the sample subjected to heat treatment (b). The concentration is esti-

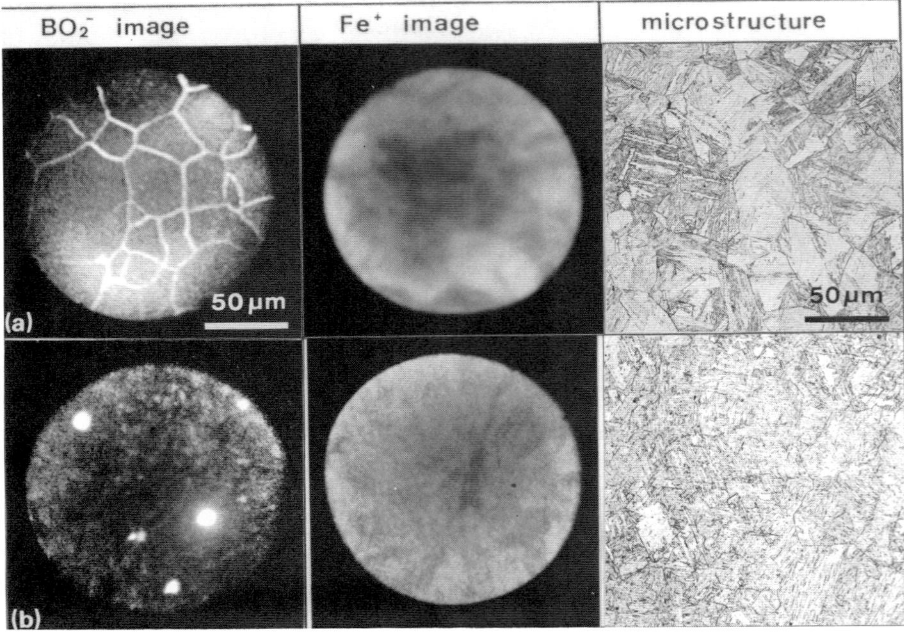

Fig. 7. BO_2^- ion and iron ion (Fe^+) images and related microstructures of the 42 wt.ppm B steel: (a) sample with heat treatment (a); (b) sample with heat treatment (b).

TABLE 3

The Vickers hardnesses of the boron steels (load, 10 kgf)

Steel	*Actual B concentration* (wt.ppm)	*Vickers hardness* (HV) *after the following heat treatments*	
		(a)	*(b)*
40 wt.ppm B steel	42	370	309
25 wt.ppm B steel	25	381	298
10 wt.ppm B steel	10	375	311

mated to be over 10%. However, this cannot be correct because the secondary ion yield is changed in the different chemical state, *e.g.* the boron in steel and boron nitride.

The secondary ion yield in SIMS depends on the incident energy [13] and angle [14], the surface morphology [15] and the amount of oxygen on a surface [5, 6]. Therefore, our measurements were carried out at a constant energy and constant angle to keep the secondary ion yield constant. The amount of oxygen on a surface was always monitored by SIMS to keep it constant when the oxygen ion beam was used for the measurement. However, it is difficult to keep the surface morphology constant for each sample, and this would influence the quantitative analysis.

It is also very important to consider the radiation effects in the ion implantation and the SIMS analysis. The knock-on and mixing effects lead to a change in the concentration of elements within certain layers influenced by radiation [15]. Radiation-induced surface enrichment of certain elements would affect the SIMS analysis [16]. These effects are all taken into consideration for the quantitative analysis using SIMS. However, the variation in our calibration curve is within 200% which is higher than that induced by the above effects.

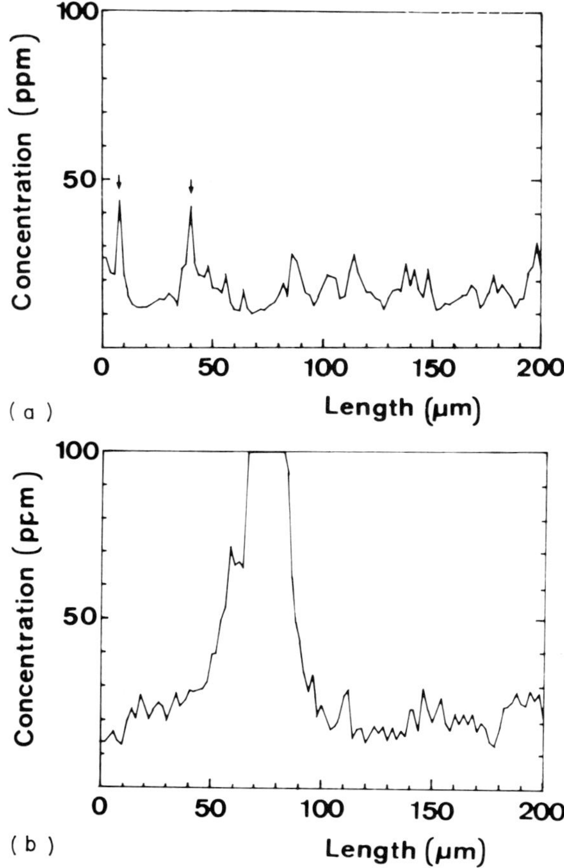

Fig. 8. The line analysis of 25 wt.ppm B steel: (a) the sample with heat treatment (a); (b) the sample with heat treatment (b).

5. SUMMARY

Quantitative analysis of trace boron in steels was performed using SIMS. Iron samples implanted with boron were used as standards because the ion implantation technique can easily control the concentration of implanted elements and produce homogeneous dilute binary alloys. Using these samples, a calibration curve ranging from the per cent to the parts per billion level was obtained. The boron distributions in 40, 25 and 10 wt.ppm B steels were observed using BO_2^- ion images. The concentration of boron segregated to the grain boundaries was estimated using the calibration curve.

The results obtained will be applied to investigate details of the mechanism of boron segregation during heat treatment although the quantitative analysis using SIMS is subject to large errors (200% obtained here). We are further studying the quantitative analysis of carbon, sulphur, hydrogen etc., in steels using ion-implanted samples as standards.

REFERENCES

1 J. E. Morral and T. B. Cameron, in *Boron in Steel*, Metallurgical Society of AIME, Warrendale, PA, 1979, p. 19.
2 T. Shiraiwa, J. Murayama and K. Fujino, *Tetsu To Haganê, 61* (1975) S664.
3 Ph. Maitrepierre, D. Thivellier and R. Tricot, *Metall. Trans. A, 6* (1975) 287.
4 S. Ueda, M. Ishikawa and N. Ohashi, in *Boron in Steel*, Metallurgical Society of AIME, Warrendale, PA, 1979, p. 181.
5 A. Benninghoven and A. Müller, *Surf. Sci., 39* (1974) 483.
6 A. Müller and A. Benninghoven, *Surf. Sci., 39* (1974) 493.
7 M. L. Yu and W. Reuter, *J. Appl. Phys., 52* (1981) 1478.
8 S. P. Holland, B. J. Garrison and N. Winograd, *Phys. Rev. Lett., 43* (1979) 220.
9 K. Tsunoyama, Y. Ohashi and T. Suzuki, *Anal. Chem., 48* (1976) 832.
10 S. Kurosawa, Y. Homma, T. Tanaka and M. Yamawaki, *Secondary Ion Mass Spectrometry 4*, Springer, Berlin, 1984, p. 107.
11 R. V. Criegern and I. Weitzel, *Secondary Ion Mass Spectrometry 5*, Springer, Berlin, 1985, p. 319.
12 S. R. Keown and F. B. Pickering, *Met. Sci., 11* (1977) 225.
13 N. Matsunami, Y. Yamamura, Y. Itikawa, N. Itoh, Y. Kazumata, S. Miyagawa, K. Morita, R. Shimizu and H. Tawara, *At. Data Nucl. Data Tables, 31* (1984) 1.
14 T. Okutani, M. Shikata, S. Ichimura and R. Shimizu, *J. Appl. Phys., 51* (1980) 2884.
15 M. A. Taubenblatt and C. R. Heims, *J. Appl. Phys., 54* (1983) 2667.
16 J. Bartella and H. Oechsner, *Surf. Sci., 126* (1983) 581.

Materials Science and Engineering, 90 (1987) 127–134

Compositional Instabilities in Aluminum-implanted Nickel*

M. AHMED and D. I. POTTER

Metallurgy Department, and Institute of Materials Science, University of Connecticut, Storrs, CT 06268 (U.S.A.)

(Received July 10, 1986)

ABSTRACT

Nickel substrates were implanted with 180 keV aluminum ions (Al^+) to fluences near 10^{18} ions cm^{-2}. The resulting aluminum concentration profiles and microstructures were examined after implantation and following annealing at temperatures below 800 °C. The phases detected after annealing were consistent with both the Ni–Al equilibrium phase diagram and the aluminum concentrations in the implanted layers. The concentration profiles changed only slightly during annealing below 650 °C for fluences less than 1.4×10^{18} Al^+ cm^{-2}. At fluences above this, however, the aluminum concentration profiles after annealing at 650 °C extended to depths approaching 8000 Å, at least 4000 Å deeper than the as-implanted profiles. The aluminum concentration near the surface decreased during annealing, consistent with conservation of total aluminum content and extension of the profiles.

The kinetics of the decrease in near-surface aluminum content were followed in situ by measuring the IR emissivities of the specimens, which correlated with their near-surface aluminum concentrations. The aluminum concentrations were constant with annealing time for specimens implanted to less than 1.4×10^{18} Al^+ cm^{-2}. For higher fluences the emissivities and aluminum contents changed by a factor of 3 over annealing time periods that increased with decreasing temperatures. The diffusion coefficients derived from the aluminum concentration dependence on time ranged from 1.1×10^{-13} to 21×10^{-13} cm^2 s^{-1} for specified annealing temperatures between 500 and 800 °C. The activation energy for this diffusion, 62 kJ mol^{-1}, is close to that ascribed to grain boundary diffusion in Ni_3Al, 50 kJ mol^{-1}. The nickel behind the implanted layers recrystallized during annealing, providing the necessary grain boundary diffusion paths.

1. INTRODUCTION

The system consisting of aluminum ions (Al^+) implanted into nickel is being investigated in order to identify the mechanisms that govern the development of implanted concentration *vs.* depth profiles [1, 2], their elevated temperature stability and the formation and stability of crystalline phases [3]. Here we particularly address the stability of the concentration profiles to annealing under conditions where volume diffusion processes would not significantly alter the profiles. The diffusion coefficient for aluminum in nickel [4] is 1.6×10^{-16} cm^2 s^{-1} at 600 °C and the 1 h r.m.s. diffusion distance is only about 100 Å. Since this distance approaches the depth resolution of the instruments used here to determine the concentration profiles, little observable change in profile shape is expected during such annealing. Although this was found to be the case for implant fluences less than 1.4×10^{18} Al^+ cm^{-2}, it was definitely not the case at higher fluences where the profiles after annealing were found to extend to depths several thousand ångströms beyond the depths of profiles measured following room temperature implantation. In this paper we characterize the kinetics exhibited by the changes in the concentration profiles and identify the probable mechanism by which these changes occur.

2. EXPERIMENTAL PROCEDURES

Electropolished nickel specimens (purity, 99.999%) were implanted with 180 keV alu-

*Paper presented at the International Conference on Surface Modification of Metals by Ion Beams, Kingston, Canada, July 7–11, 1986.

0025-5416/87/$3.50

minum ions (Al^+) to fluences between 1×10^{17} and 3×10^{18} ions cm^{-2}. The residual gas pressure was about 10^{-6} Pa and the aluminum ion flux was 1×10^{14} ions cm^{-2} s^{-1}. Further details of the implantation procedures can be found elsewhere [1]. The annealing temperatures of individual specimens, heated via electron emission furnaces, were monitored with an IR pyrometer. Concentration *vs.* depth profiles were obtained by Auger electron spectroscopy (AES) and sputter etching, using Ni–Al standards as described further elsewhere [2]. These profiles were confirmed by Rutherford backscattering spectroscopy [5]. The microstructures and phases in and below the implanted layers were identified using transmission electron microscopy (TEM) and electron diffraction.

3. RESULTS

3.1. *Instabilities in concentration vs. depth profiles*

Annealing effects were examined both in conjunction with and separate from implantation, with emphasis on the latter. Implantation at 25 °C yielded profiles consistent with calculated profiles. The calculations included components describing the range and straggling of the incident ions, the changes in range, straggling and sputtering with aluminum concentration and the expansion of the substrate by the added atoms [2]. The profiles obtained after implantation at elevated temperatures and those obtained from 25 °C implanted-then-annealed specimens varied slightly from the 25 °C implanted profiles only when the fluence was less than 1.4×10^{18} Al^+ cm^{-2}. Substantial changes occurred at higher fluences.

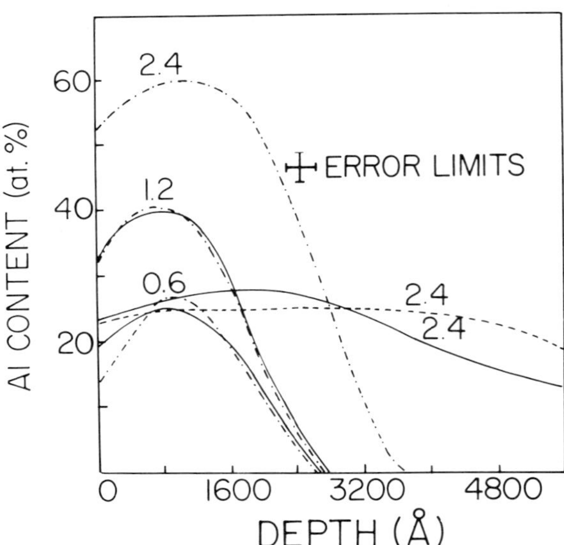

Fig. 1. Aluminum concentration determined with AES *vs.* depth, following 180 keV aluminum ion (Al^+) implantation of nickel at 25 °C (-·-·-·) and 500 °C (——) and after samples implanted at 25 °C had been annealed at 600 °C for 1 h (- - -). The fluences are indicated on curves in units of 10^{18} Al^+ cm^{-2}.

Figure 1 shows the aluminum concentration profiles measured after 25 and 500 °C implantation at three fluences. For the fluences of 0.6×10^{18} and 1.2×10^{18} Al^+ cm^{-2}, the 500 °C profiles match the 25 °C profiles closely, with a small indication (within experimental error) that aluminum is diffusing to the surface at a fluence of 0.6×10^{18} Al^+ cm^{-2}. Little transport by conventional diffusion is anticipated since the diffusion coefficient for aluminum diffusion in nickel at 500 °C is about 10^{-18} cm^2 s^{-1}. The 1 h r.m.s. diffusion distance is about 10 Å and is small even when radiation-enhanced diffusion is taken into account [6], *i.e.* about 100 Å.

Implanting 2.4×10^{18} Al^+ cm^{-2} at 25 °C (Fig. 1, uppermost profile) causes a further increase in aluminum peak concentration relative to the 1.2×10^{18} Al^+ cm^{-2} profiles. The ions come to rest at greater depths because of the higher concentrations of aluminum in regions which they pass through and the lower stopping power of aluminum compared with nickel. The concentration profiles for 25 and 500 °C implants (Fig. 1) differ greatly at 2.4×10^{18} Al^+ cm^{-2}. The aluminum peak content falls from 60 at.% for the 25 °C implant to 28 at.% for the 500 °C implant, and the 500 °C profile extends to depths two to three times greater than the 25 °C profile. The results of implanting at 25 °C and then annealing at 600 °C were similar to those obtained by implanting at 500 °C (Fig. 1). In this case also, the aluminum peak concentration decreases sharply and the depth over which aluminum is distributed increases to two to three times its as-implanted depth. The 1 h r.m.s. diffusion distance for aluminum in nickel at 600 °C is about 100 Å and is not sufficient to account for the observed increase in depth of several thousand ångströms. From the comments in the paragraph above, we note that radiation-enhanced diffusion is likewise unable to account for the changes between the 25 and 500 °C implant profiles at 2.4×10^{18} $Al^+ cm^{-2}$.

3.2. Phase transformations during annealing

Figure 2 represents a typical annealing sequence wherein the phases were identified in several specimens examined after each was annealed at 600 °C for a specified time. A more detailed description of the phases, their morphologies and the depths beneath the surface at which they occur has been given elsewhere [3]. Each specimen portrayed in Fig. 2 was implanted with 2.7×10^{18} $Al^+ cm^{-2}$ at 25 °C, leading to an initial microstructure of β'-NiAl phase and an amorphous phase [2]. The diffraction patterns at the top of Fig. 2 were used to identify the phases indicated on the aluminum concentration *vs.* annealing time curve. The data points on the curve are concentrations averaged from the surface to 1000 Å, the approximate thickness of the TEM specimens. The symbols for the data points along the curve in Fig. 2 show the phases observed during annealing and the meanings of the symbols used are given in the figure caption. These symbols are also used beneath the diffraction patterns and indicate the phases identified as present by indexing each pattern. The equilibrium phases and their homogeneity ranges at 600 °C, taken from the Ni–Al phase diagram [7], are indicated on the right-hand ordinate in Fig. 2.

We note that the amorphous phase present in the specimens from the 25 °C implantation transforms to β' during annealing for 120 s and that the composition of this β' is outside the equilibrium homogeneity range of β'. The composition is correct for δ'-Ni_2Al_3 and this phase forms during 30–80 s of further annealing. The crystal structure of δ' is similar to that of β', differing only somewhat because of the presence of ordered arrays of vacancies

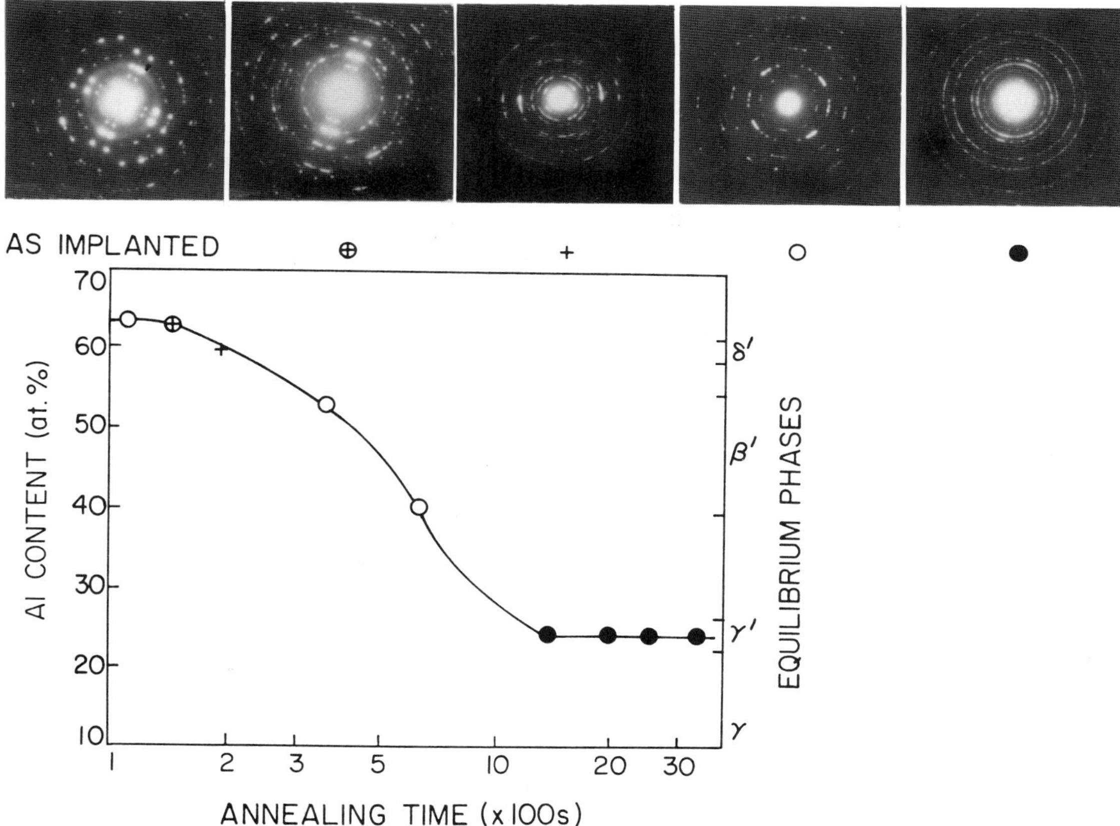

Fig. 2. Aluminum concentration plotted *vs.* annealing time at 600 °C for nickel specimens implanted at 25 °C with 2.7×10^{18} $Al^+ cm^{-2}$: +, δ'; ⊕, $\delta' + \beta'$; ○, β'; ●, γ'. The phases observed by TEM at specific times are noted along the curve and identified by the associated electron diffraction patterns above the graph.

in δ' which are not present in β'. Radiation damage accompanying implantation suppresses this vacancy ordering [8], leaving β' whose composition is that of equilibrium δ'. During further annealing (Fig. 2) the vacancies order, causing the β' to transform to δ'. The β' forms again and then transforms to γ'-Ni_3Al as the aluminum concentration decreases and enters into the corresponding homogeneity ranges of the phase diagram with still further annealing.

This series of phase transformations contrasts sharply with the absence of any transformations occurring in specimens implanted at fluences of less than 1.4×10^{18} Al^+ cm^{-2}. For example, the γ' phase that formed after 2 min at 600 °C in a sample implanted at 25 °C with 6×10^{17} Al^+ cm^{-2} remained γ' after 6 h of annealing at 600 °C. Similarly the β' phase in specimens implanted with 1.2×10^{18} Al^+ cm^{-2} at 25 °C did not transform during annealing at 600 °C. The stability of the phases in specimens implanted to these fluences is in agreement with the stability of the corresponding aluminum concentration profiles (Fig. 1).

3.3. *Near-surface aluminum concentration changes*

The rates at which the implanted concentrations changed during annealing were monitored as a function of temperature. The measurements were made *in situ* with the IR pyrometer. The device is customarily used to measure the temperatures of materials whose emissivities are known (Section 2). Alternatively, the pyrometer will display the surface emissivity at wavelengths near 0.65 μm when the temperature is known. Important here is the observation that the emissivity of an Ni–Al alloy is sensitive to the composition of the alloy. The emissivities of several specimens were measured together with their near-surface aluminum concentrations using Auger spectroscopy. The emissivities were directly proportional to aluminum contents, varying from 0.13 ± 0.01 at 64 at.% Al to 0.28 ± 0.01 at 25 at.% Al. Other experimental work [9] and straightforward absorption calculations show that the measured emissivities are controlled by material within approximately 100 Å of the specimen surface.

The response of the emissivity to annealing at 600 °C was examined for specimens implanted to different fluences. The results are shown in Fig. 3(a). The uppermost curve applies to a fluence of 6×10^{17} Al^+ cm^{-2}, but the constant emissivity with time which it portrays was shared at all fluences of less than 1.4×10^{18} Al^+ cm^{-2}. The emissivity increased with increasing annealing time for specimens implanted with greater than 1.6×10^{18} Al^+ cm^{-2}, indicating decreasing aluminum concentration. This is seen in the remaining curves shown in Fig. 3(a). The rate of emissivity increase (aluminum decrease) is fluence dependent, being greater at higher fluences (compare the 3.0 and 1.6×10^{18} Al^+ cm^{-2} curves in Fig. 3(a)). An initial decrease in emissivity is seen for fluences greater than 2.4×10^{18} Al^+ cm^{-2}. This occurred as a result of vacancy ordering and formation of δ' phase (Fig. 2).

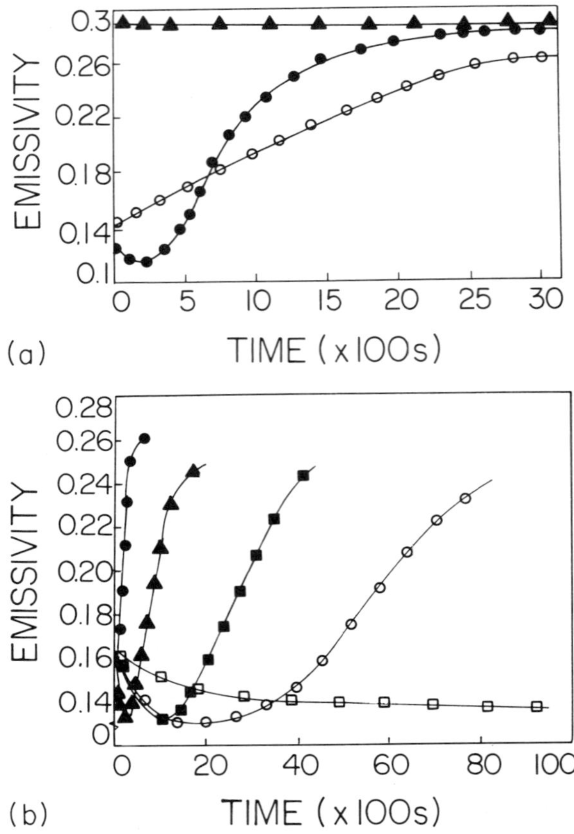

Fig. 3. Annealing kinetics as portrayed by emissivity changes with time (a) for specimens implanted at room temperature with different fluences (▲, 6×10^{17} Al^+ cm^{-2}; ○, 1.6×10^{18} Al^+ cm^{-2}; ●, 3.0×10^{18} Al^+ cm^{-2}) and subsequently annealed at 600 °C and (b) for specimens implanted at room temperature with 2.7×10^{18} Al^+ cm^{-2} and then annealed at various temperatures (●, 800 °C; ▲, 600 °C; ■, 550 °C; ○, 500 °C; □, 400 °C).

Figure 3(b) presents emissivity *vs.* time data for specimens implanted to the same fluence of 2.7×10^{18} Al^+ cm^{-2} and then annealed at different temperatures. TEM examination, after annealing was terminated, showed that the near surface of the 400 °C specimen consisted entirely of δ'-Ni_2Al_3 phase. This specimen shows the decrease in emissivity that was mentioned at the end of the last paragraph. The near-surface regions of the remaining specimens in Fig. 3(b) contained only γ'-Ni_3Al phase. The emissivities recorded after long annealing times (Fig. 3) asymptotically approach a value of about 0.28, characteristic of γ'-Ni_3Al, consistent with these TEM observations.

The emissivities in Fig. 3(b) were converted to near-surface aluminum concentrations via the proportionality between these quantities that is mentioned at the beginning of this section. These concentrations, divided by the as-implanted aluminum concentration of about 64 at.%, are plotted *vs.* annealing time for four temperatures in Fig. 4. Data points describing the initial emissivity decreases in Fig. 3(b) are omitted from Fig. 4. The curves in Fig. 4 were calculated at each of the four temperatures, as described in Section 4 and were not simply drawn to fit the data points.

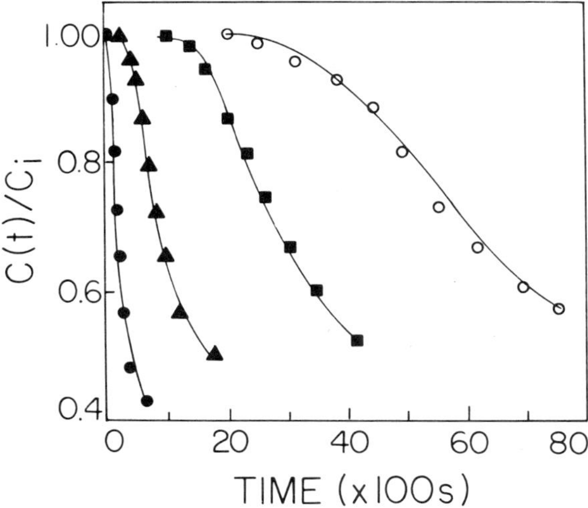

Fig. 4. Data points showing surface concentrations, divided by initial concentration, plotted *vs.* time for specimens implanted with 2.7×10^{18} Al^+ cm^{-2} and then annealed at various temperatures (●, 800 °C; ▲, 600 °C; ■, 550 °C; ○, 500 °C). The curves were calculated using eqn. (1) and the parameters listed in Table 1.

3.4. Microstructures beneath the implanted layers

The microstructures in and beneath the implanted layers were examined by TEM [3]. The microstructural developments occurring beneath the implanted layers are particularly important here and are summarized in Fig. 5. The as-implanted microstructures consist of dislocations and interstitial dislocation loops (Figs. 5(a) and 5(b)). The number density of these loops increases with increasing fluence (compare Fig. 5(a) with Fig. 5(b)). During annealing, these dislocations form lower energy networks in specimens implanted to less than 1.4×10^{18} Al^+ cm^{-2}. Recrystallization occurs at higher fluences (Fig. 5(c)) and is driven by the energy gained in eliminating the higher densities of dislocations which were generated by the higher fluences. The recrystallized layer extends from the implanted layer to a depth of about 8000 Å. The β' phase within the implanted layer is also recrystallized [2].

4. DISCUSSION

The results of this investigation have shown that the aluminum concentration profiles from implanting nickel at a fluence greater than 1.6×10^{18} Al^+ cm^{-2} change dramatically during annealing at temperatures near 600 °C. Profiles established by fluences less than this change very little during annealing. R.m.s.

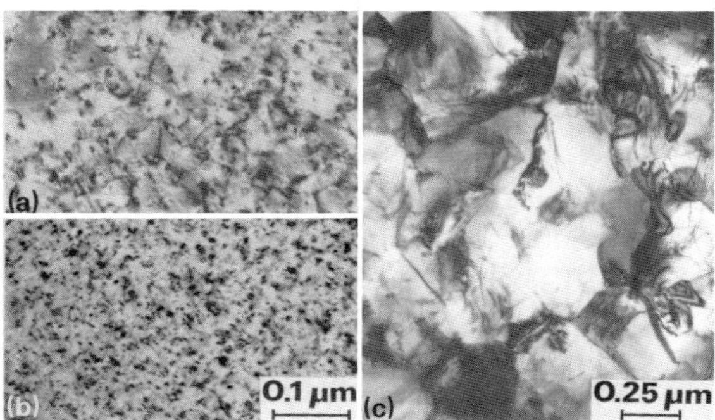

Fig. 5. Microstructures at a depth of about 5000 Å beneath the specimen surface: (a) dislocations and dislocation loops after implantation at 25 °C with 1.2×10^{18} Al^+ cm^{-2}; (b) dislocations and dislocation loops after implantation at 25 °C with 3.0×10^{18} Al^+ cm^{-2}; (c) recrystallized structure in nickel implanted with 2.4×10^{18} Al^+ cm^{-2} at 25 °C and then annealed at 600 °C for 120 s.

diffusion distances, calculated from diffusion coefficients that pertain to volume diffusion by vacancy–atom interchange, are small. For annealing times used here and temperatures near 600 °C, these r.m.s. distances suggest that little, if any, change in profiles should occur, consistent with the lower fluence results. These r.m.s. diffusion distances cannot explain the compositional redistributions in the higher fluence specimens, even if radiation-enhanced diffusion is considered.

Figure 5 shows that the nickel beneath the implanted layers of the higher fluence specimens recrystallizes during annealing. This does not occur for specimens implanted to less than 1.6×10^{18} Al^+ cm^{-2}, as noted in Section 3.4. The large-angle grain boundaries formed by the recrystallization will act as fast diffusion paths, allowing aluminum diffusion into the substrate and nickel diffusion from the substrate into the implanted layer. In order to model the diffusion process, we treated the implanted layer as an extended source, and this and the recrystallized nickel layer as a system of finite extent. Crank [10] explored the solution to this diffusion problem, including the reflection contributions that take place at the recrystallized nickel–unrecrystallized nickel interface. The contributions of these reflections to the surface concentrations under the annealing conditions of Fig. 4 are negligible. The analytical expression for the surface concentration $C_s(t)$ as a function of time t is then

$$C_s(t) = C_i \operatorname{erf}\left\{\frac{h}{2(Dt - Dt_0)^{1/2}}\right\} \tag{1}$$

Here C_i is the concentration in the implanted layer before diffusion begins, h is the original thickness of the implanted layer and D is the diffusion coefficient describing the transport process. Time is referenced to the time t_0 at which diffusion begins. This time is the time required for recrystallization of the nickel layer behind the implanted layer.

The curves in Fig. 4 describing the decrease in surface aluminum content with annealing time were calculated using eqn. (1). A value of h equal to 2400 Å was chosen in accord with Fig. 1, while C_i was taken to be 64 at.% Al (Fig. 2). The diffusion coefficients and times for recrystallization used at each temperature are presented in Table 1. Changing either D or t_0 by ±15% resulted in a poor fit of the curves to the data points in Fig. 4. These error limits are attached to the diffusion coefficients in the Arrhenius plot in Fig. 6. The activation energy for the diffusional process, found from the slope of the line in Fig. 6, is 62 kJ mol^{-1}.

The activation energy of 62 kJ mol^{-1} can be compared with those of other processes involving diffusion in nickel. These include the following: 288 kJ mol^{-1} [11] for nickel self-diffusion; 230 kJ mol^{-1}, 268 kJ mol^{-1} and 272 kJ mol^{-1} for volume diffusion coefficients

TABLE 1

Diffusion coefficients and recrystallization times

Temperature (°C)	t_0 (s)	D ($\times 10^{-13}$ cm^2 s^{-1})
500	3000	1.1
550	1500	2.3
600	300	5.0
800	30	21

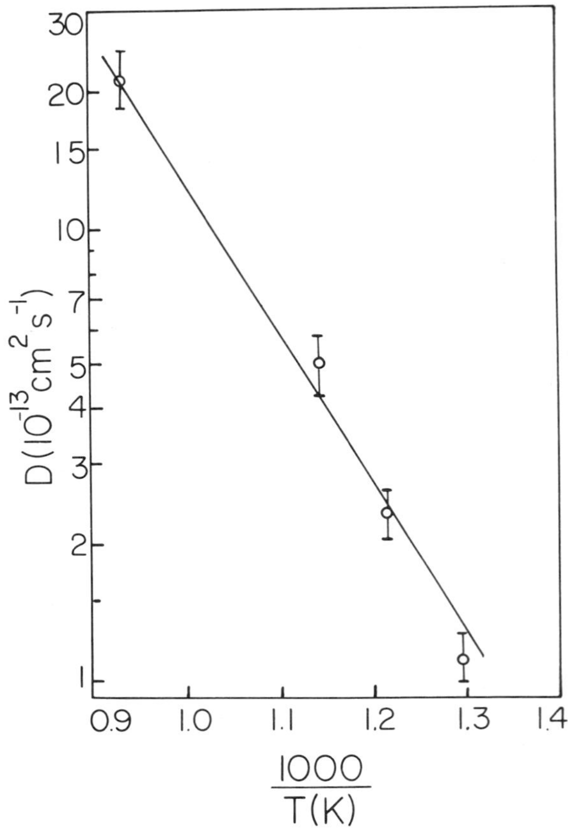

Fig. 6. Arrhenius plot of the diffusion coefficients listed in Table 1.

obtained by the Boltzmann–Montano method [12] for β'-NiAl, γ'-Ni_3Al and the solid solution of aluminum in nickel respectively; 121 kJ mol^{-1} [13] for volume diffusion in δ'-Ni_2Al_3. The activation energy determined in the present work is considerably less than any of these energies.

The activation energies for grain boundary diffusion are closer to the value of 62 kJ mol^{-1} found in this investigation. Peterson [14] suggests that the energy for grain boundary diffusion in most materials will be less than half the energy for volume diffusion, *i.e.* less than 144 kJ mol^{-1} for nickel. Upthegrove and Sinnott [15] and Wazzan [16] have measured activation energies of 109 kJ mol^{-1} and 115 kJ mol^{-1} respectively for nickel diffusing along grain boundaries in nickel. Transport by grain boundary diffusion is most noticeable at temperatures below that at which volume diffusion is appreciable, and Wazzan's measurements span the temperature interval from 475 to 650 °C. This includes part of the temperature range investigated in this work. Janssen and Rieck [17] have determined an activation energy for γ'-Ni_3Al layer growth over a somewhat higher range of temperatures, from 650 to 1000 °C. The energy value that they determined, 50 kJ mol^{-1}, is close to ours. Janssen [12] attributes the low energy found in their investigation to grain boundary diffusion of aluminum and nickel in γ'-Ni_3Al and demonstrates that this diffusion is overwhelmed by volume diffusion at temperatures above 1000 °C. Thus the evidence available from the TEM microstructural observations and the diffusion measurements supports our identification of grain boundary diffusion as the cause for the compositional instabilities observed in the specimens implanted with more than 1.6×10^{18} Al^+ cm^{-2}.

In closing, we note that, although the high fluence composition profiles changed quite rapidly during annealing, the layer of γ'-Ni_3Al that resulted persisted to much longer annealing times. As mentioned previously, for example, the γ' phase was still present after annealing for 6 h, while the aluminum content decreased from 64 to 25 at.% in less than one-tenth of this time (Figs. 2 and 3(b)). The instability of the aluminum content as it decreases from 64 to 25 at.% is caused by the grain boundary diffusion discussed above. The relative stability of the γ', as exemplified by the fact that the aluminum content of 25 at.% remains relatively constant with annealing time, results from the finite depth of the recrystallization that occurs beneath the implanted layers (Fig. 5). TEM specimens sectioned to depths beyond those in Fig. 5 showed that recrystallization did not extend to depths beyond about 8000 Å. Thus the aluminum concentration changes within the implanted and recrystallized regions in accord with the finite system error function solution to the diffusion problem [10]. This occurs relatively rapidly via grain boundary diffusion. The final result of this process would be a uniform aluminum concentration from the surface to the recrystallized–unrecrystallized nickel interface, were it not for the fact that this concentration rests in the γ'–γ solid solution phase field (Fig. 2). This uniform aluminum concentration with depth is not achieved because a layer of γ' forms during annealing, extending from the surface to a depth of beyond 5000 Å. Thus the aluminum concentration profile after annealing is nearly flat at 25 at.%, corresponding to γ'-Ni_3Al, from the surface to considerable depths (Fig. 1). Given long enough diffusion times, a second flat portion corresponding to the γ–γ' solubility limit would presumably extend from the γ' layer to the recrystallized–unrecrystallized nickel interface. This two-step aluminum concentration profile will be relatively stable to annealing at 600 °C because further aluminum transport must occur by volume diffusion into the unrecrystallized nickel, a very slow process at 600 °C compared with grain boundary diffusion.

5. CONCLUSIONS

The following conclusions can be drawn from the results and discussion of the present investigation.

(1) Concentration profiles describing specified fluences of 180 keV aluminum ions (Al^+) implanted into nickel are quite similar when the implantations are performed at room temperature and at 500 °C and when the specimens implanted at 25 °C are annealed for up to 6 h at temperatures near 600 °C. This is true for specimens implanted to fluences less than 1.4×10^{18} Al^+ cm^{-2}.

(2) Concentration profiles for specimens implanted with aluminum fluences greater than 1.6×10^{18} ions cm^{-2} are not stable to annealing at temperatures near 600 °C. These annealed profiles and those from implantation at 500 °C show lower peak aluminum concentrations, and aluminum contents extending to greater depths, than do the specimens implanted at room temperature.

(3) The near-surface aluminum concentrations in specimens implanted at room temperature with more than 1.6×10^{18} Al^+ cm^{-2} decrease during annealing. The kinetics of this decrease can be described analytically by the error function solution to the diffusion problem involving an extended source on a finite system.

(4) The activation energy governing the decrease in the near-surface aluminum concentrations during annealing, obtained using the diffusion problem solution outlined in conclusion (3), is 62 kJ mol^{-1}. This energy is close to that for grain boundary diffusion in Ni_3Al, as measured by other investigators.

(5) Recrystallization occurs behind the implanted layers in specimens implanted with more than 1.6×10^{18} Al^+ cm^{-2}. The recrystallization extends to depths approaching 8000 Å, but not much beyond this depth. The recrystallization provides the grain boundaries for the diffusion mentioned in conclusion (4), thus causing the instability in the concentration profiles. The finite depth of recrystallization leads to the relatively stable nature of the γ' layer that forms during annealing.

ACKNOWLEDGMENTS

We thank the following people at the University of Connecticut who contributed to this research: Q. Kessel and J. Gianoupolis for assistance with the Rutherford backscattering analysis; H. Hayden and C. Koch for assisting in the ion implantations; L. McCurdy for TEM assistance; J. Soracchi and T. Swol of the instrument laboratory; Lane Witherell for typing the manuscript; J. Hampikian for the Rutherford backscattering spectral analysis.

This material is based on work supported by the National Science Foundation under Grant DMR8507641. The transmission electron microscope was purchased with support from National Science Foundation Grant DMR8207266 and the State of Connecticut. The authors are grateful for the continuing support of these people and agencies.

REFERENCES

1 D. I. Potter, M. Ahmed and S. Lamond, *J. Met.*, *35* (1983) 17.
2 M. Ahmed and D. I. Potter, *Acta Metall.*, *33* (1985) 2221.
3 M. Ahmed and D. I. Potter, *Acta Metall.*, to be published.
4 J. Hirvonen, *Appl. Phys., A*, *27* (1982) 243.
5 D. I. Potter, M. Ahmed and S. Lamond, in G. K. Hubler, O. K. Holland, C. R. Clayton and C. W. White (eds.), *Ion Implantation and Ion Beam Processing of Materials, Materials Research Society Symp. Proc., Boston, MA, November 14-17, 1983*, Vol. 27, Elsevier, New York, 1984, p. 117.
6 D. Potter and A. McCormick, *Acta Metall.*, *27* (1979) 933.
7 *Metals Handbook*, Vol. 8, American Society for Metals, Metals Park, OH, 1973, p. 262.
8 L. A. Grunes, J. C. Barbour, L. S. Hung, J. W. Mayer and J. J. Ristsko, *J. Appl. Phys.*, *56* (1984) 168.
9 S. Lamond, *M.S. Thesis*, University of Connecticut, 1983.
10 J. Crank, *The Mathematics of Diffusion*, Oxford University Press, London, 2nd edn., 1979, p. 16.
11 A. Ya. Shinyaev, *Fiz. Met. Metalloved.*, *15* (1963) 100.
12 M. M. P. Janssen, *Metall. Trans.*, *4* (1973) 1623.
13 A. K. Sarkhel and L. L. Siegel, *Metall. Trans. A*, *13* (1982) 1313.
14 N. L. Peterson, *Int. Metall. Rev.*, *28* (1983) 65.
15 W. R. Upthegrove and M. J. Sinnott, *Trans. Am. Soc. Met.*, *50* (1958) 1031.
16 A. R. Wazzan, *J. Appl. Phys.*, *36* (1965) 3596.
17 M. M. P. Janssen and G. D. Rieck, *Metall. Trans.*, *239* (1967) 1372.

Materials Science and Engineering, 90 (1987) 135–142

Compounds and Microstructures of Silicon-implanted Nickel*

S. G. B. FISHMAN, M. AHMED and D. I. POTTER

Metallurgy Department, and Institute of Materials Science, University of Connecticut, Storrs, CT 06268 (U.S.A.)

(Received July 10, 1986)

ABSTRACT

During ion implantation the phase stability and implanted ion composition are strongly affected by the radiation damage inherent in the process. A study was made of the role of radiation damage processes in the implantation of silicon into nickel at temperatures from 25 to 650 °C. Radiation-induced segregation occurred, resulting in a greatly enriched silicon content on substrate surfaces. The crystalline nickel silicide phases, Ni_5Si_2, Ni_2Si and $NiSi_2$, were observed over a wide temperature range, including $T < 200$ °C. The Ni_3Si phase, however, was destabilized by radiation damage and formed only at $T >$ 500 °C.

At $T > 250$ °C and fluences near or exceeding 1×10^{18} Si^+ cm^{-2}, the implanted layer recrystallized. Penetration of implanted silicon several thousand ångströms beyond the ballistic range of the implanted ions was also observed at these fluences for $T > 350$ °C. A microcrystalline phase formed in this fluence range at $T < 200$ °C.

1. INTRODUCTION

Studies of the Ni–Si system during irradiation have added significantly to the understanding of radiation damage processes in crystalline solids. Two features of the system have attracted considerable interest. One is the presence of radiation-induced segregation which causes silicon enrichment on internal sinks such as dislocations and grain boundaries and on the external surfaces. This process is attributed to an Ni–Si mixed interstitial defect, strongly bound and mobile at room temperature [1–4]. A second area of study concerns destabilization of the $L1_2$ ordered phase Ni_3Si during irradiation [5–7]. Since the Ni–Si phase diagram (Fig. 1 [8]) consists only of ordered stoichiometric compounds beyond 10 at.% Si, the question arises whether any of these phases would form during ion implantation or would inherently be destabilized by the radiation damage. An investigation was conducted to determine what role, if any, these radiation damage processes would play in ion implantation with regard to silicon depth distribution, microstructure and phase stability. Implantation temperatures ranged from 25 to 650 °C, and implanted silicon contents ranged from 10 to

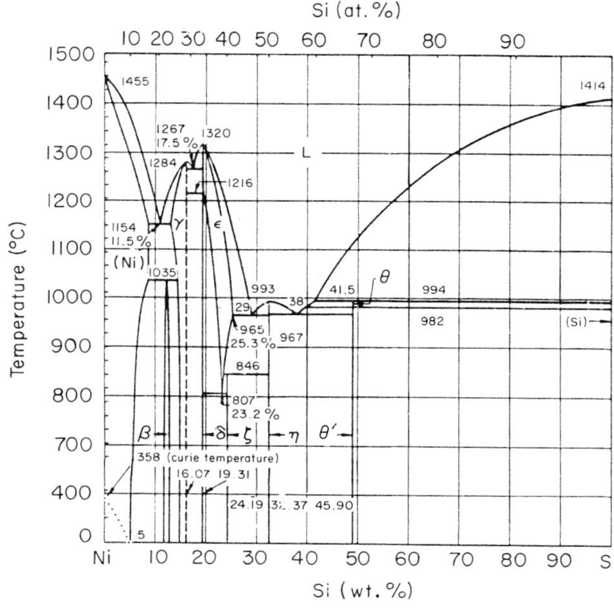

Fig. 1. Equilibrium phase diagram of the Ni–Si system [8].

*Paper presented at the International Conference on Surface Modification of Metals by Ion Beams, Kingston, Canada, July 7–11, 1986.

0025-5416/87/$3.50

90 at.%. The results of the investigation are reported here.

2. EXPERIMENTAL PROCEDURES

Specimens were in the form of discs of 3 mm diameter, approximately 0.25 mm thick. They were cut from 99.998% pure nickel, encapsulated in fused quartz at 10^{-6} Pa and annealed for 6 h at 900 °C. One surface of each disc was electropolished. The polished faces were then implanted with 175 keV silicon ions at a flux of 10^{14} ions cm^{-2} s^{-1} in a vacuum at 10^{-6} Pa, to fluences between 3×10^{17} and 1×10^{19} ions cm^{-2}. The temperature of the specimen during implantation was regulated and monitored to within 5 °C between 25 and 650 °C. Concentration profiles (depth distributions of silicon) were determined primarily by Auger electron spectroscopy (AES) in conjunction with ion milling. The ratio of the silicon-to-nickel Auger signal intensity was calibrated with respect to composition by comparison with results from energy-dispersive X-ray spectroscopy (EDXS). Two groups of specimens implanted below 200 °C were used for the calibration. One group was profiled by AES. The other group was made into thin foils, and the average silicon concentration of each foil was measured by EDXS. The results, corrected for foil thickness, were accurate to ±3 at.%. The composition obtained by EDXS was plotted against fluence. From this plot, compositions were assigned to the AES specimens based on their fluences. These compositions were plotted *vs.* the AES peak-to-peak signal ratios. The resulting calibration curve is shown in Fig. 2. The standard deviation for the calibration is ±3 at.%.

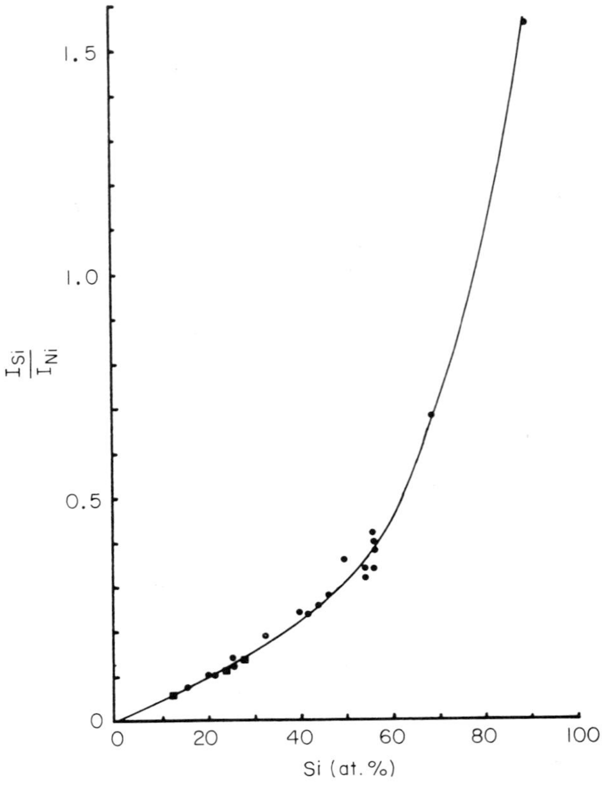

Fig. 2. Calibration of the AES peak-to-peak signal ratio with the silicon concentration: ●, implanted, average first 1000 Å; ■, bulk standard.

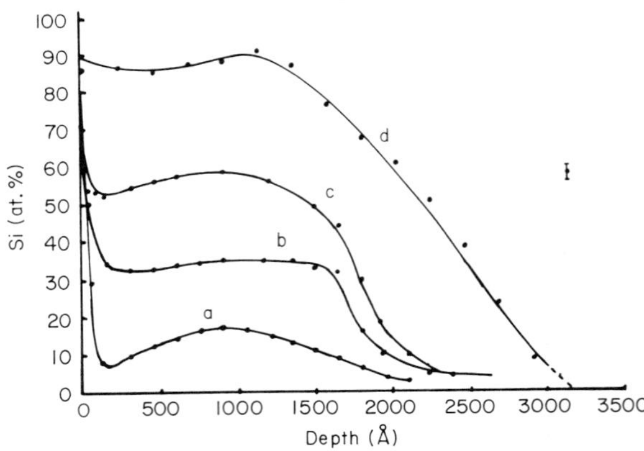

Fig. 3. Concentration profiles of silicon implanted into nickel at $T \leq 200$ °C to various fluences: curve a, 5.1×10^{17} Si^+ cm^{-2}; curve b, 1.2×10^{18} Si^+ cm^{-2}; curve c, 3.0×10^{18} Si^+ cm^{-2}; curve d, 1.0×10^{19} Si^+ cm^{-2}.

3. RESULTS AND DISCUSSION

3.1. Concentration and depth distribution of implanted silicon

Figure 3 shows the concentration profiles of silicon implanted at temperatures below 200 °C, at several fluences from 5×10^{17} to 1×10^{19} Si^+ cm^{-2}. The silicon concentration is high at the surface, falls to a minimum at about 200 Å depth and rises again to a maximum or to a level plateau at about 1000 Å depth. The concentration of silicon at the maximum or plateau level increases with increasing fluence. Thereafter, it falls and becomes negligible between 2000 and 3000 Å from the surface. These implanted depths are comparable with the ballistic ranges of silicon ions, approximately 850 Å for silicon implanted into pure nickel and up to 2500 Å for silicon implanted into pure silicon. The ranges were calculated for 175 keV silicon ions by a computer code based on the model of Biersack and Haggmark [9].

Enrichment of silicon on the surface indicates that radiation-induced segregation occurs under these conditions. A study of radiation-induced segregation below 200 °C showed that the segregation is greatest at low fluences. For the purpose of this specific study the depth interval for AES sampling was reduced to permit 11 data points to be recorded within the first 300 Å. Figure 4(a) presents AES data for the near-surface region of a specimen implanted to 5×10^{17} Si^+ cm^{-2}. These data, transformed into a concentration profile, constitute curve c in Fig. 4(b), which compares radiation-induced segregation at several fluences below 200 °C.

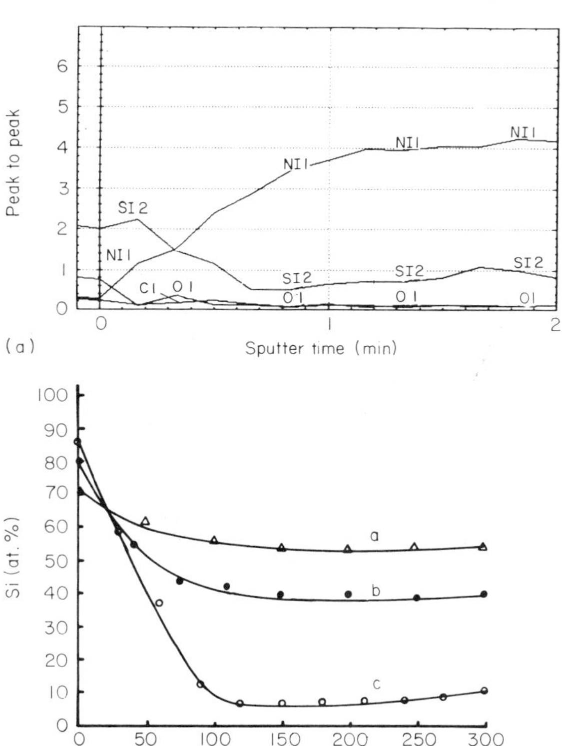

Fig. 4. Radiation-induced segregation below 200 °C: (a) AES data from a specimen implanted with 5×10^{17} Si^+ cm^{-2}; (b) radiation-induced segregation resulting from implantations below 200 °C with fluences of 3.0×10^{18} Si^+ cm^{-2} (curve a), 1.7×10^{18} Si^+ cm^{-2} (curve b) and 5.1×10^{17} Si^+ cm^{-2} (curve c) based on data in (a).

During elevated temperature implantation, increasing fluence results in deeper penetration of the silicon. Figure 5 shows the results for several fluences implanted near 500 °C. The initial portions of these curves are shown as broken because they are defined by at most three data points, in contrast with Figs. 3 and 4(b). A consequence of the deeper penetration is that silicon concentration remains essentially unchanged as fluence increases. The deeper penetration cannot be attributed to thermal diffusion alone, nor even to diffusion enhanced by the greater point defect population because of irradiation. The r.m.s. diffusion length due to thermal diffusion at 500 °C, based on the data of Swalin *et al.* [10], is only about 100 Å. The diffusion distance due to radiation-enhanced diffusion, based on the data of Macht *et al.* [11] for pure nickel, is about 1000 Å at 500 °C. To account for an increase of more than 4000 Å, we must first consider the microstructural results of implantation.

3.2. Microstructures and phases resulting from the implantation of silicon ions into nickel

The phases and microstructural results are summarized in Fig. 6. The phases are shown on axes of temperature and implanted composition, with fluence as a hidden variable. The results are from thin foils, revealing structures only within the first 1000 Å from the implanted surface. For comparison, the equilibrium phase diagram of the Ni–Si system is

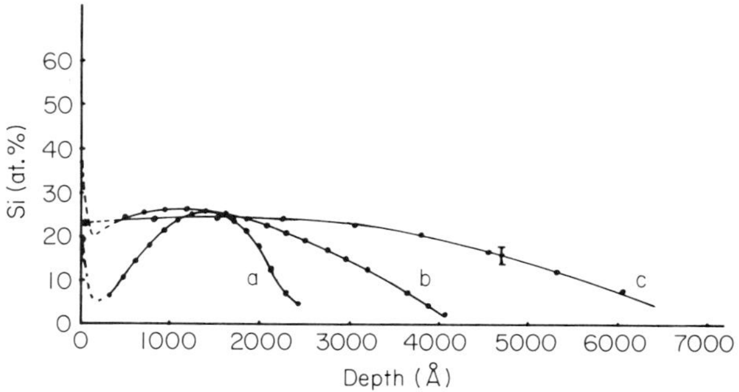

Fig. 5. Concentration profiles of silicon implanted into nickel at $T \approx 500$ °C with various fluences: curve a, 5.9×10^{17} Si^+ cm^{-2}; curve b, 1.2×10^{18} Si^+ cm^{-2}; curve c, 2.2×10^{18} Si^+ cm^{-2}. See text for an explanation of the broken portions of the curves.

also depicted in the figure. The text which follows explains these results.

3.2.1. Low fluence

At fluences below 7×10^{17} Si^+ cm^{-2}, the nickel single-crystal matrix remains intact. Radiation damage produces networks of dislocations. Silicon dissolves in the nickel until the solubility limit is reached, after which nickel silicides precipitate, oriented towards the nickel matrix.

Figure 7 shows the phase and microstructural results of implanting a low fluence at various temperatures. Figure 7(a) is from a specimen implanted with 5×10^{17} Si^+ cm^{-2} at below 200 °C. The silicon concentration in this specimen, measured by EDXS, is 16

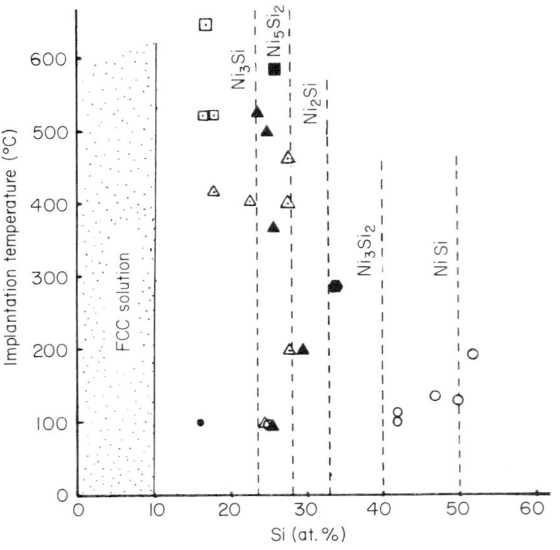

Fig. 6. Phase and microstructural results within the first 1000 Å of the surface of nickel implanted with silicon ions at the temperatures indicated: ⊡, Ni_3Si precipitate; ■, Ni_3Si polycrystal; △, Ni_5Si_2 precipitate; ▲, Ni_5Si_2 polycrystal; ⬢, Ni_2Si polycrystal; ●, f.c.c. solution; ○, diffuse ring. The equilibrium phase diagram of the Ni–Si system is depicted on the same axes for comparison.

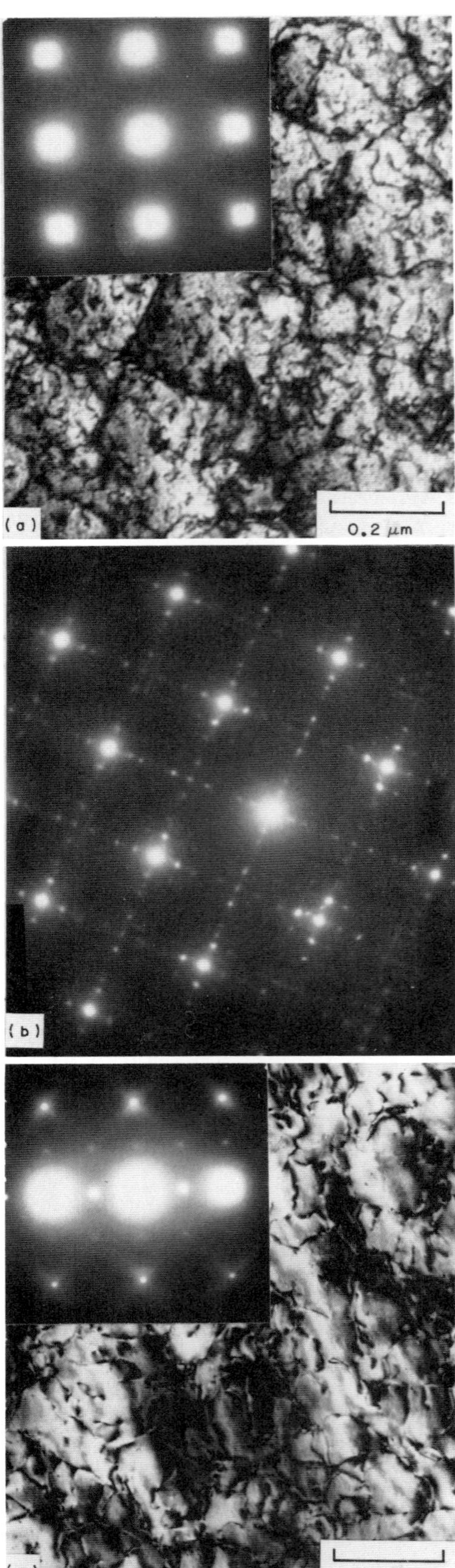

Fig. 7. Phase and microstructural results of implanting silicon into nickel with a low fluence: (a) extended solubility of silicon in f.c.c. nickel, 16 at.% Si (5×10^{17} Si^+ cm^{-2}) implanted below 200 °C; (b) Ni_5Si_2 superlattice at 14 at.% Si (3.2×10^{17} Si^+ cm^{-2}) implanted at 417 °C; (c) Ni_3Si superlattice at 17 at.% Si (4.9×10^{17} Si^+ cm^{-2}) implanted at 520 °C.

at.%. The f.c.c. diffraction pattern shows that the silicon has dissolved in the nickel, although the silicon concentration exceeds the equilibrium solubility limit, which is about 10 at.%. Figure 7(b) is from a specimen implanted with 3.2×10^{17} Si^+ cm^{-2} at 417 °C. The maximum silicon concentration in this specimen, based on AES profiles, is 19 at.%, and the average concentration in the first 1000 Å is about 14 at.%. The diffraction pattern shows the $\{00l\}$ reflections of hexagonal Ni_5Si_2 parallel to the [220] directions of the nickel matrix. The equilibrium phases at this composition are a mixture of f.c.c. solution and Ni_3Si; however, Ni_3Si was not seen below 500 °C. Figure 7(c) shows the result of implanting 4.9×10^{17} Si^+ cm^{-2} at 520 °C. The silicon concentration, measured by EDXS, is 17 at.%. Reflections from the Ni_3Si superlattice are seen together with the nickel f.c.c. single-crystal pattern.

The destabilization of Ni_3Si by irradiation has been explained by Potter [6] according to the model of Liou and Wilkes [5]. Briefly, the degree of long-range order represents a balance between disordering by irradiation and thermal reordering. When Ni_3Si is disordered, its free energy increases to that of the random f.c.c. solution. This is shown by the arrow in Fig. 8, which schematically depicts the free energies of several Ni–Si phases. It is noteworthy that Ni_5Si_2 is not destabilized by irradiation (at the present flux) and in fact was observed even in specimens implanted below 200 °C (Fig. 6). The tangent construction in Fig. 8 predicts that destabilization of Ni_3Si leads to an increase in the solubility limit, from X_1 to X_2, as is in fact observed (Fig. 7(a)).

3.2.2. At higher fluences and lower temperatures

Below 200 °C, fluences between 1×10^{18} and 3×10^{18} Si^+ cm^{-2} result in diffraction patterns consisting of diffuse rings. The matrix becomes microcrystalline. Figure 9(a) shows the diffraction pattern from a specimen implanted with 1.75×10^{18} Si^+ cm^{-2} at 137 °C. The silicon content, measured by EDXS, is 47 at.%. The bright field image (Fig. 9(b)) shows considerable structure. The dark field image (Fig. 9(c)), using electrons from the bright rings, reveals microcrystalline regions of about 20 Å diameter.

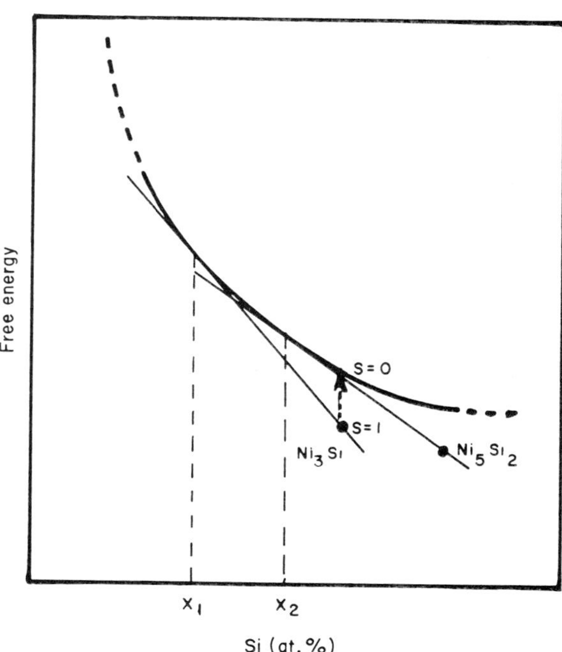

Fig. 8. Schematic diagram showing the effects of disordering by irradiation on free energies and phase stabilities in the nickel-rich region of the Ni–Si system: X_1, equilibrium solubility; X_2, irradiated solubility. See text for an explanation.

Fig. 9. Microcrystallinity results from implantation with a high fluence below 200 °C, at 47 at.% Si (1.75×10^{18} Si^+ cm^{-2}) implanted at 136 °C: (a) the diffraction pattern contains a bright diffuse ring, and a pale diffuse ring of larger diameter is also seen; (b) the bright field image; (c) the dark field image using electrons from the bright ring.

Diffraction patterns identical with that in Fig. 9(a) were obtained for average implanted concentrations between 35 and 52 at.% Si, measured by EDXS. In all cases the microstructures contained a microcrystalline phase extending to depths greater than 1000 Å from the surface. The pale outer ring in Fig. 9(a) corresponds to an interatomic distance of 1.2 Å, which matches a reflection of amorphous silicon [12]. Other amorphous silicon reflections may also be present but hidden in the brightness of the transmitted beam and the very bright diffuse ring. The range of interatomic distances giving rise to the bright ring includes those shared by several nickel silicides, Ni_5Si_2, Ni_2Si, Ni_3Si_2 and $NiSi_2$ [13]. Thus, positive unambiguous identification of near-surface microcrystalline phases is not possible. Transmission electron microscopy (TEM) sections exploring depths to 2500 Å from the surface of specimens implanted to 52 at.% Si showed recrystallization, with two phases present, Ni_2Si and $NiSi_2$, in addition to some amorphous or microcrystalline material. The nearest stoichiometric phase, NiSi, was not seen. This suggests that some higher nickel silicides are more readily disordered by irradiation than are others, an effect already noted in comparing Ni_5Si_2 and Ni_3Si.

The formation of microcrystalline phases during ion implantation may be understood by considering the relative free energies of Ni–Si phases below 200 °C. Nickel silicides richer in silicon than Ni_5Si_2 exhibit very small homogeneity ranges (Fig. 1). As noted by Brimhall *et al.* [14], the free energies of these "line compounds" will increase rapidly as the composition departs from stoichiometry. Microscopic composition fluctuations will result as the radiation-produced point defects form clusters. In the absence of a counteracting mechanism, the silicides will become unstable as their free energies exceed those of (originally) higher energy phases with wide homogeneity ranges. Phase transformations will then occur. Transformation to an amorphous phase is favored when the free-energy curve of the liquid is lower than that of the crystalline solid [15,16].

However, for the Ni–Si system, calculations [17] based on Kaufman's [18] thermochemical parameters for Ni–Si alloys show that the liquid phase is higher in free energy than is the f.c.c. phase (Fig. 10). The difference is approximately 6 kJ mol^{-1} at temperatures below 200 °C. Hence the transformation to microcrystalline rather than to amorphous material can be understood. The f.c.c. phase stability under irradiation is also favored by the fact that the f.c.c. free energy changes relatively little with increase in silicon content from 30 to 60 at.%.

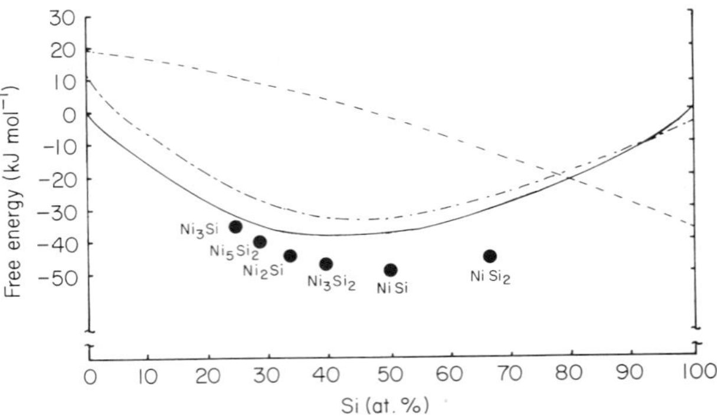

Fig. 10. Free energies of Ni–Si phases at 125 °C, based on the thermochemical parameters of Kaufman [18]: – – –, diamond cubic; —·—, liquid; ——, f.c.c. The pure elements in the f.c.c. crystal structure are taken as the reference state.

Finally, the presence of a highly mobile point defect below 200 °C has been established by others, as noted earlier, as well as through the radiation-induced segregation noted above (Fig. 4). This defect may help to stabilize the nickel silicide compounds (rather than f.c.c.) as microcrystalline phases by annealing out severely destabilizing concentration fluctuations. Alternatively, it might provide a mechanism for phase separation similar to the mixture of Ni_2Si and $NiSi_2$ seen at greater depths but forming on a much finer scale. A two-phase mixture would be stable relative to the f.c.c. solution if the sum of the compound free energies and the energies of their interfaces were lower than that of the f.c.c. solution. Such speculations are useful in connection with the problem of assigning a specific identity to the microcrystalline phases, as discussed above.

3.2.3. At higher fluences and higher temperatures

Above 250 °C, fluences near 1×10^{18} Si^+ cm^{-2} or greater cause the nickel matrix to recrystallize. Polycrystalline diffraction patterns are obtained. Figure 11(a) shows a bright field image and diffraction pattern from a speci-

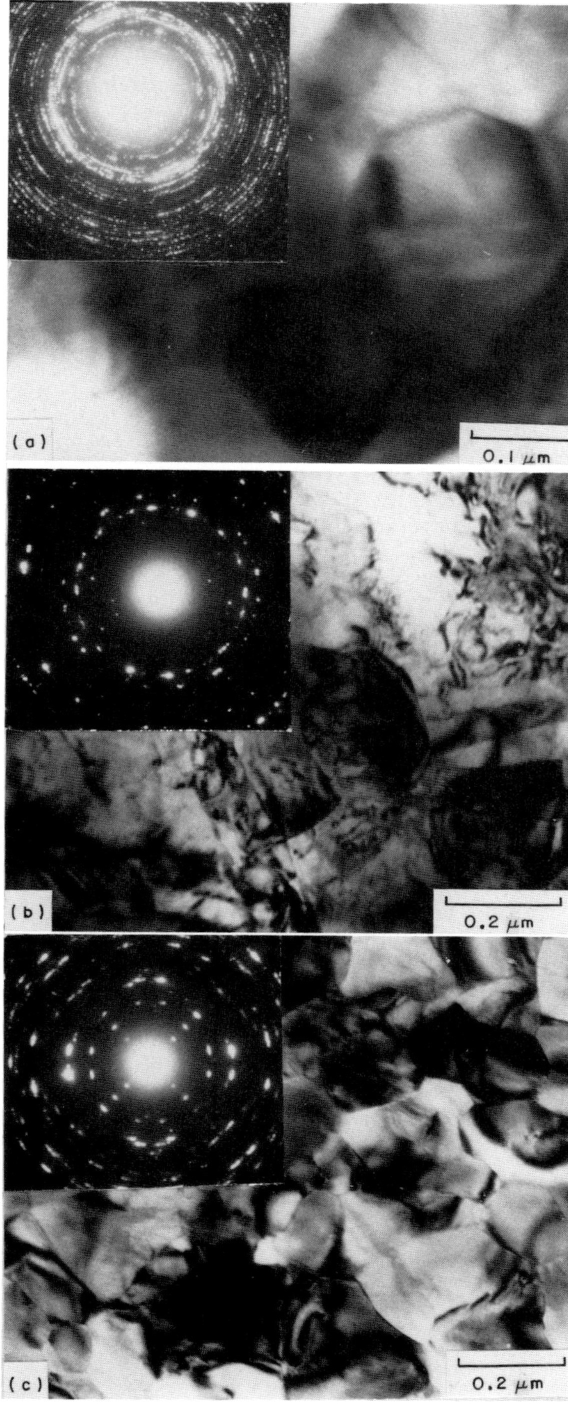

Fig. 11. Recrystallization occurs, giving polycrystalline diffraction patterns, when silicon is implanted into nickel with a high fluence at an elevated temperature: (a) Ni_5Si_2 at 28 at.% Si (2.5×10^{18} Si^+ cm^{-2}) at 370 °C; (b) Ni_3Si at 25 at.% Si (1.75×10^{18} Si^+ cm^{-2}) at 585 °C; (c) Ni_2Si at 34 at.% Si (9.2×10^{17} Si^+ cm^{-2}) at 287 °C.

men implanted with 28 at.% Si (by AES) at 370 °C (2.5×10^{18} Si^+ cm^{-2}). This particular specimen shows Ni_5Si_2. Other phases were also seen with similar microstructures, depending on the temperature and silicon concentration, namely Ni_3Si, at temperatures high enough to stabilize the phase (Fig. 11(b)) and Ni_2Si at intermediate temperatures (Fig. 11(c)). At the two higher temperatures, holes were also seen in the implanted layer, suggesting significant vacancy mobility.

The deep penetration of silicon when a high fluence is applied at an elevated temperature (Fig. 5) may be a consequence of the recrystallization seen here. Ahmed and coworkers [19, 20] have shown that, in nickel implanted with aluminum ions, dislocations due to radiation damage extend several thousand ångströms below the surface when fluences exceed 1.4×10^{18} ions cm^{-2}. It is reasonable to suppose that such a process occurs during silicon implantation also. The dislocations ultimately cause recrystallization and lead to new grain boundaries extending several thousand ångströms below the surface. The grain boundaries are high diffusivity paths and would carry silicon deep into the substrate, much as they act to transport aluminum in the Ni–Al case.

4. SUMMARY

Radiation damage processes were shown to play a significant role in the silicon depth distribution, phase stability and microstructures, resulting from silicon implantation into nickel. Over a temperature range from 25 to 650 °C, and implanted silicon contents from 10 to 90 at.%, the major observations are as follows.

(1) Radiation-induced segregation occurs, causing very high silicon concentrations within the first 100–200 Å of the implanted surface.

(2) Crystalline Ni–Si compounds form during implantation. The Ni_5Si_2, Ni_2Si and $NiSi_2$ phases were observed at room temperature and higher temperatures. The Ni_3Si phase was destabilized by radiation damage at implantation temperatures below 500 °C.

(3) Implantation of fluences below 7×10^{17} Si^+ cm^{-2} resulted in f.c.c. solutions or textured nickel silicide precipitates, depending on temperature and implanted composition.

(4) The implantation of fluences above 1×10^{18} Si^+ cm^{-2} at temperatures below 200 °C resulted in microcrystallinity which extended at least 1000 Å from the surface. The implanted ion depth was similar to the ballistic range of the ions.

(5) The implantation of fluences near or above 1×10^{18} Si^+ cm^{-2} at temperatures above 250 °C resulted in recrystallization.

(6) A high fluence implanted at temperatures above 350 °C resulted in the penetration of silicon several thousand ångströms beyond the ballistic range of the ions. This might be attributed to the formation of high diffusivity paths by recrystallized grain boundaries.

ACKNOWLEDGMENTS

We thank the following people for contributing to this research: H. Hayden and C. Koch for assisting in the implantations; L. McCurdy for TEM assistance; L. Witherell for typing.

This material is based on work supported by the National Science Foundation under Grant DMR8507641. The transmission electron microscope was purchased with National Science Foundation support under Grant DMR8207266 and with State of Connecticut funding.

REFERENCES

1 P. R. Okamoto and L. E. Rehn, *J. Nucl. Mater.*, *83* (1979) 2.
2 R. C. Piller and A. D. Marwick, *J. Nucl. Mater.*, *71* (1978) 309.
3 A. D. Marwick, R. C. Piller and P. M. Sivell, *J. Nucl. Mater.*, *83* (1979) 35.
4 D. I. Potter, P. R. Okamoto, H. Wiedersich, J. R. Wallace and A. W. McCormick, *Acta Metall.*, *27* (1979) 1175.
5 K. Y. Liou and P. Wilkes, *J. Nucl. Mater.*, *87* (1979) 317.
6 D. I. Potter, in J. R. Holland, L. K. Mansur and D. I. Potter (eds.) *Phase Stability During Irradiation*, AIME, New York, 1981, p. 521.
7 S. P. Lamond and D. I. Potter, *J. Nucl. Mater.*, *117* (1983) 64.
8 *Metallography, Structure and Phase Diagrams*, Vol. 8, *Metals Handbook*, American Society for Metals, Metals Park, OH, 8th edn., 1973, p. 325.
9 J. P. Biersack and L. G. Haggmark, *Nucl. Instrum. Methods*, *174* (1980) 257.
10 R. A. Swalin, A. Martin and R. Olson, *Trans. AIME*, *209* (1957) 936.
11 M. P. Macht, A. Muller, V. Naundorf and H. Wollenberger, *Nucl. Instrum. Methods B*, *16* (1986) 148.
12 D. J. Mazey, R. S. Nelson and R. S. Barnes, *Philos. Mag.*, *17* (1968) 1145.
13 A. Osawa and M. Okamoto, *Sci. Rep. Tohoku Univ., Ser. 1*, *27* (1939) 326.
14 J. L. Brimhall, H. E. Kissinger and L. A. Charlot, in S. T. Picraux and W. J. Choyke (eds.) *Metastable Materials Formation by Ion Implantation*, Elsevier, New York, 1982, p. 235.
J. L. Brimhall and E. P. Simonen, *Nucl. Instrum. Methods B*, *16* (1986) 187.
15 N. Saunders and A. P. Miodownik, *Calphad*, *3* (1985) 283.
16 N. Saunders and A. P. Miodownik, *Ber. Bunsenges. Phys. Chem.*, *87* (1983) 830.
17 S. G. B. Fishman, *Ph.D. Thesis*, University of Connecticut, Storrs, CT, 1986.
18 L. Kaufman, *Calphad*, *3* (1979) 45.
19 M. Ahmed, *Ph.D. Thesis*, University of Connecticut, Storrs, CT, 1985.
20 D. I. Potter, M. Ahmed and K. Ruffing, *Proc. 43rd Annu. Meet. of the Electron Microscopy Society of America*, San Francisco Press, San Francisco, CA, 1985, p. 282.
D. I. Potter and M. Ahmed, *Mater. Sci. Eng.*, *90* (1987) 127.

Materials Science and Engineering, 90 (1987) 143–148

Characterization of the Aluminium Surface Layer Implanted with Nitrogen*

SHIGEO OHIRA

Nikkei Techno-Research Company Ltd., 1-34-1 Kambara, Ihara-gun, Shizuoka-ken 421-32 (Japan)

MASAYA IWAKI

RIKEN (The Institute of Physical and Chemical Research), Hirosawa 2-1, Wako, Saitama-ken 351-01 (Japan)

(Received July 10, 1986)

ABSTRACT

Polycrystalline or single-crystal aluminium sheets with a purity of 99.99% were implanted with nitrogen ions (N_2^+) *at energies of 50, 100 and 150 keV at room temperature and fluences ranging from* 1×10^{16} *to* 1×10^{18} *ions* cm^{-2} (2×10^{16}–2×10^{18} *nitrogen atoms* cm^{-2}). *The surface structure, chemical composition and chemical bond of the implanted aluminium surface layer were investigated by Auger electron spectroscopy, X-ray photoelectron spectroscopy, X-ray diffraction, transmission electron diffraction, transmission electron microscopy and IR spectroscopy. Furthermore, the microhardness of the nitrogen-implanted aluminium surface was evaluated by measuring the penetration depth with an applied load. From the characterization of the nitrogen-implanted aluminium layer, high fluence ion implantation into aluminium can result in the formation of polycrystalline or single-crystal aluminium nitride (AlN) layers at room temperature without any thermal annealing. It was found that nitrogen implantation into aluminium was effective in enhancing the hardness of the aluminium surface, which is attributed to the formation of the crystalline AlN produced by nitrogen implantation.*

1. INTRODUCTION

Aluminium nitride (AlN) is an attractive material which is used in the electronic industry because of its electrical, optical, dielectric and acoustic properties. There are several known processes for forming AlN films including chemical vapour deposition [1], ion plating [2], sputtering [3] and molecular beam epitaxy [4]. Most of these processes involve high temperature processing during film growth and need subsequent annealing, although low temperature processing is necessary in the presence of an aluminium substrate.

Ion implantation may be used as a low temperature and well-controlled technique for changing the surface properties of metals as well as for forming metastable alloys and chemical compounds. The surface modification of aluminium by ion beams has been studied recently. For example, nitrogen ion implantation into aluminium can result in the formation of AlN [5]. However, few investigations have been made on the characterization of mechanical properties such as the microhardness of nitrogen-implanted aluminium specimens [6].

In the work presented in the present paper, high fluence implantation has been used to prepare crystalline AlN layers in aluminium. In order to confirm the formation of AlN, the characteristics of nitrogen-implanted aluminium surface layers were investigated by means of Auger electron spectroscopy (AES), X-ray photoelectron spectroscopy (XPS), X-ray diffraction (XRD), transmission electron diffraction (TED), transmission electron microscopy (TEM) and IR spectroscopy. Furthermore, the mechanical properties of the nitrogen-implanted aluminium surface were evaluated by measuring the microhardness.

2. EXPERIMENTAL DETAILS

The substrates used were polycrystalline and single-crystal aluminium sheets with a

*Paper presented at the International Conference on Surface Modification of Metals by Ion Beams, Kingston, Canada, July 7–11, 1986.

0025-5416/87/$3.50

purity of 99.99%. Nitrogen ions (N_2^+) were implanted into the substrate, over an area of 3 cm × 3 cm, at room temperature using the RIKEN 200 kV low current implanter. Acceleration energies of 50, 100 and 150 keV were selected and the fluences ranged from 1×10^{16} to 1×10^{18} N_2^+ cm^{-2} (2×10^{16}–2×10^{18} nitrogen atoms cm^{-2}). The nitrogen ion beam current density was about 0.6 μA cm^{-2}, 2.5 μA cm^{-2} and 4.0 μA cm^{-2} for 50 keV, 100 keV and 150 keV respectively, and the pressure within the target chamber was about 1×10^{-4} Pa.

The depth profiles and chemical bonding states of nitrogen, oxygen and aluminium in the nitrogen-implanted aluminium layers were investigated by means of AES and XPS combined with 5 keV argon ions for sputter etching. AES measurements were accomplished using 5 keV primary electrons at a beam current of 5 μA at the probe area which was 5 μm in diameter. The atomic concentrations were calculated at any given depth by assuming that peak-to-peak heights were proportional to the concentration using appropriate sensitivity factors, and the Auger sensitivity factor of Al LVV was assumed to be constant between aluminium and AlN. The Auger electron spectra were utilized to observe both the change in the characteristic lineshape and the aluminium concentration. In XPS measurements, electron ejection from specimens was induced by 10 kV, 40 mA Mg Kα. The chemical states of aluminium, nitrogen and oxygen were investigated from the binding energy of Al 2p, N 1s and O 1s. All peaks were referred to the position of the C KLL peak at 271 eV for AES and the C 1s line at 284.6 eV for XPS. The crystallography of nitrogen-implanted aluminium layers was determined by XRD using 1.5418 Å unresolved 50 kV Cu Kα radiation. The aluminium surface implanted with nitrogen was removed from the substrate by bromine–methyl alcohol prior to TEM observation or TED analysis with the help of selected area diffraction using a 200 keV electron beam. The IR spectra derived from the nitrogen-implanted aluminium surface were measured by an attenuated total reflection method. The annealing after the nitrogen implantation was made with a nitrogen atmosphere of 7×10^{-2} Pa at 100 and 200 °C. The IR spectra of the annealed specimens were also examined and compared with specimens without annealing.

The microhardness of the nitrogen-implanted and the unimplanted specimens was evaluated by measuring the penetration depth due to an applied load, using a Shimadzu dynamic ultramicrohardness tester (DUH-50). The penetration depth D was plotted against the applied load P (maximum, 2 gf) and the microhardness H was defined as

$$H = \frac{37.838P}{D^2}$$

where H (kg mm^{-2}), P (gf) and D (μm) are the dynamic hardness, the applied load and the penetration depth respectively. The relative hardness between nitrogen-implanted and unimplanted aluminium specimens was measured as a function of implanted nitrogen fluence.

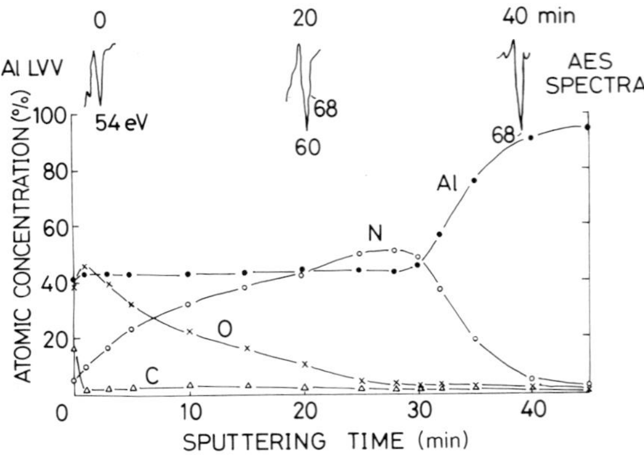

Fig. 1. AES composition depth profile for the aluminium specimens implanted with 1×10^{18} N_2^+ cm^{-2} at 150 keV (2×10^{18} nitrogen atoms cm^{-2} at 75 keV) and the corresponding AES spectra for Al LVV.

3. RESULTS AND DISCUSSION

3.1. Auger electron spectroscopy and X-ray photoelectron spectroscopy analyses

Figure 1 shows the distribution of composition in atomic per cent for aluminium implanted with 10^{18} N_2^+ cm^{-2} at 150 keV (2×10^{18} nitrogen atoms cm^{-2} at 75 keV) measured by AES combined with argon sputtering. The nitrogen depth profiles obtained for various implanted fluences differ significantly. They show a nearly gaussian-like shape at low fluences where the atomic ratio is smaller than the stoichiometric ratio (not shown in the figure) and become more rectangular at high fluences where the atomic ratio is estimated to be larger than the stoichiometric ratio by extrapolation from the low fluence case. The nitrogen concentration profile was found to deviate from a gaussian-like shape and the nitrogen atomic ratio seems to exceed the stoichiometric ratio at a deeper depth. This unsaturated feature indicates either that Al—N bonds are not readily formed or that excess nitrogen atoms do not migrate in the present implantation conditions. However, the following XPS, IR and TEM results reveal that implanted nitrogen atoms form Al—N bonds and that no excess nitrogen atoms exist. Therefore the increased nitrogen yield above aluminium in the metal depth is considered to arise as follows. In the present AES measurement, it is assumed that the Auger sensitivity factor of aluminium is always constant between the aluminium metal and nitrogen-implanted aluminium. If we use an appropriate sensitivity factor along the nitrogen-implanted aluminium region, the stoichiometric ratio of AlN is never exceeded. Furthermore, as observed in the figure, the oxygen depth profile shows a diffusion-like distribution and, the greater the amount of nitrogen implanted in the aluminium, the deeper the oxygen migrates. This mechanism of oxygen migration in aluminium during nitrogen implantation has been discussed in our previous paper [7].

AES spectra of the Al LVV energy region for different depths are also depicted in Fig. 1. It is found that the Auger spectra depend on the ratio of the nitrogen to aluminium concentration and that the line shape changes in going from the surface to the deeper region. The Al LVV spectra are oxide like near the surface region and nitride like in the deeper region. After sputtering for 20 min the main peak of Al LVV is at 60 eV, as well as 68 eV for elemental aluminium. This AES spectrum is easily distinguishable from the corresponding elemental aluminium and Al_2O_3 spectra. A chemical shift of 8 eV to lower energies and a different signal shape from that of pure aluminium are observed. This is in good agreement with the result reported in the literature [8]. The resulting Auger spectrum from our aluminium nitride is similar to that of AlN, indicating the formation of a new chemical phase of AlN buried in the aluminium metal. It is considered that these AlN layers are formed by implanting with nitrogen ions until the Auger signal ratio of nitrogen to Al LVV reached a plateau.

The binding energy spectra of Al 2p, N 1s and O 1s measured by means of XPS supported the AES results in the formation of Al_2O_3 near the surface and AlN at intermediate depth [7]. These surface analyses clearly indicate the formation of a new chemical state, *i.e.* AlN.

3.2. *Crystal structure*

Direct evidence of formation of an AlN layer was obtained by XRD studies. In Fig. 2 are shown the XRD patterns of the polycrystalline aluminium surface implanted with 1×10^{18} $N_2^+ cm^{-2}$ at 150 keV. The nitrogen-implanted aluminium layers have an h.c.p. AlN polycrystal structure with a strong preferred (002) orientation. No evidence of Al_2O_3 is provided, consistent with the presence of amorphous oxide. Thus the XRD results indicate that high fluence nitrogen ion implantation in aluminium can result in the formation of a würtzite-type h.c.p. AlN structure with a *c* axis preferred orientation.

Figure 3 consists of a TEM micrograph and the corresponding TED pattern taken from the nitrogen-implanted surface of single-crystal aluminium. The incident electron beam was chosen to be along a [011] direction. No evidence of voids or bubble formation, up to a fluence of 5×10^{17} $N_2^+ cm^{-2}$, can be observed in the irradiated area of the specimen. It is suggested from these results that most of the implanted nitrogen atoms are incorporated in the AlN structure in the region where the implanted nitrogen concentration exceeds the aluminium concentration.

It can be seen that the selected area diffraction pattern of diameter 0.5 μm reveals a

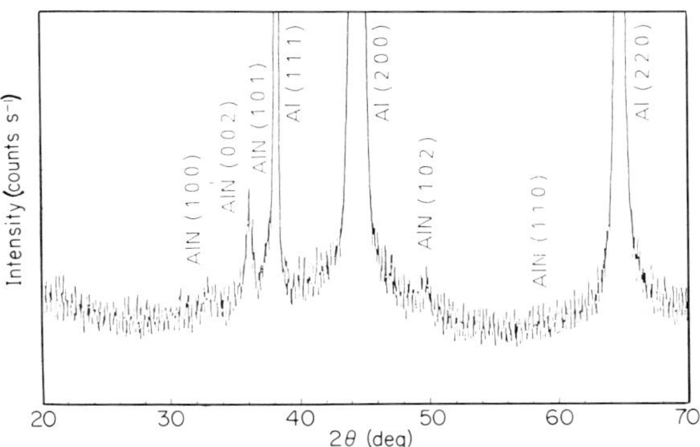

Fig. 2. XRD patterns for the aluminium specimens implanted with 1×10^{18} $N_2^+ cm^{-2}$ at 150 keV.

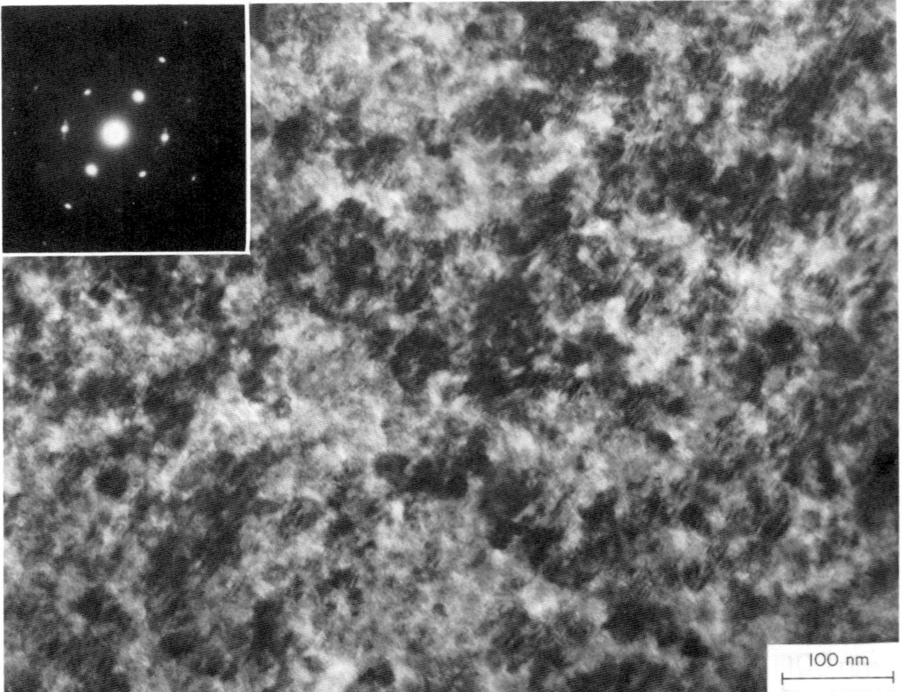

Fig. 3. TEM micrograph for the single-crystal aluminium implanted with 5×10^{17} N_2^+ cm^{-2} at 150 keV and the corresponding TED pattern.

TABLE 1

The observed lattice indices and the corresponding phase after implantation of 1×10^{18} N_2^+ cm^{-2} into single-crystal aluminium

Observed values d (Å)	*AlN from ASTM card*	
	d (Å)	(*hkl*)
2.696	2.695	(100)
—	2.490	(002)
2.348	2.371	(101)
1.805	1.829	(102)
1.567	1.559	(110)
1.405	1.413	(103)
1.348	1.348	(200)
1.196	1.185	(202)

hexagonal spotty pattern, indicating the formation of würtzite-type h.c.p. single-crystal AlN. The grain size of the single-crystal AlN was determined by dark field image observations to be about 10–100 nm. The *d* values observed after implantation of 1×10^{18} N_2^+ cm^{-2} are presented in Table 1. The values calculated from the data are 3.09 Å for *a* and 4.97 Å for *c*, which agree very well with the values for AlN in the table and in ref. 9. These results give direct evidence of the formation of single-crystal AlN by nitrogen ion implantation into single-crystal aluminium. However, the relative orientation between the AlN and the aluminium matrix was not determined, because the substrate of the aluminium matrix was perfectly resolved by bromine–methyl alcohol. Therefore, it is considered that the diffraction pattern does not arise from islands of aluminium.

3.3. *IR spectra*

The IR transmission spectra of nitrogen-implanted aluminium surfaces obtained by the attenuated total reflection method depend on the fluence of implanted nitrogen ions, as shown in Fig. 4. The IR transmission band became sharper in the spectra while there was no shift in the peak position. The first clear appearance of an absorption band near the 850 cm^{-1} wavenumber region occurs for specimens implanted with 5×10^{17} N_2^+ cm^{-2}. Its shape reflects the Al—N stretching vibration energies due to AlN compound formation. With increasing ion fluence, the sharpening of the transmission bands indicates a gradual

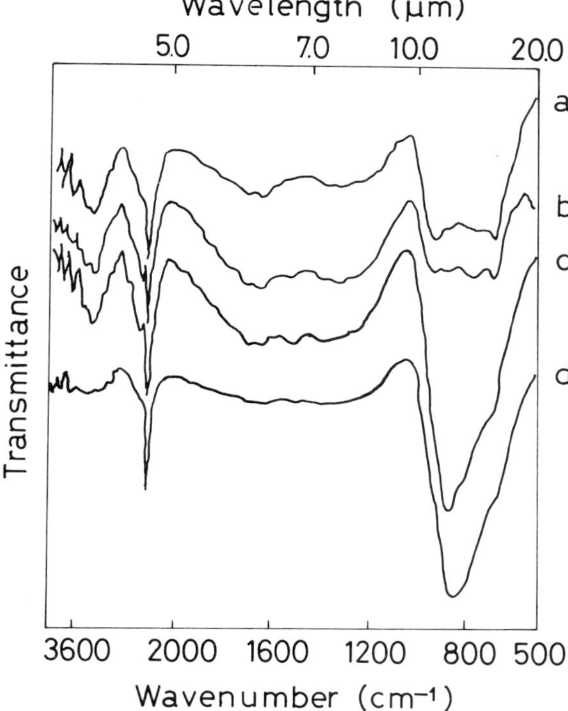

Fig. 4. IR transmission spectra of aluminium specimens implanted with various fluences of nitrogen ions (N_2^+) at 100 keV: spectrum a, 5×10^{16} ions cm^{-2}, spectrum b, 1×10^{17} ions cm^{-2}; spectrum c, 5×10^{17} ions cm^{-2}; spectrum d, 1×10^{18} ions cm^{-2}.

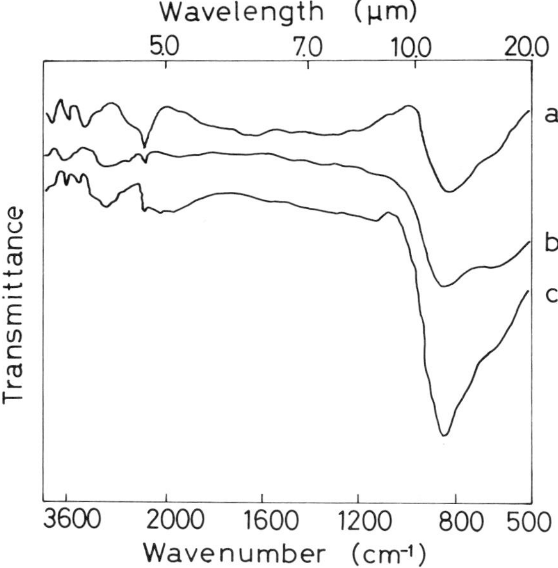

Fig. 5. IR transmission spectra of annealed aluminium specimens implanted with 5×10^{17} N_2^+ cm^{-2} at 100 keV: spectrum a, without annealing; spectrum b, 100 °C; spectrum c, 200 °C.

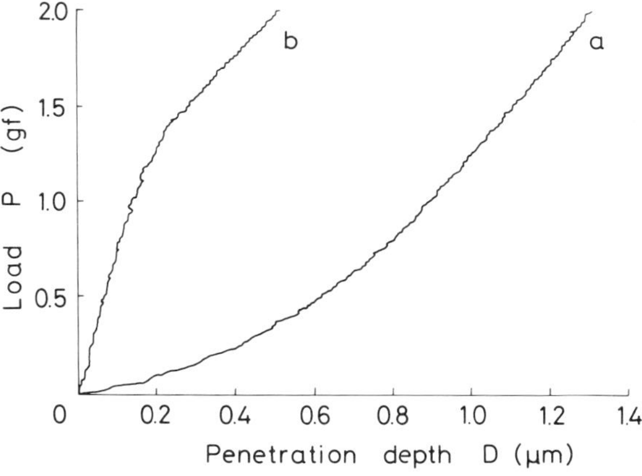

Fig. 6. Relationship between applied load and penetration depth: curve a, unimplanted pure aluminium substrate; curve b, aluminium surface implanted with 1×10^{18} N_2^+ cm^{-2} at 100 keV.

transformation of the initial AlN precipitates to aluminium metal into layers which must be close stoichiometrically to the composition AlN.

In order to verify that there is bond strain and disorder in the implanted layer due to highly energetic implantation process, studies of the effect of annealing on the IR transmission spectra were carried out. IR studies of the annealed specimens indicate that the annealing temperature has an influence on the IR transmission band. Figure 5 shows that the annealing of the nitrogen-implanted aluminium specimens causes the IR band peaks to shift towards shorter wavelengths and the shape of spectra to become sharper. This can result from a release of the bond strain and disorder which were introduced in the formation of AlN during nitrogen implantation into aluminium. It was considered that this bond strain and disorder effect might be responsible for the observed difference in the peak positions and the shapes.

3.4. *Microhardness*

Figure 6(a) and Fig. 6(b) show the relationship between the applied load and penetration depth for the unimplanted pure aluminium substrate and the nitrogen-implanted aluminium surface respectively. Comparison of the data reveals that a decrease in the penetration depth is observed in Fig. 6(b). This infers hardening of the aluminium substrate due to

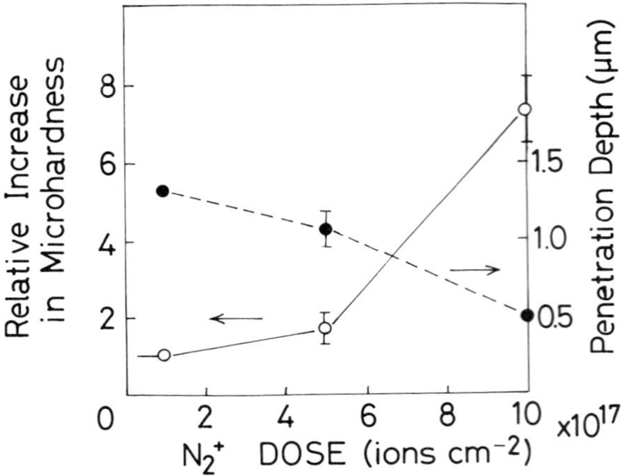

Fig. 7. Implanted fluence dependence on (a) relative increases in microhardness and (b) penetration depth.

nitrogen implantation. The steep slope from the surface to a depth of 0.2 μm may correspond to the formation of AlN.

The microhardness and penetration depth for nitrogen-implanted aluminium depend on the implanted fluence at a given energy of 100 keV, as shown in Fig. 7. As the implanted fluence increases, the microhardness increases and the penetration depth decreases. The remarkable hardening of the aluminium surface was considered to arise as a result of the formation of crystalline AlN.

4. SUMMARY AND CONCLUSIONS

A study was made on the characterization of aluminium surface layers implanted with nitrogen. The results are as follows.

(1) AlN layers in aluminium substrates can be formed by nitrogen ion implantation into aluminium sheets at room temperature without any thermal annealing.

(2) Irradiation of aluminium single crystals with nitrogen ions leads to the formation of single-crystal AlN layers.

(3) The IR spectra of annealed nitrogen-implanted aluminium specimens become sharp and shift towards a shorter wavelength.

(4) An increase in microhardness was observed when high fluence migration ions are implanted into aluminium.

From these results, it is concluded that nitrogen implantation in aluminium is effective in enhancing the hardness of the aluminium surface, which is attributed to the formation of crystalline AlN produced by nitrogen implantation.

ACKNOWLEDGMENTS

The authors gratefully acknowledge Mr. Y. Habu and Mr. S. Tanai of the Analytical Centre of the Nikkei Techno-Research Company Ltd. for their AES analyses and TED studies.

REFERENCES

1 M. Morita, S. Isogai, K. Tsubouchi and N. Mikoshiba, *Appl. Phys. Lett., 38* (1981) 50.
2 Y. Murayama and K. Kashiwagi, *J. Vac. Sci. Technol., 17* (1980) 796.
3 A. Fathimulla and A. A. Lakhani, *J. Appl. Phys., 54* (1983) 4586.
4 S. Yoshida, S. Misawa and S. Gonda, *J. Vac. Sci. Technol. B, 1* (1983) 250.
5 B. Rauschenbach, A. Kolitsch and E. Richter, *Thin Solid Films, 109* (1983) 37.
6 P. B. Madakson, *J. Phys. D, 18* (1985) 531.
7 S. Ohira and M. Iwaki, *Proc. 5th Int. Conf. on Ion Beam Modification of Materials, Catania, June 9-13, 1986*, in *Nucl. Instrum. Methods*, to be published.
8 N. Lieske and R. Hezel, *J. Appl. Phys., 52* (1981) 5806.
9 *Powder Diffraction File*, ASTM, Philadelphia, PA, 1967, Card 8-262.

Materials Science and Engineering, 90 (1987) 149-159

Effect of Temperature on High Fluence Transition Metal Implants into Polycrystalline Aluminum*

F. H. SANCHEZ†, F. NAMAVAR, J. I. BUDNICK, A. FASIHUDDIN, C. H. KOCH and H. C. HAYDEN

Department of Physics, and Institute of Materials Science, University of Connecticut, Storrs, CT 06268 (U.S.A.)

(Received July 10, 1986)

ABSTRACT

Polycrystalline aluminum was implanted with 150 keV titanium ions (Ti^+), manganese ions (Mn^+) and nickel ions (Ni^+) at 100 and 623 K, with fluences ranging from 0.6×10^{17} to 1.1×10^{18} ions cm^{-2}. At the higher temperature the implanted elements penetrated to regions much deeper than the projected range, the largest effect being observed for the manganese implants. The low temperature implants did not induce important diffusion effects, with the exception of the nickel implants at a fluence of 9×10^{17} ions cm^{-2}. In contrast, they led to higher concentrations of the transition metals. The measured depth distributions of the implanted species are discussed in terms of radiation-enhanced diffusion, radiation-induced segregation, sputtering and phase formation. In most cases the formation of the aluminum-richest transition metal–aluminum crystalline phase was observed. The production of the crystalline $NiAl_3$ and $MnAl_6$ phases by ion implantation at 100 K is reported here for the first time. These phases were previously found to be unstable under irradiation. Their formation is interpreted considering the depth distributions of damage and transition metals.

1. INTRODUCTION

The transition metal–aluminum (TM–Al) system has been previously studied for a number of reasons. Potter and coworkers [1, 2] have investigated the fabrication of "superalloys" by implanting aluminum into nickel. Studies of the lattice location of nickel implanted into aluminum as well as the amorphization of this system were pursued by Picraux and coworkers [3, 4], Follstaedt *et al.* [5] and Thomé *et al.* [6]. The stability under irradiation of nickel aluminides and the structure of the TM–Al phases produced by ion beam mixing of Ni/Al multilayers were investigated by Nastassi *et al.* [7, 8], Hung *et al.* [9] and Eridon *et al.* [10]. Budai and Aziz [11] investigated the production of the quasi-crystalline icosahedral phase by implantation of manganese ions (Mn^+) into aluminum to intermediate fluences, and Knapp and Follstaedt [12] studied the metastable Mn–Al phases fabricated by ion beam and thermal mixing of Mn/Al multilayers, also in connection with the icosahedral symmetry. Finally, the thermally induced reaction of TM–Al films was surveyed by Howard *et al.* [13, 14] in connection with the improvement of electromigration properties in electronic devices.

From these studies, some features concerning the TM–Al system are becoming clear. While there seems to be no difficulty in producing the equilibrium phases by thermally induced solid state reaction between films of a TM and aluminum under appropriate conditions, their fabrication by ion beam mixing of TM/Al layers could not be achieved in some cases. For example, mixing of Ni/Al multilayers with the average composition corresponding to $NiAl_3$ produced only amorphous material, or amorphous material plus NiAl. This remained so even when the mixing was performed at temperatures at which identical samples not exposed to the mixing beam yielded the crystalline orthorhombic $NiAl_3$ [7-9]. Most of the experiments reported in-

*Paper presented at the International Conference on Surface Modification of Metals by Ion Beams, Kingston, Canada, July 7-11, 1986.

†Permanent address: Departamento de Fisica, Universidad Nacional de La Plata, C.C. 67, 1900 La Plata, Argentina.

0025-5416/87/$3.50

dicate that ion beam processes can form only NiAl (with the simple CsCl structure) or amorphous phases when the Ni–Al system is treated [1, 3–10], while the amorphization of the system by neon, xenon and nickel ion beams was explained with different models [6, 7]. Also for the Mn–Al system, ion beam processes did not produce the crystalline $MnAl_6$ orthorhombic phase at temperatures below 473 K, but only the amorphous or the quasi-crystalline icosahedral phase [11, 12]. A systematic study of the production of TM aluminides by ion implantation of the TMs into aluminum has not yet been performed. We believe that such a study could give more insight into the questions already raised. It may further be used for comparison with the ion-beam-mixing results.

Also, it would be interesting to compare the TM–Al system with the TM–Si system which we have recently studied [15, 16]. Silicides of all the elements from titanium to nickel and also niobium were obtained by implantation of the TMs into silicon. We found that both the thermodynamic and the kinetic properties of the individual TM–Si systems strongly affected both the compound formed and the development of a uniform silicide layer. For instance, in those cases where the heat of formation of one of the silicides was much more negative than for the other silicides of the same element, this single phase was produced. Furthermore, under appropriate kinetic conditions, the silicide phase spread uniformly toward both the surface and the interior as the implantation progressed. Finally, we also observed that the implantation temperature was an important parameter for TM silicide formation by ion implantation.

Our present studies include the implantation of titanium ions (Ti^+), manganese ions (Mn^+) and nickel ions (Ni^+) at both 100 and 623 K, and with different fluences ranging from 0.6×10^{17} to 1.1×10^{18} ions cm^{-2}. With these choices, we intended to compare the behavior of the TM–Al systems for TMs at both ends (refractory and near-noble metals) and at the center of the first row in the periodic table, as well as at temperatures where vacancies are frozen (100 K) or mobile (623 K). Whenever crystalline phases were detected, they mostly corresponded to the aluminum-richest equilibrium aluminides, even when the concentration of the TM in the implanted zone was in some cases much higher than that corresponding to the observed phase. One striking result is that crystalline $NiAl_3$ was detected after implanting nickel ions at 100 K, in contrast with the ion-beam-mixing experiments [7]. Furthermore, the results indicate that, at the lower temperature, a higher concentration of the TM is achieved because of the low mobility of the implanted species while, at the higher temperature, diffusion processes play a major role, extending the transformed region to a much greater depth.

2. EXPERIMENTAL DETAILS

Cleaned but otherwise untreated aluminum foils of 99.97% purity were cut into pieces 1.5 cm $\times$ 4.0 cm from a sheet of 0.25 mm thickness. The layer of native oxide at the surface of the samples was not removed in order to facilitate the achievement of higher TM concentrations and retained fluences, since it has been reported that the sputtering rate of aluminum in its oxide is half as small as for aluminum metal [17]. Given the fact that the heat of formation of Al_2O_3 is much more negative than those of any of the implanted elements, a small probability was assigned to the eventuality of a preferential O–TM chemical interaction. The samples were mounted onto either a cold or a hot stage for implantation at 100 K and at 623 K respectively. They were fixed to the sample holder using a conducting silver paint. The sample holder was then surrounded by a cold trap kept at liquid nitrogen temperature. The implantations were carried out with the TM ions accelerated to 150 keV, using the 200 kV Varian–Extrion DF-4 implanter at the University of Connecticut. The beam was scanned both horizontally and vertically through an aperture 1 cm $\times$ 3 cm to ensure homogeneity on the implanted area, the scanning being monitored continuously with an x–y oscilloscope. For each ion species and temperature, several fluences were implanted during the same run. The different fluences were achieved by moving vertically either the samples or a mask, the higher fluences then being the result of accumulating several partial fluences. By this means, areas of typically 0.5–1.0 cm $\times$ 1.0 cm were obtained for each accumulated fluence. The temperature was moni-

TABLE 1

Fluences and fluxes for the transition metal implants into aluminum

TM ion	*Temperature* (K)	*Beam current density* (μA cm^{-2})	*Flux* ($\times 10^{13}$ ions cm^{-2} s^{-1})	*Fluence* ($\times 10^{17}$ ions cm^{-2})	*Time*[a] (min)
Ti^+	100	10	6.3	1.2	32
Ti^+	100	10	6.3	4.2	111
Ti^+	100	10	6.3	7.2	190
Ti^+	623	11	6.9	1.0	24
Ti^+	623	11	6.9	3.0	72
Ti^+	623	11	6.9	6.0	145
Ti^+	623	11	6.9	6.8	164
Mn^+	100	6	3.8	1.1	48
Mn^+	100	15	9.4	1.0	18
Mn^+	100	15	9.4	4.0	71
Mn^+	100	15	9.4	7.0	124
Mn^+	623	27	16.9	0.6	6
Mn^+	623	27	16.9	1.5	15
Mn^+	623	27	16.9	3.2	32
Mn^+	623	27	16.9	6.2	61
Mn^+	623	27	16.9	11.0	108
Ni^+	100	7.5	4.7	3.0	106
Ni^+	100	7.5	4.7	6.0	213
Ni^+	100	7.5	4.7	9.0	319
Ni^+	623	11	6.9	1.0	24
Ni^+	623	11	6.9	3.0	72
Ni^+	623	11	6.9	6.0	145
Ni^+	623	11	6.9	8.0	193

[a]The time periods needed to perform the implants are listed in the last column.

tored with a thermocouple during the implantations and the pressure varied between 2.0×10^{-4} and 4.7×10^{-4} Pa. Table 1 illustrates the fluences and fluxes for the different implants performed.

The Rutherford backscattering spectrometry analyses were performed with a 1.5 MeV helium ion (He^+) beam provided by the 2 MV Van de Graaff accelerator at the University of Connecticut. Again the samples were fastened to the sample holder using a conducting silver paint and surrounded with a cold trap kept at liquid nitrogen temperature. The angle of incidence of the beam was 10° from the normal to the surface of the samples and the detection was carried out at 170° with respect to the original beam direction. We used a detector with a resolution of better than 15 keV, and the spectra were recorded with a PDP-11 computer in 256 channels with a calibration of 5.3 keV channel^{-1}. The pressure during the analyses was better than 4×10^{-4} Pa.

For the X-ray diffraction (XRD) studies a Norelco diffractometer and a Read camera facility for surface analyses were employed. In both cases, we used the Kα radiation from copper. For the Read camera analyses the angle between the incident X-rays and the surface of the samples was varied between 1° and 15°. This procedure was followed in order to achieve some depth resolution for the structural analyses.

3. RESULTS AND DISCUSSION

Figure 1 compares the Rutherford backscattering spectra from samples implanted with similar fluences of the TMs at 100 and 623 K. It is obvious from this figure that the TM profiles depend drastically on the im-

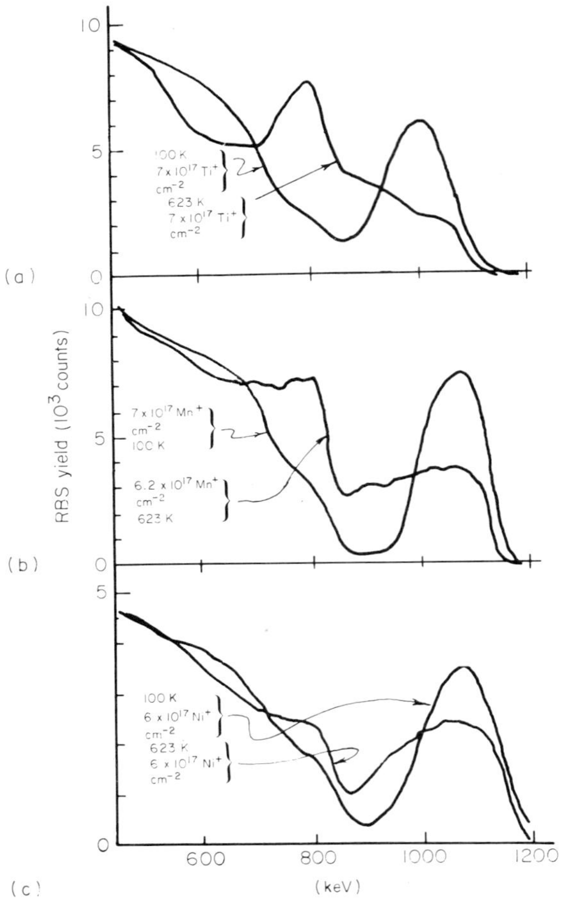

Fig. 1. 1.5 MeV helium ion Rutherford backscattering (RBS) spectra of aluminum samples implanted at 100 and 623 K to similar fluences of (a) titanium ions (Ti^+), (b) manganese ions (Mn^+) and (c) nickel ions (Ni^+).

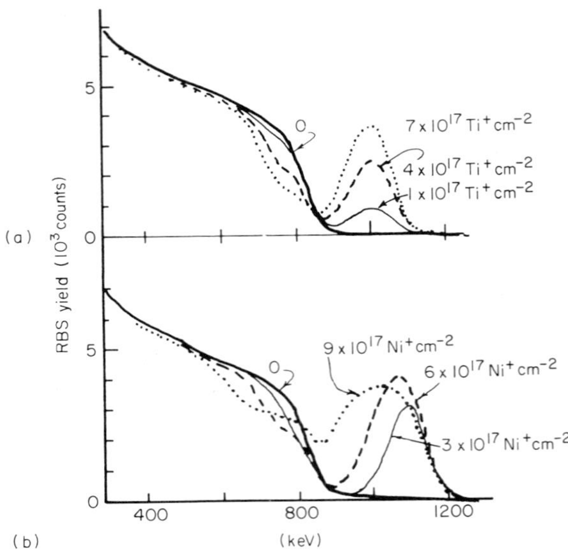

Fig. 2. 1.5 MeV helium ion Rutherford backscattering (RBS) spectra of aluminum samples implanted with several fluences of (a) titanium ions (Ti^+) and (b) nickel ions (Ni^+) at 100 K.

plantation temperature, indicating that at 623 K diffusion plays a major role. This behavior could be expected, at least for manganese and nickel. At 623 K the calculated mean displacement $x = (2Dt)^{1/2}$ of titanium, manganese and nickel atoms for $t = 1$ h are about 18 nm, 3500 nm and 250 nm respectively (using the coefficients for bulk diffusion in aluminum [18, 19]). In contrast, the large amount of defects provided by the irradiation process can facilitate the migration of the system's constituents, giving rise to so-called radiation-enhanced diffusion.

More interesting perhaps is the fact that some diffusion occurred in the samples implanted with nickel ions at the lower temperature. In Fig. 2 we compare the Rutherford backscattering spectra of the samples implanted with nickel ions at 100 K with those corresponding to the implants with titanium ions at the same temperature, where no signs of diffusion can be detected. It is unlikely that the above-mentioned effect could have been produced by a substantial temperature increase in the implanted region, because of the energy deposited by the beam. For, when a beam current density of 7.5 μA cm^{-2} is considered (see Table 1), the power deposited per surface unit by the 150 keV nickel ion beam is 1.125×10^4 W m^{-2}. Assuming that the thermal conductivity between the samples and the massive copper sample holder was 0.03 W m^{-1} K^{-1} (*i.e.* 10^4 times worse than that of aluminum, copper or silver), and taking a thickness of 50 μm for the silver paint layer, we calculate the equilibrium temperature at the sample surface to be less than 20 K higher than the sample-holder temperature, *i.e.* less than 120 K. Nevertheless, this temperature may be close to the beginning of the stage III recovery of defects in aluminum. This stage, attributed to the onset of vacancy mobility, starts at about 150 K [19]. The change in the nickel depth distribution with fluence (see also Fig. 3) does not seem to be associated with radiation-induced segregation. In this process the solute (or the solvent) atoms interact preferentially with one type of defect.

This produces an enrichment or a depletion of solute at the damage peak, depending on whether its flux has the opposite sign to or the same sign as that of the flux of point defects. The nickel distribution simply broadens as the fluence increases. However, the occurrence of radiation-enhanced diffusion may be possible if the diffusivity of nickel in the material was increased as a consequence of the development of a continuous new phase. Once the nickel concentration peak reaches a value above 25 at.%, the distribution of this element spreads inward beyond the implantation range as the fluence increases, and a flat peak top develops at about 25 at.% (see Fig. 3). In accordance with this, the formation of the orthorhombic $NiAl_3$ was detected by XRD (see below).

In Fig. 3 and Fig. 4 we present the concentration profiles of the TMs for the implants at 100 K and 623 K respectively. The depth scale was deduced using Bragg's rule for the stopping power [20], assuming for the atomic density of the material a linear combination of the aluminum and the TM atomic densities. This is a good approximation for the solid solution, elemental segregation and formation of TM–Al-rich compounds since the densities of TM–Al-rich compounds and solid solutions do not differ by more than 10% from the linear combination calculated

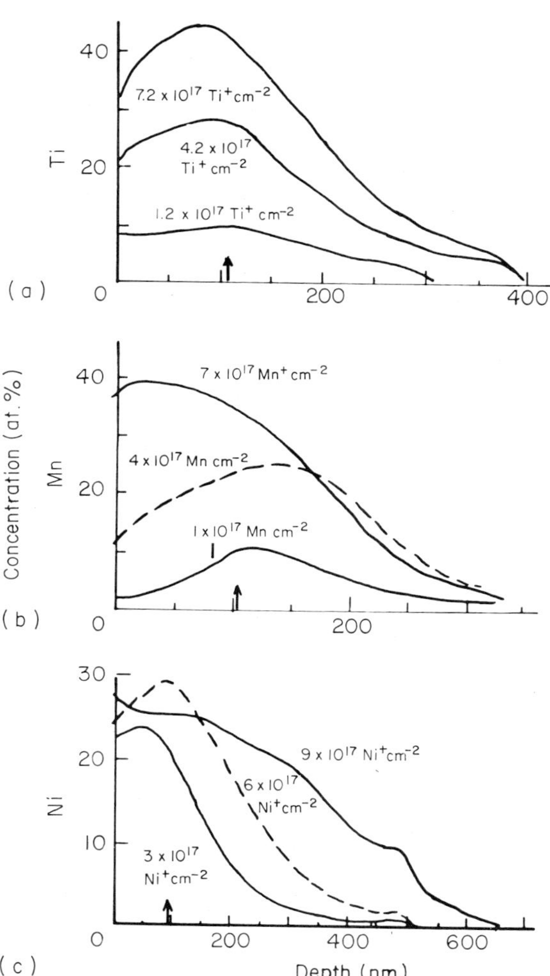

Fig. 3. Depth profiles of the transition metals ((a) titanium; (b) manganese; (c) nickel) implanted into aluminum at 100 K to several fluences. The profiles were calculated from the Rutherford backscattering spectra. The ballistic ranges for implantation into aluminum are indicated by the vertical arrows.

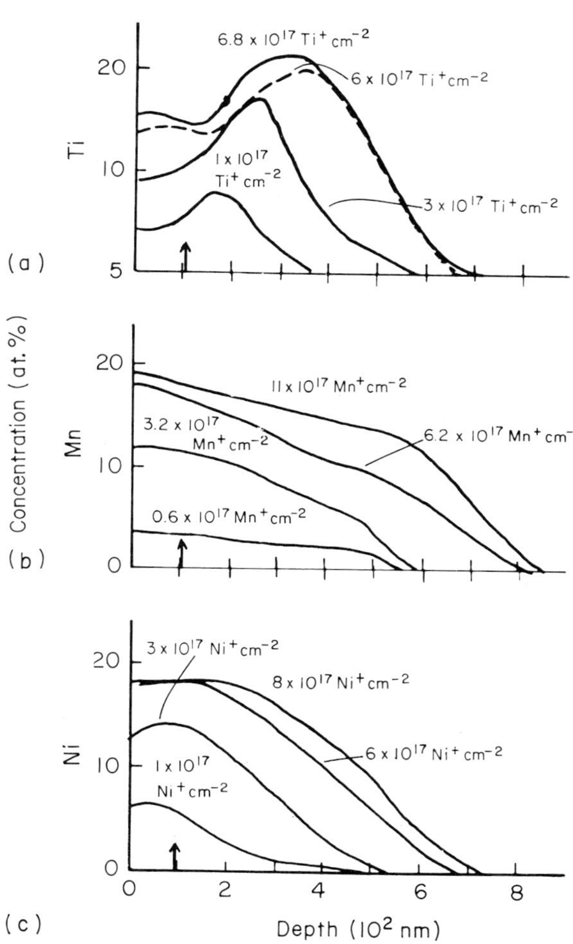

Fig. 4. Depth profiles of the transition metals ((a) titanium; (b) manganese; (c) nickel) implanted into aluminum at 623 K to several fluences. The profiles were calculated from the Rutherford backscattering spectra. The ballistic ranges for implantation into aluminum are indicated by the vertical arrows.

values. At 623 K the diffusion effects already discussed can be clearly seen. The development of almost flat manganese profiles to depths much greater than the projected range of the manganese ions should be noted. These profiles resemble those observed in the formation of SiO_2 by oxygen implantation into silicon [21]. Hence, they may indicate the production and growth of a continuous new phase. Accordingly, the XRD studies showed the formation of orthorhombic $MnAl_6$ (Table 2). Figure 4 indicates that manganese diffuses rapidly in the transformed layer under our experimental conditions. Manganese diffuses inward and produces more $MnAl_6$. Clearly, once the $MnAl_6$ has nucleated, the rate at which manganese-rich phases can nucleate is much lower than the diffusion rate of manganese in the aluminide layer already existing. Therefore, phases such as $MnAl_4$, Mn_4Al_{11} and MnAl, which have more negative heats of formation [22] (see Fig. 9), cannot be produced under these experimental conditions.

The nickel concentration profiles for the 623 K implants also show the effect of diffusion and the development of a near-surface layer, more than 200 nm thick, with a uniform concentration (about 18 at.% Ni). However, there is no equilibrium compound reported to exist with this concentration. Most probably the above-mentioned concentration limit is due to a combination of sputtering and inward diffusion. The small difference in retained fluence between the implant with 6×10^{17} ions cm^{-2} and the implant with 8×10^{17} ions cm^{-2} (see Fig. 4) indicates that the concentration is close to saturation.

The titanium concentration profiles corresponding to the 623 K implants are shown in Fig. 4. It can be seen that the titanium distributions have their peaks beyond the implantation range. The peak maximum shifts inward as the fluence increases. This depletion in the implanted element at the ballistic range could have been caused by a preferential association of one of the components of the system with one type of defect (for instance titanium–interstitial or aluminum–vacancy interactions). If this type of interaction happened, radiation-induced segregation [23] would have taken place, giving rise to the observed profiles. However, since the three TMs have similar atomic sizes, it is not clear why segregation did not occur for the manganese and nickel ion implants, if caused only by the process discussed above. Surface gettering of impurities and subsequent mixing by the ion beam should also be considered as a potential cause for the singular titanium distributions. This phenomenon would produce a two-region near-surface zone, *i.e.* a zone with high and low contaminant contents. Thus the titanium profiles would be the result of the different solubilities and diffusion parameters in the two regions. Since gettering is largely determined by the chemical properties at the surface [24], implantation with other elements should lead to different degrees of quality

TABLE 2

X-ray diffraction data from aluminum samples implanted with titanium, manganese and nickel ions at 623 K

TM ion	*Fluence* ($\times 10^{17}$ ions cm^{-2})	*Angle 2θ* (deg)	*d space* (nm)	*Phase*	*Reflection* ⟨*hkl*⟩
Ti^+	6	39.2	0.2298	$TiAl_3$	⟨113⟩, ⟨202⟩
Ti^+	6	43.2	0.2094	$TiAl_3$	⟨004⟩
Mn^+	10	35.5	0.2529	$MnAl_6$	⟨113⟩
Mn^+	10	39.85	0.2262	$MnAl_6$	⟨131⟩
Mn^+	10	40.75	0.2214	$MnAl_6$	⟨004⟩
Mn^+	10	41.4	0.2181	$MnAl_6$	⟨203⟩
Mn^+	10	42.05	0.2149	$MnAl_6$	⟨310⟩
Ni^+	7	41.35	0.2183	$NiAl_3$	⟨221⟩, ⟨031⟩
Ni^+	7	41.9	0.2156	$NiAl_3$	⟨112⟩
Ni^+	7	43.7	0.2071	$NiAl_3$	⟨131⟩
Ni^+	7	46.15	0.1967	$NiAl_3$	⟨230⟩
Ni^+	7	47.2	0.1926	$NiAl_3$	⟨202⟩, ⟨311⟩, ⟨122⟩

and quantity of the contaminants incorporated in the system, finally resulting in different concentration profiles of the implanted species. In addition to segregation, Fig. 4 shows important diffusion effects. For the more heavily implanted samples, the concentration of titanium reaches 10 at.% (half its maximum) at a depth of 500 nm. Finally, comparison between the titanium concentration profiles for the samples implanted with 6×10^{17} and 6.8×10^{17} ions cm^{-2} suggests that equilibrium among implantation, diffusion and sputtering processes has already been reached at these fluences.

The manganese profiles of the samples implanted at 100 K (Fig. 3) show little or no diffusion effects. Instead, it seems that sputtering produces a shift of the distribution toward the surface for the most heavily implanted sample.

Next, we present the results from the structural analyses of our samples. Figure 5 shows

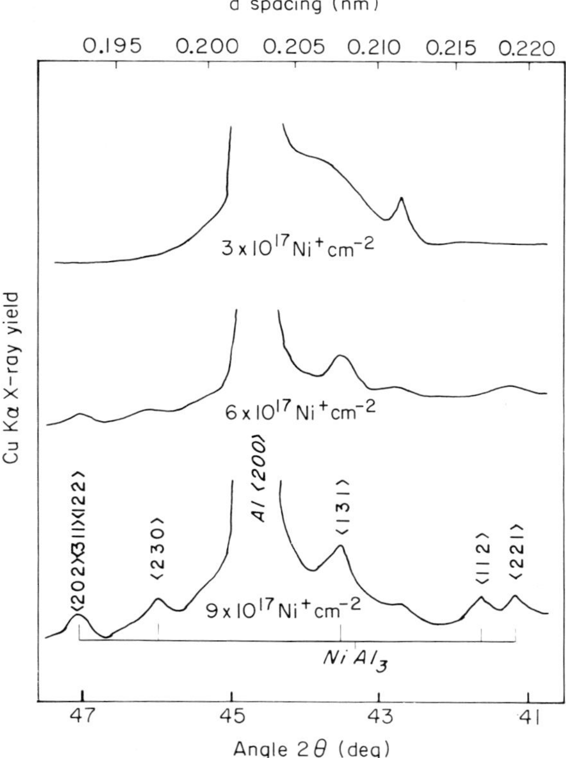

Fig. 5. Diffractometer scans of the samples implanted with nickel ions (Ni^+) to fluences of 3×10^{17}, 6×10^{17} and 9×10^{17} ions cm^{-2} at 100 K. The abscissae are indicated both in degrees (for the Bragg angle 2θ) and in nanometers (for the interplanar distance).

the diffractometer scans of the samples implanted with nickel ions at 100 K and fluences of 3×10^{17}, 6×10^{17} and 9×10^{17} ions cm^{-2}. For the lowest fluence, the only indication of a developing new phase is the broad peak appearing at the position where the ⟨131⟩ reflection of $NiAl_3$ is expected. The ⟨131⟩ reflection is the strongest corresponding to a single family of planes in $NiAl_3$. But the spectrum of the sample implanted with the highest fluence already shows all the five strongest reflections from this phase. We speculate that, after implantation with 3×10^{17} ions cm^{-2}, only short-range order exists in the developing new phase. Therefore, only the most densely populated family of planes reflects noticeably. It would produce a broad line because of the small number of planes contained in each short-range ordered region and because of the slightly different local orders. This situation may indicate either the existence of small $NiAl_3$-like isolated complexes (for instance, at the tail of the nickel distribution) or the formation of a more continuous amorphous $NiAl_3$ phase [7]. A weak peak, midway between the positions corresponding to the ⟨112⟩ and ⟨131⟩ reflections, belongs to an impurity phase already present in the aluminum employed. It also appeared in scans taken on unimplanted samples. Since the X-rays penetrate a few micrometers into the material, the intensity of this peak does not really represent the relative concentration of the impurity phase in the transformed region.

Figure 6 shows the Read camera pictures of the specimens implanted with the two highest fluences of nickel ions at 100 K, taken for the incident X-rays at an angle of 15° to the sample surface. The reflections from the $NiAl_3$ phase are also visible there. It is worth mentioning that these reflections were not seen in the Read camera pictures taken on the same samples when the angle of incidence of the X-rays was reduced to 1°. We interpret this result as an indication that the crystalline $NiAl_3$ phase did not form appreciably at the very near surface. Instead it was produced at a deeper region. Finally, the diffractometer scans and Read camera pictures of the samples implanted with nickel ions at 623 K also indicated the formation of orthorhombic $NiAl_3$ (see Table 2), even though in this case the average nickel concentration was well

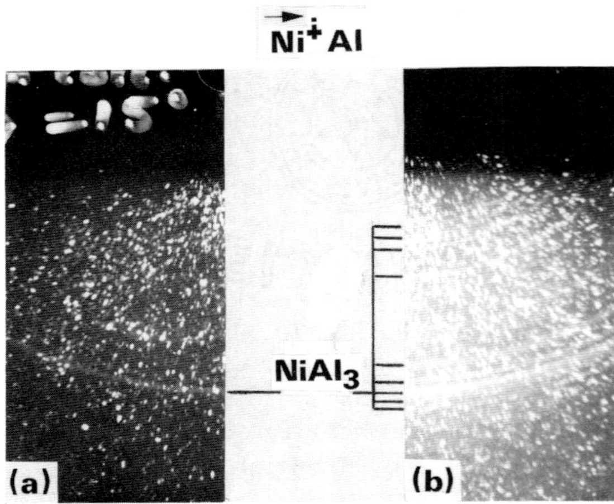

Fig. 6. Read camera pictures of the samples implanted with nickel ions (Ni^+) to fluences of (a) 6×10^{17} ions cm^{-2} and (b) 9×10^{17} ions cm^{-2} at 100 K. The incident X-rays made an angle of 15° with the surfaces of the samples.

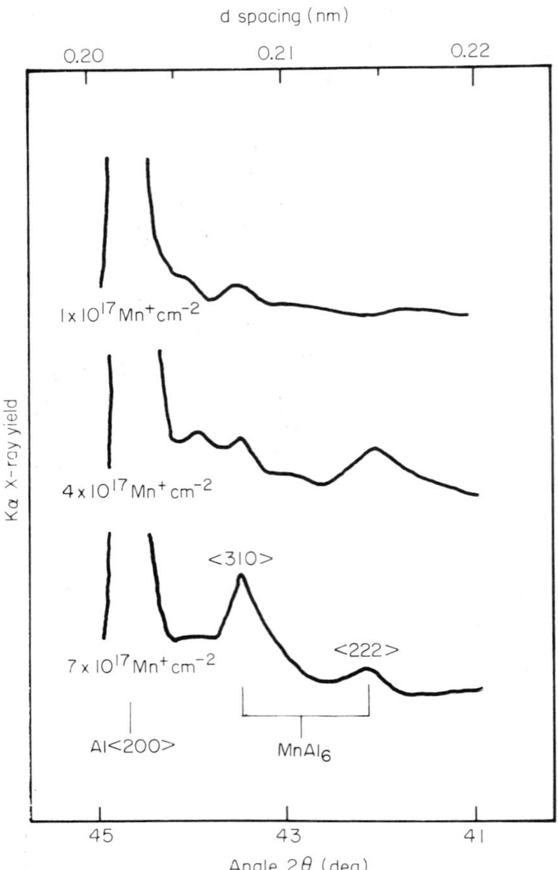

Fig. 7. Diffractometer scans of the samples implanted with manganese ions (Mn^+) to fluences of 1×10^{17}, 4×10^{17} and 7×10^{17} ions cm^{-2} at 100 K. The abscissae are indicated both in degrees (for the Bragg angle 2θ) and in nanometers (for the interplanar distance).

below 25 at.% (see Fig. 4). Hence the transformed region of these samples is concluded to be mostly a mixture of $NiAl_3$ and elemental aluminum.

Figure 7 displays the diffractometer scans of the samples implanted with manganese at 100 K with fluences of 1×10^{17}, 4×10^{17} and 7×10^{17} ions cm^{-2}. Two peaks can be seen to increase as the fluence increases, corresponding to d spacings of 0.2083 and 0.2144 nm. These lines fit the ⟨310⟩ and ⟨222⟩ reflections of orthorhombic $MnAl_6$, both in intensity and position, better than the main lines occurring in the quasi-crystalline icosahedral $Mn_{14}Al_{86}$ phase. The main reflections from the icosahedral phase are expected at about 0.2065 and 0.2170 nm, and they could be present, although weakly, in the scan taken from the sample implanted with 4×10^{17} ions cm^{-2}. Two lines, consistent with the ⟨310⟩ and ⟨222⟩ reflections from $MnAl_6$, can also be seen in the Read camera picture (taken at an incidence angle of 1°) shown in Fig. 8. However, since the method is not accurate enough for precise lattice parameter determinations at such small angles, true distinction between these reflections and the main reflections from the quasi-crystalline icosahedral phase cannot be made. The XRD patterns of the samples implanted with manganese ions at 623 K clearly showed the formation of $MnAl_6$ and are summarized in Table 2.

The XRD studies of the samples implanted with titanium ions at 623 K indicated the production of $TiAl_3$ (see Table 2). In contrast, when the samples implanted at 100 K were analyzed, the diffractometer scans and Read camera pictures showed only the appearance of a single line (as well as the f.c.c. reflections). It is very unlikely that the new phase could be identified from just one diffraction line. Its corresponding d value (0.2194 nm) agrees with both the main reflections ⟨201⟩ of Ti_2Al and ⟨021⟩ of Ti_3Al.

The Read camera pictures taken at 1° from the samples implanted at 100 K also showed several reflections from a cubic phase in the three cases (titanium, manganese and nickel implants). The intensity of these lines decreased as the angle of incidence increased,

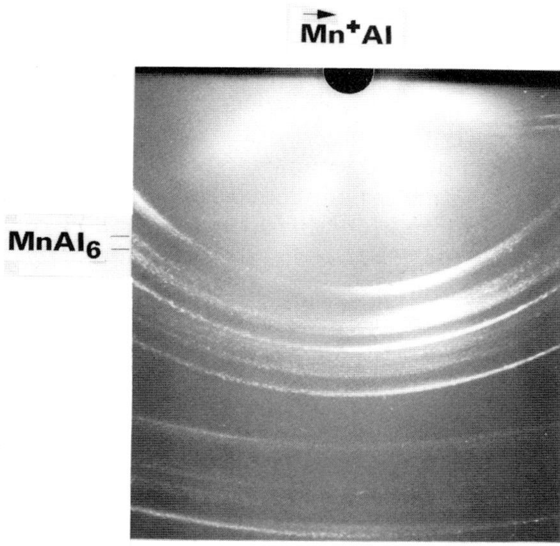

Fig. 8. Read camera picture of the sample implanted with manganese ions (Mn^+) to a fluence of 7×10^{17} ions cm^{-2} at 100 K. The incident X-rays made an angle of 1° with the surface of the sample.

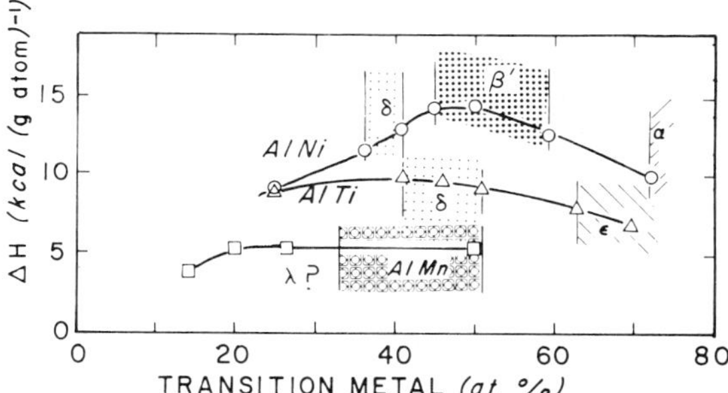

Fig. 9. Heats of formation at 298 K for the equilibrium TM–Al intermetallic compounds as a function of the atomic concentration of the TM.

indicating a smaller contribution from these phases to the Read camera patterns when greater depths were preferentially observed. For the small-incidence-angle experiments, their measured lattice parameters were close to, but did not agree with, that of pure aluminum. This situation may correspond to the formation of Al–TM solid solutions next to the surface. However, the above-mentioned inaccuracy of the method for lattice parameter determinations at such small angles prevents us from reaching a definite conclusion on the nature of these phases.

4. SUMMARY

Figure 9 shows the heats of formation at room temperature of the known equilibrium phases for the TM–Al systems studied here. From a thermodynamic point of view, the most favorable phases are TiAl, those from $MnAl_4$ to MnAl, and NiAl [22]. However, in contrast with our observations on the TM–Si system [16], only the aluminum-richest compounds were detected in our implanted samples, with the only possible exception occurring for the titanium implants at 100 K. This was so, even for the samples implanted with manganese ions at 100 K, where the manganese concentration was as high as 40 at.%, *i.e.* much greater than that corresponding to the observed $MnAl_6$. We shall discuss below the possible causes for this departure from the thermodynamic predictions.

These results raise the question of where the rest of the TM atoms are in the samples implanted at the lower temperature. We cannot discard the formation of amorphous phases. In fact, the diffractometer pattern of the sample implanted with the lowest nickel fluence at 100 K shows a broad amorphous-like peak close to the position expected for the ⟨131⟩ $NiAl_3$ reflection (Fig. 5). Amorphous phases are difficult to detect by XRD when they do not constitute a large percentage of the sample. Besides, the Read camera pictures taken at an angle of incidence of 1° show strong reflections from a cubic phase. It may correspond to an Al–TM solid solution stabilized by the presence of impurities close to the surface. This structure must exist at the very near surface, since the conditions for the Read camera experiments were selected to analyze this region. Unfortunately, the accuracy of lattice parameter measurements with the Read camera for such small grazing angles is rather poor. Thus an identification of this phase cannot be given. The samples implanted at the lower temperature may consist of a mixture of amorphous, a cubic solid solution and the above-mentioned TM–Al crystalline phases.

In the low temperature nickel-implanted samples, the crystalline $NiAl_3$ forms at a certain depth, and not at the near surface, although the concentration profiles do not show any kind of frontier between the two regions. Then, as the depth increases, the amount of the postulated solid solution should decrease while the amount of orthorhombic phase increases. This seems to be so, given the weakening of the reflections from the cubic structure in the Read camera pictures as the incidence angle of the X-rays is increased. In this context the differences between our results and those obtained in previous studies on ion-beam-induced solid reactions in the Ni–Al system [1, 6–9] can be understood as follows. Under our experimental conditions, diffusion plays an important role even at 100 K, especially for the highest nickel fluence. As a consequence, nickel diffuses far beyond the implantation range and reaches a concentration above 10 at.% at depths as great as 400 nm. Therefore, we have two different zones in our samples: the near-surface zone, where the beam produces damage while supplying nickel, and a deeper region, where the nickel enters by diffusion. We speculate that in the near-surface region the irradiation (and/or the presence of impurities) prevents the formation of orthorhombic $NiAl_3$. This phase is able to grow at a deeper zone where the conditions may be close to those occurring during thermal annealing experiments.

$MnAl_6$ has a complex orthorhombic structure with 28 atoms per unit cell. Therefore, this phase may be unstable under irradiation [7]. In fact, recent experiments seem to indicate that this is indeed the case [12, 25]. For low temperature implants, we did not see any important increase in the inward penetration of the implanted manganese as the fluence was increased (see Fig. 3). Thus the manganese concentration steadily increased as the fluence increased up to a maximum value of 40 at.%. Since the diffractometer scans seem to indicate that, as the fluence increases, the aluminide structure agrees better with that of orthorhombic $MnAl_6$, it may tentatively be concluded that this crystalline compound is more stable under irradiation in a manganese-rich environment.

It should be noted that, while manganese diffuses faster than nickel in aluminum, diffusion effects during the low temperature implants were more important in the Ni–Al than in the Mn–Al system. Our results simply indicate that diffusion in the NiAl transformed layer (mostly $NiAl_3$ according to the XRD and Rutherford backscattering data) occurs at a higher rate than in the MnAl layer, which may consist of a mixture of phases and/or short-range order regions (given the non-agreement between structural and compositional data in the last case). Thus a comparison with diffusion processes in pure aluminum does not exactly follow, especially for the most heavily implanted samples.

All the samples implanted at 623 K experienced important diffusion effects. The diffusion of the TMs toward the interior seemed to have prevented the implanted species from reaching concentration values above about 20 at.%. In the titanium- or nickel-implanted samples the transformed regions are likely to be a mixture of crystalline $TiAl_3$ or $NiAl_3$ and f.c.c. aluminum. The samples implanted with manganese produced a thick layer of orthorhombic $MnAl_6$. The excess of manganese supplied by the implantation process diffused inward toward the $MnAl_6$–Al interface where it reacted with the aluminum to form more $MnAl_6$.

As stated above, during the low temperature implants, the beam-induced damage and/or the presence of impurities seemed to have precluded the production of crystalline phases in the regions closer to the surface of the samples. Thus, crystalline phases with high heats of formation were not produced, even when the average concentrations of the TM were closer to the concentrations corresponding to these phases (see Fig. 9). At deeper regions, the beam did not produce damage. Also, there was a smaller probability of the presence of contaminants. Then, equilibrium phases more compatible with the lower concentration of the TM at these locations were formed. At high temperatures the kinetics controlled the process in the three systems. Because of the higher diffusion coefficients, the concentration of the implanted species never reached the values corresponding to the more stable phases and then only the aluminum-richest equilibrium compounds were produced.

ACKNOWLEDGMENTS

The authors would like to acknowledge Professor Q. Kessel for kindly facilitating the use of the Van de Graaff accelerator and J. Gianopoulus for his valuable technical assistance.

This work was carried out while one of the authors (F.H.S.) was on a fellowship from the Consejo Nacional de Investigaciones Científicas y Técnicas, Argentina.

REFERENCES

1 M. Ahmed and D. I. Potter, *Acta Metall., 12* (1985) 2221.
2 B. Cordts, M. Ahmed and D. I. Potter, *Nucl. Instrum. Methods, 209-210* (1983) 873.
3 S. T. Picraux, D. M. Follstaedt, J. A. Knapp, W. R. Wampler and E. Rimini, in J. F. Gibbons, L. D. Hess and T. W. Sigmon (eds.), *Laser and Electron-Beam Solid Interactions and Materials Processing, Materials Research Society Symp. Proc.*, Vol. 1, Elsevier, New York, 1981, p. 575.
4 S. T. Picraux and D. M. Follstaedt, in J. M. Poate, G. Foti and D. C. Jacobsen (eds.), *Surface Modification and Alloying*, Plenum, New York, 1983.
5 D. M. Follstaedt, in J. W. Mayer (ed.), *Proc. 4th Int. Conf. on Ion Beam Modification of Materials, Cornell University, Ithaca, NY, July 16-20, 1984*, in *Nucl. Instrum. Methods B, 7-8* (1985) 11.
6 L. Thomé, F. Pons, J. C. Pivin and C. Cohen, *Nucl. Instrum. Methods B, 15* (1986) 269.
7 M. Nastassi, H. H. Johnson, J. W. Mayer and J. M. Williams, *J. Mater. Res., 1* (1986) 268.
8 M. Nastassi, L. S. Hung and J. W. Mayer, *Appl. Phys. Lett., 43* (1983) 831.
9 L. S. Hung, M. Nastassi, J. Gyulai and J. W. Mayer, *Appl. Phys. Lett., 42* (1983) 672.
10 J. Eridon, L. Rehn and G. Was, in H. Kurz, G. L. Olson and J. M. Poate (eds.), *Beam-Solid Interactions and Phase Transformations, Materials Research Society Symp. Proc.*, Vol. 51, Materials Research Society, Pittsburgh, PA, 1986.
11 J. D. Budai and M. J. Aziz, *Phys. Rev. B, 33* (1986) 2876.
12 J. A. Knapp and D. M. Follstaedt, in H. Kurz, G. L. Olson and J. M. Poate (eds.), *Beam-Solid Interactions and Phase Transformations, Materials Research Society Symp. Proc.*, Vol. 51, Materials Research Society, Pittsburgh, PA, 1986.
13 J. K. Howard, R. F. Lever, P. J. Smith and P. S. Ho, *J. Vac. Sci. Technol., 13* (1976) 68.
14 J. K. Howard, J. F. White and P. S. Ho, *J. Appl. Phys., 49* (1978) 4083.
15 F. H. Sánchez, F. Namavar, J. I. Budnick, A. Fasihuddin and H. C. Hayden, in H. Kurz, G. L. Olson and J. M. Poate (eds.), *Beam-Solid Interactions and Phase Transformations, Materials Research Society Symp. Proc.*, Vol. 51, Materials Research Society, Pittsburgh, PA, 1986, p. 439.
16 F. Namavar, F. H. Sánchez, J. I. Budnick, A. Fasihuddin and H. C. Hayden, in S. T. Picraux, M. O. Thompson and J. S. Williams (eds.), *Beam-Solid Interactions and Transient Processes, Materials Research Society Symp. Proc.*, Vol. 74, Materials Research Society, Pittsburgh, PA, 1987, to be published.
17 G. K. Wehner, in R. Behrisch (ed.), *Sputtering by Particle Bombardment II*, Springer, Berlin, 1983, p. 57.
18 J. Askill, *Tracer Diffusion Data for Metals, Alloys and Simple Oxides*, IFI-Plenum, New York, 1970, p. 1.
19 L. F. Mondolfo, *Aluminum Alloys: Structure and Properties*, Butterworths, London, 1976, pp. 20-38.
20 W. K. Chu, J. W. Mayer and M.-A. Nicolet, *Backscattering Spectrometry*, Academic Press, New York, 1978, p. 44.
21 F. Namavar, J. I. Budnick, F. H. Sánchez and H. C. Hayden, in A. Chiang, M. W. Geis and L. Pfeiffer (eds.), *Semiconductor-on-Insulator and Thin Film Transistor Technology, Materials Research Society Symp. Proc.*, Vol. 53, Materials Research Society, Pittsburgh, PA, 1986, p. 233.
22 R. Hultgren, P. D. Desai, D. T. Hawkins, M. Gleiser and K. K. Kelly, *Selected Values of the Thermodynamic Properties of Binary Alloys*, American Society for Metals, Metals Park, OH, 1973, pp. 185-188, 191-195, 221-223.
23 N. Q. Lam and G. K. Leaf, *J. Mater. Res., 1* (1986) 251.
24 F. Namavar, J. I. Budnick and F. A. Otter, *Materials Research Society Symp. Proc.*, Vol. 36, Materials Research Society, Pittsburgh, PA, 1985, p. 55.
25 D. M. Follstaedt and J. A. Knapp, *Proc. Int. Conf. on Surface Modification of Metals by Ion Beams, Kingston, Canada, July 7-11, 1986*, in *Mater. Sci. Eng., 90* (1987).

Materials Science and Engineering, 90 (1987) 161-165

Study of a Gold-implanted Magnesium Metastable System*

M. R. DA SILVA[a], A. A. MELO[a], J. C. SOARES[a], M. F. DA SILVA[b] and R. VIANDEN[c]

[a]*Centro de Física Nuclear da Universidad de Lisboa, 1699 Lisboa Codex (Portugal)*
[b]*Departamento de Física do Instituto de Energia, Laboratório Nacional de Engenharia e Tecnologia Industrial (LNETI), 2685 Sacavém (Portugal)*
[c]*Institut für Strahlen- und Kernphysik, University of Bonn, Bonn (F.R.G.)*

(Received July 10, 1986)

ABSTRACT

Studies on the Au–Mg system formed by implanting a low fluence of gold ions in magnesium single crystals were carried out using the Rutherford backscattering–channelling technique. A distorted tetrahedral location was observed for gold. Furnace annealings in a helium atmosphere were carried out up to 525 K. The results show a fast segregation of gold to the Mg–MgO interface where it precipitates as metallic aggregates. The observed migration of gold to the surface with an activation energy of about 0.21 eV should indicate that the interstitial–vacancy mechanism is the dominant process of the impurity segregation.

1. INTRODUCTION

Ion implantation is now a well-established technique for forming metastable alloys in the near-surface region of metals [1, 2]. Moreover, being a non-equilibrium technique which is almost free from thermodynamic constraints, it allows the classical limits for diffusion or solubility to be overcome and the formation of supersaturated alloys is possible. However, the heat treatment of such metastable alloys can introduce radical changes, shifting the system towards its thermodynamic equilibrium. In addition, when ion implantation is used as a means for introducing impurities in a matrix, radiation damage must be considered. In the past few years a large amount of work has been done in this field and the basic physical mechanisms of such alloying processes are well established. However, the prediction of the behaviour of a specific implant–host system is often difficult.

One of the basic problems in describing the properties of the implanted system is knowledge of the impurity atom location within the host matrix. Previous work carried out on a number of implanted magnesium single crystals showed that a substitutional location is the preferred lattice site for all the implanted species [3–5]. However, most of these experimental data are contradictory to the predictions of the different schemes for solid solubility in magnesium [6].

Following systematic work on the lattice location and solid solubility of implanted species in magnesium [3–5], the results of a Rutherford backscattering–channelling study of an Mg–Au system formed by the implantation of a magnesium single crystal with a low fluence of gold ions are reported in this paper. The behaviour with respect to temperature of the metastable alloy obtained was studied using annealing techniques.

2. EXPERIMENTAL DETAILS

High purity magnesium single crystals spark cut approximately 5° to the $\langle 10\bar{1}0 \rangle$ axis were cleaned by chemical etching in HNO_3 (65%), rinsed with doubly distilled water and dried [4]. Despite the enhanced reactivity of the pure magnesium, this method allows stabilization of the growth of a native oxide layer with a typical thickness of about 25 nm.

The gold implantations at 160 keV and with a fluence of 5×10^{14} ions cm^{-2} were carried out at room temperature and at normal incidence using the mass separator facilities of the University of Bonn. The impurity depth distribution and the lattice location measure-

*Paper presented at the International Conference on Surface Modification of Metals by Ion Beams, Kingston, Canada, July 7–11, 1986.

0025-5416/87/$3.50

ments were carried out using the Rutherford backscattering–channelling facilities at the LNETI, Sacavém [4, 6]. The energy of the analysing helium ($^4He^+$) beam was 1.6 MeV with typical currents of 5 nA.

Isochronal annealings for 30 min in a pure helium atmosphere at 50 mbar and at temperatures ranging from 433 to 525 K and isothermal annealing at 525 K for up to 3 h were performed using a tubular furnace.

The analysis of the measured depth profiles was carried out using a least-squares fitting procedure with a set of two joined half-gaussians as the distribution function.

3. RESULTS AND DISCUSSION

Figure 1 shows random and aligned ⟨11$\bar{2}$0⟩ axial Rutherford backscattering spectra for the gold-implanted magnesium single crystal. The surface peak due to the native oxide layer present in all samples [5] has a thickness of 28 ± 5 nm. The centroid of the gold peak is located at about 77.9 ± 6.0 nm below the surface. This value is in good agreement with the theoretical mean projected range obtained using the transport and range of ions in matter (TRIM) procedure [7], taking into account the native oxide layer. The measured mean concentration of about 0.43 at.% for the implanted gold exceeds the solubility limit of gold in magnesium [8] by a factor of more than 30, indicating that the system created is supersaturated.

The lattice location of the gold impurity was obtained by complete angular scan measurements for the major axis and planes of the magnesium crystal (Fig. 2). These data definitely exclude the possibility of a substitutional or random location for the gold ions in the magnesium lattice. Also the absence of flux peaks at the centre of the gold angular scans for the ⟨11$\bar{2}$0⟩ and (0001) directions is inconsistent with an occupancy of octahedral sites in the lattice. The structure shown in the ⟨11$\bar{2}$0⟩ axis and (0001) plane scans is typical of a tetrahedral location of the impurity. However, the (1$\bar{1}$01) plane angular scan shows a substitutionality that is consistent with an off-centre location for the gold in the tetrahedral cage.

Rutherford backscattering spectra normal to the surface were obtained for the annealed samples in a helium atmosphere. Figure 3 shows the impurity depth profiles as implanted and for several annealing temperatures. The increase in the oxide layer with consecutive annealings is the same for both the implanted

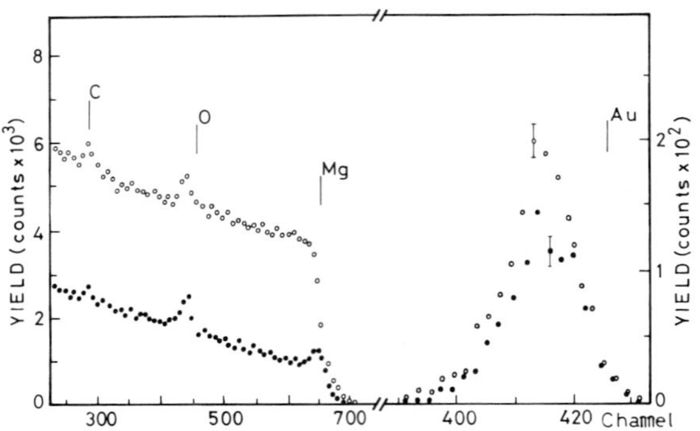

Fig. 1. Random (○) and ⟨11$\bar{2}$0⟩ aligned (●) Rutherford backscattering spectra for 1.6 MeV helium ($^4He^+$) ions obtained from a magnesium single crystal implanted at 160 keV with 5 × 10^{14} $^{197}Au^+$ ions cm^{-2}.

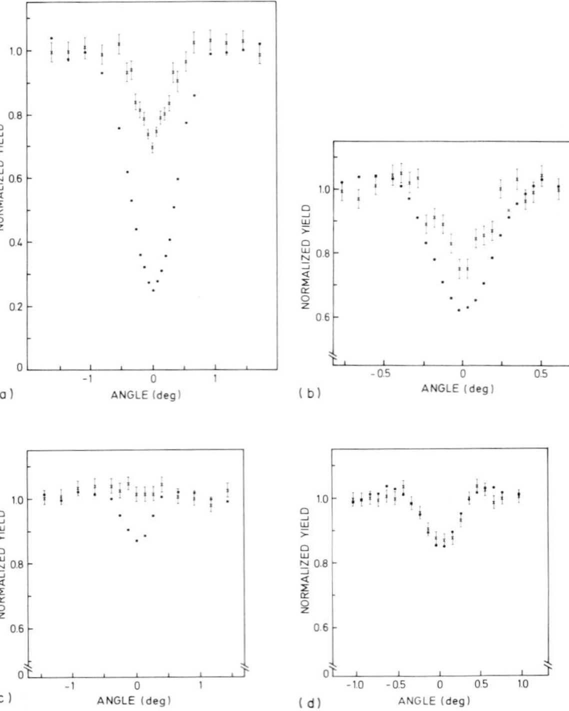

Fig. 2. Angular scan curves through (a) the ⟨11$\bar{2}$0⟩ axial direction, (b) the (0001) planar direction, (c) the (1$\bar{2}$10) planar direction and (d) the (1$\bar{1}$01) planar direction for a gold-implanted magnesium crystal (5 × 10^{14} ions cm^{-2} at 160 keV; E_{He^+} = 1.6 MeV): ●, magnesium; ×, gold.

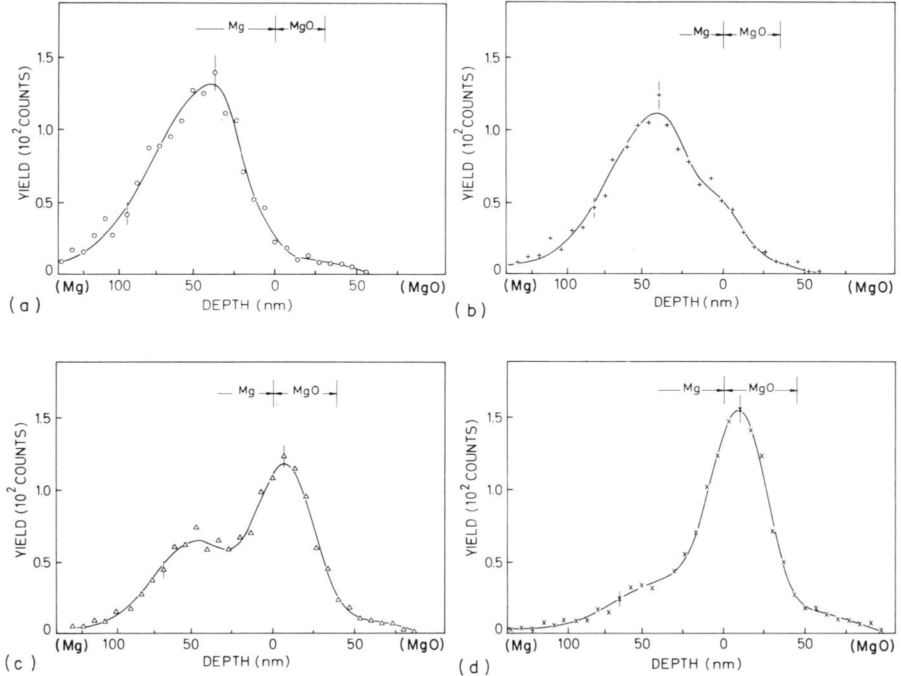

Fig. 3. Rutherford backscattering depth profiles for a gold-implanted magnesium crystal (5×10^{14} $^{197}Au^+$ ions cm^{-2} at 160 keV) (a) as implanted (○), (b) after annealing for 30 min at 433 K (+), (c) after annealing for 30 min at 495 K (△) and (d) after annealing for 30 min at 525 K (×): ——, non-deconvoluted fitted profiles.

and the unimplanted surfaces and is considered to be due to the residual oxygen present in the atmosphere of the annealing furnace. Large changes in the gold depth profile occur during the annealing treatments. However, the total amount of impurity remains constant. The gold impurity rapidly migrates towards the surface, giving rise to a peak near the surface, but does not diffuse into the bulk as shown by the small concentration of gold at greater depths, even for the highest annealing temperature (Fig. 3(d)). This peak increases both as a fast function of the temperature and as a slow function of time. The gold reaching the surface precipitates at the Mg–MgO interface, probably in the form of metallic aggregates. This is supported by the (0001) angular scan for the gold surface fraction measured after annealing at 495 K (Fig. 4) and by the results obtained by Abouchacra *et al.* [9] who found that gold implanted in an MgO crystal precipitates after annealing in stable f.c.c. metallic aggregates which do not show any long-range diffusion process up to 1370 K.

It is observed that during the annealing procedure the remaining interstitial gold does

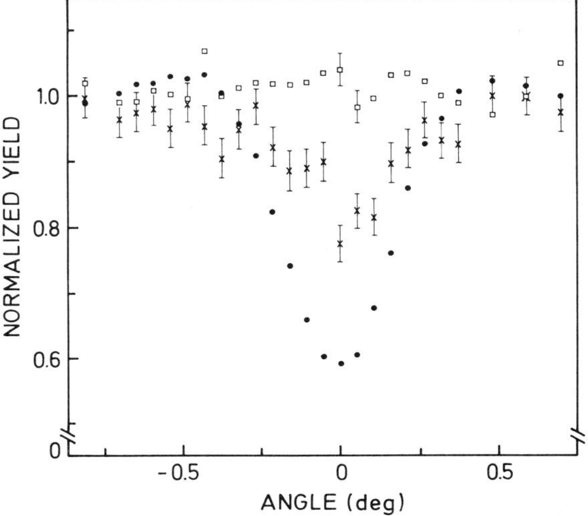

Fig. 4. Angular scan curves through the (0001) planar direction after annealing at 495 K for 30 min: ●, magnesium; ×, gold (in the magnesium bulk); □, gold (at the Mg–MgO interface).

not change position in the magnesium lattice (Fig. 4) and its profile centroid remains the same. The evolution with consecutive annealings of this interstitial fraction F_i of the im-

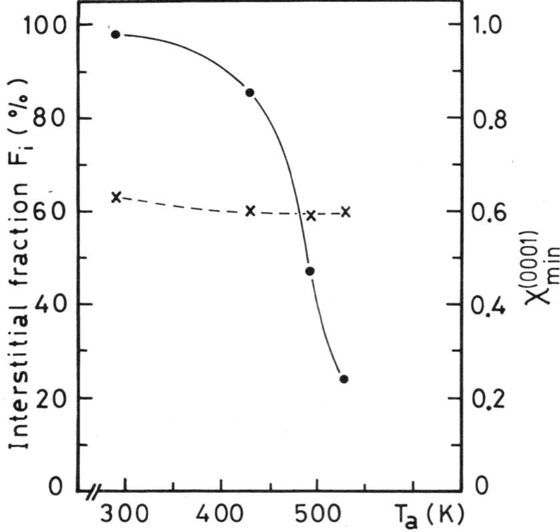

Fig. 5. Interstitial fraction F_i of implanted gold (●) and minimum yield χ_{min} of the (0001) angular scan for the magnesium host (×) as a function of the annealing temperature T.

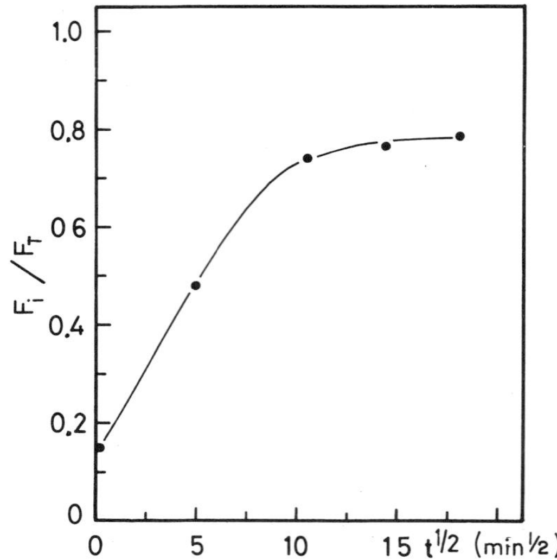

Fig. 6. Evolution of the gold precipitates at the Mg–MgO interface as a function of the annealing time $t^{1/2}$ ($T_a = 525$ K).

planted gold (in atomic per cent of the total amount) and of the magnesium minimum yield χ_{min} for the (0001) direction are shown in Fig. 5. The segregation and precipitation of gold at the Mg–MgO interface do not produce any significant change in the damage recovery rate of the host lattice as evidenced by the almost constant minimum yield. The same behaviour, within experimental error, was observed for χ_{min} evolution at the unimplanted surface. This is in good agreement with the low temperature values for the recovery stages of magnesium [10–12], indicating that at room temperature and above the defect concentration is already the equilibrium concentration.

From the flat tail of the interstitial gold (Fig. 3(d)) an estimate for the solubility limit of about $C_s = 0.016$ at.% was obtained. This result compares well with previously obtained values of $C_s = 0.1$ at.% at 834 K [8] when extrapolated to the same temperature.

The isothermal annealing at 525 K (Fig. 6) shows that up to 1.5 h there is an increase in the gold at the Mg–MgO interface with a $t^{1/2}$ time dependence, reaching an almost constant value for longer annealing times. This means that some equilibrium has been reached between the gold dissolved in the magnesium and the gold precipitates at the Mg–MgO interface. The observed time dependence indicates that the kinetics of the gold segregation might be well described by a model of the type

$$M_s = f\{(D_i, C_i, t)^{1/2}\}$$

as suggested by Okamoto *et al.* [13] where M_s is the amount of gold segregated to the surface, and D_i and C_i are respectively the diffusion coefficient and the concentration of the interstitial gold.

The activation energy of the segregation process estimated from the behaviour of the interstitial fraction F_i is about 0.21 eV. This very low value makes a striking contrast with the activation energies for magnesium self-diffusion and vacancy self-diffusion which are 1.40 eV and 1.31 eV respectively [14]. The difference between the activation energies for this segregation process and for the magnesium self-diffusion yields a low negative value of −0.99 eV typical of a fast diffusion process. When these results and the fact that gold is interstitially located in the magnesium lattice and does not diffuse into the bulk are taken into account, the segregation could be explained by a mechanism involving an interstitial impurity–vacancy complex. In fact an interstitial mechanism is usually invoked to explain an abnormal high mobility of diffusers even though the ratio of the atomic radii does not obey Hagg's [15] rule which is the case. Since the work of Miller [16] there is evidence that fast diffusion occurs via impurity–vacancy complexes for some f.c.c. metals where impurities dissolve interstitially. More recently the existence of this trapping mechanism was demonstrated for h.c.p. and tetragonal crystals [17-19].

In the present case the self-interstitials and free vacancies created during the implantation at room temperature are annealed out because of the low temperature of the damage recovery stages for magnesium [10-12]. Part of the free moving vacancies become trapped at the gold atoms, forming a solute–vacancy pair. These pair complexes migrate to the vacancy sinks located at the Mg–MgO interface, the vacancy concentration gradient providing the driving force for the complex diffusion. At the interface the migrating solute–vacancy pair is destroyed, precipitating the gold atoms in the form of metallic aggregates.

In conclusion, this study has demonstrated some unique features of the system created by gold implantation in a magnesium single crystal. It has been shown that supersaturated solutions are achievable at room temperature, exceeding the gold solubility limit by a factor of at least 30. In these solutions the gold ions occupy distorted tetrahedral sites in the magnesium lattice, in spite of a value of 0.91 for the ratio of the gold to magnesium atomic radii. These studies have also shown that the solution formed is metastable as, because of annealing, gold segregates to the Mg–MgO interface. The experimental data are consistent with an impurity–vacancy pair mechanism for the gold segregation.

ACKNOWLEDGMENTS

Thanks are due to Dr. K. Freitag (Institut für Strahlen- und Kernphysik, Bonn) for the gold implantation.

Financial support from the Gulbenkian Foundation is acknowledged by one of us (M.R.S.). This work was partially supported by Junta Nacional de Investigação Científica e Technológica (JNICT) under Contract 4268273 and by the Volkswagenwerk Foundation.

REFERENCES

1 D. K. Sood, *Radiat. Eff., 63* (1982) 141.
2 D. M. Follstaedt, *Nucl. Instrum. Methods B, 7-8* (1985) 11.
3 L. M. Howe, M. L. Swanson and A. F. Quenneville, *Radiat. Eff., 35* (1978) 227.
4 M. R. da Silva, A. A. Melo, J. C. Soares, E. J. Alves, M. F. da Silva, P. M. Winand and R. Vianden, *Port. Phys., 14* (1983) 175.
5 M. R. da Silva, A. A. Melo, J. C. Soares, M. F. da Silva and R. Vianden, *Nucl. Instrum. Methods B, 15* (1986) 344.
6 M. F. da Silva, M. R. da Silva, E. J. Alves, A. A. Melo, J. C. Soares, P. M. Winand and R. Vianden, in *Surface Engineering, NATO Adv. Study Inst. Ser., Ser. E, 85* (1984) 74.
7 J. P. Biersack and L. G. Haggmark, *Nucl. Instrum. Methods, 174* (1980) 257.
8 M. Hansen and K. Anderko, *Constitution of Binary Alloys*, McGraw-Hill, New York, 2nd edn., 1958.
R. P. Elliott, *Constitution of Binary Alloys, First Supplement*, McGraw-Hill, New York, 1965.
F. A. Shunk, *Constitution of Binary Alloys, Second Supplement*, McGraw-Hill, New York, 1969.
9 G. Abouchacra, G. Chassagne and J. Serughetti, *Radiat. Eff., 64* (1982) 223.
10 S. Takamura and M. Kobiyama, *Radiat. Eff., 49* (1980) 247.
11 B. Schönfeld and P. Ehrardt, *Radiat. Eff., 59* (1981) 93.
12 H. Maeta, in J. Takamura, M. Doyama and M. Kiritani (eds.), *Point Defects and Defect Interactions in Metals*, North-Holland, Amsterdam, 1982, p. 14.
13 P. Okamoto, L. Rehn and R. Averback, *J. Nucl. Mater., 108-109* (1982) 319.
14 P. Shewmon, *Trans. AIME, 206* (1956) 918.
15 G. Hagg, *Z. Phys. Chem. B, 8* (1930) 445.
16 J. Miller, *Phys. Rev., 181* (1969) 1095.
17 D. Yeh and P. Huntington, *Phys. Rev. Lett., 35* (1984) 1469.
18 G. Hood and R. Schultz, *Philos. Mag., 26* (1972) 329.
19 F. Dyment, *J. Nucl. Mater., 61* (1976) 271.

Materials Science and Engineering, 90 (1987) 167–171

Induced X-ray Emission from Neon Implanted into Metals*

G. DECONNINCK[a,b] and A. LEFEBVRE[a]

[a] *Laboratoire d'Analyse par Réactions Nucléaires (LARN), Facultés Universitaires de Namur, 22 rue Muzet, B-5000 Namur (Belgium)*

[b] *Université Catholique de Louvain, B-1348 Louvain-la-Neuve (Belgium)*

(Received July 10, 1986)

ABSTRACT

The physical state of neon implanted into different metals has been studied by means of soft X-ray emission spectroscopy. Metal samples (beryllium, aluminium, titanium, chromium, iron, copper, zinc, molybdenum, silver, tungsten, gold, lead and bismuth) are implanted with 50 keV neon ions ($^{20}Ne^+$) and subsequently irradiated with 4 keV electrons to excite the neon X-ray fluorescence which is analysed with a flat-crystal spectrometer. A considerable broadening of the atomic K X-ray line is observed for all samples, showing the existence of a band structure. The observed 2p band width W ranges between 1.95 and 4.07 eV depending on the host metal (compared with 1.3 eV for solid neon at atmospheric pressure). X-ray fluorescence analysis yields two observations (the total width W and the spectrum shape) which should be related to the following macroscopic parameters: the average metal shear modulus μ and the internal pressure. A plot of W vs. shear modulus shows a relation which can be approximated by the following straight line (where W is in electronvolts and μ in gigapascals):

$$W = 0.017\mu + 1.3$$

Using the relationship between the shear modulus and the pressure, it is thus possible to relate W to the pressure.

1. INTRODUCTION

There is considerable technological interest in the interaction of inert gases with metals. Much of the work on this subject relates to helium implantation into various metals in connection with nuclear and fusion reaction technology [1]. More recently a number of data on other inert gases implanted into various materials have been obtained by electron diffraction and high resolution electron microscopy, showing the presence of so-called "bubbles" of small size (1–2 nm). At concentrations of a few atomic per cent the physical state of the aggregates is solid [2–8], liquid or fluid depending on the temperature. Evans and Mazey [6] have reported a correlation between the pressure and the metal shear modulus μ, indicating that μ would be the parameter which controls the size of the bubbles and the mechanism of multiplication. During irradiation at temperature below half the melting point T_m, a loop-punching mechanism allows the bubbles to grow until the threshold for coalescence is reached at a concentration of a few atomic per cent; above this concentration the bubbles grow rapidly and the pressure decreases [7].

The mechanisms involved in this are still hypothetical, these ideas being supported by a number of measurements from transmission electron microscopy, electron diffraction [2–8] and Doppler shift analysis [7]. In a previous paper [9], we have described the application of X-ray emission spectroscopy to these problems; in particular, we outlined the observation of considerable broadening of the K X-ray lines from neon implanted into aluminium excited by 2 MeV helium (^4He) particles.

This broadening was attributed to the formation of a valence band, confirming the presence of solid or fluid neon. The aim of the present work is to establish a relation between the bandwidth and the metal shear modulus μ by measuring X-ray spectra from various

*Paper presented at the International Conference on Surface Modification of Metals by Ion Beams, Kingston, Canada, July 7–11, 1986.

0025-5416/87/$3.50

metals implanted with neon. The excitation of X-rays is produced by low energy electrons rather than by helium in order to reduce the background and to avoid additional radiation damage caused by helium bombardment [9].

2. EXPERIMENTAL PROCEDURE

The experimental set-up has already been described in ref. 9 where 2 MeV helium (^{4}He) particles were used to irradiate aluminium samples implanted with neon; a flat-crystal spectrometer was used to analyse X-rays from ionization. In the present work a 4 keV electron gun was used to produce ionization, this energy being an optimum for a maximum efficiency. The samples were disks cut in standard metal sheets supplied by Goodfellow. After polishing, the surface was implanted at 45° with a 50 keV neon ions ($^{20}Ne^+$) beam.

Range calculations indicate for each metal that the implanted ions lie in a diffuse layer below the surface which includes the maximum ionization cross-section (occurring at 2.4 keV) for 4 keV electrons travelling in the metal. Repeated measurements taken on the same sample show that there is no observable modification resulting from radiation damage or temperature-dependent effects. The implantation fluence can be estimated from backscattering using 2.4 MeV α particles on beryllium metal ($Z = 4$). In Fig. 1(a) a backscattering spectrum from neon-implanted beryllium is reproduced. The detector resolution is 9 keV and the scale is 4 keV count^{-1}. The spectrum indicates the presence of neon (^{20}Ne) and three peaks of impurities (carbon, oxygen and silicon) localized at the surface. The neon profile was obtained by simulating the backscattering curve using a computer program. The concentration curve is not constant with depth but is a typical Lindhard curve with a maximum of 2 at.%. For other metals the fluence was calculated to obtain a concentration maximum well below the bubble coalescence value.

K X-ray spectra are analysed by diffraction. The overall resolution curve has a full width at half-maximum of 0.74 eV [9]. In Fig. 2 we reproduce the K X-ray spectra (not deconvoluted) from neon gas and from a beryllium sample implanted with neon. The data in Fig. 2 clearly show that the Kα_{12} transition (na-

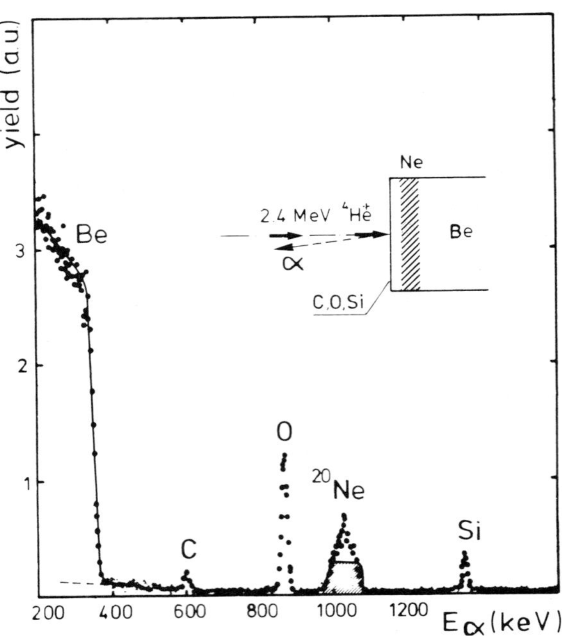

Fig. 1. Backscattering spectrum of 2.4 MeV helium ions ($^4He^+$) on a sample of beryllium implanted with 50 keV neon ions ($^{20}Ne^+$). The curve drawn is the result of a calculation based on a constant concentration (1 at.%). Carbon, oxygen and silicon peaks are due to surface contamination from the polishing procedure.

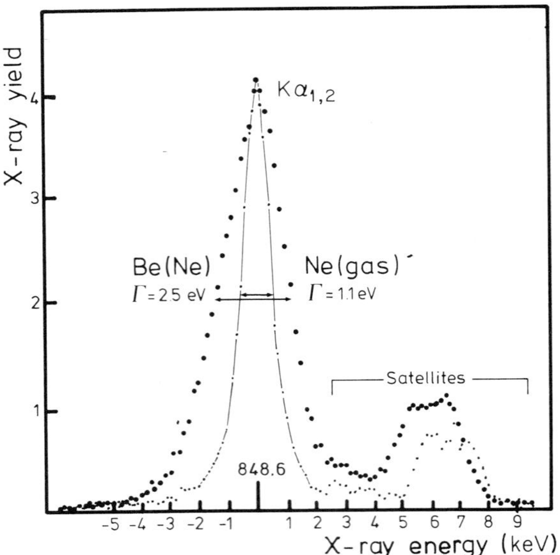

Fig. 2. A comparison between two X-ray yield curves (not deconvoluted) from neon (gas) and a sample of beryllium implanted with neon (Γ is the full width at half-maximum). The total width W will be obtained by deconvolution of the Kα_{12} component (excluding the satellite amplitudes).

tural width, 0.2 eV) is considerably broadened when neon is implanted into the metal. We attribute this to a broadening of the atomic level into a valence band, indicating the existence of condensed neon. The X-ray yield curves from different samples have been deconvoluted for the natural line shape of the $(1s)^{-1}$ hole state ($\Gamma = 0.2$ eV) and for the instrumental resolution curve ($\Gamma = 0.74$ eV).

The resulting X-ray yield curve gives a good picture of the electron density-of-states curve since the valence band in solid neon has a pure 2p character. We can reasonably assume that the total spectral width W is equal to the extremal width of the 2p electron density curve of solid or fluid neon.

3. RESULTS AND DISCUSSION

The following metals were implanted with neon: beryllium, aluminium, titanium, chromium, iron, copper, zinc, molybdenum, silver, tungsten, gold, lead, bismuth. The fluences were calculated in order to obtain concentration profiles with a maximum peak value of 2 at.% as estimated from the total number of ions, neglecting sputtering. The X-ray results are presented in Fig. 3 where the spectra are plotted for different host metals and for decreasing bandwidth. The corresponding data are also given in Table 1. The total bandwidth W shows considerable variation, ranging from 1.96 eV in titanium to 4.07 eV in beryllium. A maximum error of 10% has been attributed to the deconvolution procedure and to experimental uncertainties. The widths reported in Table 1 should be compared with the 1.3 eV value which is the electron density width measured using photoelectrons [10] on solid neon at atmospheric pressure (and 24.48 K). The shapes of the different curves also show some resemblance to the curve shown for comparison in Fig. 3. Thus, we believe that, assuming a pressure broadening mechanism, neon is present in condensed form. The case for lead or bismuth is completely different. For these soft materials the spectrum is reduced to a single line exactly at the energy of the neon gas atomic X-ray line. Atomic neon is thus present, probably included into voids at a relatively low pressure. This observation is in favour of the hypothesis that above $T_m/2$ the bubbles have been able to coalesce and the pressure to drop (from Table 1, $T_m/2 = 300$ K for lead and $T_m/2 = 272$ K for bismuth).

Superimposed on the wide band are some elements (zinc, titanium, copper and silver) which show a strong and narrow line (shaded area), corresponding exactly to the atomic neon line. This line is attributed to the neon gas contained in low pressure cavities in metals. This atomic component is observed

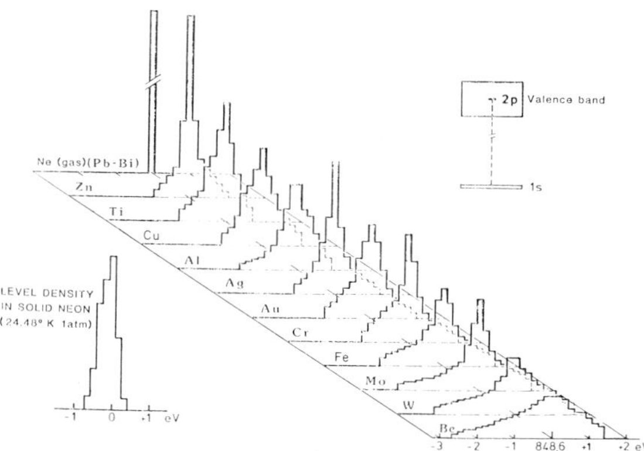

Fig. 3. X-ray spectra obtained by deconvolution of the $K\alpha_{12}$ X-ray yield curves from neon implanted into metals. The spectra are arranged according to decreasing width W. Lead and bismuth give a single line which coincides exactly with the neon gas line (atomic line). The electron density curve from solid neon at atmospheric pressure is represented for comparison.

TABLE 1

Average shear modulus [13] and $T_m/2$ values for a series of metals

Element	μ (GPa)	$T_m/2$ (K)	W (eV)
Be	150	775	4.07
Al	27.2	466	2.44
Ti	40.6	966	1.96
Cr	90.0	1065	2.61
Fe	84.7	904	3.10
Cu	46.4	678	2.12
Zn	37.9	346	1.95
Mo	122	1445	3.59
Ag	29.4	617	2.28
W	151	966	3.59
Au	28.2	669	2.44
Pb	5.7	300	—
Bi	13.1	272	—

The width W (eV) is the total width of the Ne $K\alpha_{12}$ X-ray spectrum excited by irradiation with 4 keV electrons (error, 10%).

in small proportions in all samples except beryllium.

All metals except lead and bismuth show the existence of a band spectrum which is wider than the low temperature solid neon valence band measured at 1 atm [10]. We attribute these bands to neon aggregates subjected to a very high pressure imposed by the metal matrix. The effect of this pressure is to reduce the volume of the atomic cell and to increase the interatomic overlap of valence electrons wavefunctions, hence causing valence level shift and broadening.

Solid or liquid neon at room temperature requires a very high equilibrium pressure. The *p–V* isotherms for neon at 293 K have been measured by Finger *et al.* [11] between 48.3 and 144 kbar. In this pressure range the unit cell size of solid neon decreases from 3.78 to 3.47 Å. Such compressions are probably responsible for the observed broadening. To our knowledge, there has been no valence band calculation aimed at reproducing such density-of-states widths. These calculations should take into account the interaction of the inert gas solid with the host metal since the crystallographic structure is determined by the metal lattice type [3–8].

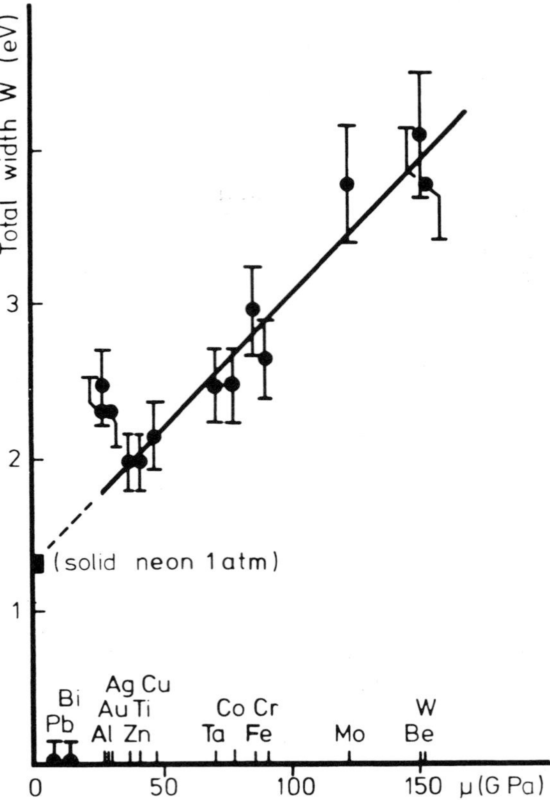

Fig. 4. The total width *W* is plotted against the shear modulus values [13]. The straight line was drawn in order to include the "zero" point value corresponding to the electron density curve value of solid neon at normal pressure. For metals with a shear modulus below this value (lead and bismuth), there is negligible width.

4. RELATION BETWEEN *W* AND THE SHEAR MODULUS

In all the proposed mechanisms the pressure required for growth is related to the metal bulk modulus [12] and essentially to the shear modulus [6]. A determination of the relaxation volume would provide a means of measuring the neon pressure [11]. However, the bandwidth *W* increases when the elementary cell size decreases. Hence *W* should be related to the shear modulus μ. For polycrystalline metals the shear modulus μ must be understood as an average over randomly distributed crystallographic directions; an extensive table of μ has been given by Köster and Franz [13] (Table 1). Of course, for individual microcrystals (inside which the bubbles grow) the stress–strain parameters depend on the orientation and on the symmetry (h.c.p., b.c.c. etc.). In Fig. 4 we plotted our data on the total width *W* (eV) against the shear modulus value μ given in ref. 13 and reported in Table 1. The total 2p bandwidth for solid neon at atmospheric pressure ($W = 1.3$ eV) is also given in this figure.

The results clearly indicate a linear correlation between *W* and μ. Moreover a straight line drawn through the points also extrapolates to the "zero" point value at 1.3 eV corresponding to solid neon at "zero" pressure. The first-order linear relation between *W* and μ can be written as

$$W = 0.017\mu + 1.3 \text{ eV} \tag{1}$$

where *W* is in electron volts and μ in gigapascals. The critical point for neon is 44 K; at room temperature, only the solid or the fluid phases can exist. The pressure threshold for solid neon is 48.3 kbar at room temperature [11]; according to the conclusions of ref. 6, this pressure would correspond to a total width of 2.6 eV (tantalum in Fig. 4). In fact the measurements of Vom Felde *et al.* [2] show that neon implanted into aluminium is in the fluid phase (well below tantalum in Fig. 4). These conclusions must be considered as preliminary. In particular the significance of W in its relation to extremal width of the electron density curve can be further refined (for relaxation effects, excimer states etc.). In any case, W can be considered as a new parameter for characterizing the physical state of neon bubbles in various host metals. More measurements made on neon-implanted beryllium at various fluences (the maximum concentration ranging between 0.5 and 10 at.%) indicate that the width W is independent of the neon concentration for low fluences on a given metal. These observations will be reported in detail in a further publication. The essential conclusions of this paper are that a quantitative relation, eqn. (1), between W and μ exists. The soft X-ray emission technique will remain a new tool for the investigation of neon gas implanted into metals.

ACKNOWLEDGMENTS

We are indebted to Professor A. Lucas, Dr. J. H. Evans and S. E. Donnelly for helpful discussions and comments on this subject and to the IISN for the financial support of this research.

REFERENCES

1 S. E. Donnelly, *Radiat. Eff.*, *90* (1985) 1-47.
A. A. Lucas, *Physica B*, *127* (1984) 225-239.
2 A. Vom Felde, J. Fink, Th.Müller-Heinzerling, J. Pflüger, B. Scheerer, G. Linker and D. Kaletta, *Phys. Rev. Lett.*, *53* (1984) 922.
3 C. Templier, R. J. Gaboriand and H. Garem, *Mater. Sci. Eng.*, *69* (1985) 63.
4 S. E. Donnelly and C. J. Rossouw, *Science*, to be published.
C. J. Rossouw and S. E. Donnelly, to be published.
5 J. H. Evans and D. J. Mazey, *J. Phys. F*, *15* (1985) L1-L6.
6 J. H. Evans and D. J. Mazey, *J. Nucl. Mater.*, *138* (1986) 176.
7 A. Luukkainen, J. Keinonen and M. Erola, *Phys. Rev. B*,*32* (1985) 4814.
8 J. H. Evans and D. J. Mazey, *Scr. Metall.*, *19* (1985) 621.
9 A. Lefebvre and G. Deconninck, *Nucl. Instrum. Methods B*,*15* (1986) 616-619.
10 N. Schwentner, F. J. Himpsel, V. Saile, M. Skibowski, W. Steinmann and E. Koch, *Phys. Rev. Lett.*, *34* (1975) 529.
11 L. W. Finger, R. M. Hazen, G. Zou, H. K. Mao and P. M. Bell, *Appl. Phys. Lett.*, *39* (1981) 892.
12 H. G. Haubold and J. S. Lin, *J. Nucl. Mater.*, *111-112* (1982) 709-714.
13 W. Köster and H. Franz, *Metall. Rev.*, *6* (1961) 1.

Characterization of Ion-beam-synthesized Boron Nitride Films*

AMARJIT SINGH[a], ROGER A. LESSARD[a] and EMILE J. KNYSTAUTAS[b]

[a] *Laboratoire de Recherches en Optique et Laser, Département de Physique, Université Laval, Quebec G1K 7P4 (Canada)*

[b] *Laboratoire de l'Accélérateur Van de Graaff, Département de Physique, Université Laval, Quebec G1K 7P4 (Canada)*

(Received July 10, 1986)

ABSTRACT

The formation of a boron nitride (BN) layer by nitrogen ion implantation into boron has been reinvestigated. Boron films were implanted with 30 keV nitrogen (N_2^+) ions to a fluence of 4.5×10^{17} ions cm^{-2} and were subsequently characterized for nitride formation by IR spectroscopy. Evidence of BN layer formation was confirmed by a strong absorption band at 1360 cm^{-1}, which is in good agreement with the known absorption band for bulk BN at 1368 cm^{-1}.

1. INTRODUCTION

The properties of the near-surface region can be modified by implanting suitable energetic ions, thereby bringing about changes in the surface composition and the properties of the material. This technique has been widely used to synthesize dielectric layers on semiconductors. Moreover, there are several beneficial effects of ion implantation on metallurgical properties [1] such as surface hardening and improvement in resistance to wear, corrosion and fatigue. Among several refractory materials, boron nitride (BN) appears to have some interesting properties, such as a high thermal conductivity, exceptional hardness and a low thermal expansion and chemical inertness at very high temperatures. These films have also been recently obtained by ion beam or ion-plating techniques [2, 3]. Very recently, Guzman *et al.* [4] have reported the formation of BN by nitrogen implantation and have characterized these films using Auger electron spectroscopy combined with argon ion etching. Since there is very little other work on the same subject, in this paper we report the reinvestigation of the formation of BN layers by nitrogen implantation. IR spectroscopy has been used to characterize the implanted films because of its extreme sensitivity to small changes in the vibrational state of the molecule and hence in its stoichiometry. Moreover, the technique is simple, inexpensive and most useful in identifying amorphous materials, when conventional techniques such as X-ray and electron diffraction have limitations. Furthermore, ion-beam-synthesized layers of SiO_2 [5], Si_3N_4, AlN [6] and TeO_2 [7] have been characterized by IR spectroscopy in recent years.

2. EXPERIMENT

Amorphous boron powder was compressed into a small disc and was evaporated from an electron-gun-heated tantalum crucible. Boron films 80 nm thick were deposited on compressed high purity KBr discs. During evaporation the pressure rose from 3×10^{-5} to 8×10^{-5} Torr. The films were then implanted with nitrogen ($^{14}N_2^+$) ions at 30 keV. Nitrogen ions, obtained from an r.f. source, were selected with a 30° analysing magnet and then accelerated to the desired energy. To maintain a clean vacuum, a turbomolecular pump and a liquid nitrogen trap were used at the target to maintain a pressure of 3×10^{-7} Torr. The current density at the target was 9–10 μA cm^{-2}. All IR spectra were obtained with a Beckmann double-beam IR spectrophotometer

*Paper presented at the International Conference on Surface Modification of Metals by Ion Beams, Kingston, Canada, July 7–11, 1986.

0025-5416/87/$3.50

model 4250A. For structural studies a Philips transmission electron microscope model EM-420 was used.

3. RESULTS AND DISCUSSION

To characterize the presence of IR bands, if any, pure amorphous boron powder was dispersed in KBr and compressed into a pellet. The IR scan of pure boron did not show any absorption band in the frequency range 4000–900 cm^{-1}. However, a similar IR scan of a boron film 80 nm thick evaporated on a KBr disc showed a weak absorption band at 1250 cm^{-1}, as indicated by curve B in Fig. 1. This is in contrast with the corresponding result for pure boron which does not have any absorption band in the IR region, thus suggesting that the film might have incorporated some oxygen atoms during evaporation. To establish the characteristic vibrational frequency of BN, an IR spectrum of bulk BN powder finely dispersed in a KBr disc was obtained. The spectrum indicated a broad well-defined absorption band at 1368 cm^{-1}, shown by Fig. 1, curve A. To observe the changes in the IR spectrum of boron films after nitrogen implantation, IR spectra of films implanted at 30 keV to a fluence ranging from 2.0×10^{17} to 4.5×10^{17} ions cm^{-2} were obtained. The IR spectrum after nitrogen implantation showed the presence of a broad absorption band in the frequency range 1500–1200 cm^{-1}, as indicated by Fig. 1, curve D. To eliminate the effect of absorption in boron films at 1250 cm^{-1} from the IR spectrum of nitrogen-implanted boron films, an IR spectrum was obtained by placing a boron film evaporated on a KBr disc in the reference beam. This resulted in an absorption band at 1360 cm^{-1} shown by Fig. 1, curve C, which is in good agreement with the absorption band at 1368 cm^{-1}. Moreover, it also establishes that the nitrogen implantation induces an absorption band which is very different in intensity and position from those in evaporated boron films.

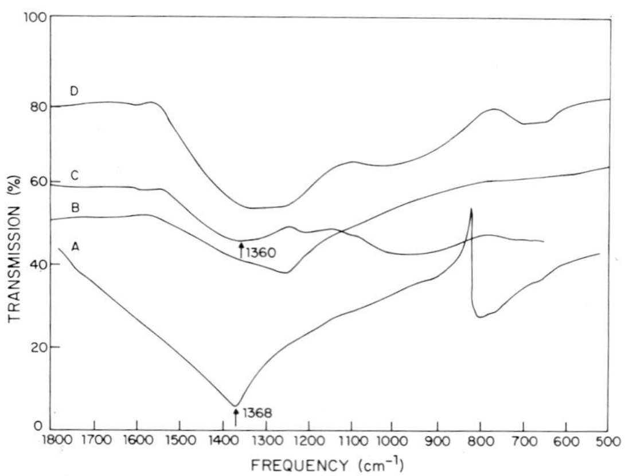

Fig. 1. The IR transmission spectra of bulk BN (curve A), electron-beam-evaporated boron films (curve B), a boron film implanted with $^{14}N_2^+$ ions at 30 keV to a dose level of 4.5×10^{17} ions cm^{-2} with a boron film in the reference beam (curve C) and a boron film after implantation at 30 keV with a dose of 4.5×10^{17} $^{14}N_2^+$ ions cm^{-2} (curve D).

The electron diffraction pattern of the electron-beam-deposited boron film consisted of broad diffuse rings characteristic of an amorphous nature, shown in Fig. 2(a). Furthermore, these films were found to be very homogeneous and featureless, as depicted by the electron micrograph in Fig. 2(b). The electron diffraction of the implanted film could not be obtained, probably because of its highly amorphous state (a faint diffuse spot could be seen). Figure 2(c) represents a typical electron micrograph of a film implanted with 4.5×10^{17} N_2^+ cm^{-2}. The effect of radiation damage resulting from a high current density of 9 μA cm^{-2} can be seen.

A well-defined absorption band observed after nitrogen implantation at 1360 cm^{-1} (Fig. 1, curve C) can be assigned to the formation of BN because of its close agreement with the characteristic band of bulk BN observed at 1368 cm^{-1} in Fig. 1, curve A. Moreover, this absorption band was obtained with an evaporated boron film placed in the reference beam so as to eliminate the effect of the vibrational nature of B—O bond from the nitrogen-implanted boron film. Thus, it is reasonable to believe that this absorption band at 1360 cm^{-1} is in fact BN. The IR bands in hexagonal BN and boron oxide (B_2O_3) have been reported by Brame *et al.* [8], while various other nitrides such as AlN and Si_3N_4 also have their characteristic band corresponding to the X—N stretching mode [9]. Similar experiments performed on silicon have revealed absorption bands at 800 cm^{-1} for Si_3N_4 formation after nitrogen implantation and at 1050 cm^{-1} for SiO_2 formation [5] after oxygen implantation, while AlN after nitrogen

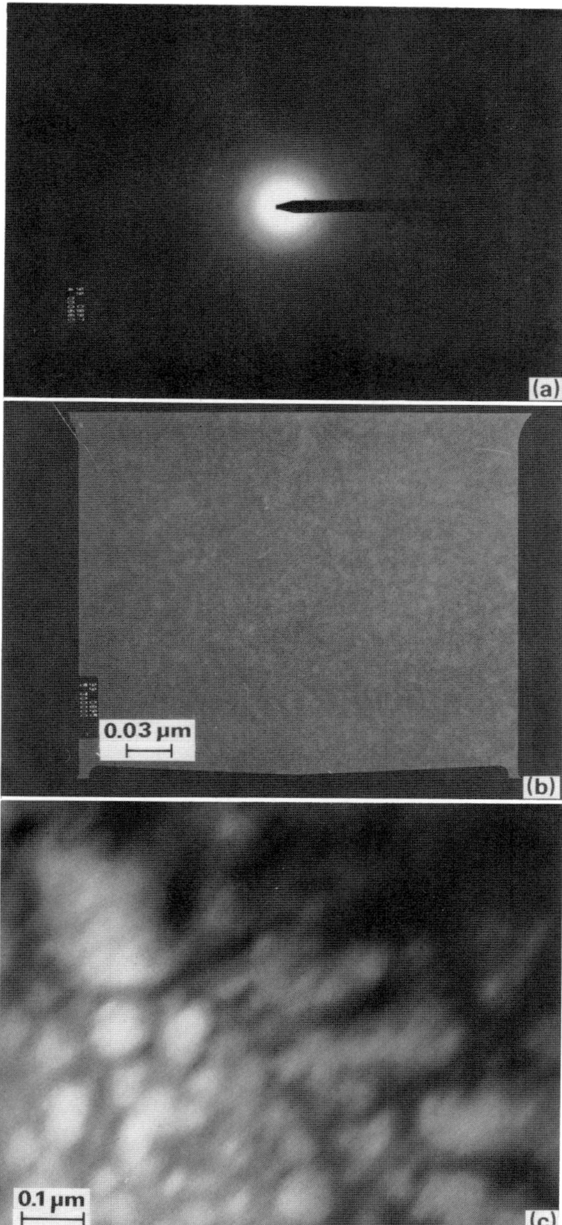

Fig. 2. (a) Electron diffraction of an evaporated boron film; (b) transmission electron micrograph of an as-deposited boron film; (c) transmission electron micrograph of a boron film after nitrogen implantation at 30 keV (dose, 4.5×10^{17} ions cm^{-2}).

implantation and TeO_2 formation after oxygen implantation have depicted the characteristic IR bands at 648 cm^{-1} and 600 cm^{-1} respectively [6, 7]. The band observed at 1360 cm^{-1} after nitrogen implantation corresponds to the B—N bending mode [10–12], while another weak band at 813 cm^{-1} (12.3 µm), shown in Fig. 1, curve A, is due to the B—N—B bending vibration [12]. Furthermore, this weak band (Fig. 1, curve D) after nitrogen implantation is observed to shift towards a lower frequency. Such a shift towards a lower frequency has been commonly observed in thin films because of amorphization due to implantation. Suitable annealing tends to restore the band towards a higher frequency.

The evaporated boron films also show an absorption band at 1250 cm^{-1} (Fig. 1, curve B) and this frequency is very close to the B—O stretching mode observed at 1200 cm^{-1} in B_2O_3 [8]. Thus, this band suggests the presence of oxygen bonded with a boron atom in evaporated films. It is very likely that the film might have absorbed some oxygen during evaporation because of the observed increase in pressure from 2×10^{-5} to 8×10^{-5} Torr. The microanalysis of these BN films obtained by other deposition techniques such as plasma, ion beam or ion-plating techniques has revealed the presence of oxygen and carbon in significant amounts [2, 3]. Thus the band at 1250 cm^{-1} observed for boron films can be assigned to the presence of oxygen atoms. Similar IR results have been obtained for BN films prepared by activated reactive evaporation [13, 14]; these films showed IR bands at 6.8 and 11.5 µm. However, it is believed in the present case that starting with a thicker boron film (about 0.5 µm) and a greater dose (about 1.0×10^{18} ions cm^{-2}) could lead to a significantly thicker BN film (about 0.15 µm), hence producing relatively higher intensity absorption bands at 6.8 and 11.5 µm.

Now, on the assumption that the nitrogen concentration necessary to form BN depends on the concentration of boron atoms in the bulk, which is 1.3×10^{23} atoms cm^{-3}, a concentration of at least 10^{23} atoms cm^{-3} or more will be required to convert boron to BN. The peak concentration of the implant can be calculated from

$$N_p = \frac{\phi}{(2\pi)^{1/2}\,\Delta R_p}$$

where N_p (atoms cm^{-3}) is the peak concentration, ϕ (ions cm^{-2}) the fluence and ΔR_p (cm) the straggling, which can be determined from range theory [15] or from depth profiles of the implant.

Thus for $\phi = 4.5 \times 10^{17}$ ions cm^{-2} and $\Delta R_p = 20$ nm the peak concentration is

$$N_p = 0.9 \times 10^{23} \text{ atoms cm}^{-3}$$

which is much less than the concentration of 1.3×10^{23} ions cm^{-3} but is sufficient for the onset of the formation of B—N bonds which can be observed in the IR spectrum.

4. CONCLUSION

From the present study, it can be concluded that BN films can be synthesized by nitrogen implantation. It has been shown that a typical dose of 4.5×10^{17} N_2^+ cm^{-2} at 30 keV into boron films tends to exhibit a characteristic IR band at 1360 cm^{-1} which is in good agreement with the band at 1368 cm^{-1} for bulk BN.

REFERENCES

1 G. Dearnaley, *J. Met., 35* (1982) 18.
2 C. Weissmantel, *J. Vac. Sci. Technol., 18* (1981) 179.
3 S. Shanfield and R. Wolfson, *J. Vac. Sci. Technol., 41* (1983) 323.
4 L. Guzman, F. Marchetti, L. Calliari, I. Scotoni and F. Ferrari, *Thin Solid Films, 117* (1984) L66.
5 S. S. Gill and J. H. Wilson, *Thin Solid Films, 55* (1978) 435.
6 A. Singh, R. A. Lessard and E. J. Knystautas, *Thin Solid Films, 138* (1986) 79.
7 A. Singh, E. J. Knystautas and R. Lapointe, *Appl. Surf. Sci., 22-23* (1985) 681.
8 E. G. Brame, J. L. Margrave and V. W. Meloche, *J. Inorg. Nucl. Chem., 5* (1957) 48.
9 R. A. Nyquist and R. A. Kagel, *Infrared Spectra of Inorganic Compounds*, Academic Press, New York, 1973.
10 T. Takahashi, H. Itoh and A. Takeuchi, *J. Cryst. Growth, 47* (1979) 245.
11 M. J. Rand and J. F. Roberts, *J. Electrochem. Soc., 115* (1968) 423.
12 A. C. Adams and C. D. Capio, *J. Electrochem. Soc., 127* (1980) 399.
13 K. L. Chopra, V. Agarwal, V. D. Vankar, C. V. Deshpandey and R. F. Bunshah, *Thin Solid Films, 126* (1985) 309.
14 T. H. Yuzuriha and D. W. Hess, *Thin Solid Films, 140* (1986) 199.
15 J. P. Gibbons, W. S. Johnson and S. W. Mylroie, *Projected Range Statistics in Semi-conductors*, Dowden, Hutchinson and Ross, Stroudsburg, PA, 1975.

The Influence of Cerium, Yttrium and Lanthanum Ion Implantation on the Oxidation Behaviour of a 20Cr–25Ni–Nb Stainless Steel in Carbon Dioxide at 900–1050°C*

M. J. BENNETT, H. E. BISHOP, P. R. CHALKER and A. T. TUSON

Materials Development Division, Atomic Energy Research Establishment, Harwell, Didcot, Oxon. OX11 0RA (U.K.)

(Received July 10, 1986)

ABSTRACT

The influence of the implantation of 10^{17} Ce^+ cm^{-2}, 10^{17} La^+ cm^{-2} and 10^{17} Y^+ cm^{-2} on the oxidation behaviour of 20Cr-25Ni-Nb stainless steel (where the composition is in weight per cent) in carbon dioxide has been examined at temperatures between 900 and 1050 °C. The oxidation kinetics during isothermal exposures of about 250 h duration were followed using a controlled-atmosphere microbalance, and spallation during subsequent cooling to room temperature was assessed. The composition and microstructure of the multicomponent scales formed were examined by a range of surface-analytical techniques, with the location of the implanted atoms being established by dynamic secondary ion mass spectrometry. The oxidation of the 20Cr-25Ni-Nb stainless steel over this temperature range is characterized by the transition from protective- to non-protective-type behaviour with scale breakdown due to the occurrence of cracking during isothermal oxidation and due to spallation on thermal cycling. By incorporation within the oxide scale, and in particular by segregation to the oxide grain boundaries, the implanted reactive elements extended the temperature range of protective-type behaviour to 1000 °C or less, depending on the element, by modifying the initial scale growth mechanism and also the resultant scale microstructure and mechanical properties. Implantation had no significant beneficial influence on the oxidation behaviour of the steel at 1050 °C.

*Paper presented at the International Conference on Surface Modification of Metals by Ion Beams, Kingston, Canada, July 7-11, 1986.

1. INTRODUCTION

A series of studies [1-7] during the last decade has concerned the influence of the implantation of various ions on the oxidation behaviour of 20wt.%Cr-25wt.%Ni-0.9wt.%-Mn-0.6wt.%Si-Nb (hereafter referred to as 20Cr-25Ni-Nb) stainless steel in carbon dioxide. The main focus has centred on the beneficial role of the so-called reactive elements, such as yttrium and rare earths (cerium, europium, ytterbium and lanthanum), one of whose important attributes is having a greater oxygen affinity than the steel constituents do. These studies have provided unique mechanistic insight into the reactive element effect, which is of crucial relevance to the behaviour of many technologically important alloys whose corrosion protection also derives from Cr_2O_3-based oxide scales.

Most of this work has been undertaken at 825 °C, where oxidation of the 20Cr-25Ni-Nb stainless steel results in the development, primarily by outward cation movement, of a multicomponent uniform protective scale, consisting of an outer (Fe, Mn, Cr) spinel, a middle Cr_2O_3 layer and an inner SiO_2 layer delineating the interface with the steel [8]. On thermal cycling to room temperature, and provided that the scale is sufficiently thick [9], discrete scale areas spall, with the primary plane of fracture being either within the SiO_2 layer or at the SiO_2-steel interface [10]. Spallation involves both cracking through the scale thickness and localized scale decohesion. This latter process is believed to control spallation of the uniform protective scale, which will occur if the strain energy within the scale equates to or exceeds the interfacial fracture energy [9]. The primary driving force for

0025-5416/87/$3.50

scale rupture is the biaxial compressive stress developed on cooling, as a result of the differential contraction between the scale and the 20Cr–25Ni–Nb stainless steel. On subsequent re-oxidation, the regions of the steel exposed, now depleted in chromium, undergo pitting-type attack characterized by the external development of iron-rich spinel oxides and internal oxidation, until slowed by the formation of a Cr_2O_3 healing layer at the pit base [10, 11].

Despite the original shallow depth (typically 500–1000 Å) of implantation into the 20Cr–25Ni–Nb stainless steel, the implanted yttrium and rare earth atoms significantly improved its oxidation and spallation resistance during extended exposures of at least 7 kh duration, in carbon dioxide, at the same temperature. Initially (less than 1 h) the rate of transient oxidation was enhanced slightly by the implanted rare earth ions but not by yttrium, which reduced it [4, 12]. Thereafter the rate of attack of the unimplanted steel gradually exceeded those for the steel implanted with rare earths, such that after 7 kh all the reactive elements reduced the extent of oxidation of the 20Cr–25Ni–Nb stainless steel by a factor of 3.

The oxidation and segregation of the implanted atoms as oxide (*e.g.* CeO_2 and Y_2O_3) grains modified the mechanism for the growth of the initially formed (Fe, Cr, (Mn, Ni)) spinel layer by enhancing inward oxidant movement. This in turn promoted the earlier completion of the transient oxidation stage, with the formation of a thinner underlying Cr_2O_3 layer than on unimplanted 20Cr–25Ni–Nb stainless steel. The reactive element oxide grains became located within the spinel near the interface with the Cr_2O_3 layer. These grains, or more probably reactive element atoms segregated from the oxide grain along adjoining spinel oxide grain boundaries [13], subsequently reduced the outward grain boundary diffusion of cations responsible for continuing growth of the multicomponent scale. The extent of the overall attack was reduced also as a consequence of the maintenance of scale adhesion and, thereby, the avoidance of pitting attack.

Although several possible mechanisms for the improved adhesion (such as the formation of a graded-seal interlayer between the scale and steel, or vacancy condensation on the reactive element or oxide rather than at the scale–steel interface) can be eliminated, the underlying cause has yet to be fully revealed. However, it undoubtedly originates from the scale microstructural changes resulting from the incorporation of the implanted atoms, possibly modifying either the growth stresses or the scale mechanical properties, which thereby enable the scale to deform rather than to crack during thermal cycling. There is some evidence [7] to suggest that another possible consequence of the changes in growth mechanism is the manner in which silicon in the implanted steel oxidizes, as a continuous SiO_2 layer does not appear to be formed at the scale–steel interface to provide a plane of weakness for scale fracture.

Recent work [14] indicated that the oxidation of the 20Cr–25Ni–Nb stainless steel in carbon dioxide at higher temperatures becomes characterized increasingly by non-protective-type behaviour. The influences of the ion implantation of cerium, yttrium and lanthanum are now being studied over the temperature range 900–1050 °C and the initial results obtained are reported in this paper.

2. EXPERIMENTAL DETAILS

2.1. *Material*

By analysis the 20Cr–25Ni–Nb stainless steel contained 19.9 wt.% Cr, 22.6 wt.% Ni, 0.9 wt.% Mn, 0.7 wt.% Nb, 0.6 wt.% Si and 0.040 wt.% C, with the balance being iron. The specimens were in the form of coupons, 10 mm × 10 mm × 500 μm. Prior to implantation they were abraded to a 1 μm diamond surface finish, degreased and finally annealed in tantalum-gettered hydrogen, for 30 min, at 930 °C.

2.2. *Implantation*

The uniform implantation of cerium, yttrium, lanthanum and krypton, at a nominal fluence of 10^{17} ions cm^{-2}, into the principal coupon faces was undertaken with the Harwell Cockcroft–Walton accelerator, using a 300 keV accelerating potential. Previous [4] secondary ion mass spectrometry (SIMS) and X-ray photoelectron spectroscopy analyses of the cerium and yttrium distributions implanted into the 20Cr–25Ni–Nb stainless steel agreed well with those predicted by the Lindhard–

Scharff–Schiøtt [15] procedure. The distributions were essentially gaussian, with a maximum reactive element concentration of 20–40 at.% at a depth of 100–500 Å. The implantation depths ranged between 1000 and 1600 Å. The implantation of krypton ions, having a comparable mass and thus energy with those of the reactive elements, was studied in order to assess possible physical effects of implantation, as krypton would not be expected to exercise any chemical influence on the oxidation behaviour of the steel.

2.3. *Oxidation and scale examination*

The implanted and unimplanted 20Cr–25Ni–Nb stainless steel specimens were oxidized isothermally at 900, 950, 1000 and 1050 °C in carbon dioxide, for periods of about 250 h, using a CI Electronics controlled-atmosphere microbalance. The specimens were contained in SiO_2 crucibles so that any scale which spalled at the completion of oxidation on cooling to room temperature would be retained for gravimetric measurement.

The scales were examined by a variety of surface-analytical techniques, including optical and scanning electron microscopy, X-ray diffraction and dynamic SIMS. The latter technique was used primarily to locate the implanted atoms within the scales.

3. RESULTS

The variations with time of the extents of oxidation (expressed as mass gain per square centimetre) of implanted and unimplanted 20Cr–25Ni–Nb stainless steel during isothermal oxidation for about 250 h, in carbon dioxide, at 900 and 950 °C are shown in Fig. 1 and at 1000 and 1050 °C in Fig. 2. At 900 and 950 °C, after times which decreased with increasing temperature, a transient occurred in the kinetics of the unimplanted 20Cr–25Ni–Nb stainless steel, such that the oxidation rate increased sharply but then decreased gradually to a slightly higher rate than that before

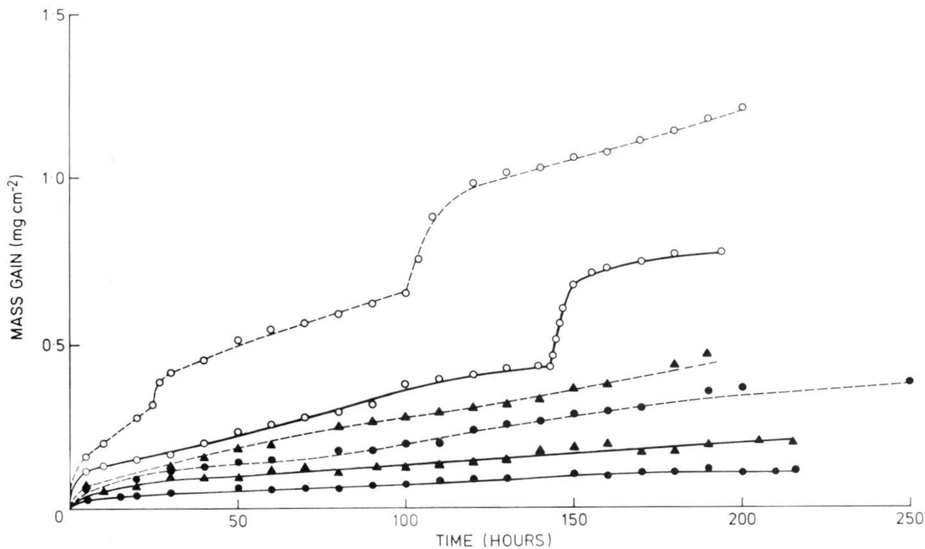

Fig. 1. Influence of the implantation of 10^{17} Ce^{+} cm^{-2} or 10^{17} Y^{+} cm^{-2} on the oxidation of 20Cr–25Ni–Nb stainless steel in carbon dioxide at 900 and 950 °C.

Symbol	*Ion implanted*	*Temperature* (°C)
—○—	None	900
- -○- -	None	950
—▲—	Ce^{+}	900
- -▲- -	Ce^{+}	950
—●—	Y^{+}	900
- -●- -	Y^{+}	950

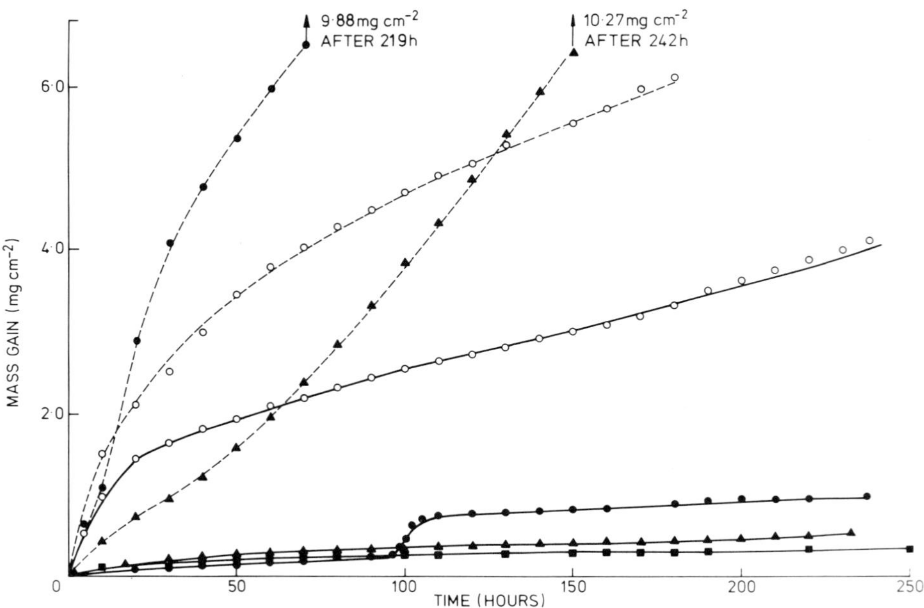

Fig. 2. Influence of the implantation of 10^{17} Ce^+ cm^{-2}, 10^{17} Y^+ cm^{-2} or 10^{17} La^+ cm^{-2} on the oxidation of 20Cr–25Ni–Nb stainless steel in carbon dioxide at 1000 and 1050 °C.

Symbol	*Ion implanted*	*Temperature* (°C)
—○—	None	1000
- -○- -	None	1050
—▲—	Ce^+	1000
- -▲- -	Ce^+	1050
—●—	Y^+	1000
- -●- -	Y^+	1050
—■—	La^+	1000

the perturbation. Cerium and yttrium ion implantation reduced significantly the rate of oxidation of the 20Cr–25Ni–Nb stainless steel at 900–1000 °C, as also did that of lanthanum implantation at 1000 °C. The overall effect after about 250 h was to reduce the extents of attack by factors of between 4 and 8. Again the kinetics were essentially parabolic but transient scale-type breakdown, resulting in an enhanced oxidation rate, was apparent during the oxidation of the yttrium-implanted steel at 1000 °C and possibly also at 950 °C. At the highest temperature (1050 °C), yttrium and cerium ion implantation increased the oxidation rate, on the former from the outset and on the latter after an initial period (about 50 h) of improved protection. By comparison with the reactive elements, the implantation of 10^{17} Kr^+ cm^{-2} had little influence on the oxidation kinetics of the 20Cr–25Ni–Nb stainless steel at 900–1000 °C (Fig. 3).

Cooling to room temperature after the completion of oxidation caused spallation of scale formed on 20Cr–25Ni–Nb stainless steel, whose extent increased with increasing temperature up to 1000 °C but was considerably smaller at 1050 °C (Table 1). There was a similar pattern to the temperature dependence of the influence of all the implanted reactive elements, namely prevention at lower temperatures, decreasing inhibition at intermediate temperatures and maintenance of scale adhesion at 1050 °C. Cerium and yttrium ion implantation became less effective at 950 °C and 1000 °C respectively. In contrast, krypton ion implantation did not exert any significant influence on scale adhesion.

Since the maximum X-ray penetration depth was about 10 μm, the majority of the scales were completely examined by X-ray diffraction. However, only components of the outer regions of some scales formed at

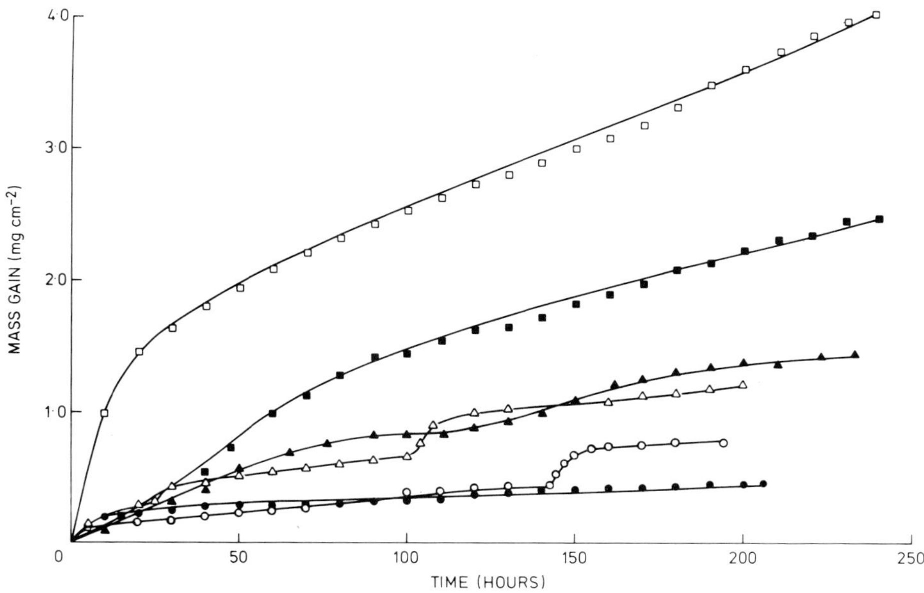

Fig. 3. Influence of the implantation of 10^{17} Kr^+ cm^{-2} on the oxidation of 20Cr–25Ni–Nb stainless steel in carbon dioxide at 900–1000 °C.

Symbol	*Ion implanted*	*Temperature* (°C)
○	None	900
△	None	950
□	None	1000
●	Kr^+	900
▲	Kr^+	950
■	Kr^+	1000

TABLE 1

Spallation on cooling to room temperature of oxide formed on unimplanted and implanted 20Cr–25Ni–Nb stainless steel during oxidation in carbon dioxide at 900–1050 °C

Temperature (°C)	*Unimplanted*			*Implanted with 10^{17} Kr^+ cm^{-2}*			*Implanted with 10^{17} Ce^+ cm^{-2}*			*Implanted with 10^{17} Y^+ cm^{-2}*		
	Time (h)	*Mass gain* (mg cm^{-2})	*Mass of spall* (mg cm^{-2})	*Time* (h)	*Mass gain* (mg cm^{-2})	*Mass of spall* (mg cm^{-2})	*Time* (h)	*Mass gain* (mg cm^{-2})	*Mass of spall* (mg cm^{-2})	*Time* (h)	*Mass gain* (mg cm^{-2})	*Mass of spall* (mg cm^{-2})
900	194	0.78	0.44	216	0.46	0.23	216	0.19	—	216	0.11	—
950	216	1.20	0.64	234	1.44	1.11	190	0.47	0.14	193	0.33	0.01
1000	238	4.03	0.70	240	2.47	1.10	233	0.54	0.12	237	0.99	0.68
1050	191	6.12	0.21				242	10.27	—	219	9.88	0.02

the higher temperatures were identified. The main constituents of all scales (Table 2) were a rhombohedral phase and spinel oxides. The former was Cr_2O_3, except on yttrium-implanted 20Cr–25Ni–Nb stainless steel oxidized at 1050 °C when Fe_2O_3 formed part of the outer scale layer. Two spinel oxides, having a larger and smaller lattice parameter, were always present in the scales that developed on the 20Cr–25Ni–Nb stainless steel. The former

TABLE 2

X-ray diffraction of oxide scales formed on unimplanted and implanted 20Cr–25Ni–Nb stainless steel during oxidation in carbon dioxide at 900–1050 °C

Temperature (°C)	*Unimplanted*			*Implanted with 10^{17} Ce^+ cm^{-2}*			*Implanted with 10^{17} Y^+ cm^{-2}*		
	Scale constituents[a]	*Spinel lattice parameter* (Å)	*Peak ratio*[b]	*Scale constituents*[a]	*Spinel lattice parameter* (Å)	*Peak ratio*[b]	*Scale constituents*[a]	*Spinel lattice parameter* (Å)	*Peak ratio*[b]
900	R $S_1 + S_3$	8.39–8.50	0.1	R S_1 CeO_2	8.51 ± 0.05	2.17	R S_1	8.53 ± 0.02	0.96
950	R $S_1 + S_2$	8.38 ± 0.02 8.46 ± 0.02	0.73	R S_1 CeO_2	8.46 ± 0.02	1.47	R S_1	8.438 ± 0.005	0.57
1000	R $S_1 + S_3$ α-quartz	8.49 ± 0.02	0.8	R S_1 CeO_2	8.421 ± 0.008	0.44	R S_1	8.393 ± 0.008	0.93
1050	R $S_1 + S_2$	8.41 ± 0.05 8.49 ± 0.01	6.97	R $S_1 + S_2$ CeO_2	8.34 ± 0.04 8.29 ± 0.04 (strain or lattice dislocation indicated)	NM[c]	R (near Fe_2O_3) S_1	8.367 ± 0.003	Strong spinel pattern

[a]R, rhombohedral phase, Cr_2O_3 unless stated otherwise; S_1 and S_2, spinels with well-resolved lattice parameters; S_3, spinels with a smaller but unresolvable lattice parameter.

[b]Peak ratio given by

$$\frac{\text{peak height} \times \text{full-width half-maximum, spinel (220) peak}}{\text{peak height} \times \text{full-width half-maximum, rhombohedral (104) peak}}$$

[c]NM, not measurable.

was probably $Mn(Fe, Cr)_2O_4$ and the latter $Fe(Fe_{2-x}Cr_x)O_4$. Only a single spinel with the larger lattice parameter was apparent in the corresponding scales on the implanted 20Cr–25Ni–Nb stainless steel, except on cerium-implanted steel oxidized at 1050 °C where again two spinels could be distinguished. The lattice parameters of the spinel constituents of the implanted 20Cr–25Ni–Nb steel scales decreased with increasing temperature, reflecting changes in spinel composition which possibly arose from iron incorporation, and thereby increasing strain in the scale and/or lattice distortion. There was a higher proportion of the spinel to the rhombohedral constituent in the scales produced on the implanted than on the unimplanted 20Cr–25Ni–Nb steel, although this difference decreased with increasing temperature. The implanted cerium was present as CeO_2 in the scales formed at all four temperatures.

SIMS profiles, using a 10.5 kv primary argon ion (Ar^+) beam, through cerium and yttrium-implanted 20Cr–25Ni–Nb stainless steel oxidized at 900 °C are given in Figs. 4 and 5. These indicate that the variations in the steel constituents were similar to those in the uniform protective-type scale which developed at this temperature on the unimplanted steel [14]. Both the cerium and the yttrium concentrations increased from the outer surface to a maximum within the spinel layer near the interface with the underlying Cr_2O_3 layer. The silicon was enriched, presumably as SiO_2, at the scale–steel interface. The broader peak does not necessarily indicate a thicker layer than the remainder of the scale, as it reflected the different sputtering rates of oxide and steel as the argon ion beam crossed that interface. With increasing temperature up to 1000 °C, and as a consequence also of scale thickness, the cerium and yttrium profiles broad-

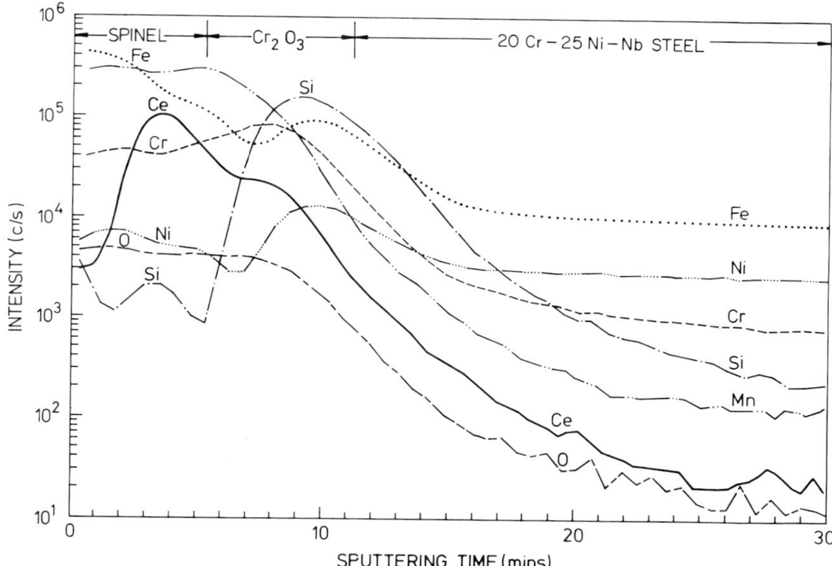

Fig. 4. SIMS profile through the oxide scale formed on 20Cr–25Ni–Nb stainless steel implanted with 10^{17} Ce^+ cm^{-2} after oxidation for 216 h in carbon dioxide at 900 °C.

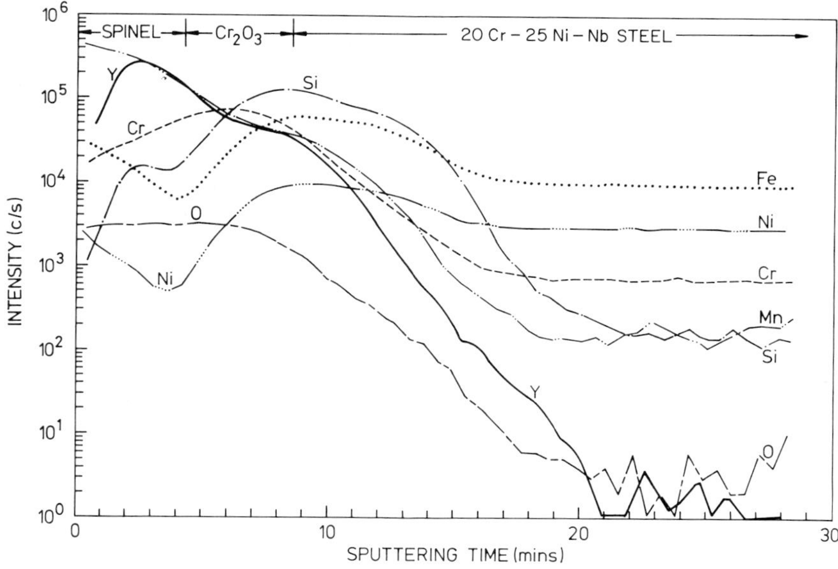

Fig. 5. SIMS profile through the oxide scale formed on 20Cr–25Ni–Nb stainless steel implanted with 10^{17} Y^+ cm^{-2} after oxidation for 216 h in carbon dioxide at 900 °C.

ened, while the maximum concentration decreased (Figs. 6 and 7). The position of the maximum concentration was maintained within the spinel layer with the possible exception of cerium, which appeared to be distributed more uniformly throughout the scale formed at 1000 °C. Other features to emerge with increasing temperature, and not shown in Figs. 6 and 7, were the broadening of the silicon and niobium distributions at the scale–steel interface and an increase in the carbon concentration in the scale, which decreased from the gas interface to peak with that of silicon. Lanthanum was located in a similar position to yttrium in the scale formed at 1000 °C (Fig. 8).

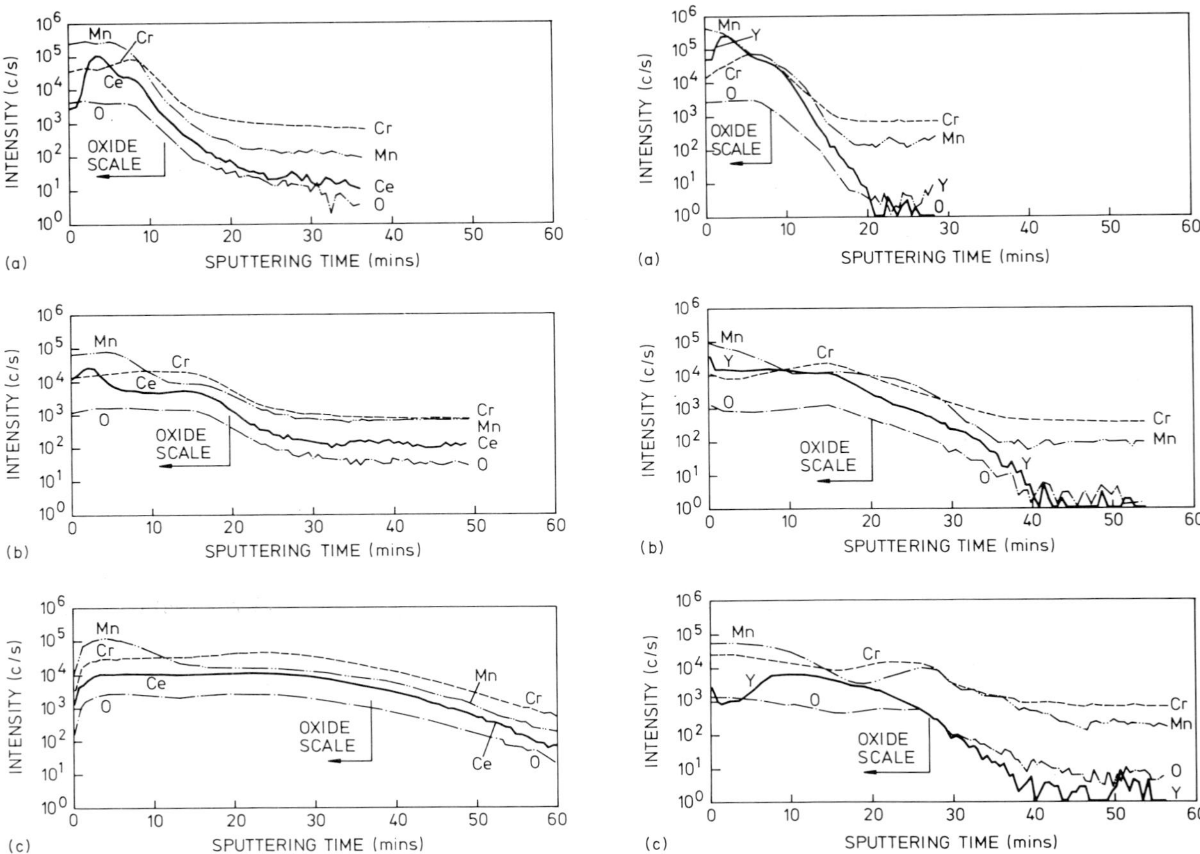

Fig. 6. Variation in SIMS profiles through the oxide scales formed on 20Cr–25Ni–Nb stainless steel implanted with 10^{17} Ce^{+} cm^{-2} after oxidation in carbon dioxide: (a) 216 h, 900 °C; (b) 190 h, 950 °C; (c) 233 h, 1000 °C.

Fig. 7. Variation in SIMS profiles through the oxide scales formed on 20Cr–25Ni–Nb stainless steel implanted with 10^{17} Y^{+} cm^{-2} after oxidation in carbon dioxide: (a) 216 h, 900 °C; (b) 193 h, 950 °C; (c) 237 h, 1000 °C.

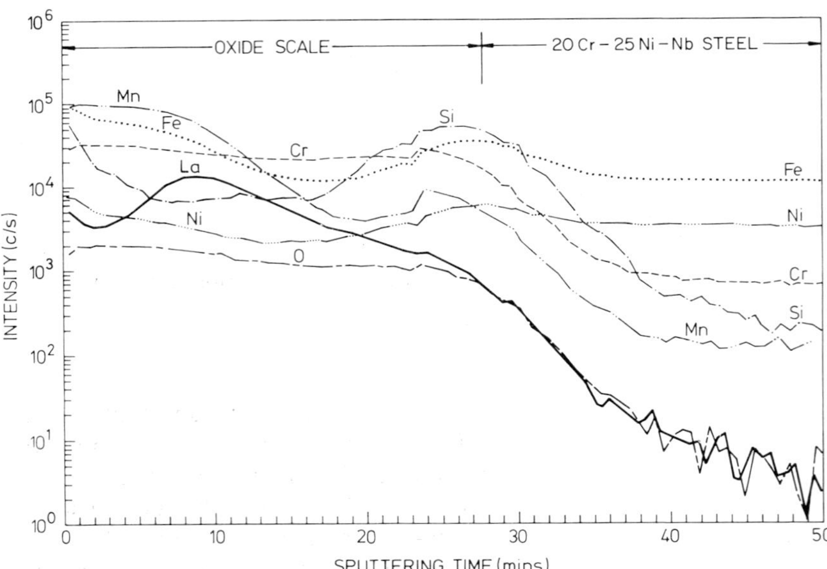

Fig. 8. SIMS profile through the oxide scale formed on 20Cr–25Ni–Nb stainless steel implanted with 10^{17} La^{+} cm^{-2} after oxidation for 311 h in carbon dioxide at 1000 °C.

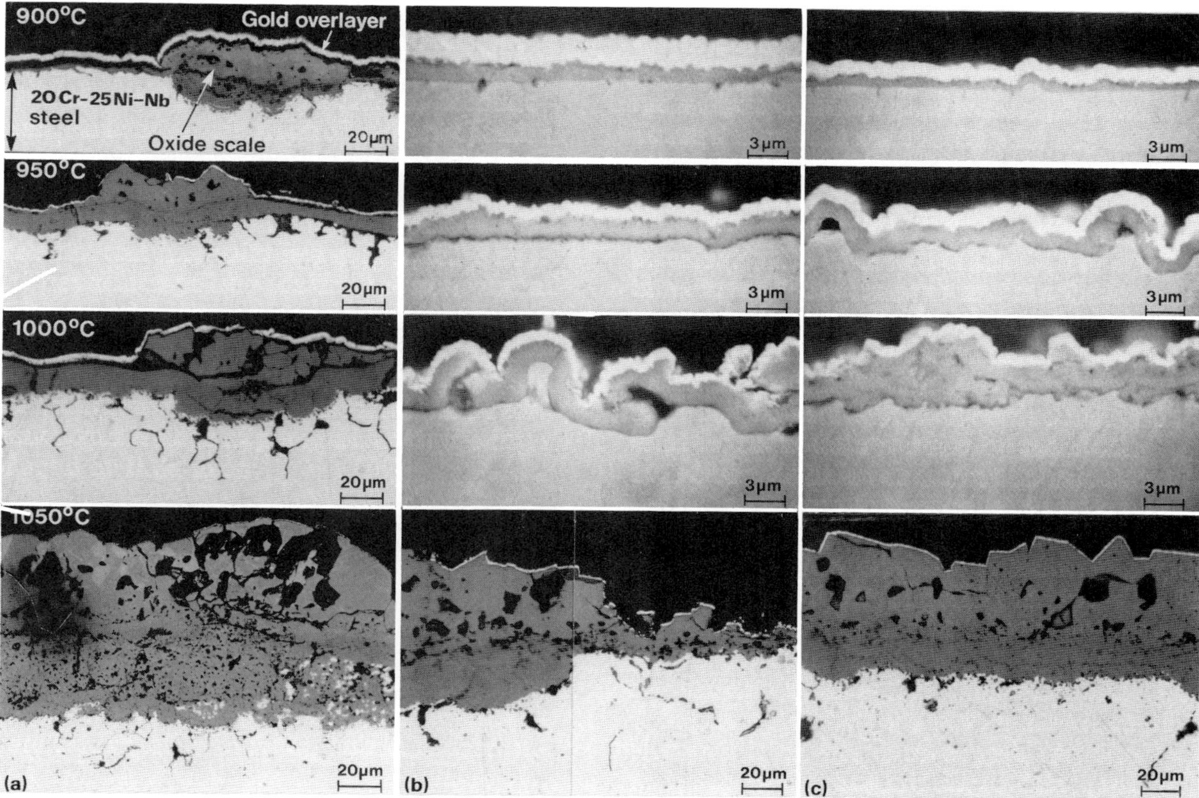

Fig. 9. Micrographs of cross-sections of oxide scales formed on (a) unimplanted, (b) cerium-implanted and (c) yttrium-implanted 20Cr–25Ni–Nb stainless steel during oxidation for 250 h in carbon dioxide at 900–1050 °C.

Micrographs of scales prepared by standard metallographic techniques are shown in Fig. 9. Before mounting, gold was sputtered onto the outer oxide scale surface to retain the oxide and to define the oxide–gas interface. The optical micrographs of the scales on the unimplanted steel at all temperatures and on the implanted steels at 1050 °C are at a lower magnification than the scanning electron micrographs of the scales formed on the implanted 20Cr–25Ni–Nb stainless steel at 900–1000 °C. On the unimplanted steel oxidized at 900 °C, a thin uniform protective-type scale is shown adjacent to a region where its fracture resulted in pitting-type attack. With increasing temperature the scales on the 20-Cr–25Ni–Nb stainless steel were characterized by more severe internal attack, in both depth and coverage, with an associated development intergranularly of SiO_2 intrusions. At 1050 °C the steel was entirely covered with a duplex scale. Thinner uniform protective-type scales formed at 900 °C on the cerium- and yttrium-implanted steel than on the unimplanted steel. With increasing temperature the scales became detached locally from the steel and formed convolutions between which steel became enveloped; finally the apex of the scale loops cracked to initiate internal pitting-type attack. The appearance of buckled scales and of scale breakdown, giving rise to a transient oxidation kinetic excursion, both appeared to occur at slightly lower temperatures (by about 50 °C) on the yttrium- than on the cerium-implanted 20Cr–25Ni–Nb stainless steel. On the latter at 1050 °C, regions of thinner protective-type scale were interspersed with the thicker duplex scale, whereas the yttrium-implanted 20Cr–25Ni–Nb steel was uniformly covered with the latter type of scale. SiO_2 intrusions were also formed on both implanted steels at this temperature.

A complementary picture emerged from scanning electron microscopy examination of the oxidized surfaces. After oxidation of the 20Cr–25Ni–Nb stainless steel at 900 °C (Fig.

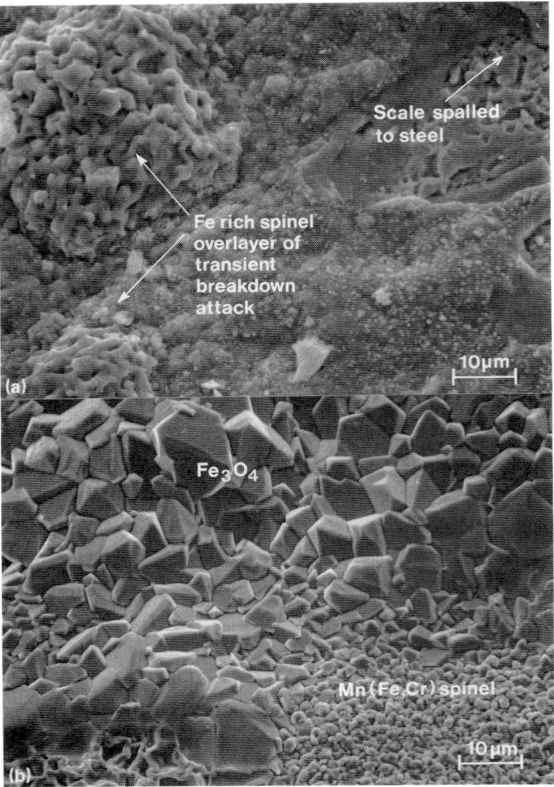

Fig. 10. Scanning electron micrographs of 20Cr–25Ni–Nb stainless steel surface after oxidation in carbon dioxide (a) for 194 h at 900 °C and (b) for 191 h at 1050 °C.

Fig. 11. Scanning electron micrographs of implanted 20Cr–25Ni–Nb stainless steel surfaces following oxidation in carbon dioxide: (a) 10^{17} Y^+ cm^{-2}, 216 h, 900 °C; (b) 10^{17} Y^+ cm^{-2}, 193 h, 950 °C; (c) 10^{17} La^+ cm^{-2}, 311 h, 1000 °C; (d) 10^{17} Ce^+ cm^{-2}, 242 h, 1050 °C.

10(a)) the outer Mn, (Fe, Cr) spinel layer identified the original protective-type scale while, at sites where it failed, iron-rich spinel mounds developed. On cooling to room temperature, discrete areas of the uniform scale spalled largely to the underlying steel but at the edges with fracture through the scale at the spinel–Cr_2O_3 layer interface. At 1050 °C (Fig. 10(b)) the scale surface was characterized by regions of large interlocking Fe_3O_4 crystals and of smaller Mn, (Fe, Cr) spinel crystallites. The outer surface of the implanted steel surfaces oxidized at 900 °C was covered with granular Mn, (Fe, Cr) spinel grains with, particularly on the yttrium-implanted steel, arrays of thicker, and possibly convoluted, grains (Fig. 11(a)). These features became more pronounced at 950 °C (Fig. 11(b)). On both the cerium- and the lanthanum-implanted steel surface oxidized at 1000 °C, cracks were apparent along the apex of some of the convoluted scale protrusions (Fig. 11(c)), which may also have been the sites for spallation. The yttrium-implanted 20Cr–25Ni–Nb steel surface oxidized at 1000 °C was covered with small iron-rich spinel oxide crystallites, which at 1050 °C had grown into larger Fe_3O_4 crystals like those found on unimplanted 20Cr–25Ni–Nb after oxidation at 1050 °C (Fig. 10(b)). In agreement with the optical micrographs, two types of scale morphology were apparent on the cerium-implanted 20Cr–25Ni–Nb steel oxidized at 1050 °C: small Mn (Fe, Cr) spinel crystallites and massive Fe_3O_4 crystals (Fig. 11(d)).

4. DISCUSSION

Over the temperature range of this study (900–1050 °C) the oxidation behaviour of the 20Cr–25Ni–Nb stainless steel was characterized by a progressive change from essentially protective- to non-protective-type attack. At 900 °C a uniform protective scale developed on the steel initially in a similar manner to that at lower temperatures, such as 825 °C [8]. The significant difference was that, whereas, at the lower temperatures, scale protection was lost primarily by spalling on thermal cycling, at 900 °C or higher it also occurred during isothermal exposure as a result of cracking [16]. The sequence of the mechanisms involved was firstly the localized detachment of regions of the scale from the steel, possibly because of differential strain between the growing oxide and the underlying steel and then the continued growth of the scale in the form of convolutions, thicker at the base and thinner at the apex, along which cracking eventually occurred. The buckled scale regions did not spall during isothermal oxidation. Access of the environment to the chromium-depleted steel beneath the scale convolution resulted in pitting-type attack until restrained by the formation of a Cr_2O_3 healing layer at the pit base. This type of scale failure was evidenced by the transient in the oxidation kinetics (Fig. 1), the scale cross-section (Fig. 9) and topography (Fig. 10). The time to breakdown of the initial protective-type scale decreased progressively with increasing temperature. At 950 °C the oxidation behaviour appeared to be similar to that at 900 °C but with more pronounced internal attack and SiO_2 intrusion development. At the highest temperatures (1000 and 1050 °C), however, failure of the initially formed protective-type scale occurred without decohesion. The subsequent broad front internal attack of the steel derived from inward oxidant movement and resulted in a non-protective duplex-type scale and enhanced intergranular SiO_2 formation.

A further consequence of the changes in the morphology and composition of the scales over the temperature range was their behaviour on cooling from the oxidation to room temperature. At temperatures up to 1000 °C, essentially two types of scale spalled. The first was uniform protective-type oxide whose decohesion proceeded by the mechanism described above for similar scales formed at 825 °C. The second type of scale was the convoluted regions, which already had become detached from the steel during oxidation and through which cracks were propagated by the tensile strain introduced during cooling. By contrast the duplex-type oxide scales formed at about 1000 °C or higher were less prone to spall so that the oxide strain energy on cooling did not exceed the energy needed to cause fracture either within the scale or at the scale–steel interface. In this context the absence of a defined fracture plane such as a continuous SiO_2 layer could be note worthy. It is also conceivable that residual stresses originating from scale growth into the steel could have reduced the overall strain on cooling. Finally, the SiO_2 intrusions could have pegged the scale to the steel.

The implantation of krypton ions, having a comparable mass with those of cerium, yttrium and lanthanum, would cause similar physical damage to the 20Cr–25Ni–Nb stainless steel surface as the implantation, using the same beam energy, of any of the reactive elements but would not be expected to exercise any chemical role. Over the temperature range 900–1000 °C, in confirmation of previous observations at 825 °C [3], krypton ion implantation did not exert any marked influence on the oxidation behaviour of the 20Cr–25Ni–Nb stainless steel. It may be concluded, therefore, that any physical effects of the implantation of the reactive elements would also be likely to be insignificant.

The overall influence of cerium, yttrium and lanthanum ion implantation was to extend the protective-type oxidation behaviour of the steel to higher temperatures, in the current exposures of about 250 h duration, typically up to 1000 °C. This is exemplified by an Arrhenius plot of the rate constants for these regions where the isothermal oxidation of cerium-implanted, yttrium-implanted and unimplanted 20Cr–25Ni–Nb stainless steel, in carbon dioxide, at 825–1050 °C, conformed to parabolic kinetics. The baseline shown for the 20Cr–25Ni–Nb stainless steel, and against which the results from the present study were plotted, represents the best estimate from all relevant data obtained within the U.K. nuclear industry. The transition from protective- to non-protective-type oxidation of 20Cr–25Ni–Nb stainless steel is measured at 925 °C by an increase in the effective activation energy from 286 to 504 kJ. For protective-type oxidation, cerium and yttrium ion implantation reduced the activation energy slightly and the pre-exponential factor markedly by about three orders of magnitude. As mentioned already, the transition temperature to non-

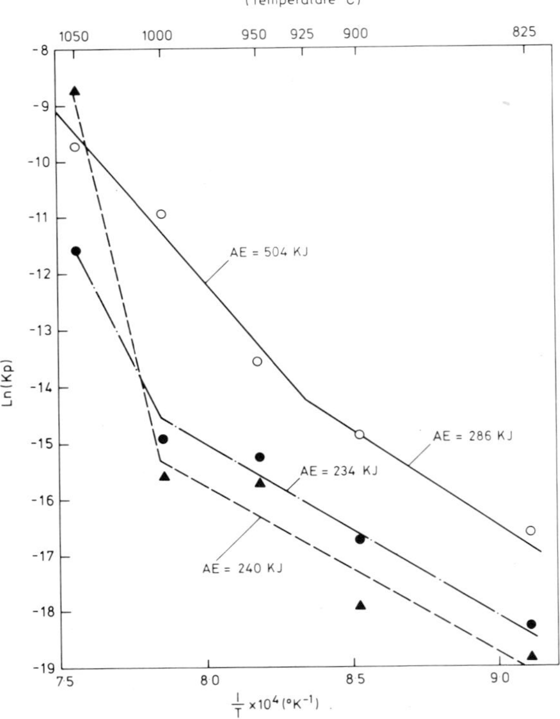

Fig. 12. Arrhenius plot of the parabolic rate constants (k_p) for the oxidation of 20Cr–25Ni–Nb stainless steel in carbon dioxide at 825–1050 °C: ○, unimplanted; ●, implanted with 10^{17} Ce^{+} cm^{-2}; ▲, and implanted with 10^{17} Y^{+} cm^{-2}; AE, activation energy.

protective oxidation was increased to about 1000 °C.

SIMS analysis indicated that the implanted atoms were located in a similar position in the scales formed at higher temperatures to that in the uniform protective scale developed at 825 °C [4, 7], *i.e.* within the outer spinel layer at the interface with the inner Cr_2O_3 layer. It is probable, therefore, that similar mechanisms applied at 900–1000 °C to those at 825 °C discussed already [4], for the incorporation of the implanted atoms into the scale and for their influence on scale growth and adhesion. Our own [5, 13] and other recent studies [17, 18] have revealed that the reactive elements probably inhibit the continuing scale growth by outward cation diffusion as a consequence of their segregation along the oxide grain boundaries, making this process energetically more difficult. The overall scale growth rate inhibition will be governed by the proportion of all the oxide grain boundaries affected and, for this particular system, the Arrhenius plot in Fig. 12 suggests that effectively at least 99.9% of the boundaries could have been blocked.

With respect to the improved scale adhesion by cerium, yttrium and lanthanum ion implantation at temperatures of 950 °C or less, a minimum thickness of scale has to develop on the 20Cr–25Ni–Nb stainless steel before spallation is initiated [9, 16]. Work in progress [19] is indicating that this crucial thickness is at least a factor of 2 higher for the cerium- and yttrium-implanted 20Cr–25Ni–Nb stainless steel. This implies that a greater strain has to be imposed by the thermal contraction differences between scale and steel to cause spallation. Possible causes could be the necessity to offset growth stresses due to the initial inward scale development or the modification of the scale morphology at the steel interface (the primary fracture plane for scales formed on the 20Cr–25Cr–Nb steel), in that the SiO_2 layer might not be continuous [7]. Alternatively the incorporated dispersed reactive element oxide grains and segregated atoms along scale grain boundaries could enhance the mechanical properties (*e.g.* strength) of the scale or increase the required energy for through-scale crack initiation and propagation. The combined effect of the implanted reactive elements, increasing the minimum scale thickness to spall and reducing the scale growth rate, was that exposure of the implanted 20-Cr–25Ni–Nb stainless steel could be significantly more onerous (with respect to both temperature and duration) than for the unimplanted steel before any oxide spalled.

As on unimplanted 20Cr–25Ni–Nb stainless steel, the transition between protective and non-protective oxidation of the cerium- and yttrium-implanted steel was marked by the formation of buckled scales. This again stemmed from consequences of the outstanding adhesion of the scales and also the recrystallization and grain growth of the underlying 20Cr–25Ni–Nb stainless steel. Sliding of the oxide and steel grains at the scale interface with emerging steel grain boundaries initiated scale detachment. On account of its increased strength, because of the incorporated reactive elements, the scale accommodated the differential strain with the steel at higher temperatures than on unimplanted 20Cr–25Ni–Nb stainless steel by buckling. At probably a critical scale thickness, however, cracking occurred along the apex of the convolutions. The breakdown in the protection afforded by the scale led inevitably to internal attack, as during the isothermal oxidation of yttrium-implanted 20Cr–25Ni–Nb stainless steel at 1000 °C. The implanted reactive elements had little beneficial influence on the faster diffusion processes at 1050 °C, probably dominated by inward oxidant movement.

5. CONCLUSIONS

(1) The isothermal oxidation of the 20Cr–25Ni–Nb stainless steel, in carbon dioxide, changes between 900 and 1050 °C from protective- to non-protective-type behaviour. Two main contributory causes are the breakdown of scale protection by cracking at decreasing times with increasing temperature and the progressively increasing role of inward oxidant movement in scale growth, again with increasing temperature. The non-protective duplex scale is less prone to spall, possibly because of the absence of a defined fracture plane or pegging by the underlying SiO_2 intrusions.

(2) Ion implantation of cerium, yttrium and lanthanum provides continuing protection to the 20Cr–25Ni–Nb stainless steel to higher temperatures (about 1000 °C or less). Initial scale development is modified by the promotion of inward oxidant movement. The implanted reactive elements become incorporated as oxide grains and segregants along grain boundaries in the outer spinel layer of the scale adjacent to the inner Cr_2O_3 layer. Cation transport through these boundaries is energetically less favourable so that continuing scale growth is inhibited by an extent governed by the proportion of the boundaries not affected.

(3) The incorporated reactive elements necessitate a greater strain to be imposed because of the thermal contraction differences between the scale and steel on cooling to room temperature to cause spallation than for the corresponding scale formed on unimplanted 20Cr–25Ni–Nb stainless steel. This could derive from an enhancement either of the mechanical properties (*e.g.* strength) of the scale or of the energy for through-scale crack initiation and propagation.

ACKNOWLEDGMENTS

We are grateful to Dr. G. Dearnaley, Dr. B. J. Laundy and the Cockcroft–Walton accelerator staff for carrying out the implantations, to B. A. Bellamy and M. D. Fones for the X-ray diffraction examinations, to D. J. Savage for oxidizing some of the specimens and to Dr. H. E. Evans and Dr. R. C. Lobb of the Central Electricity Generating Board, Berkeley Nuclear Laboratories, for the baseline in Fig. 12.

REFERENCES

1 M. J. Bennett, J. E. Antill, R. F. A. Carney, G. Dearnaley, F. H. Fern, P. D. Goode, B. L. Myatt, J. F. Turner and J. B. Warburton, *Corros. Sci.*, *16* (1976) 729.
2 M. J. Bennett, G. Dearnaley, M. R. Houlton, R. W. M. Hawes, P. D. Goode and M. A. Wilkins, *Corros. Sci.*, *20* (1980) 73.
3 M. J. Bennett, G. Dearnaley, M. R. Houlton and R. W. M. Hawes, in V. Ashworth, W. A. Grant and R. P. M. Procter (eds.), *Proc. Conf. on Ion Implantation into Metals, Manchester, 1981*, Pergamon, Oxford, 1982, p. 264.
4 M. J. Bennett, B. A. Bellamy, C. F. Knights, N. Meadows and N. J. Eyre, in G. K. Wolf, W. A. Grant and R. P. M. Procter (eds.), *Proc. Int. Conf. on Surface Modification of Metals by Ion Beams, Heidelberg, September 17–21, 1984*, in *Mater. Sci. Eng.*, *69* (1984) 359.
5 M. J. Bennett, J. A. Desport and P. A. Labun, in G. W. Bailey (ed.), *Proc. 43rd Annu. Meet. of the Electron Microscopy Society of America*, San Francisco Press, San Francisco, CA, 1985, p. 270.
6 M. J. Bennett, B. A. Bellamy, G. Dearnaley and M. R. Houlton, *Proc. 9th Int. Congr. on Metallic Corrosion, Toronto, June 3–7, 1984*, Vol. 2, National Research Council of Canada, Ottawa, 1984, p. 416.
7 C. H. Yang, P. A. Labun, G. Welsch, T. E. Mitchell and M. J. Bennett, *UKAEA Rep. AERE-R12088*, 1986 (Atomic Energy Research Establishment); *J. Mater. Sci.*, *22* (1987) 449.
8 M. J. Bennett, J. A. Desport and P. A. Labun, *UKAEA Rep. AERE-R11034*, 1984 (Atomic Energy Research Establishment); *Proc. R. Soc. London, Ser. A*, to be published.
9 H. E. Evans and R. C. Lobb, *Proc. 9th Int. Congr. on Metallic Corrosion, Toronto, June 3–7, 1984*, Vol. 2, National Research Council of Canada, Ottawa, 1984, p. 46.
10 M. J. Bennett, M. R. Houlton and R. W. M. Hawes, *Corros. Sci.*, *22* (1982) 111.
11 H. E. Evans, D. A. Hilton, R. A. Holm and S. J. Webster, *Oxid, Met.*, *14* (1980) 235.
12 A. Galerie, M. Caillet and M. Pons, in G. K. Wolf, W. A. Grant and R. P. M. Procter (eds.), *Proc. Int. Conf. on Surface Modification of Metals by*

Ion Beams, Heidelberg, September 17–21, 1984, in *Mater. Sci. Eng., 69* (1984) 329.
13 M. J. Bennett, J. A. Desport, P. A. Labun and J. T. Titchmarsh, unpublished results, 1986.
14 M. J. Bennett, J. B. Price and D. J. Savage, *UKAEA Rep. AERE-R11228*, 1984 (Atomic Energy Research Establishment).
15 J. Lindhard, M. Scharff and H. E. Schiøtt, *Mat.-Fys. Medd. K. Dan. Vidensk. Selsk., 33* (1963) 14.
16 J. Asher, S. Sugden, M. J. Bennett, R. W. M. Hawes, D. J. Savage and J. B. Price, *UKAEA Rep. AERE-R12196*, 1986 (Atomic Energy Research Establishment); *Werkst. Corros.*, to be published.
17 K. Przybylski, A. J. Garratt-Reed and G. J. Yurek, in G. W. Bailey (ed.), *Proc. 44th Annu. Meet. of the Electron Microscopy Society of America*, San Francisco Press, San Francisco, CA, 1986, p. 518.
18 D. P. Moon, *D.Phil. Thesis*, University of Oxford, 1987.
19 M. J. Bennett and A. T. Tuson, unpublished results, 1987.

Materials Science and Engineering, 90 (1987) 191-196

A Rutherford Backscattering–Channelling Study of Yttrium-Implanted Stainless Steel Before and After Oxidation*

G. DEARNALEY[a], T. LAURSEN[b] and J. L. WHITTON[b]

[a]*Atomic Energy Research Establishment, Harwell, Didcot, Oxon. OX11 0RA (U.K.)*

[b]*Department of Physics, Queen's University, Kingston, Ontario K7L 3N6 (Canada)*

(Received July 10, 1986)

ABSTRACT

A single crystal of austenitic stainless steel has been implanted with 200 keV 2×10^{16} Y^+ cm^{-2} at room temperature. Random and aligned Rutherford backscattering spectra were obtained along the ⟨100⟩ and ⟨110⟩ axes before and after oxidation. The oxidation was made by a residual oxygen component in an argon atmosphere for 1 h at 500 °C.

The implanted yttrium ions occupy a specific non-substitutional site in the stainless steel crystal and it is suggested that the yttrium atoms at shallow depths interact with oxygen atoms from the native oxide. No specific site could be assigned after oxidation and the yttrium ions redistributed from a range distribution centred around 44 nm to a narrow distribution centred around the metal–oxide interface at about 110 nm from the surface. In addition, increased dechannelling is observed at this interface, a feature which is absent in the unimplanted case.

1. INTRODUCTION

Implantation of yttrium is an established method of treating steels for oxidation protection and is now used on an industrial scale at the Atomic Energy Research Establishment, Harwell, U.K. The same implantation treatment also has beneficial effects on the wear properties of steels, some of which may be related to oxidative wear [1]. The Rutherford backscattering–channelling technique is very useful for characterizing ion-implanted and oxidized surfaces providing, as it does, information regarding the elemental composition and the microstructure. Channelling studies require the use of single crystals; these we have used in earlier studies of nitrogen-implanted and laser-alloyed surfaces of austenitic stainless steel [2–4].

2. EXPERIMENTAL DETAILS

The single crystal of f.c.c. austenitic stainless steel was purchased from Cristal-Tec, Grenoble, France. It has a quoted composition of 70wt.%Fe–17wt.%Cr–13wt.%Ni and was cut by spark machining perpendicular to the ⟨100⟩ axial direction. The crystal was mechanically polished followed by electropolishing in a solution consisting of 1 part perchloric acid to 20 parts acetic acid.

A circular area 8 mm in diameter was implanted with 200 keV 2×10^{16} Y^+ cm^{-2} at room temperature using the Harwell ion implanter. The oxidation was done by a heat treatment at 500 °C for 1 h in an argon atmosphere. The residual oxygen component was sufficient to grow an oxide layer about 100 nm thick on both implanted and unimplanted parts of the crystal.

The Rutherford backscattering–channelling analysis was carried out using a well-collimated 2 MeV helium ion (He^+) beam impinging on the crystal mounted on a manually controlled two-axis goniometer. The beam was 1–2 mm in diameter and beam currents were less than 1 nA. The spectrum of backscattered ions was measured with a surface barrier detector at a scattering angle of 160°.

*Paper presented at the International Conference on Surface Modification of Metals by Ion Beams, Kingston, Canada, July 7–11, 1986.

0025-5416/87/$3.50

3. RESULTS

The untreated stainless steel crystal showed good channelling behaviour; the minimum yield χ_{min} was 0.04 along the ⟨100⟩ axial direction. The aligned spectrum showed that the chromium surface peak is resolved from the combined Fe–Ni peak, which is measured at a slightly higher energy (see also ref. 2). Nuclear reaction analysis using the $^{16}O(d, p_1)^{17}O$ reaction with a 900 keV deuterium beam showed that the native oxide had about 3×10^{16} O cm^{-2} with a thickness of about 5 nm.

3.1. Yttrium implantation

Two sets of Rutherford backscattering–channelling spectra of the yttrium-implanted sample are shown in Fig. 1. They are obtained for the ⟨100⟩ and ⟨110⟩ axes. Since the host atoms differ by only a few mass units, the substrate edge appears within a relatively narrow energy range around channel 400. The implanted yttrium atoms, however, have a significantly higher atomic mass and the resulting signal has a higher energy. Most of the yttrium signal is separated from the substrate signal to the right of the substrate edge. The maximum yttrium concentration derived from the spectra in Fig. 1 is 2.6 at.% at a depth of 44 nm. This depth value is in agreement with the theoretical mean projected range obtained using the TRIM code [5].

The substrate signal shows a pronounced channelling effect for both directions, indicating a high degree of crystallinity, but the channelling behaviours are different. In both cases, dechannelling from extended defects occurs in a surface layer affected by the implantation; however, the dechannelling close to the surface is much higher in the ⟨110⟩ direction than in the ⟨100⟩ direction. It can also be observed that the distribution of extended defects goes much deeper than the yttrium distribution.

The yttrium signal also shows channelling effects and, to show this more clearly, it is enlarged and shown in Fig. 2. The detailed channelling behaviour of the impurity signal is interesting since it is related to its atomic location in the lattice. Figure 2(a) shows that the relative channelling yield is depth dependent; at the surface, there is virtually no dip in the ⟨100⟩ aligned direction, while a significant dip is seen at greater depth. In order to study this depth effect in more detail and to understand it in terms of the lattice location, detailed angular scans were measured at different depth intervals; each interval is ten channels wide (33 nm). The result of this study is presented in Fig. 3. There is only a small indication of a channelling dip at shallow depths (Figs. 3(a) and 3(b)), while a large dip is seen at greater depth (Fig. 3(c)). This is not the channelling behaviour expected for substitutional impurities, since the dip is substantially narrower (full width at half-maximum, 0.6°) than that of the host (full width at half-maxi-

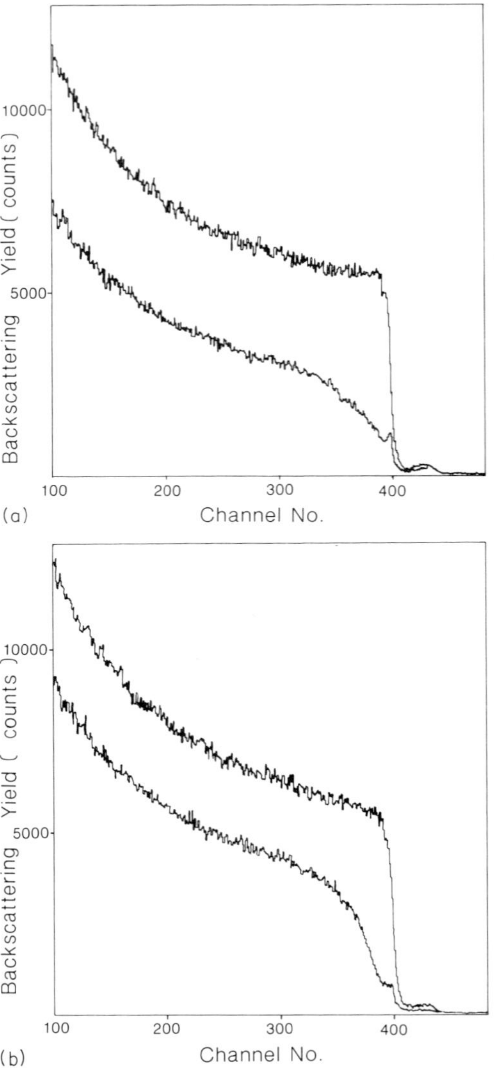

Fig. 1. Channelling spectra of 2 MeV helium ions from yttrium-implanted stainless steel before oxidation: (a) ⟨100⟩; (b) ⟨110⟩.

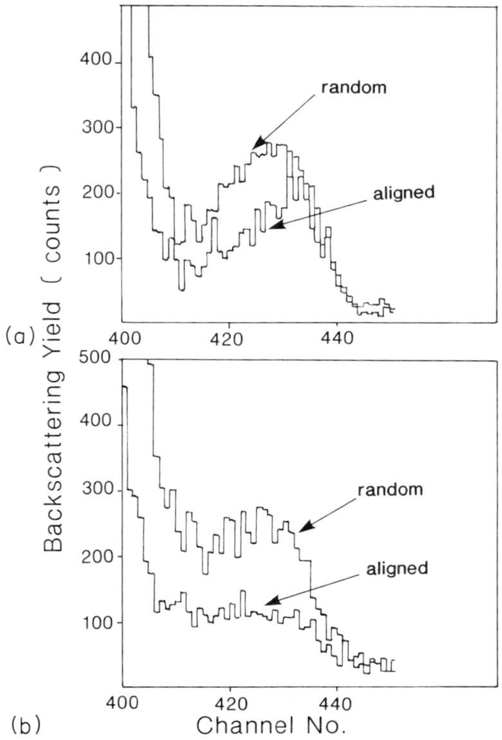

Fig. 2. Enlarged view of the yttrium signal from Fig. 1: (a) ⟨100⟩; (b) ⟨110⟩.

mum, 1.35°). Angular scans were also carried out for the ⟨110⟩ direction, but planar effects were too severe for these to reveal information in addition to what is provided by the spectra in Figs. 1(b) and 2(b).

3.2. *Oxidation*

The visual appearance of the oxidized surface was a purple colour within the implanted area and a less intense colour on the unimplanted part. Examination by optical microscopy showed a varying spotted appearance within the implanted area, while the unimplanted part appeared relatively uniform.

The Rutherford backscattering–channelling spectra of a moderately oxidized unimplanted sample are shown in Fig. 4. The oxygen has a low atomic mass and is thus seen at a low energy on top of the dominating substrate background. The high concentration of oxygen at the surface affects the substrate signal by reducing the yield. The thickness of the oxide layer is best determined from the substrate signal, and on the assumption of a metal-to-oxygen ratio of 2 to 3 the thickness obtained from Fig. 4 is 96 nm.

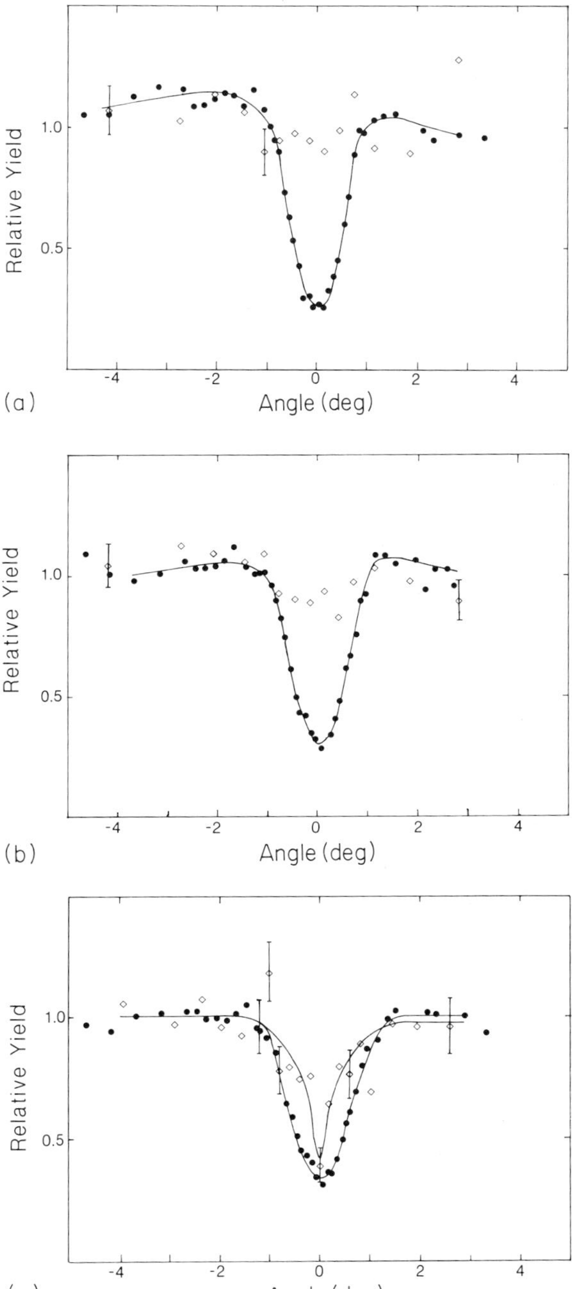

Fig. 3. Angular scan along the ⟨100⟩ direction of yttrium-implanted stainless steel at depths of (a) 0–33 nm, (b) 33–66 nm and (c) 66–99 nm: ◇, yttrium; ●, stainless steel.

The aligned spectra show that the oxide has not grown epitaxially and, therefore, shows a random yield within the oxide. The channelling is very good in the substrate, but the random fraction is elevated (the values of χ_{min} are 0.28 and 0.27 for ⟨100⟩ and ⟨110⟩

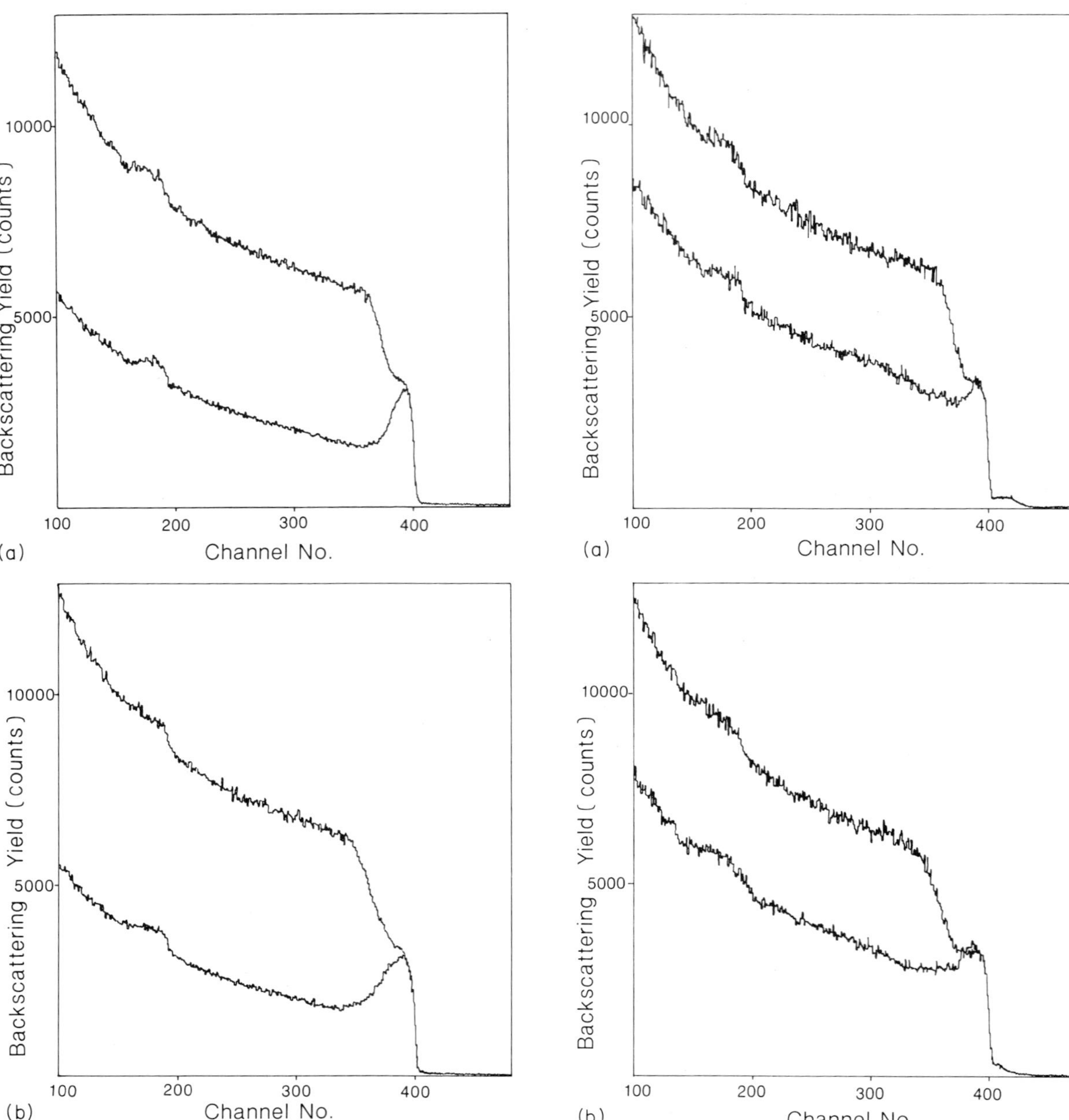

Fig. 4. Channelling spectra of 2 MeV helium ions from unimplanted stainless steel after oxidation: (a) ⟨100⟩; (b) ⟨110⟩.

Fig. 5. Channelling spectra of 2 MeV helium ions from yttrium-implanted stainless steel after oxidation: (a) ⟨100⟩; (b) ⟨110⟩.

respectively) because of spreading of the beam in the oxide.

Figure 5 shows the Rutherford backscattering–channelling spectra of the oxidized yttrium-implanted crystal. The oxide is slightly thicker (110 nm) and the metal–oxide interface appears sharper than in the previous case, which suggests that the oxide has a more uniform thickness. The random fraction behind the oxide is substantially higher than before ($\chi_{min} = 0.5$). Additional multiple scattering in the thicker oxide is not sufficient to explain this high value, and the implantation damage, which mainly consists of extended defects (*cf.* Fig. 1), is expected to contribute to the random fraction at a greater depth. In order to analyse the increased random fraction in more detail, Fig. 6 shows the result of subtracting the aligned spectra in Figs. 4 and 5. Two very different slopes are

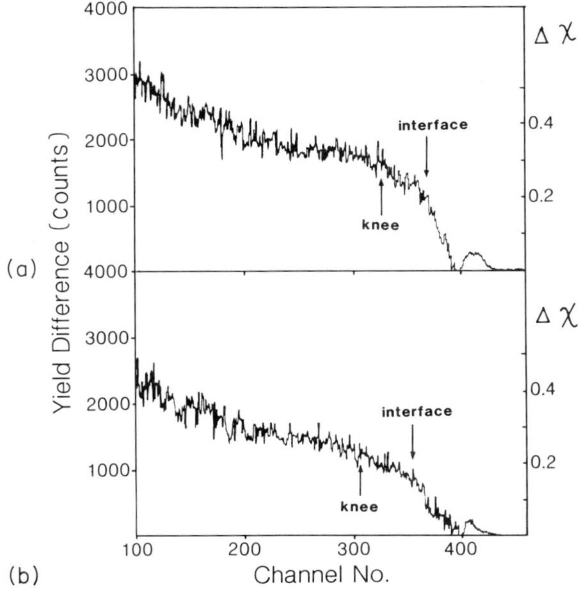

Fig. 6. Difference spectra obtained by subtracting the aligned spectra in Figs. 4 and 5: (a) ⟨110⟩; (b) ⟨100⟩.

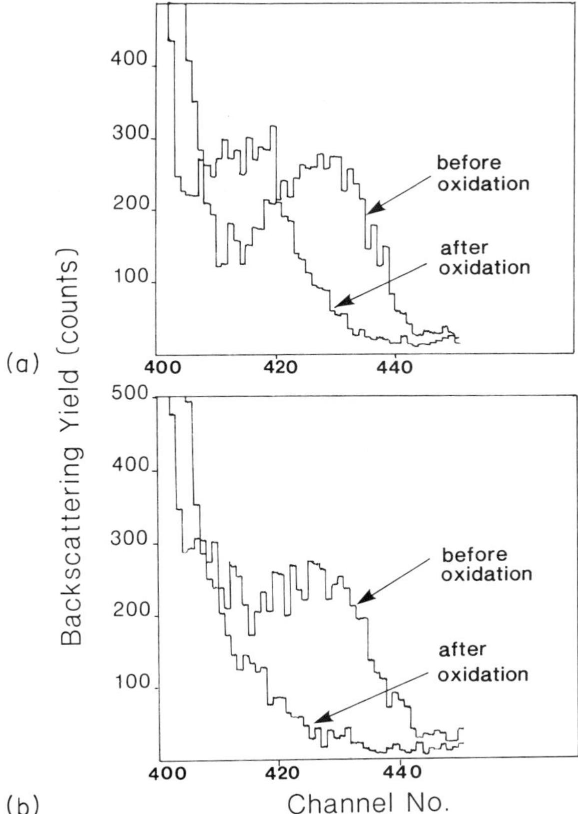

Fig. 7. Rutherford backscattering spectra of yttrium-implanted stainless steel obtained using a 2 MeV helium ion beam. The yttrium signal is shown before and after oxidation at two different tilt angles: (a) 6°; (b) 45°.

seen before and after the interface. The effect is most evident in Fig. 6(a), where a steep gradient is seen before the interface, and a moderate gradient due to implantation damage is seen between the interface and the characteristic knee.

There is also interesting information in the yttrium spectra even though no channelling effects are observed (Fig. 5). This is best seen when comparing them with the random spectra of a non-oxidized sample (Fig. 2). This has been done in Fig. 7, and the effect of oxidation is seen to be large and different from the signal of the host atoms where the backscattering signal is moderately depressed; instead the yttrium ions have migrated away from the surface and their distribution is centred around the metal–oxide interface. A 4 MeV helium ion analysing beam provides a spectrum in which the yttrium signal is fully separated from the signal of the host atoms (Fig. 8).

4. DISCUSSION

4.1. *Yttrium lattice location*

The present results show that the lattice location is depth dependent. Yttrium has a

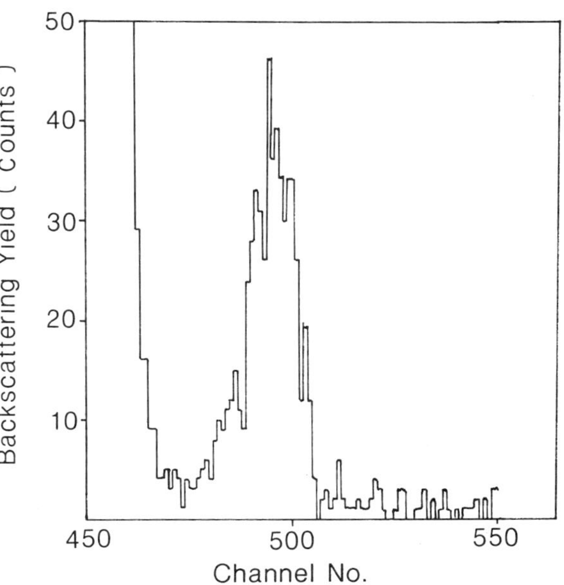

Fig. 8. Rutherford backscattering spectrum of yttrium-implanted stainless steel after oxidation obtained using a 4 MeV helium ion beam at normal incidence. The yttrium signal appears as a peak separated from the host signal.

high affinity for oxygen, and it is possible that the implantation process has moved oxygen atoms from the 5 nm native oxide to depths of the order of 30 nm. Therefore, the lattice site seen at shallow depths could be associated with an oxygen impurity.

The lattice location observed at greater depths is expected to be free of impurity interactions. The narrow yttrium dip compared with that of the host suggests that the yttrium impurity is displaced a few tenths of an ångström from a substitutional site. It is not surprising that the yttrium atom is non-substitutional because of its large size (the ratio of the yttrium to iron atomic radii is 1.4). It is also known that implanted yttrium in iron is associated with one or more vacancies [6]. It may, therefore, be likely that the yttrium location is stainless steel is determined by the interaction with neighbouring vacancies.

4.2. Effect of oxidation

At first glance it may be surprising that the yttrium-implanted surface has a thicker oxide than the unimplanted surface but this may be due to the low (500 °C) oxidation temperature. It should therefore be emphasized that the present oxidation treatment is not a test for oxidation resistance since the degree of oxidation does not reveal which oxide is protective against further oxidation. It has been established in other experiments that the yttrium-implanted surface is oxidation resistant, at least at temperatures above 800 °C [1]. The difference between the two types of oxide layer must be sought amongst specific microstructural components.

There is a clear difference between the channelling spectra, and the high random fraction observed in the implanted case can only be explained by dechannelling in the substrate due to crystalline damage. One explanation for the damage in the substrate is lattice strain due to the growth of the oxide. This would indicate a strong adherence between the oxide and the substrate, which is in contrast with the unimplanted case, where no substrate strain is detectable. Another explanation could be that dislocations introduced by the implantation have migrated during oxidation.

The redistribution of yttrium atoms suggests that these impurities play an essential role in forming an interface region, thus providing an oxidation-resistant scale.

5. CONCLUSION

The yttrium implantation (200 keV 2×10^{16} Y^+ cm^{-2}) of stainless steel is associated with a high degree of crystalline damage. The implantation damage consists mainly of extended defects and exhibits a surprising directional dependence. A strong component is seen within the first 100 nm in the ⟨110⟩ direction, which is 45° from the surface normal.

The yttrium lattice location is complicated by being depth dependent. At shallow depths, interactions with oxygen impurities may play a role while, at larger depths, the yttrium atoms are suggested to be slightly displaced from a substitutional site as the result of vacancy interactions.

The effect of oxidation on the implanted crystal is to concentrate yttrium atoms and extended defects at the metal–oxide interface. It is not clear whether the high density of defects at the interface is implantation-induced damage which has migrated during oxidation or is due to lattice strain brought about by an improved adherence facilitated by the presence of yttrium atoms.

REFERENCES

1 G. Dearnaley, *Nucl. Instrum. Methods, 182–183* (1981) 899.

2 J. L. Whitton, G. T. Ewan, M. M. Ferguson, T. Laursen, I. V. Mitchell, H. H. Plattner, M. L. Swanson, A. V. Drigo, G. Celotti and W. A. Grant, in G. K. Wolf, W. A. Grant and R. P. M. Procter (eds.), *Proc. Int. Symp. on Surface Modification of Metals by Ion Beams, Heidelberg, September 17–21, 1984*, in *Mater. Sci. Eng., 69* (1985) 111.

3 J. L. Whitton, T. Laursen, J. A. Nilson, Wing Nip, I. V. Mitchell, H. H. Plattner and M. L. Swanson, in J. B. Roberto, R. W. Carpenter and M. C. Wittels (eds.), *Advanced Photon and Particle Techniques for the Characterization of Defects in Solids, Materials Research Society Symp. Proc.*, Vol. 41, Materials Research Society, Pittsburgh, PA, 1985, p. 313.

4 T. Laursen, J. L. Whitton, J. A. Nilson, Wing Nip, I. V. Mitchell, H. H. Plattner and M. L. Swanson, to be published.

5 J. P. Biersack and L. G. Haggmark, *Nucl. Instrum. Methods, 174* (1980) 257.

6 S. M. Myers, S. T. Picraux and R. E. Stoltz, *Appl. Phys. Lett., 37* (1980) 168.

Materials Science and Engineering, *90* (1987) 197-203

Oxidation Inhibition of Iron by Deposited and Ion-implanted Boron*

F. GIACOMOZZI, L. GUZMAN, F. MARCHETTI, A. MOLINARI, M. SARKAR† and A. TOMASI

Istituto per la Ricerca Scientifica e Tecnologica, 38050 Povo (Trento) (Italy)

(Received July 10, 1986)

ABSTRACT

Fe-B and Fe-B-N surface alloys were produced by several deposition and ion-mixing procedures including ion bombardment of multilayer configurations and ion-beam-enhanced deposition. The surface treatments examined were boron and BN deposited on iron. Fe/B/Fe/B... multilayers deposited on iron and analogous samples implanted with krypton ions (Kr^+) or nitrogen ions (N_2^+) at an energy of 160 keV.

The influence of these layers on the oxidation of iron in air at 500 and 650 °C was examined. The samples were characterized by scanning electron microscopy. Auger electron spectroscopy, thermogravimetry, secondary ion mass spectrometry and X-ray diffractometry. The aim of the investigation was to clarify the mechanisms by which boron so effectively inhibits the oxidation rate of iron.

1. INTRODUCTION

Boron inhibition of the oxidation of iron and of iron alloys is well established [1, 2]. Several ion implantation studies have been performed to clarify the reasons for this behaviour [3, 4]. Boron-ion-implanted iron shows excellent oxidation resistance at temperatures of less than 600 °C [2]. The lack of solubility of boron into iron and its consequent segregation to grain boundaries and other fast diffusion paths are thought to inhibit the oxidation. In order to extend the protection to higher temperatures, however, it may be necessary to turn to other mechanisms such as blocking by particle agglomeration or by a diffusion barrier.

It is clear that ion-implanted surfaces do not contain sufficient dopant to form a coherent protective layer during oxidation. Moreover, the implantation of boron ions at fluences of 1×10^{17} ions cm^{-2} or more could be constrained by practical limitations in the life of the ion sources. Therefore an alternative treatment to direct implantation, known as ion beam mixing, has been employed. This consists of ion bombarding (with an easily obtained gaseous beam) a thin layer or multilayer previously deposited on the substrate to form an alloy surface. This method bypasses the limits imposed by sputtering and allows the formation of strongly adherent interface-free coatings.

The effect of nitrogen ion implantation across boron deposits was the formation of a boron nitride (BN) layer just below a surface layer of "untransformed" boron [5]. Adhesion between the coating and the substrate was enhanced because of ion beam mixing at the B-Fe interface. BN appears to be extremely interesting in view of its unique physical properties: its high thermal conductivity, exceptional strength, low thermal expansion and high chemical inertness, even at very high temperatures. BN is not easily oxidized, at least up to 1000 °C. The inhibition of the high temperature oxidation rate resulting from nitrogen ion (N^+) implantation across boron deposits was substantial [6] but the mechanism involved needs further clarification. It could result from the presence of both the BN barrier layer and the B-Fe ion-beam-mixed layer.

In this work the oxidation behaviour of iron specimens treated by different deposition and ion-mixing techniques is reported in order to discriminate between the possible contributions from the respective layers.

*Paper presented at the International Conference on Surface Modification of Metals by Ion Beams, Kingston, Canada, July 7-11, 1986.

†Permanent address: Department of Physics, University of Dacca, Bangladesh.

0025-5416/87/$3.50

2. EXPERIMENTAL PROCEDURES

Single boron (15 nm) (Fig. 1, sample 2) or BN (100 nm) (Fig. 1, sample 4) layers or thin multilayer films (consisting of 31 nm iron and 5 nm boron layers making up a total thickness of about 100 nm) with a predetermined composition $Fe_{80}B_{20}$ (Fig. 1, sample 3) were deposited on both sides of a rolled sheet of 99.95% pure iron (supplied by Goodfellow Metals Ltd.) using the magnetron r.f. sputtering technique. The multilayer deposited samples were then ion implanted, either with 160 keV krypton ions (Kr^+) to a fluence of 2×10^{16} ions cm^{-2} (Fig. 1, sample 5) or with 160 keV nitrogen ions (N_2^+) to a fluence of 3×10^{17} ions cm^{-2} (Fig. 1, sample 6).

Ion-assisted coatings were also produced on iron using four sequential steps of 27 nm boron deposition followed by nitrogen ion (N_2^+) implantation to a fluence of 1.4×10^{17} ions cm^{-2} (Fig. 1, sample 7). This technique has been called reactive ion-beam-enhanced deposition (RIBED) [7].

The specimens were cut into 1 cm $\times$ 0.5 cm pieces and suspended in a recrystallized alumina reaction tube, which was part of a Mettler thermobalance. Oxidation was performed in air flowing at 10 l h^{-1} and the specimen attained the test temperature (500 or 650 °C) at a rate of 10 °C min^{-1}. At the end of the run, the specimen was allowed to cool at the same rate. Mass changes during oxidation were monitored continuously with an accuracy of ±0.1 mg and checked after the oxidation with an analytical Mettler balance (accuracy, ±0.01 mg); loss of oxide during cooling was only observed occasionally. Characterization of the specimen surfaces prior to and after oxidation was undertaken using the techniques given in Table 1.

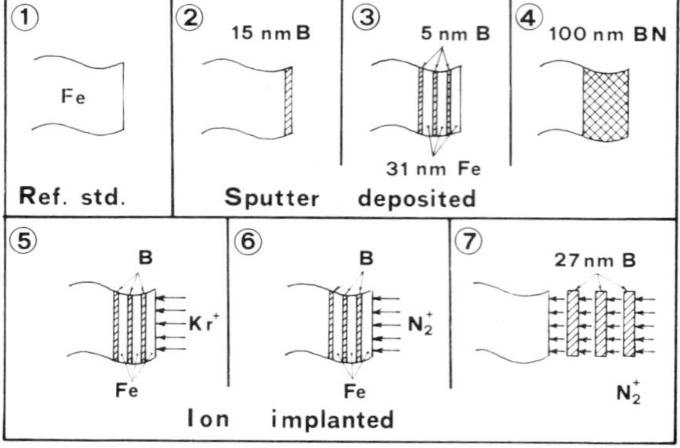

Fig. 1. Diagrammatic representation of the samples used in this investigation. The numerals indicate the sample numbers.

3. RESULTS AND DISCUSSION

3.1. *Characterization of ion-bombarded Fe–B multilayers deposited on iron*

The AES depth profiles (Figs. 2(a)–2(c)) show the distinct succession of the as-deposited iron and boron multilayers and their interdiffusion consequent to krypton and nitrogen ion implantation respectively. The morphologies of the implanted samples were very different; the surface of the former was apparently undamaged, while that of the latter appeared full of holes. In contrast, the boron and BN deposited on iron substrates did not show an appreciable number of defects, the films being fine grained and coherent. The films obtained by RIBED also appeared to be homogeneous. Their Auger depth concentration profiles (Fig. 2(e)) are similar to those of previously studied samples, deposited with a single boron layer and implanted with 100 keV nitrogen ions (N^+) in which a buried BN layer formed with enhanced adhesion to the substrate [6]. However, the uniformity of the present samples was better than the previous samples.

SIMS analysis of the krypton-implanted surface of sample 5 gives evidence of the following species: krypton, FeB and Fe_2B. Similar mass spectra for the nitrogen-implanted surface of sample 6 indicates the presence of the following species: BN, FeB, FeN, Fe_2N and Fe_3N. Finally, the mass spectra of the RIBED obtained by film (sample 7) indicate the predom-

TABLE 1

Techniques used to characterize surface layers both prior to and after oxidation

Property	*Method*
Thickness	Profilometer
Crystalline structure	X-ray diffraction (XRD)
Morphology	Scanning electron microscopy
Composition	Auger electron spectroscopy (AES); secondary ion mass spectrometry (SIMS)

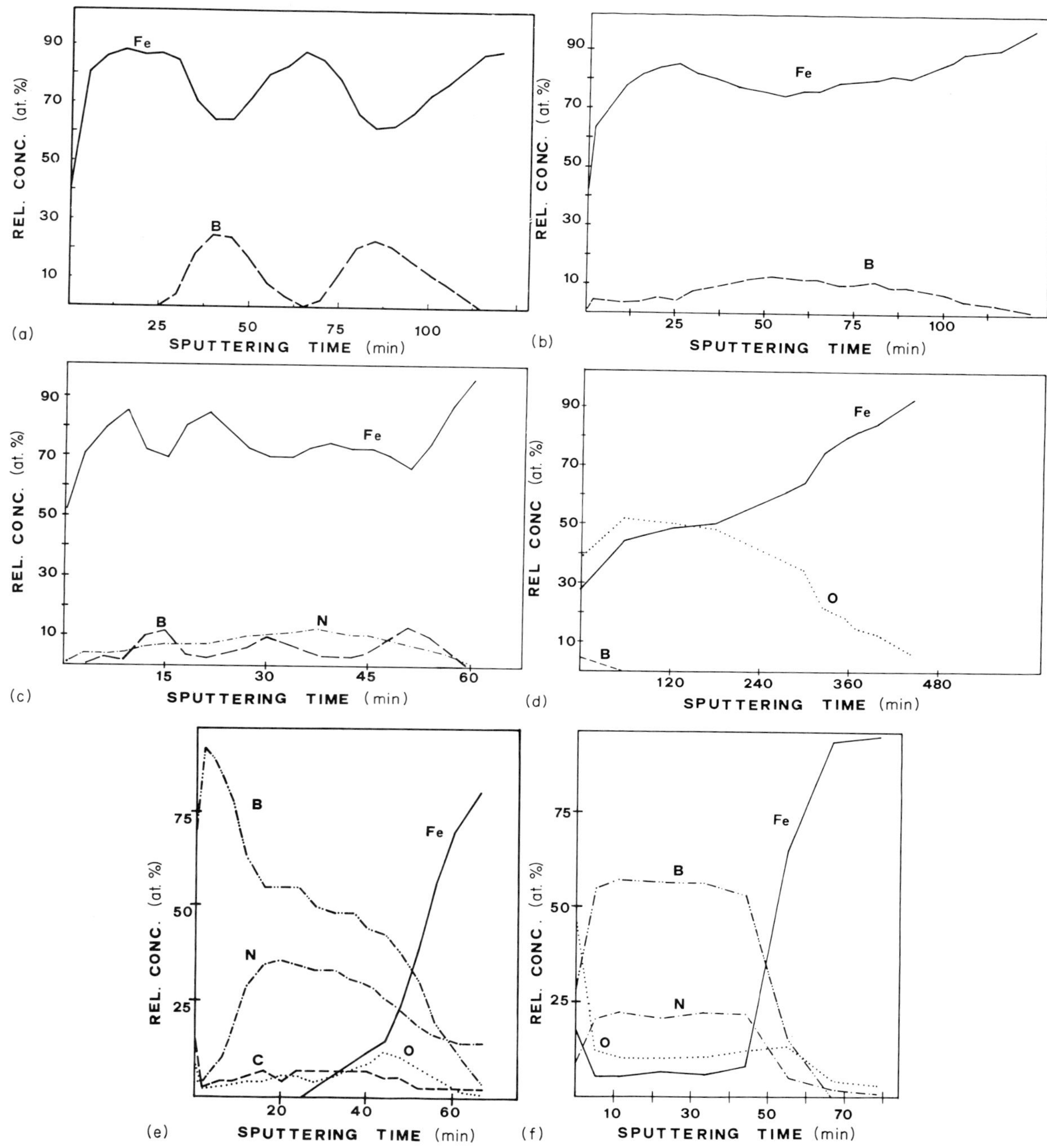

Fig. 2. Auger depth concentration profiles: (a) sample 3; (b) sample 5; (c) sample 6; (d) sample 6 after oxidation at 500 °C; (e) sample 7; (f) sample 7 after oxidation at 500 °C.

inant formation of BN, with some evidence of FeB, Fe_2B, FeN, Fe_2N and Fe_3N.

3.2. *Oxygen uptake data*

Plots of mass gain *vs.* time for unimplanted iron and for the different boron-treated samples during oxidation at 500 and 650 °C are presented in Fig. 3. The kinetics can be fitted generally by two parabolic regimes: the first is valid for oxidation times of less than 3 h, while the second applies for times in excess of 6 h.

In all cases, oxidation inhibition is observed. Thus, even in pure iron (sample 1), a single

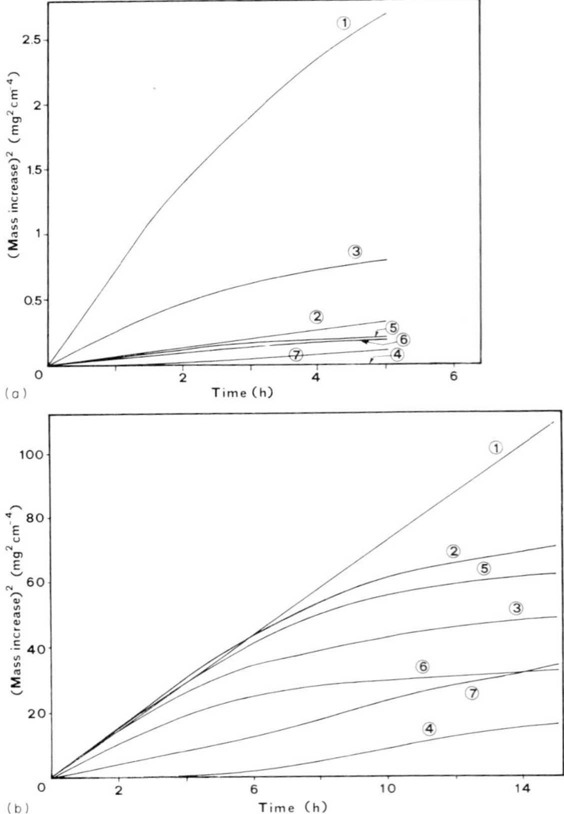

Fig. 3. Square of the mass increase *vs.* time for samples 1–7, during oxidation in air at (a) 500 °C and (b) 650 °C. The numerals on the curves indicate the sample numbers.

parabolic law is not observed and this is attributed to contamination in the oxidation chamber. In fact, SIMS analysis was able to find boron traces at the surface of oxidized iron.

Boron addition in the form of Fe/B multilayers on iron (sample 3) gives rise to an improvement visible after reaction for only a few hours. The oxidation is then no longer parabolic and shows a slower dependence than $t^{1/2}$ on time, which indicates that boron atoms enter progressively into the process. In contrast, boron deposited on the surface of iron (sample 2) is immediately effective against oxidation. A substantial reduction in the oxidation rate at 500 °C is also observed for the krypton-ion-mixed surface (sample 5), as well as for that bombarded with nitrogen ions (sample 6). At this temperature the effectiveness of both implants is similar. The BN coatings obtained by RIBED (sample 7) behave even better and are almost comparable with the sputter-deposited BN coatings (sample 4) which only react with oxygen within the measurement errors.

At 650 °C the situation changes radically; the surface boron layer (sample 2) is clearly inadequate, while the krypton-ion-mixed boron layer (sample 5) appears to be far less efficient than the Fe/B multilayer (sample 3), which probably is converted into an alloy layer at this temperature. A radiation damage effect can probably explain this trend inversion. For the nitrogen-implanted multilayer (sample 6), the BN phase is still very effective at this temperature. However a coherent BN barrier layer (sample 7 or 4) performs better than a simple dispersed phase.

3.3. *Oxide morphologies and composition*

Scanning electron micrographs of the topographies of the oxide scales formed on pure iron and the different boron-treated samples after oxidation at 500 °C for 5 h are shown in Fig. 4. On iron the oxide growth is characterized by the presence of whiskers, which increase the total surface area of the specimens [8]. In contrast, on samples 3 and 5, globular oxide morphologies are interposed with fine acicular structures. The microstructure is still finer for the scales formed on samples 6 and 7. The BN deposited on iron (sample 4) shows hardly any sign of oxidation. At 650 °C, after oxidation for 15 h, the morphology is rather similar for all boron-treated samples, *i.e.* a more or less fine globular structure.

The samples oxidized at 500 °C were also examined with SIMS and AES. The following species were found on profiling sample 5 (krypton-implanted multilayer): FeO, Fe_2O, B_2O_3, Fe_2O_3 and FeO·B_2O_3, predominantly concentrated in the near-surface region. The borides FeB and Fe_2B present in the original layer have disappeared. In sample 6 (nitrogen-bombarded multilayer), the following oxide species appear: FeO, Fe_2O, B_2O_3 and FeO·B_2O_3 although less abundant than in sample 5 and more homogeneously distributed in depth. The nitrides FeN, Fe_2N and Fe_3N have disappeared completely, but BN is still present. In sample 4, as well as in sample 7 (BN layers on iron), the following phases are present: BN, FeB and Fe_2N, while new phases (FeO, B_2O_3 and FeO·B_2O_3) appear as a result of oxidation.

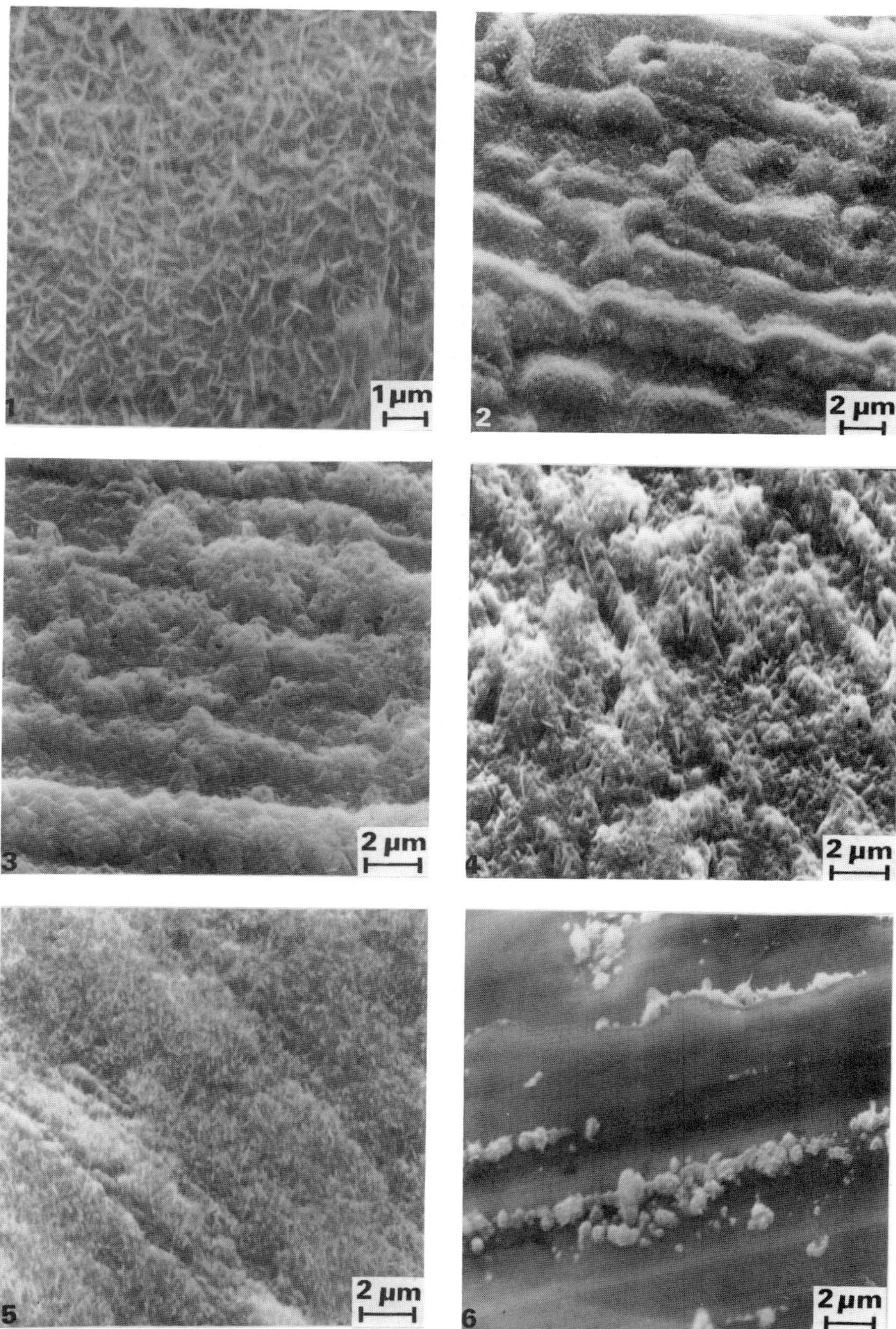

Fig. 4. Scanning electron micrographs of oxide topographies on samples 1, 3, 4, 5, 6 and 7, after oxidation for 5 h in air at 500 °C. The numerals on the micrographs indicate the sample numbers.

Iron is present on the surface of all oxidized samples, as confirmed by Auger depth concentration profiles (Figs. 2(e) and 2(f)). On sample 6, an oxide scale 1.1 nm thick is present with little boron being left at the surface, while nitrogen has desorbed. In sample 7, the boron-rich BN layer still exists and only the near-surface region appears oxidized. Nevertheless, it is important to note that, in spite of the presence of oxygen, the Auger line shape analysis never showed any evidence of oxidized iron at the BN–Fe interface.

TABLE 2

Peak intensities of phases present after oxidation at 500 and 650 °C

Sample	*500 °C*		*650 °C*	
	Fe_2O_3	*Fe_3O_4*	*Fe_2O_3*	*Fe_3O_4*
1	w	s	m	s
2	m	s	m	m
3	m	w	m	m
4	—	—	m	m
5	m	w	m	s
6	m	w	m	s
7	m	m	m	s

s, strong; m, medium; w, weak.

XRD analysis of the different oxidized samples, summarized in Table 2, reveal only a mixture of magnetite (Fe_3O_4) and haematite (Fe_2O_3), except for sample 4, where no oxide is present after oxidation at 500 °C, in good agreement with the mass gain data. Figure 5 shows, as an example, diffractograms of samples 1, 6 and 7 oxidized at 500 °C. At this lower oxidation temperature, two different trends are clearly observed. In the first, in samples 3, 5 and 6 (the multilayer-coated series), haematite predominates, which suggests that the ion-induced phases act as a barrier against cationic diffusion. In the second group, in samples 2, 4 and 7 (boron- or BN-coated samples), the inhibition mechanism derives from the sacrificial oxidation of boron. At 650 °C the barrier mechanisms still act for all samples but are less effective.

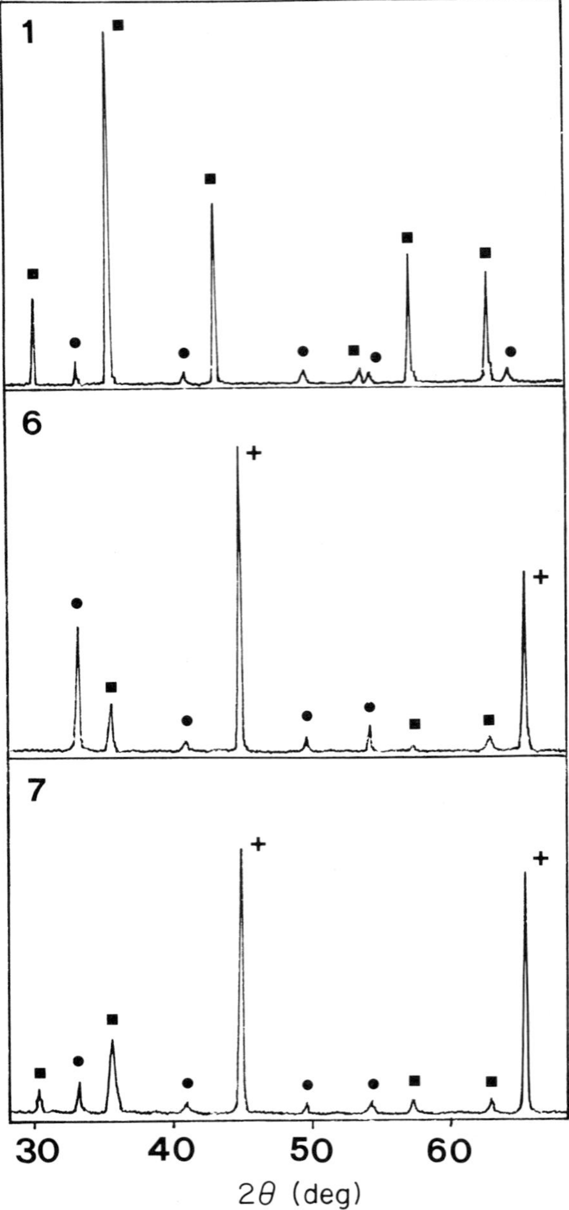

Fig. 5. X-ray diffractograms of sample 1 (pure iron), sample 6 (nitrogen-bombarded Fe/B/Fe/...) and sample 7 (BN coating obtained by RIBED) surfaces after oxidation at 500 °C: +, iron; ■, Fe_3O_4; ●, Fe_2O_3. The numerals on the diffractograms indicate the sample numbers.

4. CONCLUSIONS

Boron has been shown to confer a good oxidation resistance to iron, probably by means of several different mechanisms.

(1) The action of boron deposited on the surface as an oxidation inhibitor has been mentioned previously in the literature [9]. Boron would prevent oxygen from entering into the oxide film by the introduction of networks of —O—B—O— covalent bonds in the oxide structure, reducing the growth of stress in the oxide.

(2) The current oxidation data at 500 °C indicates that Fe-B alloys, formed by ion beam mixing, are effective in inhibiting oxidation. However, the same Fe–B alloys may be formed thermally at 500 °C or lower and result in analogous protection. The inhibition mechanism in samples treated with boron seems to be a blocking effect by particle agglomeration of ternary oxide $FeO \cdot B_2O_3$, also known as FeB_2O_4, as already shown by Pons *et al.* [10].

(3) At 500 °C, the krypton-ion-mixed Fe–B alloys perform better than the as-deposited alloys. At higher temperatures, however, the reverse applies, probably because of radiation damage induced by krypton ion bombardment. This can induce modifications in scale morphology and is deleterious to the oxidation behaviour, as has been observed by Goode [11] for other substrates and implants.

(4) Nitrogen ion implantation in iron does not bring about any improvement because massive nitrogen loss occurs on heating to temperatures as low as 200 °C. However, nitrogen ion implantation across boron deposits is extremely effective in inhibiting oxidation at 500 °C and higher. BN formation is thought to be responsible for the significant effect observed. The presence of second-phase particles exercises some influence on the motion of defects across the scale. The greatest inhibition is provided by a continuous BN layer which provides a diffusion barrier. The magnitude of its influence depends on the coherency of the BN layer and also on its adhesion to the iron substrate, which can be enhanced if the coating is produced by an ion-assisted technique.

ACKNOWLEDGMENTS

We are indebted to Mr. S. Pedrotti for technical assistance.

One of the authors (M.S.) is grateful to the International Centre for Theoretical Physics of Trieste for its financial support under the Programme for Training and Research in Italian Laboratories.

REFERENCES

1 C. Lea, *Met. Sci., 5* (1979) 301.
2 M. Pons, M. Caillet and A. Galerie, *Mater. Tech., 12* (1985) 699.
3 M. Pons, M. Caillet and A. Galerie, *Nucl. Instrum. Methods, 209–210* (1983) 1011.
4 A. Galerie, M. Caillet and M. Pons, in G. K. Wolf, W. A. Grant and R. P. M. Procter (eds.), *Proc. Int. Conf. on Surface Modification of Metals by Ion Beams, Heidelberg, September 17–21, 1984,* in *Mater. Sci. Eng., 69* (1985) 329.
5 L. Guzman, F. Marchetti, L. Calliari, I. Scotoni and F. Ferrari, *Thin Solid Films, 117* (1984) L63.
6 F. Giacomozzi, L. Guzman, A. Molinari, A. Tomasi, E. Voltolini and L. M. Gratton, in G. K. Wolf, W. A. Grant and R. P. M. Procter (eds.), *Proc. Int. Conf. on Surface Modification of Metals by Ion Beams, Heidelberg, September 17–21, 1984,* in *Mater. Sci. Eng., 69* (1985) 341.
7 L. Guzman, F. Giacomozzi, B. Margesin, L. Calliari, L. Fedrizzi, P. M. Ossi and M. Scotoni, *Proc. Int. Conf. on Surface Modification of Metals by Ion Beams, Kingston, July 7–11, 1986,* in *Mater. Sci. Eng., 90* (1987).
8 H. Howarth, in L. L. Shreier (ed.), *Corrosion*, Vol. 1, Butterworths, London, 1979, p. 18.
9 P. Hendy, B. Kent, G. O. Lloyd, J. E. Roades-Brown and S. R. J. Saunders, *Proc. Eur. Congr. on Corrosion, 1977,* Society of Chemical Industry, London, 1977, p. 105.
10 M. Pons, M. Caillet and A. Galerie, in V. Ashworth, W. A. Grant and R. P. M. Procter (eds.), *Proc. Conf. on Ion Implantation into Metals, Manchester, 1981*, Pergamon, Oxford, 1982, p. 201.
11 P. D. Goode, in G. Carter, J. S. Colligon and W. A. Grant (eds.), *Proc. Int. Conf. on Applications of Ion Beams to Metals, Warwick, 1975*, in *Inst. Phys. Conf. Ser., 28* (1976) 154.

Materials Science and Engineering, 90 (1987) 205–212

Effects of Tin Ion and Nitrogen Ion Implantation on the Oxidation of Titanium*

PETER MADAKSON

IBM Thomas J. Watson Research Center, Yorktown Heights, NY 10598 (U.S.A.)

(Received July 10, 1986)

ABSTRACT

The high temperature oxidation of ion-implanted titanium was studied using the scanning electron microscopy, X-ray photoelectron spectroscopy and Rutherford back-scattering techniques. The specimens were implanted with either tin ions (Sn^+) or nitrogen ions (N_2^+) or a combination of the two, to doses of between 10^{15} and 10^{17} ions cm^{-2}. The implantation energies were 200 keV for tin ions and 200 or 400 keV for nitrogen ions. The specimens, treated and untreated, were oxidized in air at 500 °C for 100 h. The combination of nitrogen and tin ions was found to increase the thickness of the nitride–oxide film on the titanium surface by four orders of magnitude and to increase the hardness significantly. Nitrogen ions alone produced a more uniform oxide layer but the thickness of the layer was unchanged. The implantation of tin ions alone led to a reduction in the oxidation of titanium by the formation of SnO_2. The results are discussed in terms of well-established mechanisms of the oxidation of titanium, in which oxygen diffuses into the metal along anion vacancies of the rutile lattice.

1. INTRODUCTION

Titanium and its alloys are important in engineering applications where lightness, corrosion resistance and strength are needed. They are, therefore, widely used in the aerospace industry. The metal is fairly resistant to oxidation because of the formation of an adherent and compact oxide film but, at high temperatures, significant in-diffusion of oxygen takes place. At about 600 °C the metal displays parabolic oxidation kinetics which are believed to involve inward migration of oxygen atoms along short-circuit diffusion paths. Such a mechanism does not, therefore, require the generation of vacancies at the oxide–metal interface.

The need to control the high temperature oxidation of titanium without affecting the mechanical properties and dimensions of components has led to the use of ion implantation. The process is known to modify greatly the rate of gas–metal reactions. Several research workers [1] have shown that the oxidation of metals is significantly reduced by ion implantation. For titanium a selected number of ion species have been found to modify the oxidation characteristics [2–6]. Dearnaley *et al.* [2] first demonstrated that implanted calcium ions reduce the high temperature oxidation and that the reduction is dose dependent. A 70% decrease in oxidation was observed at a moderate dose of about 2×10^{16} ions cm^{-2}. Pons *et al.* [4] implanted a number of elements but they found only phosphorus to be effective in inhibiting the high temperature oxidation of titanium.

The general view about the effectiveness of ion implantation in inhibiting the oxidation of metals is that the electronegativity of the ion species is very important and that other atomic parameters such as valency or ionization potential are irrelevant [3]. Hence, ions with a low electronegativity would be most effective because they would "inhibit oxidation by providing traps for migrating oxygen ions at grain boundaries, which then serve to block further migration". However, it has recently become apparent that other parameters such as ionic radius, heat of formation and solubility are also very important [4, 6].

*Paper presented at the International Conference on Surface Modification of Metals by Ion Beams, Kingston, Canada, July 7–11, 1986.

0025-5416/87/$3.50

There are no available data on the study of the role of nitrogen on the oxidation of titanium but tin has been studied, both as an alloy addition in titanium and as an implanted ion species [4, 7, 8]. Tin increases the strength of the α phase of titanium and it has been observed to increase high temperature oxidation. The purpose of using both nitrogen and tin in the present work was to try to form an SnO_2 layer on top of a TiN layer on the surface. It was hoped that this would lead not only to a significant reduction in high temperature oxidation but also to an improvement in the mechanical properties of titanium.

2. EXPERIMENTAL PROCEDURE

2.1. *Ion implantation and Rutherford backscattering analysis*

The titanium alloy was supplied by Rolls Royce Limited and it contained the following: 5.5 wt.% Al, 3.5 wt.% Sn, 3 wt.% Zr, 1 wt.% Nb and 0.3 wt.% Si. Four polished specimens, each 1 in in diameter and $\frac{1}{8}$ in thick were used. Specimen A was not implanted but specimen B was implanted with 0.97×10^{17} N_2^+ cm^{-2} at 200 keV, specimen C with 1.06×10^{16} Sn^+ cm^{-2} at 200 keV and specimen D with 0.97×10^{17} N_2^+ cm^{-2} at 400 keV plus 0.97×10^{16} Sn^+ cm^{-2} at 200 keV. They were all heated at 500 °C, in air, for 100 h and allowed to cool slowly. They were then analysed by Rutherford backscattering using 1.5 MeV helium ions (He^+) from a 2 MeV Van de Graaff accelerator.

2.2. *X-ray photoelectron spectroscopy*

The specimens were examined by X-ray photoelectron spectroscopy (XPS) to produce compositional depth profiles. The analysis was carried out using a Vacuum Generator Scientific ESCA 3 Mark II electron spectrometer operating under the following conditions: analyser pass energy, 50 eV; X-ray, Al Kα, 1486.6 eV; vacuum pressure, 5×10^{-10} Torr; ion bombardment, 6 kV argon ions (Ar^+). For each specimen a compositional depth profile was produced by sequential XPS with argon ion bombardment. After each etch step, C 1s, O 1s, Ti 2p and N 1s spectra were recorded, together with a survey spectrum after five or so etch steps.

2.3. *Scanning electron microscopy and microhardness measurement*

The specimens were first cleaned by ultrasonic irradiation in acetone and air dried. The microscope was operated at 15 keV, giving an electron penetration of about 1 μm below the surface on normal incidence. The working distance was 20 mm. The hardness was measured with a Vickers microhardness tester at a load of 5 gf applied for 60 s.

3. RESULTS

Figure 1 shows the Rutherford backscattering spectra from the unimplanted specimen (specimen A) and the nitrogen-implanted specimen (specimen B). There is an oxygen peak and a reduced titanium yield on the surface. This suggests the formation of a TiO_2 film, which is about 30 channels (1000 Å) thick. There is no evidence of nitrogen or carbon on either surface. On comparison of the spectrum from the unimplanted specimen (specimen A) with that from the tin-implanted specimen (specimen C) (Fig. 2), an increased surface concentration of tin on specimen C can be observed. The tin is about 250 Å below the surface. The thickness of the TiO_2 is 26 channels (800 Å), about four channels less than that on the unimplanted and the nitrogen-implanted samples.

The Rutherford backscattering spectrum from the specimen implanted with tin plus nitrogen ions (specimen D) is compared with that from the unimplanted specimen (speci-

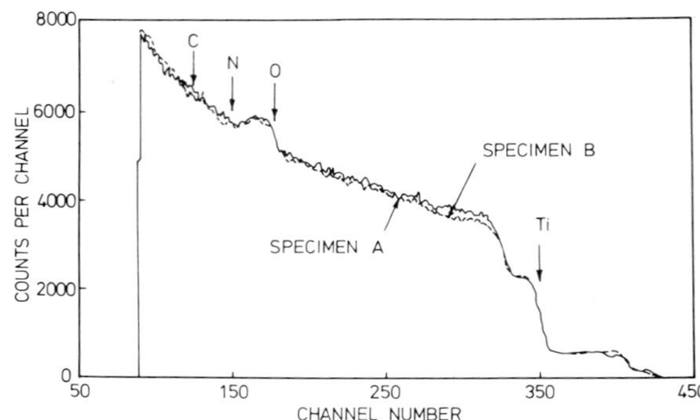

Fig. 1. Rutherford backscattering spectra of 1.5 MeV helium ions (He^+) from the unimplanted specimen (specimen A) and the nitrogen-implanted specimen (specimen B).

men A) in Fig. 3. Interestingly, there is no tin on the surface of specimen D. There is a large dip below the surface. This corresponds to increased concentrations of nitrogen and

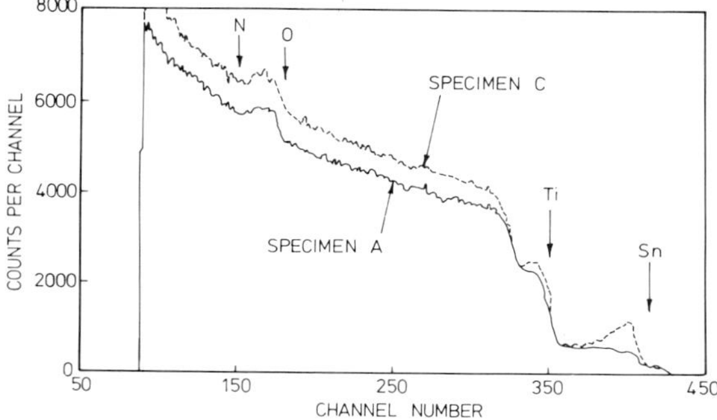

Fig. 2. Rutherford backscattering spectra of 1.5 MeV helium ions (He^+) from the unimplanted specimen (specimen A) and the tin-implanted specimen (specimen C).

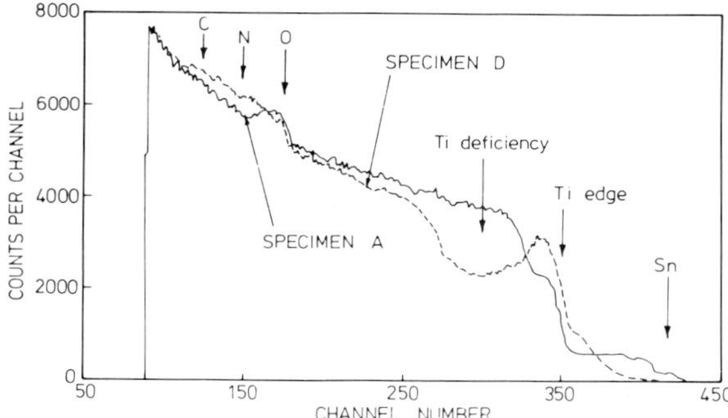

Fig. 3. Rutherford backscattering spectra of 1.5 MeV helium ions (He^+) from the unimplanted specimen (specimen A) and the tin-plus-nitrogen-implanted specimen (specimen D).

oxygen. The film is about 80 channels (about 2700 Å) thick.

By using energy analysis with scanning electron microscopy the impurity concentration within the top 1 μm layer was obtained. The results are given in Table 1. It can be seen that the concentrations of aluminum and tin on specimens C and D have increased very slightly. The concentration of aluminum was slightly reduced after the nitrogen ion implantation (specimen B) and the concentration of silicon on both specimen B and specimen C was almost halved.

A sample spectrum from the XPS analysis is given in Fig. 4. This is from specimen C. It shows large concentrations of carbon and oxygen on the surface and small amounts of aluminum and tin. The ion gun in the XPS analysis was operated at a 3 kV focus potential, with 6 kV argon ions (Ar^+). This was estimated to give an etch rate of approximately 2.5 nm min^{-1}. On the assumption that this is constant, the etching time was converted to depth below the surface. The plots of surface impurities with depth are given in Figs. 5–8.

Figure 5 shows the variations in the concentrations of titanium, oxygen, carbon and nitrogen with depth. The top layer on the surface (about 25 Å thick) consists mainly of carbon. Below this are fluctuating film thicknesses of oxygen, the first of them being about 0.07 μm. These oxide islands are similar to those reported by Pons *et al.* [4]. At the present, there is no explanation for why the oxidation process occurs in such a manner. The fluctuations in the titanium concentration with depth correspond to those of carbon and oxygen, *i.e.* a decrease in the titanium curve

TABLE 1

Surface concentration of impurities measured by scanning electron microscopy in implanted and unimplanted titanium after oxidation at 500 °C for 100 h

Specimen	*Implantation conditions*		*Impurities (average concentration* (wt.%) *in a layer 1 μm thick)*			
	Energy (keV)	*Dose* (ions cm^{-2})	*Al*	*Si*	*Zr*	*Sn*
A	—	—	4.44	0.27	3.28	1.76
B	200	0.97×10^{17} N_2^+	3.80	0.16	3.43	1.55
C	200	1.06×10^{17} Sn^+	4.75	0.15	3.40	2.32
D	400	0.97×10^{17} N_2^+	4.67	0.26	3.84	2.62
	200	0.97×10^{16} Sn^+				

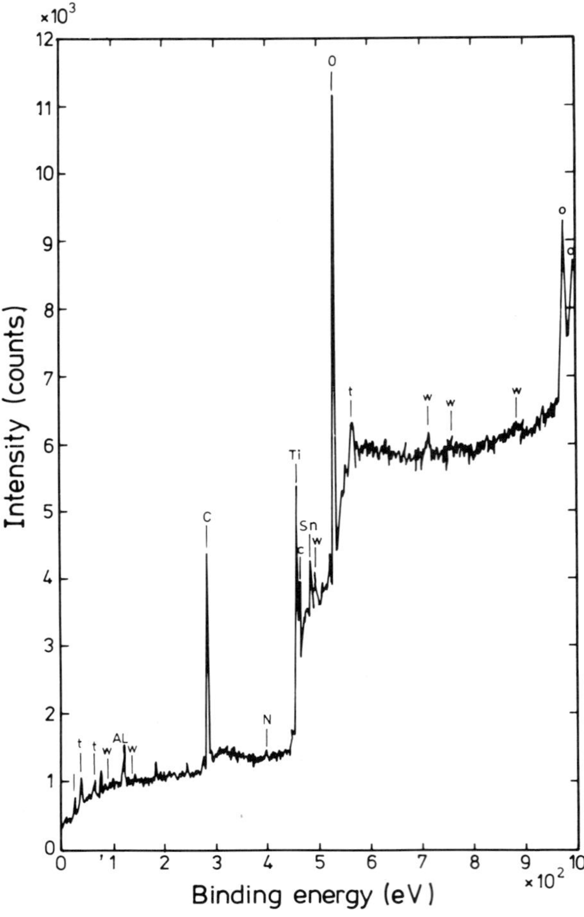

Fig. 4. X-ray photoelectron spectrum from the tin-implanted specimen (specimen C; 1.0×10^{16} Sn^{+} cm^{-2}; 200 keV): c, carbon; o, oxygen; w, tin; t, titanium.

corresponds to an increase in the oxygen curve and a decrease in the carbon curve. The displacement of titanium is probably due to the formation of carbides, nitrides and oxides.

The oscillations in the oxygen and carbon concentrations with depth, observed for the unimplanted specimen, did not appear for the nitrogen-implanted specimen (specimen B) (Fig. 6). After an initially high value the concentration of oxygen decreased gradually with depth for the first 0.17 μm layer. Beyond this, a rapid decrease in oxygen concentration is observed. As with the unimplanted specimen (specimen A), there is a high concentration of carbon in the top surface layer (less than 25 Å thick). After this, the carbon concentration decreases almost exponentially with increasing depth to a depth of about 0.17 μm, after which there is a gradual increase. It should be noted that the decreases in the oxygen and carbon concentrations with depth corresponds to an increase in nitrogen concentration, suggesting that the nitride layer forms below the oxide and the carbide layers.

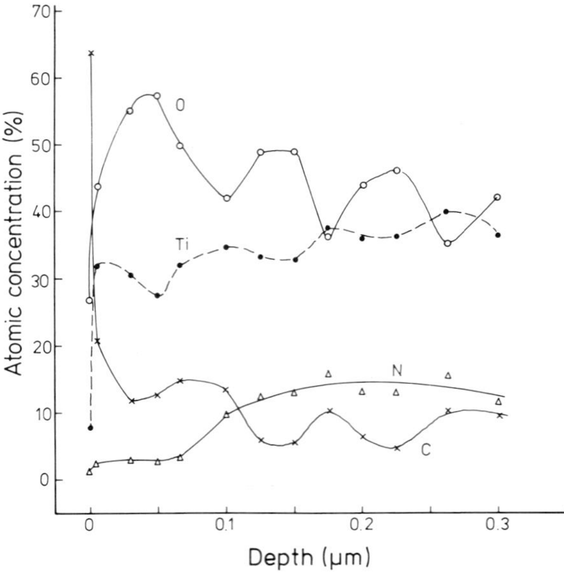

Fig. 5. The variations in the concentrations of oxygen, titanium, nitrogen and carbon with the depth for the unimplanted specimen (specimen A).

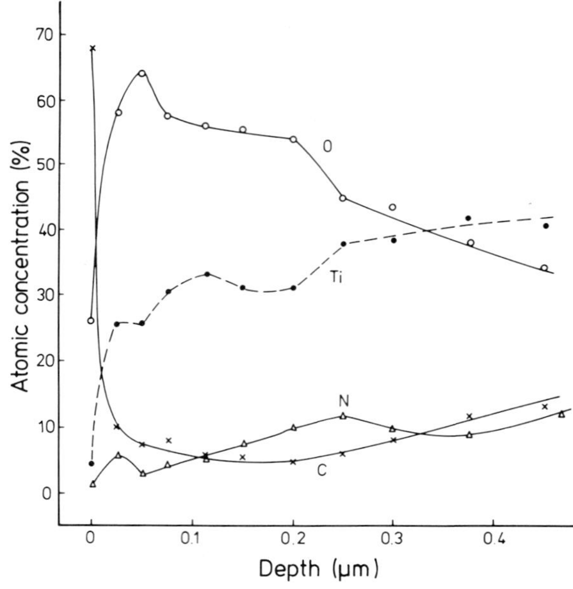

Fig. 6. The variations in the concentrations of oxygen, titanium, nitrogen and carbon with the depth for the nitrogen-implanted specimen (specimen B).

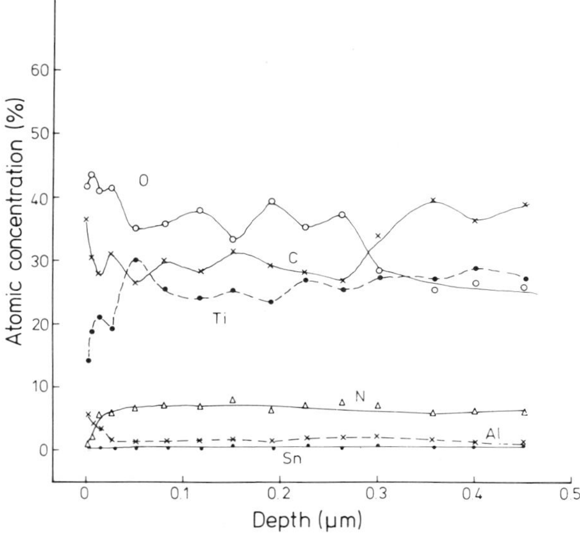

Fig. 7. The variations in the concentrations of oxygen, tin, aluminum, carbon, nitrogen and titanium with the depth for the tin-implanted specimen (specimen C).

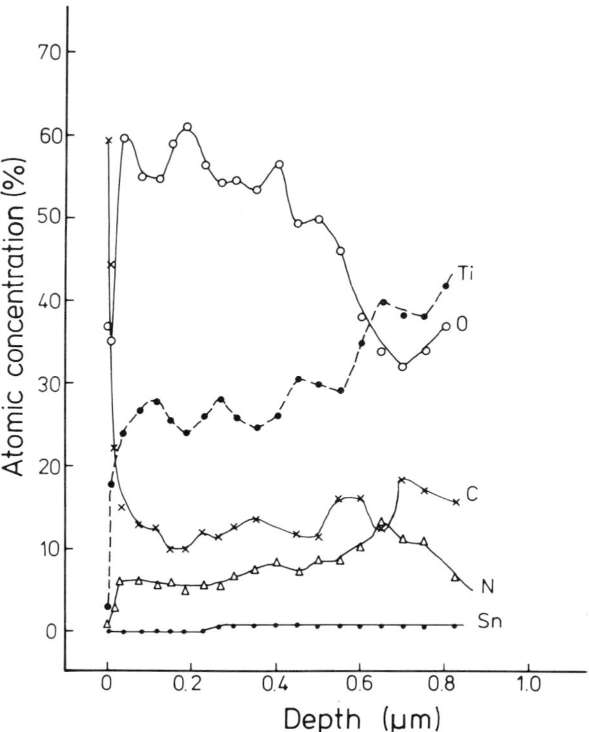

Fig. 8. The variations in the concentrations of oxygen, titanium, nitrogen, carbon and tin with the depth for the tin-plus-nitrogen-implanted specimen (specimen D).

In the tin-implanted specimen (specimen C), there was a fairly constant distribution of impurities with depth (Fig. 7). The concentration of oxygen was lower and that of carbon was higher than on any of the other specimens. The surface concentration of aluminum was also higher. There was less titanium in the implanted region, probably because of the displacement of titanium by tin, carbon and aluminum atoms. The tin-plus-nitrogen-implanted specimen (specimen D) showed a very high concentration of oxygen which extended to a depth of about 0.4 μm (Fig. 8). On top of this very thick oxide layer was a thin carbide film. Both tin and nitrogen seem to be evenly distributed throughout the oxide layer.

Table 2 indicates that specimen D produced the largest increase in hardness, an increase of about seven times before oxidation and 53 times after oxidation. Nitrogen implantation (specimen B) also produced a significant increase in hardness but there was no change in the tin-implanted specimen.

TABLE 2

Average increase in hardness

Specimen	*Factor of improvement*	
	As implanted	*After oxidation*
A	1	11
B	3	22
C	1	11
D	7	53

4. DISCUSSION

The Rutherford backscattering data are considered to be more reliable and so most of the discussion will be based on these. However, although it seems that there was surface contamination of all the specimens during the XPS analysis, the results support the formation of thick TiO_2 films, as observed in the Rutherford backscattering data, on all the specimens. Each specimen shows a significant concentration of carbon with depth (Figs. 5–8), with the tin-implanted specimen having about 30 wt.% C. This is probably associated with vacuum contamination.

Nitrogen implantation has no effect on the oxidation of the titanium alloy investigated.

The thickness of the TiO_2 film was 1000 Å, the same as that of the unimplanted specimen. This is probably because of the presence of nitride-forming elements such as zirconium, silicon and aluminum. Si_3N_4 and ZrN are more stable than TiN. So, if the implanted nitrogen is taken up by silicon and zirconium, titanium will still be free to take up oxygen. The mechanism of oxidation would, therefore, remain unchanged and the thickness of the TiO_2 film would remain unchanged. The situation might be different if nitrogen were implanted to a much shallower depth. The theoretical mean depth of penetration of 200 keV nitrogen ions (N_2^+) is about 1700 Å, with most of the nitrogen concentrated in a layer about 1000 Å thick, and 500 Å below the surface. The tail of the nitrogen distribution extends to about 3000 Å below the surface. So the ions have a large area in which to move about, to give a thinly dispersed concentration of nitrides that would have no effect on the oxidation of the titanium alloy.

Tin was implanted to a mean depth of about 620 Å, producing a layer 700 Å thick about 150 Å below the surface. This resulted in two TiO_2 layers: one on the surface and the other below the tin layer (see Fig. 2, specimen C). It should be noted that the tin edge is below the surface and that there are two slopes in the titanium spectrum: one at the edge and the other after the dip. The tin is most probably in the form SnO_2, which is about 500 Å thick. There is no significant change in the thickness of the oxide layer, but the layer is not predominantly TiO_2 as in the unimplanted or nitrogen-implanted specimens. Oxygen might have diffused through the tin layer because of insufficient tin atoms to trap it; a higher dose of tin ions is, therefore, needed.

Implantation with tin plus nitrogen ions produces a thick layer (of about 0.3 μm thickness) that consists of TiO_2 and TiN, below the surface. Tin was implanted after the nitrogen implantation. The purpose was to form an SnO_2 layer on a nitride layer (SnO_2/TiN/Ti), the two layers acting as double barriers to the diffusion of oxygen to the titanium surface. Tin would trap oxygen to form SnO_2 and the compressive stress generated by this in the titanium lattice would help to block inward diffusion of oxygen and outward diffusion of nitrogen. This would enable a TiN layer to form below the SnO_2 layer. Nitrogen implantation alone would have no effect on oxidation because, as suggested above, nitrogen atoms would be free to move to the surface so that only dispersed nitride precipitates would form. Homogeneous and continuous nitride and oxide layers are therefore needed to produce a significant effect on the high temperature oxidation of the implanted titanium.

Microscopic analysis of the specimens revealed the formation of colors on the surfaces. Patches of yellow on top of a white background were observed on the unimplanted and the nitrogen-implanted specimens. Yellow, gold, brown and green on a white background were observed on the tin- and the tin-plus-nitrogen-implanted specimens (Table 3). The observed colors appeared to nucleate at isolated points on the specimen and then to spread outwards on the surface. The formation of colors on oxidized titanium was first reported by Cotton and Hayfield [9] in 1967 and it is now well known and used in industry. They found that the colors depend on film thickness and whether titanium is heated in air or in, say, an oxygen environment [10].

The mechanism of the oxidation of titanium is thought to be largely associated with oxygen passing inwards, probably along vacant anion sites in the lattice [11]. Enhanced anion vacancy concentration, therefore, increases oxygen diffusion. When the oxygen reaches the metal–oxide interface, "part of it enters the metal in solid solution and part goes into forming fresh oxide". The oxide film subsequently formed is under stress. Any inhibition of further oxidation, therefore, depends largely on the compressive stresses

TABLE 3

Colors and corresponding compounds after oxidation

Color	Yellow	White	Brown	Green
Compound	TiN	TiO, TiO_2, SnO_2	Not known, possibly, Ti_3Sn	Not known

within the oxide film [12]. Ion implantation can produce large compressive stresses within the treated material [13, 14]. The stresses increase linearly with increasing dose and stress relief occurs at high doses (above 5×10^{16} ions cm^{-2} for metals) [14].

However, whether implanted ions decrease or increase oxidation depends on other factors also. Solid solubility and the electronegativity of the ion species have been considered to be important [3, 15]. The results presented here show that only the tin and tin-plus-nitrogen implantation produced a significant effect on oxidation. Perhaps, by precipitating at grain boundaries the implanted tin forms "gates" which trap oxygen to form SnO_2. The formation of SnO_2 further increases the compressive stress to result in tighter "gates" which inhibit any inward migration of oxygen.

The effect of radiation damage does not seem to be important. Dearnaley [16] observed that bombarding titanium with titanium ions (Ti^+) "had no measurable effect on the subsequent oxidation, showing that radiation damage, sputtering etc. do not influence the process". Antilla *et al.* [17] think that this radiation insensitivity of titanium is because "the defects, at least vacancy-interstitial pairs, arising from the bombardment will disappear at rather low temperatures", owing to the low activation energy for the self-diffusion of titanium. Hence, if titanium recovers very rapidly from radiation damage because of its low activation energy (1.6 eV) [12, 18], then implanted ions with a higher activation energy of migration, *e.g.* nitrogen (with an activation energy of 2.3 eV), would not have any influence on the oxidation of the metal. The results from the nitrogen implantation support this point. Collins *et al.* [19] indicated that ions with a low diffusivity in the implanted metal (*e.g.* tin in titanium) offer the best protection against high temperature oxidation because they produce the largest residual stress.

5. CONCLUSIONS

The implantation of nitrogen has little effect on the oxidation of titanium. Implanted tin is thought to act as a "gate" that traps oxygen, but a sufficient amount of tin is needed to block the inward migration of oxygen effectively. The combination of tin and nitrogen seems promising, especially where titanium is exposed to a very high temperature and long periods of oxidation. The formation of two oxygen diffusion barriers, SnO_2 and TiN, should be very effective under these conditions. More work is, therefore, needed on the Sn–N system. Nitrogen should be implanted at high temperatures to promote the formation of a TiN layer in the vacuum before a high dose (5×10^{16} ions cm^{-2} or more) of tin is implanted.

ACKNOWLEDGMENTS

The author would like to thank the U.K. Science and Engineering Research Council for the award of a grant for this research, Mr. G. Syers, Rolls Royce Limited, for the provision of specimens, Dr. C. Jynes, University of Surrey, for the implantation and Rutherford backscattering analysis, Professor J. C. Anderson, Imperial College, London, for the scanning electron microscopy analysis and Dr. J. F. Watts, University of Surrey, for the XPS analysis. The work was done at King's College, University of London.

REFERENCES

1 *Ion Implantation and Ion Beam Analysis Techniques in Corrosion, Corros. Sci., 20* (1980).
2 G. Dearnaley, P. D. Goode, W. S. Miller and J. F. Turner, in B. L. Crowder (ed.), *Ion Implantation in Semiconductors and Other Materials*, Plenum, New York, 1973, p. 405.
3 G. Dearnaley, in S. T. Picraux, E. P. EerNisse and F. L. Vook (eds.), *Proc. Int. Conf. on Applications of Ion Beams to Metals, Albuquerque, NM, 1974*, Plenum, New York, 1974, p. 63.
4 M. Pons, M. Caillet and A. Galerie, *Nucl. Instrum. Methods, 209–210* (1983) 1001.
5 G. H. Gleaves, R. A. Collins and G. Dearnaley, *J. Electroanal. Chem., 137* (1982) 51.
6 J. D. Benjamin and G. Dearnaley, in G. Carter, J. S. Colligon and W. A. Grant (eds.), *Proc. Int. Conf. on Applications of Ion Beams to Metals, Warwick, 1975*, in *Inst. Phys. Conf. Ser., 28* (1976) 141.
7 B. de Gelas, R. Molinier and L. Seraphin, in J. C. Williams and A. F. Below (eds.), *Titanium and Titanium Alloys*, Plenum, New York, 1982, p. 2121.
8 D. V. Ignator, Z. I. Kornilova and E. M. Lazarev, *Russ. Metall. (Engl. Transl.), 4* (1975) 182.

9 J. B. Cotton and P. C. S. Hayfield, *Trans. Inst. Met. Finish.*, *45* (1967) 48.

10 A. McQuinllan and M. McQuillan, *Metallurgy of the Rarer Metals–Titanium*, Butterworths, London, 1956.

11 U. R. Evans, *The Corrosion and Oxidation of Metals*, Edward Arnold, London, 1960.

12 A. E. Jenkins, *J. Inst. Met.*, *82* (1953) 213.

13 A. E. EerNisse and S. T. Picraux, *J. Appl. Phys.*, *48* (1977) 9.

14 P. B. Madakson, *J. Phys. D*, *18* (1985) 531.

15 A. Galerie, in V. Ashworth, W. A. Grant and R. P. M. Proctor (eds.), *Proc. Int. Conf. on Ion Implantation into Metals, Manchester, 1981*, Pergamon, Oxford, 1981, p. 190.

16 G. Dearnaley, in J. H. Hirvonen (ed.), *Ion Implantation Treatise Mater. Sci. Technol.*, *18* (1980) 276.

17 A. Antilla, J. Ráisánen and J. Keinonen, *Appl. Phys. Lett.*, *42* (1983) 500.

18 L. E. Modolfo, *Aluminum Alloys, Structure and Properties*, Butterworths, London, 1979.

19 R. A. Collins, S. Muhl and G. Dearnaley, *J. Phys. F*, *9* (1979) 1245.

Aqueous Corrosion Behaviour of Ion-Implanted Metals*

H. FERBER and G. K. WOLF

Physikalisch-Chemisches Institut, Universität Heidelberg, Heidelberg (F.R.G.)

(Received July 10, 1986)

ABSTRACT

We can now look back at 10 years of application of ion beams in corrosion studies. Therefore, after the introduction, we first attempt to give an overview of what has been accomplished during this period in the field of aqueous corrosion, with emphasis on developments in more recent years. Then we present a more detailed discussion of some particular examples of research which make use of different types of corrosion protection mechanism as well as applications of different types of ion beam technique to metal surfaces. These examples include the application of ion beam mixing and ion-beam-assisted vapour deposition to (i) the prevention of localized corrosion, (ii) the reduction of hydrogen uptake by metals (the formation of "migration barriers"), (iii) corrosion protection by means of ion-beam-mixed monolayers and multilayers of aluminium and boron and (iv) ion-beam-modified carbon layers and their influence on the corrosion of mild steel. Following these examples, we attempt to deduce recommendations for the future application of ion beams in corrosion science.

1. INTRODUCTION

The usefulness of ion implantation in the field of basic research and understanding of corrosion mechanisms has been stressed by many researchers (see for example ref. 1). None the less, there exists only a relatively small number of papers in which use is made of the method in basic electrochemical research. This is obviously because only a few electrochemists have access to accelerators, and the scientists running these machines are not normally interested in electrochemical research. However, the few publications in existence in which basic corrosion research is dealt with reflect strikingly how well the two methods go together (see for example refs. 2-5). The possibility of combining most elements in either stable or metastable alloys over a wide range of compositions should present a tempting invitation to any electrochemist involved in basic research. This is particularly so since one of the major disadvantages of implanted surface layers for technical application, namely their shallowness (usually not exceeding 100 nm), is not detrimental to basic research experiments.

The number of papers containing application-oriented work is also still limited. Unlike the modification of mechanical properties (namely friction and wear), ion beams have still to be applied industrially to corrosion problems, although some work has suggested that this could happen in the near future. In this paper, we have attempted to compile some of the basic as well as the application-oriented work of recent years which is concerned with the corrosion properties of ion-implanted metal surfaces.

2. WHAT HAS BEEN DONE SO FAR?

2.1. *Basic corrosion studies*

Most of the basic research in the past has been on iron and different types of steel. Such substrates as aluminium [6], copper [7] and titanium [8] were also dealt with. Early work was concerned with the application of rare gas (mainly argon) ions to examine the contribution of physical effects (lattice damage) to the observed electrochemical behaviour

*Paper presented at the International Conference on Surface Modification of Metals by Ion Beams, Kingston, Canada, July 7-11, 1986.

0025-5416/87/$3.50

of implanted samples. Next was the production of conventional alloys such as Fe–Cr and metastable alloys such as Fe–Au, Fe–Pb, Fe–Ta and Fe–Hg for basic electrochemical research. Molybdenum and titanium ion beams were used to improve the pitting behaviour of steels [9].

In some papers the formation of an amorphous surface layer has been reported. The ions used in these cases were boron and phosphorus which, after implantation into iron or steel at high fluences (a maximum concentration of more than 10 wt.%), showed the ability to render a superficial layer of the metal amorphous. Chen *et al.* [10] showed that such layers on stainless steel exhibited outstanding corrosion resistance in acid solution. Table 1 contains further examples of work dealing with amorphous metals [9, 12, 13].

In the majority of papers, potentiodynamic recording of voltammograms served as the method of investigating the corrosion behaviour of implanted metals. This method has the advantage of being rapid but also has some disadvantages. Changing the potential of the metal electrode under investigation over a wide range from the hydrogen evolution region up to the onset of oxygen evolution yields a valuable overall impression of the corrosion and passivation behaviour of the electrode. However, it must be borne in mind that these results can only be interpreted in a rather qualitative way by means of comparison of the critical current density, the passive current density and the corrosion potential measured on different samples. They cannot normally be a substitute for long-time immersion tests under free-corrosion (open-circuit) conditions and therefore usually fail to impress any potential user.

Most of the publications mentioned so far were concerned with anodically controlled corrosion of metals and consequently the

TABLE 1

The effect of ion implantation on aqueous corrosion (selected examples)

Substrate	*Ion*	*Result*	*Test solution*	*Corrosion mechanism*	*Effect*	*Reference*
Fe	Ar	Radiation damage	Acetic acid buffer	Anodic	Slight	Ashworth *et al.* [5]
Fe	Cr	Stable alloy	Acetic acid buffer	Anodic	Considerable inhibition	Ashworth *et al.* [5]
Fe	Ta	Metastable alloy	Acetic acid buffer	Anodic	Considerable inhibition	Ferber [11]
Fe	Pb	Metastable alloy	1 N H_2SO_4	Anodic	Inhibition	Ferber and Wolf [3]
Fe	Hg	Metastable alloy	1 N H_2SO_4	Anodic	Inhibition	
Fe	Au	Metastable alloy	1 N H_2SO_4	Anodic	Promotion	
AISI 52100	Ti	Amorphous Fe–Ti–C alloy	1 N H_2SO_4	Anodic	Moderate inhibition	Hubler and McCafferty [8]
Type 316L stainless steel	B, P	Amorphous	1 N H_2SO_4	Anodic	Inhibition	Chen *et al.* [10]
Type 304 stainless steel	P	Amorphous	1 N H_2SO_4	Anodic	Inhibition	Wang *et al.* [9]
Fe	B + N mixing	FeB + BN amorphous?	1 N NaCl, pH 4	Anodic	Inhibition	Marchetti *et al.* [12]
AISI 316 and 440 stainless steel	B	Amorphous	1 N H_2SO_4	Anodic	Inhibition	Kim *et al.* [13]
Ti	Pd	Stable alloy	1 N H_2SO_4	Cathodic	Inhibition	Munn and Wolf [14]
Stainless steel	Pt	?	20% H_2SO_4	Cathodic	Inhibition	Ensinger *et al.* [15]

researchers were looking for lowered critical current densities as well as passive current densities in the recorded voltammograms of ion-beam-treated electrodes relative to those of untreated electrodes. Such findings indicate a potential slow-down of the active corrosion rate.

Table 1 summarizes some of the basic research mentioned so far. It also contains an example of work on the Ti–Pd system where the corrosion is under cathodic control [14]. Through stimulation of the hydrogen evolution reaction, titanium electrodes can be rendered passive; the corrosion potential is shifted into the passive region of titanium and the electrode remains passive. This method had previously been successfully applied in the use of conventionally produced, low percentage Ti–Pd bulk alloys; the application of a palladium ion beam to titanium can therefore also be regarded as an example of application-oriented work.

A summary of the evaluation of the basic corrosion research carried out in recent years shows that the formation of surface alloys by ion implantation produced in most cases beneficial effects on aqueous corrosion in various electrolytes. This holds for both uniform corrosion attack and pitting; however, little has so far been accomplished in the fields of stress corrosion cracking and hydrogen embrittlement.

With respect to the aforementioned beneficial effects, it has to be stressed that the corrosion protection exerted by the implanted layers was, in most cases, shown to deteriorate rather quickly, especially in active corrosion of the implanted substrates. It is therefore now a common assumption that surface modification by pure ion implantation of metal substrates is unsuitable for long-term corrosion protection against active corrosion. This is due to the restricted penetration depth of the metal ions implanted with energies that are usually in the region 20–400 keV. Fortunately, such new developments in ion beam applications as ion beam mixing (interface mixing), multilayer mixing and ion-assisted vapour deposition are now at hand and present a means for producing thicker protective layers than does ion implantation on its own. Some research groups have recently produced evidence that these new methods of surface modification can be successfully applied to aqueous corrosion and, in Section 3, recent examples of such investigations are presented.

2.2. *Application-oriented corrosion studies*

In Table 2, we have listed examples of work on systems which could be useful for technical application. Although this list is far from complete it is none the less true that, to date, comparatively little has been achieved in the practical application of ion-beam-modified metals in corrosive environments. However, the wider attention that the method has recently attracted because of its considerable success in surface hardening and wear reduction of metals will almost certainly influence the number of applications in the field of corrosion in a positive way. One reason for such a trend could be the fact that metals with modified wear properties may also, depending on the field of application, have to be checked for possible changes in their corrosion behaviour.

The examples of corrosion research given in Table 2 cover a wide area of different corrosion phenomena and include ion-beam-mixing experiments as well as multilayer mixing. One of the most elaborate examples of ion beam modification of a material for practical application is given by the work of the Naval Research Laboratory who studied the prevention of the pitting of an M50 ball-bearing steel in a sea water spray environment [16]. This research has now reached the phase of technical maturity. The only other work for which this can be said is concerned with the ion implantation of Ti–6Al–4V surgical alloy [24] which is used for prostheses such as hip joints and knee caps. Recently, Sioshansi [25] has reported that ion implantation for improved wear and corrosion resistance has now found its way into the actual production process of such surgical implants.

So far, stress corrosion cracking has not been shown to be influenced by ion implantation although various metal ion species have been implanted into various steels [18]. Implantation of silver into a special aluminium alloy had a slight beneficial effect but the reproducibility of the experiments was poor [19]. The remaining examples in Table 2 are discussed in Section 3 in more detail.

TABLE 2

Selected examples of application-oriented corrosion studies

Substrate	*Ion*	*Corrosion type*	*Corroding agent*	*Effect*	*Reference*
M50 steel	Cr (+ P)	Pitting	NaCl solution	Improvement	Hubler *et al.* [16]
Type 304 stainless steel	P	Pitting	H_2SO_4	Improvement	
Ti	Pd (implanted and mixed)	Crevice corrosion	$MgCl_2$, 142 °C	Long-term improvement	Munn and Wolf [17]
Various steels	Ti, Mo, Ni	Stress corrosion cracking	Na_2CO_3	No effects	Craig and Parkins [18]
Al alloys	Ag	Stress corrosion cracking	3% NaCl solution	Small effects	Wolf and Buhl [19]
Fe–steel	O, P	Hydrogen embrittlement	Aqueous solutions	Permeation barrier	Reiß *et al.* [20]
Ta	Pt	Hydrogen embrittlement	Aqueous solutions	Permeation promoter	Ensinger and Wolf [21]
Low carbon steel	Al (ion-assisted deposited)	Uniform corrosion and pitting	Aqueous	Improved adhesion and corrosion	Andoh *et al.* [22]
Low carbon steel + C (+ H) layer	Rare gas interface mixing	Uniform corrosion	Acetate buffer solution	Improved adhesion and corrosion	Ferber *et al.* [23]

3. SOME DETAILED EXAMPLES OF RECENT RESEARCH

3.1. *Crevice corrosion of titanium*

Titanium, which is known for extremely good corrosion resistance in most environments, is susceptible to crevice corrosion attack under extreme conditions. Munn and Wolf [17] have shown that the implantation of palladium or the ion beam mixing of previously deposited palladium layers into titanium protects the metal against this form of corrosion.

Test specimens with narrow crevices were immersed in boiling $MgCl_2$ solution (42%) and long-time corrosion tests were carried out. Untreated titanium showed pitting attack inside the crevice after only 5 days under test. Implanted samples (10^{16} Pd^+ cm^{-2} at 200 keV), however, remained passive for 4 weeks under the same test conditions. Inspection under the optical microscope confirmed the total absence of pits inside the crevices of the implanted samples. Following these experiments, Munn and Wolf produced Ti–Pd surface layers by means of ion beam mixing. Palladium layers of 40 nm thickness were deposited on titanium samples and irradiated with a rare gas ion beam (4×10^{16} Kr^+ cm^{-2} at 380 keV). Corrosion tests were carried out as described above. It was shown that such specimens had still not been attacked after 4 months in boiling $MgCl_2$ solution. The improved resistance was explained as follows: the palladium on the titanium surface catalyses the hydrogen evolution reaction which is part of the total corrosion process. When this reaction does not take place, the only possible cathodic reaction counterbalancing the anodic dissolution of titanium is oxygen reduction. Since mass transfer inside the narrow crevice is restricted, oxygen is depleted and the pH value inside the crevice falls, rendering the titanium active and thus vulnerable to pitting attack. In the presence of the catalytically active palladium on the titanium surface, hydrogen ion reduction takes place, a decrease in the pH value inside the crevice is avoided and the titanium consequently remains passive. As in all cases of cathodic protection, the thickness and uniformity of the palladium-implanted or palladium-intermixed layers are not crucial, as long as enough palladium remains on the surface to stimulate sufficiently the hydrogen evolution reaction. For protection against crevice corrosion, ion implantation or ion beam mixing of palladium could obviously be a useful alternative to bulk alloying titanium with palladium.

3.2. *Influence of ion implantation on the permeation of hydrogen through iron*

The hydrogen uptake of some metals corroding in aqueous solutions with hydrogen evolution as the cathodic reaction can be considerable and, as a consequence, hydrogen embrittlement can occur. This is effectively described as the weakening of the metal substructure by a build-up of high pressure hydrogen gas inside small voids and along grain boundaries in the metal. The formation of hydrides can also have a detrimental effect in this context. The problem is especially severe with high strength steels. Reiß [26] has studied hydrogen uptake, permeation and diffusion through iron foils (thickness, 0.5 mm) by means of an electrolytic double cell which allowed hydrogen to be evolved under galvanostatic control at one face of an iron foil and to be oxidized under potentiostatic control on the opposite face, after diffusion through the foil. This set-up allowed measurement of both the hydrogen permeation kinetics through the foil (by detecting the hydrogen oxidation current on the exit face) and the diffusion coefficient (through evaluation of the current transients caused by stepwise increases in the cathodic overpotential at the hydrogen entry surface of the iron foil). The measurements were carried out in 0.1 N NaOH.

Implantation of many elements, amongst them oxygen and phosphorus, into the entrance side of iron foils was carried out (2×10^{17} P^+ cm^{-2} at 150 keV; 1×10^{18} O_2^+ cm^{-2} at 100 keV). The choice of these two species derived from the idea that an "internal oxide layer" and an amorphous layer produced by phosphorus implantation could present effective barriers to hydrogen migration into and within the iron lattice respectively. These effects should show up as a decrease in the measurable hydrogen permeation current; Fig. 1 shows the appropriate results.

The graph shows the dependence of the hydrogen permeation current density on the charging current density for unimplanted, oxygen-implanted and phosphorus-implanted foils. For the same charging current, both phosphorus- and oxygen-implanted samples yield a lower hydrogen permeation current; this is attributable to the above-mentioned effects of oxygen and phosphorus after implantation, although in the phosphorus implantation the formation of an amorphous surface layer could not be convincingly confirmed.

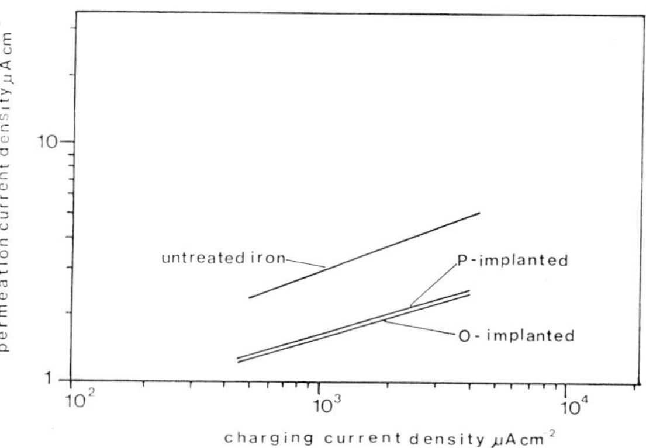

Fig. 1. The influence of oxygen ion (O_2^+) and phosphorus ion (P^+) implantation on the hydrogen permeation through an iron foil.

The decrease measured in the permeation current seems to be small but is in fact astonishingly large when it is borne in mind that the thickness of the ion-beam-modified layer is almost negligible compared with the overall thickness of the foil through which permeation took place. We believe that these results hold some promise for further attempts to suppress the hydrogen uptake, and thereby the hydrogen embrittlement, of metals using ion beams; however, very little attention has been given to this possibility so far.

3.3. *Protective surface layers produced by ion beam mixing*

We have already mentioned that the relatively new development of ion beam mixing promises to help to overcome some of the intrinsic limitations of direct ion implantation, namely the shallow depth of the modified surface layers which therefore cannot produce long-time corrosion protection of metal surfaces as long as measurable corrosion rates prevail. Since this is normally the case on modified iron and steel substrates, first efforts have been made to overcome these limitations by depositing non-corroding elements such as boron or aluminium on iron substrates and irradiating these layers with gas ions. Multilayers and sandwich layers have also been used to produce thicker, homogeneously intermixed surface films. By choosing an appropriate element for the bombarding beam, additional chemical effects can be produced by formation of inert quasi-stoichiometric chemical compounds such as nitrides or carbides. The mixing process ensures the necessary good adhesion of these compound layers. Recently, Elena *et al.* [27] have reported the latest results of ion-beam-mixing experiments with boron layers on Armco iron. They implanted krypton and nitrogen ions into previously deposited multilayers of iron and boron on iron. Voltammograms of these intermixed layers were recorded and a positive shift in the pitting potential was found (particularly for nitrogen implantation) which, in the opinion of Elena *et al.*, led to the formation of a very stable BN film with excellent adhesion to the iron substrate. The tests were carried out in deaerated 1 M NaCl solution of pH 4 and in Fig. 2 some of their results are presented.

As can be seen, the unimplanted multilayer pits rather quickly after going through a repassivation phase and implantation of rare gas ions does not change this behaviour much. Nitrogen implantation, however, leads to a substantial shift in the pitting potential to more positive values. Stationary potentiostatic current density *vs.* time plots also confirmed this finding. However, it should not be forgotten that this type of experiment cannot accurately predict long-time corrosion resistance under free-corrosion conditions.

A different ion beam technique, closely related to ion beam mixing, should not be left out at this stage, namely simultaneous vapour deposition and ion implantation (ion-beam-assisted deposition). The equipment necessary for this method is a vacuum chamber containing a gas ion source and an electron gun evaporator (with one or more crucibles). A prototype of such a machine was built by Andoh *et al.* [22] and it contained a bucket-type ion source for the production of relatively low energy (2–40 keV) beams with high beam currents (100 mA for nitrogen ions). Using such high currents, the sample holder has to be kept at low temperatures by water cooling. Although more machines of this type have now been built by other research groups, little work in the field of corrosion protection has so far been performed. Andoh *et al.* used the method for the production of strongly adherent aluminium coatings (evaporated and subsequently treated with an argon ion beam

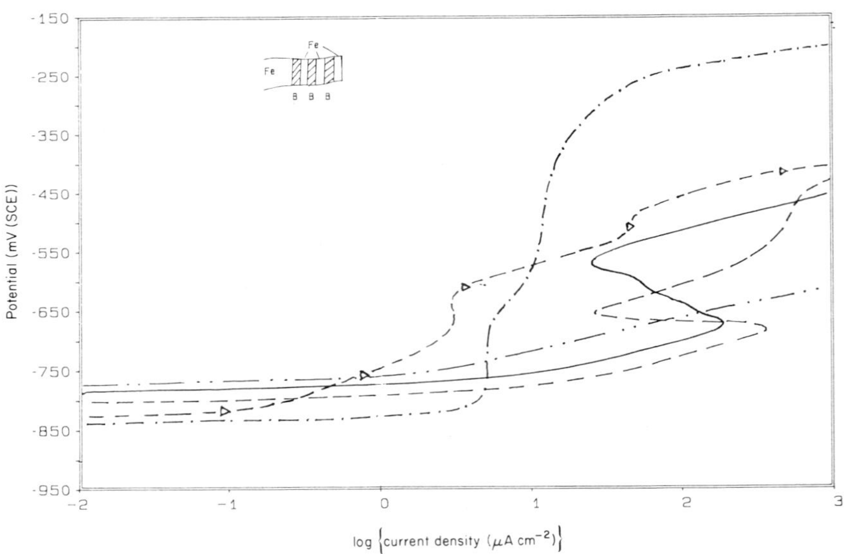

Fig. 2. The effect of mixing Fe/B multilayers with rare gas and nitrogen on the polarization behaviour of Armco iron: —·—, Fe/B/N multilayer; - -△- -, Fe-B-N; —··—, Armco iron; - - -, Fe/B multilayer; ——, Fe/B/Kr multilayer.

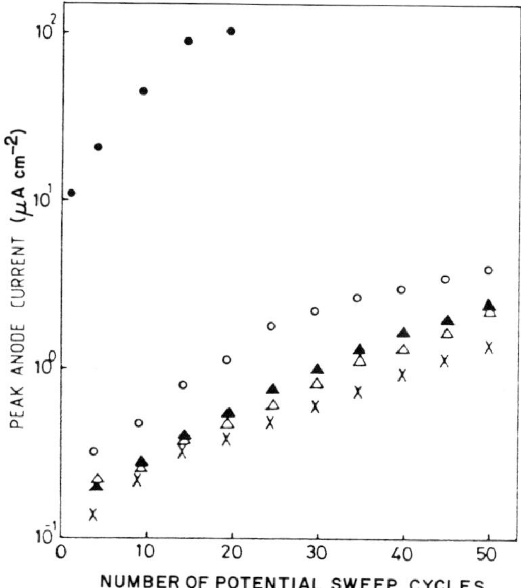

Fig. 3. Corrosion behaviour of ion-beam-mixed aluminum coatings on a low carbon steel: ●, uncoated; ○, as coated at room temperature; ×, as coated at 200 °C; ▲, coated at room temperature, 1×10^{16} Ar^+ cm^{-2}; △, coated at room temperature, 1×10^{17} Ar^+ cm^{-2}.

at 30 keV) on a low carbon steel. Apart from the adhesion and bending properties, they also investigated the corrosion behaviour of the layers by means of cyclic volammetry in an acetic buffer solution of pH 5.

Figure 3 shows the results plotted as the critical current density (anodic peak density) *vs.* the number of sweep cycles, comparing uncoated, coated and mixed samples. Unfortunately the additional improvement after implantation in these preliminary experiments is not substantial but, none the less, ion-beam-assisted deposition looks very suitable for engineering corrosion protection layers.

3.4. *Ion-beam-modified carbon layers for corrosion protection*

As a last example of the work on corrosion reduction by ion beam applications, we present some more recent results of our own research group in Heidelberg dealing with ion-bombarded carbon layers on low carbon steel [28]. The early results of this project have been reported previously [23]. In this work, use was made of the corrosion properties of carbon-deposited and subsequently irradiated mild steel samples to monitor the adhesion and corrosion properties of these very inert layers. Carbon layer formation was by means of plasma decomposition of hydrocarbons (ethylene). Such layers are known to contain up to 40 wt.% H. Application of ion beams to carbon layers had previously been shown to render them more dense and, through a loss of hydrogen, the carbon can be transformed into a diamond-like modification (sp^3 hybridization) which exhibits excellent corrosion resistance and very good mechanical properties.

Whereas some other researchers have simply applied a relatively low bias voltage to the samples during the deposition, we have tried to modify the carbon (plus hydrogen) layers in a more defined way by energy- and fluence-controlled ion implantation subsequent to the evaporation step. We concentrated on layers about 50 nm and 100 nm thick and implanted them with krypton ions with energies high enough to achieve mixing through the Fe–C interface (200 keV and 400 keV respectively). The ion fluences varied between 2×10^{15} and 6×10^{16} Kr^+ cm^{-2}.

Again the test method applied was potentiostatic polarization in an acetic acid–acetate buffer of pH 5.6. Although multisweep measurements were carried out as a routine, for comparison between samples we used only the first anodic-going sweep.

Our results can be summarized as follows (Fig. 4).

(i) The corrosion properties of implanted samples were greatly improved compared with those of unimplanted samples, and the estimated corrosion rate decreased by several orders of magnitude. Both the anodic and the cathodic reaction kinetics were beneficially influenced. For high fluence implantation into 100 nm carbon layers the prevailing hydrogen evolution and iron dissolution currents were hardly measurable with our electronic set-up and we believe that they were in fact only due to imperfections in the evaporated layers already present prior to implantation.

(ii) The beneficial effects could only be produced with beam energies high enough for the ions to penetrate the carbon layers and were obviously predominantly attributable to improved adhesion of the carbon films after implantation. Unimplanted carbon layers and layers bombarded with low energy ions could be quantitatively removed by cathodic pre-polarization (as a result of spalling caused by hydrogen evolution) and the

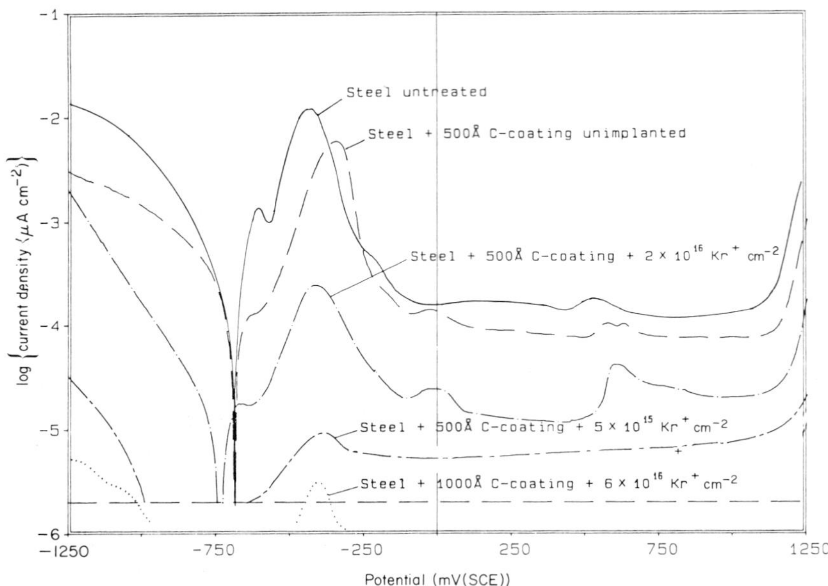

Fig. 4. Influence of rare-gas-implanted carbon (plus hydrogen) layers on the corrosion behaviour of a low carbon steel.

samples thereafter behaved like untreated samples (see Fig. 4).

(iii) Changes in the implantation fluence had a strong influence on the resulting behaviour, especially on thinly coated (50 nm) samples. The good protection produced by 5×10^{15} Kr^+ cm^{-2} was strongly reduced if we increased the fluence to more than 10^{16} Kr^+ cm^{-2}; this was due to carbon removal from the surface through sputtering. At a fluence of above 5×10^{16} Kr^+ cm^{-2} the specimens were effectively sputtered clean and all protection was lost. For layers 100 nm thick, sputtering caused no major problem up to high fluences (10^{17} Kr^+ cm^{-2}); in fact the best corrosion results were received with relatively high fluences (6×10^{16} Kr^+ cm^{-2} at 400 keV) on 100 nm carbon layers.

We have recently started long-time immersion tests with samples treated as described above and first inspections confirmed the trends shown in the voltammograms. However, none of the samples performed as well as was hoped. This is mainly due to the evaporation step in the production of the test specimens. Since the underlying material is very susceptible to anodic dissolution in the acetate buffer, the layers have to be absolutely flawless for good performance. Pinholes or shadows from dust grains on a specimen cannot be "repaired" by the subsequent ion beam irradiation. At such sites, corrosion sets in immediately after immersion and, starting from such points of attack, the protected layer is undercut by the solution, the underlying iron dissolves and the layer collapses after some time. This process leaves shallow circular depressions which grow in all directions until the surface layer has been removed completely. Since different samples were protected more or less effectively according to their treatment, the progressive removal of the carbon layers could be detected under the optical microscope. As a consequence of these results we believe that the evaporation step in the process has to be performed under "clean-room" conditions in order to render the method satisfactorily applicable.

4. FUTURE RECOMMENDATIONS

The present authors believe that the application of ion beams to practical aqueous corrosion problems still makes sense. In the field of electrochemistry and basic corrosion research the usefulness of the method is unquestionable. With respect to future practical applications, appropriate systems have to be selected for investigation having regard to the following criteria: in general, ion implantation is only applicable successfully in systems where very low corrosion rates can be achieved; suppressing the initiation of local attack seems to be a promising field for application; in most cases, cathodically controlled corrosion is easier to handle than anodic dissolution because of the high susceptibility of thin surface layers to inhomogeneities and damage such as scratches; anodically controlled corrosion requires much thicker modified layers than ion implantation alone can yield and this could be a field for the application of the more advanced ion beam techniques such as ion beam mixing, interface mixing and ion-beam-assisted deposition of protective surface layers; finally the application of ion beam techniques to the problem of hydrogen embrittlement deserves more attention.

REFERENCES

1 G. Dearnaley, in E. McCafferty, L. R. Clayton and J. Oudar (eds.), *Proc. Int. Symp. on Fundamental Aspects of Corrosion Protection by Surface Modification, Washington, DC, October 9–14, 1983*, in *Spec. Publ. 84-3*, 1984 (Corrosion Division, Electrochemical Society, Pennington, NJ).
2 V. Ashworth, R. P. M. Procter and W. A. Grant, in J. K. Hirvonen (ed.), *Ion Implantation, Treatise Mater. Sci. Technol., 18* (1980) 175.
3 H. Ferber and G. K. Wolf, in V. Ashworth, W. A. Grant and R. P. M. Procter (eds.), *Proc. Conf. on Ion Implantation into Metals, Manchester, 1981*, Pergamon, Oxford, 1982, p. 1.
4 G. K. Wolf and H. Ferber, *Nucl. Instrum. Methods, 209–210* (1983) 197.
5 V. Ashworth, W. A. Grant, R. P. M. Procter and T. C. Wellington, *Corros. Sci., 16* (1976) 393.
6 A. H. Al-Saffar, V. Ashworth, W. A. Grant and R. P. M. Procter, *Proc. Eur. Congr. on Corrosion, 1977*, Society of Chemical Industry, London, 1977, p. 13.
7 R. O. Toivanen, J. Hirvonen and V. K. Lindner, *Nucl. Instrum. Methods B, 7–8* (1985) 200.
8 G. K. Hubler and E. McCafferty, *Corros. Sci., 20* (1980) 103.
9 Y. F. Wang, C. R. Clayton, G. K. Hubler, W. H. Lucke and J. K. Hirvonen, *Thin Solid Films, 63* (1979) 11.
10 Q. M. Chen, H. M. Chen, X. D. Bai, J. Z. Zhang and H. H. Wang, *Nucl. Instrum. Methods, 209–210* (1983) 867.
11 H. Ferber, *Thesis*, Universität Heidelberg, 1981.
12 F. Marchetti, L. Fedrizzi, F. Giacomozzi, L. Guzman and A. Borgese, in G. K. Wolf, W. A. Grant and R. P. M. Procter (eds.), *Proc. Int. Conf. on Surface Modification of Metals by Ion Beams, Heidelberg, September 17–21, 1984*, in *Mater. Sci. Eng., 69* (1985) 289.
13 H. J. Kim, W. B. Carter, R. F. Hochman and E. I. Meletis, in G. K. Wolf, W. A. Grant and R. P. M. Procter (eds.), *Proc. Int. Conf. on Surface Modification of Metals by Ion Beams, Heidelberg, September 17–21, 1984*, in *Mater. Sci. Eng., 69* (1985) 297.
14 P. Munn and G. K. Wolf, *Nucl. Instrum. Methods, 7–8* (1985) 205.
15 W. Ensinger, A. Meger and G. K. Wolf, *Proc. 8th Eur. Congr. on Corrosion, Nice, November 19–21, 1985*, Société de Chimie Industrielle et Centre Français de la Corrosion, Paris, 1985, p. 71.
16 G. K. Hubler, J. K. Hirvonen, C. R. Gossett, J. Singer, C. R. Clayton, Y. F. Wang, H. E. Munson and G. Kuhlman, *NRL Memo. Rep. 4481*, 1981 (Naval Research Laboratory).
17 P. Munn and G. K. Wolf, in G. K. Wolf, W. A. Grant and R. P. M. Procter (eds.), *Proc. Int. Conf. on Surface Modification of Metals by Ion Beams, Heidelberg, September 17–21, 1984*, in *Mater. Sci. Eng., 69* (1985) 303.
18 I. H. Craig and R. N. Parkins, cited in V. Ashworth, R. P. M. Procter and W. A. Grant, in J. K. Hirvonen (ed.), *Ion Implantation, Treatise Mater. Sci. Technol., 18* (1980) 175.
19 G. K. Wolf and H. Buhl, in G. K. Wolf, W. A. Grant and R. P. M. Procter (eds.), *Proc. Int. Conf. on Surface Modification of Metals by Ion Beams, Heidelberg, September 17–21, 1984*, in *Mater. Sci. Eng., 69* (1985) 317.
20 G. Reiß, G. Frech and G. K. Wolf, in D. Hirschfeld (ed.), *Gase in Metallen*, Deutsche Gesellschaft für Metallkunde, Darmstadt, 1984.
21 W. Ensinger and G. K. Wolf, *Proc. Int. Conf. on Surface Modification of Metals by Ion Beams, Kingston, July 7–11, 1986*, in *Mater. Sci. Eng., 90* (1987).
22 Y. Andoh, Y. Suzuki, K. Matsuda, M. Satou and F. Fujimoto, *Nucl. Instrum. Methods, 6* (1985) 111.
23 H. Ferber, G. K. Wolf, H. Schmiedel and G. Dearnaley, in G. K. Wolf, W. A. Grant and R. P. M. Procter (eds.), *Proc. Int. Conf. on Surface Modification of Metals by Ion Beams, Heidelberg, September 17–21, 1984*, in *Mater. Sci. Eng., 69* (1983) 261.
24 J. M. Williams and R. A. Buchanan, in G. K. Wolf, W. A. Grant and R. P. M. Procter (eds.), *Proc. Int. Conf. on Surface Modification of Metals by Ion Beams, Heidelberg, September 17–21, 1984*, in *Mater. Sci. Eng., 69* (1985) 237.
25 P. Sioshanshi, *Proc. 5th Int. Conf. on the Ion Beam Modification of Materials, Catania, June 9–13, 1986*, in *Nucl. Instrum. Methods*, to be published.
26 G. Reiß, *Thesis*, Universität Heidelberg, 1985.
27 M. Elena, L. Fedrizzi, V. Zanini, A. Cavelli, M. Sarkar, L. Guzman and P. L. Bonora, *Proc. 5th Int. Conf. on the Ion Beam Modification of Materials, Catania, June 9–13, 1986*, in *Nucl. Instrum. Methods*, to be published.
28 H. Ferber, G. K. Wolf and H. Wirth, *Radiat. Eff.*, to be published.

Electrochemical Properties of Chromium-implanted Fe-Cr Alloys*

YOSHIO OKABE†, MASAYA IWAKI and KATSUO TAKAHASHI

RIKEN (The Institute of Physical and Chemical Research) 2-1 Hirosawa, Wako, Saitama 351-01 (Japan)

(Received July 10, 1986)

ABSTRACT

Chromium ions (Cr^+) were implanted in Fe–10Cr, Fe–15Cr, and Fe–20Cr binary alloys (where the compositions are in approximate weight per cent) at an energy of 150 keV with fluences of 1×10^{16}–1×10^{17} ions cm^{-2}. The polarization curves were measured by means of multisweep cyclic voltammetry in 0.5 (mol Na_2SO_4) dm^{-3} solution containing 10^{-2} (mol H_2SO_4) dm^{-3}. The anodic peak current in the transpassive region of the polarization curves was found to be proportional to the chromium concentration in the binary bulk alloys. For chromium-ion-implanted Fe–Cr alloys the anodic peak current on change in the number of potential sweep cycles showed that the chromium concentration depended on the depth from the specimen surface. The concentration–depth profile obtained by this electrochemical method was in good agreement with that obtained by secondary ion mass spectrometry.

1. INTRODUCTION

Surface modifications of metals by ion implantation [1] have been carried out to improve surface properties such as corrosion inhibition, hardness, friction and wear. It has been reported that chromium ion (Cr^+) implantation into iron and Fe–Cr alloys can cause a corrosion-resistant layer to be formed [2–4]. This effect results from the chromium richness in the passive film. Surface analyses of chemical composition play an important role in the understanding of the effects of ion implantation on the various surface properties. Secondary ion mass spectrometry (SIMS) and Rutherford backscattering spectroscopy have often been employed to analyse the chemical composition of material surfaces.

In this paper an electrochemical method is proposed for the analysis of surface compositions of Fe–Cr binary alloys, and in particular for concentration–depth profile measurements of chromium-implanted Fe–Cr alloys. This electrochemical method is simpler and more convenient than techniques such as SIMS and Rutherford backscattering. Furthermore, it is expected that the method will be applied to analyse the oxidation state of implanted chromium and the electrochemical dissolution mechanism of the Fe–Cr alloys.

2. EXPERIMENTAL DETAILS

Fe–Cr binary alloys used as target specimens were mechanically mirror polished down to about 1 μm alumina paste with a buffing wheel. The nominal compositions of the specimens used are shown in Table 1.

Chromium ion (Cr^+) implantations into Fe–Cr alloys were carried out using the RIKEN 200 kV Low Current Implanter. The acceleration energy was 150 keV and the fluences were 1×10^{16}, 3×10^{16}, 6×10^{16} and 1×10^{17} Cr^+ cm^{-2}. The current density of the chromium ion beam was 1–2 μA cm^{-2}.

SIMS was performed using an ion beam surface mass analyser (Commonwealth Science Co. Ltd.) in order to determine the concentration–depth profiles of chromium for implanted and unimplanted specimens. Oxygen ions (O_2^+) at 7 keV were used as the primary ions for ion beam sputter etching. The primary ion beam current was about 30 μA at the target

*Paper presented at the International Conference on Surface Modification of Metals by Ion Beams, Kingston, Canada, July 7–11, 1986.

†Permanent address: Department of Electronics, Faculty of Engineering, Saitama Institute of Technology, 1690 Fusaiji, Okabe-machi, Oosato-gun, Saitama 369-02, Japan.

0025-5416/87/$3.50

TABLE 1

Nominal compositions of the Fe–Cr alloys

Alloy	*Amount* (wt.%) *of the following elements*							
	C	*Si*	*Mn*	*P*	*S*	*Cr*	*Al*	*O*
Fe–10Cr	0.002	0.017	<0.001	0.002	0.002	10.11	<0.001	0.0165
Fe–15Cr	0.002	0.014	0.001	0.001	0.002	15.00	<0.001	0.0130
Fe–20Cr	0.003	0.018	0.001	0.002	0.002	20.22	<0.001	0.0173

and the area for SIMS was about 0.5 cm^2. The relative concentration of chromium was determined as the ratio of the intensity of the secondary ions of chromium to that of iron. The depth was estimated assuming a constant sputtering rate of about 2 nm min^{-1}.

A conventional voltammetric system with a three-electrode type of cell was employed for multisweep cyclic voltammetry the sweep rate for which was 100 mV s^{-1}. The electrical charge flow with anodic current was also measured during cyclic voltammetry. The electrolytic solution used was deaerated 0.5 (mol Na_2SO_4) dm^{-3} containing 10^{-2} (mol H_2SO_4) dm^{-3}. All measurements were made at 25 °C.

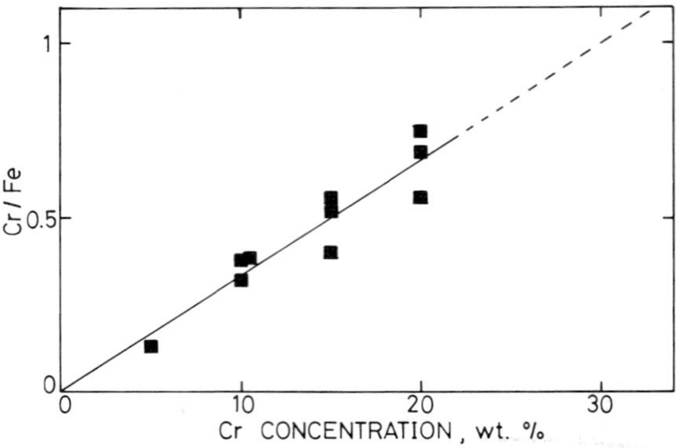

Fig. 1. Calibration curve of chromium for Fe–Cr alloys by means of SIMS.

3. RESULTS AND DISCUSSION

3.1. Concentration–depth profiles of chromium measured by secondary ion mass spectrometry

Figure 1 shows the calibration curve of chromium obtained by means of SIMS. The vertical axis of the figure is the ratio of the secondary ion current of chromium to that of iron. The specimens used for the calibration are Fe–Cr alloys whose chromium concentrations are known. The figure shows that the chromium-to-iron current ratio is proportional to the chromium concentration of Fe–Cr alloys.

The concentration–depth profiles of chromium for Fe–15Cr alloys implanted with 3×10^{16} and 6×10^{16} Cr^+ cm^{-2} and unimplanted Fe–15Cr alloys are shown in Fig. 2. Gaussian-like distribution curves are observed for the depth region up to 100 nm from the surface.

For chromium ion implantation in iron, chromium concentrations at the average projected range $\bar{R}_p$ calculated by the Lindhard–Scharff–Schiøtt (LSS) range theory are 7 wt.%

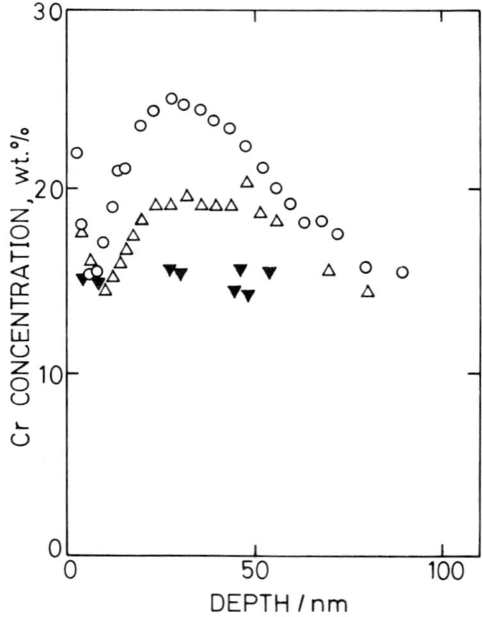

Fig. 2. Concentration–depth profiles of chromium for unimplanted and chromium-implanted Fe–15Cr alloys measured by SIMS: △, implanted with 3×10^{16} Cr^+ cm^{-2}; ○, implanted with 6×10^{16} Cr cm^{-2}; ▼, unimplanted.

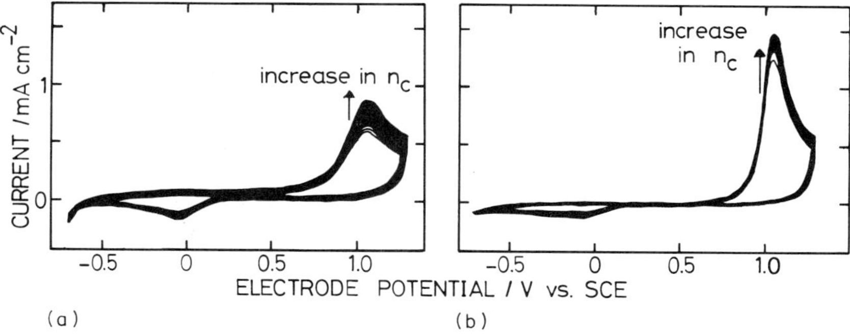

Fig. 3. Polarization curves of unimplanted (a) Fe–10Cr and (b) Fe–20Cr alloys.

and 14 wt.% for the fluences of 3×10^{16} Cr^+ cm^{-2} and 6×10^{16} Cr^+ cm^{-2} respectively implanted into iron. The maximum chromium concentrations observed for the chromium-implanted Fe–15Cr alloys, which are about 20 and 25 wt.%, approximately agree with the sum of the bulk chromium concentration and the chromium concentration predicted by the LSS range theory.

The measured $\bar{R}_p$ values were about 38 nm and about 30 nm for the alloys implanted with 3×10^{16} Cr^+ cm^{-2} and 6×10^{16} Cr^+ cm^{-2} respectively. These $\bar{R}_p$ values are less than the theoretical $\bar{R}_p$, 47 nm; this may be due to sputtering effects during ion implantation.

3.2. *Electrochemical method for the concentration–depth profiling*

Figure 3 shows polarization curves of unimplanted Fe–20Cr and Fe–10Cr alloys measured by means of cyclic voltammetry. Anodic current peaks appeared near 1.05 V, which correspond to the dissolution behaviour of chromium as Cr^{VI} [5]. The peak current density i_p near 1.05 V was approximately constant after a few potential sweeps. The stationary peak current densities for Fe–10Cr, Fe–15Cr and Fe–20Cr alloys were 1 mA cm^{-2}, 1.5 mA cm^{-2} and 2 mA cm^{-2} respectively. This indicates that the peak current density is proportional to the chromium concentration.

The peak current densities i_p were plotted against the number n_c of potential sweep cycles for chromium-implanted Fe–15Cr alloys as shown in Fig. 4. The curves denoted by open triangles and open circles, corresponding to the specimens implanted with fluences of 3×10^{16} Cr^+ cm^{-2} and 6×10^{16} Cr^+ cm^{-2} respectively, have a gaussian-like shape which suggests that the i_p–n_c curves indicate the

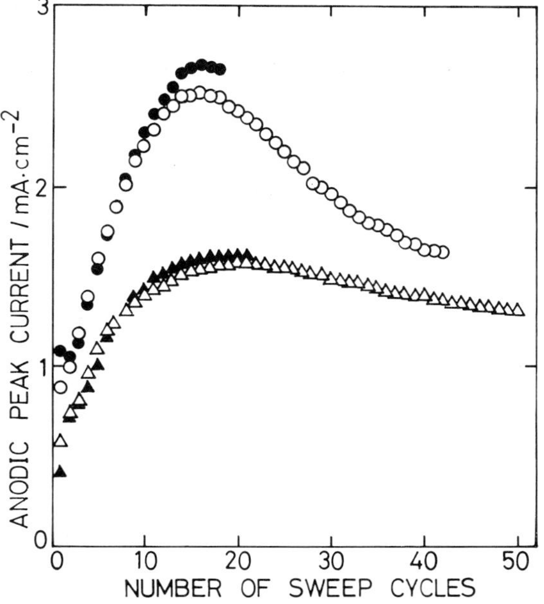

Fig. 4. i_p–n_c curves for chromium-implanted Fe–15Cr alloys: △, implanted with 3×10^{16} Cr^+ cm^{-2}, $n_{cs} = 50$; ○, implanted with 6×10^{16} Cr^+ cm^{-2}, $n_{cs} = 42$; ▲, implanted with 3×10^{16} Cr^+ cm^{-2}, $n_{cs} = 21$; ●, implanted with 6×10^{16} Cr^+ cm^{-2}, $n_{cs} = 18$.

concentration–depth relationship of chomium, where the surface is etched by an electrochemical dissolution process with multipotential sweeps.

In order to prove the relationship between i_p and the surface concentration of chromium, the surface concentrations of chromium were measured by SIMS for the specimens treated by cyclic voltammetry with the number n_{cs} of potential sweeps.

For example, the i_p–n_c curve for the Fe–15-Cr alloy implanted with 6×10^{16} Cr^+ cm^{-2} with $n_{cs} = 18$ is shown in Fig. 4 as full circles, and the concentration-depth profile measured by SIMS for this specimen is shown in Fig. 5 as open triangles. The surface concentration in this specimen determined by SIMS is about 26 wt.%, which agrees with the concentration of about 26 wt.% estimated from i_p at $n_{cs} = 18$ for the specimen.

The surface concentration measured by SIMS for various specimens treated by cyclic voltammetry with n_{cs} is plotted against i_p at n_{cs} as shown in Fig. 6. The unimplanted specimens are indicated by full circles and the ion-implanted specimens by open circles (concentration peak region, $n_{cs} = 14$-22) and open squares (deeper region, $n_{cs} = 35$-50). All these points show linear correlation between i_p and the surface concentration of chromium determined by SIMS. The correlation can be written by the following empirical equation:

$$i_p \approx 0.1C_{Cr}$$

where i_p (mA cm^{-2}) is the anodic peak current density and C_{Cr} (wt.%) is the chromium concentration.

In order to correlate n_c with the depth, the dissolved thickness δ_d of the specimens after cyclic voltammetry measurement with n_{cs} was measured by a stylus method. In Fig. 7, δ_d is plotted against electrical charge density q_d where q_d is obtained by integration of the anodic current in cyclic voltammetry with n_{cs}. A linear relationship between δ_d and q_d can be seen in the figure, which suggests that the q_d measured with cyclic voltammetry at n_c gives the thickness for the profiling of the

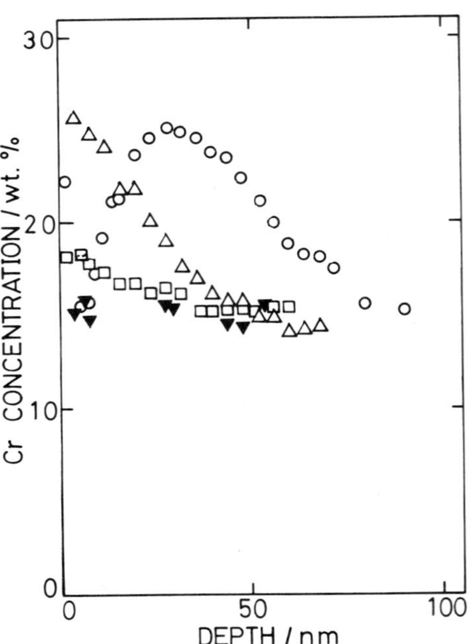

Fig. 5. Concentration-depth profiles of Fe-15Cr alloys measured by SIMS after cyclic voltammetry: □, implanted with 6×10^{16} Cr^+ cm^{-2}, $n_{cs} = 42$; △, implanted with 6×10^{16} Cr^+ cm^{-2}, $n_{cs} = 18$; ○, as implanted with 6×10^{16} Cr^+ cm^{-2} (before cyclic voltammetry); ▼, unimplanted.

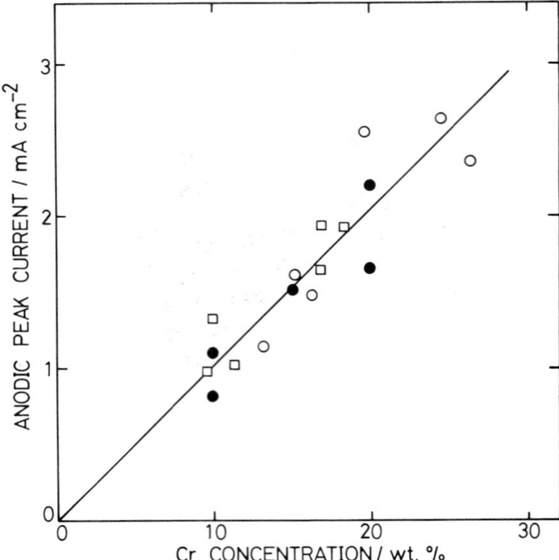

Fig. 6. Relationship between i_p and chromium concentration: ●, unimplanted specimen; ○, implanted specimen, $n_{cs} = 14$-22; □, implanted specimen, $n_{cs} = 35$-50.

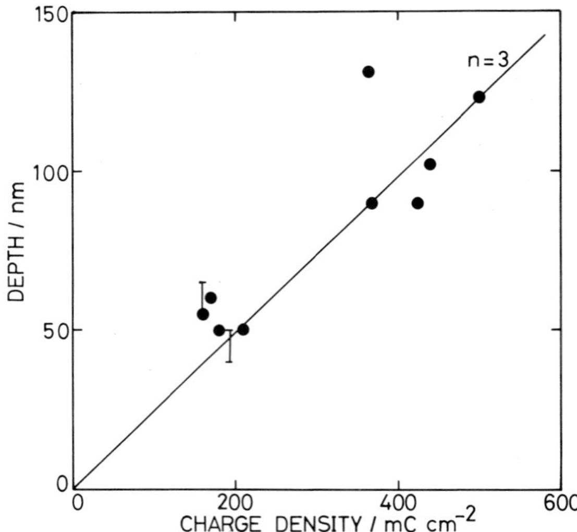

Fig. 7. Dissolved thickness as a function of electrical charge: n, the number of electrons concerned in the anodic dissolution reaction.

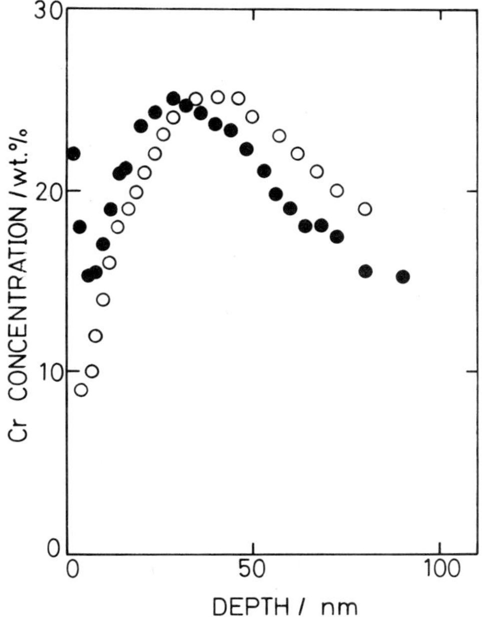

Fig. 8. Comparison between concentration–depth profiles obtained by SIMS and the electrochemical method: ●, SIMS; ○, electrochemical method.

chromium concentration by the electrochemical method.

Figure 8 shows the concentration–depth profile obtained by the electrochemical method; this is analogous to that obtained by SIMS.

3.3. *Future trends*

The electrochemical profiling method discussed above is at present limited to a determination of the chromium concentration in the Fe–Cr system. However, we have found that the polarization curves observed in cyclic voltammetry with $n_c = 1$–5 are affected by the oxidation state of chromium [6]. The anodic current peak appears near 1.13 V as a result of chromium oxidation near the original surface of specimens. The oxidation effect of chromium on the polarization curve has been examined for Fe–Cr alloys with the implantation of oxygen ions. The anodic dissolution current peak appeared at 1.13 V for $n_c = 20$–110. These results suggest that the oxidation state of chromium may be detected by the electrochemical method as a function of depth in the specimen surface layer.

REFERENCES

1 J. K. Hirvonen (ed.), *Ion Implantation, Treatise Mater. Sci. Technol., 18* (1980).
2 V. Ashworth, W. A. Grant and R. P. M. Procter, *I&EC Prod. Res. Dev., 17* (1978) 176.
3 B. S. Covino, Jr., B. D. Sartwell and P. B. Needham, Jr., *J. Electrochem. Soc., 125* (1978) 366.
4 Y. Okabe, M. Iwaki, K. Takahashi, H. Hayashi, S. Namba and K. Yoshida, *Surf. Sci., 86* (1979) 257.
5 J. M. West, *Electrodeposition and Corrosion Processes*, Van Nostrand, Princeton, NJ, 1965.
6 Y. Okabe, M. Iwaki, S. Namba and K. Yoshida, *Nucl. Instrum. Methods, 182–183* (1981) 231.

Materials Science and Engineering, 90 (1987) 229-236

Surface Modification of Electrodeposited Chromium Films by Ion Implantation*

KEIICHI TERASHIMA[a], TOMOYA MINEGISHI[a], MASAYA IWAKI[b] and KENICHI KAWASHIMA[c]

[a] *Chiba Institute of Technology, 17-1 Tsudanuma, Narashino-shi, Chiba 275 (Japan)*
[b] *The Institute of Physical and Chemical Research, 2-1 Hirosawa, Wako-shi, Saitama 351 (Japan)*
[c] *Marubeni Hytech Co. Ltd., 3-1-1 Higashi-Ikebukuro, Toshima-ku, Tokyo 170 (Japan)*

(Received July 10, 1986)

ABSTRACT

Electrodeposited chromium films implanted with nitrogen and argon ions were characterized by surface analysis, surface hardness, wear, and corrosion. Depth profiles were examined by Auger electron spectroscopy and the chemical bonding state was estimated by X-ray photoelectron spectroscopy. The near-surface hardness of nitrogen-implanted chromium was measured with a Vickers microhardness tester and a dynamic ultramicrohardness tester. Wear tests were carried out using an Ohgoshi rapid wear-testing machine and a Taber abrasion test apparatus. The corrosion resistance was investigated by the electrochemical method of cyclic voltammetry in 0.05 M H_2SO_4. The following results were obtained.

(1) Nitrogen implantation caused the surface hardness to increase.

(2) Nitrogen-implanted chromium showed a remarkable wear resistance.

(3) The corrosion resistance of chromium was improved markedly by nitrogen implantation.

(4) Nitrogen depth profiles showed definitely that there were two peaks.

(5) Depth profiles and measurement of chemical bonding showed that the implanted nitrogen combined with chromium to form Cr_2N.

1. INTRODUCTION

Ion implantation has received attention in recent years as a method for metal surface modification and has progressed from the stage where it was used to "produce some sort of effect or other" [1] to the point where it may soon constitute a practical technique. It has been suggested [2] that the most fertile area for the application of ion implantation techniques is thin film materials since the depth of penetration of ions is normally in the range 0.1-0.5 μm. If it were possible to implant any desired element at any desired depth in single- or multiple-layer films, it would be possible to obtain film materials with properties and functions that are undreamed of at present.

Electrodeposited chromium films, because of their hardness and resistance to wear, are used to make a wide variety of mechanical components for industry, but they do not provide good corrosion protection because of the tendency for cracks to appear in the films.

In the present study, therefore, some fundamental considerations relative to the application of ion implantation electrodeposited chromium films are presented as a first step in research to modify surface-finishing films using this technology; this has led to the discovery that, in terms of the electrochemical properties of films under nitrogen implantation, there was a drastic reduction in anodic dissolution, producing a surface film with corrosion-resistant properties.

In this paper the implantation of nitrogen and argon ions into electrodeposited chromium films and pure chromium sheet is reported and the relationship between the changes produced by implantation in the organization of the surface film and its electrochemical properties is investigated. The physical properties of the surface film such as hardness and wear resistance were also evaluated, in order to provide material on which to base future statistical studies on

*Paper presented at the International Conference on Surface Modification of Metals by Ion Beams, Kingston, Canada, July 7-11, 1986.

0025-5416/87/$3.50

ion implantation to modify surface-finishing films.

2. EXPERIMENTAL DETAILS

2.1. Materials

The electrodeposited chromium samples used in the experiments were prepared from a Sargent bath of the type normally used in industrial chromium plating, which contained 250 (g CrO_3) l^{-1} and 2.5 (g H_2SO_4) l^{-1}. Deposits were obtained at a bath temperature of 40 °C and current densities of 20–50 A dm^{-2}. Plating substrates were primarily mild steel bars 12 mm in diameter by 17 mm long, with the upper and lower end surfaces electroplated to a depth of 50–100 μm and buffed to a mirror finish. To provide comparison with the electrodeposited films, pieces of commercial grade chromium sheet (purity, 99.8%; arc melted; 4 mm × 1.5 mm × 30 mm) were also used as samples for implantation.

2.2. Ion implantation method

We used a nitrogen ion beam (consisting of 40% of N^+ ions plus 60% of N_2^+ ions) generated by a Zymet model Z-100 ion implantation system. The acceleration energy was 90 keV, the beam current was 3 mA and the fluence was 3×10^{17} ions cm^{-2}. Some samples were subjected to argon implantation at an acceleration energy of 90 keV and a fluence of 1×10^{17} ions cm^{-2}.

2.3. Measurement of the surface hardness

To determine the hardness near the surface of the electrodeposited chromium and pure chromium sheet samples after ion implantation, we used a Vickers microhardness tester at loads of 2–20 gf. The results of these measurements were tabulated to contrast the average Vickers microhardness values obtained before and after ion implantation at each load level.

Some of the samples were subjected to ultramicrohardness tests using a pyramid indenter having an angle of 100° between the pressure surfaces and a loading speed of 0.03 gf min^{-1} [3]; the depth of penetration was plotted against the load.

2.4. Cyclic voltammetry

Electrolysis was carried out in a 0.05 M H_2SO_4 bath (pH 1.3) deoxidized by argon, and cyclic voltammetry was performed in a static solution at a constant temperature of 28 °C.

Using a potentiostat and function generator at a constant scanning speed of 20 mV min^{-1}, measurements were made with repeated potential sweeps from −0.9 to 0.0 V. Anodic polarization curves were plotted using an x–y recorder. In measuring the potential, a reference saturated calomel electrode and a platinum counterelectrode were used.

2.5. Wear tests

Two types of test were carried out to evaluate wear resistance. These were (1) an improved version of the Taber abrasion resistance test in which a CS-10 abrasion wheel was loaded at 1 kgf, and the mass of the sample was measured after every 1000 rev and (2) a test using an Ohgoshi rapid wear-testing machine in which wear was measured under oil-lubricated conditions with a sliding speed of 1.4 m s^{-1} and a final load of 18.9 kgf, with sliding distances of 200, 400 and 600 m. Hard metal was used for the counterposed rotating sample in the wear tests.

2.6. Depth profiles and chemical bonding

The distribution profile of the implanted ions in the depth direction was measured by Auger electron spectroscopy (AES) accompanied by 6 kV argon sputter etching, and the degree of chemical bonding between the ions and the electrodeposited chromium was investigated by measuring the bonding energy between Cr 2p and N 1s using X-ray photoelectron spectroscopy (XPS) with an Mg Kα X-ray source (VG ESCA-5). The sputtering rate of metallic chromium film under the conditions applied in these experiments was 2.4 nm min^{-1}.

3. RESULTS AND DISCUSSION

3.1. Surface observations

The mirror-bright surfaces of electrodeposited chromium and chromium sheet samples suffered yellowish discolouration after nitrogen implantation. Samples were marked using a Vickers microhardness tester so that the same points could be observed by scanning electron microscopy (SEM) before and after implantation, and these were also subjected to surface observation.

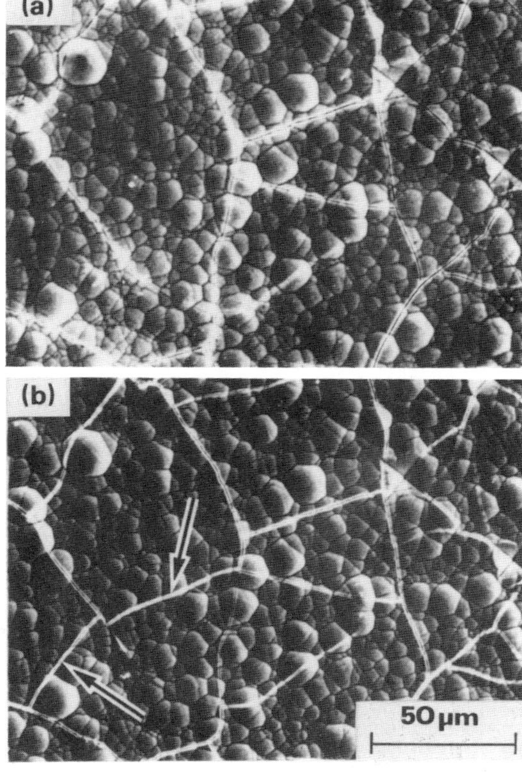

Fig. 1. SEM photographs of an electrodeposited chromium surface: (a) unimplanted; (b) nitrogen implanted.

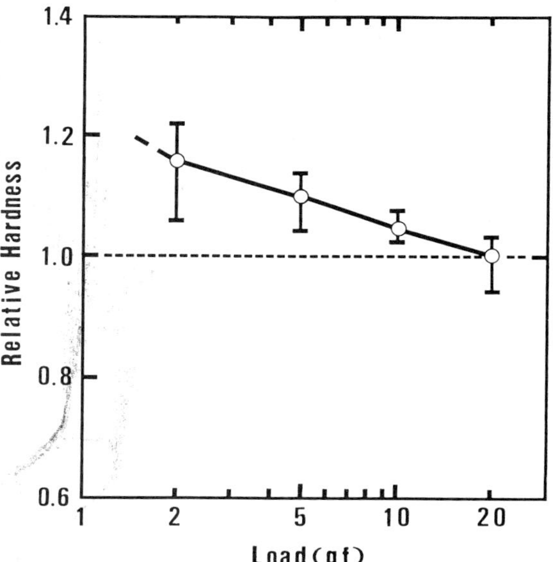

Fig. 2. Relationship between the applied load and the relative change in the Vickers microhardness for electrodeposited chromium samples (—o—): -----, unimplanted.

Figure 1 consists of SEM photographs of unimplanted and nitrogen-implanted samples. In Fig. 1(a) (unimplanted sample), the nodular protrusions characteristic of electrodeposited chromium films were seen while, in Fig. 1(b) (nitrogen-implanted sample), newly formed cracks (shown by the arrows) were detected. The facts observed in these experiments were at variance with the previously obtained results [4] in which the formation of Cr_2N as a result of ion implantation caused a volumetric expansion of approximately 25% so that the cracks in the electrodeposited chromium film disappeared.

3.2. *Surface hardness*

Figure 2 shows the relationship between the applied load and the relative change in the Vickers microhardness after nitrogen implantation of electrodeposited chromium samples formed at current densities of 30–50 A dm^{-2}. The change in relative hardness was clear at low loads, reaching a hardness that was a factor of 1.2 greater at a load of 2 gf. At a load of 20 gf the depth of indentation was approximately 1.2 μm, a value that was 12 times greater than the projected range of values for R_p calculated by the Lindhard–Scharff–Schiøtt (LSS) theory for the conditions used in these experiments.

The indentation depth at a load of 2 gf was about 0.2 μm, around twice the thickness of the ion implantation layer. This leads us to believe that the actual relative change in surface hardness is even greater.

These findings concerning the relationship between the Vickers microhardness and the load have been corroborated by Iwaki *et al.* [5].

Figure 3 shows the results of measurements on a pure chromium sheet sample of the relationship between the depth of penetration and the load for unimplanted and nitrogen-implanted samples using a Vickers microhardness tester. At a load of 0.1 gf the penetration was 0.3 μm for unimplanted samples and 0.1 μm for nitrogen-implanted samples. This decrease in the depth of penetration shows that the hardness increased. The results for electrodeposited chromium films showed a similar trend.

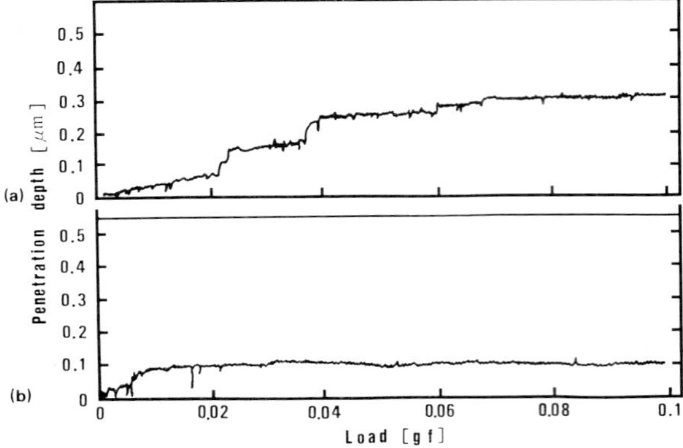

Fig. 3. Relationship between the applied load and the depth of penetration in ultramicrohardness tests on pure chromium samples: (a) unimplanted pure chromium; (b) nitrogen-implanted pure chromium.

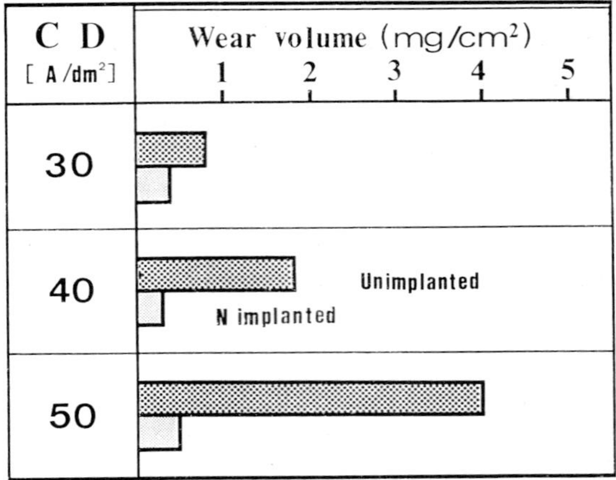

Fig. 5. Variations in the wear volume in the Taber abrasion tests for chromium films electrodeposited at different current densities CD.

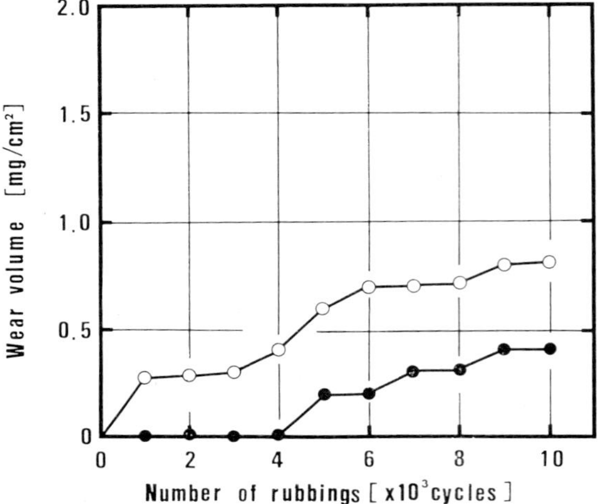

Fig. 4. Results of the Taber abrasion resistance tests of electrodeposited chromium films: ○, unimplanted; ●, nitrogen implanted.

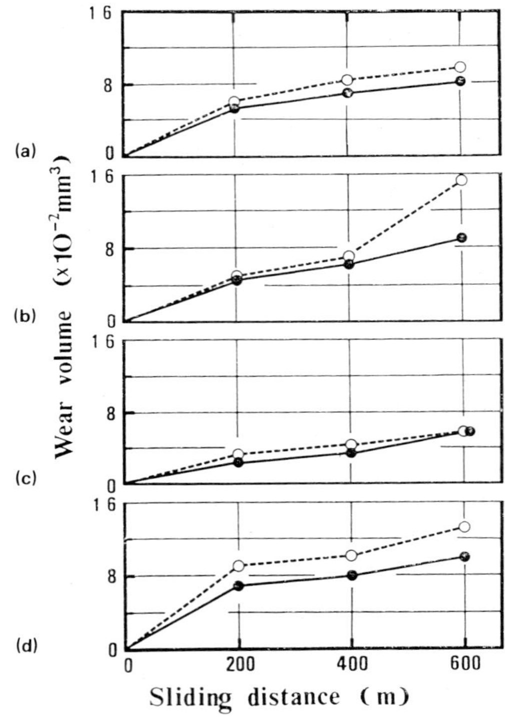

Fig. 6. Variation in the wear volume in the Ohgoshi rapid wear tests for chromium films electrodeposited at current densities of (a) 20 A dm^{-2}, (b) 30 A dm^{-2}, (c) 40 A dm^{-2} and (d) 50 A dm^{-2}: ---○---, unimplanted; —●—, nitrogen implanted.

3.3. Wear tests

Figure 4 shows typical results of a Taber abrasion resistance test of chromium films electrodeposited at a bath temperature of 40 °C and a current density of 30 A dm^{-2}. Nitrogen-implanted samples did not show wear for less than 3000 rev and at 10 000 rev showed only half the wear volume of unimplanted samples, a clear indication of improved abrasion resistance. Figure 5 shows the results of abrasion up to the 10 000 rev (the maximum number of revolutions used in these experiments), with current density as a parameter. It can be seen that the wear volume decreased with nitrogen implantation for all samples tested.

Figure 6 shows variations in wear volume in Ohgoshi rapid wear tests for electrodeposited chromium at different current densities.

Except for samples deposited at 40 A dm^{-2}, differences in the values of wear volume in the initial wear region appeared after nitrogen implantation, showing that nitrogen implantation had the effect of increasing wear resistance.

It can thus be seen that, despite the fact that the layer affected by ion implantation is only about 0.1 μm in thickness, it has a considerable effect on the wear process. For example, in the Taber abrasion tests, the results of which are shown in Fig. 4, the reduction in the thickness of the chromium film was equivalent to 0.6 μm at a wear volume of 0.25 mg cm^{-2}, and to 1.4 μm at a wear volume of 1 mg cm^{-2}. Thus the implantation effect was present even at depths greater than the thickness of the implantation layer. This supports the reasoning based on measurements obtained through nuclear fission analysis for the amount of nitrogen after wear [6].

3.4. Cyclic voltammetry

Figure 7 shows the polarization curves obtained by cyclic voltammetry for electrodeposited chromium films. When unimplanted samples (Fig. 7(a)) are subjected to potential scanning from −0.9 V in the noble direction, an active dissolution passivation reaction current was observed in the range from −0.65 to −0.58 V, and at more noble potentials a passivation holding current and a transpassivity reaction current were observed. When the sweep was repeated from the starting potential of −0.9 V, the anode current increased.

Nitrogen-implanted chromium films (Fig. 7(b)) show an anodic dissolution current at around −0.7 V, and this is small even after 20 potential sweeps. Further, it was found that the passivation holding current was extremely minute. Furthermore, in contrast with the pre-implantation films, which showed black discolouration after ten potential sweeps, the surfaces of nitrogen-implanted films retained their mirror finish. The cyclic voltammetry curves for unimplanted pure chromium sheet samples were of approximately the same shape but the passivation holding current was larger for the pure chromium sheet than for the electrodeposited films.

These results, as well as the processes of anodic dissolution and passivation were closely observed by SEM, and the values of peak anode current observed at this time provide a yardstick for determining electrode reactivity and ease of dissolution.

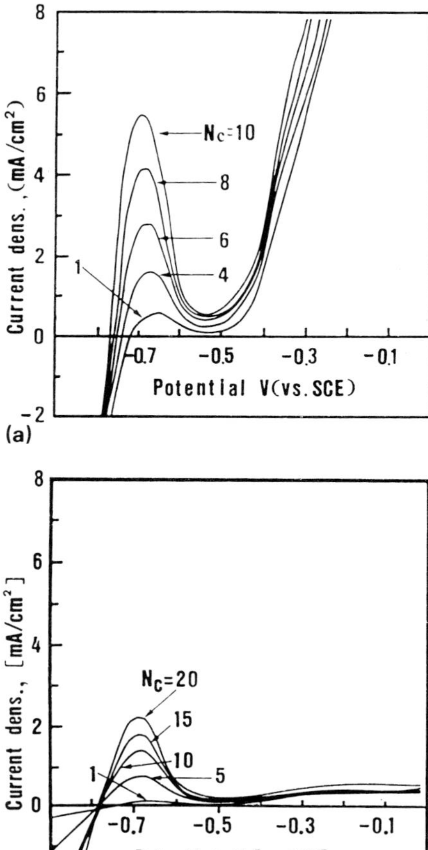

Fig. 7. Cyclic voltammograms for electrodeposited chromium films: (a) unimplanted; (b) nitrogen implanted.

Next, Fig. 8(a) shows the relationship between the anodic peak current density and the number of potential sweeps for electrodeposited chromium films. On comparison of the unimplanted and nitrogen-implanted samples, the increase in current as the number of potential sweeps increased was seen to undergo a marked suppression due to chromium dissolution.

In argon-implanted films, however, the implantation layer dissolved more easily than in unimplanted films, and for this reason a large anodic current was observed from the beginning of cyclic voltammetry.

Figure 8(b) shows the same relationship between the anodic peak current density and the number of potential sweep cycles for pure chromium sheet samples. In nitrogen-implanted pure chromium sheet the current increased in very small increments as the number of potential sweeps increased, but surface dissolution scarcely occurred, and even after 20 sweeps the anodic current density was 0.3 mA cm^{-2}, only about one-fifteenth of the dissolution current for pre-implantation samples. It can be seen that argon-implanted pure chromium sheet samples were more easily dissolved than unimplanted samples. Up to seven sweeps, the anodic peak current was observed to increase, but it stabilized thereafter. This differed from the tendency in electrodeposited films, shown in Fig. 8(a), for the anodic peak current to continue to increase with increases in the number of sweeps. After 20 sweeps, crystalline boundaries appeared on the sample surfaces, and corrosion was observed.

3.5. Depth profiles

Figure 9 shows the depth profiles plotted against the normalized ratio C_N/C_{Cr} of the AES peak voltage for nitrogen to that for chromium. The C_N/C_{Cr} ratio showed two peaks. The first peak was near the surface, and the second, caused by nitrogen ions (N_2^+), was measured at a depth of approximately 70 nm. This second peak is considerably deeper than the average penetration of 47 nm calculated from the LSS theory. Figure 9(b) shows depth profiles after 20 cyclic voltammetry potential sweeps. Definite changes were observed in the C_N/C_{Cr} distribution after cyclic voltammetry, and it can be seen that a surface layer of approximately 12 nm was dissolved. The C_N/C_{Cr} ratio measured for Cr_2N was 0.35 and, since the height of the second peak in Fig. 9(a) was 0.42, it is con-

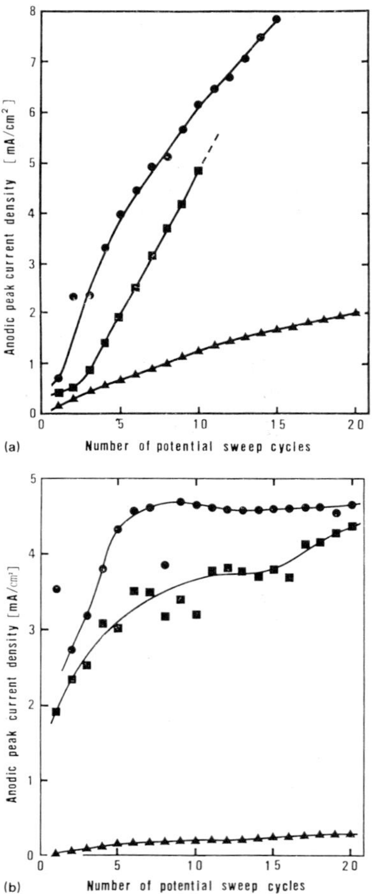

Fig. 8. Relationship between the anodic peak current density and the number of potential sweeps for (a) electrodeposited chromium films and (b) pure chromium sheet: ■, unimplanted; ●, argon implanted; ▲, nitrogen implanted.

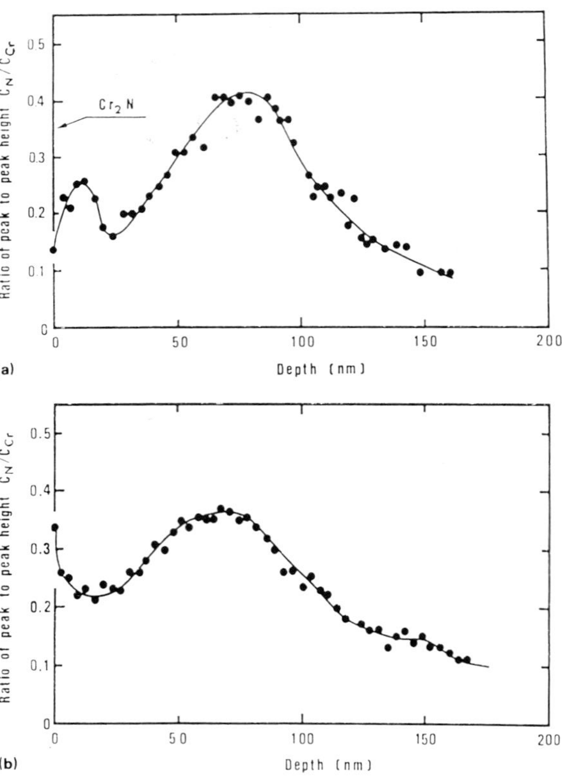

Fig. 9. Depth profiles normalized with the AES peak-to-peak height ratio for electrodeposited chromium films: (a) before cyclic voltammetry; (b) after 20 sweeps.

sidered that the region of Cr_2N formation is from 60 to 90 nm.

Figure 10 shows the depth profile for nitrogen in electrodeposited chromium films. The ion beam used in these experiments consisted of 60% N^+ ions and 40% N_2^+ ions in an atomic ratio of 3 to 1. The LSS theory indicates that implanted ion depth profiles should be represented by one gaussian distribution centred on the average implantation depth R_{p1} and another centred on R_{p2}. The first peak R_{p1} is the average implantation depth of the N_2^+ ions, and the second peak is the average depth corresponding to the N^+ ions. The line joining the full squares in Fig. 10 combines the gaussian distributions for N_2^+ ions (45 keV) and N^+ ions (90 keV). The values calculated for the average implantation depths in chromium of N_2^+ ions is 47 nm and of N^+ ions is 94 nm. As can be clearly seen by comparing Figs. 9 and 10, the average of actual measured depth values is 67 nm, which is approximately midway between the theoretical values for the N_2^+ and N^+ ions.

The nitrogen peak near the surface in Fig. 9, however, was not predicted by the LSS theory. This phenomenon has been explained by Iwaki *et al.* [5] and others as due to the migration of nitrogen atoms to the surface as a result of increases in substrate temperature during the implantation process.

3.6. Chemical bonding

Figure 11 shows the changes in bonding state in the depth direction through X-ray photoelectron spectra for Cr 2p and N 1s in nitrogen-implanted chromium as a function of sputtering time. The bonding energy of N 1s, 396.9 eV, is equivalent to that of Cr_2N.

The Cr 2p spectrum shows that Cr_2O_3 is present at the very surface and, as sputtering proceeds, the bonding energy approaches that of chromium and Cr_2N; so close, in fact, are the two energy levels (only 0.2 eV apart) that they are indistinguishable. These results of AES and XPS analyses demonstrate that the nitrogen implanted in the chromium bonds with it to form Cr_2N.

4. CONCLUSIONS

AES and XPS techniques were used to measure the nitrogen distribution and bonding in

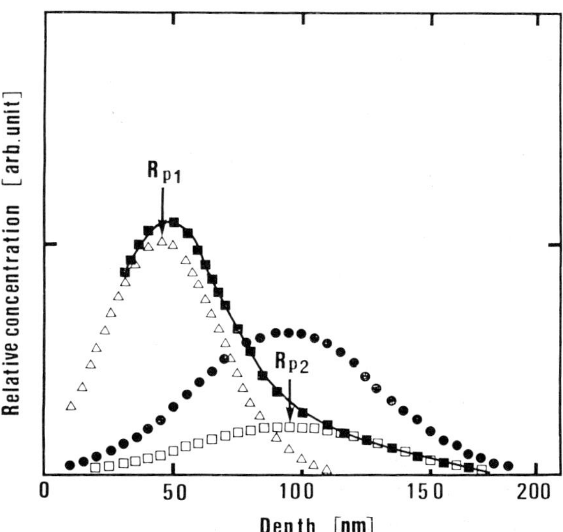

Fig. 10. Distribution of implanted nitrogen in chromium predicted by the LSS theory: □, N^+ ions (90 keV) (reduced by a factor of $\frac{1}{3}$); ●, N^+ ions (90 keV); △, N_2^+ ions (90 keV) (equivalent to N^+ ions (45 keV)); ■, N_2^+ ions (45 keV) plus N^+ ions (90 keV).

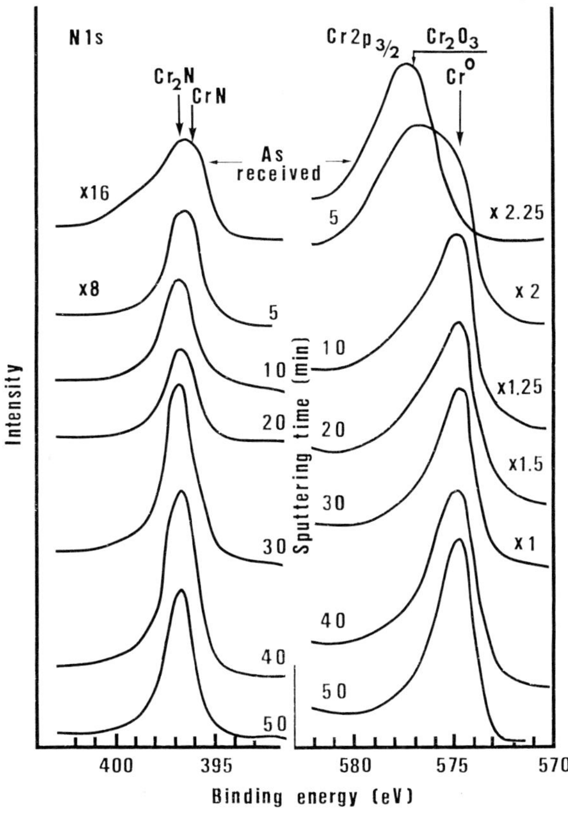

Fig. 11. X-ray photoelectron spectra for Cr 2p and N 1s in a nitrogen-implanted electrodeposited chromium film as a function of sputtering time.

the chromium surface layer after nitrogen ion implantation in electrodeposited chromium films, and attention was given to the mechanical properties, including the changes in hardness and wear resistance. The corrosion resistance was also investigated using electrochemical techniques. The results obtained may be summarized as follows.

(1) Nitrogen implantation caused the surface layer to harden, and the extent of the change depended on the load applied.

(2) Taber abrasion tests revealed that nitrogen implantation of electrodeposited chromium films resulted in a decrease in the wear volume. Tests using an Ohgoshi machine showed that nitrogen implantation improved the wear resistance of electrodeposited chromium films.

(3) Cyclic voltammetry revealed that nitrogen implantation brought about a clear suppression of the dissolution of chromium, and an anticorrosion effect was also seen. Argon-implanted samples dissolved more easily than unimplanted samples, indicating that argon implantation produced an effect.

(4) The nitrogen distribution in the chromium showed two peaks, one near the surface and another approximately 20 nm deeper than the point at which it could be predicted by the LSS theory.

(5) Analysis of N 1s by XPS techniques showed that nitrogen implanted in the chromium reacted chemically to form Cr_2N.

REFERENCES

1 M. Iwaki, *Jpn. J. Appl. Phys.*, *53* (1984) 700 (in Japanese).

2 K. Saito, *Hyomen*, *22* (1984) 361 (in Japanese).

3 K. Kanazawa, A. Khono, T. Sada and S. Shoma, *Proc. J. Jpn. Soc. Prec. Eng.*, (October 1985) 149 (in Japanese).

4 W. C. Oliver, R. Hatching and J. B. Pechica, *Metall. Trans. A*, *15* (1984) 222.

5 M. Iwaki, T. Fujihana and K. Okitaka, in G. K. Wolf, W. A. Grant and R. P. M. Procter (eds.), *Proc. Int. Conf. on Surface Modification of Metals by Ion Beams, Heidelberg, September 17–21, 1984*, in *Mater. Sci. Eng.*, *69* (1985) 211.

6 N. E. W. Hartley, *Thin Solid Films*, *64* (1979) 177.

Materials Science and Engineering, 90 (1987) 237-241

The Influence of Platinum Implantation on the Hydrogen Embrittlement of Tantalum

W. ENSINGER and G. K. WOLF

Physikalische Chemie der Universität Heidelberg, Im Neuenheimer Feld 500, 6900 Heidelberg (F.R.G.)

(Received July 10, 1986)

ABSTRACT

The present study is an attempt to investigate the hydrogen embrittlement of tantalum in order to explore the possibility of modifying the surface of refractory metals by ion beam techniques with respect to improved resistance against this form of degradation. Tantalum foils were implanted in the as-delivered state and also after anodizing in order to prepare a thicker defined oxide on the metal with 5×10^{16} *Pt^+ cm^{-2} at 80 keV. These specimens were charged cathodically with hydrogen and subjected to a bending test until fracture. The ratio of the number of bends until fracture with charging to the number of bends until fracture without charging was used as a measure of the degree of embrittlement. It turned out that specimens implanted in the as-delivered state were less resistant to hydrogen than were unimplanted samples with undamaged oxide films. However, implanted specimens with oxide layers thicker than 100 nm showed an increased resistance to embrittlement compared with that of unimplanted specimens. This behaviour is discussed in terms of desorption catalysis which is only effective if the implanted platinum has no direct contact with metallic tantalum. A good long-term protective effect is expected for the systems studied.*

1. INTRODUCTION

During the last 10 years, many studies on the applications of ion implantation in corrosion science have been performed [1, 2]. Mostly the oxidation behaviour or the influence of ion bombardment on active corrosion or passivity has been studied [3, 4]. In the field of localized corrosion, interest has mainly focused on pitting [5, 6]. The problem of hydrogen-induced degradation of metals, however, has been nearly completely omitted. The only exceptions have been the investigation of stress corrosion cracking of steel [7] and aluminium alloys [8]. In these systems, hydrogen may be involved in the cracking mechanism. Until now, no indication of the positive effects of implanted ions has been reported.

The present study is an attempt to investigate the hydrogen embrittlement of tantalum and the possibility of modifying the tantalum surface in order to improve its resistance to hydrogen degradation. Tantalum is used in the chemical industry as a construction material which is exposed to strong acids. The corrosion of tantalum in these solutions involves the formation of hydrogen, which may enter the metal and cause embrittlement. A conventional method of protection is the attachment of a platinum foil to the surface of the tantalum. Because these foils must have a minimum area in order to be effective, we tried to conserve noble metals by implantation of platinum into the tantalum surface.

2. EXPERIMENTAL DETAILS

2.1. Sample preparation and implantation

Tantalum foils, 0.125 mm thick with an area of 30 mm $\times$ 7.5 mm and a purity of 99.9%, were implanted with 5×10^{16} Pt^+ cm^{-2} at an energy of 80 keV using the 300 keV ion accelerator at Gesellschaft für Schwerionenforschung, Darmstadt. Prior to implantation the specimens were degreased in hot ace-

*Paper presented at the International Conference on Surface Modification of Metals by Ion Beams, Kingston, Canada, July 7-11, 1986.

0025-5416/87/$3.50

tone and washed with hot water. For some measurements the samples were anodized prior to implantation in order to produce an oxide layer of reproducible thickness. These samples were etched before anodizing in a solution consisting of 5 parts (by weight) of sulphuric acid, 2 parts of nitric acid and 1 part of hydrofluoric acid and electropolished at 100 mA cm^{-2} in a solution of 9 parts of sulphuric acid and 1 part of hydrofluoric acid. The anodizing was done in a solution of 0.1 M Na_2SO_4 at an elevated temperature, first galvanostatically at 1 mA cm^{-2} until the appropriate voltage was reached and then potentiostatically. This procedure resulted in oxide layers up to 700 nm thick with good adhesion and without incorporation of electrolyte. The oxide thickness was measured by recording the faradaic current during anodizing and controlled by observation of the capacitance and the colour of the surface layers. In addition, the depth profile of the implanted platinum and the composition of the near-surface regions of the tantalum specimens were analysed by Auger electron spectroscopy.

2.2. *Hydrogen charging*

In order to cause embrittlement of the tantalum specimens in a reasonable time, cathodic charging was applied. Tantalum served as the cathode of an electrochemical cell with separate anode and cathode regions; carbon was used as the anode because of its good resistance to the chlorine generated during the process. The cell was filled with 5 M HCl and 1 M H_2SO_4 and the charging current was usually 100 mA cm^{-2} at 25 °C. The potential was measured with respect to a calomel reference electrode connected to a voltmeter and an x–t plotter. The anodized specimens had to be predamaged before hydrogen charging. Because of the rather thick and dense oxide layer, no embrittlement took place without predamage. Therefore the oxide surface was scratched in a reproducible way with hardened steel tips moving across the surface under a constant load of 100–200 gf. Usually, six parallel equidistant scratches of 50–100 μm width were made.

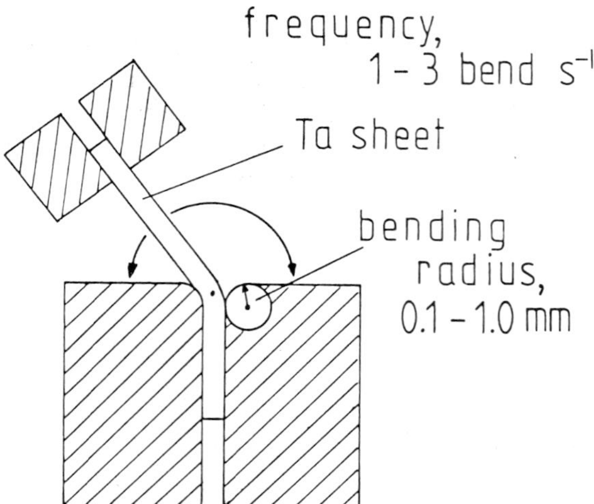

Fig. 1. Schematic representation of the bending test apparatus. The apparatus is driven by an E motor.

2.3. *Determination of embrittlement*

The degree of embrittlement was determined by means of a bending test [9]. The apparatus for this is shown in Fig. 1. The specimens were bent by means of a motor around a metal block with a defined and variable radius. Generally, the test was performed with one bend (180°) per second around a radius of 0.1 mm. The ratio of the number B of bends until fracture with charging to the number B_0 of bends until fracture without charging was recorded as a function of the charging time.

For future experiments an additional and more precise test for embrittlement was developed. In this test the conductivity of the specimen is measured by the four-point technique. Hydrogen uptake causes an additional resistance which is proportional to the charging time [10]. Figure 2 shows the results of preliminary experiments with a tantalum foil anodized to an oxide thickness of 200 nm; curve a represents the undamaged specimens not undergoing embrittlement and curve b the predamaged sample degrading increasingly with increasing charging time. Experiments with implanted samples are being performed at present, but no results are available yet.

3. RESULTS AND DISCUSSION

3.1. *Tantalum foils with thin oxide layers*

The first series of tantalum foils was irradiated and tested in the as-delivered state, with

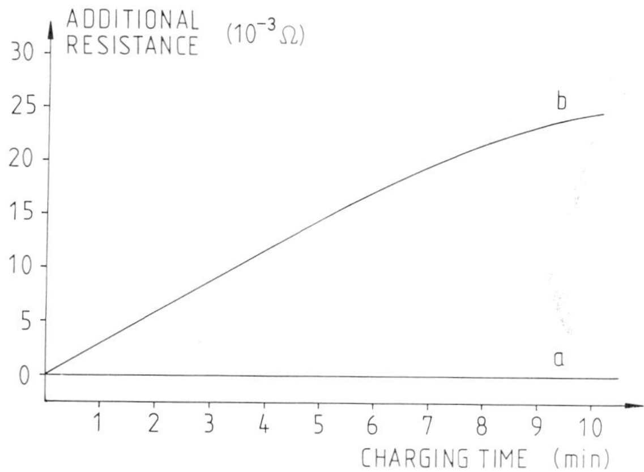

Fig. 2. Resistivity measurements with the four-point technique: curve a, the additional resistance of a tantalum sample anodized to a surface oxide thickness of 200 nm as a function of the hydrogen-charging time; curve b, the same for a similar sample pre-damaged by scratches through the oxide layer.

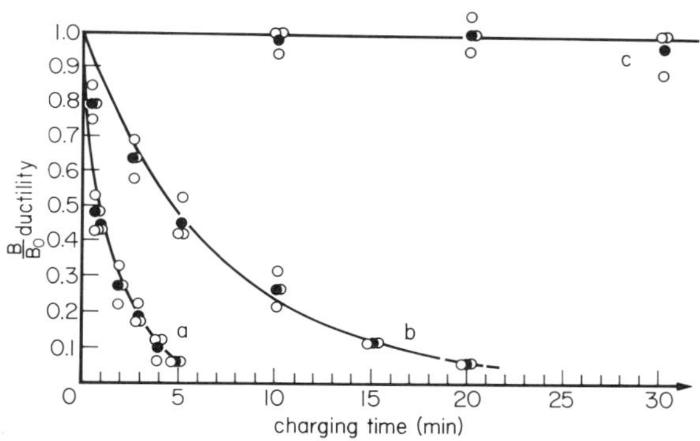

Fig. 4. Results of bending tests performed with tantalum foils in the as-delivered state (B is the number of bends until fracture of hydrogen-charged samples; B_0 is the number of bends until fracture of uncharged samples; $\bar{B}$ is the mean value of three measurements): curve a, tantalum foil etched with hydrofluoric acid directly before charging; curve b, tantalum foil bombarded with 5×10^{16} Pt^+ cm^{-2}; curve c, tantalum foil with an undamaged oxide film, as delivered; ●, $\bar{B}/B_0$.

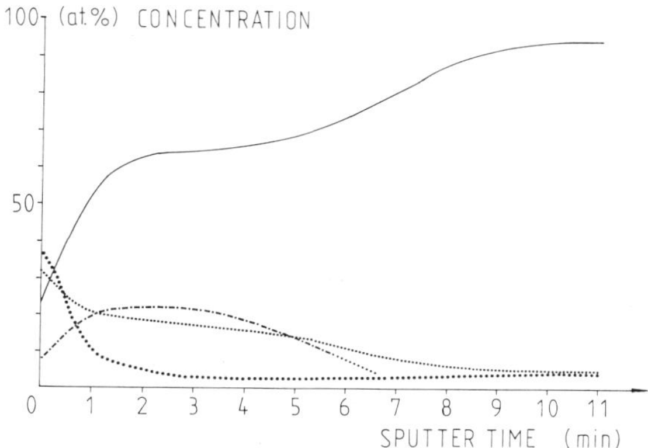

Fig. 3. Auger depth profile of a tantalum foil in the as-delivered state after bombardment with 5×10^{16} Pt^+ cm^{-2} at 80 keV: ——, tantalum; —·—·—, platinum; -----, carbon; · · ·, oxygen.

an oxide layer of approximately 5 nm thickness. Figure 3 shows the Auger depth profile of an implanted sample. The surface concentration of platinum is 5–10 at.% and the peak concentration 20 at.%. Considerable carbon contamination of the implanted region is obvious. The concentrations are calculated from the sensitivity factors and may contain rather large errors. The results of the bending tests are displayed in Fig. 4. Curve a represents a sample for which the oxide layer was etched away with hydrofluoric acid before the hydrogen charging. The result is that it has a very strong susceptibility to hydrogen embrittlement. Curve b is for a sample implanted with 5×10^{16} Pt^+ cm^{-2}. The resistance against embrittlement is better than that of curve a but still much worse than that of curve c, which represents a sample in the as-delivered state with an undamaged oxide layer. These samples do not undergo any embrittlement during charging for 30 min in 5 M HCl with a current density of 100 mA cm^{-2}. These results show that platinum implantation into tantalum impairs its resistance to hydrogen embrittlement. The reason is probably a deterioration in the quality of the oxide layer due to incorporated platinum clusters. The effect of the platinum as a hydrogen desorption catalyst is counterbalanced by the presence of additional hydrogen diffusion paths along the oxide–platinum phase boundary.

3.2. *Tantalum foils with thick oxide layers*

Because of the unsatisfactory results for implantations into tantalum foils with thin oxide layers, where the platinum concentration profile extends deep into the bulk metal (see Fig. 3), the influence of the oxide thickness was studied in more detail. It turned out that ion implantation into the oxide layer only had a detrimental effect below a certain thickness of the oxide. In Fig. 5 the dependence of the B/B_0 ratio on the thickness of the anodized oxide layer is displayed for samples

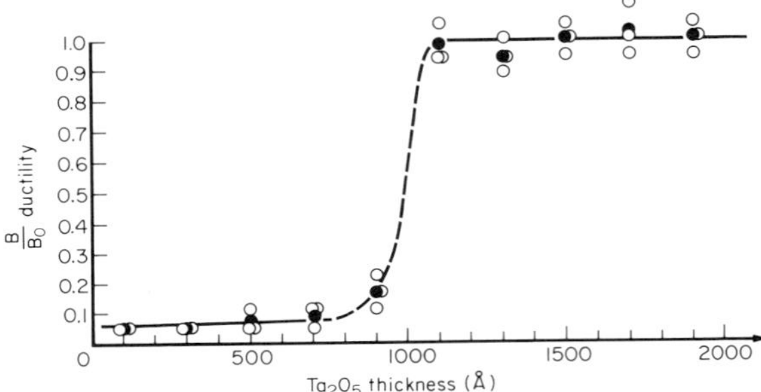

Fig. 5. Results of bending tests showing the dependence of the B/B_0 ratio, as defined in Fig. 4, on the oxide thickness for anodized tantalum foils bombarded with 5×10^{16} Pt$^+$ cm^{-2}: ●, $\bar{B}/B_0$ (charging time, 30 min).

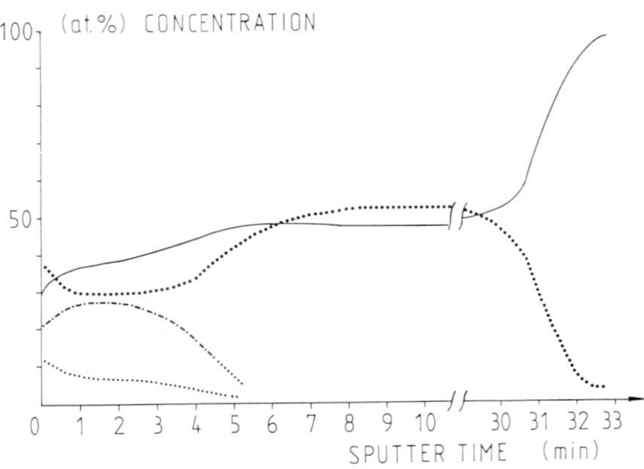

Fig. 6. Auger depth profile of a tantalum foil anodized to an oxide thickness of 200 nm and bombarded with 5×10^{16} Pt$^+$ cm^{-2} at 80 keV: ——, tantalum; —·—·—, platinum; -----, carbon; . . ., oxygen.

implanted with 5×10^{16} Pt$^+$ cm^{-2}. Hydrogen embrittlement is strong for an oxide thickness of less than 80 nm. For an oxide thickness greater than 100 nm the samples do not become embrittled at all and behave like unimplanted specimens in the as-delivered state. 80 nm corresponds roughly to the range of the platinum ions after consideration of sputter erosion of the surface. This leads to the conclusion that the presence of a very thin undisturbed oxide layer between the implanted layer and the bulk metal is necessary for protection against embrittlement.

The next question is whether platinum implantation in a thick oxide layer would have a beneficial effect. For this purpose, tantalum specimens with 200 nm Ta_2O_5 films were prepared and bombarded with 1×10^{16} and 5×10^{16} Pt$^+$ cm^{-2}. Because these samples do not become brittle under normal hydrogen-charging conditions, they were predamaged in a reproducible way according to the procedure described in Section 2. Figure 6 shows the Auger depth profile of the sample bombarded with 5×10^{16} Pt$^+$ cm^{-2}, indicating a near-surface region about 30 nm thick containing up to 25 at.% Pt, an intermediate pure oxide layer approximately 150 nm thick and the oxide–metal interface.

After hydrogen charging, the predamaged samples were subjected to the bending test described in Section 2. The results are shown in Figs. 7 and 8. In contrast with undamaged specimens, the damaged specimens became brittle during the first 10 min of hydrogen charging (curves a). In Fig. 7, curve b, the behaviour of a sample bombarded with 1×10^{16} Pt$^+$ cm^{-2} is displayed. To reach the same

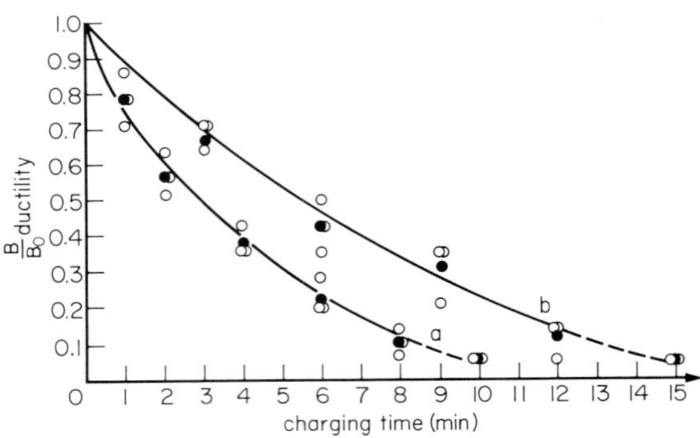

Fig. 7. Results of bending tests performed with tantalum foils anodized to an oxide thickness of 200 nm and predamaged by equidistant parallel scratches: curve a, B/B_0 ratio for the unimplanted specimens; curve b, B/B_0 ratio for specimens implanted with 1×10^{16} Pt$^+$ cm^{-2}; ●, $\bar{B}/B_0$.

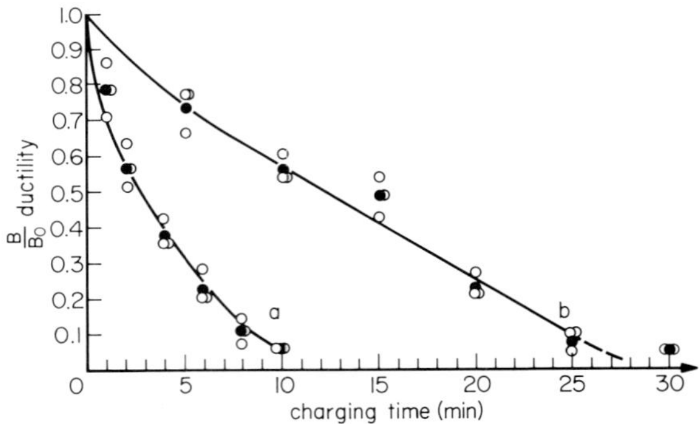

Fig. 8. Results of bending tests as described in Fig. 7: curve a, B/B_0 ratio for unimplanted specimens; curve b, B/B_0 ratio for specimens implanted with 5×10^{16} Pt$^+$ cm^{-2}; ●, $\bar{B}/B_0$.

degree of embrittlement, double the charging time was required. A sample bombarded with 5×10^{16} Pt^+ cm^{-2} (Fig. 8, curve b) required four times the charging time of unimplanted tantalum. These results show that platinum implantation into the surface oxide layer of tantalum, without direct Pt–Ta contact, is an effective method of inhibiting hydrogen embrittlement. The platinum acts partly as a desorption catalyst for hydrogen and partly by accelerating the repassivation of the damaged oxide layer by stimulating the cathodic partial reaction of the corrosion of tantalum in acids; as a consequence of the faster cathodic reaction, anodic passivation is facilitated.

4. CONCLUSIONS

The results of the study shows that it is possible to protect metals such as tantalum, niobium, titanium and hafnium from hydrogen embrittlement in acids by implantation of metals such as platinum which catalyse the hydrogen evolution reaction. However, care must be taken to implant only into the oxide layer and to avoid direct contact between platinum and the bulk metal. In the latter case the quality of the oxide film is reduced and embrittlement may proceed faster than in the unimplanted case. Good long-term stability of the protective implanted layer can be expected because the reported experiments were carried out under much more severe conditions than exist in normal corrosion processes. Realistic cases of embrittlement are often caused by damaged metal surfaces. As shown, these are conditions where implantation is especially helpful. Because the presence of platinum on the surface of refractory metals is expected to reduce the corrosion rate [11], the loss of platinum during dissolution of the oxide should be low; this conclusion will be examined in long-term experiments. Also the effect of the ion beam mixing of thin evaporated platinum layers is at present under study; without ion bombardment, these layers have been shown to have no adhesion to the Ta_2O_5 surface. All future experiments will use conductivity measurements (see Section 2) for determining embrittlement because they are more accurate and reproducible than the bending tests used in the present study.

REFERENCES

1 V. Ashworth, R. P. M. Procter and W. A. Grant, *Treatise Mater. Sci. Technol., 18* (1980) 176.

2 H. Ferber and G. K. Wolf, *Proc. Int. Conf. on Surface Modification of Metals by Ion Beams, Kingston, Canada, July 7–11, 1986*, in *Mater. Sci. Eng., 90* (1987).

3 A. Galerie, M. Caillet and M. Pons, in G. K. Wolf, W. A. Grant and R. P. M. Procter (eds.), *Proc. Int. Conf. on Surface Modification of Metals by Ion Beams, Heidelberg, September 17–21, 1984*, in *Mater. Sci. Eng., 69* (1985) 329.

4 Q. M. Chen, H. M. Chen, X. D. Bai, J. Z. Zhang, H. H. Wang and H. D. Li, *Nucl. Instrum. Methods, 209–210* (1983) 867.

5 W. K. Chan, C. R. Clayton, R. G. Allas, C. R. Gosset and J. K. Hirvonen, *Nucl. Instrum. Methods, 209–210* (1983) 857.

6 H. J. Kim, W. B. Carter, R. F. Hochman and E. I. Meletis, in G. K. Wolf, W. A. Grant and R. P. M. Procter (eds.), *Proc. Int. Conf. on Surface Modification of Metals by Ion Beams, Heidelberg, September 17–21, 1984*, in *Mater. Sci. Eng., 69* (1985) 297.

7 I. H. Craig and R. N. Parkins, in V. Ashworth, R. P. M. Procter and W. A. Grant, *Treatise Mater. Sci. Technol., 18* (1980) 245.

8 G. K. Wolf and H. Buhl, in G. K. Wolf, W. A. Grant and R. P. M. Procter (eds.), *Proc. Int. Conf. on Surface Modification of Metals by Ion Beams, Heidelberg, September 17–21, 1984*, in *Mater. Sci. Eng., 69* (1985) 317.

9 L. A. Gypen, M. Brabers and A. Deruyttere, *Werkst. Korros., 35* (1984) 37.

10 E. A. Garcia, *Z. Metallkd., 63* (1972) 561.

11 P. Munn and G. K. Wolf, *Nucl. Instrum. Methods B, 7–8* (1985) 205.

Materials Science and Engineering, 90 (1987) 243-251

Hydrogen Permeation Behavior in Polycrystalline Nickel Implanted with Helium, Argon, Nickel, Yttrium and Platinum*

R. NISHIMURA[a]†, R. M. LATANISION[a] and G. K. HUBLER[b]

[a] *The H.H. Uhlig Corrosion Laboratory, Massachusetts Institute of Technology, Cambridge, MA 02139 (U.S.A.)*

[b] *U.S. Naval Research Laboratory, Washington, DC 20375 (U.S.A.)*

(Received July 10, 1986)

ABSTRACT

The entry and transport of hydrogen in helium-, argon-, nickel-, yttrium- and platinum-implanted polycrystalline nickel specimens with a fluence range 1×10^{14}-2×10^{17} ions cm^{-2} have been investigated using an electrochemical permeation technique. It was found that the effective diffusion coefficient of hydrogen was lower and the effective solubility was larger in nickel-, yttrium- and platinum-implanted nickel with a fluence of more than 1×10^{15} ions cm^{-2} than in unimplanted nickel. Increased solubility caused the steady state permeation current in nickel- and platinum-implanted nickel to increase. The steady state permeation current and the effective hydrogen solubility of nickel-implanted nickel with a fluence of 1×10^{14} ions cm^{-2} were small compared with the corresponding values for unimplanted nickel, and those of helium- and argon-implanted nickel decreased with increasing fluence. Furthermore, the permeation current of platinum-implanted nickel treated with aqueous 0.2% HF decreased by a factor of 2 from that of platinum-implanted nickel. The results are explained by considering the presence of gas bubbles, compressive stress, shallow and deep hydrogen traps and the electrocatalytic properties of the implanted element.

*Paper presented at the International Conference on Surface Modification of Metals by Ion Beams, Kingston, Canada, July 7-11, 1986.

†Present address: The Government Industrial Research Institute, Chugoku, 15000 Hiromachi Kure City, Hiroshima, Japan.

1. INTRODUCTION

Many researchers have investigated the interrelation between hydrogen embrittlement and metalloid segregation at grain boundaries in metals and alloys [1-10]. Segregated metalloids may act as preferential sites for the absorption of hydrogen, which may decrease the cohesive strength of the material. However, it is clear that not all metalloid elements have the same effect on the hydrogen embrittlement of nickel [8,10] and nickel-base alloys [6,7], once hydrogen is absorbed.

Ion implantation may be the best method to prepare surfaces for segregation studies without affecting bulk physical or mechanical properties because any element can be implanted and the concentration of segregated elements may be controlled over several orders of magnitude. At present, ion implantation is being used as a technique for the beneficial modification of surfaces, including resistance to corrosion, wear and fatigue [11,12]. Most studies on ion implantation in the field of corrosion and electrochemistry have been concerned with improving the corrosion resistance [13-15] and enhancing the hydrogen evolution reaction [16]. However, Zamanzadeh *et al.* [17] have studied hydrogen permeation transients in helium-, iron- and platinum-implanted iron while Myers *et al.* [18,19], Besenbacher *et al.* [20] and Wampler and Myers [21] have performed detailed studies of damage and impurity trapping of hydrogen in iron and nickel.

The hydrogen permeation transients of polycrystalline nickel implanted with inert elements (helium or argon), the substrate element (nickel), metalloids (phosphorus, sulfur,

0025-5416/87/$3.50

antimony or arsenic), a catalytic element (platinum) and other elements (lead, bismuth, boron, yttrium or tin) have been investigated as a function of fluence using an electrochemical permeation method. In the present paper the entry and transport of hydrogen in helium-, argon-, nickel-, yttrium- and platinum-implanted nickel were investigated to sort out effects caused by the process of ion implantation. These elements do not produce a crystalline-to-amorphous phase transition at high fluences [22, 23]. The results obtained are interpreted in terms of damage and/or gas bubbles, produced during the process of ion implantation, and the electrocatalytic property of the implanted element. In subsequent papers, metalloid-implanted nickel will be treated, and relationships between segregants and hydrogen embrittlement will be inferred.

2. EXPERIMENTAL DETAILS

2.1. *Specimen preparation*

Strips of Ni 270 (99.97% Ni) were cold rolled, annealed at 1000 °C for 10 min and then water quenched. Electropolishing of the specimens was performed in an aqueous 60% H_2SO_4 solution (bath voltage, 5 V) at room temperature using a platinum electrode as the cathode. The specimens were not coated with palladium.

2.2. *Ion implantation*

The ions of helium and argon were chosen because previous work indicated that helium forms deep hydrogen traps in iron and nickel [19-21] and argon may be expected to behave similarly. Nickel ions were selected to assess the effects of radiation damage without the presence of chemical effects, the physical effects of gas bubbles [21] or extreme compressive stress [22]. Platinum was chosen because of expected surface catalytic effects on the hydrogen evolution reaction [17] and yttrium because it is likely to interact strongly with defects [18].

The electropolished specimens were implanted on the cathodic side (the side where hydrogen evolution occurs) by clamping the edge of a rectangular (5 cm × 2.5 cm) foil along the long dimension with razor blade masks. The masks provided a thermally conductive path to remove heat from the foils by providing a pressure contact with a water-cooled aluminum block. The sharp edges on the razor blades minimized the sputtering of mask material onto the nickel surface during ion implantation. The cryopumped vacuum was kept at pressures between 3×10^{-7} and 2×10^{-6} Torr and the ion beam current density was held below 1 $\mu A\ cm^{-2}$ in order to minimize specimen heating. The foil temperature during implantation was measured to be less than 100 °C by means of an optical pyrometer. The ions were implanted at normal incidence to the surface and the beam was raster scanned to produce a uniform fluence measured to better than ±2.0%. The fluences and energies of implantation are given in Table 1 which also shows the calculated projected range R_p of implanted ions and the standard deviation ΔR_p. R_p was almost constant, given the various ion implantation conditions, with the exception of helium. Different fluences were used in an attempt to separate progressively the effects of only radiation damage at low fluences, bulk alloying effects at intermediate fluences (*e.g.* a bulk concentration of a few atomic per cent and a surface concentration of about 0 at.%) and surface chemical effects at high fluences (*e.g.* surface and bulk concentrations of about 20 at.%). The 10^{17} ions cm^{-2} fluence for argon, nickel and yttrium are all near the sputter saturation limit for retained dose.

2.3. *Permeation measurements*

The method developed by Devanathan and Stachurski [24] was used for the permeation measurements. The specimens were mounted between the two half-cells, giving a 0.95 cm^2 area of exposure. The cathodic and anodic compartments contained 0.1 N H_2SO_4 and 0.1 N NaOH solutions respectively. Both solu-

TABLE 1

Implantation parameters and ion range statistics

Ion	*Energy* (keV)	*Fluence* (ions cm^{-2})	R_p (nm)	ΔR_p (nm)
He^+	25	10^{17}, 2×10^{17}	87	14
Ar^+	60	10^{15}, 10^{16}, 10^{17}	24	12
Ni^+	90	10^{14}, 10^{15}, 10^{16}, 10^{17}	25	12
Y^+	125	10^{17}	25	11
Pt^+	180	10^{15}, 3×10^{15}, 10^{16}	25	8.6

tions were deaerated with prepurified nitrogen prior to the experiments. Hydrogen charging in the cathodic compartment was controlled by galvanostatic cathodic polarization with a current density of 1.47 mA cm^{-2}. The anodic side of the specimens was potentiostatically polarized at +0.1 V with respect to a saturated calomel electrode (SCE) with a Wenking 66TAI potentiostat. When the anodic current became negligible (less than 10 nA cm^{-2}), hydrogen charging was started. Hydrogen permeation was monitored with a Keithley 602 electrometer and an Allen 2100 series strip chart recorder. All experiments were carried out at 30 ± 0.1 °C. The steady state permeation current P_∞ was measured twice on two different areas of the same implanted foil and the values averaged. Good reproducibility was obtained between each side both for the shape of the transient and for P_∞. The errors are estimated to be ±10% on the absolute values of the transient data.

Permeation data are frequently analyzed for D_{lag}, the diffusion coefficient (proportional to the rise time of the transient) and C_H, the concentration of hydrogen adsorbed just beneath the surface (proportional to P_∞ [25, 26]). The analysis works well for unimplanted nickel and yields a D_{lag} value of 3.8×10^{-10} cm^2 s^{-1} and a C_H value of 2.5×10^{-4} mol cm^{-3}. P_∞ is thickness dependent. Values of C_H and D_{lag} are not emphasized here because of the lack of physical significance of the derived quantity D_{lag} which is an "effective" diffusion coefficient for the entire sample. Its value always decreased for implanted samples although not by more than a factor of 2. Because the implantation depth constitutes only 1 part in 4000 of the total thickness, D would have to decrease by four orders of magnitude in the implanted layer in order to be responsible for the observed decrease in D_{lag}. This order is physically unreasonable. Also, in the analysis of hydrogen trapping and detrapping data by Wampler and Myers [21] the bulk diffusion constant of hydrogen in helium-ion-damaged nickel gave good agreement with experimental results, indicating that D for ion-damaged nickel is not greatly different from that for pure annealed nickel. Values of C_H in the simple analysis of our data followed closely the trends in P_∞ but usually underwent larger percentage changes in the absolute value.

2.4. Background

A number of different mechanisms may be expected to have an effect on the permeation transient in nickel. In a review of hydrogen trapping in metals, Besenbacher *et al.* [20] identified a number of traps in nickel with small activation energies that may be considered shallow traps at room temperature. These are as follows: substitutional impurity, $E_B \leqslant 0.1$ eV; dislocation, $E_B = 0.1$ eV; interstitial impurity, $E_B = 0.1$–0.2 eV; self-interstitial, $E_B = 0.24$ eV. Deep traps that were identified include the following: vacancy, $E_B = 0.43$ eV; voids and bubbles, $E_B = 0.55$ eV; yttrium–vacancy complex in iron, $E_B = 1.27$ eV. The effect of these traps introduced by ion implantation is to delay the development of the permeation transient while the traps fill, and to increase the solubility for hydrogen in the surface. Deep traps increase the solubility but do not contribute to the mobile fraction of C_H that serves as the driving force for hydrogen diffusion whereas shallow traps which fill and empty readily at room temperature do contribute to hydrogen diffusion. Another bulk effect can come from decreasing the solubility in the surface. The effective medium theory states that the solution sites for hydrogen in the lattice are those where there is a minimum in the electron density [21]. Compressive stress which is the usual state of implanted surfaces (exceptions are for small interstitial species) increases the average electron density and decreases hydrogen solubility. A high concentration of interstitial impurities can also remove solution sites. The final bulk effect to consider is a change in the diffusion coefficient and, as already discussed, we believe that this is an unlikely mechanism for the data presented here.

Surface effects that can alter the permeation transient are the catalytic action on hydrogen absorption or surface recombination rates. Another possibility is a change in active surface area induced by topographical changes, caused by sputtering, by alloying of an inactive element at the surface or by strong hydride formers which may trap hydrogen at the surface, removing that site from chemical activity. These mechanisms are summarized in Table 2. For a given combination of ion and ion fluence, more than one of these mechanisms may be responsible for changes observed in the permeation transient.

TABLE 2

Summary of expected effects of ion implantation on permeation transients

Bulk effects	
(1)	Deep traps and shallow traps contribute to delay in breakthrough
(2)	C_H increases by introduction of shallow traps
(3)	C_H decreases by compressive stress or occupied interstitial sites
(4)	Decrease in diffusion coefficient (not likely)
Surface effects	
(1)	Increase or decrease rate of H adsorption
(2)	Increase or decrease H recombination rate
(3)	Alter chemically active surface area

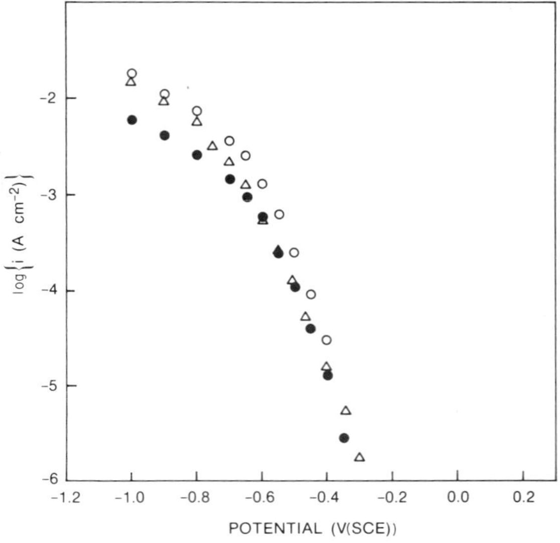

Fig. 1. Cathodic potentiostatic polarization curves for pure nickel (△), and nickel implanted with 10^{16} Ni^+ cm^{-2} (○) and nickel implanted with 10^{17} Ni^+ cm^{-2} (●) in deaerated 0.1 M H_2SO_4.

3. RESULTS

3.1. *Cathodic polarization*

Figure 1 shows cathodic potentiostatic polarization scans for nickel-implanted nickel for the two highest fluences. There is a shift in open-circuit potential of +40 mV(SCE) for the highest fluence and 0 mV(SCE) for the 10^{16} Ni^+ cm^{-2} fluence. The largest shifts for the other elements were smaller (−35 mV(SCE) for 1×10^{16} Pt^+ cm^{-2}; −20 mV(SCE) for 1×10^{16} Ar^+ cm^{-2}). At the highest argon fluence of 10^{17} ions cm^{-2} the shift was +10 mV. We attribute these small shifts in open-circuit potential chiefly to experimental drift and uncertainty in the measurement of the surface voltage. Carbon build-up on the surface induced by implantation is also a possibility which creates a less cathodically active surface. Since occasionally the shift was positive, this explanation seems less likely. In all cases the Tafel slope of 120 mV $decade^{-1}$ was identical with the slope for unimplanted nickel, indicating that the mechanism of hydrogen exchange on the surface was unchanged by all the surface treatments in the as-implanted condition.

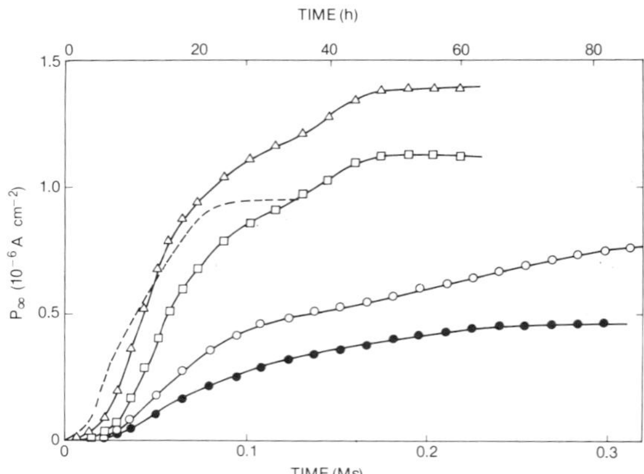

Fig. 2. Hydrogen permeation transients for unimplanted nickel (– – –), nickel implanted with 10^{14} Ni^+ cm^{-2} (L = 135 μm) (●), nickel implanted with 10^{15} Ni^+ cm^{-2} (L = 110 μm) (○), nickel implanted with 10^{16} Ni^+ cm^{-2} (L = 90 μm) (△) and nickel implanted with 10^{17} Ni^+ cm^{-2} (L = 100 μm) (□).

3.2. *Nickel-implanted nickel*

Figure 2 shows the hydrogen permeation transients of nickel implanted with fluences of 1×10^{14}–1×10^{17} Ni^+ cm^{-2}. It is found that the steady state permeation currents tend to become larger for nickel implanted with fluences of 1×10^{16} and 1×10^{17} Ni^+ cm^{-2} and smaller for nickel implanted with fluences of 1×10^{14} and 1×10^{15} Ni^+ cm^{-2} than for unimplanted nickel, even if the difference in membrane thickness (shown in the figure) is taken into consideration (the thinner the membrane thickness, the larger is the steady state permeation current). As was the case for all the implanted ions, there was a sizable time delay in the transient response. Increased surface roughness due to sputtering would only be possible for the highest fluence as an explanation of the increase in permeation current. Since there is no increase in P_∞ between 10^{16} and 10^{17} Ni^+ cm^{-2}, surface roughness is not a factor in these measurements.

3.3. *Argon-implanted nickel*

Figure 3 shows the hydrogen permeation transients of argon-implanted nickel with a fluence of 1×10^{15}–1×10^{17} ions cm^{-2}. The broken line in the figure indicates a typical permeation transient for unimplanted nickel (pure nickel) as a reference. It is evident that the breakthrough time of the transients is larger than that of unimplanted nickel. All the steady state permeation currents were small in comparison with that of unimplanted nickel and the permeation current of argon-implanted nickel decreased with increasing fluence.

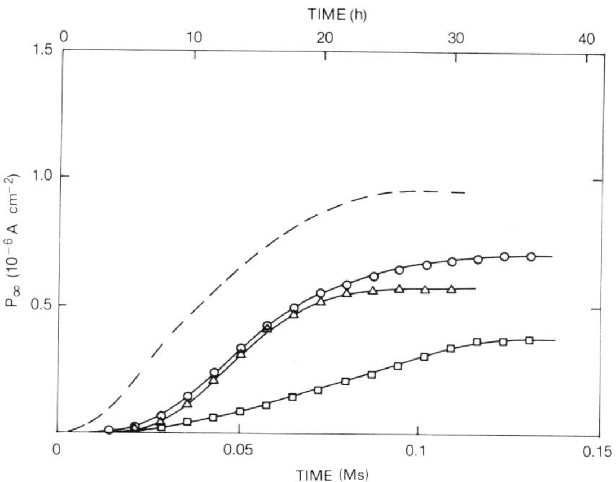

Fig. 3. Hydrogen permeation transients for unimplanted nickel (— —), nickel implanted with 10^{15} Ar^+ cm^{-2} (○), nickel implanted with 10^{16} Ar^+ cm^{-2} (△) and nickel implanted with 10^{17} Ar^+ cm^{-2} (□) ($L = 105\,\mu m$).

3.4. *Platinum-implanted nickel*

The range of fluences for platinum-implanted nickel was only 1×10^{15}–1×10^{16} ions cm^{-2} because of the difficulty in obtaining high beam currents of platinum ions. Figure 4 shows the permeation transients of the platinum-implanted nickel. Similar to nickel-implanted nickel, the breakthrough time was much longer than for unimplanted nickel and P_∞ was found to increase with increasing fluence.

3.5. *Helium- and yttrium-implanted nickel*

Figure 5 shows the permeation transients of nickel implanted with helium (fluences of 1×10^{17} and 2×10^{17} ions cm^{-2}) and yttrium (a fluence of 1×10^{17} ions cm^{-2}). The permeation transients of helium- and argon-implanted nickel are similar. For yttrium-implanted nickel the breakthrough time is much longer than that for unimplanted nickel, whereas the permeation current is the same as that for unimplanted nickel.

The value of P_∞, normalized to that for an unimplanted nickel foil 100 μm thick, is shown

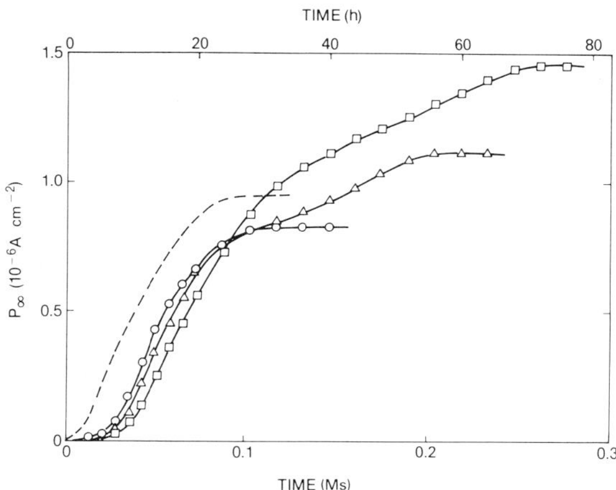

Fig. 4. Hydrogen permeation transients for unimplanted nickel (- - -), nickel implanted with 10^{15} Pt^+ cm^{-2} (○), nickel implanted with 3×10^{15} Pt^+ cm^{-2} (△) and nickel implanted with 10^{16} Pt^+ cm^{-2} (□) ($L = 105\,\mu m$).

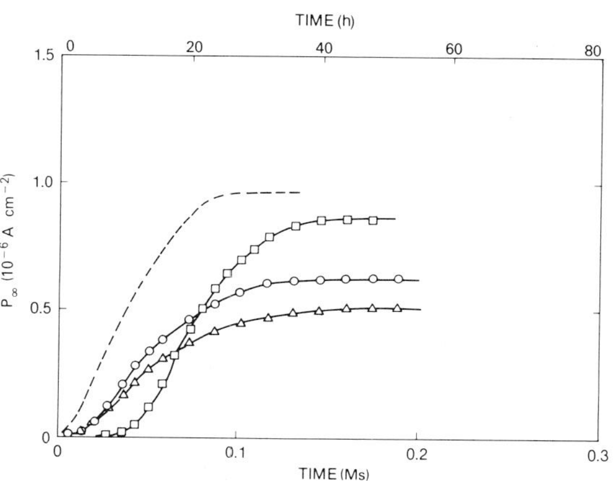

Fig. 5. Hydrogen permeation transients in unimplanted nickel (- - -), nickel implanted with 10^{17} He^+ cm^{-2} (○), nickel implanted with 2×10^{17} He^+ cm^{-2} (△) and nickel implanted with 10^{17} Y^+ cm^{-2} (□) ($L = 105\,\mu m$).

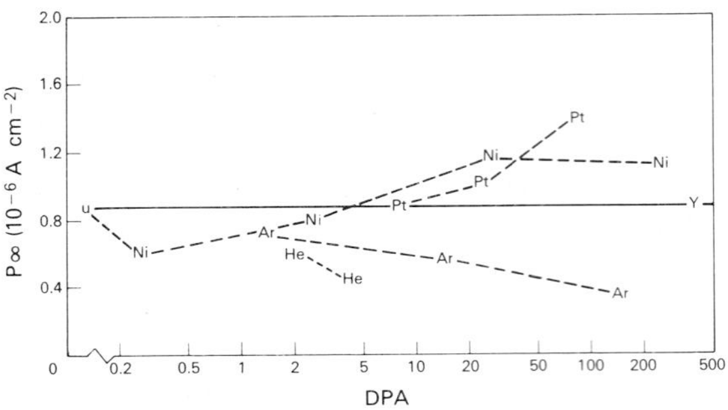

Fig. 6. The permeation current P_∞ for each implanted element *vs.* displacements per substrate atom: u, unimplanted.

for all the ions *vs.* displacements per atom (the abbreviation for which is dpa), a unit of radiation damage in Fig. 6. A unit of 1 dpa means that, on the average, each atom in the implanted region has been displaced from its equilibrium lattice site once. The range parameters R_p and ΔR_p and the total energy deposited in displacement damage used in the calculation of displacements per atom were obtained using the code E-DEP-1 [27] and a simple Kinchin and Pease definition of displacements per atom [28]. From Fig. 6 it is clear that there are three types of behavior. At low displacements per atom, all the data group together and P_∞ is less than for unimplanted samples. At high fluences, P_∞ for the rare gases decreases, the P_∞ values for nickel and platinum increase, while that for yttrium remains unchanged.

4. DISCUSSION

4.1. *Nickel- and platinum-implanted nickel*

In the as-implanted condition the nickel- and platinum-implanted nickel permeation transients shown in Figs. 2 and 4 are very similar. For example, the plot of P_∞ *vs.* displacements per atom in Fig. 6 indicates that 10^{16} Pt^+ cm^{-2} produces equivalent radiation damage to 10^{17} Ni^+ cm^{-2} for foil thicknesses of 105 μm and 110 μm respectively, and the two permeation transients for these ions and fluences are in good agreement. Both ions introduce a delay in the transient. Because of the similarities, the effects of both platinum and nickel can be considered to be caused by

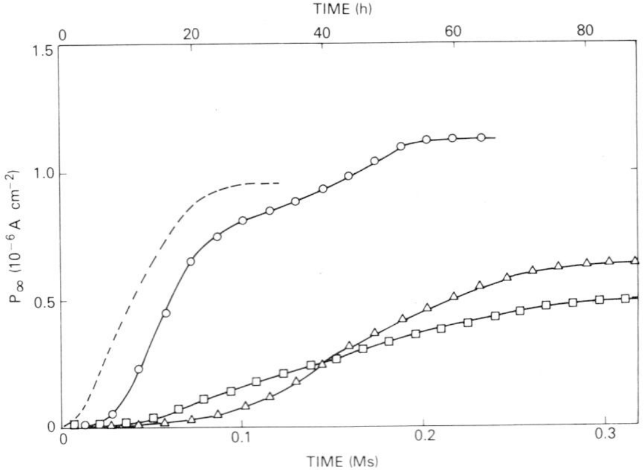

Fig. 7. Hydrogen permeation transients of nickel implanted with a fluence of 3×10^{15} Pt^+ cm^{-2} before and after treatment with aqueous 0.2! HF: ○, $t = 0$ s; △, $t = 60$ s; □, $t = 180$ s.

radiation damage with no surface or bulk chemical effects present. Platinum is near to nickel in atomic size and would be expected to be substitutional in the nickel lattice and, at the highest fluence, the surface concentration of platinum is only about 3 at.% as measured by Rutherford backscattering. This platinum surface concentration evidently is not high enough to exert a chemical influence on the surface. In order to stimulate the chemical effect of platinum, a sample was etched in aqueous 0.2% HF for several time periods to dissolve some of the surface and to expose additional platinum. Figure 7 shows that, after this treatment, the hydrogen permeation transient exhibited a large delay and a much lower P_∞ which can be explained by the well-known catalytic stimulation by platinum of the hydrogen recombination reaction on the surface. More of the hydrogen atoms absorbed on the surface during hydrogen charging leave the surface as H_2 gas, creating a lower H_2 partial pressure and a reduction in atomic hydrogen entry.

Returning to Figs. 2 and 4, it is noted that for fluences of 10^{16} and 10^{17} Ni^+ cm^{-2} and of 3×10^{15} and 10^{16} Pt^+ cm^{-2}, the permeation current is larger than for unimplanted nickel. This behavior can be understood when the mechanism of trap formation in the bulk described in Section 2.4 is considered; deep traps produce a delay in the transient and shallow traps increase the solubility C_H which increases P_∞.

Implantation damage (in the absence of phase formation) takes the form of simple point defects (*e.g.* interstitials, di-interstitials, vacancies and divacancies) which at higher fluences coalesce into vacancy clusters, vacancy loops, voids and dislocations. Bright field transmission electron microscopy cannot resolve the point defect structures so that only dislocations, voids (when formed) and large point defect clusters are observed. In heavily damaged material (greater than 1–100 dpa) a dense tangled network of dislocations is observed which resembles the micrographs of cold-worked material. The permeation behavior of nickel-implanted nickel to a fluence of 1×10^{14} ions cm^{-2} (corresponding to about 0.3 dpa) is much different from that of the high fluence samples as seen in Fig. 2. It is reasonable to assume that the development of the permeation transients in Fig. 2 with increasing fluence follows the build-up and saturation of disorder which occurs for values of 0.3 dpa, 3 dpa, 30 dpa and 300 dpa for fluences of 10^{14} Ni^+ cm^{-2}, 10^{15} Ni^+ cm^{-2}, 10^{16} Ni^+ cm^{-2} and 10^{17} Ni^+ cm^{-2} respectively. It should be noted that saturation of P_∞ occurs between 3 and 30 dpa. At the lowest fluence the effects on permeation are predominantly those caused by simple defect formation and vacancy loops, while at high fluences the effects are caused by simple defects plus extended defects. At the higher fluences, shallow traps are introduced which raise C_H and P_∞ compared with the values for unimplanted material. At low fluences the behavior of P_∞ is not understood and the effect was unexpected. In view of the well-behaved open-circuit voltages and Tafel slopes described earlier, carbon or oxygen contamination during implantation do not appear to be responsible.

Finally, the permeation transient for platinum and nickel-implanted nickel shows a dual-rise-time effect with an initial rapid rise followed by a second slower rise prior to reaching steady state. This effect may be caused by a dynamic filling and emptying of a distribution of traps both in depth and in trap energy. It is interesting to note that the effect is specific to the ionic species and that it may only occur when there is no chemical effect between the impurity and point defects as for nickel- and platinum-implanted nickel. The influence of deep and shallow traps might be clarified by precharging samples in order to fill the deep traps prior to recording the permeation transient. These experiments were not attempted in this work.

4.2. Argon- and helium-implanted nickel

Figures 3 and 5 show that the effect of argon and helium implantation is to decrease P_∞. At the lowest fluence for each case, the effect can be considered to be caused by damage alone because of the similarity of the values in the plot of P_∞ *vs.* displacements per atom in Fig. 6. The departure from the nickel and platinum values as the displacements per atom increase means that the effect of increasing C_H with shallow traps is overcome by another effect(s) that decreases C_H, since as for nickel-implanted nickel we expect no surface chemical effect with helium and argon. Indeed, the helium is buried beneath the surface at the highest fluence, and the argon has a surface concentration of 20 at.% at the highest fluence; yet they behave similarly.

There are probably two effects acting to reverse the effect of damage alone. It is well known that rare gases are repelled by the nickel lattice [21] and therefore they introduce strain. Hartley [29] measured a very large strain in argon-implanted iron foils at a fluence of 10^{16} ions cm^{-2} and nickel would be expected to behave similarly. This compressive stress decreases the solubility for hydrogen and the presence of interstitial argon also removes interstitial solution sites for the hydrogen. At a fluence of 10^{16} Ar^+ cm^{-2}, where the peak concentration of argon is about 3 at.%, these effects combine to overcome increases in C_H caused by shallow traps and P_∞ is smaller than for nickel-implanted nickel. At the highest fluence, gas bubbles form which introduce additional deep traps and increase the stress further so that the delay increases and P_∞ again decreases. Trapping caused by helium gas bubbles is known to occur in nickel under these conditions [21] and we assume that argon implantation produces bubbles as well. It should be noted that the transients for argon at the two highest fluences and for helium do not appear to have the dual-rise-time behavior that is characteristic of nickel- and platinum-implanted samples.

4.3. Yttrium-implanted nickel

The permeation transient for yttrium-implanted nickel in Fig. 5 shows the same shape as that for unimplanted nickel, a sizable delay and a P_∞ essentially equal to that for unimplanted nickel. The apparent effect of the yttrium is only to create deep traps which, after filling, do not alter P_∞. However, the value of P_∞ for yttrium implantation probably arises from a cancellation of the effect of increasing C_H by damage on the one hand and the effect of decreasing C_H by compressive stress on the other hand. The atomic radius of yttrium is much larger than that of nickel, making it likely that yttrium is non-substitutional as it is known to be in iron where it forms an yttrium–vacancy complex that is a very deep trap for hydrogen. If yttrium behaves similarly in nickel, then the above explanation is plausible for the permeation behavior of yttrium-implanted nickel.

The surface concentration of yttrium for a 1×10^{17} ions cm^{-2} fluence is 20 at.%. Since yttrium has a strong chemical affinity for hydrogen, it is possible that hydrogen is trapped by yttrium at surface sites and that these are removed as possible hydrogen entry sites (*e.g.* 20% reduction in surface area).

As described above, it is clear that the hydrogen permeation behavior of nickel is influenced by radiation damage, even at low levels (0.3 dpa), by compressive stress, gas bubbles and the electrochemical property of the implanted element.

4.4. *Prevention of hydrogen entry*

It is well known that many metals and alloys are susceptible to embrittlement from hydrogen supplied by gaseous and aqueous environments. It is, therefore, easily understood that, if hydrogen does not enter, hydrogen embrittlement is prevented or does not occur. At present, there are three approaches to decreasing or inhibiting hydrogen entry without modification of the bulk alloy: (i) coat the surface of an element such as platinum which catalytically stimulates H_2 evolution rather than atomic hydrogen adsorption [30], (ii) apply a dense continuous coating of an element which has a low hydrogen solubility and/or diffusivity compared with those of the substrate [31, 32] or (iii) add an inhibitor to the solution which shows a low overpotential for the hydrogen evolution reaaction [33, 34].

However, in the first two methods ((i) and (ii)), it is difficult to obtain a uniform coating layer without pores and a strong cohesion between coating element and substrate. The coated materials may enhance galvanic corrosion or pitting-like corrosion owing to pores and cracks which exist in the coating layer, especially when they are subjected to a stress field. The third method (iii) can be applied only to closed aqueous systems.

In contrast, the ion implantation technique does not affect bulk physical or mechanical properties and has been successfully applied to various problems involving semiconductors, thermal oxidation, corrosion and wear [11, 12, 15]. Furthermore, ion implantation is not restricted by the laws of thermodynamics governing equilibrium processes since it is essentially an athermal process involving individual atoms [35] and therefore any atom can be implanted to a high concentration into a metal.

On the basis of the present results, it seems that hydrogen entry can be reduced with the ion implantation technique incorporating (i) low fluences (*e.g.* nickel implanted with a fluence of 1×10^{14} Ni^+ cm^{-2}), (ii) ion implantation of inert elements (helium- and argon-implanted nickel) and (iii) ion implantation of catalytic elements (platinum-implanted nickel treated with aqueous 0.2% HF).

5. CONCLUSIONS

The following conclusions may be drawn from this work.

(1) Values of D_{lag}, the effective diffusion coefficient for the entire sample (derived from the rise time of the transient), decreased for all the implanted elements at all fluences, but by no more than a factor of 2. Values of C_H, the effective hydrogen solubility in the near surface, followed closely the trend of the permeation current P_∞.

(2) The hydrogen permeation transients in nickel- and platinum-implanted nickel in the fluence range 1×10^{15}–1×10^{17} ions cm^{-2} are chiefly influenced by the damage resulting from the process of ion implantation. The steady state permeation currents P_∞ in nickel- and platinum-implanted nickel increase because of the contribution of hydrogen in shallow traps which increases the effective solubility C_H of hydrogen. However, P_∞ in yttrium-implanted nickel is almost the same

as that in unimplanted nickel. This implies that the increase in C_H caused by the hydrogen in shallow traps balances the decrease in C_H caused by compressive stress accompanying the dissolved oversized yttrium atoms or that the presence of yttrium alters the defect structure.

(3) For nickel implanted with a fluence of 1×10^{14} Ni^+ cm^{-2}, D_{lag}, C_H and P_∞ become small compared with those for unimplanted nickel. This effect is not understood and deserves further investigation. The dual-rise-time behavior of platinum- and nickel-implanted nickel transients is likewise not at present explained.

(4) With regard to helium- and argon-implanted nickel with a fluence 1×10^{15}–2×10^{17} ions cm^{-2}, P_∞ and C_H both decrease with increasing fluence. This is explained in terms of compressive stress and the formation of gas bubbles in the implanted layers.

(5) The permeation transient of treated platinum-implanted nickel treated with aqueous 0.2% HF shows that P_∞ becomes much lower than that for unimplanted nickel. Thus, platinum has a catalytic effect on the hydrogen evolution reaction.

(6) The use of multiple ion species and different ion fluences is a useful means to separate surface, impurity, radiation damage and bulk alloying effects in permeation transient experiments.

ACKNOWLEDGMENTS

We gratefully acknowledge stimulating discussions with G. Wolf and J. A. Sprague.

REFERENCES

1 R. M. Latanision and H. Opperhauser, Jr., *Metall. Trans. A, 5* (1974) 483.
2 K. Yoshino and C. J. McMahon, Jr., *Metall. Trans. A, 5* (1974) 363.
3 B. J. Berkowitz and R. D. Kane, *Corrosion, 36* (1980) 24.
4 R. D. Kane and B. J. Berkowitz, *Corrosion, 36* (1980) 29.
5 R. H. Jones, S. M. Bruemmer, M. T. Thomas and D. R. Baer, *Metall. Trans. A, 13* (1982) 241.
6 A. W. Funkenbusch, L. A. Heldt and D. F. Stein, *Metall. Trans. A, 13* (1982) 611.
7 M. Cornet, C. Bertrand and M. D. Cunha Bela, *Metall. Trans. A, 14* (1983) 223.
8 S. M. Bruemmer, R. H. Jones, M. T. Thomas and D. R. Baer, *Metall. Trans. A, 14* (1983) 223.
9 B. D. Craig, *Metall. Trans. A, 15* (1984) 565.
10 Y. Ogino and T. Yamasak, *Metall. Trans. A, 15* (1984) 519.
11 S. T. Picraux, *Annu. Rev. Mater. Sci., 14* (1984) 335.
12 P. Sioshansi, *Thin Solid Films, 118* (1984) 61.
13 V. Ashworth, D. Baxter, W. A. Grant and R. P. M. Proctor, *Corros. Sci., 16* (1976) 775; *17* (1977) 947.
14 V. Ashworth, W. A. Grant, R. P. M. Proctor and E. J. Wright, *Corros. Sci., 18* (1978) 681.
15 C. R. Clayton, *Nucl. Instrum. Methods, 182–183* (1981) 875.
16 S. T. Picraux, E. P. Eer Nisse and F. L. Vook (eds.), *Proc. Int. Conf. on Applications of Ion Beams to Metals, Albuquerque, NM, 1974*, Plenum, New York, 1974.
17 M. Zamanzadeh, A. Allam, H. W. Pickering and G. K. Hubler, *J. Electrochem. Soc., 127* (1980) 1688.
18 S. M. Myers, S. T. Picraux and R. E. Stoltz, *Appl. Phys. Lett., 37* (1980) 168.
19 S. M. Myers, S. T. Picraux and R. E. Stoltz, *J. Appl. Phys., 50* (1979) 5710.
20 F. Besenbacher, S. M. Myers and J. K. Norskov, *Nucl. Instrum. Methods B, 7–8* (1985) 55.
21 W. R. Wampler and S. M. Myers, *Nucl. Instrum. Methods B, 7–8* (1985) 76.
22 J. M. Poate, *J. Vac. Sci. Technol., 15* (1978) 1634.
23 W. A. Grant, *J. Vac. Sci. Technol., 15* (1978) 1644.
24 M. A. Devanathan and Z. Stachurski, *Proc. R. Soc. London, Ser. A, 270* (1962) 90.
25 H. A. Dayness, *Proc. R. Soc. London, Ser. A, 97* (1920) 286.
26 P. L. Chang and W. D. G. Bennett, *J. Iron Steel Inst., London, 170* (1952) 205.
27 I. Manning and G. P. Mueller, *Comput. Phys. Commun., 7* (1974) 85.
28 G. H. Kinchin and R. S. Pease, *Rep. Prog. Phys., 18* (1955) 1.
29 N. E. W. Hartley, *J. Vac. Sci. Technol., 12* (1975) 485.
30 S. S. Chatterjee, B. G. Ateya and H. W. Pickering, *Metall. Trans. A, 9* (1978) 389.
31 M. A. Devanathan and Z. Stachurski, *J. Electrochem. Soc., 110* (1963) 887.
32 M. Zamanzadeh, A. Allam, K. Kato, B. Ateya and H. W. Pickering, *J. Electrochem. Soc., 129* (1982) 284.
33 D. L. Dull and K. Nobe, *Corrosion, 35* (1979) 535.
34 K. Nobe, I. M. Pearson and Y. Saito, *Werkst. Korros., 31* (1980) 763.
35 *Encyclopedia of Chemical Technology*, Vol. 13, Wiley, New York, 3rd edn., 1981, p. 712.

Materials Science and Engineering, 90 (1987) 253-262

Electrochemical Studies of TiO_2 Films with Various Palladium Implantation Profiles*

J. W. SCHULTZE[a], L. ELFENTHAL[a], K. LEITNER[a] and OTTO MEYER[b]

[a]*Institut für Physikalische Chemie II der Universität Düsseldorf, Universitätsstrasse 1, D-4000 Düsseldorf (F.R.G.)*

[b]*Kernforschungszentrum, Institut für Nukleare Festkörperphysik, Karlsruhe, (F.R.G.)*

(Received July 10, 1986)

ABSTRACT

To investigate the radiation damage as well as the doping effect of ion implantation, oxide-covered titanium electrodes were implanted with various concentration profiles of palladium ions (Pd^+). The comparison of the electrochemical data allows separation of the influences of doping and Frenkel defects.

Measurements of repassivation, capacity and photocurrent show a dominant influence of defect production increasing in the following order of the profiles: out < in < tot < through. For repassivation up to 2 V, defects with a high mobility migrate out of the film and form new oxide of up to 0.2 nm thickness. In capacity measurements, donor production of up to 4×10^{19} cm^{-3} can be detected, but less than 40% of the donors can be eliminated during repassivation. The photocurrent spectrum shows amorphization of the oxide, a decrease in the mobility gap and the production of localized electronic states in the mobility gap.

The rate of electron transfer reactions increases in the order out < in ≈ through < tot by several orders of magnitude. This shows that the production of defects enhances electron transfer reactions but that doping with palladium ions is most effective. For the out profiles, the unchanged inner layer represents an additional barrier for anodic electron transfer. When the data for the various profiles are compared and summarized, various defects with different concentrations, energies and mobilities can be distinguished, and the potential distributions can be estimated for the various profiles.

1. INTRODUCTION

The ion implantation of oxide films yields interesting systems for electrocatalysis, corrosion protection, solar cells and production of microelectronic circuits. The electronic changes within the film can be detected best by electrochemical methods. The analysis of capacity *vs.* potential data yields the charge distribution within the film, the donor concentration N and the dielectric constant D. Measurements of the rate of electron transfer reactions are very sensitive to the electronic processes at a constant electron energy in the vicinity of the Fermi level. Finally, photoelectrochemical experiments yield information on the crystallinity or amorphicity of the oxide film and the electronic band structure. Recent experiments have been carried out with rectangular concentration profiles in the oxide film which reveal a drastic change in the electronic properties of the oxide film [1-4]. In these experiments the effects of doping and radiation damage were observed simultaneously.

To separate these two effects, we carried out electrochemical experiments with various concentration profiles: (i) 0 profiles, control experiment without implantation; (ii) tot profiles, homogeneous implantation of the total film; (iii) in profiles, implantation of the inner half of the film; (iv) out profiles, implantation of the outer half of the film; (v) through profiles, penetration of high energy ions through the film. Figure 1(a) shows schematically the various concentration profiles. If

*Paper presented at the International Conference on Surface Modification of Metals by Ion Beams, Kingston, Canada, July 7-11, 1986.

0025-5416/87/$3.50

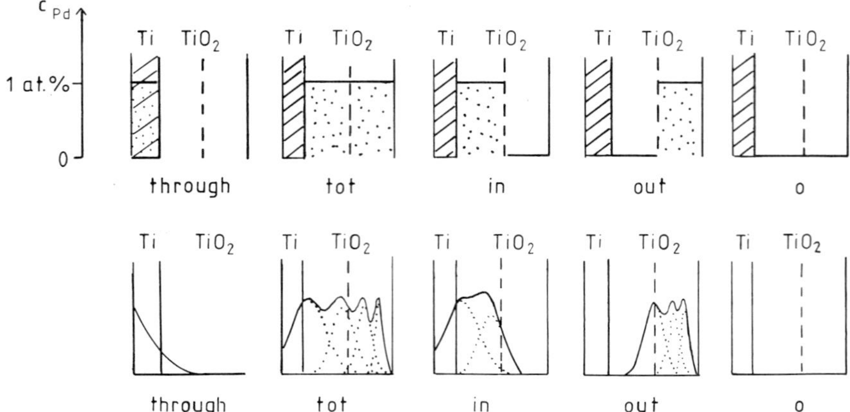

Fig. 1. (a) Schematic and (b) calculated palladium distributions in the TiO_2 film for the various implantation profiles.

radiation effects dominate, then the behaviour of the tot, through and in profiles should be similar and the out and 0 profiles should be comparable. However, if doping effects dominate, then we expect the through and 0 profiles to be similar, the out and in profiles to be similar, and the most marked difference for the tot profile. Figure 1(b) shows the profiles calculated from the theory of Lindhard, Scharff and Schiøtt. The deviations from the profiles (tails instead of steps) are almost negligible since implantation effects below 0.3% are not detectable [2]. Small discrepancies between the assumed and realized concentration profiles (*e.g.* $c_{Pd^+} \approx 0$ in the outer 3 nm layer) are unimportant, since radiation damage gives sufficient conductivity in the outer layer. The results of electrochemical measurements with these different implantation profiles are described and discussed in this paper. It is the aim to evaluate the details (E_g, N and D) of the band structure model which will be given later in Fig. 12.

2. EXPERIMENTAL DETAILS

Experiments were carried out with titanium (purity, 99.6%) electrodes anodized in 1 N H_2SO_4 up to 20 V, yielding a TiO_2 film thickness of 40 nm. These electrodes have a blue colour due to interference and absorption of the oxide. Then the electrodes were implanted with palladium ions (Pd^+) with different energies, according to the calculated profiles shown in Fig. 1(b). The highest energies were applied first to avoid radiation damage of the outer implantation layer. The total implantation fluences and calculated total deposited energies are given in Table 1. The local con-

TABLE 1

Implantation data and resulting parameters of the implanted oxide films for the various palladium concentration profiles

Profile type	*Pd fluence* ΣN_{impl} ($\times 10^{15}$ cm^{-2})	*Total implantation energy* ΣEN_{impl} ($\times 10^{20}$ eV cm^{-2})	*Dielectric constant* D	*Donor concentration* N ($\times 10^{19}$ cm^{-3})	*Flat-band potential* ϵ_{fb} (V)	*Anodic transfer coefficient* α_+	*Cathodic transfer coefficient* α_- $\epsilon = 0.1$ V	*Cathodic transfer coefficient* α_- $\epsilon = 0.3$ V	*Exchange current density* $\log j_0$ (A cm^{-2})
through	2.6	5.2	452	4.2	−0.12	0.1	0.6	0.2	−6.0
tot	3.4	2.98	255	1.45	−0.1	0.1	i_{diff}	0.2	−5.2
in	2.5	2.83	217	1.13	−0.06	0.1	0.6	0.2	−6.0
out	1.7	0.73	97	0.32	−0.06	0.1	0.6	0.3	−7.0
0	0	0	40	0.15	−0.2	0	0.6	—	−12

centration of palladium is about 8×10^{20} ions cm^{-3}. Implantation caused a profile-dependent shift of the colour to blue-yellow, increasing in the order out < in < tot ≈ through. After electrochemical measurements the electrodes were even brighter.

Electrochemical experiments were carried out potentiodynamically in 1 N H_2SO_4 with a sweep rate $d\epsilon/dt$ of 10 mV s^{-1}. For the electron transfer reaction measurements the Fe^{2+}–Fe^{3+} redox couple was added with a concentration of 0.05 M. Electrode potentials are referred to the standard hydrogen electrode (SHE). The current densities for the repassivation and electron transfer reactions were measured simultaneously with the electrode capacity C using the impedance method with an a.c. voltage of 0.5 mV and a frequency of 1 kHz. The photoelectrochemical equipment has been described elsewhere [5].

3. RESULTS AND DISCUSSION

3.1. Repassivation

After ion implantation the electrodes are in an active state which changes during electrochemical measurements depending on polarization. To obtain reproducible results, we repassivated the electrodes first by a potentiodynamic cycle between 0 and 2 V. Figure 2 shows typical repassivation currents of the first cycle for the five types of implantation profile. While the electrodes of types 0 and out show only capacitive currents, the electrodes of types in, tot and through show a repassivation current i_{rep}, increasing at a potential of 1.2 V to 8 $\mu A\ cm^{-2}$. The dissolution of TiO_2 in 1 N H_2SO_4 is negligible [6]. Figure 3 shows the corresponding charge q obtained by integration of the current over the time. We see that the repassivation charge is almost negligible for film types 0 and out but it increases up to 0.4 mC cm^{-2} for the film type through. It will be shown in a subsequent paper that, during this repassivation process, new oxide is formed at the oxide interfaces. Calculating the thickness d_{rep} of the new oxide by means of the expression

$$d_{rep} = \frac{qV_{TiO_2}}{4Fr} \qquad (1)$$

where q is the charge, V the molar volume, r the roughness factor and F the Faraday constant, we see that, even for films of the type through, the film thickness formed at the interfaces is less than about 0.2 nm. This means that oxide formation by repassivation is much smaller than the original oxide formation of 2 nm V^{-1}.

Figures 2 and 3 show that repassivation starts at potentials exceeding 1 V only. This is reasonable since the electrodes are in contact with air after the implantation and, therefore, a potential of about 1.2 V may be built up by oxygen adsorption at the electrode surface according to the oxygen evolution potential.

During the film formation the field strength E within the growing oxide is 20 V/40 nm = 5×10^8 V m^{-1}. During repassivation the potential difference and the resulting field strength of 5×10^7 V m^{-1} are only one-tenth

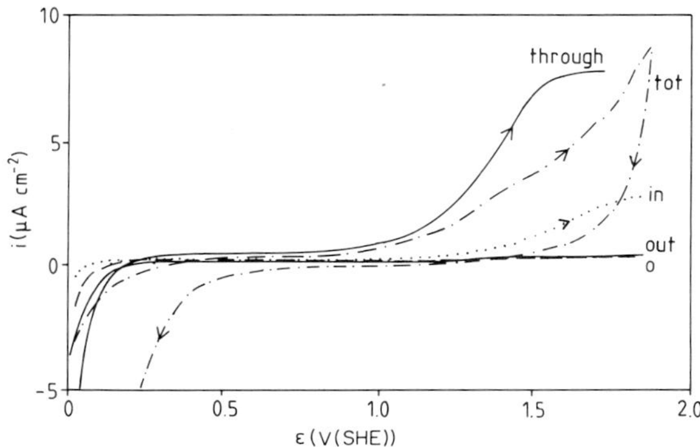

Fig. 2. Repassivation of palladium-implanted TiO_2 electrodes in 1 N H_2SO_4 for the profiles in Fig. 1 during the first potentiodynamic cycle (sweep rate $d\epsilon/dt$ = 10 mV s^{-1}).

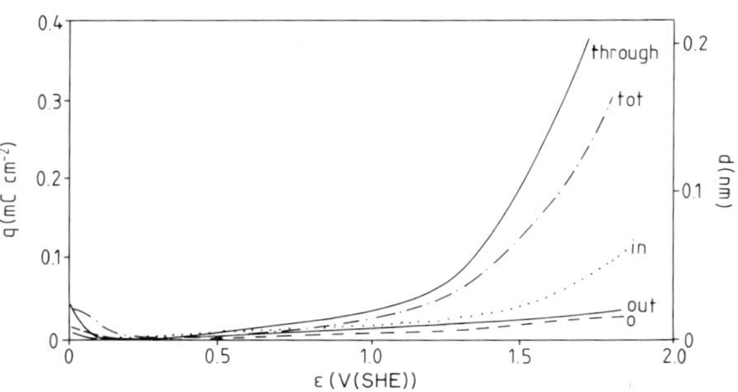

Fig. 3. Anodic repassivation charge and corresponding oxide growth during the first potentiodynamic cycle (data in Fig. 2).

of the values during film formation. In spite of the low field strength, defects move out of the film since they have a much lower migration enthalpy in the damaged oxide than in an anodically grown undisturbed oxide film.

Comparison of the curves in Figs. 2 and 3 shows very clearly that the repassivation cannot be due to the migration of palladium ions out of the oxide. If so, the current should be largest for out and tot and much smaller for in and through. The experiments, however, show the strongest repassivation currents for type through which can be correlated with the largest radiation damage caused by the largest total implantation energy and the field distribution within the oxide (see Fig. 12).

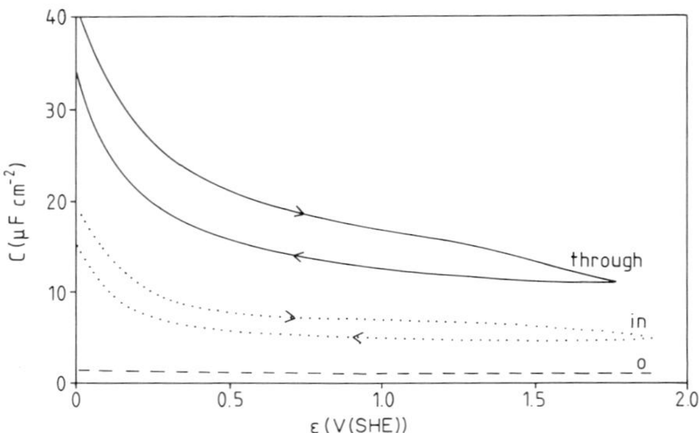

Fig. 4. Differential capacity of TiO_2 electrodes with three different palladium concentration profiles during the first repassivation cycle (sweep rate $d\epsilon/dt$ = 10 mV s^{-1}; U_{ac} = 0.5 mV; f = 1 kHz).

3.2. Capacity

The charge distribution in the oxide is given by the Poisson equation, the integration of which yields the potential dependence of the capacity $C(\epsilon)$. The analysis of $C(\epsilon)$ of thin oxide films has to distinguish between two different situations. At high potentials, all donors in the semiconducting n-type oxide film are exhausted and cannot contribute to C. Then the electrode capacity becomes potential independent since the thickness of the space charge layer exceeds that of the whole oxide and the simple condenser equation is valid;

$$C = \frac{DD_0}{d} \tag{2}$$

where D_0 is the dielectric constant of the vacuum and d the thickness of the oxide film. At low potentials under depletion layer conditions, we can use the Schottky–Mott equation

$$C^{-2} = \frac{2(\epsilon - \epsilon_{fb} - kT/e)}{eDD_0N} \tag{3}$$

where k is the Boltzmann constant, T the absolute temperature and e the electron charge to determine the flat-band potential ϵ_{fb} and the product of the donor concentration N and the dielectric constant D for the n-type semiconducting oxide film.

During the first cycle after ion implantation the electrode capacity decreases continuously with increasing potential which is caused by repassivation. Figure 4 shows such capacity measurements during the anodic and cathodic cycle for the films of types through, in and 0. For the films of through and in the electrode capacity decreases by about 20% which could be due to an increase in thickness or a decrease in D. Model calculations were carried out for the possible contribution of a new oxide layer with a capacity $C_{rep} = DD_0/d_{rep}$ in series with the implanted oxide film. Only the new layer on the film of type through with 0.2 nm thickness could explain the capacity decrease, but it is too small for the other films. Therefore, we favour the second explanation, a decrease in D.

The decrease in C is similar for films of type tot but very small for films of type out. This may be combined with the fact that the field distribution differs; for films of types in, through and tot the field strength decreases continuously from the oxide surface to the inner film, but for the type out the potential drop in the outer implanted film may be very small because of the high dielectric constant compared with the potential drop in the inner film which is the same as in an unimplanted film.

After repassivation the dependence of the electrode capacity on the potential follows the theoretical curves for n-type semiconductors [7]. Figure 5 shows the capacity curves for all types of concentration profiles. The capacity values increase in the order 0 < out < in < tot < through at all electrode potentials. At high potentials the capacity increases by a factor of 2 for films of type out; this can be explained by an unchanged dielectric constant of the inner layer and a

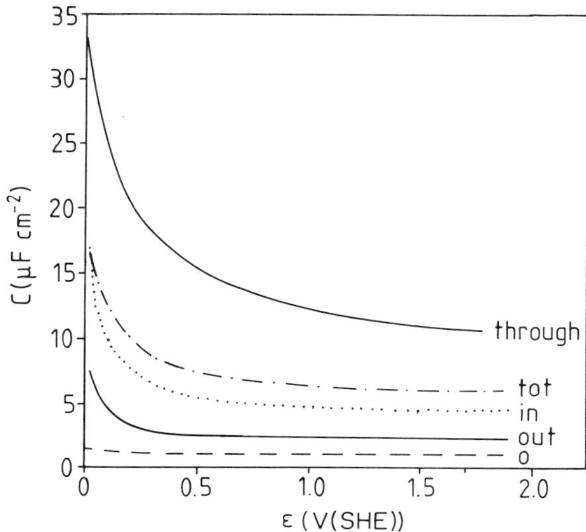

Fig. 5. Potential dependence of the capacity of all palladium implantation profiles after repassivation up to 2 V (cathodic sweep; sweep rate $d\epsilon/dt = 10$ mV s^{-1}).

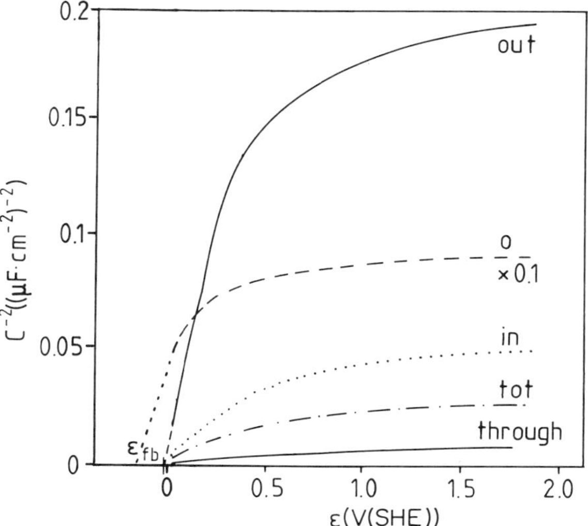

Fig. 6. Schottky–Mott plot for the data in Fig. 5. Scale for curve 0 is reduced by a factor of 10.

marked increase in the dielectric constant of the outer film. Since the layers are in series, with the film of type out we observe mainly the smaller capacity of the inner film.

For the film of type tot the dielectric constant of the total film is increased by a factor of 6. A comparison with the films of types in and through shows clearly that this cannot be attributed to the doping of palladium ions but must be explained by the radiation damage which is almost the same for types in and tot and even larger for type through because of the higher energy of the implanted ions.

The quantitative analysis is given by the Schottky–Mott diagram (Fig. 6). It can be seen clearly that the flat-band potential is about 0 V for all implanted films, but it is about −0.2 V for the control experiment 0. The slope of the Schottky–Mott lines yields the product ND. Using the dielectric constant from the asymptotic values and eqn. (2), we get donor concentrations increasing from about 0.15×10^{19} cm^{-3} for type 0 up to 4.2×10^{19} cm^{-3} for type through (Table 1). A comparison of these values with the concentration of implanted palladium ions ($c_{\mathrm{impl}} \approx 8 \times 10^{20}$ cm^{-3}) shows that c_{impl} is about two orders of magnitude higher than N, which demonstrates that the palladium ions do not act as donors, and we conclude that all occupied electronic levels of the implanted ions lie deep in the band gap or are localized by opposite charges in their vicinity so that they could not be charged during anodization [2]. This means that the marked increase in D and N produced by implantation must be explained by the formation of Frenkel defects. The increase in D after implantation indicates increased polarizability of the ions which can be correlated to an increase in the disorder since the ions are not fixed strongly in lattice positions.

The pronounced bend in the Schottky–Mott line at about 0.4–0.6 V is in the same potential range for all films, indicating an exhaustion of mobile charge in the oxide film at the same field strength and potential.

Surprisingly the slopes of the Schottky–Mott lines for types 0 and out differ. Assuming that the pure TiO_2 films are homogeneous, we believe that the potential drop for type out would be located mainly at the inner unchanged layer. Then, however, the slopes of the Schottky–Mott lines should be the same but the inflection point, indicating the exhaustion of charge, should appear at a smaller potential for type out. The discrepancy between expectation and experiments means that even for type out the band bending starts at the oxide–electrolyte interface. At 0.6 V, however, the outer film is exhausted, and the potential drop extends over the whole film at about 1.5 V.

3.3. Electron transfer reactions

Electronic processes can take place in thin oxide films by transport of electrons at a constant electronic energy E and by excitation at constant coordinates. The former can be investigated by measurement of the current *vs.* potential curves for redox systems, while the latter can be investigated by photoelectrochemical experiments.

First we investigated the outer sphere electron transfer reaction Fe^{2+}–Fe^{3+} which has an equilibrium potential of 0.68 V in H_2SO_4 solution. For pure films, it is well known that the reaction can take place on the cathodic side only with large overvoltages [8]. The anodic reaction is blocked because of the increasing band bending in the n-type semiconducting TiO_2 with increasing anodic potential. For the ion-implanted oxide films the electron transfer reaction is enhanced. Since the electrode potential has an exponential influence on the rate of electron transfer reaction, the results of redox measurements are shown in Fig. 7 in the Tafel plot of log j *vs.* potential. The anodic transfer coefficient α_+ and cathodic transfer coefficient α_- can be calculated from the slopes b_+ and b_- of the Tafel curves and are related to the conductivity of the oxide. On the cathodic side the current densities increase in the order 0 < out < in ≈ through < tot. On the anodic side the pure film blocks the electron transfer reaction completely while the implanted films show increasing anodic current densities and Tafel lines with an anodic transfer coefficient of about 0.1:

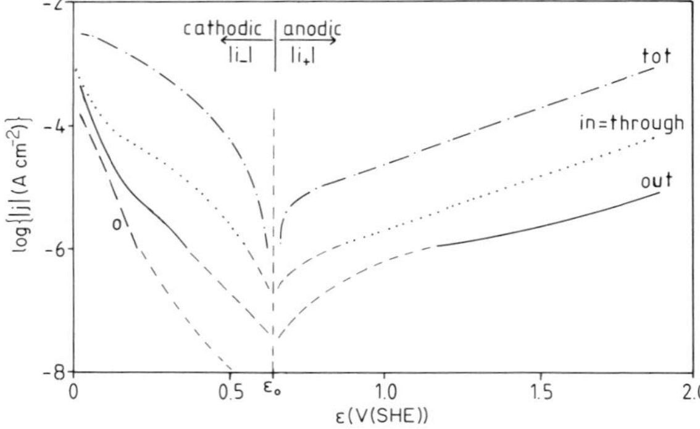

Fig. 7. Tafel curves of the Fe^{2+}–Fe^{3+} redox reaction (0.05 N) in 1 N H_2SO_4 on TiO_2 electrodes for various implantation profiles (cathodic sweep; sweep rate $d\varepsilon/dt = 10$ mV s^{-1}).

$$\eta_+ = a_+ + b_+ \log i$$

$$\eta_- = a_- - b_- \log|i|$$

$$b_+ = \frac{2.3RT}{\alpha_+ zF} \tag{4}$$

and

$$b_- = \frac{2.3RT}{\alpha_- zF}$$

where η_+ and η_- are the overvoltages

$$a_+ = -b_+ \log i$$

$$a_- = b_- \log i$$

z is the number of exchanged electrons and j_0 is the exchange current density. The results of these measurements can be explained by the Gurney–Gerischer theory of electron transfer reactions [8–10]. The anodic and cathodic current densities are given by the distribution functions of occupied and vacant states in the oxide and in the solution respectively and the tunnel probability which depends exponentially on the tunnel distance. Implantation causes a higher number of localized states in the band gap because of the implanted palladium ions and also because of the formation of defects in the oxide film. Therefore the rate-determining tunnel distance decreases and multiple resonance tunnel processes can take place via these states. Moreover, the energy of the electron transfer reaction is changed. In a low doped crystalline semiconductor the electron transfer reaction takes place near the band edges. With increased doping and amorphization the rate-determining electron transfer reaction is shifted to energies near E_F. Current densities increase and the transfer coefficients become potential dependent. This is observed here especially for α_-. Model calculations by Schmickler and Leiva [11], using the percolation theory, take these multiple processes into account. Small transfer coefficients and $\alpha_+ + \alpha_- < 1$ are calculated. This prediction is in line with the experimental results given in Table 1.

The enhancement of the anodic current density is most pronounced for the tot profile and much less for the out profile. The in and through profiles yield similar results, lying between the tot and out profiles.

This sequence can be explained in the following way. Since the rates of electron transfer reaction for films of types through and in are much lower than for type tot, the defects produced by implantation damage are of smaller efficiency than those obtained by doping with palladium ions. This may be caused by the energy position of the states with respect to the band gap; the palladium ions lie at deep levels near the Fermi level while the smaller number of donors detected in the capacity measurements lie near the conduction band and so the palladium ion states can contribute more effectively to the electron transfer reaction [10].

The small electron transfer reaction rate of type out compared with the complementary film of type in can be explained by the presence of the inner layer which was not changed by implantation. Hence, in addition to space charge layers at the oxide surface, pure TiO_2 may represent an additional barrier to electron transfer.

3.4. Photoelectrochemistry

The anodic photocurrent i_{ph} (ϵ, hν) gives a great deal of information on the band structure and the amorphization produced by irradiation. Figure 8 shows i_{ph} (ϵ) at a constant wavelength of 300 nm. Implantation can cause an increase as well as a decrease in i_{ph} depending on the implanted species and concentration. For this reason the absolute photocurrents will not be discussed, but the shapes of the curves reveal much information. According to the Gärtner [12] model for crystalline semiconductors, the photocurrent should show a square root dependence on the potential. The control experiment shows a linear increase at higher potentials which demonstrates that even the unimplanted film is not crystalline, but the implanted films show an almost exponential increase, especially for films of types out, tot and through. This exponential increase indicates the large contribution from the Poole–Frenkel effect [13]. Electron excitation can occur from and to the localized electronic states produced by implantation. The photoexcited electron on an ionized donor experiences a coulombic force which results in an energy barrier. If an external electric field E is superimposed, the escape barrier is modified by $E^{1/2}$ and so the photocurrent increases exponentially with the square root of the electric field.

This effect is commonly observed in amorphous semiconductors for fields E greater than 10^6 V m^{-1}. At very high field strengths, tunnelling through the barrier also has to be taken into account. The fact that the Poole–Frenkel effect is also observed at photon energies higher than the band gap energy E_g indicates a localization of the upper part of the valence band or the lower part of the conduction band due to amorphization.

Figure 9 shows the spectral response of the implanted electrodes at a constant electrode potential of 1.5 V. All electrodes show a constant increase in photocurrent at photon energies exceeding 3.5 eV. It should be mentioned here that the absolute values of the quantum efficiency given in Fig. 9 are not reliable enough for a quantitative interpretation. However, for all implanted films, we obtain larger quantum efficiencies at low photon energies. The dependence of the

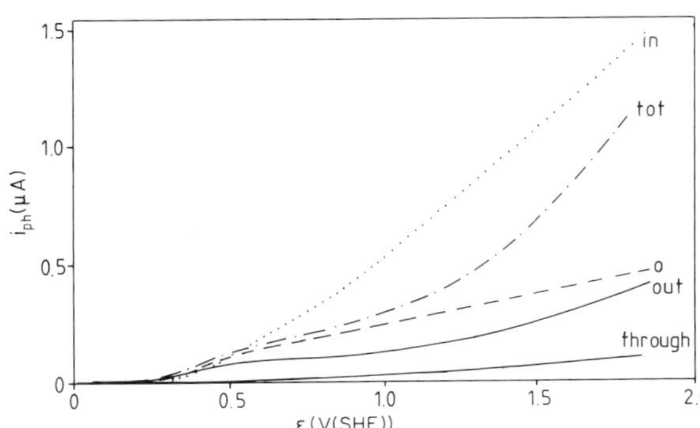

Fig. 8. Potential dependence of the photocurrent at palladium-implanted TiO_2 electrodes with various implantation profiles (λ = 300 nm).

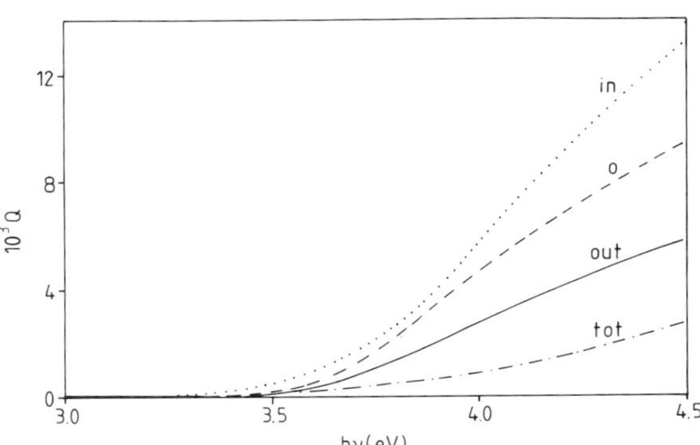

Fig. 9. Wavelength dependence of the quantum efficiency of TiO_2 electrodes with various implantation profiles (ϵ = 1.5 V).

photocurrent on the photon energy is given by

$$i_{ph}h\nu = \text{constant} \times (h\nu - E_g)^n \quad (5)$$

where E_g is the band gap or mobility gap, assuming that the photocurrent is proportional to the absorption coefficient A [14].

According to the selection rules in crystalline materials for allowed transitions, $n = \frac{1}{2}$ denotes direct and $n = 2$ denotes indirect transitions. In amorphous materials the selection rules are believed to be relaxed, but eqn. (5) is still observed and must be considered as an empirical relation. The quantitative evaluation of the indirect transitions is shown in Fig. 10. Almost straight lines are obtained for all films but the mobility gap obtained by extrapolation decreases from about 3.45 eV for type 0 to type out to type in finally to type tot with 3.2 eV. The sequence indicated here equals that obtained by the other electrochemical measurements and demonstrates the strong influence of radiation damage and the small influence of doping. At photon energies close to the band gap energy in strongly disordered materials, the so-called Urbach tail is often found. In this case, an exponential decrease in electronic states below the band gap has the following spectral dependence for the absorption coefficient A [15]:

$$A = A_0 \exp\{\gamma(h\nu - E_g)\} \quad (6)$$

with the disorder energy γ^{-1}. Figure 11 shows that the slopes of the Urbach lines for the implanted films are smaller than those for type 0, which can be clearly correlated to the higher amorphization of the implanted films.

4. CONCLUSIONS

The results of the experiments described above are summarized in Table 1. The investigations of passive films with different ion implantation profiles again show the marked sensitivity of electrochemical investigations to disordered systems. According to the expectations mentioned in Section 1, we obtained strong evidence of the influence of radiation damage on the electronic properties for ion implantation with heavy metal ions in oxide films. This is shown clearly by the increase in activation in the profile order $0 < \text{out} < \text{in} < \text{tot} < \text{through}$, which is observed for i_{rep}, C and the band gap energy in i_{ph} measurements. This order parallels that of the total implantation energy $\Sigma\, EN_{impl}$. The capacity measurements show that the radiation damage causes the production of defects of displaced oxygen and titanium ions as well as an increase in the dielectric constant. The number of detectable defects, however, is smaller than that of the implanted palladium ions. This is easily understood since the palladium ions produce electronic states lower than the usual donor level detected in Schottky–Mott measurements. Measurements of electron transfer reactions, in contrast, are sensitive to the radiation damage too, but they also show the dominance of palladium doping, yielding the order $0 < \text{out} < \text{in} \approx \text{through} < \text{tot}$. The photoelectrochemical experiments show partial localization of the valence band or conduction band and the creation of localized states in the gap close to the band edge as a result of implantation.

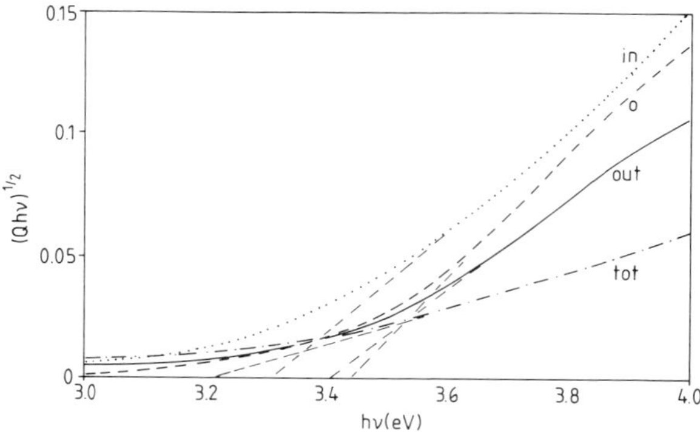

Fig. 10. Plot according to eqn. (5) with $n = 2$ for the various profiles. Extrapolation of the curves yields the mobility gap for indirect transitions.

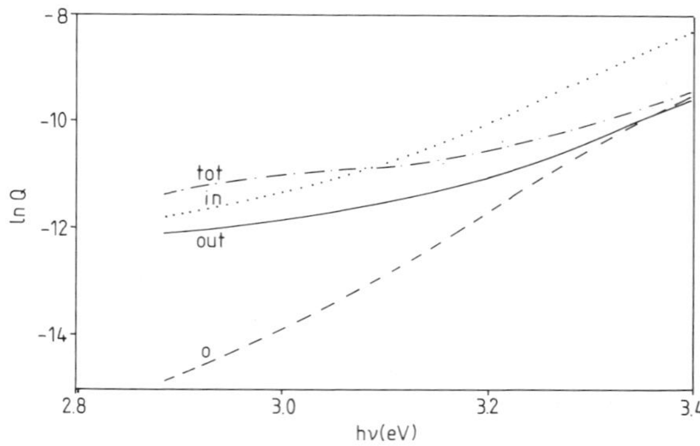

Fig. 11. Urbach tail plot for the various implantation profiles.

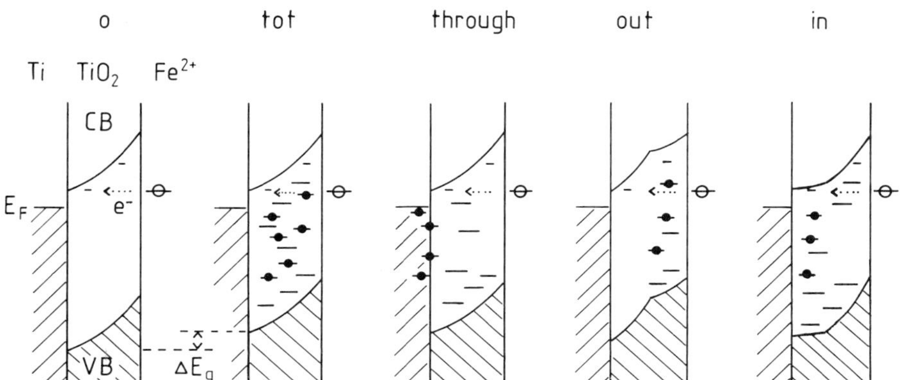

Fig. 12. Schematic diagram of electronic states of the implanted films and possible electron transfer reactions: CB, conduction band; VB, valence band; E_F, Fermi level; ΔE_g, decrease in mobility gap energy; -•-, states of palladium; —, states of Frenkel defects; —, states of original oxygen vacancies; ⊖, electronic state of Fe^{2+} (representative of a broad distribution function).

Finally, it is interesting to compare the charges measured in various experiments. The total charge of palladium (Pd^+) ions (calculated with $z = 1$) is about 1 mC cm^{-2}. We conclude from other measurements that the ions become mobile only at larger field strengths ($E > 10^8$ V m^{-1}). The maximum repassivation charge q_{rep} of 0.4 mC cm^{-2} (through) is much smaller. Therefore, it probably corresponds to mobile defects of TiO_2.

In the capacity, we detect a maximum donor concentration of 4×10^{19} cm^{-3} corresponding to a charge q_d of 25 μC cm^{-2}, *i.e.* only a few per cent of q_{Pd} and q_{rep}. q_d is probably the charge of oxygen vacancies. After repassivation,

$$\int C \, d\epsilon = q_c$$

decreases by about 10 μC cm^{-2}, *i.e.* about 40% of q_d. Probably, oxygen vacancies move to the oxide surface and are filled up there. All these data refer to the film of type through; smaller values are obtained for the other types (0 < out < in < tot) but the relations between q_{rep}, q_d, q_{Pd} etc. are similar.

Because of the different electronic data for the various film types, the distributions of electric potential differ. From the Schottky–Mott lines, we must conclude that the space charge layer extends into the film with increasing ϵ in the usual way. For types tot and through the field strength will decrease monotonically. For types in and out, however, the field strength in the unimplanted part (with radiation damage in the outer part of type in) will be larger than in the implanted part.

This leads to the model of the electronic band structure given in Fig. 12: type 0, homogeneous layer, only donor states due to intrinsic defects, no palladium; type tot, homogeneous layer, donor states due to intrinsic and radiation-induced defects and to palladium, high conductivity; type through, homogeneous layer, donor states due to intrinsic and radiation-induced defects, no palladium within the oxide; type out, inhomogeneous layer, palladium and radiation-induced defects only in the outer half, high electric field strength E in the inner half and low E in the outer half; type in, inhomogeneous layer, palladium only in the inner half, radiation-induced defects in the complete oxide, high E in the outer half and low E in the inner half.

ACKNOWLEDGMENT

We gratefully acknowledge the support of this work by the Bundesministerium für Forschung und Technologie.

REFERENCES

1 C. Bartels, B. Danzfuss and J. W. Schultze, in M. Froment (ed.), *Passivity of Metals and Semiconductors*, Elsevier, Amsterdam, 1984, p. 35.
2 J. W. Schultze, B. Danzfuss, O. Meyer and U. Stimming, in G. K. Wolf, W. A. Grant and R. P. M. Procter (eds.), *Proc. Int. Conf. on Surface Modifi-*

cation of Metals by Ion Beams, Heidelberg, September 17–21, 1984, in *Mater. Sci. Eng., 69* (1985) 273.

3 B. Danzfuss, J. W. Schultze, U. Stimming and O. Meyer, in M. Froment (ed.), *Passivity of Metals and Semiconductors*, Elsevier, Amsterdam, 1984, p. 503.

4 B. Danzfuss, J. W. Schultze and U. Stimming, in J. W. Schultze (ed.), *Grundlagen von Elektrodenreaktionen, Dechema Monogr., 102* (1986) 465.

5 B. Danzfuss, *Dissertation*, University of Düsseldorf, 1985.

6 K. D. Allard, M. Ahrens and K. E. Heusler, *Werkst. Korros., 26* (1975) 694.

7 K. Bohnenkamp and H. J. Engell, *Z. Elektrochem., 61* (1957) 1184.

8 J. W. Schultze and L. Elfenthal, *J. Electroanal. Chem., 204* (1986) 153.

9 R. W. Gurney, *Proc. R. Soc. London, Ser. A, 134* (1931) 137.

10 H. Gerischer, *Z. Phys. Chem. N.F., 26* (1960) 223.

11 W. Schmickler and H. Leiva, to be published.

12 W. Gärtner, *Phys. Rev., 116* (1959) 84.

13 A. R. Newmark and U. Stimming, *J. Electroanal. Chem., 204* (1986) 197.

14 R. H. Bube, *Photoconductivity of Solids*, Wiley, New York, 1967.

15 F. Urbach, *Phys. Rev., 92* (1953) 1324.

Tribological Properties of Ion-implanted Steels*

MASAYA IWAKI

RIKEN (Institute of Physical and Chemical Research), Hirosawa 2-1, Wako, Saitama-ken 351-01 (Japan)

(Received July 10, 1986)

ABSTRACT

The tribological properties such as surface hardness, friction and wear have been studied for low carbon steels and tool steels implanted with many types of ion including metallic elements. The hardness measured by Vickers or Knoop hardness testers as a function of normal load is dependent on the implanted species, fluence and substrate. The friction coefficients measured by Bowden–Leben type of friction tests or detected during wear tests also depend on the implantation conditions. The improvement in the wear resistance, which is most important for industrial use of implanted materials, has been investigated for AISI H13 prehardened and tool steels implanted with nitrogen and boron ions. The relationship between hardness, friction and wear is discussed in comparison with the microcharacteristics such as composition and chemical bonding states measured by means of secondary ion mass spectrometry and X-ray photoelectron spectroscopy. It is concluded that the increase in hardness and/or the decrease in friction coefficient play(s) an important role in improving the wear resistance, and the relationship between relative wear volume and relative hardness is correlated for boron and nitrogen implantation.

1. INTRODUCTION

For 10 years, ion implantation in metals has been applied to fundamental studies of physical, chemical and mechanical properties in surface layers such as composition, structure, hardness, wear and corrosion. In order to perform these research studies, many types of element have been implanted in a variety of metals, and many results have shown that ion implantation in metals results in significant modification of the surface properties [1]. Some technological areas where such a process can be applied with commercial justification have already been reported (*e.g.* tribology) [2].

Tribology is the study of the friction, lubrication and wear of engineering surface layers including the theories of surface interactions; it is therefore interdisciplinary, covering physics, chemistry, materials science and mechanics. The majority of the investigations on ion-implanted surface layers have concerned hardness, friction and wear properties, which will be closely related to each other [3].

Most of the time the element implanted to improve wear resistance has been nitrogen in order to form nitrides, which seems to be similar to a conventional process for nitriding, but in reality the ion implantation process without thermal equilibrium is wholly different from it. Anyway, the formation of compounds such as nitrides, carbides and borides near the surface results in hardening at the surface layers, which will reduce the friction coefficient and improve the wear resistance.

It is also found that the formation of a metastable alloy plays a role in improving the wear resistance [4]; titanium implantation into iron causes a Fe–Ti–C metastable and amorphous alloying layer to form, which improves the wear resistance. Additional carbon implantation is more effective for the production of deeper amorphous layers [5] and more drastic improvement in the wear resistance of AISI 52100 steel than is implantation with titanium ions alone [6]. The implantation of both titanium and carbon

*Paper presented at the International Conference on Surface Modification of Metals by Ion Beams, Kingston, Canada, July 7-11, 1986.

0025-5416/87/$3.50

ions with a very high fluence above 3×10^{17} ions cm^{-2} is considered to be most useful for modification of semicritical materials with a thermal equilibrium limit. For the available stainless steels and tool steels, however, the influence of titanium implantation on hardness and wear resistance is complicated [7].

In this paper the effect of ion implantation in low carbon steels and tool steels (commercially available materials) on hardness, friction and wear properties is studied. The implanted elements were boron, nitrogen, argon, chromium, nickel and copper; the hardnesses and friction coefficients for all the implanted steels were measured and their relationship is discussed. To investigate the practical use of implanted steels, boron and nitrogen implantations into prehardened steels and alloy tool steels (AISI H13) were carried out and the correlation between hardness and wear volume is indicated.

2. HARDNESS

Hardness is one of the most important surface properties of materials, and we can measure it simply by means of conventional Vickers or Knoop hardness testers. The indentation made by the hardness testers, however, penetrates through most implanted layers even at lower normal loads because these implanted layers are of submicron thickness (*i.e.* they have a thickness of the order of the ion range). Therefore, it is difficult to measure precisely the hardness of implanted layers; the hardness measured as a function of normal loads will give us an average hardness, and it approaches the value of the implanted layer at lower loads.

Figure 1 shows the Vickers hardness as a function of the normal load for nitrogen-, argon-, chromium-, nickel- and copper-implanted low carbon steels. Ion implantation was performed with a fluence of 10^{17} ions cm^{-2} at an energy of 150 keV. Even for the unimplanted specimen, a decrease in the normal applied load results in an apparent increase in hardness. The hardening phenomenon at relatively low loads may be explained by the existence of work-hardened layers near the surface.

No change in hardness at a high load such as 100 gf could be seen between implanted and unimplanted specimens. Specimens implanted with argon, nickel and copper ions

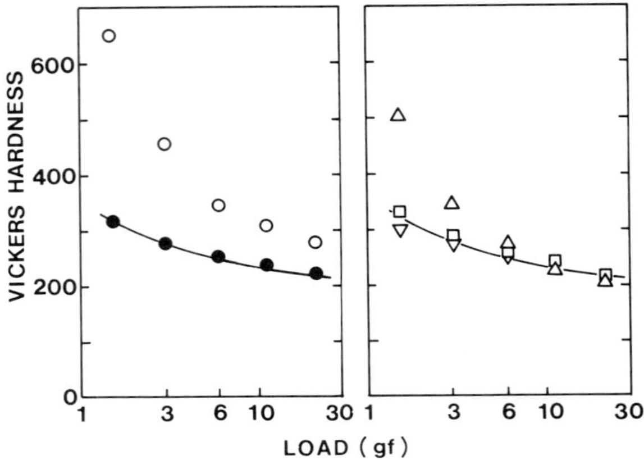

Fig. 1. The Vickers hardness as a function of load for nitrogen (○), argon (●), chromium (△), nickel (▽) and copper (□) implantations into low carbon steels: ——, unimplanted.

do not show a change in hardness, even at lower loads. No change in hardness at lower loads may imply a decrease in hardness because the measurement of the surface hardness is markedly affected by a harder bulk if the surface layers become softer as a result of ion implantation.

It is clear that nitrogen- and chromium-implanted specimens are harder than unimplanted specimens when measured at lower loads. The depth of indentation is several times deeper than the thickness of the implanted layer, even at the lower loads. This observation indicates that even a thin layer with an amorphous structure or containing new compounds induced by ion implantation will strongly affect the hardness.

X-ray photoelectron spectroscopy (XPS) measurements showed that nitrogen implantation in irons and steels formed some nitrides. For chromium implantation in steels, the enhancement of oxidized chromium occurs near the surface. Figure 2 shows the depth profile and XPS spectra of Cr 2p for chromium-implanted low carbon steel with a fluence of 5×10^{16} Cr^{+} cm^{-2} at an energy of 150 keV, as measured by XPS combined with argon sputtering. The chromium depth profile has a gaussian-like distribution except near the surface [8]. The Cr $2p_{3/2}$ spectra obtained by XPS measurements indicate that implanted chromium atoms combine with oxygen to form Cr_2O_3, enriched near the surface, and are in a metallic state in the interior region. It is considered that the enrichment of chro-

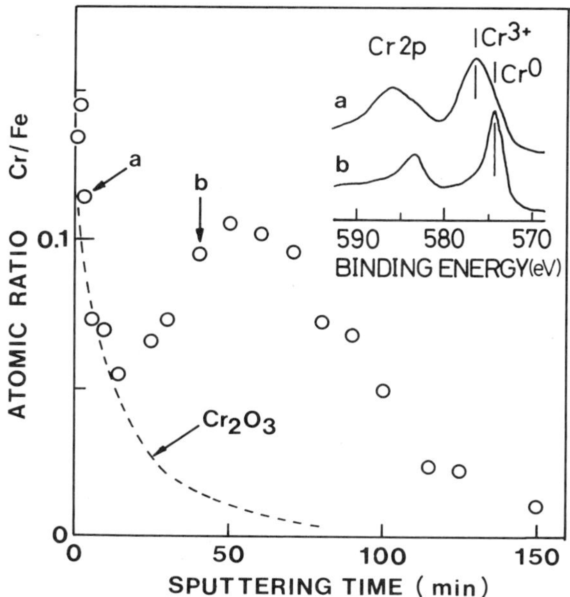

Fig. 2. Depth profile and Cr 2p spectra for chromium-implanted low carbon steels measured by means of XPS. The ion implantation was carried out with 5×10^{16} Cr^+ cm^{-2} at 150 keV.

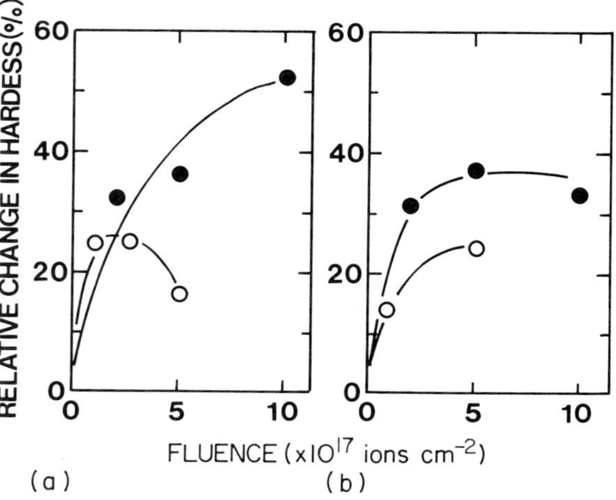

Fig. 3. Dependence of the relative change in hardness on fluence for (a) prehardened H13 and (b) H13 alloy tool steels implanted with nitrogen (●) and boron (○) at 75 keV.

mium atoms occurs as a result of the attraction of oxygen atoms in iron oxide layers because chromium has a stronger affinity for oxygen than iron has.

Titanium implantation in iron results in the in-diffusion of carbon atoms from the surface to form an Fe–Ti–C amorphous alloy. For titanium implantation into iron to 10^{17} ions cm^{-2}, XPS measurements showed that in-diffusing carbon combined with iron to form iron carbide [9], and consequently titanium-implanted iron showed an increase in hardness [7, 10].

These results suggest that the formation of compounds such as nitrides, oxides and carbides ensures the increase in hardness.

The hardness of implanted layers at a given applied load appears to depend on the fluence. Figure 3 shows the dependence of Knoop hardness on fluence at an applied load of 5 gf for nitrogen- and boron-implanted H13 prehardened and alloy tool steels (the alloy tool steel is hardened by quenching). As the fluence increases, the relative hardness change increases gradually, and at higher fluences it saturates. It is difficult to understand the mechanism of saturation of hardness because the sputtering of nitrogen implantation in iron was low [11], and consequently it may not be simply due to sputter-related saturation of retained atoms. We think it is due to formation of a brittle structure by a higher fluence implantation, which will reduce the effect of ion implantation on hardness. Anyway, nitrogen-implanted H13 steel becomes harder than boron-implanted specimens; nitrogen implantation is more effective in increasing the hardness than boron implantation is.

3. FRICTIONAL PROPERTIES

Friction behaves in a complicated way under the influence of such physical properties as adhesion and such engineering operations as ploughing. The friction coefficient is defined as the ratio of the resistance to the plastic flow of the weaker of the contacting materials in shear and in compression [3]. According to this relationship, it is expected that an increase in hardness may lead to a decrease in the friction coefficient.

Hartley *et al.* [12] carried out the first published friction test on ion-implanted surfaces using a comparatively simple slow speed sliding apparatus based on a Bowden–Leben type of friction machine. Ion implantation with various ions (tin, indium, silver, lead and molybdenum) at fluences in excess of 10^{16} ions cm^{-2} and energies around 120 keV resulted in macroscopic changes in the friction coefficient. The majority of the implanted ions produced a decrease in the friction coefficient, and Hartley *et al.* assumed that the dominant behaviour of an ion-implanted species weakens the shear strength of the implanted layers by promoting the formation of oxide films.

However, we implanted many types of element into low carbon steels and investigated the relation between hardness and friction coefficient. The friction coefficients of implanted steels were measured under dry sliding at a low speed. The loads applied to an AISI 1025 steel pin ranged from 50 to 200 gf. The friction coefficients of all the unimplanted specimens were 0.5–0.6 and differed slightly from specimen to specimen. In order to estimate the effect of ion implantation on friction, the relative change in friction is defined as

$$\Delta\mu = \frac{\mu_i - \mu_u}{\mu_u}$$

where μ_i and μ_u are the friction coefficients of implanted specimens and unimplanted specimens respectively.

The relative change in friction for copper- and chromium-implanted low carbon steels is shown as a function of the total fluence at energies of 50, 100 and 150 keV in Fig. 4 [13, 14]. Copper implantation causes the friction coefficient to increase, and almost the same result was obtained for nickel implantation. In contrast, chromium implantation resulted in a decrease in friction coefficient. Because the masses of both ions are almost the same, the change in friction is due to the

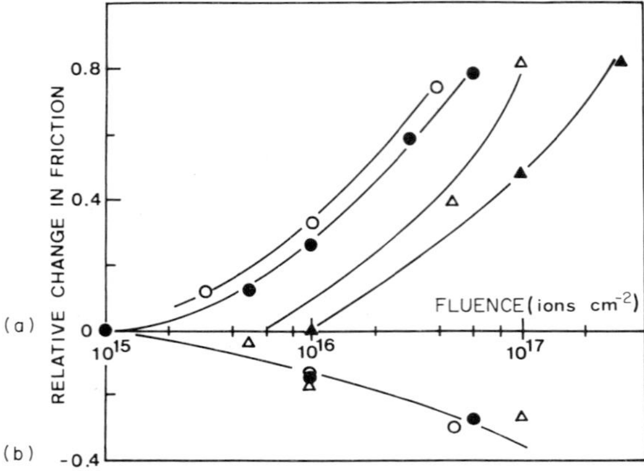

Fig. 4. The relative change in friction as a function of fluence for (a) copper and (b) chromium implantations into low carbon steels: ○, 50 keV; ●, 100 keV; △, 150 keV; ▲, 200 keV.

characteristics of implanted elements and not to radiation damage.

Copper implantation shows the effect on friction clearly over a fluence of 10^{15} ions cm^{-2} at a low energy and 10^{16} ions cm^{-2} at a high energy. The friction coefficient shows a tendency to increase as the total fluence at each energy increases. It also increases as the acceleration energy at a given fluence decreases. The phenomenon can be explained by the depth profile of implanted ions as follows. The distribution of implanted ions is affected by the acceleration energy, and the depth at the peak in the profile moves towards the surface as the energy decreases. When ion implantation was performed at the same fluence, the element concentration which may contribute to frictional properties is large if the energy is low. For this contributed concentration, three quantities are considered: the first is the surface concentration, the second is the maximum concentration and the third is the amount of implanted ions accumulated from the surface to a certain depth.

To investigate the change in frictional force which may be caused by the concentration at the surface or at the peak, the implanted specimens were removed from the surface to near the peak by sputtering with an oxygen beam. The sputtering removal effect causes an increase in concentration at the surface and no change in the maximum concentration. The friction coefficients of all the sputtered specimens were lower than those of the untreated specimens. These experimental results show that the concentration at the surface or at the peak is not of prime importance in the frictional properties.

An investigation of the way in which concentration contributes to the frictional properties has demonstrated that the most important parameter is the amount of implanted ions in the surface layers. The relative change in friction for copper-implanted steels varies according to the amount of implanted ions down to 40 nm from the surface, which is defined as the effective fluence, as shown in Fig.5. These data include the friction coefficient for specimens sputtered after ion implantation and for specimens implanted with two implantation conditions using the same ions (double implantation). The results suggest that the amount of implanted ions remaining in the surface layers (0–40 nm) is of prime

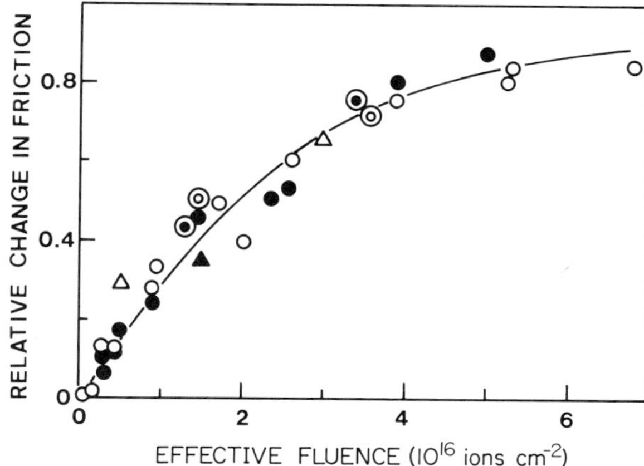

Fig. 5. The relative change in friction as a function of effective fluence for copper and nickel implantations: ○, as-implanted copper; △, sputtered copper; ◎, double copper; ● as-implanted nickel; ▲, sputtered nickel; ⊙, double nickel.

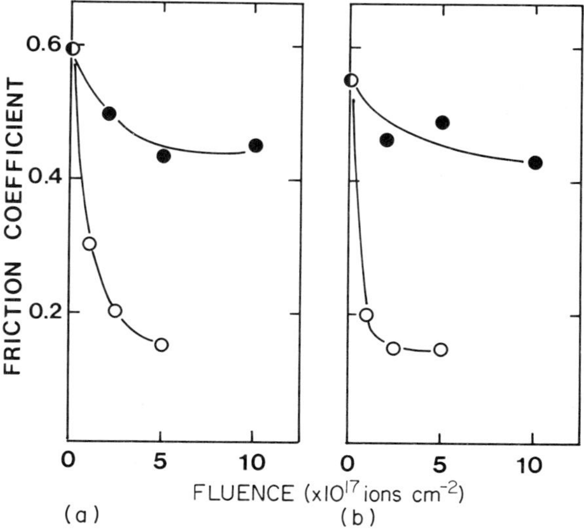

Fig. 6. Dependence of the friction coefficient on fluence for (a) prehardened H13 and (b) H13 alloy tool steels implanted with nitrogen (●) and boron (○) at 75 keV: ◐, unimplanted.

importance in the frictional properties of copper implanted steels. The frictional properties of nickel-implanted steels, also included in Fig. 5, were almost the same as those of copper-implanted steels. The results suggest that the surface layers, and not just the surface itself, play an important role in the frictional properties.

For chromium implantation into low carbon steels, an increase in the total fluence results in a decrease in the friction coefficient, which appears to be independent of the acceleration energy. The reduction in friction due to chromium implantation is considered to be a result of surface hardening or adhesive force weakening caused by the presence of chromium-based oxide particles. The chromium-implanted layers, as mentioned above, have indeed been hardened as determined from Vickers hardness tests and showed enrichment of Cr_2O_3 near the surface. The oxide layer is 10–20 nm thick [14], independent of the acceleration energy, which is shallower than the effective thickness for an increase in friction coefficient. The results of chromium and copper implantation suggest that the formation of compounds near the surface layers will increase the hardness and consequently the frictional force will be reduced compared with that without compound formation at metal–metal interfaces.

To assess the practical use of implanted materials, the frictional coefficients were measured during wear tests described later for nitrogen- and boron-implanted H13 prehardened and alloy tool steels, which indicated the increase in hardness. High fluence implantation shows a reduction in the friction coefficients, as in Fig. 6. The progress of hardness, however, does not correspond to the decrease in friction coefficients; although boron-implanted specimens appear to be softer than those implanted with nitrogen, they have lower friction coefficients than nitrogen-implanted specimens. These results suggest that the friction coefficient of boron-implanted specimens is markedly affected by such physical properties as adhesion rather than by such engineering operations as hardness. The friction coefficient is also affected by wear properties.

4. WEAR PROPERTIES

Wear is a complex process involving parameters that can be altered by several orders of magnitude depending on the dominant mechanisms and their interactions with the environment. Ion implantation has a particular role to play in the science and technology of wear. Unexpected benefits which result from the shallow depths of treatment not only can improve the wear resistance but also may lead to a new understanding of the wear process.

Hartley [15] and Hirvonen [16] have reported significant changes in the wear rate of steels implanted with relatively high fluences of nitrogen. Many experimental results have shown that the dependence of the relative improvement in wear on fluence becomes sensitive to fluences in excess of about 10^{17} ions cm^{-2} and it has been proposed that it is presumably necessary to achieve complete saturation of the surface microstructure in order to ensure that every contact event is altered by the presence of the implanted ions [17]. However, no simple relationship between the nitrogen fluence and the improvement in wear resistance has been found.

Figure 7 shows the relationship between the relative change in wear volume and fluence for H13 prehardened and alloy tool steels implanted with boron and nitrogen. Wear tests were carried out under unlubricated conditions at room temperature using pin-on-disc (pin, AISI 1025 steel; disc, H13 steel) configuration test equipment. The relative change in wear volume was defined as the ratio of the wear volume for the nitrogen- and boron-implanted steel to that for the unimplanted steel, which was measured after wear tests at a sliding distance of 100 m at a velocity of 210 mm s^{-1} with a load of 2.11 kgf.

The wear volume is reduced by nitrogen and boron implantation. In the nitrogen implantation described in our previous paper [18], for maraging steels the reduction in wear volume tends to increase gradually as the nitrogen fluence increases. For stainless steels the wear is drastically reduced up to a fluence of 2×10^{17} N^+ cm^{-2} and then saturates beyond this dose. The wear resistance is reduced again, however, beyond a fluence of 10^{18} N^+ cm^{-2}. These results suggest that the improvement in wear resistance induced by nitrogen implantation depends on the history of the specimens (*e.g.* heat treatment) before ion implantation.

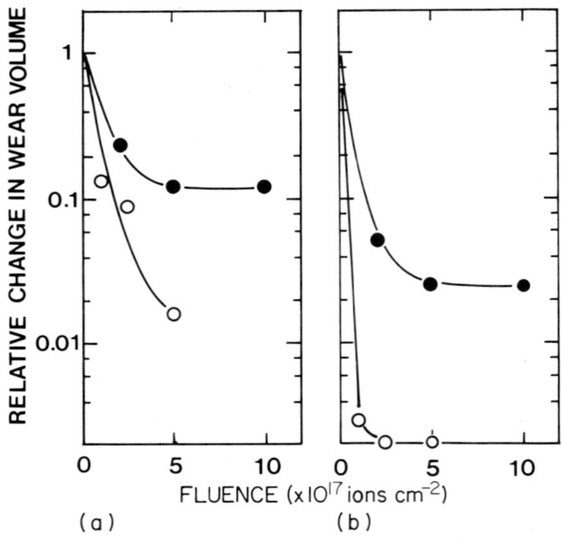

Fig. 7. Dependence of the relative change in wear volume on fluence for (a) prehardened H13 and (b) H13 alloy tool steels implanted with nitrogen (●) and boron (○) at 75 keV.

In many wear tests, measurements of the wear track on the ion-implanted disc showed that the depth of the track far exceeded the range of nitrogen implantation. The nitrogen content at the base of wear tracks was measured by means of nuclear reaction analysis [19], by means of Auger electron spectroscopy (AES) combined with argon sputtering [20] and by means of secondary ion mass spectrometry (SIMS) [21]. In order to understand the wear mechanisms in the experiment as shown in Fig. 7, the nitrogen depth profile outside and within the wear tracks after wear tests was measured by means of AES combined with argon sputtering. The results show that the presence of 10%–20% of the total fluence was observed at a depth about ten times deeper than the ion range. The migration of nitrogen atoms inwards may be explained by the generation of dislocations.

However, the depth profiles of nitrogen implanted into steels at a target temperature lower than 100 °C are different as a function of substrate material. Figure 8 shows the depth profiles of FeN^- secondary ions measured by means of SIMS, using caesium (Cs^+) ions as the primary ion, outside and within the tracks after wear tests on the steels. This technique is a new way of measuring nitrogen depth profiles which has a higher sensitivity than nuclear reaction analysis, AES and SIMS using oxygen (O^+) ions as the primary ion. The profiles of FeN^- are considered to correspond to nitrogen depth profiles, because the depth profiles related to nitrogen measured by this technique, AES etc. were almost the same in higher concentration regions.

Three types of depth profile can be seen: Fig. 8(a) is for H13 alloy tool steel and indicates most improvement in the wear resistance, Fig. 8(b) is for maraging steel and is an

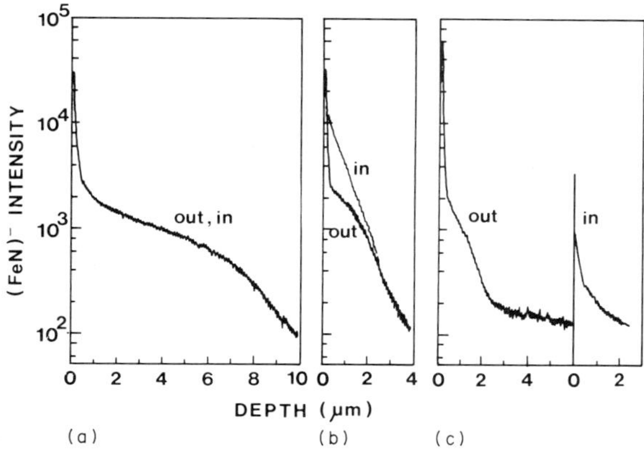

Fig. 8. Depth profiles of FeN^- intensity for (a) an alloy tool steel, (b) a maraging steel and (c) prehardened steel (including nickel) measured by means of SIMS. Nitrogen was implanted at 2.5×10^{17} ions cm^{-2} and 75 keV. The profiles correspond to those of nitrogen.

intermediate case with respect to its improvement in wear resistance and Fig. 8(c) is for prehardened steel with little improvement in the wear resistance. The depth profile of nitrogen implanted in the H13 alloy tool steel shows an unexpected deeper penetration of nitrogen up to 10 μm at a lower concentration, and this profile does not change even after wear tests (within the track). For a maraging steel, implanted nitrogen migrates into the deep region to about 4 μm; this is shallower than the depth for the H13 alloy tool steel, but it is also an unexpectedly high value. Moreover, in the high concentration region, nitrogen atoms migrate into the deeper region and the high concentration region becomes thicker after wear tests. The phenomenon indicates the migration of nitrogen during wear tests. For the prehardened steel, most of the implanted nitrogen forms a gaussian-like distribution, which is similar to nitrogen implantation into pure iron. The depth profile within the track, the depth of which is 6 μm, shows that very little implanted nitrogen remains. It is considered that the unexpected deeper penetration of nitrogen atoms in as-implanted specimens is due to grain boundary diffusion. These results suggest that important factors in improving the wear resistance are not only the migration of nitrogen during wear tests but also the deeper penetration of implanted nitrogen in as-implanted specimens.

The concentration of nitrogen atoms after the wear test could be seen at depths beyond the original implanted layers. Moreover, the fact that oxygen atoms have a higher concentration and deeper penetration than nitrogen atoms in the wear tracks could be seen [18, 22] and the concentration and penetration are considered to play an important role in the wear mechanisms. From the analyses of wear tracks and wear particles in our experiment, it is suggested that the main wear mechanism is adhesive wear in unimplanted steels and oxidative wear in nitrogen-implanted steels. If surface oxidation plays a role in the wear process, the initial stage of the wear tests is important. In the initial stage, if the oxidation rate is higher than the wear rate, the wear rate will be low. Nitrogen implantation produces nitrides which cause a decrease in the adhesive force and an increase in the hardness. Therefore the increase in hardness induced by nitrogen implantation may aid surface oxidation during the initial stage of a wear test and this will cause the wear volume to decrease.

Figure 9 shows the relationship between the relative wear volume and the relative hardness. The relative values were obtained by normalizing the wear volume and hardness of all the steels by those of the unimplanted H13 alloy tool steel. As shown in the figure, it appears that wear resistance and surface hardness are related for both boron and nitrogen implantations. This correlation for nitrogen implantation supports the suggestion that the increase in hardness due to nitrogen implantation enhances surface oxidation during wear tests and thus improves the wear resistance of steels.

However, boron implantation is more effective in improving the wear resistance than nitrogen implantation is; the wear volume decreases drastically for a small increase in hardness and for such a large decrease in friction coefficient (lower than 0.2). The result suggests that the improvement in wear resistance due to boron implantation may be affected more by the friction coefficient than by the hardness. It is proposed that in boron implantation the mechanism for improving the wear resistance therefore will be a reduction in the adhesive force.

Therefore the relationship between the wear volume and hardness is very important in the industrial use of boron and nitrogen implantation into steels. The amount of

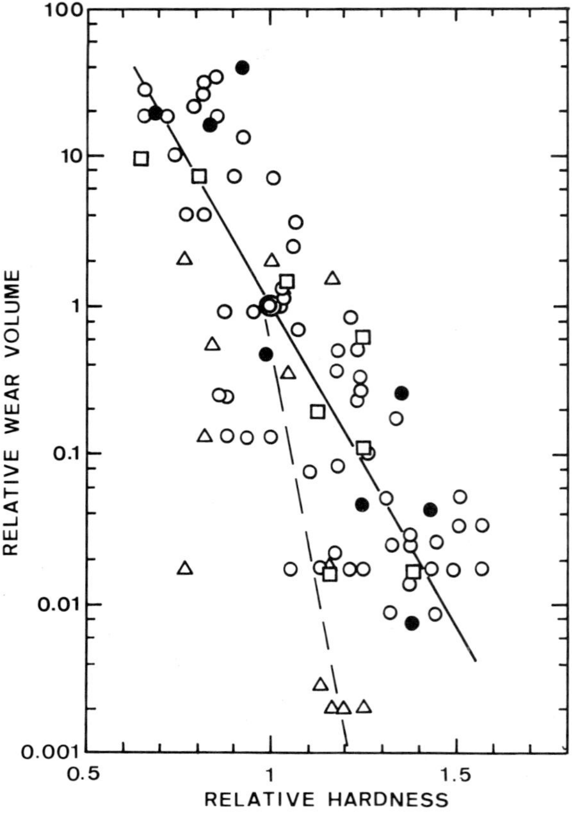

Fig. 9. Relationship between the relative wear volume and relative hardness obtained for boron and nitrogen implantations: ○, nitrogen (N_2^+); ●, NO^+; □, nitrogen (N^+ + N_2^+); △, boron (B^+).

wear from these results can be anticipated by measuring the changes in hardness after implantation.

5. SUMMARY

The tribological properties of low carbon steels and tool steels implanted with various types of species including metallic ions have been investigated by measuring the hardness, friction and wear. The near-surface hardness of implanted steels was measured with Vickers hardness and Knoop hardness testers. The friction coefficients on implanted layers were measured using a Bowden–Leben type of friction-testing machine without lubrication at atmospheric pressure and room temperature. Wear tests were carried out under unlubricated conditions at atmospheric pressure and room temperature using pin-on-disc test equipment. The relationship between the tribological properties and surface analyses obtained by SIMS and XPS is discussed.

The surface hardness of steels implanted with species such as boron, nitrogen and chromium (apparently forming hard compounds near the surface) appears to increase compared with that of the original substrate. These specimens also indicate a decrease in the friction coefficient. In contrast, copper and nickel implantations in steels show no change in hardness and cause the friction coefficient to increase. The surface analyses suggest that the surface layers of a few nanometers (and not just the surface) are of prime importance in the frictional properties.

The wear resistance of H13 steels is improved by high fluence boron and nitrogen implantation. The improvement is considered to be due to the increase in hardness for nitrogen implantation and to the decrease in friction coefficient for boron implantation. The relationship between wear volume and hardness was determined and will be useful in the practical use of ion implantation.

ACKNOWLEDGMENTS

I would like to thank Mr. K. Okitaka and Mr. T. Fujihana of Canon Inc. for the wear tests, Dr. A. Ishitani, Mr. F. Soeda, Mr. K. Okuno, Mr. A. Karen and Mr. K. Yoshida of Toray Research Center Inc. for the SIMS measurements, and Dr. K. Yabe and Mr. O. Nishimura of the Government Industrial Development Laboratory, Hokkaido, for the XPS measurements.

REFERENCES

1 J. K. Hirvonen (ed.), *Ion Implantation, Treatise Mater. Sci. Technol., 18* (1980).
2 P. Sioshansi, *Thin Solid Films, 118* (1984) 61.
3 F. P. Bowden and D. Taber, *The Friction and Lubrication of Solids*, Part II, Oxford University Press, London, 1964.
4 I. L. Singer, *Appl. Surf. Sci., 18* (1984) 28.
5 J. A. Knapp, D. M. Follstaedt and B. L. Doyle, *Nucl. Instrum. Methods B, 7-8* (1985) 38.
6 P. Sioshansi and J. J. Au, in G. K. Wolf, W. A. Grant and R. P. M. Procter (eds.), *Proc. Int. Conf. on Surface Modification of Metals by Ion Beams, Heidelberg, September 17-21, 1984*, in *Mater. Sci. Eng., 69* (1985) 161.
7 T. Fujihana, T. Kuwabara and M. Iwaki, *Proc.*

17th Symp. on Ion Implantation and Submicron Fabrication, RIKEN (Institute of Physical and Chemical Research), Wako, 1986, p. 209.
8 M. Iwaki, S. Namba, K. Yoshida, N. Soda, T. Sato and K. Yukawa, *J. Vac. Sci. Technol.*, *15* (1978) 1089.
9 M. Iwaki, K. Yabe, M. Suzuki and O. Nishimura, submitted to *Nucl. Instrum. Methods.*
10 Y. Okabe, S. An, M. Iwaki and K. Takahashi, submitted to *Nucl. Instrum. Methods.*
11 M. Iwaki, *Proc. Int. Ion Engineering Congr., Kyoto, 1983*, Institute of Electrical Engineers, Tokyo, 1983, p. 1793.
12 N. E. W. Hartley, G. Dearnaley and J. F. Turner, in B. L. Crowder (ed.), *Ion Implantation in Semiconductors and Other Materials*, Plenum, New York, 1973, p. 423.
13 M. Iwaki, H. Hayashi, A. Kohno and K. Yoshida, *Jpn. J. Appl. Phys.*, *20* (1981) 31.
14 M. Iwaki, H. Hayashi, A. Kohno, S. Namba, K. Yoshida and N. Soda, *Proc. 1st Conf. on Ion Beam Modification of Materials*, Central Research Institute of Physics, Budapest, 1978, p. 1981.
15 N. E. W. Hartley, *Wear, 34* (1975) 427.
16 J. K. Hirvonen, *J. Vac. Sci. Technol.*, *15* (1978) 1662.
17 G. Dearnaley and N. E. W. Hartley, *Thin Solid Films, 54* (1978) 215.
18 M. Iwaki, T. Fujihana and K. Okitaka, in G. K. Wolf, W. A. Grant and R. P. M. Procter (eds.), *Proc. Int. Conf. on Surface Modification of Metals by Ion Beams, Heidelberg, September 17–21, 1984*, in *Mater. Sci. Eng.*, *69* (1985) 211.
19 N. E. W. Hartley, *Thin Solid Films, 64* (1979) 177.
20 R. N. Bolster and I. L. Singer, *Appl. Phys. Lett.*, *36* (1980) 208.
21 G. F. Zhai, L. H. De and Z. X. Zhong, *Nucl. Instrum. Methods, 209–210* (1983) 881.
22 B. L. Doyle, D. M. Follstaedt, S. T. Picraux, F. G. Yost, L. E. Pope and J. A. Knapp, *Nucl. Instrum. Methods B, 7–8* (1985) 166.

Materials Science and Engineering, 90 (1987) 273-280

Nitrogen Implantation of Steels: A Treatment which can Initiate Sustained Oxidative Wear*

E. B. HALE,[a] R. REINBOLD[a] and R. A. KOHSER[b]

[a] *Materials Research Center, and Physics Department, University of Missouri-Rolla, Rolla, MO 65401 (U.S.A.)*

[b] *Metallurgical Engineering Department, University of Missouri-Rolla, Rolla, MO 65401 (U.S.A.)*

(Received July 10, 1986)

ABSTRACT

Falex wear tests on mild (SAE 3135) steel samples treated by either nitrogen implantation (2.5×10^{17} N_2^+ cm^{-2} at 180 keV) or low temperature (about 315 °C) oxidation are reported. The results show that both treatments lead to about an order-of-magnitude reduction in the long-term wear rate of the steel. In addition to the wear rate measurements, the wear member asymmetry behavior, scanning electron microscopy studies, Auger spectra and sputter profiles all indicate that the wear modes induced by both treatments are the same and are oxidative wear. These results confirm the previously proposed initiator–sustainer wear model in which implanted nitrogen simply acts as an initiator of favorable oxidative wear but is not directly involved in maintaining the sustained wear resistance. Possible mechanisms for both the initiation process and the sustained wear process are reviewed and discussed.

1. INTRODUCTION

It has been known for more than a decade that nitrogen implantation of various steels can cause significant (often by factor of 10) reduction in their wear rates [1]. Many tools and medical devices are now routinely implanted to improve their wear lifetimes. Nitrogen implantation works but, ironically, there are still many cases in which the scientific causes for the wear improvements are not clear. This paper shows that one effect of nitrogen implantation is to act as an initiator of favorable oxidative wear, which is then self-sustained, and provides a desirable low wear rate. It is believed that this initiation role of implanted nitrogen is not an uncommon occurrence and is important in applications involving sliding wear, especially if lubricated.

The role of implantation as an initiator was pointed out at the last conference [2] and the strong possibility that the sustained wear mode was one of favorable oxidative wear was conjectured by us [2, 3]. Recently, low temperature oxidation was found also to initiate much improved wear [4] and further confirmed the initiator–sustainer model concept. In this paper, more details are presented on the effects of both initiation treatments and the sustained oxidative wear mode. In the latter section of this paper, an overview is presented on the differing views of wear mechanisms in nitrogen-implanted steels, especially those related to the initiator–sustainer concept.

2. EXPERIMENTAL PROCEDURES

The samples studied can be classified into three types: untreated samples, implanted samples and oxidized samples. The untreated samples were mechanically polished to a final stage in which 600 grit paper was used on the pins and 400 grit paper on the blocks. The implanted samples were similarly polished and then exposed to a beam of 180 keV nitrogen ions (N_2^+) to a fluence corresponding to 5×10^{17} N atoms cm^{-2}. The oxidized samples were heated in the range 315 ± 35 °C in still air for 75 min and then cooled slowly in the oven. It should be noted that samples oxidized

*Paper presented at the International Conference on Surface Modification of Metals by Ion Beams, Kingston, Canada, July 7–11, 1986.

0025-5416/87/$3.50

outside this temperature range did not yield significantly improved wear rates [5].

All the wear tests were run on a modified Falex testing machine in a manner which has been described previously [6]. During testing, a cylindrical pin is positioned between two V-grooves which have been machined in opposing blocks. Wear occurs when the applied load squeezes the blocks against the rotating pin. The samples used in this cylinder-in-groove wear geometry were the standard Falex pin and grooved block set, in which the pins are a low concentration Ni–Cr steel (SAE 3135) and the blocks are a free-machining steel (AISI 1137). Both are medium carbon steels with a tempered martensitic structure. All tests were run in Duo-seal oil with an initial 200 lbf applied load and were run until a 10% load drop occurred. Thus, this cut-off criterion resulted in an average wear run time for untreated pins of 35 min, for implanted pins of 60 min and for oxidized pins of 105 min.

3. COMPARISON OF TREATMENT METHODS

Since nitrogen implantation initiated favorable wear and oxygen was involved in the sustained wear stage, it was decided to investigate the influence of direct oxidation of the samples. Thus, oxide films were grown as described above and wear tests were conducted. The results are shown in Table 1. These wear rate data clearly established that low temperature oxidation can also induce favorable wear rates and, since the wear rate improvement was about a factor of 10 for both the implantation and the oxidation treatments, the data suggested that the wear rate improvement mechanism might be the same.

As a test of the similarity of the sustained wear rate mechanisms, measurements involving wear member asymmetry behavior were made. Previous observations in implanted samples [7, 8] showed that implanting the member of the wear couple which had continuous wear contact (*i.e.* an 100% duty cycle) had little effect on the wear rate, but implanting the low duty cycle member of the wear couple was very effective in reducing the wear rate. To test whether the oxidation treatment also caused this wear member asymmetry, blocks were oxidized and run against unoxidized pins. The results of these wear tests are shown in Table 2 which, for comparison purposes, also contains the results on other untreated pins run against both untreated and implanted blocks. The results indicate that there is no major improvement in either the pin or the block wear rates as a result of oxidizing the blocks. Thus the oxidation treatment also causes the wear member asymmetry behavior. It seems rather unlikely that this "unusual" result would occur unless the mechanism underlying the sustained wear modes for both the implantation and the oxidation treatments was the same.

Further evidence of wear mode similarity was provided by scanning electron microscopy (SEM). Figure 1 shows a series of micrographs taken of pins and blocks treated in a variety of ways. The left-hand micrographs in Figs. 1(a) and 1(b) show the as-polished conditions, while the right-hand micrographs in Figs. 1(a) and 1(b) show the severe wear observed when the samples were not treated. Clearly, wear-created groove depths of several

TABLE 1

Average wear rates of pins

	Wear rate ($ng\ s^{-1}$)
Untreated pins	900 ± 300 (5)
Implanted pins	60 ± 30 (5)
Oxidized pins	70 ± 35 (19)

The numbers of pins evaluated are given in parentheses.

TABLE 2

Average wear rates

	Wear rate ($ng\ s^{-1}$)	
	Pins	*Blocks*
Untreated blocks	900 ± 300 (5)	240 ± 600 (3)
Implanted blocks	2000 (1)	450 (1)
Oxidized blocks	670 ± 200 (6)	190 ± 20 (5)

The numbers of samples evaluated are given in parentheses.

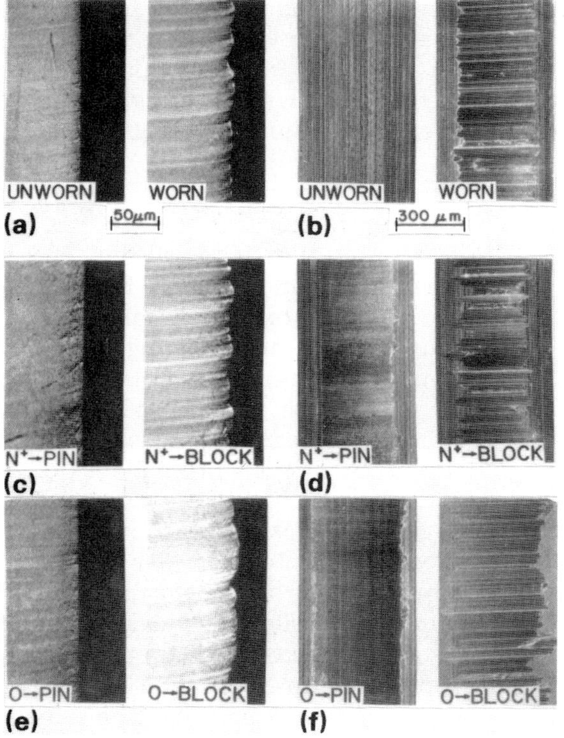

Fig. 1. SEM micrographs of various samples: (a), (b) untreated; (c), (d) implanted (worn); (e), (f) oxidized (worn). The worn horizon profiles of the pins are all shown with the same magnifications in the two left-hand columns ((a), (c), (e)). In the two right-hand columns ((b), (d), (f)) the wear scars on the blocks are all shown with the same magnifications (which are four times smaller than the magnifications used on the pins). In (a) and (b), neither member of the wear couple was treated. In (c) and (d), worn regions are shown for the case when only the pin was nitrogen implanted or when only the blocks were nitrogen implanted. In (e) and (f), worn regions are shown for the case when only the pin was oxidized or when only the blocks were oxidized. In the treated pin cases, the wear runs lasted about three times longer than in the other cases.

microns are seen on the pin and similar groove structures also appear on the blocks. These severe groove features were seen when only the blocks were implanted, as in the right-hand micrographs in Figs. 1(c) and 1(d), or were oxidized, as in the right-hand micrographs in Figs. 1(e) and 1(f). Thus, treating only the blocks had no apparent effect, in agreement with the wear data presented in Table 2. However, radical changes in the wear features were observed when only the pin was treated. These are shown in the left-hand micrographs in Figs. 1(c) and 1(d) for implanted pins and in the left-hand micrographs in Figs. 1(e) and 1(f) for oxidized pins. Despite the fact that the wear tests of the treated pins were longer, the surface roughness of the pins remained about the same as that on unworn pins (and appeared somewhat smoother under an optical microscope). However, although the block wear scars were not changed much in width, they were radically changed in appearance, appearing smooth with some porosity under the scanning electron microscope and burnished under the optical microscope.

The major transition in wear rates, seen as a result of treating the pins, suggest that the wear mode is changed from severe to mild, a transition commonly found in tribological studies [9, 10]. Such transitions usually occur as a result of a change in load or temperature but have also been found when a change of material composition was made [9]. The SEM micrographs support this view of a severe-to-mild transition since it appears that wear in the untreated case occurs by the mutual plowing action of the pin and block on each other with major mass loss on the pin. When the pin is treated, the pin has a smooth and burnished appearance with little evidence of plowing and sizable reduction in mass loss. However, on the blocks a worn layer has formed and the mass loss of the blocks has increased about 50% compared with the untreated case. Results from both energy-dispersive analysis and Auger measurements (see below) show that this layer is mainly iron, although the front part of the first micron of this layer is largely iron oxide. It does not appear to be formed from wear debris since there is very little evidence of graininess and no evidence of wear debris on the edges of the wear scar, and little oxygen is found in Auger results at depths comparable with the layer thickness, which is estimated from the photographs to exceed several microns at the edges. Therefore, it appears that this layer occurs by plastic deformation since the resulting surface is so smooth. The formation of irregularly shaped regions on the exit edge (Figs. 1(b), 1(d) and 1(f)) of the wear scar also suggests plastic deformation since these regions appear to be formed by extrusion and such regions are not seen on the entrance edge of the scar. Wear on the pin by this plastically deformed layer is quite favorable since it is definitely mild and occurs with little mass loss.

As is clear from Fig. 1, the SEM results indicate a marked similarity in the wear scars produced by both treatments of the pins and again suggest a similarly sustained wear mode. This similarity was especially noticeable because of the radical change in wear mode from that observed in the untreated samples or treated block cases. The SEM studies also confirm the wear member asymmetry behavior observed in the wear rate data, *i.e.* the effect of treating the blocks made little change, if any, in the wear scars, while treating the pins caused major changes.

Auger measurements have also been made on the worn samples. The excitation beam was 3 kV and had a diameter of about 30 μm with an ion sputtering beam diameter of about 1 mm. Typical spectra found in the worn region of oxidized pins after various sputtering times are shown in Fig. 2. Unsputtered samples (top spectrum) show a strong oxygen line near 510 eV, a triplet line structure in the 600–750 eV range as well as other structures below 100 eV which are characteristic of non-metallic iron, and the carbon line near 275 eV. The carbon is a surface contaminant and disappears after sputtering for less than 1 min. However, the oxygen line remains strong after sputtering for many minutes. Finally, after much further sputtering, the oxygen signal reaches a minimal (background) intensity.

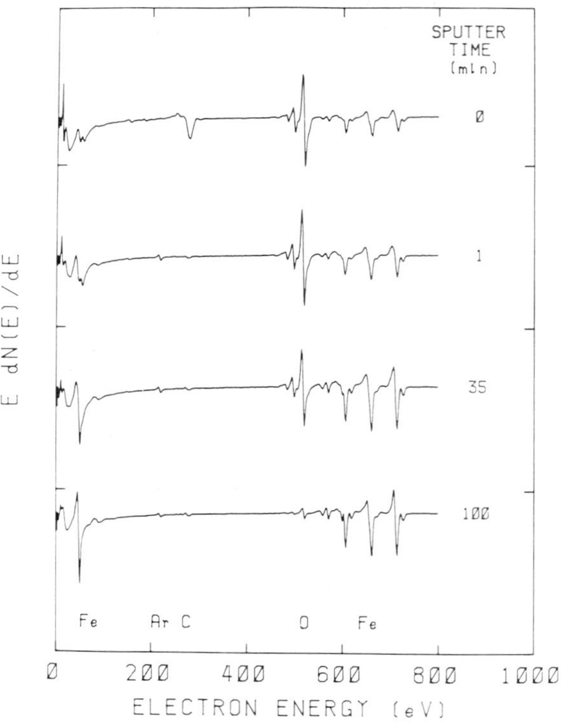

Fig. 2. Auger spectra taken in a worn region of an oxidized pin after various sputter times.

Oxygen sputter time profiles are shown in Fig. 3. Data from the unworn region of untreated pins showed that there was only a very thin surface oxide layer, formed in the polishing stage, and that the oxygen signal reached its background level after sputtering for about 3 min. The unworn regions on implanted pins had slightly more oxide, but on the scale of Fig. 3 the profile was basically the same as for the untreated pins. The unworn regions on oxidized pins show that an oxide layer of rather well-defined thickness was formed by the oxidation, as expected, and was sputtered away in about 15 min. In the worn regions, both untreated and treated pins yielded similar results with above background oxygen signals seen after sputtering for more than 50 min.

It is common practice and tempting to interpret the sputter time scale in Fig. 3 as a depth scale; however, this is often fraught with grave uncertainty. Electron beam damage can cause complicating effects [11], especially in oxides. To obtain very rough depth estimates, we determined that our sputter rate in Ta_2O_5 films was 8 nm min^{-1}. Assuming this rate, we estimate the grown oxide thickness as roughly 100–150 nm.

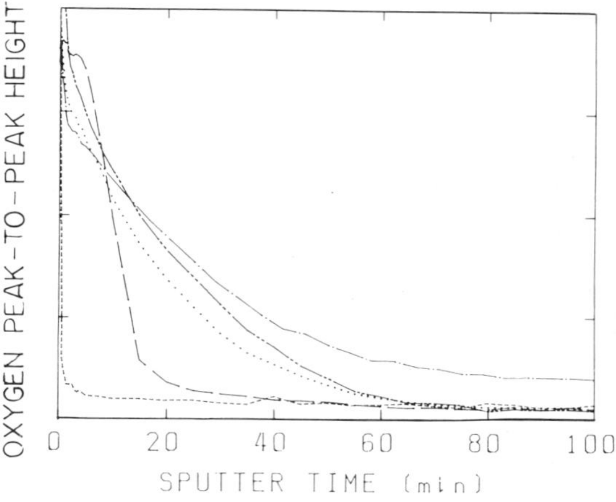

Fig. 3. Sputter time profiles of oxygen obtained from pins exposed to a variety of treatments: -----, unworn, untreated; — —, unworn, oxidized; — · —, worn, untreated; — - - —, worn, implanted; · · · , worn, oxidized. The sputter rate was 8 nm min^{-1} in Ta_2O_5.

The slowly decaying profiles obtained from the worn pins indicate that a uniform oxide layer was not formed. This lack of uniformity may arise from lateral non-uniformities (such as patch formation of the oxide), longitudinal non-uniformities (such as patch formation of the oxide), longitudinal non-uniformities (such as variations in the film thickness), density non-uniformities (such as porous film regions), compositional non-uniformities or any combination of the above. Thus the film thickness is difficult to define. Since oxygen was still observed after sputtering for 50 min, it nominally appears as if oxygen extended at least several tenths of a micron into the surface.

More importantly, these Auger results did not indicate any major difference between worn regions of treated pins. Unexpectedly, they also did not show any major compositional or bonding differences between the worn surfaces of treated and untreated specimens. Nevertheless, because of their different wear performances there must be some structural or chemical differences between these worn surfaces. Singer [12] believed that such similar Auger results simply indicate that only the underlying metal was deformed, and this may be the case. However, there is a major difference in surface morphology, as seen in Fig. 1, which is not detected in Auger measurements. Also, there could be differences in oxide stoichiometry not seen in the Auger data because of beam damage effects [11].

Auger measurements were also made on blocks worn against pins which had been treated in various ways. The sputter time profiles obtained from these samples are shown in Fig. 4. In unworn regions the blocks had a thin oxide layer just like the pins (compare the broken (short dashes) curves in Figs. 3 and 4). However, in sharp contrast with the Auger results from the pins, major differences were found in the worn regions depending on whether the pin was untreated or treated. The blocks worn against untreated pins showed some oxide, but the strength of the oxygen signal was rather weak and the background oxygen level was obtained after sputtering for 20 or 30 min. If the pins were treated by either method, strong oxygen signals were still seen after sputtering for many minutes and the background oxygen level was not obtained until after sputtering for 100 or more min

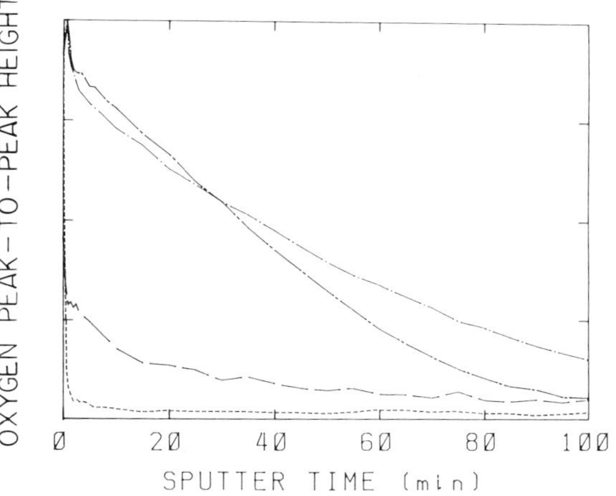

Fig. 4. Sputter time profiles of oxygen obtained from blocks worn against pins which were exposed to a variety of treatments: -----, unworn, untreated; — —, worn, untreated; —·—, worn implanted; —-—, worn, oxidized. The sputter rate was 8 nm min^{-1} in Ta_2O_5.

(see Fig. 4). These results suggest that blocks worn against untreated pins did not become heavily oxidized, with possibly a few local exceptions and even there the oxide thickness was not thick. However, the blocks worn against treated pins became heavily oxidized and yielded Auger profiles similar to those seen on the pins. As is now expected, these Auger results also showed great similarity in the sustained wear produced by the two dissimilar treatments.

4. WEAR MECHANISMS IN IMPLANTED STEELS

At this point a brief synopsis of work dealing with wear in ion-implanted steels will be given so that the above results can hopefully be placed in perspective. Review articles on wear improvements in implanted steels [3, 12-17] have basically focused on the role of nitrogen in hardening the steel (*e.g.* by solution hardening, nitride formation or dislocation pinning [18]) or in stabilizing a favorable surface layer (*e.g.* as in austenitic steels [19]). Such mechanisms could be reasonable in shallow wear cases if nitrogen remains in the specimen's surface and hence can have a continuing influence.

However, such explanations do not appear to be valid in commonly observed deep wear situations, where the favorable wear persists to depths which are often much greater (at least an order of magnitude deeper) than the range of the implanted ion. Wear induced in-diffusion of nitrogen has been discussed as a solution to this depth scale problem [18], but no convincing data have been shown to substantiate it [12, 14, 16]. In addition, two groups [2, 20] have presented data which convincingly showed that favorable wear persisted long after virtually all the nitrogen has worn away. It is also noteworthy that both groups used significantly different wear tests, one being a lubricated Falex sliding wear test [2] and the other an abrasive wear test involving vibratory polishing in a slurry [20].

This leaves little doubt that there is at least one and possibly several important wear mechanisms in nitrogen-implanted steels which are not understood and have basically been overlooked. Most of the discussions and models proposed to date have direct nitrogen involvement in the steady state wear process [12–17]. It was pointed out several years ago [2], however, that any viable model for deep wear must involve nitrogen simply as an initiator of a non-nitrogen-involving sustained wear mode.

The role of nitrogen as an initiator should not be confused with the fact that nitrogen can also cause reduced wear during run-in, especially in hard steels [16, 21]. In such cases the implantation has been found to cause a much reduced mass loss during break-in (attributed to hardening involving nitrogen [16]), but the steady state wear rate was not changed. In our initiated and then sustained wear case, there is a major change in the steady state wear rate. This major change can be seen from the accumulated mass loss data presented in Fig. 5. The untreated wear rate can be obtained from the linear slope, and the untreated wear data and the sustained wear rate from the slope of the fit to the implanted pin data. There is a difference of about a factor of 10 between these slopes, in agreement with the data in Table 1, since the run-in mass loss is relatively small.

The run-in mass losses for the differently treated pins can be estimated by extending the linear fits of the data in Fig. 5 to obtain the y axis intercepts. These intercepts are 0.8 mg for the untreated pin and 0.15 mg for the implanted pins. Thus, there is a definite reduction in the run-in mass loss. This dif-

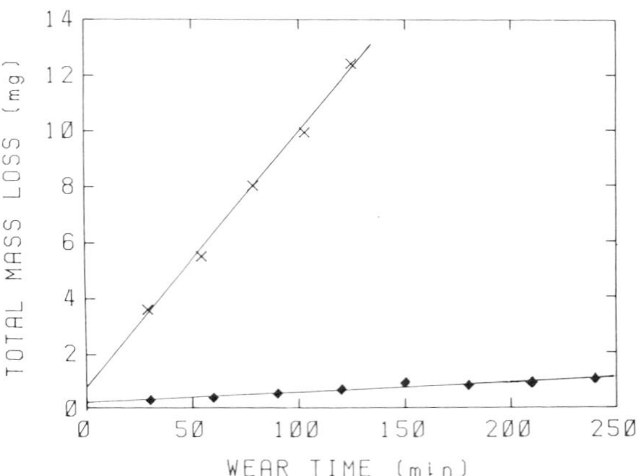

Fig. 5. Mass loss *vs.* wear time data of an untreated pin (×) and an implanted pin (♦). The much reduced mass loss for the implanted pin arises mainly from the favorably sustained wear mode.

ference between the two cases was to some extent surprising since other lubricating wear tests in mild steel found no run-in differences [16, 21]. However, the important point is that the real improvement produced by the implantation is in the steady state wear rate since the run-in losses in each case are small compared with the total mass lost in the respective tests.

It is of prime importance to determine the mechanisms which occur in both the initiation and the sustained wear processes. Once the initiation process is understood, other surface treatments may become attractive initiators or, conversely, use of other favorable wear-inducing treatments may provide additional insight. Using our wear test, it is difficult to learn much about the initiation process since it occurs so early in the run. The initiation process works rapidly, less than a minute in our case, to set up the deformed layer on the blocks. Perhaps a suitable initiation process is any one which prevents severe wear from starting, in which case mild wear occurs and results in the formation of the favorably wearing oxidized surfaces. If this is the case, then nitrogen implantation might basically just provide initial hardening and possibly a resulting temperature rise. In any case, a different oxide forms from that on the untreated surface.

The sustained wear process is oxidative wear since the Auger data show extensive oxygen involvement in the worn, but not unworn, surfaces of both the pin and the block.

Oxidative wear seems rather common in other lubricated sliding wear studies on implanted steels (see below). The fundamental importance of sustained oxidative wear needs to be emphasized. Often, the differences between the untreated and the treated wear cases are referred to using terms such as adhesive and abrasive wear. Although many tribologists also classify wear using such terms, in practice it is difficult to make this distinction [22]. One classification scheme [23, 24] simply considers all wear as either severe, in which case it is delaminational wear, or mild, in which case it is oxidative wear. Wear improvements are commonly seen as a result of a transition between these two types of wear mode [9, 10].

In the early lubricated sliding wear tests of implanted samples, no studies emphasized the role of oxygen. Recently, more researchers have been attributing effects to oxygen. Goode *et al.* [25] used SEM to observe the mild wear in their implanted samples and attributed it to oxidative wear enhanced by the presence of nitrogen. Doyle *et al.* [26] noted unusual behavior of oxygen in implanted worn samples. Pollock *et al.* [27] studied oxygen involvement in implanted steels and tentatively proposed that the nitrogen implantation may improve the formation of an oxide layer during run-in and cause the sustained wear. However, Dearnaley [3] believed that the improved sustained wear found in his abrasive wear experiments [20] was related to self-propagation of nitrogen-initiated strain-hardening processes and did not involve oxygen.

We have also observed the importance of the subsurface metal in treated worn pins. When these pins are dipped briefly in aqueous HCl, the acid dip dissolves most of the surface oxide layer. If the dipped pins are then reworn, they resume wearing with a low wear rate, *i.e.* in the sustained wear mode. Thus a thick oxide layer is not required and the dipped worn surface is able to reinitiate itself. The ability for reinitiation appears to be the key mechanism in sustaining favorable wear.

Further experimentation is needed to sort out the various factors associated with the sustaining mechanism. Still unknown is the fundamental difference between the sustained damaged layer and an untreated damaged layer. Smoothness does appear to be one important factor since rough (120 grit) treated pins do not wear in the sustained mode. However, a thick oxide layer seems to be simply a by product rather a critical factor. In addition, the wear member asymmetry behavior suggests that off-contact interactions may also play an important role in the sustaining process.

Finally, a statement about the generality of these results seems appropriate. There are certainly cases of wear improvements induced by ion implantation which are not examples of initiated and sustained oxidative wear. However, the above results and discussion suggest that the initiator–sustainer mechanism should be considered as a possible cause of implantation-induced wear rate improvements, especially in cases where oxygen is involved and the wear depth greatly exceeds the ion range. In addition, such favorable oxidative wear is probably the cause of improvements if a treated mild steel experiences low duty cycle wear under lubricated conditions. From a practical point of view, thermal oxidization should be seriously considered as a means of imparting wear resistance, before ion implantation, because it is a significantly cheaper treatment.

5. CONCLUSION

The wear rate data on the pins, the member asymmetry of the wear rates, the SEM results and the Auger results all indicate that the wear modes induced by either nitrogen ion implantation or thermal oxidization treatments are the same. This substantiates the initiator–sustainer model since it shows that various treatments can be used to obtain the same good wear resistance. In the present study the sustained mode is one of oxidative wear. However, the precise role of oxygen in the sustaining mechanism is not completely clear since a favorably damaged underlying metal layer may also be important. Thus, some of the factors which influence the improved wear rates observed in treated steels are becoming clear, but more work is required to determine precisely the wear mechanisms responsible for both the initiation and the sustained processes.

REFERENCES

1 N. E. W. Hartley, G. Dearnaley, J. F. Turner and J. Saunders, in S. T. Picraux, E. P. EerNisse and

F. L. Vook (eds.), *Applications of Ion Beams to Metals*, Plenum, New York, 1974, p. 123.
2 T. J. Sommerer, E. B. Hale, K. W. Burris and R. A. Kohser, in G. K. Wolf, W. A. Grant and R. P. M. Procter, *Proc. Int. Conf. on Surface Modification of Metals by Ion Beams, Heidelberg, September 17-21, 1984*, in *Mater. Sci. Eng., 69* (1985) 149.
3 G. Dearnaley, in G. K. Wolf, W. A. Grant and R. P. M. Procter, *Proc. Int. Conf. on Surface Modification of Metals by Ion Beams, Heidelberg, September 17-21, 1984*, in *Mater. Sci. Eng., 69* (1985) 139.
4 E. B. Hale, R. A. Kohser and R. A. Reinbold, *Appl. Phys. Lett., 49* (1986) 447.
5 R. A. Reinbold, R. A. Kohser and E. B. Hale, to be published.
6 E. B. Hale, C. P. Meng and R. A. Kohser, *Rev. Sci. Instrum., 53* (1982) 1255.
7 E. B. Hale, M. M. Muehlemann, W. Baker and R. A. Kohser, in R. Kossowsky and S. C. Singhal (eds.), *Proc. Conf. on Surface Engineering, Les Arcs, July 1983*, in *NATO Adv. Study Inst. Ser. E, No. 85*, (1984) 158.
8 J. K. Hirvonen, *J. Vac. Sci. Technol., 15* (1978) 1662.
9 N. C. Welsh, *Philos. Trans. R. Soc. London, Ser. A,1077* (1965) 31.
10 T. S. Eyre, in D. Scott (ed.), *Wear, Treatise Mater. Sci. Technol., 13* (1979) 363-442.
11 C. G. Pantano and T. E. Madey, *Appl. Surf. Sci., 7* (1981) 115.
12 I. L. Singer, in G. K. Hubler, O. W. Holland, C. R. Clayton and C. W. White (eds.), *Ion Implantation and Ion Beam Processing of Materials, Materials Research Society Symp. Proc., Boston, MA, November 14-17, 1983*, Vol. 27, Elsevier, New York, 1984, p. 585.
13 N. E. W. Hartley, in J. K. Hirvonen (ed.), *Ion Implantation, Treatise Mater. Sci. Technol., 18* (1980), Chapter 8.
14 G. K. Hubler, in S. T. Picraux and W. J. Choyke (eds.), *Metastable Materials Formation by Ion Implantation, Materials Research Society Symp. Proc.*, Vol. 7, Elsevier, New York, 1982, p. 341.
15 I. L. Singer, *Appl. Surf. Sci., 18* (1984) 28.
16 G. K. Hubler and F. A. Smidt, *Nucl. Instrum. Methods B,7-8* (1985) 151.
17 G. Dearnaley, *Nucl. Instrum. Methods B,7-8* (1985) 158.
18 G. Dearnaley and N. E. W. Hartley, *Proc. 4th Conf. on Scientific and Industrial Applications of Small Accelerators, Denton, TX, 1975*, IEEE, New York, 1976, p. 20.
19 R. N. Bolster and I. L. Singer, *Appl. Phys. Lett., 17* (1980) 327.
20 G. Dearnaley, F. J. Minter, P. K. Rol, A. Saint and V. Thompson, *Nucl. Instrum. Methods B, 7-8* (1985) 188.
21 K. Yu, H. D. Li, X. Z. Zhang and J. H. Tian, *Nucl. Instrum. Methods, 209-210* (1983) 1063.
22 E. Rabinowicz, *Mater. Sci. Eng., 25* (1976) 23.
23 J. F. Archard and W. Hirst, *Proc. R. Soc. London, Ser. A, 236* (1956) 397.
24 T. F. J. Quinn, in N. P. Suh and N. Saka (eds.), *Fundamentals of Tribology*, Massachusetts Institute of Technology Press, Cambridge, MA, 1980, pp. 447-492.
25 P. D. Goode, A. T. Peacock and J. Asher, *Nucl. Instrum. Methods, 209-210* (1983) 925.
26 B. L. Doyle, D. M. Follstaedt, S. T. Picraux, F. G. Yost, L. E. Pope and J. A. Knapp, *Nucl. Instrum. Methods B, 7-8* (1985) 166.
27 J. T. A. Pollock, M. D. Scott and M. J. Kenny, *Appl. Surf. Sci., 22-23* (1985) 128; *Proc. 5th Int. Conf. on the Ion Beam Modification of Materials, Catania, June 9-13, 1986*, to be published.

Materials Science and Engineering, 90 (1987) 281-286

Enhanced Endurance Life of Sputtered MoS_x Films on Steel by Ion Beam Mixing*

K. KOBS[a], H. DIMIGEN[a], H. HÜBSCH[a], H. J. TOLLE[a], R. LEUTENECKER[b] and H. RYSSEL[c]

[a]*Philips G.m.b.H. Forschungslaboratorium Hamburg, Vogt-Koelln-Strasse 30, D-2000 Hamburg 54 (F.R.G.)*

[b]*Fraunhofer Institut für Festkörpertechnologie, Paul-Gerhardt-Allee 42, 8000 München 60 (F.R.G.)*

[c]*Fraunhofer Arbeitsgruppe für integrierte Schaltungen, Artilleriestrasse 12, 8520 Erlangen (F.R.G.)*

(Received July 10, 1986)

ABSTRACT

R.f. sputter-deposited MoS_x films (x = 1.6–1.9) show excellent lubrication properties. The coefficient of friction under sliding conditions in an inert gas or in vacuum is very low (0.02–0.03), but the endurance life even for optimized sputter parameters depends on the internal mechanical stress and is often insufficient. The method of ion beam mixing was used to increase the film–substrate adherence and, therefore, to obtain a life enhancement of the lubricating layers. The tribological tests were carried out on a ball-on-disc tribometer under oscillating conditions in a dry nitrogen atmosphere. The mixing experiments were performed with argon and nitrogen ions at various energies (50–400 keV) and fluences ((0.3–5) × 10^{16} ions cm^{-2}). The film thicknesses ranged from 0.20 to 0.47 μm.

A considerable enhancement of the endurance life without any deterioration of the friction behaviour was obtained, depending on the energy and the film thickness. For example, an MoS_x layer of 0.36 μm thickness yielded a six-fold increase in the endurance life by implantation with 180 keV nitrogen and 400 keV argon ions. For a thinner film thickness the required energy can be lowered. This indicates a mixing of the interface since the penetration depth and the film thickness have to overlap so that the projected range of the ions falls at the interface. Profilometer traces across the wear tracks indeed show a considerable reduction in flaking after implantation, and with secondary ion mass spectrometry analysis we found a distinctly broadened interface. Additionally, we observed an increased film density of 20%–40% due to nitrogen and argon implantation, depending on the energy and film thickness.

1. INTRODUCTION

In the last few years the interest in the application of thin films for reducing wear and friction has increased. It has been shown that r.f. sputtered MoS_2 layers have excellent lubrication properties in an inert gas or in vacuum [1–4]. In previous work the influence of sputtering conditions on the composition of MoS_x films and their frictional behaviour was investigated [5]. For optimized sputtering conditions a coefficient of friction of about 0.02 has been obtained, but a wide application of this particular behaviour is restricted by an insufficient sliding life. The lifetime depends on the degree of cohesiveness between the MoS_2 crystallites and on the film–substrate adherence [6].

It has been reported that enhanced adhesion of metal films to various substrate materials occurs after bombardment with high energy ions [7–10]. In our work, we intended to improve the adherence of MoS_x films on steel substrates by ion beam mixing without any deterioration in the excellent lubrication properties. It should be pointed out that the coefficient of friction depends strongly on the structure of the MoS_x film, *e.g.* amorphous MoS_2 films show no lubrication effect, the coefficient of friction increasing to 0.4 [11]. Therefore, amorphization especially of the superficial layers should be avoided by ion bombardment, and it would be appropriate to

*Paper presented at the International Conference on Surface Modification of Metals by Ion Beams, Kingston, Canada, July 7–11, 1986.

0025-5416/87/$3.50

mix only the interface; this is possible by ion implantation, as the displacement damage shows a typical gaussian error distribution law. This means that the ion energy and fluence have to be optimized with respect to the film thickness.

2. EXPERIMENTAL TECHNIQUE

The MoS_x films ($x = 1.6$–1.9) were prepared by r.f. diode sputtering in the non-reactive mode with an MoS_2 target under an argon pressure of 2.6 Pa. The power density was 5 W cm^{-2}, the residual gas pressure below 6×10^{-4} Pa, and the temperature below 60 °C. The substrates consisted of 100 Cr6 (AISI 52100) bearing steel and of S 6-5-2 (AISI M2) high speed steel with a diamond-polished surface. Before the deposition process was started, the substrates were cleaned by ion etching to avoid influencing the adherence by contamination of the substrate surface. The implantation was carried out using a model 200 Varian–Extrion implanter with a mass-separated ion beam. The beam current density was 5 μA cm^{-2}, and the maximum temperature 150 °C. As ions we used atomic nitrogen (singly charged) and argon (doubly charged).

The tribological investigations were performed with a ball-on-disc tester under oscillating conditions [12] in a dry nitrogen atomosphere (relative humidity, less than 0.5%). The lifetime is taken to be indicated by a drastic increase in the coefficient of friction to a value of 0.6 within a time period of only a few seconds; this gives a well-defined endurance life [13].

Each lifetime was measured five times with an oscillation frequency of 7 cycles s^{-1} and an applied load of 5.1 N. The ball material was 100 Cr6 (AISI 52100) bearing steel.

To characterize the interface, secondary ion mass spectrometry (SIMS) was performed using 7 keV oxygen and argon ions at a primary ion current of 0.5 μA and a scanning area of 500 μm $\times$ 500 μm. The sputter rates were 1 μm h^{-1} for MoS_x and 0.3 μm h^{-1} for steel. The depth profiles were corrected in depth corresponding to the different sputter rates, assuming a linear relation between the reduction in the sputter rate and the iron content [14].

3. RESULTS

3.1. *Tribological investigations*

MoS_x films (thickness, 0.43 ± 0.03 μm) were implanted with nitrogen and with argon ions at various energies and a fluence of 1×10^{16} ions cm^{-2}. Figure 1 gives the result obtained for the lifetime. For both ion species a considerable increase occurs at higher energies, whereas at lower energies a slight reduction was observed. The lifetimes are nearly identical for nitrogen-implanted and for argon-implanted films, taking into consideration that, for argon ions, double the energy of the nitrogen ions is needed. This indicates that the range of the ions is important in the lifetime improvement due to interface mixing. This is in accordance with the results of films of smaller film thickness (0.31 and 0.20 μm) in which an enhancement has already been observed for 50 keV nitrogen ions (Fig. 2).

The dose dependence of the sliding life is shown in Fig. 3 at a constant energy of 150 keV for nitrogen ions and 300 keV for argon ions. The thickness was 0.43 μm, as used for the energy variation. The maximum lifetime was observed at a fluence of 1×10^{16} ions cm^{-2}, but almost the same improvement occurred at 3×10^{15} ions cm^{-2}.

It is very important for practical applications to examine whether the enhancement of the sliding lifetime remains at higher loads. The lifetime dependence on the applied load

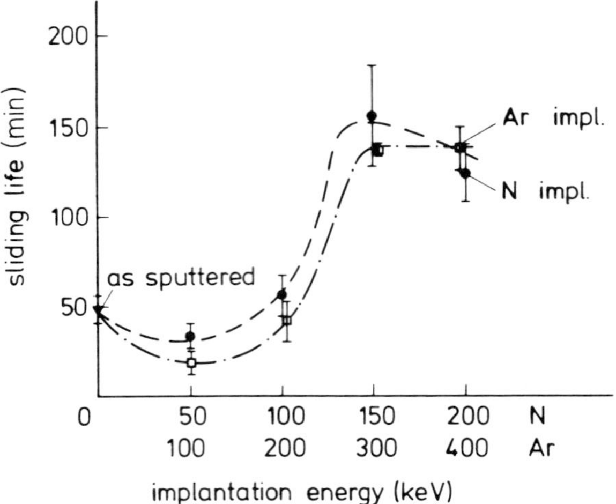

Fig. 1. Sliding life of as-sputtered and nitrogen-implanted or argon-implanted MoS_x films (thickness, 0.47 μm) on steel at various energies.

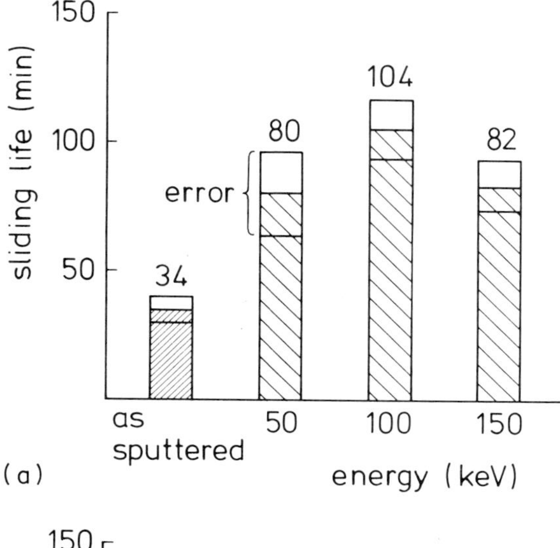

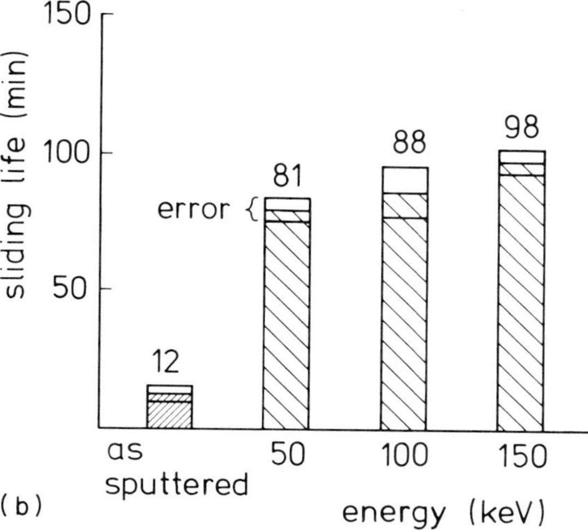

Fig. 2. Sliding life of as-sputtered and nitrogen-implanted MoS_x films ((a) thickness, 0.31 μm; (b) thickness, 0.20 μm) on steel at various energies (nitrogen (N^+) fluence, 1×10^{16} ions cm^{-2}): ▧, mean values.

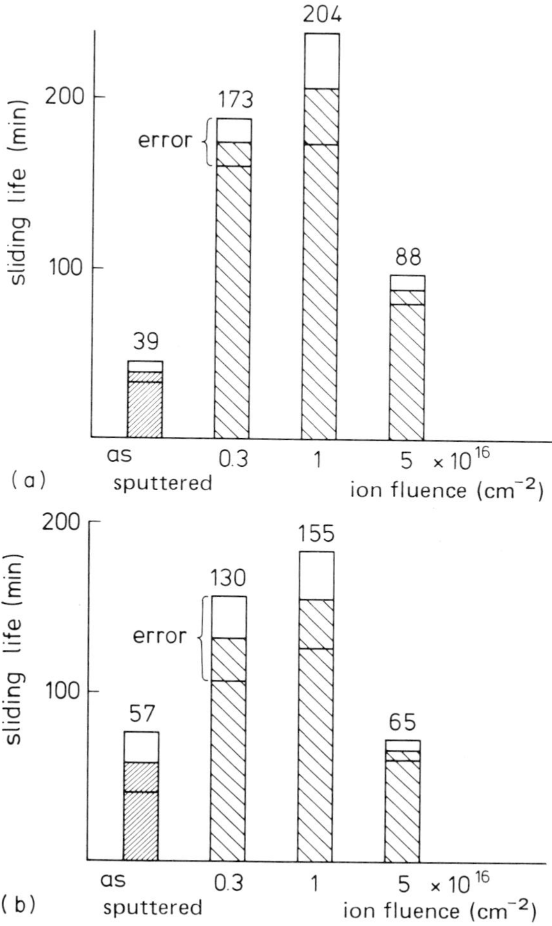

Fig. 3. Sliding life of as-sputtered and MoS_x films (thickness, 0.43 μm) implanted with (a) 150 keV nitrogen (N^+) ions and (b) 300 keV argon (Ar^+) ions on steel at various fluences: ▧, mean values.

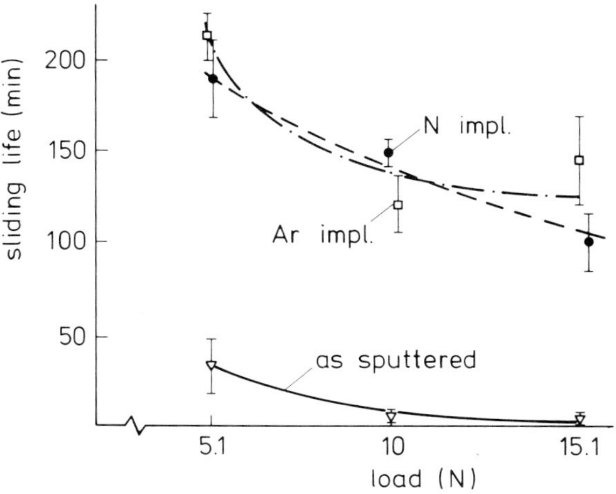

Fig. 4. Dependence of sliding life on the applied load for as-sputtered, nitrogen-implanted (100 keV) and argon-implanted (250 keV) MoS_x films on S 6-5-2 (AISI M2) steel (implanted fluence, 1×10^{16} ions cm^{-2}).

for an MoS_x film 0.29 μm thick on S6-5-2 (AISI M2) high speed steel is shown in Fig. 4. With increasing load, the lifetime of the unimplanted film decreased drastically to a value of a few minutes, whereas both the nitrogen-implanted and the argon-implanted films yielded only a slight decrease, so that at higher loads a more than 30-fold improvement in the lifetime was obtainable.

It should be pointed out that for all tribological experiments the friction behaviour, even for the highest energies and fluences, did not deteriorate, as described in more detail in ref. 13. This is an indication that amorphization due to ion bombardment did not occur.

3.2. Topographical investigations

Clarification of the enhanced lifetime was found from a study of the wear tracks. Optical microscopy investigation accompanied by profilometer measurements of wear tracks after various numbers of cycles yielded quite a different appearance. The unimplanted film showed that even after 1 cycle the MoS_x film had been partially removed up to the substrate; after 10 cycles, most parts of the wear track and the adjacent regions are damaged, indicating that flaking has occurred. This was confirmed by the profilometer measurements. The profiles had a rectangular shape. For the implanted films (400 keV argon ions; fluence, 1×10^{16} ions cm^{-2}; film thickness, 0.36 μm) the first wear occurred after 1000 cycles without any flaking due to an improved film–substrate adherence. The profile of the wear track has a typical ditch profile. Improved adhesion of sputtered MoS_2 films was also found by Chevallier *et al.* [15], who used a combination of r.f. sputtering and ion implantation (argon, krypton and xenon ions) in the 100 keV region.

Also from profilometer measurements we found that the film thickness is reduced as a result of the implantation (Fig. 5). A step was observed when going from the implanted region to the unimplanted region. The height of the step increases with increasing ion energy. This effect is not due to sputtering of the film; this was confirmed by electron probe microanalysis. We obtained the same intensities for the molybdenum and sulphur signals in both the implanted and the unimplanted regions within a deviation of 1%–2%. Only a slight decrease in the molybdenum intensity (4%–6%) was found at the lower implantation energies of 50 and 100 keV. This means that the reduction in the film thickness is due to a density enhancement of up to 40% at the higher implantation energies. The density enhancement may lead to a higher degree of cohesiveness between the MoS_2 crystallites and to a more effective film thickness and is probably caused by structural changes, indicated by the distinct increase in the reflectance of the bombarded MoS_x films, which, for example, increases from 15% to 55% after 400 keV argon implantation. In ref. 16, Spalvins described the change in the reflectance of MoS_2 films due to reorientation of the MoS_x platelets. Platelets with basal planes perpendicular to the plane of the film reorient horizontally. Also, recrystallization in the MoS_x film due to implantation can occur, as Hirano and Miyake [17] found in sputtered WS_2 films.

3.3. Secondary ion mass spectrometry investigations

In Fig. 6 the depth profiles of sulphur, molybdenum and iron for 50 keV and 200 keV nitrogen-implanted MoS_x films on 100 Cr6 steel are presented. The implantation fluence was 1×10^{16} ions cm^{-2}. We observed no broadening of the interface for 50 keV ions compared with the as-sputtered film shown in Fig. 7. A markedly broadened inter-

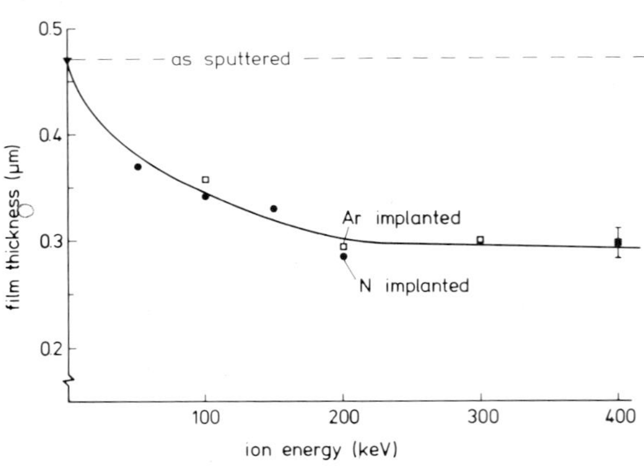

Fig. 5. Reduction in the film thickness of an MoS_x film 0.47 μm thick after nitrogen and argon implantation at various energies.

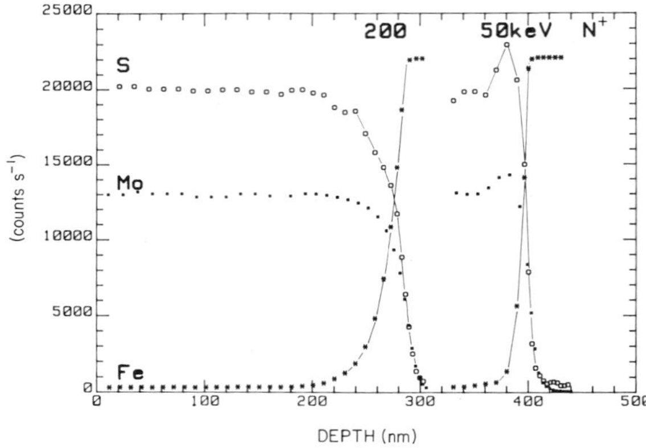

Fig. 6. Molybdenum, sulphur and iron depth profiles of 50 keV and 200 keV nitrogen-implanted MoS_x films on steel, obtained by SIMS (film thickness, 0.47 μm; 100 Cr6 (AISI 52100) steel; implanted fluence, 1×10^{16} ions cm^{-2}).

face of about 80 nm was detected after 200 keV implantation (and also for 150 keV implantation), revealing a strong correlation to the lifetime enhancement. The shift of the interface to the surface is due to the reduction in the film thickness as described before. A further SIMS profile of the same sample implanted with 400 keV argon ions is shown in Fig. 7. The broadening of the interface is more significant than for the nitrogen-mixed films at the same fluence because of the stronger mixing effect of the argon ion. Almost the same profiles were obtained with a nitrogen fluence of 5×10^{16} ions cm^{-2} that was five times larger. At this increased fluence,

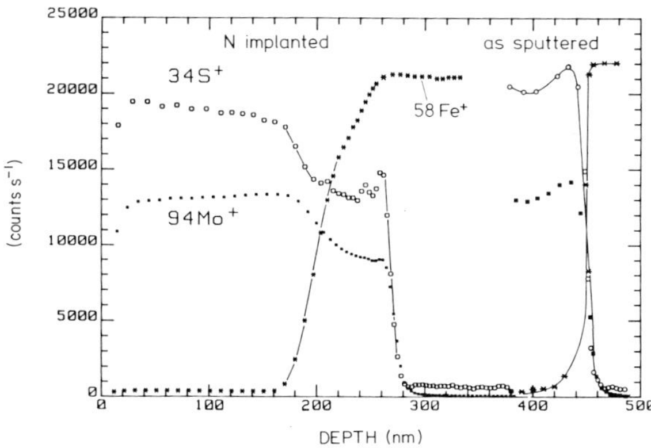

Fig. 7. Molybdenum, sulphur and iron depth profiles of a 300 keV argon-implanted MoS_x film on steel (film thickness, 0.47 μm; 100 Cr6 (AISI 52100) steel; implanted fluence, 1×10^{16} ions cm^{-2}).

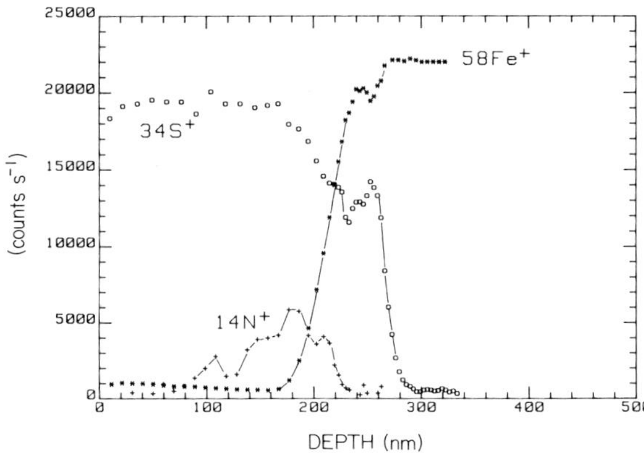

Fig. 8. Sulphur, iron and nitrogen depth profile of a 150 keV nitrogen-implanted MoS_x film on steel at a fluence of 5×10^{16} ions cm^{-2} (film thickness, 0.43 μm).

it was possible to measure the nitrogen depth profile (Fig. 8), because of the detection limits of SIMS [14]; this confirmed that the maximum nitrogen concentration is located at the interface and that the ion range is in accordance with the film thickness.

4. CONCLUSIONS

With the method of ion beam mixing, considerable enhancement of the sliding life of sputtered MoS_x (x = 1.6–1.9) films on steel was observed, especially at higher loads. The best results were obtained for a film thickness of 0.4–0.5 μm corresponding to ion energies of 150–200 keV for nitrogen ions and 300–400 keV for argon ions at a fluence of 1×10^{16} ions cm^{-2} with no deterioration in the excellent friction behaviour.

This enhancement is due to improved film–substrate adherence which has been proved by topographical and SIMS investigations. Additionally, a structural change occurs during ion bombardment, indicated by a density enhancement of up to 40% and a distinct increase in the coefficient of reflection.

ACKNOWLEDGMENTS

The authors are indebted to Mr. W. Mohwinkel for performing the tribological tests and to Dr. P. Willich and Mr. D. Obertop for the electron probe microanalysis investigations.

Part of the work was supported by the Bundesministerium für Forschung und Technologie under Grants 03 T 0002 E6, 03 T 0002 A5 and 13 N 5353/0.

REFERENCES

1 T. Spalvins, *NASA Tech. Note TND 7169*, 1973, (National Aeronautics and Space Administration).
2 M. Gardos, *Lubr. Eng.*, *32* (1976) 463.
3 R. J. Christy, *Thin Solid Films, 64* (1979) 223.
4 R. J. Christy, *Thin Solid Films, 73* (1980) 299.
5 H. Dimigen, H. Hübsch, P. Willich and K. Reichelt, *Thin Solid Films, 129* (1985) 79.
6 T. Spalvins, *Thin Solid Films, 118* (1984) 375.
7 J. E. E. Baglin and G. J. Clark, *Nucl. Instrum. Methods B, 7–8* (1985) 881.
8 R. A. Kant, B. D. Sartwell, I. L. Singer and R. G.

Vardiman, *Nucl. Instrum. Methods B, 7-8* (1985) 915.
9 D. K. Sood, W. M. Skinner and J. S. Williams, *Nucl. Instrum. Methods B, 7-8* (1985) 893.
10 A. E. Berkowitz, R. E. Benenson, R. L. Fleischer, L. Wielunski and W. A. Landford, *Nucl. Instrum. Methods B, 7-8* (1985) 877.
11 T. Spalvins, *Thin Solid Films, 73* (1980) 291.
12 H. Dimigen, K. Kobs, R. Leutenecker, H. Ryssel and P. Eichinger, *Mater. Sci. Eng., 69* (1985) 181.
13 K. Kobs, H. Dimigen, H. Hübsch, H. J. Tolle, R. Leutenecker and H. Ryssel, *Appl. Phys. Lett., 49* (1986) 496.
14 H. J. Tolle, K. Kobs, submitted to *Fresenius' Z. Anal. Chem.*
15 J. Chevallier, S. Oleson, G. Sørensen and B. Gupta, *Appl. Phys. Lett., 48* (1986) 876.
16 T. Spalvins, *Thin Solid Films, 96* (1982) 17.
17 M. Hirano and S. Miyake, *Appl. Phys. Lett., 47* (1985) 683.

Materials Science and Engineering, 90 (1987) 287–290

Mechanical Properties of Nitrogen-implanted 18W–4 Cr–1V Bearing Steel*

PETER MADAKSON

IBM Thomas J. Watson Research Center, Yorktown Heights, NY 10598 (U.S.A.)

(Received July 10, 1986)

ABSTRACT

Nitrogen was implanted into 18W–4Cr–1V bearing steel (where the composition is in approximate weight per cent) to doses of 10^{14}–10^{17} *ions* cm^{-2} *at 400 keV. The specimens were analyzed using the Rutherford backscattering and the X-ray photoelectron spectroscopy techniques and then tested on a pin-on-disc wear machine. There was a thick non-metallic film on the high dose specimens which was shown by the X-ray photoelectron spectroscopy examination to be graphite. Friction and wear were significantly decreased and hardness was increased by a factor of 4. These changes were observed to depend on the nitrogen dose and the applied load. The formation of a graphite surface layer is considered to be due to vacuum carburization and the increase in wear resistance is thought to be associated with nitride precipitation hardening.*

1. INTRODUCTION

Previous work on titanium and boron implantation of 18W–4Cr–1V bearing steel (where the composition is in approximate weight per cent) has already been reported [1, 2]. In titanium-implanted 18W–4Cr–1V the hardness, friction and wear were significantly improved. However, the changes were independent of ion dose because they were sensitive to vacuum carburization that occurred during the implantation process. With boron implantation, wear resistance was observed to increase linearly with increasing ion dose, but friction and hardness did not. The results were considered to be governed by the effectiveness of boron in blocking the migration of dislocations during the wear process. Negative wear was observed during the early stages of the wear process, but this is now thought to be due largely to the thermal drift of the measuring instruments and not to oxidation of the wear surface as reported.

Much research work has been done on the effects of nitrogen implantation on the mechanical properties of metals [3–5]. Improvements in wear resistance were observed to persist well beyond the implanted region. It was concluded that the implanted nitrogen migrated ahead of the wear surface [6], but it has been observed by the present author and many other research workers that wear resistance does persist even after the implanted nitrogen can no longer be traced in the wear track. Improvements in wear resistance have also been observed, for other ion species, to extend far beyond the implanted region without evidence of inward migration of the implanted ions. There must therefore be secondary effects of ion implantation that play an important role during the wear process, especially after the implanted layer has been worn away, to account for the improvements observed. Madakson [7] suggested that the implanted ions create dislocation networks beyond their depth of penetration in the treated material which, together with those generated during the wear process, continue to entangle with each other after the implanted layer has been worn off, to result in work hardening of the wear surface.

2. EXPERIMENTAL METHOD

The chemical composition of the bearing steel and details of the wear and microhardness tests have been given elsewhere [1].

*Paper presented at the International Conference on Surface Modification of Metals by Ion Beams, Kingston, Canada, July 7–11, 1986.

0025-5416/87/$3.50

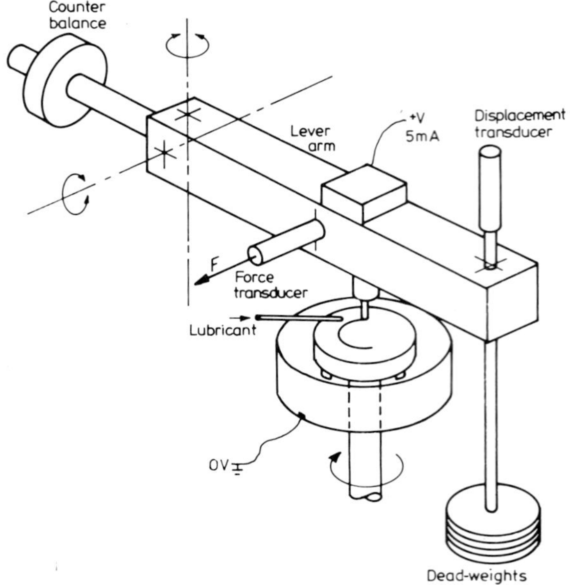

Fig. 1. Pin-on-disc wear machine, showing the measuring points for friction and wear.

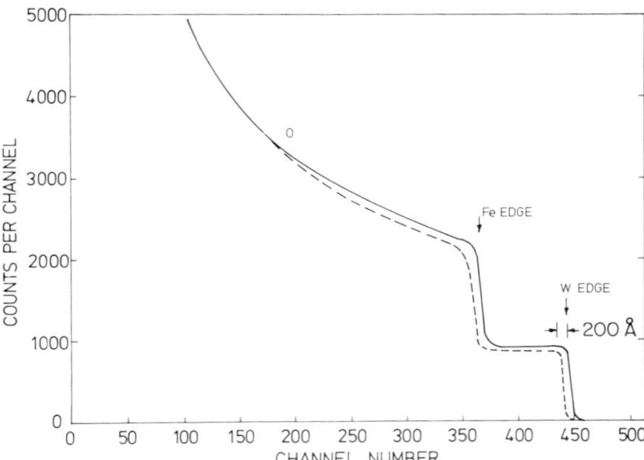

Fig. 2. Rutherford backscattering spectra of 1.5 MeV helium ions backscattered from a specimen implanted with 10^{17} N_2^+ cm^{-2} (- - -) and an unimplanted specimen (——).

Friction and wear were investigated using a pin-on-disc machine (Fig. 1). The pin was 1.5 mm in diameter and the applied loads were 1.0 kgf and 3.0 kgf, making the hertzian contact pressures 0.57 MN m^{-2} and 1.7 MN m^{-2} respectively. The lubricant, white spirit, was circulated at 4 l min^{-1}. Hardness measurements were done at a load of 10 gf applied for 60 s.

The nitrogen implantation was performed on the 500 keV Science and Engineering Research Council accelerator at the University of Surrey. The vacuum pressure was 10^{-6} Torr. Rutherford backscattering analysis was carried out on a Van de Graaff accelerator, also in the Electrical Engineering Department at the University of Surrey, using 1.5 MeV helium ions. Detailed examination of the specimen implanted with 10^{17} N_2^+ cm^{-2} was carried out using X-ray photoelectron spectroscopy on a Vacuum Generator ESCA 3 Mark II electron spectrometer, in the Materials Science Department at the University of Surrey. The operating analyzer pass energy was 50 eV and the vacuum pressure was 10^{-10} Torr; the argon ions were accelerated to 6 kV.

3. RESULTS AND DISCUSSION

Figure 2 gives the Rutherford backscattering spectra of the unimplanted specimen and the specimen implanted with 10^{17} $N_2{}^+$ cm^{-2}. It can be seen that there is no oxygen on the surface and that nitrogen implantation has resulted in a shift of the iron and tungsten edges to the left by about 200 Å. This suggests the formation of a non-metallic film on the surface. It is not an oxide film since there is no detectable oxygen. The X-ray photoelectron spectra in Figs. 3 and 4 indicate the surface to consist mainly of carbon; the low friction results at this dose suggest that it is in the form of graphite.

There was very little change in the hardness of specimens implanted with less than 10^{15} N_2^+ cm^{-2} (Fig. 5), but there was a fourfold increase in hardness at the highest dose. Heat treatment at 200 °C for 5 h after the implantation produced no further change. This temperature was chosen so as not to affect the other mechanical properties, such as toughness, of the steel. It is also fairly close to the temperature reached during the implantation process. So, the fact that annealing had no effect on hardness indicates that an equilibrium phase was reached during the nitrogen implantation. The large scatter in the hardness measurements at 10^{17} N^+ cm^{-2} is probably associated with large carbide precipitates on the surface, but more work is needed to verify this.

Figure 6 shows that only the highest dose reduced friction by (20 ± 5)%. This might be due to the graphite on the specimen. Graphite is well known to have a low friction [8]. However, the reduction in friction and the large

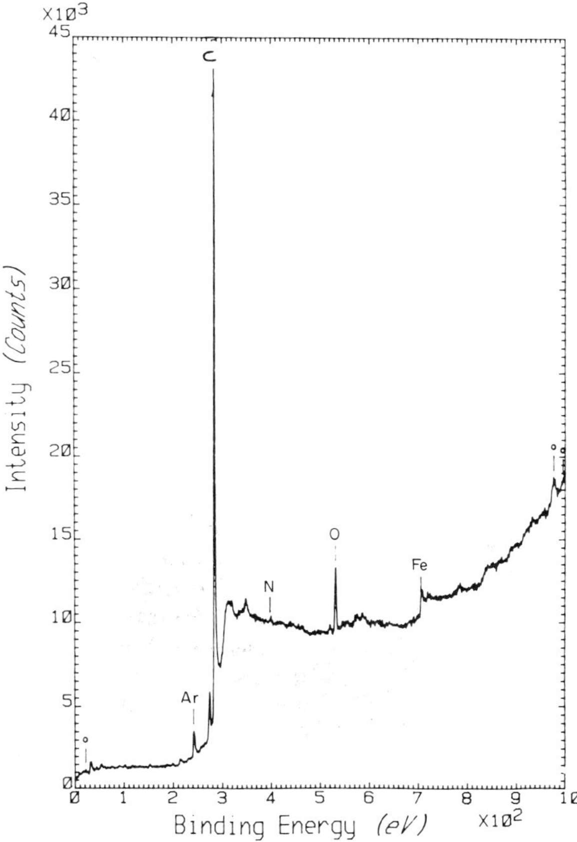

Fig. 3. X-ray photoelectron spectrum from the specimen implanted with 10^{17} N_2^+ cm^{-2} (obtained using aluminium radiation).

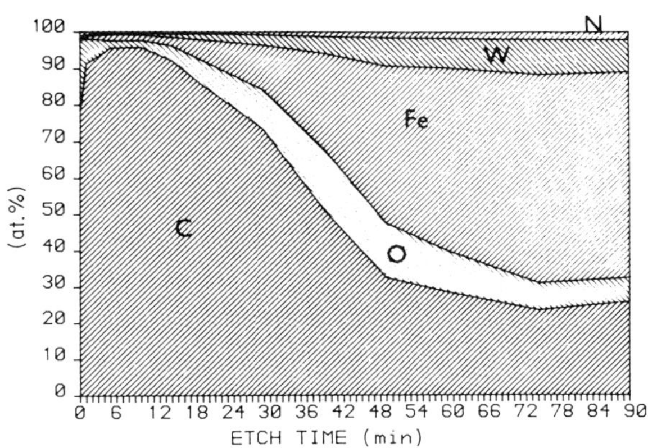

Fig. 4. X-ray photoelectron depth profile of the specimen implanted with 10^{17} N_2^+ cm^{-2}.

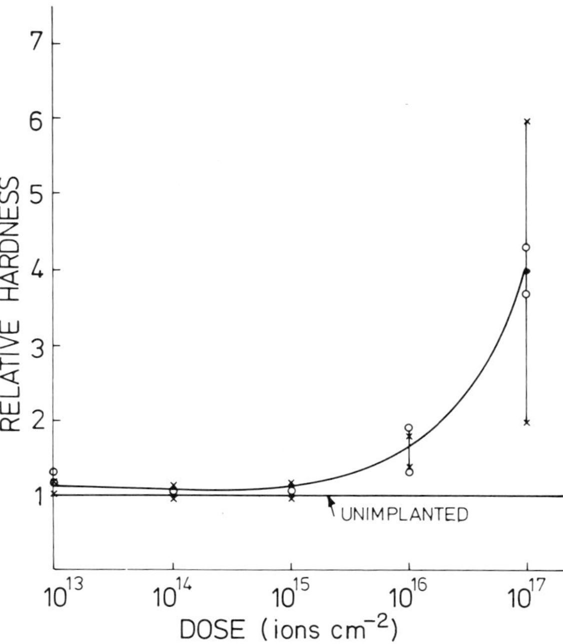

Fig. 5. Variations in hardness with nitrogen ion dose for 400 keV nitrogen ions (N_2^+) implanted into 18W–4Cr–1V steel: ×, as implanted; ○, after heat treatment at 200 °C for 5 h.

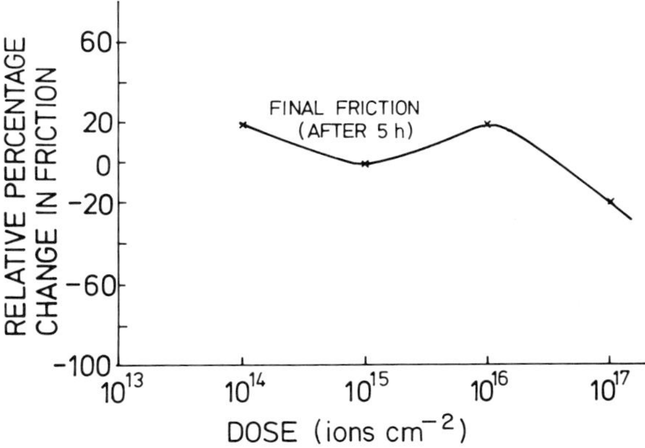

Fig. 6. Changes in friction with nitrogen ion dose for nitrogen ions (N_2^+) implanted into 18W–4Cr–1V steel.

increase in hardness only increases wear resistance by a factor of 2.4, for an applied load of 1 kgf (Fig. 7). As suggested above, the carbon on the surface was in the form of large carbides which, because they were loosely held to the steel, would come off easily. This would increase the abrasive component of wear and hence reduce the wear resistance. At 3.0 kgf, no improvement was observed; the benefits of nitrogen implantation were probably removed in the early stages of the wear process.

It would be wrong to suggest that the observed increase in wear resistance of the im-

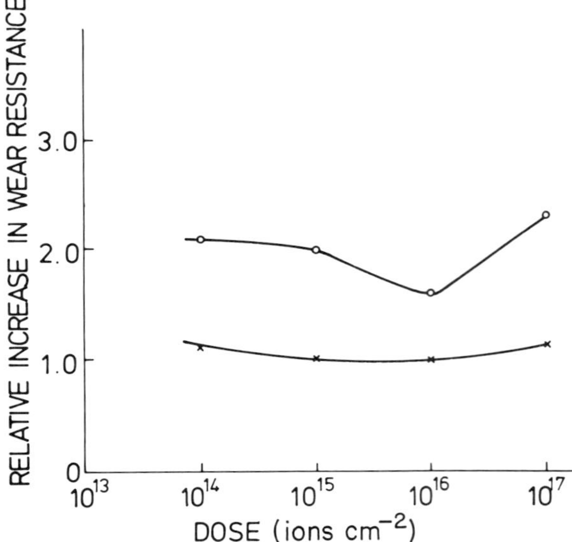

Fig. 7. Plots of relative change in wear resistance as a function of nitrogen ion dose for 400 keV nitrogen ions (N_2^+) implanted into 18W–4Cr–1V steel, showing the effect of the applied load: ○, load of 1 kgf; ×, load of 3 kgf.

planted surface, which persisted beyond the depth of penetration of the ions (about 0.5 μm), is due to inward migration of the nitrogen ions. The steel investigated contains iron, vanadium and tungsten, all of which have a strong affinity for nitrogen. There is a very strong likelihood that they would trap mobile nitrogen atoms to form stable nitrides which would strengthen the steel by precipitation hardening. The general picture is that the graphite layer on the surface would initially resist wear, principally because of the low friction. When this layer is worn away, resistance to wear would be maintained by the implanted region, which has been hardened. The network of dislocations created by the implantation process and, as a result of the precipitation and the wear process, would continue to harden the wear track after the implanted region has been removed to give a continuous resistance to wear. However, all this is possible only if the applied load is light. Heavy loads would generate large shear stresses, well above those that could be withstood by the implanted region, to give rise to severe plastic deformation on and below the wear surface. This explains why the effects of ion implantation could only be observed for the lowest contact stress.

4. CONCLUSIONS

Nitrogen implantation into 18W–4Cr–1V bearing steel increases the hardness by four times and the wear resistance by a factor of more than 2; it reduces friction by 20%. These changes are considered to be associated with the formation of nitride precipitates which strengthen the steel by dislocation pinning and with work hardening which results from the entanglement of dislocations generated during ion implantation and wear.

ACKNOWLEDGMENTS

The work reported in this paper was done at King's College, University of London. The author is grateful to Dr. C. Jeynes and Dr. J. Watts, University of Surrey, for the nitrogen implantation and Rutherford backscattering and X-ray photoelectron spectroscopy analyses, to RHP Limited for the provision of the specimens and to the Science and Engineering Research Council (U.K.) for the award of a grant for the research.

REFERENCES

1 P. B. Madakson and G. Dearnaley, in G. K. Wolf, W. A. Grant and R. P. M. Procter, *Proc. Int. Conf. on Surface Modification of Metals by Ion Beams, Heidelberg, September 17–21, 1984*, in *Mater. Sci. Eng., 69* (1985) 155.

2 P. B. Madakson, in G. K. Wolf, W. A. Grant and R. P. M. Procter, *Proc. Int. Conf. on Surface Modification of Metals by Ion Beams, Heidelberg, September 17–21, 1984*, in *Mater. Sci. Eng., 69* (1985) 167.

3 G. Dearnaley and N. E. W. Hartley, *Proc. 4th Conf. on the Scientific and Industrial Applications of Small Accelerators, Denton, TX, 1975*, IEEE, New York, 1976, p. 20.

4 S. Lo Russo, P. Mazzoldi, I. Scotoni, C. Tosello and S. Tosto, *Appl. Phys. Lett., 34* (1979) 627.

5 N. E. W. Hartley, in J. K. Hirvonen (ed.), *Ion Implantation Treatise Mater. Sci. Technol., 18* (1980) 321.

6 G. Dearnaley, F. J. Minter, P. K. Rol, A. Saint and V. Thompson, in J. W. Mayer (ed.), *Proc. 4th Int. Conf. on Ion Beam Modification of Materials, Cornell University, Ithaca, NY, July 16–20, 1984*, in *Nucl. Instrum. Methods, B 7/8* (1985) 188.

7 P. B. Madakson, *Ph.D. Thesis*, University of London, 1983.

8 F. P. Bowden and D. Tabor, *The Friction and Lubrication of Solids*, Part 1, Clarendon, Oxford, 1950; *The Friction and Lubrication of Solids*, Part 2, Clarendon, Oxford, 1964.

The Improvement in the Wear Properties of GCr15–QBe2 Wear Pairs by Ion Implantation*

ZHAO JIE[a], SU YAWEN[a], LU GUANGYUAN[a], LIU FU RUN[a], YE WEIYI[a] and LI WANG[b]

[a]*Department of Physics, Tianjin Normal University, Tianjin (China)*
[b]*North China Research Institute, Tianjin (China)*

(Received July 10, 1986)

ABSTRACT

We have studied the influence of overlapping nitrogen ion implantation and recoil implantation on the wear resistance of GCr15 bearing steel and QBe2 beryllium bronze wear pairs.

GCr15 bearing steel specimens were implanted with nitrogen ions (N^+) at 120, 70 and 30 keV in sequence at the same fluence of 1×10^{17} ions cm^{-2}. For the recoil case, after the surface of the GCr15 steel specimens had been coated with 300 Å of chromium, nitrogen implantation was performed at energies of 30 and 70 keV at a fluence of 2×10^{17} ions cm^{-2}.

A series of wear measurements were carried out on the specimens under two different loads applied with a pin-on-disk wear tester. During the wear testing, an unimplanted QBe2 beryllium bronze pin was fixed on the tester as the pin and the implanted GCr15 bearing steel specimen was used as the disk, as usual. In the recoil implantation of GCr15 steel specimens, the improvement in wear resistance is remarkable for both GCr15 and QBe2. The wear resistance of GCr15 steel specimens subjected to overlapping implantation, however, did not change significantly.

In order to investigate the microstructure, Auger electron spectroscopy was used to measure the depth profile of the implanted nitrogen and electron spectroscopy for chemical analysis was also performed. On the surface of the specimens, Cr_2O_3 and various nitrides were found. The existence of these compounds is thought to be the main reason for the improvement in wear properties.

1. INTRODUCTION

Several papers on the modification of the wear of steel–steel and bronze–bronze contact by ion implantation have been published in recent years [1–4]. Sometimes, however, the steel–bronze wear pair is used in technological processes. Therefore, it is very important to improve the wear resistance of the steel–bronze system. Ion implantation was shown to be an effective way of modifying the wear resistance for many wear pairs.

Some results of the wear tests using GCr15 bearing steel implanted with nitrogen ions (N^+) and unimplanted QBe2 beryllium bronze are reported in the present paper. These results are compared with the results of wear tests between unimplanted steel and QBe2 beryllium bronze.

In the present experiment the GCr15 steel specimens were bombarded with an ion beam in two different ways. In one case the specimens were directly implanted with nitrogen ions (N^+); in the other case the specimens were first coated with 300 Å of chromium and then implanted with nitrogen ions (N^+) (recoil implantation).

In order to obtain a uniform distribution, the nitrogen ion implantation was performed using three different energies, namely 120, 70 and 30 keV sequentially. After nitrogen ion implantation, the wear test, Auger electron spectroscopy (AES) and electron spectroscopy for chemical analysis (ESCA) were carried out to investigate the influence of the surface modification on the wear resistance and its mechanism.

*Paper presented at the International Conference on Surface Modification of Metals by Ion Beams, Kingston, Canada, July 7–11, 1986.

0025-5416/87/$3.50

2. EXPERIMENTAL METHODS

(1) For overlapping nitrogen ion (N^+) implantation the GCr15 bearing steel specimens were implanted with nitrogen ions at 120, 70 and 30 keV in sequence with the same fluence of 1×10^{17} ions cm^{-2} to obtain a relatively uniform nitrogen ion distribution over the penetrative depth of the ions into the surface of the specimens.

(2) For the recoil implantation the GCr15 steel specimens were coated with 300 Å of chromium and then implanted with nitrogen ions (N^+) at energies of 30 and 70 keV with a fluence of 2×10^{17} ions cm^{-2}. The reason for choosing these energies in the recoil implantation case is as follows.

(a) In order to obtain good mixing between the chromium coating and the iron matrix, the energy chosen should correspond to the maximum $d\epsilon/d\rho$ in the universal curve of $d\epsilon/d\rho$ *vs.* ϵ, to reach a high nuclear stopping power.

(b) The mean projected range of nitrogen ions (N^+) should be nearly equal to the thickness of the chromium film coating, to obtain a mixing layer between the chromium and iron.

(c) Because of the effect of sputtering, some chromium is lost from the surface. The thickness of the chromium film coating therefore has to be thick enough.

When the above three factors (a)-(c) were taken into account, a suitable energy was found to be about 30 keV in accordance with the Lindhard-Scharff-Schiøtt theory [5]. To compare the effect of implantation at different energies, a 70 keV ion beam was also used.

(3) Next let us consider the conditions under which the wear tests were carried out. They were performed with a pin-on-disk wear tester [6]. The pin of QBe2 beryllium bronze was 2 mm in diameter, with a taper of 120° and an initial diameter of 0.20 mm. The pins were loaded with 400 gf in the axial direction and for some measurements the load was increased to 1200 gf. For the disk the speed of rotation was 640 rev min^{-1}. A light machine oil lubricant such as sewing machine oil was used in all the tests.

(4) All specimens were analysed using a multifunctional electron spectrometer (PHI-550) by means of AES and ESCA.

3. RESULTS

3.1. *Wear test results*

The wear curve between the GCr15 steel disk implanted with nitrogen and the unimplanted QBe2 beryllium bronze is shown in Fig. 1 and indicates that the wear of the QBe2 beryllium bronze pin was obviously reduced by implantation of the steel disk. Under a metallographic microscope, it could be seen that there was some adhesion of QBe2 on the unimplanted GCr15 disk, whereas there was no obvious adhesion on the nitrogen-implanted disks. From this, we concluded that nitrogen ion (N^+) implantation improved the adhesive wear.

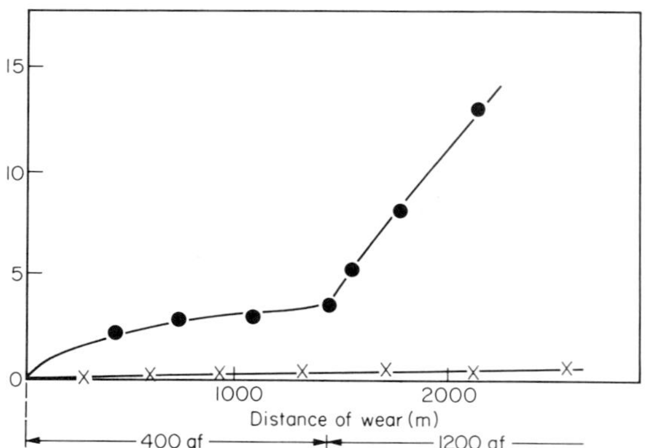

Fig. 1. Wear curve of a QBe2 pin on a disk after overlapping nitrogen ion (N^+) implantation: ●, pin on an unimplanted disk; ×, pin on an overlapped implanted disk.

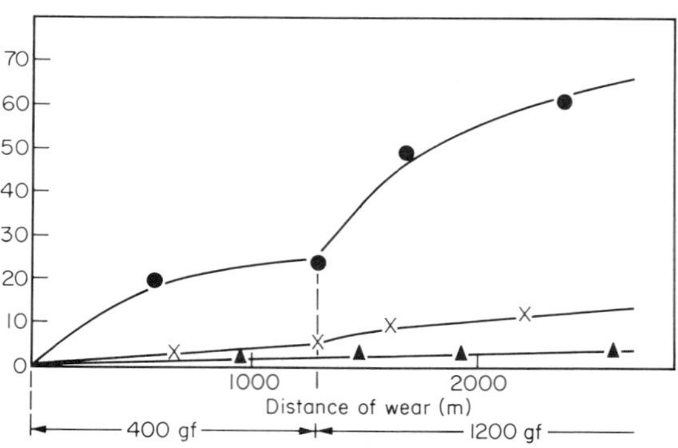

Fig. 2. Wear curve of a QBe2 pin on a disk after nitrogen ion (N^+) recoil implantation: ●, pin on an unimplanted disk; ×, pin on a disk coated with chromium and implanted with 30 keV nitrogen ions; ▲, pin on a disk coated with chromium and implanted with 70 keV nitrogen ions.

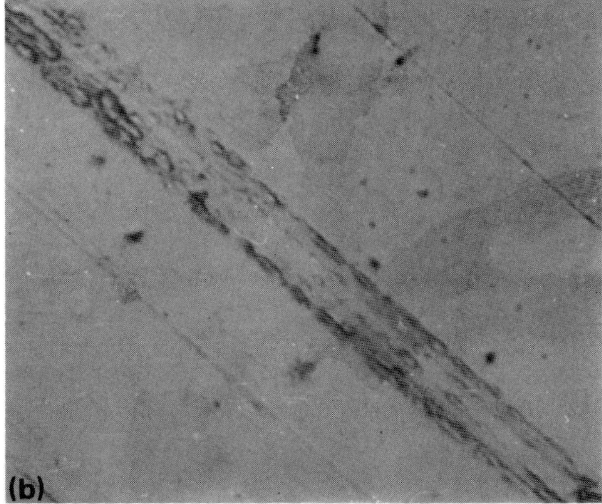

Fig. 3. Micrographs of the wear trace on the surfaces of specimens: (a) a wear trace on the surface of a specimen coated with chromium and implanted with nitrogen ions (N^+) at 30 keV; (b) a wear trace on an unimplanted specimen.

Figures 2 and 3 show that the wear resistances of both QBe2 beryllium bronze pins and GCr15 steel disks coated with chromium were improved significantly. It can be observed from the micrographs in Fig. 3 that the wear trace on unimplanted disks is very obvious with many trace lines; however, it is very difficult to find any wear lines on the recoil-implanted disks.

The wear modification coefficients of the pin for various implantation conditions are listed in Table 1, where $\tan\alpha_{un}$ and $\tan\alpha_{im}$ stand for the slopes of the straight-line part of curves for the unimplanted specimens and implanted specimens respectively.

TABLE 1

The wear modification coefficient *K* for various specimens

Implantation conditions				*Load* (gf)	*K*
Ion	*Coating*	*Energy* (keV)	*Fluence* ($\times 10^{17}$ ions cm^{-2})		
N^+	Cr	30	2	400	5.0
N^+	Cr	30	2	1200	2.1
N^+	Cr	70	2	400	4.3
N^+	Cr	70	2	1200	11.0
N^+		120, 70, 30	1 (each)	400	8.0
N^+		120, 70, 30	1 (each)	1200	14.5

$K = (\tan\alpha_{un})/(\tan\alpha_{im})$ is calculated in accordance with the slope of the straight-line part of the curves.

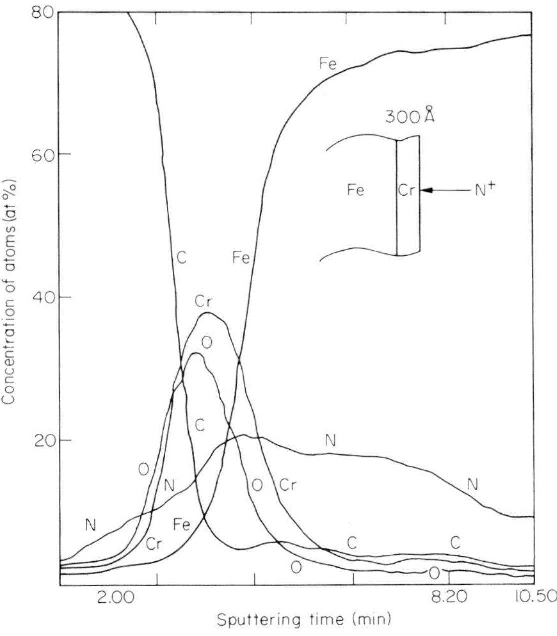

Fig. 4. AES profile analysis of GCr15 steel specimens after nitrogen ion (N^+) recoil implantation.

3.2. *The results obtained by Auger electron spectroscopy and electron spectroscopy for chemical analysis*

The AES profile results for specimens coated with chromium and bombarded with nitrogen ions (N^+) at 30 keV are shown in Fig. 4, from which the extent of mixing of chromium and iron can be found; however, because of the light mass of nitrogen, the mixing of chromium and iron is only slight. In Fig. 5 the profile for the overlapping nitrogen ion implantation can be seen. It indicates that the distribution of nitrogen is relatively uniform over a wide region. Therefore the overlapping implantation can be used in cases in which a uniform distribution of an implanted elements is required [7].

In addition, ESCA indicated that there is Cr_2O_3, CrN, (graphitic) carbon and iron in various states in the implanted layer of the specimen [8], as shown in Fig. 6.

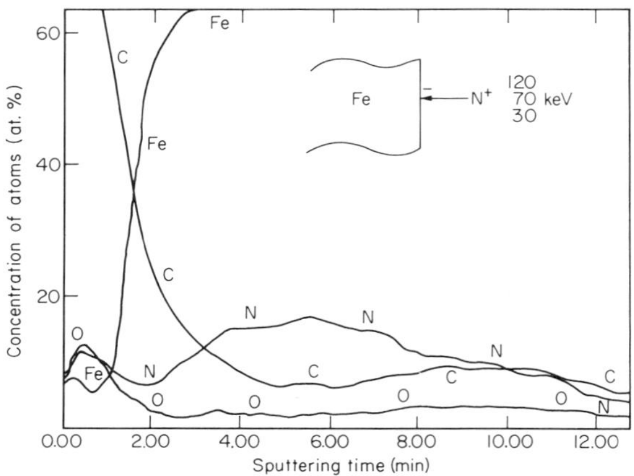

Fig. 5. AES profile analysis of GCr15 steel specimens after overlapping nitrogen ion (N^+) implantation at energies of 120 and 30 keV.

4. DISCUSSION AND CONCLUSIONS

(1) After recoil implantation of GCr15 steel specimens, the improvement in wear resistance was remarkable for both GCr15 steel and QBe2 beryllium bronze. However, for the GCr15 steel subjected to overlapping nitrogen ion implantation, the improvement in the wear resistance of the QBe2 pin slipping on the disk is obvious, but the wear resistance of the steel disk has not changed. In both cases the adhesive wear was significantly improved.

(2) With increasing ion energy (from 30 to 70 keV) the wear resistance of GCr15 steel specimens was hardly changed, but the depth for surface modification at 70 keV may be deeper than that at an implantation energy of 30 keV.

(3) The wear behaviour of GCr15–QBe2 wear pairs after nitrogen ion implantation varies with load. From Figs. 1 and 2, it can be observed that the wear resistance of the pin under a load of 400 gf is better than that under a load of 1200 gf.

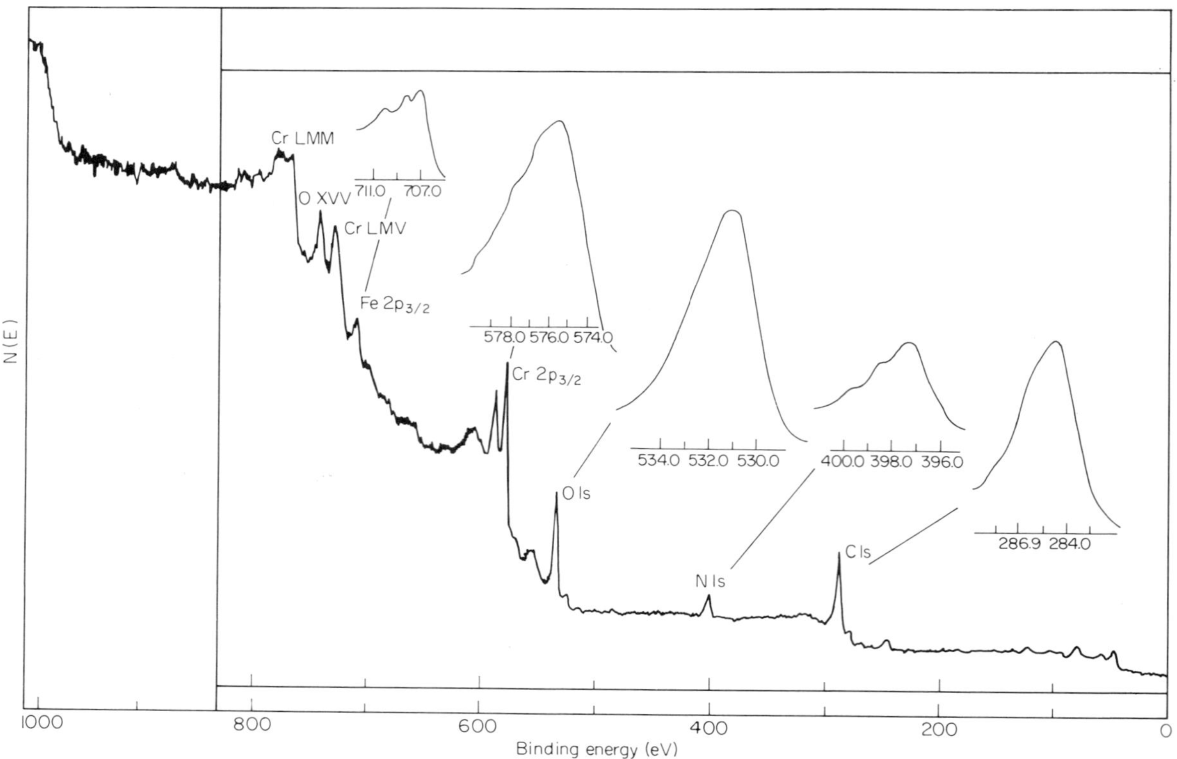

Fig. 6. ESCA of GCr15 steel specimen coated with chromium and bombarded with nitrogen ions (N^+).

(4) According to ESCA, there may be various reasons for the modification of specimens coated with chromium and implanted with nitrogen ions. Graphitic carbon could play the role of lubricant, and the existence of hard phases such as Cr_2O_3, CrN and FeN on the surface is probably the main reason for the improvement [9].

Since the load and lubricant conditions in our experiments were more severe than in normal practice, we believe that the wear resistance of GCr15–QBe2 wear pairs could be greatly improved for practical uses by nitrogen ion recoil implantation.

ACKNOWLEDGMENTS

We wish to thank Mr. Li Yongcan for his skill in the preparation of specimens and Mr. Luo Yingming for the ion implantation.

REFERENCES

1 C. Fuzhai, L. Hengde and Z. Xiaozhong, *Nucl. Instrum. Methods, 209–210* (1983) 881.
2 S. Yawen, L. Guangyuan and L. Furun, *Proc. Natl. Conf. on Electron Beam, Ion Beam and Laser Beam Methods*, China, 1984.
3 A. Kujore, S. B. Chakrabortty, E. A. Starke and K. O. Legg, *Nucl. Instrum. Methods, 182–183* (1981) 949.
4 S. Saritas, R. P. M. Procter, V. Ashworth and W. A. Grant, *Wear, 82* (2) (1982) 233.
5 P. D. Townsend, J. C. Kelly and N. E. W. Hartley, *Ion Implantation, Sputtering and their Applications*, Academic Press, New York, 1976.
6 J. K. Hirvonen, *J. Vac. Sci. Technol., 15* (5) (1978) 1662.
7 J. K. Hirvonen, *Proc. 1st Int. Conf. on Ion Beam Modification of Materials*, Central Research Institute of Physics, Budapest, 1978, p. 1753.
8 C. D. Wagner, *Handbook of ESCA*, Perkin Elmer, 1979.
9 I. L. Singer, C. A. Carosella and J. R. Reed, *Nucl. Instrum. Methods, 182–183* (1981) 923.

Materials Science and Engineering, 90 (1987) 297–306

Use of Ion Implantation to Modify the Tribological Properties of Ti–6Al–4V*

[a], R. P. M. PROCTER[b] and W. A. GRANT[c]

[a] *Department of Mechanical Engineering, Middle East Technical University, Gaziantep (Turkey)*

[b] *Corrosion and Protection Centre, University of Manchester Institute of Science and Technology, Manchester (U.K.)*

[c] *Department of Electronic and Electrical Engineering, University of Salford, Salford (U.K.)*

(Received July 10, 1986)

ABSTRACT

The effects of boron, carbon, nitrogen and oxygen ion implantation on the friction properties and wear behaviour of Ti-6Al-4V have been investigated under low and high contact stress conditions. With low contact stresses, ion implantation results in some increase in the incubation period for measurable wear mass loss. However, the effects are less than those produced by conventional surface alloying with oxygen and the subsequent steady state wear rates are not altered. The effects are attributed to an increase in the surface microhardness produced by ion implantation rather than to a change in the wear mechanism, which remains predominantly abrasive. With high contact stresses, however, boron implantation increases the incubation period for measurable wear from a sliding distance of 10 m to a sliding distance of 90 m and is far more effective than other implant species and conventional surface treatments. There is again no change in the steady state wear rate or the wear mechanism although in this case the effect is clearly not attributable to an increase in the surface microhardness.

1. INTRODUCTION

Titanium and its alloys have good corrosion resistance and an attractive strength-to-mass ratio but their wear resistance is relatively poor. There are conventional surface treatments available to improve the wear resistance of titanium and its alloys but these generally impair the fatigue resistance [1–14]. Ion implantation, however, offers the possibility both of improving the wear resistance and of improving, or at least not impairing, the fatigue resistance [15–31].

The objective of the present work, therefore, was to identify a range of implant species and parameters that affect the wear behaviour. In a subsequent study, which is currently under way, the effects of the same implant species and parameters on fatigue behaviour are being examined. Eventually these studies will make it possible to select an implantation treatment that optimizes both wear and fatigue resistance. The material chosen for this research was Ti-6Al-4V, a widely used high strength titanium alloy.

2. EXPERIMENTAL PROCEDURE

The wear tests reported in this paper were complemented by friction and microhardness measurements, surface profilometry, scanning electron microscopy and Auger electron spectroscopy (AES).

Two unlubricated wear regimes have been studied: low stress wear obtained with a cylindrical titanium alloy pin 2.5 mm in diameter wearing on a titanium alloy disc under a contact stress of 7.6 MN m^{-2} and high stress wear obtained with a cemented WC ball 6.4 mm in diameter wearing on a titanium alloy disc under a contact stress of 1070 MN m^{-2}. In both cases the relative linear velocity was 10.5

*Paper presented at the International Conference on Surface Modification of Metals by Ion Beams, Kingston, Canada, July 7–11, 1986.

0025-5416/87/$3.50

mm s^{-1} during the friction measurements and 105 mm s^{-1} during the wear measurements. The wear was measured by stopping the tests every 5 min and determining the masses of the pins and the discs.

The pins and discs were machined from as-received (mill-annealed) Ti–6Al–4V rod. The microstructure of this material, which had a Vickers hardness of 370 HV, is shown in Fig. 1. One face of each disc was polished to a mirror finish using alumina paste and then implanted separately with boron, carbon, nitrogen and oxygen ions. In all cases the implantation energy was 40 keV and the fluence was generally 2×10^{17} ions cm^{-2} (Table 1). The vacuum in the target chamber was better than 10^{-5} Torr and the specimens were cooled with a liquid nitrogen cold-finger during implantation.

Discs were tested in the as-received and as-implanted conditions. In addition, discs were examined after oxidizing in air at 600 °C for 5 h and after annealing in a vacuum of 10^{-6} Torr for 5 h.

Microhardness measurements were made using a 10 gf load which gave indentations with about 5 μm diagonals. Argon ion beam milling was used during the AES analyses to obtain depth profiles of the implanted species.

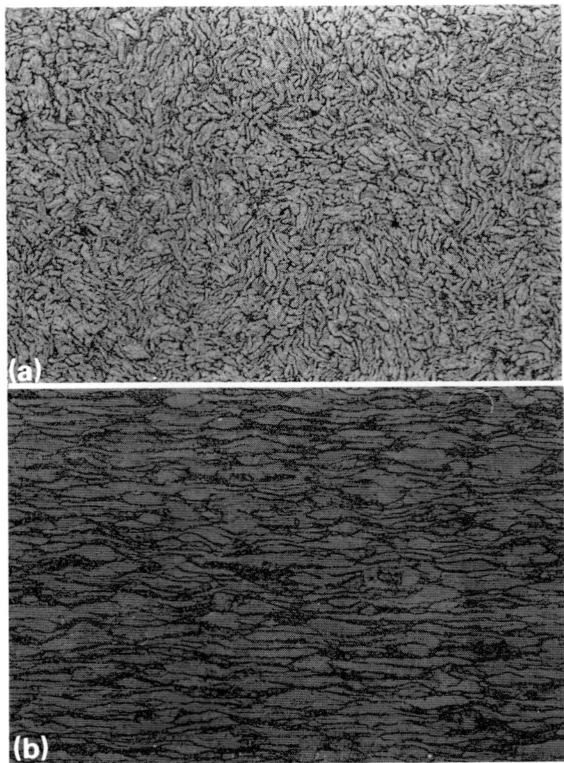

Fig. 1. Microstructure of as-received Ti–6Al–4V which had been polished and etched with Krolls reagent: (a) transverse section; (b) longitudinal section. (Magnifications, 400×.)

TABLE 1

Implantation and microhardness data

Implant species	*Implant energy* (keV)	*Fluence* (ions cm^{-2})	*Heat treatment*	*Mean projected range* (nm)	*Peak concentration* (at.%)	*Microhardness* (HV)	*Surface appearance*
Unimplanted	—	—	As received	—	—	376 ± 9	Mirror finish
Unimplanted	—	—	Annealed	—	—	704 ± 10	Straw colour
Unimplanted	—	—	Oxidized	—	—	957 ± 11	Bluish purple colour
O^+	40	2×10^{17}	None	20	40	504 ± 14	As polished
N^+	40	2×10^{17}	None	20	35	444 ± 12	As polished
N^+	40	2×10^{17}	Annealed	20	35	504 ± 11	Straw colour
N^+	40	5×10^{17}	None	Not measured		525 ± 11	Straw colour
C^+	40	2×10^{17}	None	20	30	500 ± 14	As polished
C^+	40	2×10^{17}	Annealed	20	30	638 ± 12	Straw colour
C^+	40	5×10^{17}	None	Not measured		579 ± 13	Straw colour
B^+	40	2×10^{17}	None	20	15	417 ± 8	As polished
B^+	40	2×10^{17}	Annealed	20	15	527 ± 12	Straw colour

3. RESULTS AND DISCUSSION

3.1. *Auger electron spectroscopy analyses*

Figure 2, Fig. 3, Fig. 4 and Fig. 5 show the AES depth profile analyses for implanted oxygen, nitrogen, carbon and boron respectively. The following points emerge from these data. The thickness of the air-formed oxide layer on Ti–6Al–4V is much less than the mean projected range of the implant species (of the order of 20 nm (Table 1)) and is not affected by ion implantation. Vacuum annealing of as-received Ti–6Al–4V results in some inward diffusion of oxygen but this process is apparently reduced by the presence of implanted boron and prevented by the presence of implanted carbon and nitrogen. Vacuum annealing results in no significant diffusion of implanted carbon and nitrogen and only slight inward diffusion of implanted boron. The oxidation treatment results in the growth of a relatively thick (of the order of micrometres) rutile (TiO_2) scale. Finally, there was significant inward diffusion of oxygen into a specimen which inadvertently was not cooled during implantation (Fig. 4). AES

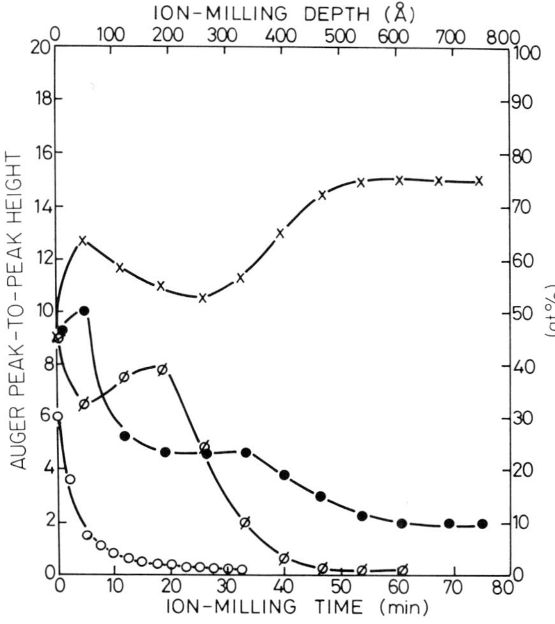

Fig. 2. Auger oxygen peak-to-peak height against argon ion-milling time for various Ti–6Al–4Y specimens: ○, as received; ϕ, oxygen implanted; ×, oxidized in air; ●, vacuum annealed.

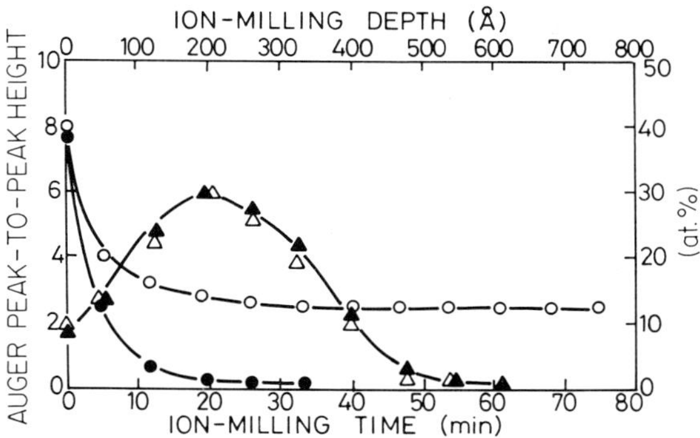

Fig. 4. Auger peak-to-peak height against argon ion-milling time for C-implanted specimens: ○, oxygen, uncooled; ●, oxygen, annealed; △, carbon, as implanted; ▲, carbon, annealed.

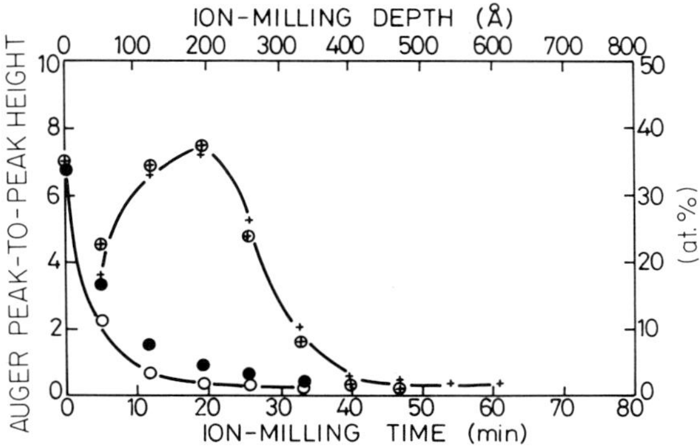

Fig. 3. Auger peak-to-peak height against argon ion-milling time for N-implanted specimens: ○, oxygen, as implanted; ●, oxygen, annealed; +, nitrogen, as implanted; ⊕, nitrogen, annealed.

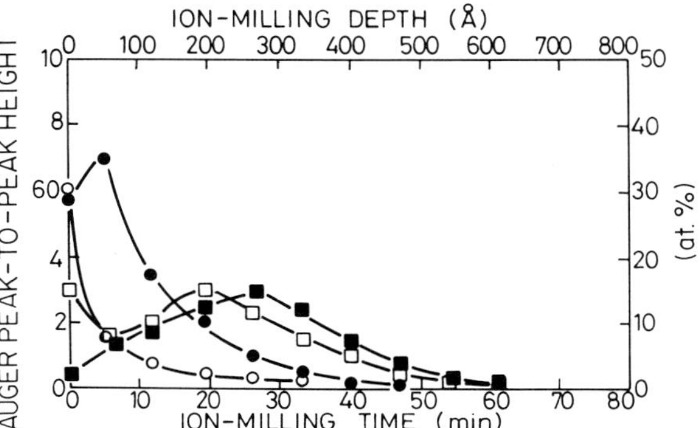

Fig. 5. Auger peak-to-peak height against argon ion-milling time for B-implanted specimens: ○, oxygen, as implanted; ●, oxygen, annealed; □, boron, as implanted; ■, boron, annealed.

analyses also showed that carbon contamination during implantation was slight or shallow. No attempt was made to quantify the AES analyses but calculations show that the peak concentrations of all the implanted species significantly exceed the respective equilibrium solid solubilities of the implant species in titanium.

Figure 6 shows the Ti(418 eV)-to-Ti(386 eV) relative Auger peak-to-peak height ratio as a function of depth for the various implant species. The significance of these curves derives from the fact that the Ti(418 eV) peak is sensitive to the chemical state of the titanium, being sharp and narrow with unbound titanium and broad and diffuse with bound titanium; the Ti(386 eV) peak height, however, is not affected by the chemical state. The Ti(418 eV)-to-Ti(386 eV) peak height ratio therefore provides a measure of the degree of chemical bonding of the titanium. Figure 6 shows that the peak height ratio for as-received titanium rises rapidly to a value of 1.5, which is characteristic of metallic titanium. In contrast, the ratio for oxidized Ti–6Al–4V remains constant with depth at 0.8; this value is characteristic of titanium in the oxide TiO_2. The most important point to be noted in Fig. 6 is that even in the as-implanted condition there appears to be considerable chemical binding between the implant species and the titanium lattice atoms. Specifically the location and depth of the minima on the various peak height ratio curves correlate well with the mean projected range and the estimated peak concentrations of the corresponding implant species. Furthermore the degree of chemical binding becomes more pronounced after post-implantation vacuum annealing.

3.2. Coefficient of friction

The variation in the coefficient of friction with sliding distance for the various surface treatments is shown in Fig. 7 and Fig. 8 (and also in Table 2) for low contact stress conditions and high contact stress conditions respectively. During low stress pin-on-disc testing the coefficient of friction, with two exceptions, increases within two turns (a sliding distance of about 150 mm) from an initial relatively low value to a value of about 0.48. However, with oxidized and as-boron-implanted specimens the coefficient of friction remains relatively low (about 0.2) for a sliding distance of 3 m while the coefficient of friction on boron-implanted-plus-vacuum-annealed specimens does not increase to a value of 0.48 until a sliding distance of 14 m.

During high stress ball-on-disc testing the beneficial effect of boron implantation in maintaining a low coefficient of friction and delaying the onset of adhesion is even more pronounced. As shown in Fig. 8, the coefficient of friction generally increases to a value of about 0.48 within 1 m and even for oxidized

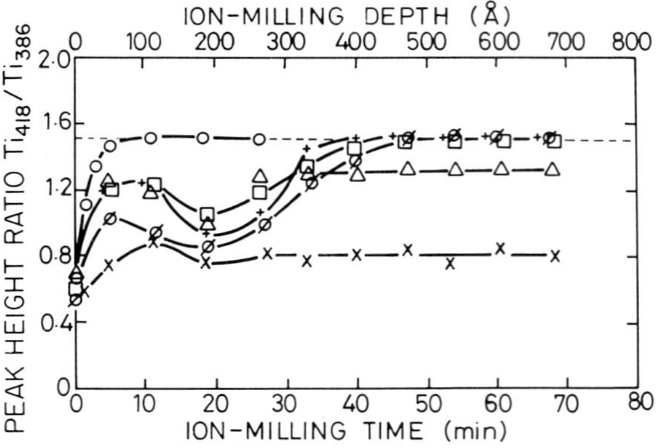

Fig. 6. Ti(418 eV)-to-Ti(386 eV) Auger peak-to-peak height ratio against argon ion-milling time for various Ti–6Al–4V specimens: ○, as received; ×, oxidized; ⌀, implanted with 2×10^{17} O^+ cm^{-2}; +, implanted with 2×10^{17} N^+ cm^{-2}; △, implanted with 2×10^{17} C^+ cm^{-2}; □, implanted with 2×10^{17} B^+ cm^{-2}.

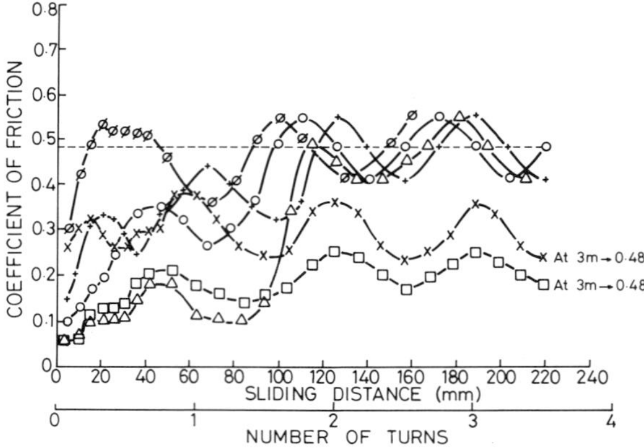

Fig. 7. Coefficient of friction against sliding distance for various Ti–6Al–4V specimens under low contact stress conditions: ○, as received; ×, oxidized; ⌀, implanted with 2×10^{17} O^+ cm^{-2}; +, implanted with 2×10^{17} N^+ cm^{-2}; △, implanted with 2×10^{17} C^+ cm^{-2}; □, implanted with 2×10^{17} B^+ cm^{-2}.

discs within 5 m. However, boron implantation delays the increase in the coefficient of friction for a sliding distance of up to 78 m.

3.3. *Wear*

Figures 9–12 and Table 2 show the pin wear under low contact stress conditions as a function of sliding distance for the various surface treatments. The following conclusions may be drawn from these results. Generally, measurable pin wear is only observed after an incubation period which varies from a sliding distance of 40 m to a sliding distance of 150 m, depending on the surface treatment. However, oxidized discs show immediate mass losses due to adhesion and spalling of the thick rutile scale. The most important conclusions to be drawn from Figs. 9–12 are that, although ion implantation treatments can delay the onset of measurable wear from a sliding distance of about 40 m for as-received Ti–6Al–4V up to a sliding distance of 100 m for carbon-implanted discs, the subsequent wear rate is not changed.

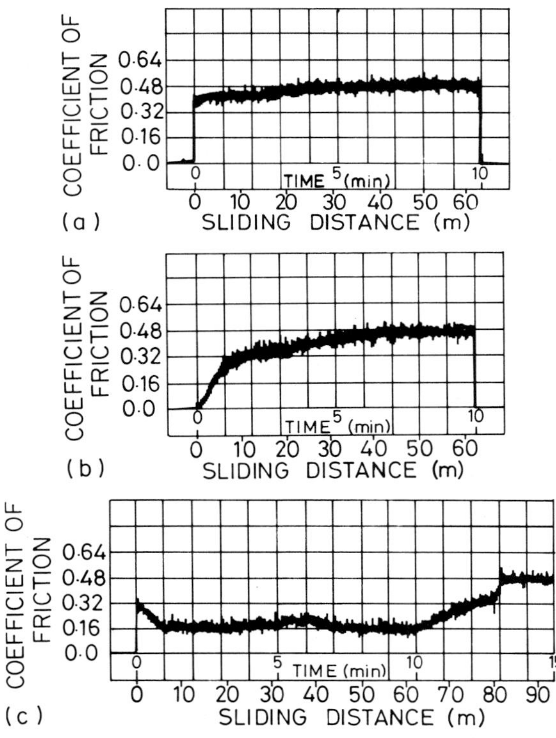

Fig. 8. Coefficient of friction against sliding distance under high contact stress conditions: (a) as-received and carbon-, nitrogen- and oxygen-implanted specimens; (b) oxidized specimens; (c) boron-implanted specimens.

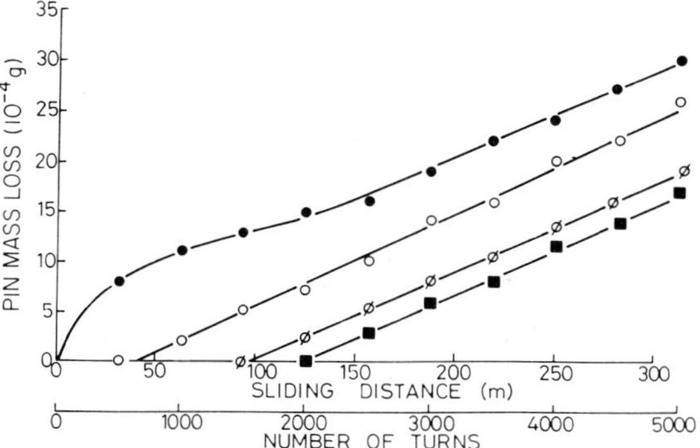

Fig. 9. Pin wear against sliding distance for various Ti–6Al–4V discs under low contact stress conditions: ○, as received; ●, oxidized; ■, annealed; ∅, implanted with 2×10^{17} O^+ cm^{-2}.

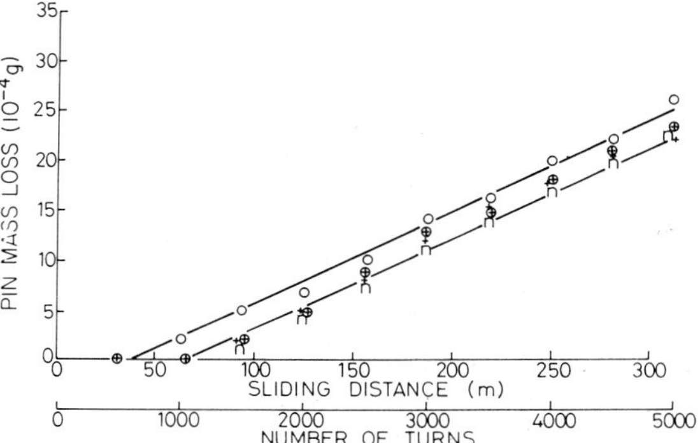

Fig. 10. Pin wear against sliding distance for various Ti–6Al–4V discs under low contact stress conditions: ○, as received; +, implanted with 2×10^{17} N^+ cm^{-2} as implanted; ⊕, implanted with 2×10^{17} N^+ cm^{-2}, annealed; n, implanted with 5×10^{17} N^+ cm^{-2}, as implanted.

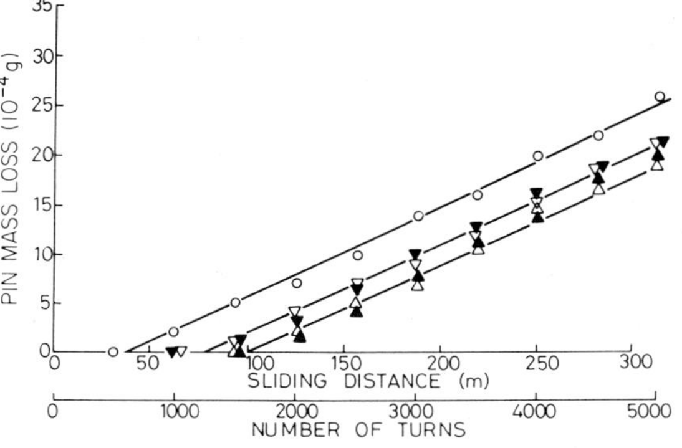

Fig. 11. Pin wear against sliding distance for various Ti–6Al–4V discs under low contact stress conditions: ○, as received; △, implanted with 2×10^{17} C^+ cm^{-2}, as implanted; ▽, implanted with 2×10^{17} C^+ cm^{-2}, as implanted, uncooled; ▲, implanted with 2×10^{17} C^+ cm^{-2}, annealed; ▼, implanted with 5×10^{17} C^+ cm^{-2}, as implanted.

Furthermore, none of the implantation treatments is as effective as vacuum annealing of as-received Ti–6Al–4V, which results in the formation of a hard titanium alloy–oxygen surface layer and delays the onset of measurable wear for a sliding distance of up to about 130 m. Again, however, the subsequent sliding-distance-based gravimetric wear rate is not affected. In summary, therefore, there is no evidence in this system of a persistent effect or that the implanted species are transported into the substrate ahead of the wear damage, as has been suggested for other systems.

Figure 13 shows the disc wear under high contact stress conditions as a function of sliding distance for the various surface treatments. In this case, measurable mass losses are observed with all surface treatments except boron

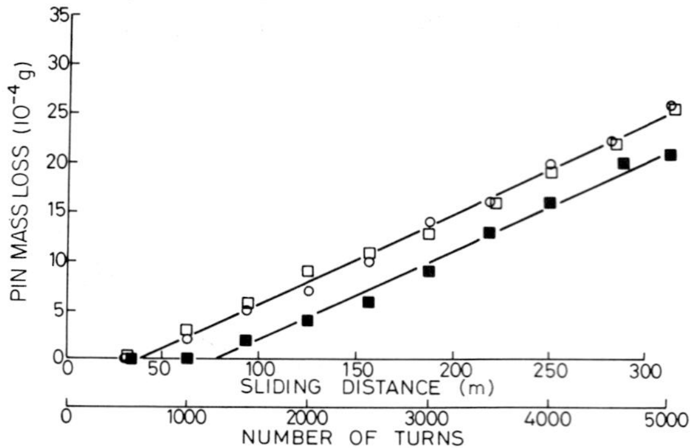

Fig. 12. Pin wear against sliding distance for various Ti–6Al–4V discs under low contact stress conditions: ○, as received; □, implanted with 2×10^{17} B^+ cm^{-2}, as implanted; ■, implanted with 2×10^{17} B^+ cm^{-2}, annealed.

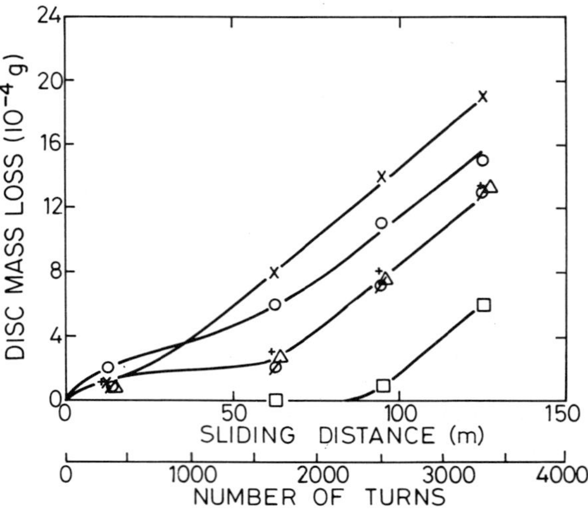

Fig. 13. Disc wear against sliding distance for various discs under high contact stress conditions: ○, as received; ×, oxidized; ϕ, implanted with 2×10^{17} O^+ cm^{-2}; +, implanted with 2×10^{17} N^+ cm^{-2}; △, implanted with 2×10^{17} C^+ cm^{-2}; □, implanted with 2×10^{17} B^+ cm^{-2}.

TABLE 2

Pin-on-disc friction and wear data under 7.6 MPa Hertzian stresses

Implant species	*Fluence at 40 keV* (ions cm^{-2})	*Heat treatment*	*Initial coefficient of friction*	*Number of turns to onset of wear*	*Wear of pin at a sliding distance of 300 m* ($\times 10^{-4}$ g)	*Reduction in wear at 300 m* (%)
Unimplanted	—	As received	0.10	660	26	—
Unimplanted	—	Annealed	0.13	2000	16	38.4
Unimplanted	—	Disc oxidized	—	60	29	11.5 increase
O^+	2×10^{17}	None	0.30	1600	18	30.7
N^+	2×10^{17}	None	0.15	1060	21	19.2
N^+	2×10^{17}	Annealed	0.18	1060	23	11.5
N^+	5×10^{17}	None	—	1060	21	19.2
C^+	2×10^{17}	None	0.08	1250	21	19.2
C^+	2×10^{17}	Not cooled	0.08	1600	19	26.9
C^+	2×10^{17}	Annealed	0.23	1600	20	23.0
C^+	5×10^{17}	None	—	1250	21	19.2
B^+	2×10^{17}	None	0.08	660	26	0.0
B^+	2×10^{17}	Annealed	0.09	1250	21	19.2

implantation after a sliding distance of only 10 m. This decrease in the incubation period with increased contact stress is not unexpected. What is surprising, however, is that boron implantation increases the incubation period under high stress conditions from a sliding distance of 10 m to a sliding distance of about 90 m. This very significant increase is likely to be of greatest practical importance in service applications in which little or no wear can be tolerated because the eventual steady state wear rate is again not affected by any of the surface treatments. As would be expected, the steady state wear rate measured under high contact stress conditions is considerably greater than that observed under low contact stress conditions.

While mass loss measurements were reasonably sensitive for the pins, they were less so for the much heavier (100 times) discs. In the high contact stress tests, therefore, the gravimetric wear measurements were supplemented by surface profilometry. The results of these measurements, which are presented in Fig. 14, confirm that boron has a marked beneficial effect in delaying the onset of wear and show that oxidation also increases the wear rate under high contact stress conditions.

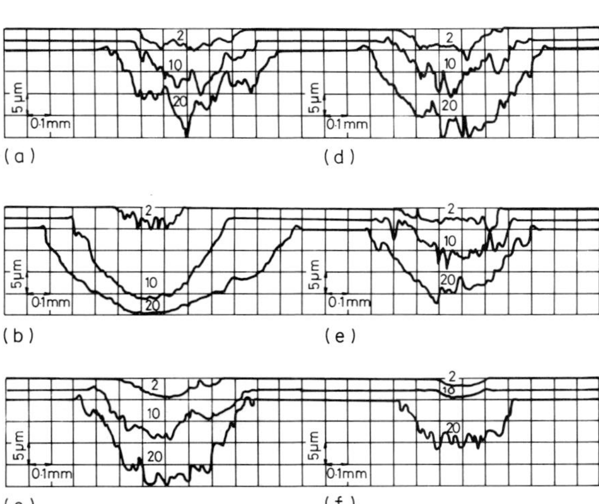

Fig. 14. Disc wear scar profiles under high contact stress conditions after 2 min, 10 min and 20 min of wear testing (after 340 turns, 1700 turns and 3400 turns respectively equivalent to sliding distances of 13 m, 65 m and 130 m respectively): (a) as-received disc; (b) oxidized disc; (c) oxygen-implanted disc; (d) nitrogen-implanted disc; (e) carbon-implanted disc; (f) boron-implanted disc.

3.4. Microhardness

Figures 15 and 16 show that there is generally a direct correlation, under low contact stress test conditions, between the microhardness of the Ti–6Al–4V discs after the various surface treatments, the incubation period for measurable wear and the wear mass loss after a sliding distance of 300 m. The only exception to this general trend is oxidized Ti–6Al–4V which shows a very high surface microhardness because of the presence of the rutile scale but which also shows a relatively poor wear resistance owing to spalling of this scale. These results suggest that the changes in the wear behaviour are merely due to the changes in the surface microhardness and that under low contact stress test conditions the basic wear mechanism is not affected by ion implantation. This conclusion is supported by scanning electron microscopy (SEM) studies of the worn surfaces. In all cases the wear is predominantly (greater than 80%) abrasive, as shown in Fig. 17; however, there is also some contribution from adhesive wear (Fig. 18) and delamination (Fig. 19). The most

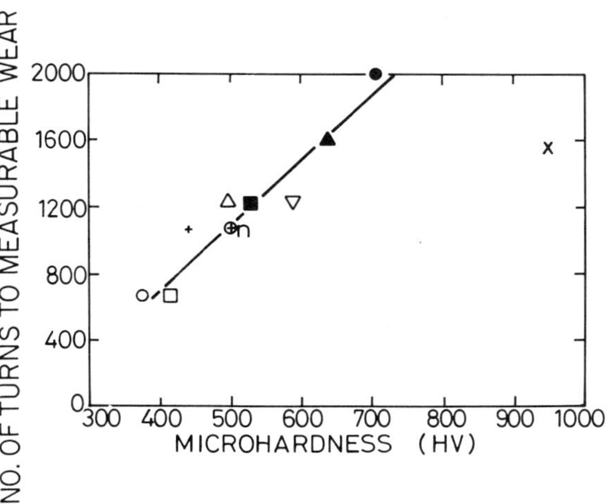

Fig. 15. Incubation period for measurable pin wear under low contact stress conditions against surface microhardness for the surface conditions investigated: ○, as received; ×, oxidized; ●, vacuum annealed; +, implanted with 2×10^{17} N^+ cm^{-2}, as implanted; n, implanted with 5×10^{17} N^+ cm^{-2}, as implanted; △, implanted with 2×10^{17} C^+ cm^{-2}, as implanted; ▲, implanted with 2×10^{17} C^+ cm^{-2}, annealed; ▽, implanted with 2×10^{17} C^+ cm^{-2}, as implanted, uncooled; □, implanted with 2×10^{17} B^+ cm^{-2}, as implanted; ⊕, implanted with 2×10^{17} N^+ cm^{-2}, annealed; ■, implanted with 2×10^{17} B^+ cm^{-2}, annealed.

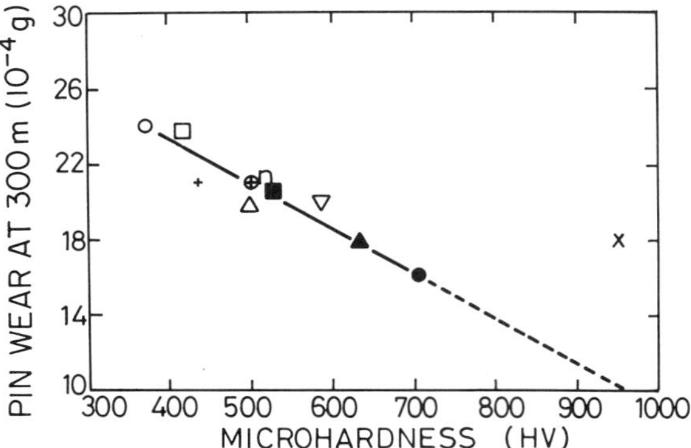

Fig. 16. Pin mass loss after a sliding distance of 300 m under low contact stress conditions against surface microhardness for the surface conditions investigated: ○, as received; ×, oxidized; ●, vacuum annealed; +, implanted with 2×10^{17} N^+ cm^{-2}, as implanted; n, implanted with 5×10^{17} N^+ cm^{-2}, as implanted; △, implanted with 2×10^{17} C^+ cm^{-2}, as implanted; ▲, implanted with 2×10^{17} C^+ cm^{-2}, annealed; ▽, implanted with 2×10^{17} C^+ cm^{-2}, as implanted, uncooled; □, implanted with 2×10^{17} B^+ cm^{-2}, as implanted; ■, implanted with 2×10^{17} B^+ cm^{-2}, annealed; ⊕, implanted with 2×10^{17} N^+ cm^{-2}, annealed.

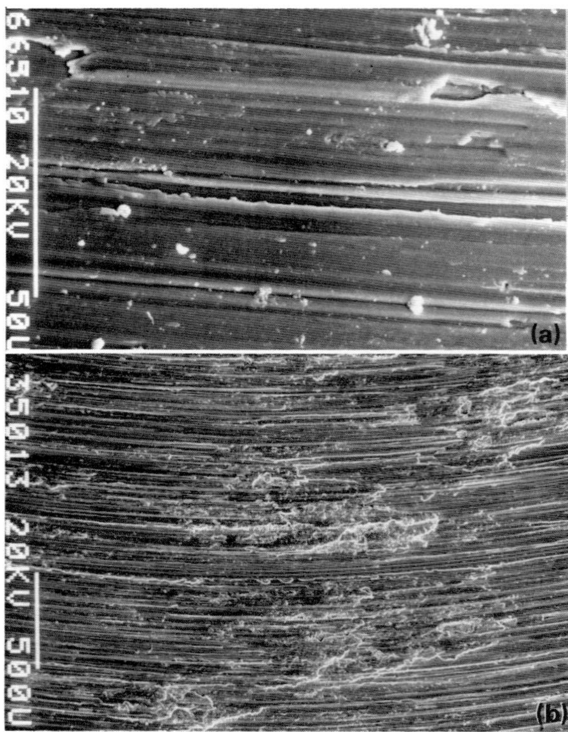

Fig. 17. (a) Scanning electron micrograph of the disc wear scar after a sliding distance of 300 m under low contact stress conditions, oxygen-implanted disc, predominantly abrasive wear; (b) scanning electron micrograph of the worn pin after a sliding distance of 300 m under low contact stress conditions, boron-implanted disc, abrasive wear. (Magnifications (a) 35×; (b) 860×.)

important point, however, is that no morphological differences were apparent between as-received, vacuum annealed and implanted specimens.

By contrast, in high contact stress wear, as shown in Fig. 20, there is no clear correlation between surface microhardness and disc wear after a sliding distance of 125 m. In this case the data fall into three groups. Firstly, oxidized Ti–6Al–4V specimens, although being very hard, again show particularly severe wear because of oxide spalling. Secondly, for the as-received, vacuum-annealed and most implanted specimens there is no real correlation between microhardness and wear mass loss. Finally, as-boron-implanted Ti–6Al–4V specimens, although showing relatively little increase in surface microhardness (from 376 to 417 HV), nevertheless show significantly less wear. This suggests that the mechanism of wear may be different in the latter case. How-

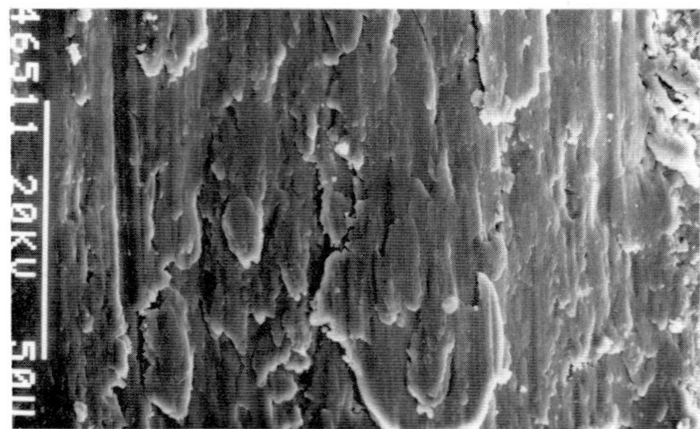

Fig. 18. Scanning electron micrograph of the worn pin after a sliding distance of 300 m under low contact stress conditions, nitrogen-implanted disc, adhesive wear. (Magnification, 860×.)

Fig. 19. Scanning electron micrograph of the worn pin after a sliding distance of 300 m under low contact stress conditions, carbon-implanted disc, delamination. (Magnification, 860×.)

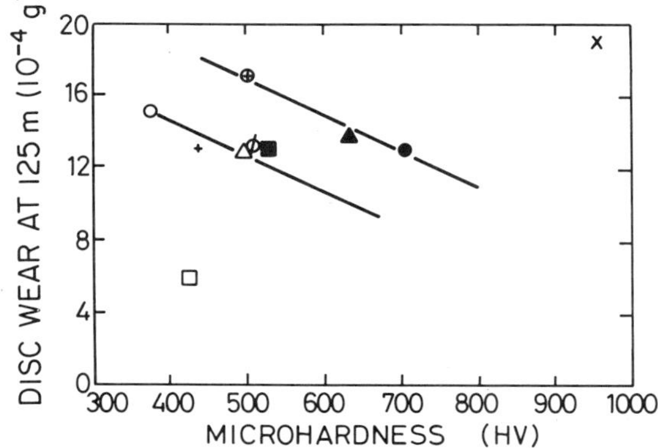

Fig. 20. Disc mass loss after a sliding distance of 125 m under high contact stress conditions against surface microhardness for the surface conditions investigated: ○, as received; ×, oxidized; ϕ, implanted with 2×10^{17} O^+ cm^{-2}, as implanted; ●, vacuum annealed; +, implanted with 2×10^{17} N^+ cm^{-2}, as implanted; △, implanted with 2×10^{17} C^+ cm^{-2}, as implanted; ▲, implanted with 2×10^{17} C^+ cm^{-2}, annealed; □, implanted with 2×10^{17} B^+ cm^{-2}, as implanted.

ever, this hypothesis is not supported by SEM examination of the worn surfaces. Again, the oxidized specimens show severe abrasive wear as a result of spalling of the oxide scale while all the other specimens show a mixture of predominantly abrasive plus some adhesive wear. The effect of boron implantation therefore appears to be primarily to delay the onset of adhesion.

4. CONCLUSIONS

(1) The effect of ion implantation on the wear of Ti–6Al–4V depends not only on the implant species and parameters but also on the wear regime.

(2) Under relatively mild wear conditions (low contact stresses), ion implantation delays somewhat the onset of wear. However, the effects are less than those produced by vacuum annealing and do not persist; the eventual steady state wear rates are similar in all cases.

(3) The effects of the various surface treatments on the incubation period are directly related to changes in the surface microhardness.

(4) Under moderately severe wear conditions (high contact stresses) a more significant and specific effect is observed. Boron ion implantation maintains a low coefficient of friction and increases the incubation period for measurable wear from a sliding distance of 10 m to a sliding distance of 90 m.

(5) In both cases the wear mechanism is predominantly abrasive and is not affected by ion implantation.

REFERENCES

1 A. E. Jenkins, *J. Inst. Met., 82* (1954) 213.
2 J. E. Reynolds, H. R. Ogden and R. I. Jaffee, *Trans. Am. Soc. Met., 49* (1956) 327.
3 K. C. Anthony, *J. Mater., 1* (1966) 456.
4 R. W. Hansel, *Met. Prog., 65* (1954) 90.
5 E. Mitchell and P. J. Brotherton, *J. Inst. Met., 93* (1965) 381.
6 R. D. Weltzin and G. Koves, *ASTM Spec. Tech. Publ. 432*, 1968, p. 283.
7 A. G. Lucas, *Light Met. Age, 14* (1956) 11.
8 A. Takamura, *Trans. Jpn. Inst. Met., 3* (1962) 10.
9 J. L. Wyatt and N. J. Grant, *Iron Age, 173* (1954) 124.
10 J. R. Cuthill, W. D. Hayes and R. E. Seybold, *J. Res. Natl. Bur. Stand., Sect. A, 64* (1964) 119.
11 R. H. Shoemaker, in R. I. Jaffee and H. M. Burke (eds.), *Proc. 2nd Int. Conf. on Titanium Science and Technology, Cambridge, MA, 1972*, Plenum, New York, 1973, p. 2501.
12 K. T. Rie and T. H. Lampe, in G. K. Wolf, W. A. Grant and R. P. M. Procter, *Proc. Int. Conf. on Surface Modification of Metals by Ion Beams, Heidelberg, September 17–21, 1984, Mater. Sci. Eng., 69* (1985) 473.
13 T. S. Eyre and H. Al-Salim, *Proc. Int. Conf. on Wear of Materials*, American Society of Mechanical Engineers, New York, 1977, p. 344.
14 N. E. W. Hartley, *Tribol. Int., 8* (1975) 65.
15 G. Dearnaley, *Mater. Eng. Appl., 1* (1978) 28.
16 S. R. Sherpard and N. P. Suh, *J. Lubr. Technol., 104* (1982) 29.
17 J. K. Hirvonen, in G. K. Hubler, O. W. Holland, C. R. Clayton and C. W. White (eds.), *Ion Implantation and Ion Beam Processing of Materials, Materials Research Society Symp. Proc., Boston, MA, November 14–17, 1983*, Vol. 27, Elsevier, New York, 1984, p. 621.
18 I. L. Singer, in G. K. Hubler, O. W. Holland, C. R. Clayton and C. W. White (eds.), *Ion Implantation and Ion Beam Processing of Materials, Materials Research Society Symp. Proc., Boston, MA, November 14–17, 1983*, Vol. 27, Elsevier, New York, 1984, p. 585.
19 H. Dimigen, K. Kobs, R. Leutenecker, H. Ryssel and P. Eichinger, in G. K. Wolf, W. A. Grant and R. P. M. Procter, *Proc. Int. Conf. on Surface Modification of Metals by Ion Beams, Heidelberg, September 17–21, 1984*, in *Mater. Sci. Eng., 69* (1985) 150.
20 P. B. Madakson, in G. K. Wolf, W. A. Grant and R. P. M. Procter, *Proc. Int. Conf. on Surface Modification of Metals by Ion Beams, Heidelberg, September 17–21, 1984*, in *Mater. Sci. Eng., 69* (1985) 167.
21 S. Saritas, R. P. M. Procter, V. Ashworth and

W. A. Grant, *Wear, 82* (1982) 233.
22 T. Bridge, R. P. M. Procter, V. Ashworth and W. A. Grant, in R. Kossowsky and S. C. Singhal (eds.), *Proc. Conf. on Surface Engineering, Les Arcs, July 1983*, in *NATO Adv. Study Inst. Ser. D.* (1983) p. 275.
23 H. Herman, *Nucl. Instrum. Methods, 182* (1981) 887.
24 R. Hutchings and W. C. Oliver, *Wear, 92* (1983) 143.
25 J. M. Williams, G. M. Beardsley, R. A. Buchanan and R. K. Bacon, in G. K. Hubler, O. W. Holland, C. R. Clayton and C. W. White (eds.), *Ion Implanta-ation and Ion Beam Processing of Materials, Materials Research Society Symp. Proc., Boston, MA, November 14-17, 1983*, Vol. 27, Elsevier, New York, 1984, p. 735.
26 W. C. Oliver, R. Hutchings, J. B. Pethica, E. L. Paradis and A. J. Shuskus, in G. K. Hubler, O. W. Holland, C. R. Clayton and C. W. White (eds.), *Ion Implantation and Ion Beam Processing of Materials, Materials Research Society Symp. Proc., Boston, MA, November 14-17, 1983*, Vol. 27, Elsevier, New York, 1984, p. 705.
27 R. Martinella, G. Chevallard and C. Tosello, in G. K. Hubler, O. W. Holland, C. R. Clayton and C. W. White (eds.), *Ion Implantation and Ion Beam Processing of Materials, Materials Research Society Symp. Proc., Boston, MA, November 14-17, 1983*, Vol. 27, Elsevier, New York, 1984, p. 711.
28 R. G. Vardiman, in G. K. Hubler, O. W. Holland, C. R. Clayton and C. W. White (eds.), *Ion Implantation and Ion Beam Processing of Materials, Materials Research Society Symp. Proc., Boston, MA, November 14-17, 1983*, Vol. 27, Elsevier, New York, 1984, p. 699.
29 R. Martinella, S. Giovanardi, G. Chevallard, M. Villani, A. Molinari and C. Tosello, in G. K. Wolf, W. A. Grant and R. P. M. Procter, *Proc. Int. Conf. on Surface Modification of Metals by Ion Beams, Heidelberg, September 17-21, 1984*, in *Mater. Sci. Eng., 69* (1985) 247.
30 J. M. Williams and R. A. Buchanan, in G. K. Wolf, W. A. Grant and R. P. M. Procter, *Proc. Int. Conf. on Surface Modification of Metals by Ion Beams, Heidelberg, September 17-21, 1984*, in *Mater. Sci. Eng., 69* (1985) 237.
31 R. G. Vardiman and R. A. Kant, *J. Appl. Phys., 53* (1982) 690.

Friction and Wear of Ion-beam-modified Ceramics for use in High Temperature Adiabatic Engines*

WILLIAM WEI[a], JAMES LANKFORD[a] and RAM KOSSOWSKY[b]

[a] *Department of Materials Sciences, Southwest Research Institute, San Antonio, TX (U.S.A.)*

[b] *Applied Research Laboratory, State College, Pennsylvania State University, PA (U.S.A.)*

(Received July 10, 1986)

ABSTRACT

An experimental program has been conducted to investigate the friction and wear behavior of ceramic materials being considered for use in high temperature adiabatic engines. Pin-on-disk-type tests were conducted in a simulated diesel environment at 800 °C using ceramic-ceramic couples made up of SiC, TiC and TiC-Ni-Mo pins tested against Si_3N_4 *and partially stabilized* ZrO_2 *disk surfaces modified by ion beam mixing with cobalt, chromium, nickel or Ti-Ni. The coefficients of friction for each couple were determined from these tests, and the amount of wear was measured by profilometry of the wear surfaces. Scanning electron microscopy, Auger electron spectroscopy and energy-dispersive spectroscopy were used to characterize the wear surfaces.*

The results of these tests show that most non-ion-mixed ceramic-ceramic couples have relatively high wear rates and coefficients of friction above about 0.25. Ion beam mixing of disk material surfaces with Ti-Ni produced coefficients of friction as low as 0.06, values close to that for liquid-lubricated metal couples run at much lower temperatures. Surface modification with cobalt or nickel produced some promising results, while chromium did not improve the wear characteristics of the disk materials. Auger analysis suggests that the formation of lubricious oxide layers is responsible for the superior friction and wear behaviour of the Ti-Ni-ion-implanted disks.

*Paper presented at the International Conference on Surface Modification of Metals by Ion Beams, Kingston, Canada, July 7-11, 1986.

1. INTRODUCTION

Improvements in low heat rejection engines and the development of a new generation of lighter, more energy-efficient engines and power systems has seen the increased application of ceramic materials to their construction, materials capable of withstanding the high operating temperatures and stresses required in such engines. The high temperature mechanical and chemical stability of many types of ceramic, most notably Si_3N_4 and ZrO_2, has made them candidates for use not only in stationary structural and insulating parts but also, with the recent consideration of all-ceramic engines, in moving parts [1, 2].

One of the major difficulties in the application of ceramics to the moving parts of high temperature engines has been the poor friction and wear performance of these materials. This problem is especially critical to applications such as cylinder liners and piston rings for high temperature adiabatic engines, or high temperature bearings in gas turbine engines, where engine design requirements include high temperatures, uncooled operation, and friction and wear properties of key parts comparable with those of conventional lubricated engines. In the last few years, numerous investigators have explored the friction and wear of ceramic-ceramic couples [3-5]. Although couples exhibiting minimum wear could be identified, wear was never negligible, and the unlubricated sliding friction coefficient μ_F was usually discouragingly high, *i.e.* about 0.2 or more [3-11]. In fact, it has been generally concluded by some investigators that ceramic components will not be used unlubricated in sliding contact engine applications at either low [3] or elevated [4] temperatures.

0025-5416/87/$3.50

While this is a discouraging conclusion, especially in the light of the breakdown of conventional lubricants at elevated temperatures, certain observations made during these tests and other related investigations provide some insight as to how the wear properties of these materials might be improved to acceptable levels. In particular, the test environment has been shown to play a major role in the friction and wear behavior of ceramics. Coefficients of friction and wear rates have been found to be reduced when tests are conducted in oxidizing environments (air or water vapor) compared with tests conducted in inert environments (vacuum or inert gas) [12–14]. Surface analysis of the wear surfaces produced in these environments has shown that the formation of a thin oxide layer during testing in the oxidizing environments apparently provides a form of solid lubrication between the ceramic surfaces and is responsible for the improvement in the fricton and wear behavior of the ceramics tested [12–16]. An especially good example of this behavior is demonstrated by the titanium-based ceramics TiC and Ni–Mo-bonded TiC cermet. When one of these materials is used as a member of a ceramic–ceramic couple, it has been shown that the coefficients of friction of these pairs often are reduced compared with those of other ceramic–ceramic combinations. This reduction is apparently the result of the formation and transfer of a TiO_2 layer from the titanium-based ceramic member to the other member of the couple [5, 17–22].

Oxygen does not necessarily have to be the lubrication-causing species for improved ceramic friction and wear behavior, however. Myristic acid, for instance, enhances surface plasticity and surface cracking during the sliding of steel on LiF [23], while the increased surface segregation of carbon as a graphite film on SiC has been shown to lead to a dramatic decrease in the coefficient of friction at temperatures above 800 °C [24].

From the results of these studies, it appears that one promising direction toward the improvement of the friction and wear behavior of ceramics is to try to enhance these apparent self-lubricating properties. This could be achieved by a modification of the material surfaces which would, for example, enhance the growth of a lubricating oxide film. One such surface modification technique which has shown initial success is the ion beam mixing of certain metals into ceramic surfaces and, in particular, into metals which are known to form stable continuous metal oxides and might induce this type of lubricating behavior [21, 22]. It has been shown that the surface ion mixing of a double layer of titanium and nickel in Si_3N_4 or partially stabilized ZrO_2 has improved the friction and wear properties of these materials in simulated diesel environments (at 800 °C) to levels approaching that of conventional lubricated engines run at lower temperatures [21]. Auger electron spectroscopy (AES) indicates that the modified surface layer is apparently oxidized and provides a lubricating layer between the ceramic surfaces [22]. The objective of this paper is to discuss further results on this preliminary investigation as well as to report on results of the ion beam mixing of other metals including cobalt, chromium and nickel.

2. EXPERIMENTAL PROCEDURE

Details of the friction and wear experiments and subsequent surface analysis have been given in detail elsewhere [21, 22] and are summarized briefly here. Si_3N_4 and partially stabilized ZrO_2 disks were surface modified by the ion beam mixing of cobalt, chromium, nickel or a Ti–Ni double layer. The choices of metal ions to be ion beam mixed were based on the results of unimplanted ceramic–ceramic tests (titanium and nickel) as discussed previously, and on work reported in the literature [25] which seemed to indicate possible lubricating properties for cobalt oxides and Cr_2O_3. The metals were vapor deposited on the ceramic disk substrates and then mixed into the surface by bombardment with argon ions. The argon ions were accelerated using a beam voltage of 140 keV, with a fluence of 10^{17} ions cm^{-2}, and a flux of about 10^{12} ions cm^{-2} s^{-1}. The thicknesses of the modified layers were estimated to be of the order of about 400 nm or less, based on subsequent Auger analysis. The modified disks, as well as unmodified disks, were tested in sliding contact in a three-pin-on-disk arrangement where the pin materials were TiC or an Ni–Mo-bonded TiC cement. Tests were run in a simulated diesel environment or an inert argon atmosphere at 800 °C. The results of these tests are reported primarily in terms of the coefficients of friction of the various couples.

Surface analysis of the wear surfaces of the pins and disks were conducted using scanning electron microscopy (SEM) and AES. The morphology of the wear surfaces was analyzed using SEM, while the chemistry of various features of the wear surfaces was analyzed using AES. The results of these analyses were then correlated with the results of the mechanical friction and wear testing to provide a preliminary mechanism for the lubricating properties of these modified surfaces.

3. RESULTS

3.1. Friction and wear testing

Results of the friction and wear testing of the unmodified and modified ceramic–ceramic couples run in the simulated diesel environment are summarized in Fig. 1 in terms of the steady state coefficient μ_F of friction. It is clear from these results that, while the coefficient of friction for pins run against unmodified (bare) disks range above 0.2, surface modification of the disks by the ion beam mixing of metals reduced the coefficients of friction to values below 0.1 for certain specific cases.

Values of the coefficients of friction for the modified disks are shown in Table 1. The metal ions showing the most promise for use in the improvement of friction and wear behaviour by ion beam mixing appear to be Ti–Ni and cobalt, while chromium and nickel appear to have no effect on the friction and wear behavior of the ceramic couples. The most promising pin–disk combinations include combinations 1 and 4 involving the Ti–Ni modification of Si_3N_4 and partially stabilized ZrO_2, where $\mu_F = 0.09$, and combination 5 involving cobalt modification of partially stabilized ZrO_2, where $\mu_F = 0.06$.

Comparison of the combinations involving the Ti–Ni modification indicates that pin composition may also play a role in the friction and wear behavior of these materials. TiC appears to improve the coefficient of friction of the Ti–Ni-modified Si_3N_4 disk, while the Ni–Mo-bonded TiC cermet appears to improve that of the Ti–Ni-modified ZrO_2. The latter is also the case for the cobalt-modified ZrO_2. However, there was no improvement in the coefficient of friction for either of the cobalt-modified Si_3N_4 disks.

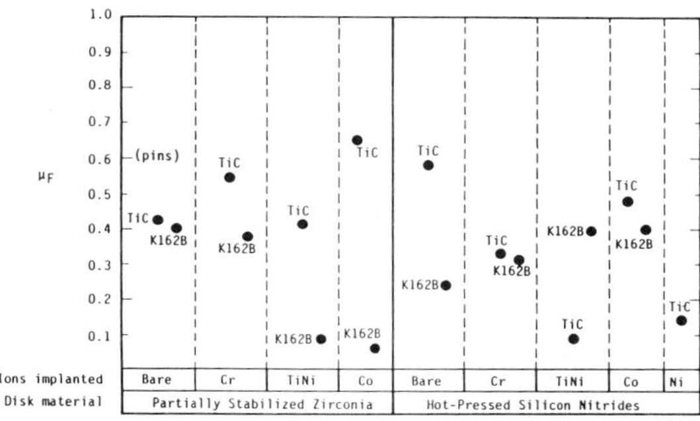

Fig. 1. Steady state coefficient of friction at 800 °C in a diesel environment.

TABLE 1

Coefficients of friction of pin-modified disk combinations tested in a diesel (oxygen, nitrogen, CO_2 and H_2O) environment at 800 °C

Combination	*Disk*[a,b]	*Pin*[c]	μ_F
1	Si_3N_4 (Ti, Ni)	TiC	0.09
2	Si_3N_4 (Ti, Ni)	TiC–5Ni–5Mo	0.22
3	Partially stabilized ZrO_2 (Ti, Ni)	TiC	0.25
4	Partially stabilized ZrO_2 (Ti, Ni)	TiC–5Ni–5Mo	0.09
5	Partially stabilized ZrO_2 (Co)	TiC–5Ni–5Mo	0.06
6	Partially stabilized ZrO_2 (Co)	TiC	>0.25
7	Si_3N_4 (Ni)	TiC	0.14
8	Si_3N_4 (Ni) (tested at room temperature)	TiC	>0.28

[a] Disk diameter, 7.62 cm; disk thickness, 0.95 cm.
[b] The elements given in parentheses are those ion beam mixed into the surfaces of the disks.
[c] Pin diameter, 0.64 cm; pin length, 1.27 cm.

Another major improvement resulting from the surface modification of the ceramic disks by ion beam mixing was the reduction in wear of the pin–disk combinations to the extent that in many cases the amount of wear could not be measured either by mass change or by surface profilometry. For combinations 1–4 involving the Ti–Ni modification, and combinations 7 and 8 involving the nickel modification, the wear track on the disks was barely discernible by eye or in the electron microscope. The morphology of the wear process will be discussed in Section 3.2.

3.2. Surface analysis

Typical results of the morphological and chemical analyses of the wear surfaces of the modified ceramic couples are shown in Figs. 2–5. A typical wear surface from a pin is shown in Fig. 2, while wear surfaces from modified disks are shown in Figs. 3 and 4. Typical Auger elemental depth profiles from the modified disk surfaces are shown in Fig. 5.

The pins from pin–disk combinations 1–6 showed a wear mark similar to that in Fig. 2(a) for the Ni–Mo-bonded cermet pin of combination 4. At a higher magnification (Fig. 2(b)), it appears that this mark is a layer of material which has transferred from the disk to the pin surface during sliding contact. AES analysis shows that this is indeed the case, the transfer layer containing nickel (and possibly titanium) transferred from the Ti–Ni modified disks [22], and cobalt transferred from the cobalt-modi-

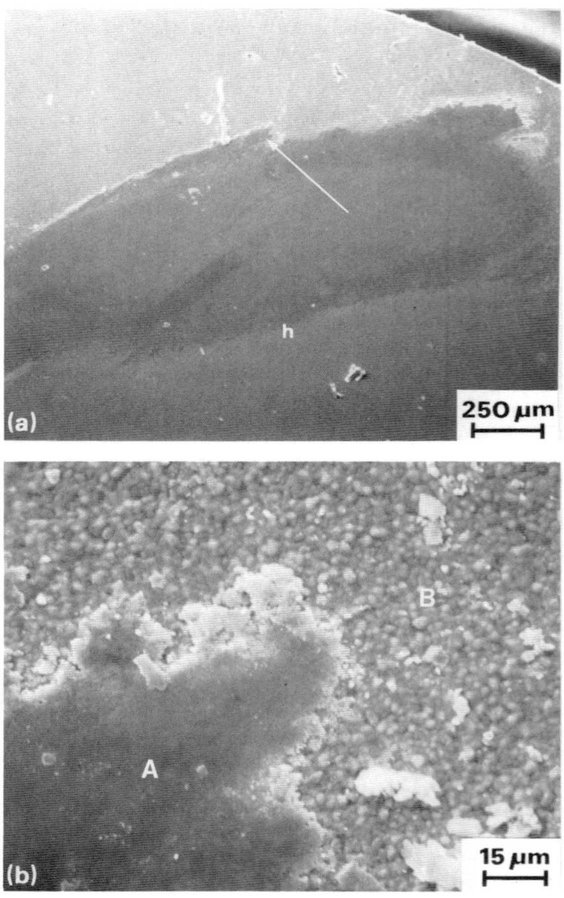

Fig. 2. Wear surface of an Ni–Mo-bonded TiC cermet pin run against a Ti–Ni-modified partially stabilized ZrO_2 disk (diesel environment; 800 °C): (a) the arrow indicates the region magnified in (b); (b) magnification of (a). A indicates the transferred layer of nickel and titanium; B indicates the original pin material. (Magnifications: (a) 40×; (b) 800×.)

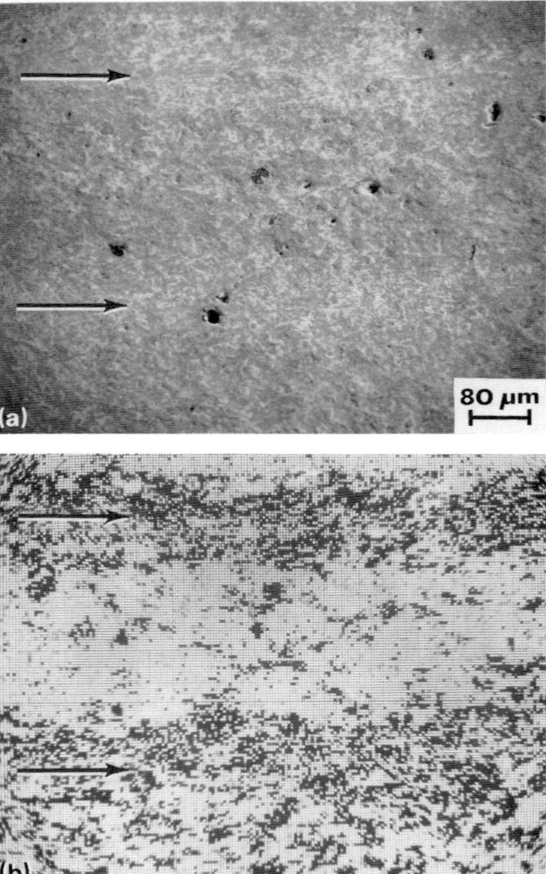

Fig. 3. Wear surface of a Ti–Ni-modified partially stabilized ZrO_2 disk (diesel environment; 800 °C): (a) scanning electron micrograph; (b) Ti(387 eV) elemental map. The dark areas correspond to the exposed ZrO_2 substrate shown by the arrows in (a) and (b).

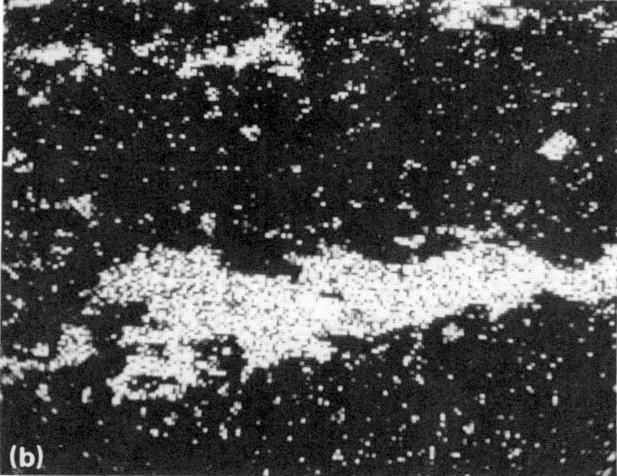

Fig. 4. Wear surface of a cobalt-modified partially stabilized ZrO_2 disk (diesel environment; 800 °C): (a) scanning electron micrograph; (b) Co(53 eV) elemental map. The light areas correspond to cobalt laminates left after friction and wear testing shown by the arrows in (a) and (b).

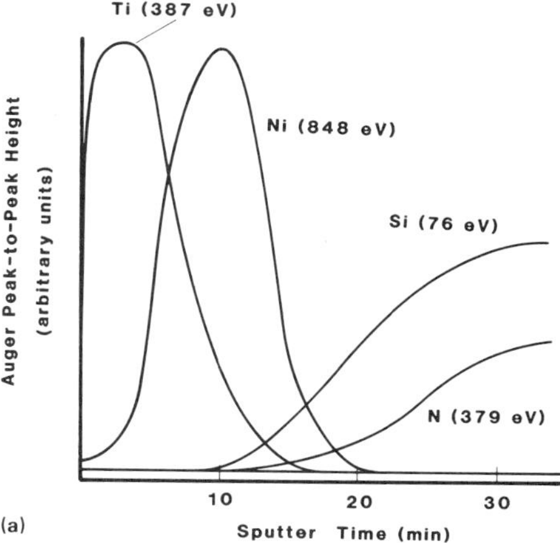

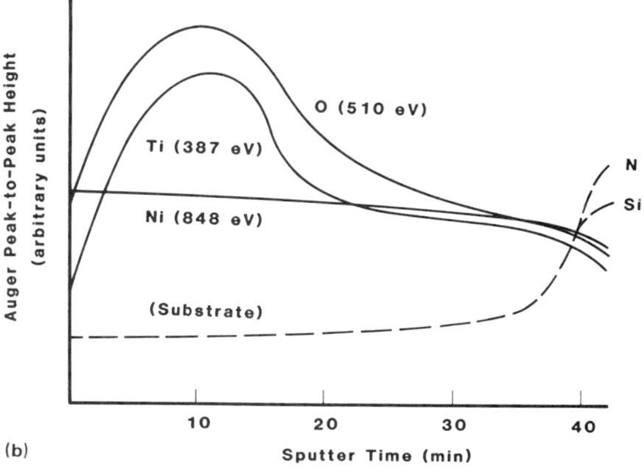

Fig. 5. Typical elemental depth profiles taken (a) before and (b) after friction and wear testing of a Ti-Ni-modified Si_3N_4 disk (diesel environment; 800 °C). Because of inhomogeneity of the preliminary mixing process and preferential sputtering, profiles are meant to show qualitative changes in the surface-modified layer (see ref. 22).

fied disks. Pins run against the nickel-modified combinations 7 and 8, as well as those run against chromium-modified disks (not given in Table 1), show no such transfer layer, instead evidencing a large amount of delamination-type wear as described by Suh [26].

Although many of the wear surfaces on the modified disks were difficult to observe optically, scanning electron micrographs show that wear of the modified surfaces did indeed occur to varying degrees. Figure 3 shows a typical wear surface from a Ti–Ni-modified ZrO_2 disk, and Fig. 4 shows a wear surface from a cobalt-modified ZrO_2 disk. The previous study on the Ti–Ni-modified disks indicated that the modified layer showed evidence of delamination wear [21, 26], with the disks which had higher coefficients of friction (combinations 2 and 3) showing larger amounts of wear than those which had lower coefficients of friction. Elemental Auger maps of the wear surfaces such as that in Fig. 3(b) showed that delamination occurred along the modified layer–substrate interface. Auger depth profile analysis of the modified layers indicated that poor mixing of the modified layer with the substrate may have been a cause for the delamination of the layer (see Fig. 5(a)). However, subsequent mixing and oxidation of the modified layer at the high temperatures expected at the contact surfaces during testing apparently led to the improved friction and wear behavior observed in the simulated diesel environment, especially for the modified Si_3N_4 disks (Fig. 5(b)).

The results of the tests on cobalt-modified disks show behavior in some ways similar to that of the Ti–Ni-implanted disks. The cobalt-modified ZrO_2 disks have a low coefficient of friction when run against the TiC–Ni–Mo pin, and a much higher value when run against the TiC pin. However, unlike the Ti–Ni case, the Si_3N_4 disks showed no improvement in the coefficient of friction when ion beam mixed with cobalt.

Auger analysis of the cobalt-modified ZrO_2 disks and the corresponding pins indicates that cobalt transfers from the disk surface to the pins. Elemental depth profiles indicate that both the pin and the disk cobalt layers are oxidized, implying the presence of a lubricating cobalt oxide film analogous to the Ti–Ni case. However, the physical characteristics of the wear tracks, especially for the low coefficient of friction disk, are disturbing. Both wear tracks show delamination of the cobalt layer, leaving exposed substrate. The wear track of the disk with a low coefficient of friction varies in width and amount of wear, ranging from the width of the pin with wear as heavy as that visible on the other three disks (see Fig. 4) to areas where the wear track is faint or almost non-existent (similar to Fig. 3). While it is thus possible that the ZrO_2 and Ni–Mo-bonded TiC results were influenced by the alignment of the pins and disk, the low coefficient of friction obtained indicates the transfer and possible lubricating effect of a cobalt oxide layer.

AES analysis of the surface of the Ni–Mo-bonded TiC pin run against the cobalt-modified disk also shows the strong presence of nickel. This indicates that nickel segregates to the pin surface and may play a role in the friction and wear behavior of the ceramic couples. This also implies that nickel may have segregated from the pins in the Ti–Ni modification tests but would not be observable because of the nickel component of the modified layer.

The results of tests involving the Ti–Ni- and the cobalt-modified disks suggest that nickel may play a particularly critical role in the friction and wear behavior of the ion-beam-modified ceramics, either as a part of the modified layer or because of its presence in the pins. Therefore, two Si_3N_4 disks were ion beam modified with a layer of nickel and tested in a diesel environment at 800 °C and at room temperature.

The results of the friction and wear tests indicate that nickel improved the coefficient of friction of Si_3N_4, but not to the same degree as did the Ti–Ni modification or the cobalt in the ZrO_2 modification (see Table 1). Analysis of the wear surfaces of the disks and the pins showed that the nickel layer was completely worn off the disk and that no nickel transferred to the pins. The wear track shows cracking, especially near the edge of the crack, typical of the wear tracks of unmodified ceramics. The pin surfaces show clear signs of wear, as opposed to the adherence of a transfer layer seen in the Ti–Ni and cobalt cases. AES analysis of the unworn areas after testing shows, however, that the nickel layer was much better mixed into the substrate than in the two previous cases.

4. DISCUSSION

The results of this investigation indicate that solid lubrication of ceramic materials at high temperatures is possible through the use of the technique of surface modification by ion beam mixing of oxide-forming metal ions. In order to optimize the conditions for the application of this technique to real systems, several factors must be considered. These include the characteristics and adhesive properties of the modified surface layer as governed by the operating environment, the ion-beam-mixing conditions, the choice of material substrates, the mechanism of lubrication by the modified layer (including its chemical state, the mechanical and structural properties and the possibility of its transfer from one material to the other) and the choice of the material to be run against the modified material.

The adhesion at the ion-beam-modified layer–ceramic boundary is clearly important in the delamination-type wear observed, since this will govern the useful lifetime of the component. While this investigation was conducted on a screening basis, emphasizing short-term tests to find promising ceramic–ceramic couples for immediate use in an experimental engine [21], the results of the analyses of the wear surfaces suggest that much longer lifetimes may be attainable through an optimization of the ion-beam-mixing conditions and/or the choice and condition of the substrate materials and metal ions to be mixed. The previous study [21] showed that the Ti–Ni modification appeared to mix the best in the Si_3N_4 substrate. While the amount of wear on the cobalt-modified disks was not satisfactory, the surface morphology of the wear surface of the cobalt-modified ZrO_2 indicates that the degree of initial surface roughness or initial microstructure in terms of grain size also may play an important role in the adhesion of the modified layer. As can be seen in Fig. 6, traces of cobalt remain at pores, asperities and grain boundaries of the substrate surface, after argon sputtering has removed the main part of the modified surface.

The results of the tests conducted on disks modified only with nickel indicate that, while nickel alone does not improve the friction and wear behavior of ceramics as well as other species, it may help to provide better bonding of those species and/or their lubricating oxides. For the Ti–Ni modification, TiO_2 appears to be responsible for the improved friction and wear behavior, but the nickel layer appears to be the bonding species which mixes into the substrate, especially for Si_3N_4. For the cobalt modification, segregation of nickel from the Ni–Mo-bonded TiC pin may be responsible for the adherence of the transferred cobalt to the pin surface at least temporarily, thus delaying the degradation of the friction and wear properties caused by removal of the cobalt layer from the disk surface. Further work is required to determine the influence of nickel on the bonding and/or friction and wear behavior of ion ceramics. Also of interest would be to determine those species which can be coupled with nickel in a modified layer to provide these beneficial results. For example, the ion beam mixing of chromium alone apparently had no beneficial influence on the friction and wear behavior of either Si_3N_4 or ZrO_2, even when run against Ni–Mo-bonded TiC pins. The ion beam mixing of chromium with nickel may provide, however, an improvement similar to the Ti–Ni modification.

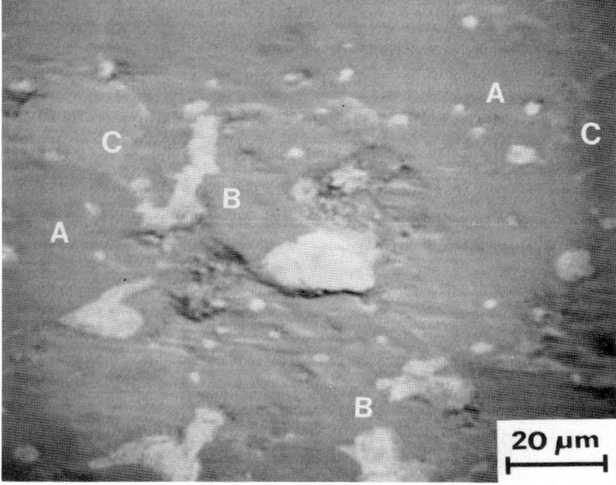

Fig. 6. Wear surface of the cobalt-modified ZrO_2 disk shown in Fig. 4 after sputter removal of most of the cobalt-modified layer. The remains of the cobalt layer are shown adhering to pores A, asperities B and grain boundaries C in the substrate surface.

The transfer of material or the formation of a surface oxide has been shown to be beneficial to the wear characteristics of the Ti–Ni-modified ceramics, as well as to many other material systems in general [21, 22, 27, 28]. From the results of the present investigation, nickel appears to have been transferred from the disks to the pins, or vice versa, as indicated by the results obtained from the pins run against Ti–Ni- or cobalt-modified disks. Although for the Ti–Ni modification, it is highly likely that titanium also transferred to the pins, it was not possible to determine definitely whether or not this was the case; a radioactive tracer technique would be useful for this determination for either nickel or titanium transfer. There appears to be no doubt that the modified and transferred layers had been oxidized and that this oxide contributed to the improvement in the friction and wear properties of the modified ceramics.

Further investigations are under way to determine the characteristics of these oxides under the extreme conditions expected at the contact surfaces in order to determine the mechanism of their lubricating effect. Characteristics such as oxide chemistry, mechanical or flow properties, and amorphous *vs.* crystallographic structure would be expected to play important roles in the lubricating phenomenon. Some clues may be derived from phase diagrams [29]. The oxides of nickel show a strong dependence of their melting points on stoichiometry. Whereas NiO melts at close to 1600 °C, there is a sharp continuous decrease in melting temperature with increase in the fraction of oxygen; Ni_2O_3 melts at less than 400 °C. It is quite plausible, therefore, that the tribological system of the sliding ceramics was lubricated by a film of fluid Ni_xO_y where y/x is greater than unity. The relationships in the Co–O system are less defined. It has been shown, however, that the stoichiometric CoO is the only form stable above 1000 °C. The higher oxygen compounds show significant instabilities at high temperatures where Co_3O_4, for example, dissociates at 900 °C. In contrast with the above, the oxides of chromium or titanium display high stabilities across the entire range of oxygen potentials. All titanium oxides are stable to at least 1600 °C. Similar observations hold for chromium oxides [30].

Related to the material transfer question is the choice of material to be run against the modified ceramic. This choice would be made on the basis of not only the actual chemical processes discussed here but also the specific applications. The current investigation was conducted to screen ceramic–ceramic couples for use as cylinder liners and piston rings in high temperature adiabatic engines. As such, the material simulating the piston ring (the pin) would see constant sliding contact and thus would be susceptible to more wear than any particular point on the cylinder liner. Thus, for these tests, it was believed that the use of the harder modified ceramic which had some oxidizing capability (titanium) as the pin material would be more advantageous than modifying the pin surface. Other types of sliding application could benefit, however, from the use of modified ceramics for both members of the couple.

From the previous investigation the choice of pin material played a confusing role in combinations 1–4 involving the Ti–Ni modification. Nickel was clearly transferred to all four pins [22], but the two pin materials had opposite effects with respect to disk material. The role of molybdenum comes into question but, because of its volatility at high temperatures, it was not detected on any wear surfaces using AES. In contrast, nickel segregated from the Ni–Mo-bonded TiC pins may have played a role in the cobalt-modified disks.

5. CONCLUSIONS

Pin-on-disk-type friction and wear tests have been conducted on a series of ceramic–ceramic couples in the unmodified state, and surface modified by ion beam mixing, in a simulated environment typical of high temperature adiabatic diesel engines. Studies of the morphology and chemistry for the resulting wear surfaces have been conducted and correlated with the friction and wear behavior of these materials.

It has been found that the surface modification of ceramics by the ion beam mixing of Ti–Ni or cobalt reduces the coefficient of friction (for certain specific pin–disk combinations) to levels falling in the upper end of the range of values considered acceptable for conventional liquid-lubricated engines. In contrast, the ion beam mixing of chromium or nickel does not appear to have a beneficial effect on the friction and wear behavior of the materials tested. The improvement in friction and wear behavior of the ceramic couples modified with Ti–Ni and cobalt is apparently the result of the formation of a lubricating oxide of these species. Significant differences among all the material combinations examined, however, indicate that several factors must be investigated before this surface modification technique can be applied to real systems. These include (1) the characteristics and adhesion properties of the modified surface layer under operating conditions, (2) the effect of ion-beam-mixing conditions on the surface-modified layer and its ability to provide lubricating properties, (3) the mechanism of lubrication by the oxide layer and (4) the choice of materials and metals to be ion mixed for optimal lubricating behavior.

ACKNOWLEDGMENTS

This work was conducted under a program sponsored by the U.S. Department of Energy and technically monitored by the Lewis Research Center, National Aeronautics and Space Administration, under Contract DEN3-352.

REFERENCES

1 J. J. Burke, E. M. Lenoe and R. N. Katz (eds.), *Proc. 2nd Int. Conf. on Ceramics for High Performance Applications, Newport, RI, March 21–25, 1977*, Brook Hill Publishing, Chestnut Hill, MA, 1978.

2 R. Kamo and W. Bryzik, Cummins/TACOM advanced adiabatic engine, in V. Danner (ed.), *Adiabatic Engines: Worldwide Review*, in *SAE Spec.*

Publ. SP-571, 1984, pp. 21–34 (Society of Automotive Engineers, Warrendale, PA).

3 J. Breznak, E. Breval and N. H. Macmillan, *J. Mater. Sci., 20* (1985) 4657.

4 T. Shimachi, T. Murakami, T. Nakagaki, Y. Tsuya and K. Umeda, in V. Danner (ed.), *Adiabatic Engines: Worldwide Review*, in *SAE Spec. Publ. SP-571*, 1984, pp. 35–45 (Society of Automotive Engineers, Warrendale, PA).

5 D. J. Boes, *Proc. 22nd CCM*, Society of Automotive Engineers, Warrendale, PA, 1984, p. 323.

6 E. Rabinowicz, *Friction and Wear of Materials*, Wiley, New York, 1965.

7 K. Miyoshi and D. H. Buckley, in *Tribology in the '80s*, Vol. 1, in *NASA Conf. Publ. CP-2300*, 1984, p. 291 (National Aeronautics and Space Administration).

8 O. O. Adewoye and T. F. Page, *Wear, 70* (1981) 37.

9 D. W. Richerson, L. J. Lindberg, W. D. Carruthers and J. Dahn, *Ceram. Eng. Sci. Proc., 2* (1981) 578.

10 J. R. Smyth and D. W. Richerson, *Ceram. Eng. Sci. Proc., 4* (1983) 663.

11 M. B. Peterson and S. F. Murray, *Met. Eng. Qt., 7* (2) (1967) 22.

12 D. J. Barnes and B. D. Powell, *Wear, 32* (1975) 195–202.

13 P. Sutor, Tribology of silicon nitride and silicon nitride–steel sliding pairs, in F. D. Gac (ed.), *Proc. 9th Annu. Conf. on Composites and Advanced Ceramic Materials, Cocoa Beach, FL, January 20–23, 1985*, American Ceramic Society, Columbus, OH, 1985, pp. 460–469.

14 T. E. Fischer and H. Tomizawa, *Wear, 105* (1985) 29–45.

15 R. P. Steijn, *Wear, 7* (1964) 48–66.

16 H. Shimura and Y. Tsuya, Effect of atmosphere on the wear rate of some ceramics and cements, *Proc. Int. Conf. on the Wear of Materials, St. Louis, MO, April 1977*, American Society of Mechanical Engineers, New York, 1977, pp. 452–461.

17 E. F. Finkin, S. J. Calabrese and M. B. Peterson, *ASLE Prepr. 72LC-7C-2*, 1972 (American Society of Lubrication Engineers).

18 L. B. Sibley and C. M. Allen, *Wear, 5* (1962) 312–329.

19 Advanced Mechanical Technology Inc., Evaluation of improved materials for stationary diesel engines operating on residual and coal based fuels, *Rep.*, 1980 (U.S. Department of Energy Contract DE-AC-03-79ET15444).

20 S. Gray, in F. D. Gac (ed.), *Proc. 9th Annu. Conf. on Composites and Advanced Ceramic Materials, Cocoa Beach, FL, January 20–23, 1985*, American Ceramic Society, Columbus, OH, 1985, pp. 965–975.

21 J. Lankford, W. Wei and R. Kossowsky, Friction and wear behavior of ion beam modified ceramics, submitted to *J. Mater. Sci.*

22 W. Wei and J. Lankford, Characterization of ion beam modified ceramic wear surfaces using Auger electron spectroscopy, submitted to *J. Mater. Sci.*

23 D. H. Buckley, *Ceram. Bull., 51* (1972) 884–905.

24 K. Miyoshi and D. H. Buckley, *Appl. Surf. Sci., 10* (1982) 357–376.

25 W. J. Lackey and D. P. Stinton, *Proc. 22nd Department of Energy Contractors Coordinating Meeting*, Society of Automotive Engineers, Warrendale, PA, 1984, p. 445.

26 N. P. Suh, *Wear, 25* (1973) 111.

27 D. H. Buckley, *Surface Effects in Adhesion, Friction, Wear, and Lubrication, Tribology Ser. 5*, Elsevier, Amsterdam, 1984.

28 I. L. Singer, *Appl. Surf. Sci., 18* (1984) 28–62.

29 *Phase Diagrams for Ceramists*, Vols. 3–5, American Society for Metals, Metals Park, OH, 1975–1980.

30 R. P. Elliott, *Constitution of Binary Alloys, First Supplement*, McGraw-Hill, New York, 1965, p. 327.

The Effect of Nitrogen and Boron Ion Implantation on Cyclic Deformation Response in Ti–24V Alloy*

JEON G. HAN and R. F. HOCHMAN

Fracture and Fatigue Research Laboratory and School of Material Engineering, Georgia Institute of Technology, Atlanta, GA 30332 (U.S.A.)

(Received July 10, 1986)

ABSTRACT

The surface structure change and its effect on fatigue response produced by nitrogen and boron implantation were studied in a Ti-24V alloy where the composition is in approximate weight per cent. The surface structure change by ion implantation was identified using transmission electron microscopy, Auger electron spectroscopy and X-ray photoelectron spectroscopy. Nitrogen and boron implantation produced irradiation defects as well as fine titanium nitride and titanium boride precipitates. A dense precipitation was found as a result of the boron implantation. Surface deformation and crack development features were altered at the implanted surface under strain-controlled fatigue conditions and hence changed the fatigue life. Nitrogen implantation slightly improved the strain-controlled fatigue life while boron implantation was detrimental. Both nitrogen implantation and boron implantation enhanced the stress-controlled fatigue life and limit. A particularly marked improvement was observed for boron implantation.

1. INTRODUCTION

Fatigue failure generally proceeds by microcrack nucleation, microcrack growth to macrocrack growth, and then macrocrack growth to final fracture. It is well known that microcrack nucleation occurs through strain localization or a dislocation avalanche process at localized regions near the surface. Both the strain localization and the dislocation avalanche processes are controlled by surface deformation behavior, yield strength and residual stresses at the near-surface region. All these properties are closely associated with the surface physical state including microstructure and damage structure [1-4].

Ion implantation has emerged as an effective process to alter the near-surface region and to produce a variety of physical states such as solid solution formation, formation of second phases, formation of amorphous and/or microcrystalline alloys, development of a wide range of damage structures, production of surface residual stresses etc. [5-10]. As a result, surface modification by ion implantation can influence the deformation behavior of the implanted surface region which will allow the fatigue initiation properties to be controlled [11-15].

The present paper illustrates the surface structure changes produced by nitrogen and boron implantation and the effect of the surface modification on the fatigue behavior of a metastable and β phase Ti-24V alloy.

2. EXPERIMENTAL DETAILS

The Ti-24V alloy used for the study was supplied by Titanium Metals Co., Henderson, NV. The chemical composition of this alloy is given in Table 1. Two types of specimen were prepared, one with a cylindrical gauge

TABLE 1

Chemical composition of Ti-24V alloy

Element	V	Fe	N	O	Ti
Amount (wt.%) *of element*	24.5	0.046	0.006	0.102	Balance

*Paper presented at the International Conference on Surface Modification of Metals by Ion Beams, Kingston, Canada, July 7-11, 1986.

0025-5416/87/$3.50

section of 11 mm and a diameter of 4 mm, and the other was a hourglass-type specimen which was 3 mm at its minimum diameter and approximately 3 mm long. The specimens were solution heat treated at 900 °C and quenched in iced brine water. The gauge section of each specimen was then mechanically fine polished followed by electropolishing.

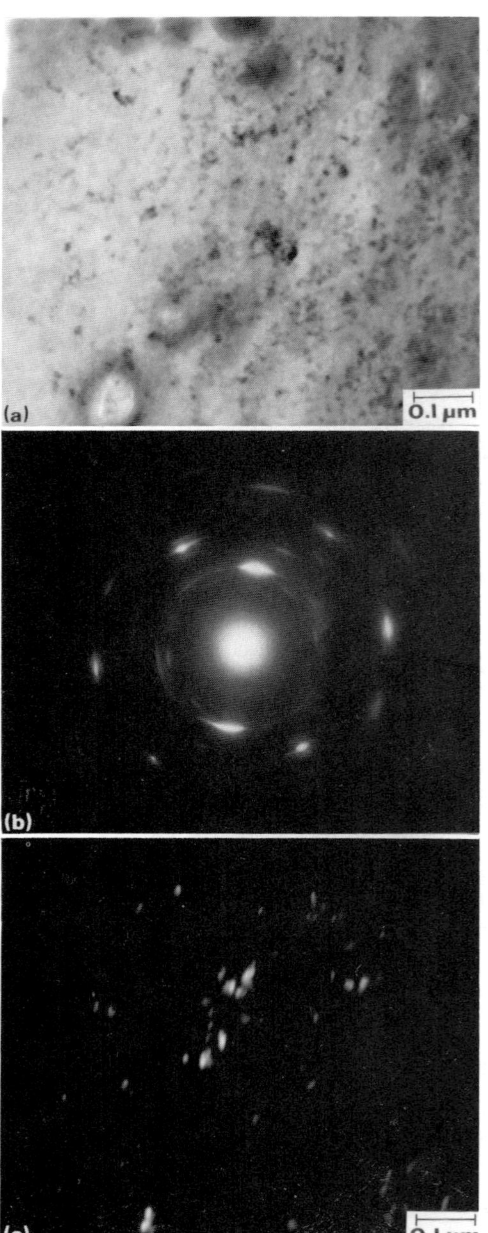

Fig. 1. (a) Bright field TEM micrograph, (b) electron diffraction pattern and (c) dark field TEM micrograph of the nitrogen-implanted Ti–24V.

Implantation with nitrogen and boron ions was carried out with the specimens held at room temperature. An accelerating voltage of 100 keV was used to achieve a total surface dose of 2×10^{17} ions cm^{-2}. All specimens were tilted and rotated during implantation for homogeneous implantation. In addition to the fatigue samples, thin foils for structure analysis utilizing transmission electron microscopy (TEM) were implanted under the same conditions.

Fatigue tests for both unimplanted specimens and ion-implanted specimens were conducted in laboratory air under strain- and stress-controlled conditions. Strain-controlled fatigue tests were conducted at a constant strain rate of 10^{-3} s^{-1} in the total strain amplitude range 10^{-3}–10^{-2}. The specimens utilized in these tests were those with the cylindrical gauge section. During the strain-controlled fatigue testing, *in situ* replication of the gauge section was carried out to determine the surface deformation mode testing as well as crack nucleation and development features. The stripped replicas were shadowed with Au–Pt and carbon at a shadowing angle of 30° to a thickness of approximately 200 Å. They were then examined by optical microscopy and scanning electron microscopy (SEM). The stress-controlled fatigue tests were performed using the hourglass-shaped specimens that were carried out at a constant frequency of 10 Hz and in a stress amplitude range 0.8–1.1 of the monotonic yield stress of the fully reversed mode ($R = -1$).

The structure changes produced by ion implantation were studied using Auger electron spectroscopy, X-ray photoelectron spectroscopy (XPS) and scanning transmission electron microscopy. SEM was also used to examine the crack origin features after stress-controlled fatigue fracture.

3. RESULTS AND DISCUSSION

3.1. *Microstructure of implanted surface*

Figure 1 and Fig. 2 illustrate the surface microstructures and corresponding electron diffraction pattern resulting from nitrogen implantation and boron implantation respectively. The bright field micrographs (Figs. 1(a) and 2(a)) show that the implanted surface was heavily damaged because of high energy

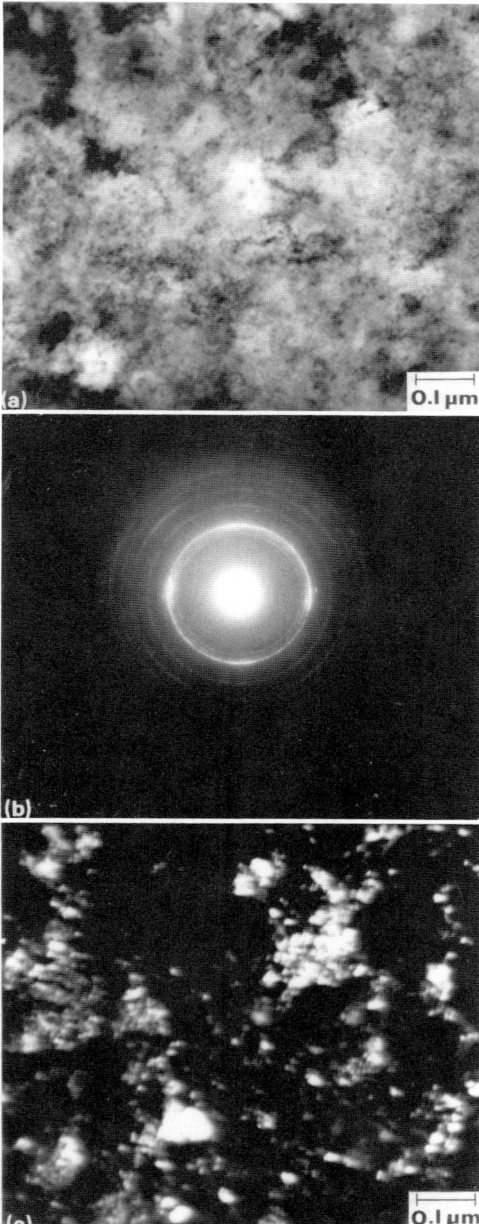

Fig. 2. (a) Bright field TEM micrograph, (b) electron diffraction pattern and (c) dark field TEM micrograph of the boron-implanted Ti–24V.

ion irradiation for both nitrogen implantation and boron implantation. The damage is apparently in the form of vacancies, defect clusters and dislocation loops as would be expected from a theoretical calculation of the threshold energy for atomic displacement [16] in these systems. Because of the limit on resolution using TEM, precise identification of the total defect structure could not be made.

Electron diffraction patterns of the implanted surface (Figs. 1(b) and 2(b)) illustrate the microstructural changes which resulted from the nitrogen and boron implantation. The electron diffraction pattern for the nitrogen-implanted surface shows extra rings in addition to the (110) reciprocal lattice reflections for the b.c.c. Ti–24V alloy. Dark field imaging, using a section of the inner rings (Fig. 1(c)), shows the formation of a fine precipitate phase. The particle sizes of these precipitates were mostly in the 50–100 Å range. Extra well-defined rings were also found in the diffraction pattern for the boron-implanted surface (Fig. 2(b)). These rings were found to originate from a dense concentration of fine precipitates shown in the dark field micrograph (Fig. 2(c)).

From a depth profile through quantitative analysis of the Auger electron peak-to-peak amplitude of each element, the maximum concentration of boron was found to be about 32 at.% at a depth of about 2000 Å below the surface. This result is very close to the theoretical estimates for this system based on a calculation using the Lindhard–Scharff–Schiøtt (LSS) theory [17]. However, the distribution of nitrogen could not be well defined because the Auger peak energy of nitrogen, N 1s, overlaps the secondary peak of titanium. However, the maximum concentration for nitrogen using the LSS theory was estimated to be about 24 at.% at a depth of about 1450 Å which would be possible in the possible resolved stage for the Auger results. When the solubility limits of nitrogen (about 6.5 at.%) and boron (about 0.5 at.%) are considered [18], it is obvious that the implanted region is highly supersaturated for both nitrogen and boron. The result would be the formation of titanium nitride and titanium boride precipitates during implantation, which is in keeping with the observations of the TEM and selected area diffraction results. The denser precipitates resulting from boron implantation can be directly attributed to the much lower solubility limit and higher implanted concentration of boron compared with the nitrogen implantation.

XPS analysis was carried out to identify the precipitates. Figure 3 illustrates the narrow scan spectra for Ti $2p_{3/2}$ for unimplanted,

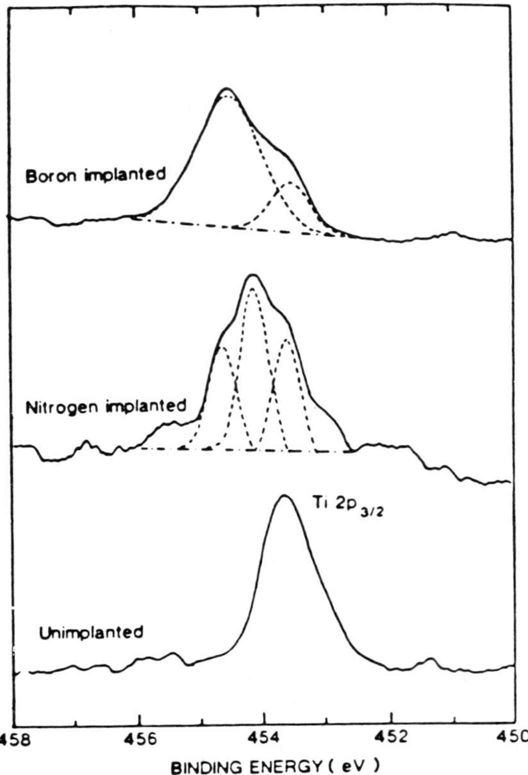

Fig. 3. X-ray photoelectron narrow-scan spectra of Ti $2p_{3/2}$ recorded from unimplanted, nitrogen-implanted and boron-implanted surfaces at a depth of about 300 Å.

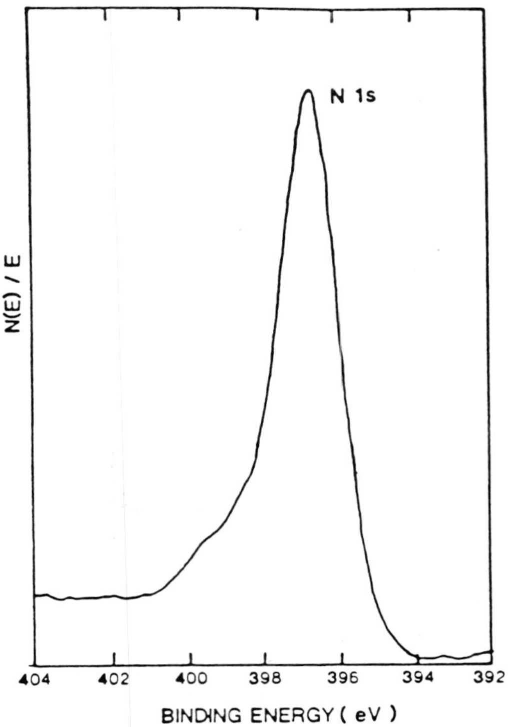

Fig. 4. X-ray photoelectron narrow-scan spectrum of N 1s recorded from a nitrogen-implanted surface at a depth of about 300 Å.

nitrogen-implanted and boron-implanted surfaces. This was taken at a depth of 300 Å from the surface to assure a typical sample of the implanted region. A distinct shape change and shift of Ti $2p_{3/2}$ peaks are observed as a result of both nitrogen implantation and boron implantation. For nitrogen implantation the binding energies from the fitted peaks for various valence states were found to be 453.9, 454.4 and 454.9 eV, which are very close to the binding energy reported for elemental titanium (453.8 eV) [19] and that for Ti $2p_{3/2}$ obtained from XPS studies of $TiN_{x\approx 0.5}$ (454.8 eV) and $TiN_{x\approx 1}$ (455.1 eV) [21]. Boron implantation also induced the shift of the binding energy of Ti$2p_{3/2}$. Peaks were found at 453.8 eV and 454.7 eV which corresponds to the binding energies of elemental titanium and titanium boride (116) respectively.

In addition, the N 1s and the B 1s core peaks (Figs. 4 and 5) indicate the binding

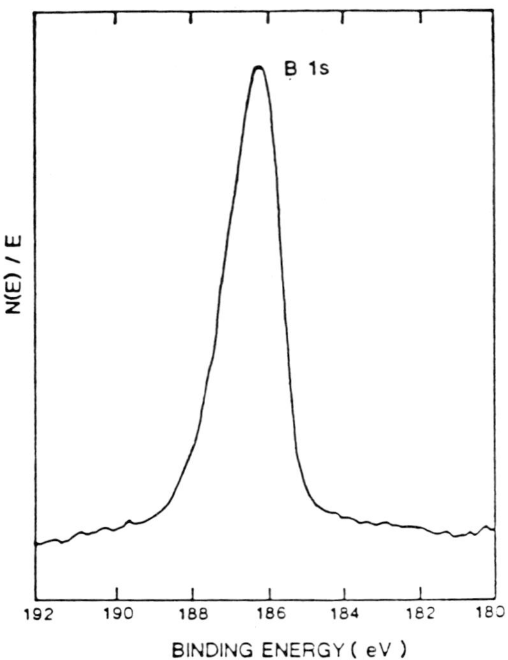

Fig. 5. X-ray photoelectron narrow-scan spectrum of B 1s recorded from a boron-implanted surface at a depth of about 300 Å.

energy to be 396.9 eV for N 1s and 186.2 eV for B 1s electrons. These electron binding energies are in keeping with the N 1s and B 1s core peaks reported for titanium nitride and titanium boride [19, 20].

In addition to these structural changes, implantation of nitrogen and boron ions introduces stresses in the implanted region caused by volume expansion resulting from the lattice disruption and the athermal effects of implanted ions. This leads to compressive residual stresses in the implanted region.

3.2. Strain-controlled fatigue response

Figure 6 shows cyclic strain *vs.* fatigue life curve for unimplanted and ion-implanted specimens following the Coffin–Manson relationship [22]. Nitrogen implantation slightly extended the fatigue life while boron implantation reduced it at all the strain amplitudes tested. Since ion implantation only modifies the near-surface region, the changes in fatigue life resulting from nitrogen and boron implantation will be related to changes in surface deformation and the crack nucleation process. These phenomena are clearly illustrated in the *in situ* replica micrographs in Figs. 7–9. On comparison of the number of cycles to the crack appearance in the replica micrographs (Figs. 7(a), 8(a) and 9(a)) with the number of cycles to the maximum stress amplitude in

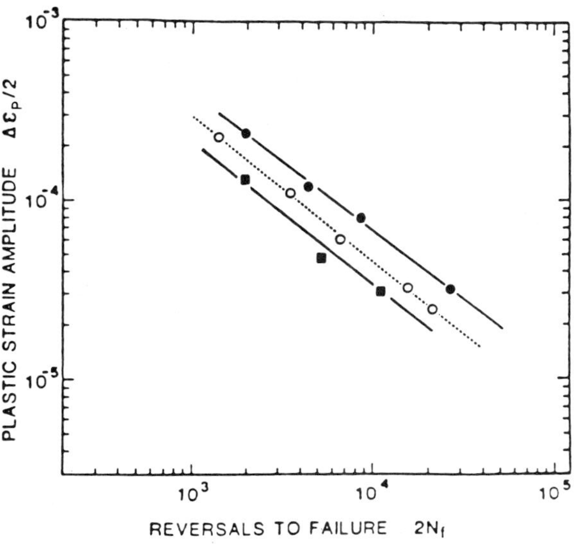

Fig. 6. Cyclic strain–life relationship for unimplanted (. . .) and ion-implanted Ti–24V: ○, unimplanted; ●, nitrogen (N^+) implanted; ■, boron (B^+) implanted.

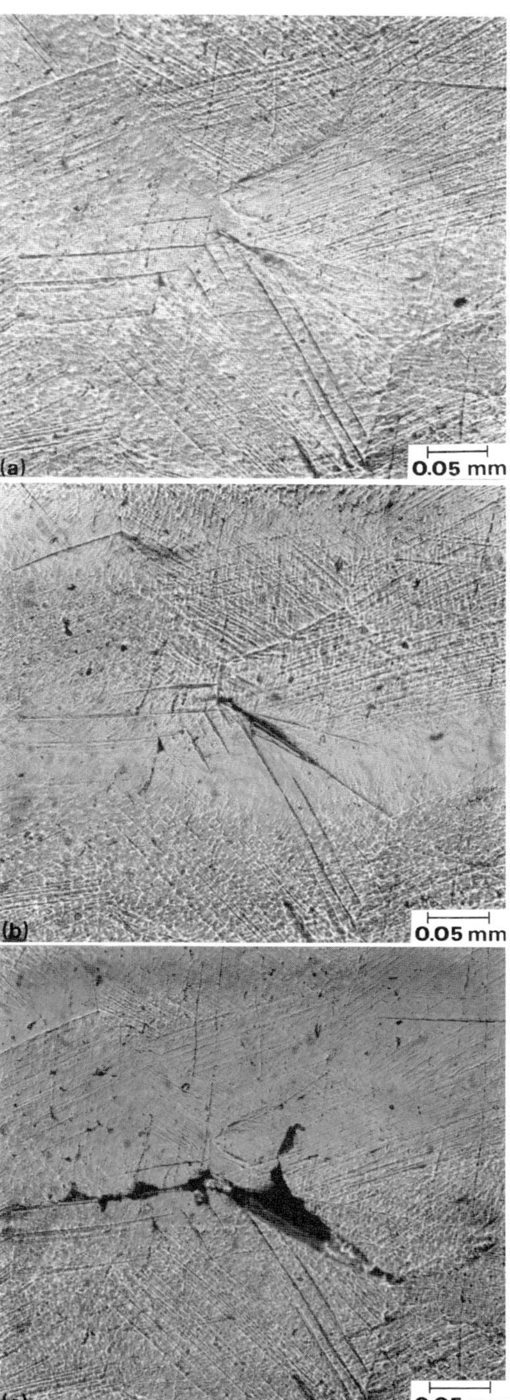

Fig. 7. Replica micrographs showing surface deformation and crack development features during strain-controlled fatigue cycling at a plastic strain amplitude of 2.3×10^{-4} for unimplanted Ti–24V: (a) 200 cycles; (b) 300 cycles; (c) 500 cycles.

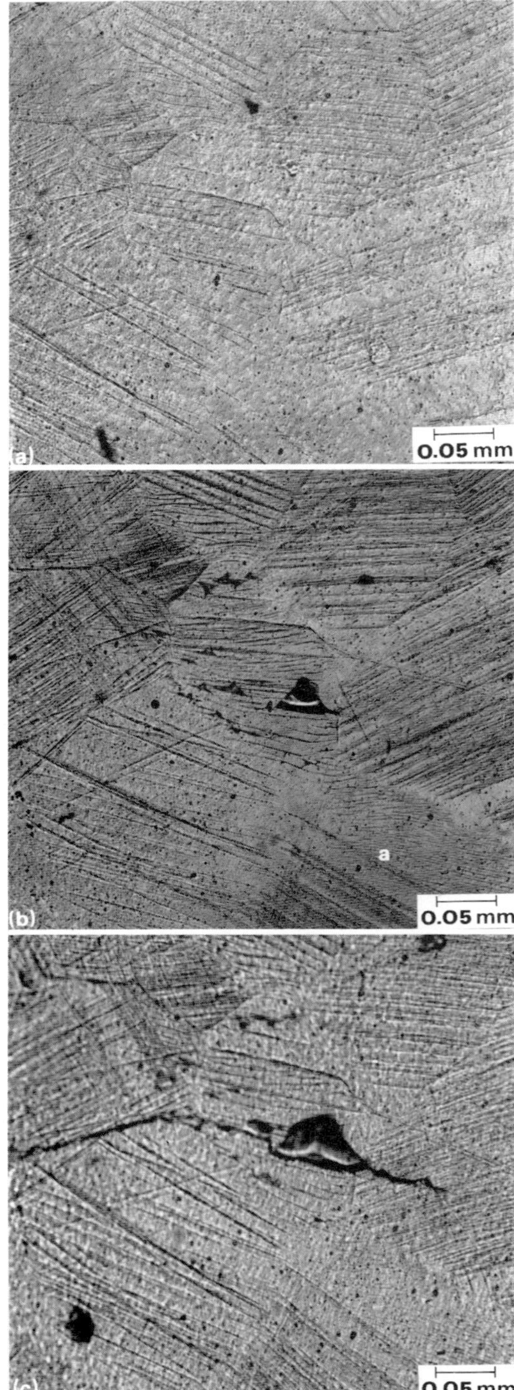

Fig. 8. Replica micrographs showing surface deformation and crack development features during strain-controlled fatigue cycling at a plastic strain amplitude of 1.2×10^{-4} for nitrogen-implanted Ti–24V: (a) 300 cycles; (b) 500 cycles; (c) 800 cycles.

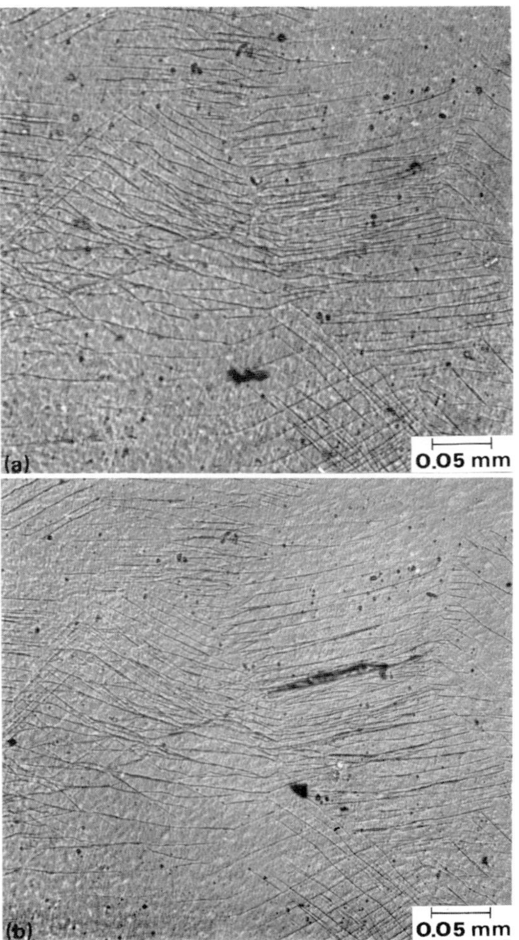

Fig. 9. Replica micrographs showing surface deformation and crack development features during strain-controlled fatigue cycling at a plastic strain amplitude of 1.3×10^{-4} for boron-implanted Ti–24V: (a) 100 cycles; (b) 200 cycles.

Fig. 10 for unimplanted and implanted specimens, there is no doubt that cracks nucleated before the stress amplitude reached the maximum. This indicates that cyclic hardening continues beyond crack initiation until the microcracks have grown to an appreciably larger size up to the maximum stress amplitude. Such trends were observed to be similar at all strain amplitudes for both unimplanted cases and implanted cases. However, it was found that nitrogen implantation extended the number of cycles to the maximum stress while boron implantation decreased it. These results were closely associated with the change in surface deformation and crack development behavior. It is observed for an unimplanted specimen that most of the surface grains were deformed in a fairly homogeneous manner and major cracks originated at a grain boundary, as shown in Fig. 7(b). The cracks then propagated along the grain boundary until they finally coalesced with a coarse deformation band, as shown in Fig. 7(c). For the nitrogen-implanted surface the discontinuous deformations appeared together with homogeneous deformation lines, as shown in grain A in Fig. 8(b). Nitrogen implantation also induced a change in crack development behavior. Cracks were found to originate at several slip bands, as can be seen in Fig. 8(b). The crack propagation proceeded by coalescence of these microcracks with further cycling, as illustrated in Fig. 7(c). This suggests that strain is not accumulated at few localized slip bands or grain boundaries but is uniformly distributed at overall surface slip bands. It is believed that the strain homogenization enhanced the resistance to crack development, thereby improving fatigue life. Boron implantation produced a significant change in the surface deformation and crack origin features. In Fig. 9 it is seen that surface deformation partially occurred in a river pattern with coarse slip. This is considered to be a result of inhomogeneous destruction of the structural order in the boron-implanted surface. These slip phenomena are thought to inhibit the structural relaxation process (*i.e.* structural reordering) during cyclic deformation. As a consequence, slip reversibility is markedly reduced and hence strain would be significantly increased at these slip bands. The result is earlier crack development.

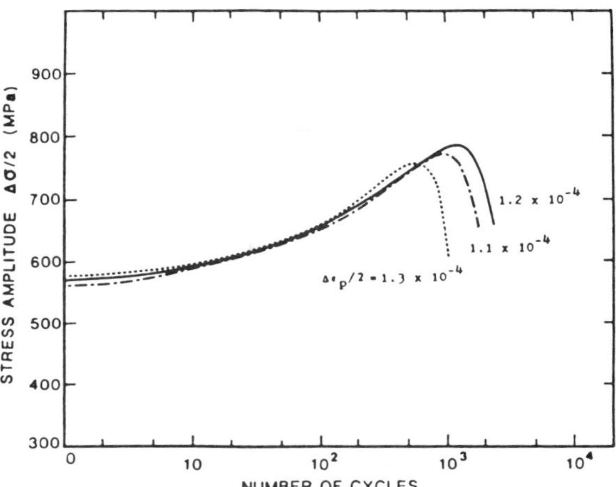

Fig. 10. Cyclic stress response for the unimplanted and implanted Ti–24V alloy during strain-controlled fatigue cycling: —·—, unimplanted; ——, nitrogen implanted; -----, boron implanted.

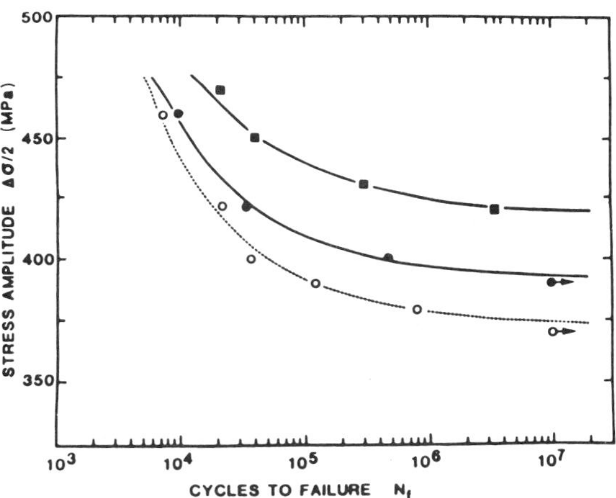

Fig. 11. Stress *vs.* life curves for stress-controlled fatigue tests of unimplanted (. . .) and ion-implanted Ti–24V alloy: ○, unimplanted; ●, nitrogen (N^+) implanted; ■, boron (B^+) implanted.

3.3. Stress-controlled fatigue response

Figure 10 illustrates the fatigue response under stress cycling for the unimplanted and implanted Ti–24V alloy. Both nitrogen implantation and boron implantation enhanced the fatigue life as well as the fatigue limit (Fig. 11). Particular improvement was observed for the boron-implanted specimens.

As discussed earlier, nitrogen and boron implantation produced non-deformable precipitates and irradiation defects. These modified structures increase the flow stress of the implanted surface region through interaction with the mobile dislocations during cyclic loading. The increase in surface flow stress reduced the plastic deformation in the surface region and hence reduced the localized strain accumulation that led to crack nucleation. The result is an improved fatigue life and a high endurance limit. The greater improvement resulting from boron implantation is undoubtedly a result of the extensive surface strengthening associated with denser precipitates which were observed. This effect is seen in crack origin features examined after fatigue failure, as shown in Fig. 12. Significant surface extrusion is observed for the unimplanted specimen at 400 MPa (about 90% of the yield

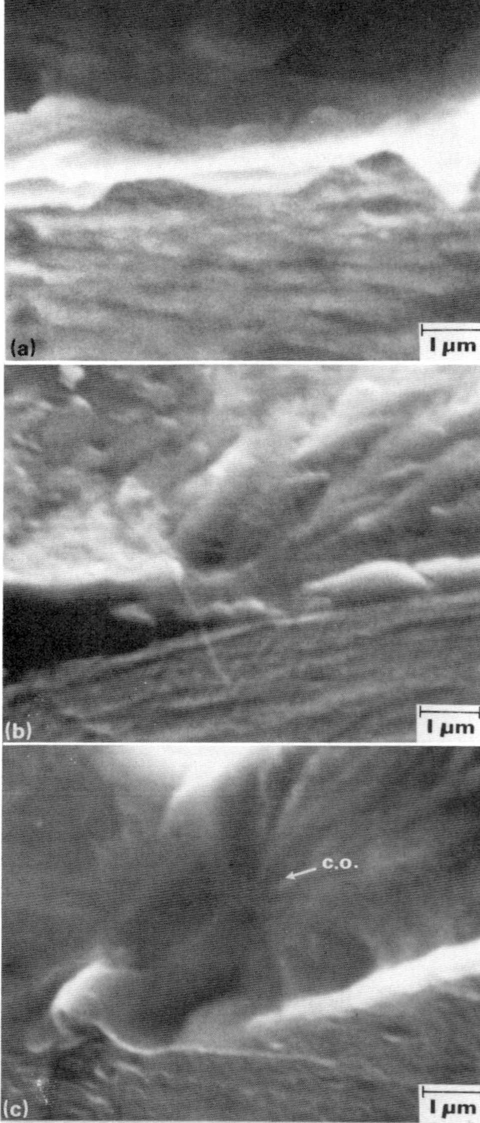

Fig. 12. SEM micrographs of the crack origins (C.O.) in the fracture surface after stress-controlled fatigue failure: (a) unimplanted (400 MPa); (b) boron implanted (400 MPa); (c) boron implanted (450 MPa).

stress) in Fig. 12(a). However, no extrusions occurred for the nitrogen- and boron-implanted surfaces shown in Figs. 12(b) and 12(c). In addition, the crack origin for the boron-implanted specimens was shifted to a subsurface area below the implanted region, even at stress levels above the bulk yield stress. This is illustrated in Fig. 12(c). Vardiman and Kant [12] found a similar phenomenon in nitrogen- and carbon-implanted Ti–6Al–4V.

Finally, compressive surface residual stresses, which are normally introduced during ion implantation, will also produce a beneficial effect on the fatigue life following nitrogen and boron implantation.

4. SUMMARY

(1) Nitrogen and boron implantation produced fine titanium nitride (TiN_x) and titanium boride (TiB) in the implanted surface region. Particularly dense fine precipitates were observed in the boron-implanted surface. This is believed to be due to the much lower solubility of boron in the b.c.c. titanium.

(2) Surface deformation and crack development features under strain-controlled fatigue conditions were altered by ion implantation. Discontinuous slip bands appeared at the nitrogen-implanted surface where cracks originated at various slip bands. This suggests that the applied strain is homogeneously dispersed at the implanted surface, leading to enhancement of the resistance to crack development. The result is an improvement in the fatigue life.

(3) During strain cycling, boron-implanted surfaces deformed in an inhomogeneous river pattern. This induced earlier crack nucleation, resulting in a shorter strain-controlled fatigue life.

(4) Both nitrogen implantation and boron implantation were beneficial to the stress-controlled fatigue life and limit. A particularly marked improvement resulted from boron implantation. This improvement is believed to be mainly due to the enhancement of surface flow stress resulting from the formation of dense fine titanium boride in the implanted surface layer.

ACKNOWLEDGMENTS

The authors wish to acknowledge the partial support of NIDR under Training Grant DE-07054 and the National Aeronautics and Space Administration under Contract NAS 8-35048 as well as institutional support from the Fracture and Fatigue Laboratory.

We would also like to thank Dr. W. B. Carter for his surface analysis work and Dr. Stephen D. Antolovich for his useful discussions. In addition, we would like to thank Dr. E. A. Starke, now Dean of Engineering, University of Virginia, who suggested some of the early possibilities of this work.

REFERENCES

1 A. S. Argon, in O. F. Devereux, A. J. McEvily and R. W. Staehl, (eds.), *Corrosion Fatigue, Chemistry Mechanisms and Microstructure*, National Association of Corrosion Engineers, Houston, TX, 1972, p. 176.

2 D. Kuhlmann-Wilsdorf and C. Laird, *Mater. Sci. Eng., 21* (1977) 137.

3 M. E. Fine, *Metall. Trans. A, 11* (1980) 365.

4 P. Lukas, M. Klesnil and J. Kreja, *Phys. Status Solidi, 27* (1968) 545.

5 J. A. Borders, *Annu. Rev. Mater. Sci., 9* (1979) 313.

6 A. Ali, W. A. Grant and P. J. Grundy, *Philos. Mag. B, 37* (1978) 353.

7 S. M. Meyers, *J. Vac. Sci. Technol., 15* (1978) 1650.

8 J. M. Poate, *J. Vac. Sci. Technol., 15* (1978) 1636.

9 G. Cater and J. S. Colligon, *Ion Bombardment Technology*, Elsevier, New York, 1968, p. 204.

10 S. Spooner and K. O. Legg, in C. M. Preece and J. K. Hirvonen (eds.), *Ion Implantation Metallurgy, Proc. Conf.*, Metallurgical Society of AIME, Warrendale, PA, 1980, p. 162.

11 S. B. Chakrabortty, A. Kujore and E. A. Starke, Jr., *Thin Solid Films, 73* (1980) 209.

12 R. G. Vardiman and R. A. Kant, *J. Appl. Phys., 53* (1) (1982) 690.

13 W. W. Hu, C. R. Clayton and H. Herman, *Scr. Metall., 12* (1978) 697.

14 H. Herman, *Nucl. Instrum. Methods, 182-183* (1981) 887.

15 J. Kumar, J. Han, E. A. Starke, Jr., and K. O. Legg, *Scr. Metall., 17* (1983) 479.

16 J. G. Han, *Ph.D. Thesis*, Georgia Institute of Technology, 1985.

17 J. Lindhard, M. Scharff and H. E. Schiøtt, *Mat.-Fys. Medd. K. Dan. Vidensk. Selsk., 33* (1963) 1.

18 A. E. Palty, H. Margolin and J. P. Nielsen, *Trans. Am. Soc. Met., 46* (1954) 312.

19 G. D. Wagner, W. M. Riggs, L. E. Davis, J. F. Moulder and G. E. Muilenberg (eds.), *Handbook of X-ray Photoelectron Spectroscopy*, Perkin–Elmer, Eden Prairie, MN, 1979.

20 L. Porte, L. Roux and J. Hanns, *Phys. Rev. B, 28* (1983) 3214.

21 L. Ranquist, K. Hamrin, G. Johanson, A. Fahlman and C. Nordling, *J. Phys. Chem. Solids, 30* (1969) 1835.

22 B. I. Sandor, *Fundamentals of Cyclic Stress and Strain*, University of Wisconsin Press, Madison, WI, 1972.

Materials Science and Engineering, 90 (1987) 327-332

Structure and Fatigue Properties of Ion-plated Nickel Films on Steel*

K. THOMA, R. FÄRBER†, TH. WIEDER and H. GÄRTNER

Arbeitsgruppe Metallphysik, Fachbereich Physik, Gesamthochschule Kassel, Postfach 101380, D-3500 Kassel (F.R.G.)

(Received July 10, 1986)

ABSTRACT

Nickel layers were ion plated on steel in order that their behaviour under the bending fatigue test could be examined. X-ray measurements showed that no residual stresses occurred in the layers. No substantial reduction in the fatigue limit was observed for samples in air. However, in salt water the coated specimens performed worse than the uncoated specimens. For these working conditions, denser layers are required, which may be produced by changing the parameters of the ion-plating process.

1. INTRODUCTION

The fatigue strength of materials under time-varying loads is an important parameter for mechanical engineering. This is investigated, for example, in the well-known reversed bending fatigue test. Evidently, surface conditions are a decisive factor since the largest stresses occur on the surface and here cracks will form predominantly. In aqueous or gaseous environments, corrosive substances will attack the surface and will influence the fatigue strength.

A number of treatments have been applied to strengthen the surface, *e.g.* shot peening, rolling, nitriding and carburizing. Also, for corrosion protection, ion implantation is used [1] and surface coatings are applied, for example, by galvanic or chemical deposition, plasma spraying and chemical or physical vapour deposition. Many galvanically deposited coatings tend to decrease the fatigue strength substantially. Nickel layers are known to reduce the fatigue limit by as much as 45% [2].

It is, therefore, of interest to investigate other methods of producing metallic coatings, which will protect against corrosion and are favourable to the fatigue strength at the same time. In this paper, reports are given of the ion plating of nickel layers on steel samples in a d.c. diode apparatus using either argon or nitrogen as the working gas. The surface topography was examined by scanning electron microscopy (SEM). X-ray diffractometry was used to measure the residual stresses in the coating and underlying substrate material. The fatigue strength was determined by producing Wöhler curves (curves of stress against the number of cycles to failure) from the rotary bending test. Finally the influence of corrosive media was studied on samples in air and salt water.

2. EXPERIMENTAL DETAILS

The basic material was a low carbon (0.45 wt.% C) non-alloyed steel (German designation, Ck 45). The rods (15 mm in diameter) were heat treated in air at 860 °C for 30 min, quenched in oil and annealed at 660 °C for 60 min. The resulting structure (Fig. 1) was a mixture of about 50% ferrite and 50% pearlite which, because of the annealing treatment, commences to segregate globularly. The material had an ultimate tensile strength R_m of 690 N mm^{-2}, an upper yield point R_{eH} of 496 N mm^{-2}, a lower yield point R_{eL} of 478 N mm^{-2} and a fracture elongation of 27%, *i.e.* the material is rather soft. This again is reflected in a Vickers hardness of 200 HV 1. The specimens were produced by turning and

*Paper presented at the International Conference on Surface Modification of Metals by Ion Beams, Kingston, Canada, July 7-11, 1986.

†Present address: Volkswagenwerk AG., 3507 Baunatal, F.R.G.

0025-5416/87/$3.50

grinding them to a central diameter of 7.52 mm (Fig. 2).

The ion plating was performed in a conventional d.c. diode apparatus, which was evacuated by a diffusion pump [3]. The cathode has an area of 120 cm^2; currents of 100 mA and high voltages of 6.5 kV can be reached. Because of the peculiar geometry of the samples below the cathode, the current distribution among the samples and cathode plane is not known, and thus the current density flowing into the samples cannot be stated. The surface material, nickel (purity, 99.9%), was evaporated from a resistively heated Al_2O_3 ceramic boat; the distance between the evaporation source and the specimens was 150 mm. Three specimens were mounted horizontally below the cathode at a distance of 21 mm. During the process, they were rotated slowly on their long axis to ensure uniform deposition on their circumference. As the specimens were held only at their ends, dissipation of heat is poor. A temperature check at the samples core proved that the temperature did not rise above 220 °C during the process so that the structure of the bulk metal was not affected. The temperature at the surface could not be measured directly.

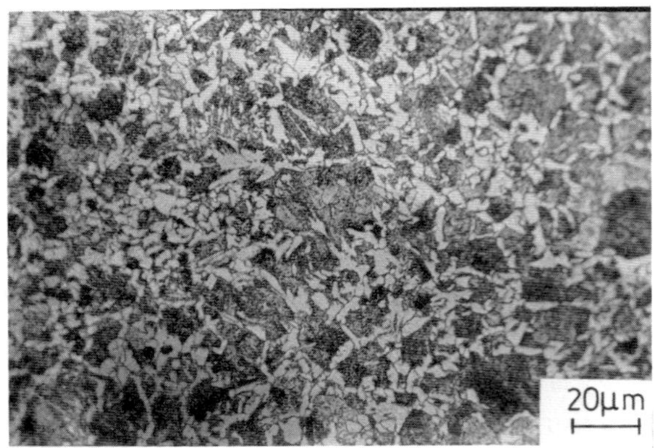

Fig. 1. Structure of steel Ck 45. (Magnification, 500.)

Prior to introduction into the vacuum chamber, the specimens were cleaned ultrasonically with trichloroethylene. Starting at a total pressure of 5×10^{-5} mbar, the samples were sputter cleaned for 15 min in argon at a pressure of 1.7×10^{-2} mbar and a voltage of 5 kV. During this time the evaporation source was shuttered. The ion plating took place in an argon or nitrogen atmosphere at a pressure of 2.2×10^{-2} mbar, and at a voltage of 4.5 kV for a duration of 15–20 min. A total of about 5 g of nickel was evaporated per run. The resulting surface structures were examined by SEM, the elemental distribution was checked by X-ray microanalysis, and the residual surface stresses were analysed by X-ray diffractometry. The goniometer was of the usual θ–2θ type (Fig. 3). It was employed in the ω tilt mode, *i.e.* the surface of the sample was rotated additionally on the goniometer axis to provide the necessary angle ψ between the surface normal and the incoming and outgoing radiation. Although this method has some material drawbacks compared with the ψ tilt mode [4], *e.g.* loss of Bragg–Brentano focusing, it is still to be preferred for cylindrical samples. The specimens can be mounted horizontally or vertically; this provides the necessary degree of freedom for stress analysis. Both the nickel layers and the underlying substrate material can be examined as the intensity peaks are well separated (for Cr Kα radiation the Bragg angles are ideally 2θ =

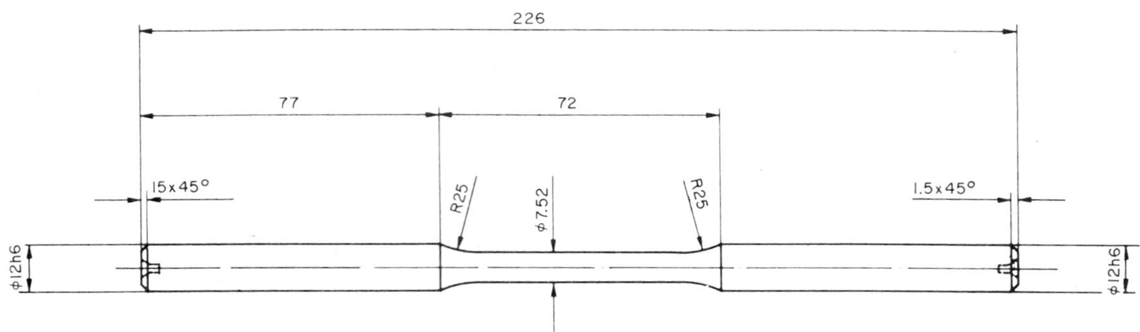

Fig. 2. Specimen for the bending fatigue test. All dimensions are in millimetres.

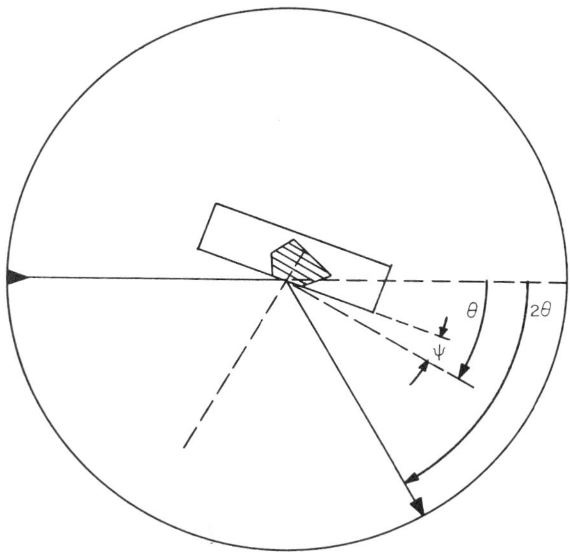

Fig. 3. Goniometer circle showing the incoming and outgoing radiation and the tilt of the scattering lattice planes.

Fig. 4. Cross-section of the nickel layer coated under nitrogen before the fatigue test.

133.53° for the {220} reflection in nickel and $2\theta = 156.07°$ for {211} in iron).

The fatigue strengths of coated and uncoated specimens were determined on a reversed bending fatigue test machine. The samples rotated at 50 rev s^{-1} approximately, and they experience a uniform bending moment along the central shaft of 7.52 mm diameter. All points on the circumference pass from a state of compression to one of tension, the stress varying sinusoidally in time. The stress is inhomogeneous in that it grows linearly in the radial direction from the centre to the surface. Two stress values in the high stress regime and two values near the fatigue limit were employed. For each stress value, six samples were run. The bending tests were performed for the uncoated-but-heat-treated material, and for coated samples under an argon and under a nitrogen atmosphere, so that a total of 72 samples was employed.

For corrosion studies a special cell was built in which the samples are guided through shaft-sealing rings and run in salt water (3% NaCl in water). A comparison between coated and uncoated surfaces was performed on a set of samples prepared from normalized Ck 45. A set of five samples was tested for each of the following: uncoated material, argon-treated material and nitrogen-treated material.

3. RESULTS AND DISCUSSION

The ion-plated nickel layers were homogeneous along the axis of the specimens and around the circumference, when inspected optically. They were sufficiently adherent to pass the well-known Scotch tape test.

The microhardness of the ion-plated layers was measured on flat samples of steel Ck 45 which were partially exposed to the gas discharge. A small increase in Vickers hardness was observed from 132 HV 0.01 on the substrate material to 156 HV 0.01 on the nickel layers, for both argon and nitrogen as process gases.

The surface roughness was measured similarly with a stylus instrument before the ion-plating process, after sputter cleaning and after plating. No significant change in roughness could be observed. The thickness of the nickel layers was determined by SEM pictures of the cross-sections (Fig. 4). Originally the determination of the thickness had been attempted by Tolanski interferometry using flat samples additionally mounted below the cathode. These samples turned out to be too thin compared with the round samples. This was due to the different geometry and position in the gas discharge. On average the thickness of the nickel layers was 1.3 μm for layers plated in argon and 4.6 μm for layers plated in nitrogen. The difference is due to the stronger sputtering rate of argon ions than that of nitrogen ions. Figure 4 shows the dense columnar structure of the nickel layers, which is a function of temperature and working gas pressure [5, 6].

For X-ray determination of the residual stresses [7] the shift in the Bragg peak due to a change in the distance between the scattering lattice planes is measured according to Bragg's law:

$$\frac{\Delta d}{d} = -\frac{1}{2} \cot \theta \ \Delta(2\theta)$$

This lattice strain is attributed to a two-dimensional stress state of the surface

$$\epsilon_{\phi,\psi} = \frac{s_2}{2} \sigma_\phi \sin^2\psi + s_1(\sigma_{ax} + \sigma_{tg})$$

where ψ is the angle between the normal to the lattice planes and the normal to the sample surface (Fig. 3), σ_{ax} and σ_{tg} are the stresses in the axial and the tangential directions respectively and ϕ is the rotation of the scattering plane with respect to a surface coordinate system in these directions. The elastic constants s_1 and s_2 for iron and nickel can be taken from tables [8]. According to the above formula, ϵ as a function of $\sin^2\psi$ should be a straight line. For fixed ϕ the stress σ_ϕ is given by a combination of axial and tangential stress according to

$$\sigma_\phi = \sigma_{ax} \cos^2\phi + \sigma_{tg} \sin^2\phi$$

Therefore, in our experimental set-up, σ_ϕ can be chosen to be σ_{ax} and σ_{tg}. If the cylindrical sample is positioned horizontally, σ_{ax} is measured. In this position the experimental errors are larger than in the vertical position, since the X-ray tube has a vertical line focus of about 1 mm × 5 mm.

Figure 5 shows examples of the peak shifts measured for the nickel layers. The error in the determination of peak position is estimated to be 0.05°, which is equivalent to an error in the stress of about 30 N mm^{-2}. This explains the large scatter of data in Fig. 5(a). The large peak shift in Fig. 5(b) and the systematic difference in the curves before and after testing are not understood at present. It was verified that no texture effects, which

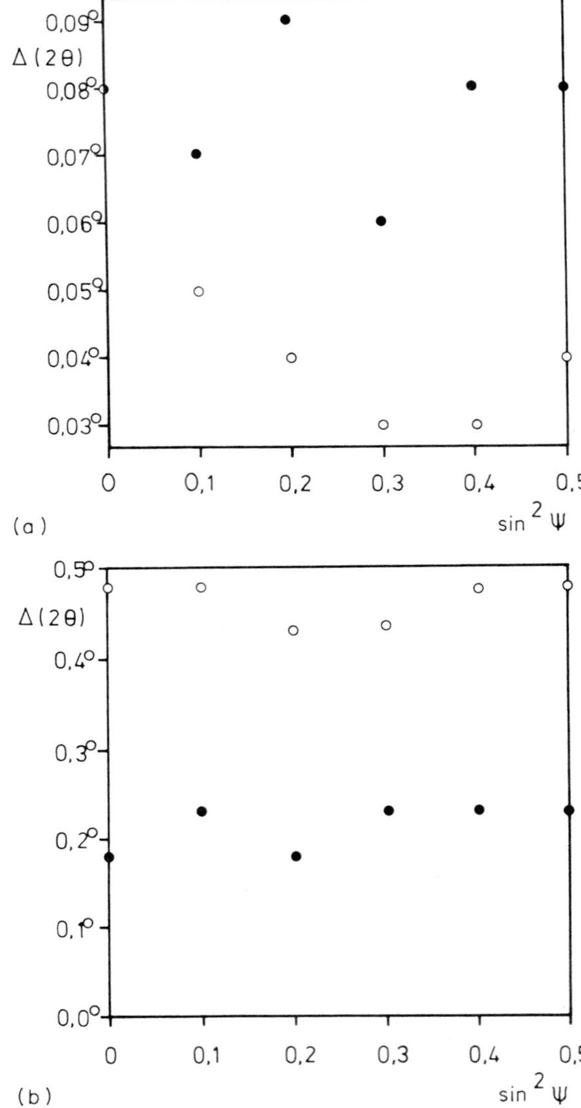

Fig. 5. Peak shifts in degrees for (a) steel nickel coated under nitrogen ($\sigma_\phi = \sigma_{ax}$) and (b) steel nickel coated under argon ($\sigma_\phi = \sigma_{tg}$): ●, before bending test; ○, after bending test.

could cause deviations from a straight line in the plot, affected the nickel reflection in question. The absence of a recognizable slope in Figs. 5(a) and 5(b) led to the conclusion that the residual stresses in the nickel layers were zero to within ±15 N mm^{-2}.

The exact crystallographic orientations of the nickel columns on the surface (Fig. 4) have not been analysed; both ⟨111⟩ and ⟨110⟩ orientations would be consistent with the peak positions measured so far. Similarly the tangential stresses were measured in the steel.

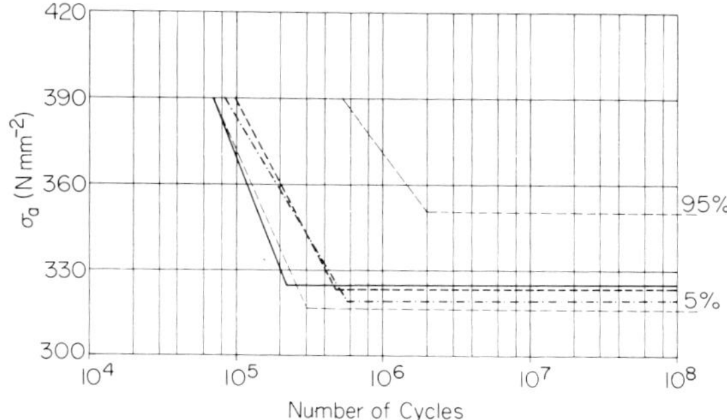

Fig. 6. Fatigue test Wöhler curves for a fracture probability of 16.6%: ——, uncoated steel; —·—, steel nickel coated under argon; - - -, steel nickel coated under nitrogen. For illustration, the curves for steel nickel coated under nitrogen for 95% and 5% fracture probabilities are also shown to illustrate the large spread in data.

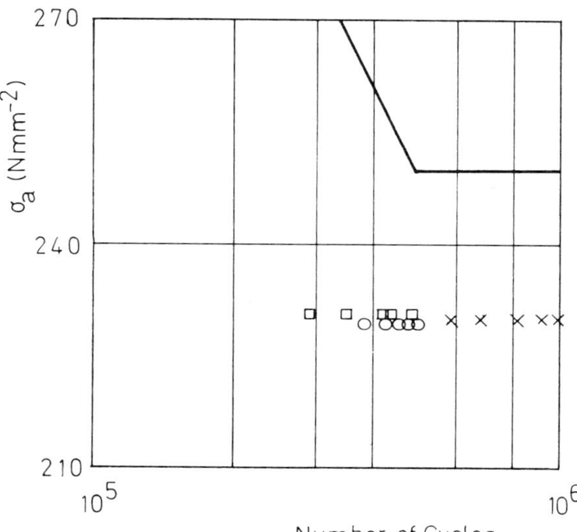

Fig. 7. Bending test under salt water at 230 N mm^{-2}: ×, uncoated steel; □, steel nickel coated under argon; ○, steel nickel coated under nitrogen.

Compressive stresses of −260 N mm^{-2} before the bending test that reduced to −130 N mm^{-2} after bending were found. No influence of the ion plating on these stress values could be detected.

The results of the bending fatigue test in air are summarized in Fig. 6. In constructing these curves the arcsin $P^{1/2}$ method [9] is employed, which assumes a statistical hypothesis for the fracture of the sample. When a number N of samples which are at a fixed stress value above the fatigue limit and which fail after a number N_i of cycles are considered, the method proceeds by assigning probabilities P_i (equal to i/N for $i = 1, \ldots, N-1$ and equal to $1 - 1/2N$ for $i = N$) to these events. The assumed relation between the logarithm of the cycle number and the fracture probability is

$$\log N_i = a + b \arcsin P_i^{1/2}$$

The constants a and b are least squares fitted to the data P_i and $\log N_i$. The experiments were carried out at two stress levels and yielded the slanting lines in Fig. 6. Similarly, in the vicinity of the fatigue limit, the fracture probability is given as the number of broken samples divided by the total number of samples at a fixed stress level and the assumed relation is

$$\sigma_i = c + d \arcsin P_i^{1/2}$$

c and d being least squares fitted. In the experiments, two stress values were employed to determine the horizontal lines. Figure 6 compares the results for uncoated steel and for steel nickel coated in an argon and a nitrogen atmosphere for the 16.6% fracture probability. The large data spread is exemplified by the 5% and 95% probability curves for the samples coated in nitrogen.

When this data spread is considered, the resulting curves show no significant differences between coated and uncoated samples. In conclusion, the formation of fatigue cracks in air is not influenced very much by the nickel layers. This should be compared with results reported in the literature on nickel layers deposited galvanically on Ck 45. For 30 μm layers, Wiegand [10] noted a slight (6%) decrease in the fatigue limit. In contrast, Broszeit [2] reported substantial decreases depending on the galvanizing baths used. The beneficial action of compressive stresses in the nickel layers is recognized in these papers.

Figure 7 shows the results of the bending test in salt water. These tests were performed on steel samples of a lower strength than those for the data in Fig. 6. The inset shows part of the Wöhler curve (50% fracture probability) for the uncoated material in air. The samples were tested at $\sigma = 230$ N mm^{-2}; this is below the fatigue limit for the material in air.

The coated samples perform worse in salt water than the uncoated samples do. The 50% fracture probability is given by $N = 7.42 \times 10^5$ for uncoated samples, $N = 3.68 \times 10^5$

for samples nickel coated in argon and $N = 4.35 \times 10^5$ for samples nickel coated in nitrogen, tested on five samples each. This is unexpected since nickel layers are used to improve the fatigue strength, and nickel layers deposited galvanically in general achieve this [10].

A possible explanation is that the ion-plated layers are not dense enough to prevent chlorine ions from attacking the steel surface. The strains occurring during each stress cycle are large enough to open the gaps between the nickel columns (Fig. 4). This leads to localized galvanic corrosion, pitting and crevice formation.

4. CONCLUSION

Although the ion-plated nickel layers described in this paper adhere and perform well under bending fatigue tests in air, their performance in corrosive media, however, is not satisfactory. A remedy would be to introduce compressive stresses into the layers by using larger gas pressures for the ion-plating process. Furthermore, according to the well-known Movchan–Demshishin–Thornton diagram [5, 6] an increase in substrate temperature leads to denser surface structures. To effect this, the process parameters will have to be changed.

ACKNOWLEDGMENTS

We thank the late Dipl.-Phys. Th. Neikes for providing the SEM pictures, and Professor Dr. H. Wohlfahrt for laboratory assistance.

REFERENCES

1 G. K. Wolf, W. A. Grant and R. P. M. Procter (eds.), *Proc. Int. Conf. on Surface Modification of Metals by Ion Beams, Heidelberg, September 17–21, 1984*, in *Mater. Sci. Eng., 69* (1985).
2 E. Broszeit, *Galvanotechnik, 75* (1984) 2, 164.
3 H. Gärtner, F. R. Hucke, K. Krug, K. Thoma, K. H. Tampe and H. W. Wagener, *Int. Tagung Verschleiß- und Korrosionsschutz durch Ionen- und Plasmagestützte Vakuum beschichtungstechnologien*, in *Tech. Hochsch. Darmstadt, Schriftenr. Wiss. Tech., 20* (1983) 199. R. Färber, K. Thoma, Th. Wieder and H. Gärtner, *2. Int. Tagung PVD'86, Verschleiß- und Korrosionsschutz durch ionen- und plasmagestützte Vakuum-Beschichtungstechnologien*, in *Tech. Hochsch. Darmstadt, Schriftenr. Wiss. Tech., 30* (1986) 306.
4 H. Neff, in V. Hauk and E. Macherauch (eds.), *Eigenspannungen und Lastspannungen*, Hanser, Munich, 1982, p. 19.
5 B. A. Movchan and A. V. Demshishin, *Fiz. Met. Metalloved., 28* (1969) 653.
6 J. A. Thornton, *J. Vac. Sci. Technol., 11* (1974) 666.
7 R. Glocker, *Materialprüfung mit Röntgenstrahlen*, Springer, Berlin, 1971.
8 V. Hauk and E. Macherauch, in V. Hauk and E. Macherauch (eds.), *Eigenspannungen und Lastspannungen*, Hanser, Munich, 1982, p. 1.
9 D. Dengel, *Z. Werkstofftech., 6* (1975) 253.
10 H. Wiegand, *Jahrbuch Oberflächentech., 15* (1959) 25.

Materials Science and Engineering, 90 (1987) 333-338

Effects of Titanium Implantation on Cavitation Erosion of Cobalt-based Metal–Carbide Systems*

N. V. H. GATELY and S. A. DILLICH

Worcester Polytechnic Institute, Worcester, MA 01609 (U.S.A.)

(Received July 10, 1986)

ABSTRACT

Cavitation erosion tests were performed on unimplanted and titanium-implanted samples of a cobalt-based superalloy and a 6wt.%Co–WC carbide. Erosion of unimplanted superalloy samples began with crack propagation through the carbides and debonding at the carbide–matrix interfaces, after which material loss from both matrix and carbide phases took place. Implantation of the alloy resulted in diminished carbide–matrix debonding and matrix phase erosion. Consequently, the erosion resistance of the alloy was significantly increased by implantation, as shown by comparison of cumulative mass loss data for unimplanted and implanted samples. Implantation also produced an improvement in the erosion resistance of the cemented carbide material, again because of increased durability of the matrix phase.

1. INTRODUCTION

Cavitation erosion is material loss of a solid surface in a fluid due to the repeated growth and collapse of bubbles near the surface. This type of wear is a severe problem in many common engineering systems, *e.g.* hydraulic turbines and pumps, pipes and valves, mining drills, diesel engine cylinders, steam turbine blades and ship propellers. Cobalt-based superalloys, consisting of hard chromium and tungsten carbides (primarily M_7C_3 and M_6C) in softer more ductile cobalt-rich solid solutions, are among the most erosion-resistant materials commercially available [1-4]. However, the high cost of these materials as well as the uncertainties in the cobalt market are incentives for further study and, if possible, improvement of the erosion resistance of these alloys.

In a previous study, high fluence titanium implantation of a cobalt-based superalloy (Stoody 3; Co–31wt.%Cr–12.5wt.%W–2.2wt.%C; Rockwell C hardness, 54–58 HRC) was found to produce significant improvements in the tribological behavior of the alloy, as shown by dry sliding friction measurements and abrasive wear tests [5, 6]. However, these tests yielded little information on the wear mechanisms active in the alloy, other than that they can be altered by ion implantation. In this work, cavitation erosion tests were performed on unimplanted and implanted Stoody 3 samples in order to investigate further the wear modes in metal–carbide composite alloys and the manner in which they are influenced by implantation. Of particular interest were the specific contributions of the different matrix and carbide phases to the erosion of the alloy.

Cavitation erosion tests were also performed on another technologically important class of metal–carbide composites: 6wt.%Co–WC cemented carbides (WC grain size, $\frac{1}{2}$–3 μm; Rockwell C hardness, 78 HRC).

2. EXPERIMENTAL PROCEDURE

Cavitation erosion test equipment (vibrating horn configuration) was assembled and calibrated according to ref. 4. A schematic diagram of the test apparatus can be seen in Fig. 1. Tests were performed under standard test conditions (frequency, 20 kHz; peak-to-peak amplitude displacement of the vibrating tip, 0.05 mm). The test liquid, distilled water, was

*Paper presented at the International Conference on Surface Modification of Metals by Ion Beams, Kingston, Canada, July 7–11, 1986.

0025-5416/87/$3.50

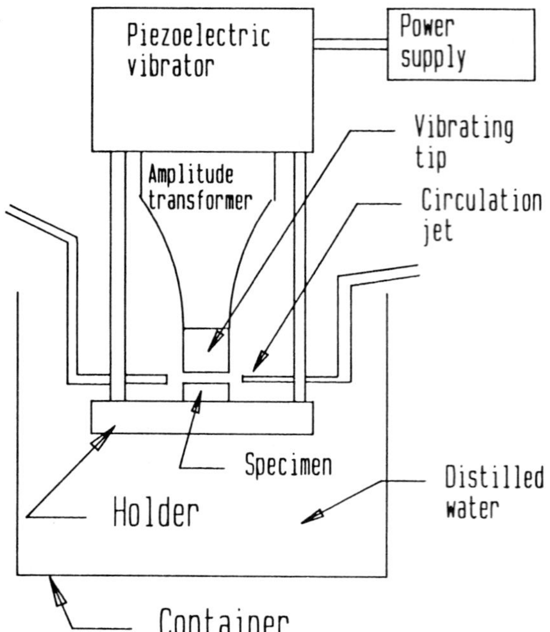

Fig. 1. Apparatus for cavitation erosion testing. The test sample is held stationary below a horn oscillating at 20 kHz with an amplitude of 0.051 mm. Air bubbles nucleated in the test liquid below the vibrating horn tip collapse on the sample surface, causing surface damage and material loss.

kept at 20 ± 2 °C by means of a cooling coil immersed in the test beaker. Additional cooling of the test surfaces was provided by water jets (pump circulated test water) through the sample–vibrating tip gap (Fig. 1).

Test conditions deviated from the standard in one respect: because of the extreme hardness and non-machinability of the alloy, samples were not attached to the horn by means of a threaded end. Rather, test samples were held stationary below the vibrating horn, at a stationary tip–sample separation distance of 0.5 mm. In addition to being convenient, this configuration also avoided the disadvantages of the moving sample geometry, *e.g.* stress imposed by the longitudinal vibrations of the horn even in the absence of cavitation [7] and rimming (non-uniform erosion of the sample due to an undamaged annular region around the perimeter of the sample).

Test samples were Stoody 3 disks (1.27 cm in diameter and 0.6 cm thick) and 6wt.%Co–WC flats (1.27 cm square and 0.48 cm thick). Before testing, the sample surfaces were diamond polished to a 1 μm diamond finish and cleaned with organic solvents. Titanium implantations were made to a fluence of 5 × 10^{17} Ti^+ cm^{-2} at 190 keV.

The losses in mass of the samples due to cavitation erosion were monitored during testing by periodic determination of the mass of the samples. Erosion depths of the samples were calculated using the mass loss measurements, bulk densities (8.63 g cm^{-3} and 14.9 g cm^{-3} for the Stoody 3 alloy and the 6wt.%-Co–WC carbide respectively) and surface areas of the samples. Mass loss measurements for the Stoody 3 samples were made with a digital scale readable to 0.01 mg. A scale accurate to ±0.1 mg proved sufficient to measure the mass loss of the carbide samples. Periodic scanning electron microscopy (SEM) inspection of the samples was made during testing to monitor the initiation and progression of damage to the surfaces.

The total test time per sample varied between 15 and 35 h. Except when noted, tests were interrupted every 1–3 h for intervals of at least 5 min, to allow the surface temperatures of the samples to equilibrate.

3. RESULTS

Cumulative erosion depths of unimplanted and implanted Stoody 3 samples, run to total test times of between 20 and 35 h, are shown in Fig. 2. SEM micrographs showing the damage to one region of unimplanted sample 4 after testing for 3, 10 and 20 h can be seen in Fig. 3. A similar series of micrographs for implanted sample 3 is shown in Fig. 4. A comparison of the surfaces of these samples after testing for 20 h is provided in Fig. 5.

The incubation period (the test time before mass loss was observed) was between 5 and 10 h for the unimplanted samples. Increases in the rates of erosion (*i.e.* increases in the cumulative depth of erosion per unit time) were noticed after 20 h for samples 2 and 4, and again after 25 h for sample 2 (the break in the ordinate axis should be noted). As was mentioned previously, samples were interrupted during testing at intervals of the order of an hour. Unimplanted sample 1 was an exception in that it was tested for 20 h without interruption before determination of its mass. Also, titanium-implanted sample 2 was tested without interruption between 5 and 15

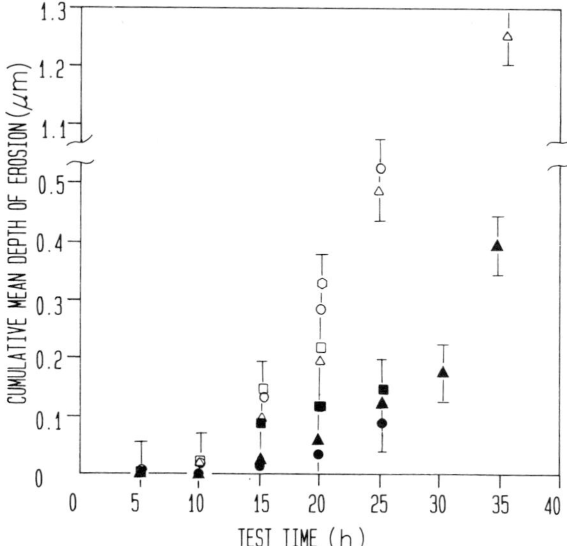

Fig. 2. Mean depth of erosion in micrometers (10^{-6} m) *vs.* test time for unimplanted and titanium-implanted samples of Stoody 3: ○, unimplanted sample 1; △, unimplanted sample 2; □, unimplanted sample 3; ○, unimplanted sample 4; ▲, titanium-implanted sample 1; ■, titanium-implanted sample 2; ●, titanium-implanted sample 3.

h. The higher erosion rates observed for these samples during these time periods compared with the other test samples suggests that the erosion rate increases with increasing length of the exposure interval, as well as with increasing cumulative exposure time.

Damage to the surfaces, namely crack propagation through the carbides and debonding at carbide–matrix interfaces (Fig. 3(a)), could be observed under SEM examination after the first hour of testing. After testing for about 5–10 h, some damage to the matrix phase could also be detected (Fig. 3(b)). Material loss from both carbides and matrix occurred after about 10 h, so that by 20 h microstructural features previously seen on the surfaces were barely recognizable (Figs. 3(c) and 5(a)).

The cumulative depths of erosion were much lower for the titanium-implanted samples (Fig. 2). The incubation period for these samples was 15 h, as opposed to less than 10 h for the unimplanted samples, after which material loss progressed steadily at erosion rates about one-third of that of the unimplanted samples. The titanium-implanted sample tested for 35 h, sample 1 (full triangle in Fig. 2), however, did show an increase in erosion rate at 30 h.

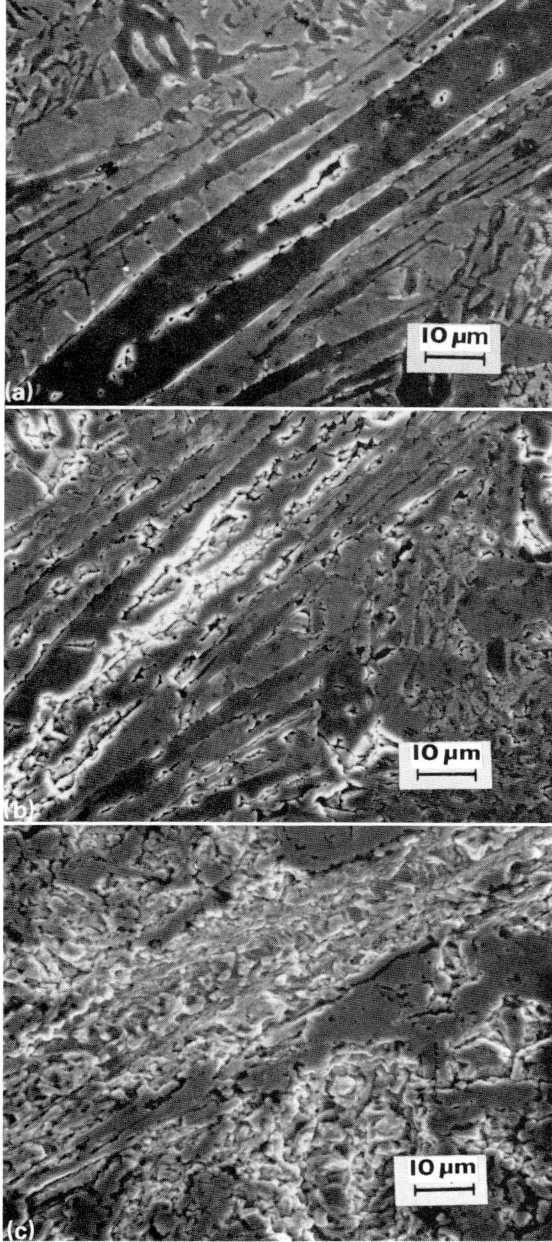

Fig. 3. SEM micrographs of a region on Stoody 3 sample 4 after testing for (a) 3 h, (b) 10 h and (c) 20 h.

As in the unimplanted sample tests, surface damage began with crack nucleation and propagation through the larger carbides (Fig. 4(a)). Initial material loss occurred primarily from carbide crack sites (Figs. 4(b), 4(c) and 5(b)). Cohesion at the carbide–matrix interfaces appeared to be enhanced relative to that in the unimplanted samples, although some

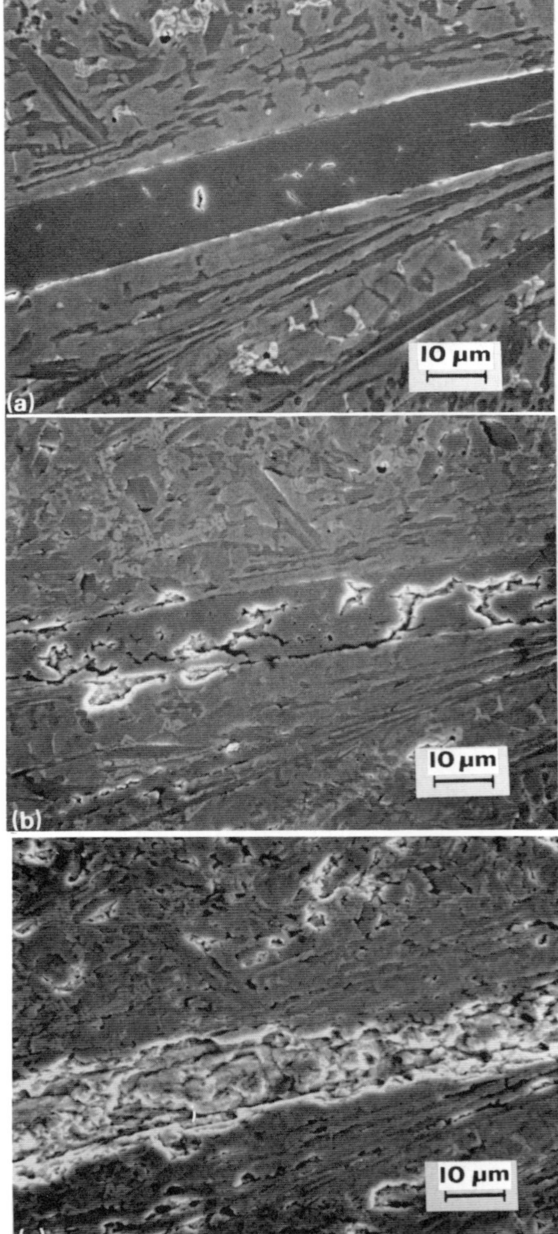

Fig. 4. SEM micrographs of a region on Stoody 3 titanium-implanted sample 3 after testing for (a) 3 h, (b) 10 h and (c) 20 h.

debonding at these interfaces was again observed. Only after testing for about 15 h was material loss from the matrix phase observed, and then to a lesser extent than in the unimplanted samples (compare Figs. 3(c) and 4(c) and Figs. 5(a) and 5(b)). However, as was the case for unimplanted samples, surface damage

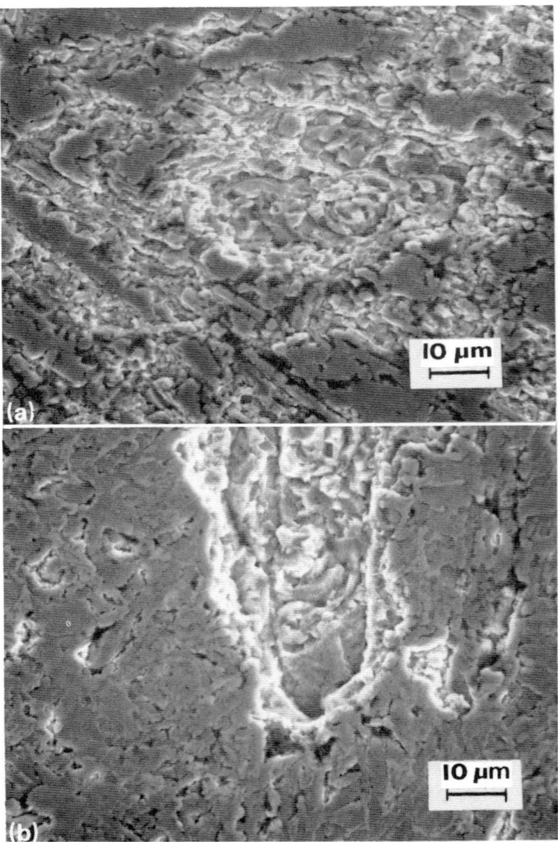

Fig. 5. SEM micrographs of (a) an unimplanted sample and (b) an implanted sample of Stoody 3 after testing for 20 h. Severe damage occurred to both matrix and carbide phases on the unimplanted sample. Less damage to the matrix phase was observed on the implanted sample.

progressed rapidly after the appearance of matrix phase erosion.

The cumulative depths of erosion of the unimplanted and implanted 6wt.%Co–WC samples are shown in Fig. 6. The incubation periods for these samples were less than 1 h. Not only did material loss begin earlier in the carbide samples than in the Stoody alloy, but also it progressed at a rate which was an order of magnitude higher (compare Figs. 2 and 6). Erosion initiated at voids originally present on the surfaces. After testing for 5 h, little of the cobalt binder phase remained on the carbide surfaces which, as shown by SEM and energy-dispersive X-ray examinations, consisted almost entirely of WC particles (Fig. 7(a)).

The implanted carbide sample showed an erosion rate about 50% lower than that of the

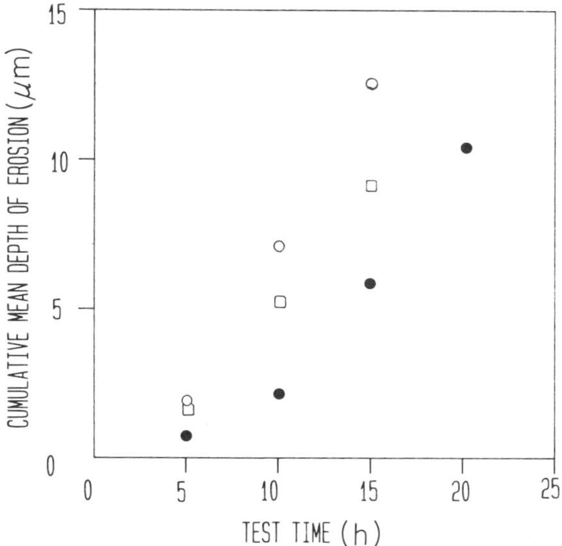

Fig. 6. Mean depth of erosion in micrometers (10^{-6} m) *vs.* test time for unimplanted and titanium-implanted samples of a 6wt.%Co–WC carbide: ○, unimplanted sample 1; □, unimplanted sample 2; ●, titanium-implanted sample.

Fig. 7. SEM micrographs of (a) an unimplanted sample and (b) a titanium-implanted sample of 6wt.%Co–WC after testing for 5 h.

unimplanted samples (Fig. 6). Again, surface voids acted as erosion initiation sites. In this case, however, the cobalt binder phase was removed more slowly. As Fig. 7(b) shows, much of the original surface remained intact after testing for 5 h. By 15 h, the surface binder phase material had been almost entirely removed from the implanted sample, and there was no longer any distinguishable difference between the appearance of this surface and that of the unimplanted samples.

4. DISCUSSION

The results described above indicate that under high impact stress conditions, such as are encountered during cavitation, the wear resistance of the Stoody 3 alloy is determined by the endurance of the cobalt-based matrix phase. Erosion of this phase is preceded by debonding of the matrix from the larger of the carbides, presumably because of differences between the elastic moduli and plasticities of the phases [8]. Although material loss from the second-phase carbides was observed on both unimplanted and implanted Stoody 3 samples, measurable sample mass loss coincided with the appearance of damage to the matrix phase.

Material loss and damage to the carbides on implanted Stoody 3 surfaces appeared, under SEM examination, comparable with or perhaps slightly less than that on the unimplanted samples. Carbide–matrix debonding and matrix phase erosion were inhibited on the implanted samples, yielding a longer incubation period and lower erosion rates relative to the unimplanted samples.

Significantly lower cumulative depths of erosion for implanted surfaces, relative to those of unimplanted surfaces, may persist long after the loss of the implant layer, as is suggested by the data for implanted sample 1 (full triangle in Fig. 2). As was previously described, an increase in the erosion rate of this sample was observed after testing for 30 h and a mean depth of erosion of 0.17 μm, *i.e.* approximately the depth of the original implant layer [5]. Since the erosion rates of the unimplanted samples were not constant but rather increased with time, the cumulative depth of erosion of the implanted sample at 35 h was still only one-third of that of the corresponding unimplanted sample (open triangle in Fig. 2). It can be expected therefore that, despite the removal of the implant layer, the ratio of the cumulative wear loss of initially implanted samples to that of unimplanted samples will increase only slowly with extended exposure to cavitation.

The superior erosion resistance of the titanium-implanted Stoody 3 samples is consistent with chemical and microstructural changes observed in the alloy after implantation. High fluence titanium implantation of the Stoody alloy has been found to produce a carburized surface layer [5] with an amorphous matrix phase and recrystallized carbides [9]. A corresponding toughening of the matrix phase and softening of the carbides can account for the observed increased erosion resistance of the matrix and enhanced carbide–matrix cohesion.

The 6wt.%Co–WC samples eroded much more quickly than did the Stoody 3 alloy samples, a consequence, most likely, of the higher porosity and carbide–metal interface density of these samples. Debonding at the carbide–metal interfaces and the subsequent rapid loss of the cobalt binder phase resulted in attrition rather than erosion of the carbide particles.

A lower erosion rate, together with increased erosion resistance of the binder phase, was observed for the implanted carbide sample. Structural changes in this material due to implantation have not been determined. However, it seems probable that, as was the case for the Stoody alloy, toughening of the cobalt binder and enhanced carbide–metal cohesion were the results of implantation-induced amorphization of the binder phase.

5. SUMMARY

Cavitation erosion testing provided a high impact stress non-contact accelerated wear situation by which the wear resistance contributions of separate components of multiphase alloys and carbides, as well as the changes produced by implantation, could be evaluated. In both materials tested, a cobalt-based superalloy (Stoody 3) and 6wt.%Co–WC, resistance to erosive wear was controlled by the durability of the matrix phase. Titanium implantation improved the erosion resistance of both materials as a result of implantation-enhanced cohesion at carbide–metal interfaces and toughening of the cobalt-based matrix and binder phases.

ACKNOWLEDGMENTS

We thank the GTE Research Laboratories in Waltham, MA, for the donation of the cemented carbide samples and the Surface Modification and Materials Analysis Group at the Naval Research Laboratory for the sample implantations.

This work was supported by the Office of Naval Research.

REFERENCES

1 K. C. Antony, *J. Met., 25* (1973) 52.
2 K. J. Bhansali and A. E. Miller, *Wear, 75* (1982) 241–252.
3 C. J. Heathcock, A. Ball and B. E. Protheroe, *Wear, 74* (1981–1982) 254–258.
4 *ASTM Stand. G-32-85*, in *ASTM Standards*, Vol. 03.02, ASTM, Philadelphia, PA, 1985, pp. 187–194.
5 S. A. Dillich and I. L. Singer, *Thin Solid Films, 73* (1981) 219–227.
6 S. A. Dillich, R. N. Bolster and I. L. Singer, in G. K. Hubler, O. W. Holland, C. R. Clayton and C. W. White (eds.), *Ion Implantation and Ion Beam Processing of Materials, Materials Research Society Symp. Proc., Boston, MA, November 14–17, 1983*, Vol. 27, Elsevier, New York, 1984, pp. 661–666.
7 C. M. Preece, *Cavitation Erosion, Treatise Mater. Sci. Technol., 16* (1979) 249–308.
8 C. J. Heathcock, B. E. Protheroe and A. Ball, *Wear, 81* (1982) 311–327.
9 S. A. Dillich and R. R. Biederman, *Proc. Int. Conf. on Surface Modification of Metals by Ion Beams, Kingston, July 7–11, 1986*, in *Mater. Sci. Eng., 90* (1987).

Ductile-to-brittle Transition Induced by Nitrogen Ion Implantation into WC Single Crystals*

S. BARTOLUCCI LUYCKX, Y. CASSUTO and J. P. F. SELLSCHOP

Wits-Council for Scientific and Industrial Research Scholand Research Centre for Nuclear Sciences, University of the Witwatersrand, Johannesburg (South Africa)

(Received July 10, 1986)

ABSTRACT

Vickers indentation patterns were studied before and after nitrogen ion implantation (N^+) on (0001) planes of WC single crystals. Before implantation the patterns consisted of only slip lines and after implantation they consisted of slip lines and cracks. At low ion fluences the cracks were dislocation-induced cleavage cracks; at high ion fluences the cracks followed paths of maximum tensile stress and were unrelated to cleavage planes. The material was clearly embrittled by ion implantation.

1. INTRODUCTION

Indentation patterns can provide simultaneous information on the deformation and fracture properties of a material. The size of the indentation depends on the resistance to plastic deformation, *i.e.* on the hardness; the amount of cracking depends on the resistance to fracture, *i.e.* on the toughness. The relative susceptibility of a material to plastic deformation and fracture determines whether the material is ductile or brittle.

In this paper a study of the susceptibility to plastic deformation and fracture of WC single crystals before and after nitrogen ion implantation is presented.

2. EXPERIMENTAL DETAILS

Indentation patterns were produced on (0001) faces of WC single crystals. The crystals were triangular prisms grown from a solution of 60wt.%Co-40wt.%WC. The sides of the prisms ranged from 1 to 3 mm in length.

The (0001) faces were mechanically polished. The crystals were then annealed at 10^{-7} Torr and 800 °C for 2 h. The (0001) faces were indented before and after implantation with 100 keV nitrogen ion (N^+) fluences ranging from 3×10^{16} to 1.5×10^{17} ions cm^{-2}. The depth of ion penetration was of the order of 0.2 μm. The indentations were produced with a standard Vickers microindenter at a load of 300 gf. The depth of the indentations was about 4.4 μm and the work done in producing an indentation was approximately 2.2×10^{-6} J [1]. The indentation patterns were studied by means of scanning electron microscopy.

3. RESULTS

Figures 1-3 show 300 gf Vickers indentations on (0001) faces of WC single crystals before and after implantation. Within the resolution of the scanning electron microscope, the sizes of the three indentations are the same.

Figure 1 shows a 300 gf indentation on a (0001) face of an unimplanted crystal. The slip lines visible around the indentation belong to the $\{01\bar{1}0\}\langle 2\bar{1}\bar{1}3\rangle$ slip system [2]. No cracks were observed around the indentations on the (0001) faces of unimplanted crystals, in agreement with the results of earlier investigations [3, 4].

Figure 2 shows a 300 gf indentation on a (0001) face of a crystal which was implanted to a fluence of 3×10^{16} N^+ cm^{-2}. The slip lines around the indentation belong to the same slip system as in Fig. 1. The indentation pattern includes cracks, which were nucleated

*Paper presented at the International Conference on Surface Modification of Metals by Ion Beams, Kingston, Canada, July 7-11, 1986.

0025-5416/87/$3.50

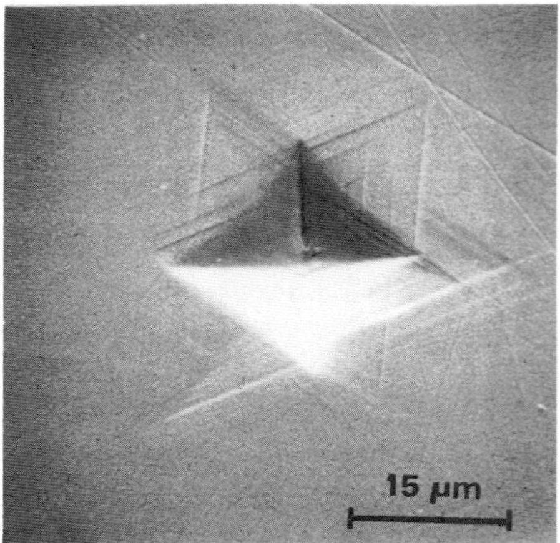

Fig. 1. 300 gf indentation on an unimplanted (0001) face of a WC single crystal. The indentation pattern consists only of slip lines.

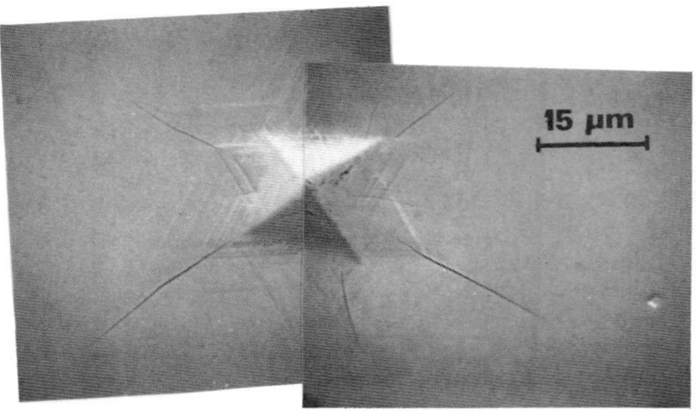

Fig. 2. 300 gf indentation on a (0001) face of a WC crystal implanted to a fluence of 3×10^{16} N^+ cm^{-2}. The indentation pattern consists of slip lines and dislocation-induced cleavage cracks.

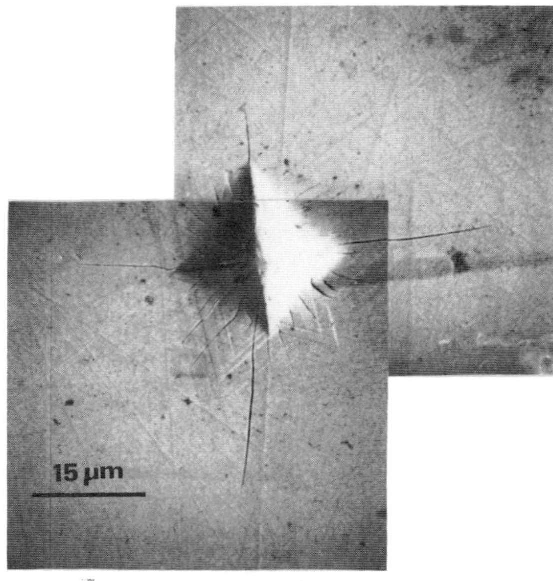

Fig. 3. 300 gf indentation on a (0001) face of a WC crystal implanted to a fluence of 1.5×10^{17} N^+ cm^{-2}. The indentation pattern consists of slip lines, Palmqvist cracks and "star" cracks, unrelated to cleavage planes.

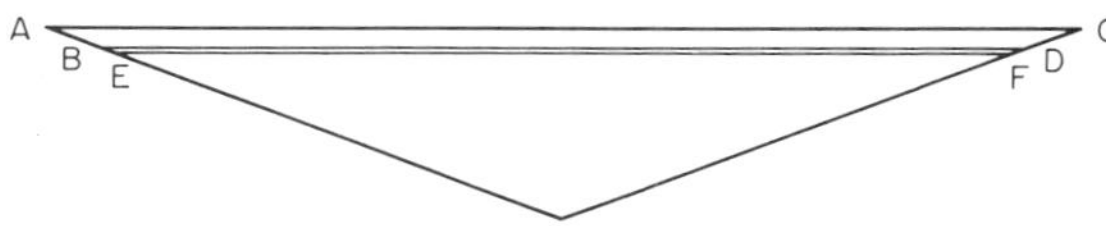

Fig. 4. Schematic representation of a section through an indentation in an ion-implanted crystal. AC represents the free surface of the crystal; BEFD represents the subsurface layer (not in scale) in which the ions came to a stop.

through the coalescence of glide dislocations on intersecting slip planes. These appear to be cleavage cracks propagating along $\{2\bar{1}\bar{1}0\}$ planes, at about 60° to each other.

Figure 3 shows a 300 gf indentation on a (0001) plane of a crystal which was implanted to a fluence of 1.5×10^{17} N^+ cm^{-2}. Slip lines are still visible, as well as cracks. These cracks, however, are unrelated to slip plane intersections. They did not propagate along cleavage planes but along paths of maximum tensile stress. The cracks which emerge from the corners of the indentation in Fig. 3 are Palmqvist cracks [5], which are typical of Vickers indentation patterns in hard brittle materials. The shorter cracks in Fig. 3, which emerge from the indentation sides, are also typical of hard brittle materials but are usually found around conical or spherical indentations. They may be called radial [6] or "star" [7] cracks.

4. DISCUSSION

The following discussion is based on the results reported above as well as on the results reported elsewhere [8] which were obtained from transmission electron microscopy of thin (about 20 nm) WC crystals. By transmission electron microscopy it was found that nitrogen ion implantation increases the dislocation density in WC to the point that the dislocation mobility is reduced, as occurs in work-hardened material.

Figure 4 is a schematic representation of a section normal to the free surface of the crystal, through a diagonal of an indentation such as those in Figs. 1–3. Lines BD and EF in Fig. 4 delimit (not in scale) the subsurface where most of the implanted ions came to a stop. The thickness of the layer BEFD in Fig. 4 is very small compared with the depth of the indentation, which accounts for the equal sizes of the indentations in Figs. 1–3.

When a crystal is indented, the material below and around the indenter may deform plastically or elastically and cracks may be formed. The elastic energy stored during indentation is released when the indenter is removed, which may produce additional plastic deformation and/or fracture.

Before ion implantation the indentation pattern on (0001) faces of WC crystals (Fig. 1) consisted only of slip lines, which were produced by dislocations gliding on $\{01\bar{1}0\}$ planes along $\langle 2\bar{1}\bar{1}3\rangle$ directions. Thus the indenting work was transformed totally into plastic work.

After implantation to a fluence of 3×10^{16} N^+ cm^{-2}, slip lines were still visible (Fig. 2) because the plasticity of zone ABDC (Fig. 4) was not affected by the implanted ions. In zone BEFD, however, the dislocation density was increased by ion implantation [8] and the dislocations were less mobile; thus the energy that could not be released by dislocation motion was stored as elastic energy. When the indenter was removed, this energy was released by the formation of the dislocation-induced cleavage cracks visible in Fig. 2. Similar cracks have been observed in the past on WC fracture surfaces [9] where they were also produced by elastic stress relaxation.

After implantation to a fluence of 1.5×10^{17} N^+ cm^{-2} the dislocations in zone ABDC (Fig. 4) could still glide (Fig. 3) but zone BEFD was so hardened by the formation of dislocation tangles that almost none of the energy supplied by the indenter could be released by plastic deformation. Cracking, therefore, occurred in the typical manner of brittle materials: by Palmqvist cracks and by "star" cracks (Fig. 3). Both Palmqvist cracks and "star" cracks usually initiate during the loading of the indenter but continue to propagate during the unloading. They also, therefore, release stored elastic energy.

The changes in the indentation patterns observed before and after implantation can thus be explained in terms of the hardening of the subsurface layer in which the implanted ions came to a stop. Before implantation the indenting work was totally transformed into plastic work, while after implantation it was also transformed into fracture surface energy. The subsurface layer BEFD, therefore, underwent a ductile-to-brittle transition. From the point of view of practical applications of implanted material, it will be important to determine at which ion fluence the transition occurs.

5. CONCLUSION

By studying indentation patterns on the (0001) faces of WC single crystals before and after implantation it was found that the subsurface layer of the crystals in which the ions come to a stop undergoes a ductile-to-brittle transition.

ACKNOWLEDGMENTS

The authors are grateful to Dr. T. E. Derry for assistance in the ion implantation and to Boart International Ltd., Sandton, South Africa, for financial support.

REFERENCES

1 S. B. Luyckx, in S. R. Valluri, D. M. Taplin, P. R. Rao, J. F. Knott and R. Dubey (eds.), *Advances in Fracture Research, Proc. 6th Int. Conf. on Fracture, New Delhi, December 4–10, 1984*, Pergamon, Oxford, p. 2665.
2 S. B. Luyckx, *Acta Metall., 18* (1970) 233.
3 T. Takahashi and E. J. Freise, *Philos. Mag., 12* (1965) 1.
4 J. Corteville, J. C. Monier and L. Pons, *C.R. Acad. Sci., 260* (1965) 2773.
5 S. B. Luyckx, *Specialty Steels and Hard Materials*, Pergamon, Oxford, 1983, p. 369.
6 C. J. Studman and J. E. Field, *J. Mater. Sci., 12* (1977) 215.
7 B. Lawn and R. Wilshaw, *J. Mater. Sci., 10* (1975) 1049.
8 S. B. Luyckx, Y. Cassuto, T. E. Derry and J. P. F. Sellschop, *Proc. 5th Int. Conf. on the Ion Beam Modification of Materials, Catania, June 9–13, 1986*, in *Nucl. Instrum. Methods*, to be published.
9 S. Bartolucci and H. H. Schlossin, *Acta Metall., 14* (1966) 337.

Materials Science and Engineering, 90 (1987) 343-347

Characterization of Ion-beam-induced Carbon Deposition on WC–Co Hard Metal by Microhardness, Scratch and Abrasive Wear Tests*

J.-P. HIRVONEN[a]†, J. KOSKINEN[a], A. ANTTILA[a], D. STONE[b] and C. PASZKIET[b]

[a] *Department of Physics, University of Helsinki, SF-00170 Helsinki (Finland)*

[b] *Department of Materials Science and Engineering, Cornell University, Ithaca, NY 14853 (U.S.A.)*

(Received July 10, 1986)

ABSTRACT

Diamond-like ion-beam-deposited carbon (i-C) layers were obtained on WC–Co cemented carbide using a mass-separated ^{12}C beam at an energy of 500 eV and a deposition rate of 3 Å s^{-1}. The mechanical properties of these layers were probed using microhardness and scratch tests and abrasive wear measurements. All these tests revealed that the depositions possess an extremely high hardness and good adhesion to the substrate. In particular, a hardness of 75 GPa was obtained, which is considerably higher than that found on i-C films involving hydrogen.

1. INTRODUCTION

Energetic ion beams provide a novel technique for producing metastable phases. This is easily understandable based on the atomistic picture of irradiation effects in material. Although the quantitative theory is still far from complete, molecular dynamics simulations allow some general conclusions to be drawn about the low energy cascade in pure materials [1]. Accordingly, the radiation cascade can be divided into three different periods of time. During the first, within about 10^{-13} s after the arrival of a primary ion, a high defect density is created by two-body collisions. By the end of this time, concentrations of self-interstitials and vacancies may reach a few atomic per cent. Also, the average kinetic energy of atoms in a cascade has increased far above the value at thermal equilibrium. This first period is followed by a second period, lasting approximately 10^{-12} s, during which the majority of the defects are removed by mutual spontaneous recombination resulting from high defect concentrations. The defect concentrations after this period are in fair agreement with the Kinchin and Pease model. These two first phases of the cascade can be considered as ballistic and are followed by the thermal spike period. During this period, point defects migrate by more or less diffusion-like mechanisms. The energy has become locally equipartitioned between different atoms, which gives a collective character to this concluding regime. In less than 10^{-11} s after the arrival of the primary ion the equilibrium with the surrounding crystal has been obtained.

The temperature inside the radiation cascade and thermal spike may reach values of several 10^3 K [2, 3]. It has also been estimated that the strong agitation of atoms increases the pressure at the cascade surface considerably [4,5]. Under such conditions the material naturally is far from the most stable equilibrium phase. Consequently, the production of ion-beam-deposited phases of such superhard materials as diamond and cubic boron nitride by depositing energetic ions has been documented by several researchers [6-9]. It is not yet known, however, whether the formation of these phases is due to the effect of the ion bombardment on the mobility and disorder of atoms or to some type of chemical activa-

*Paper presented at the International Conference on Surface Modification of Metals by Ion Beams, Kingston, Canada, July 7-11, 1986.

†Temporary address: Department of Materials Science and Engineering, Cornell University, Ithaca, NY 14853, U.S.A.

0025-5416/87/$3.50

tion. Although the structure of these phases is not clear and involves disordered characteristics, many properties correspond to those of their crystalline counterparts. Among others, they are chemically inert and possess a high electrical resistivity and an extremely high hardness. This alone, particularly for ion-beam-deposited carbon (i-C), justifies the use of the name diamond-like material.

A wide variety of ion-assisted techniques have been used to prepare ion-beam-deposited phases [6–8]. For most of them, the incorporation of a high hydrogen concentration is a common feature. Hydrogen is said to be necessary to tailor optical properties and to obtain good transparency [10]. However, in mechanical applications, high hydrogen concentrations cause problems such as the presence of high internal stresses [11]. Furthermore, the effects of hydrogen on mechanical properties such as hardness and adhesion are not yet clear, although it is expected that the production of diamond-like depositions without hydrogen is desirable. Such depositions require high purity conditions which are at present only possible using mass-analyzed ion beams. Accordingly, we have begun to study the physical properties of ion-beam-deposited diamond-like carbon. In this work we have used microhardness, scratch and abrasive wear tests to characterize diamond-like i-C on WC–Co cemented carbide.

2. EXPERIMENTAL METHODS AND MEASUREMENTS

Commercial grade WC–Co cemented carbide discs of 2 mm nominal thickness were employed as substrates. These were cut from a rod 15 mm in diameter and then were mechanically polished with diamond paste. The deposition of the i-C was carried out using a 100 kV isotope separator of the Scandinavian type. The deposition energy was 500 eV. The ion source used was a small duoplasmatron equipped with a permanent magnet. The mass separation was performed with a magnet of 90°. Because of the relatively low ion current (20–30 μA), no sweeping system was used. Accordingly, the deposited area was only a few square millimeters and the corresponding current density and deposition rate were 1 mA cm^{-2} and 3 Å s^{-1} respectively. The maximum thicknesses of the depositions varied from 5 to 7.5 μm. The deposition arrangement was nearly the same as that of Miyazawa *et al.* [12] except that no trap for neutral ions was used.

To measure hardness, a special apparatus was utilized [13]. A Berkovitch indenter of tip radius less than 0.1 μm was pushed into the surface of the specimen at a constant displacement rate (0.025 μm s^{-1}) until the desired load was reached. The indenter was then removed at the same rate. The load and position were measured continuously by an apparatus having resolutions of 30 mg and 2.5 nm respectively. The impressions were examined in a scanning electron microscope and the indentation areas were measured. The hardness was calculated by dividing the maximum load by the area of the indent.

The arrangements for the abrasive wear test are shown schematically in Fig. 1. An SiC ball 3.13 mm in diameter was employed as a wear pin. The sample performed a reciprocating motion with an amplitude of 7 mm at a sliding speed of 15 mm s^{-1}. The sliding motion carried the SiC pin over both the substrate and the deposited layer. During the experiments, the sample was mounted on the bottom of a small vessel filled with a slurry of paraffin oil and SiC grinding powder (400 mesh). A load of 32 gf was used and the measurements were extended to 20×10^3 passes.

A scratch test was also performed using a reciprocating wear apparatus. A diamond of a pyramidal shape was used as the stylus. As in the abrasive wear measurements, the test

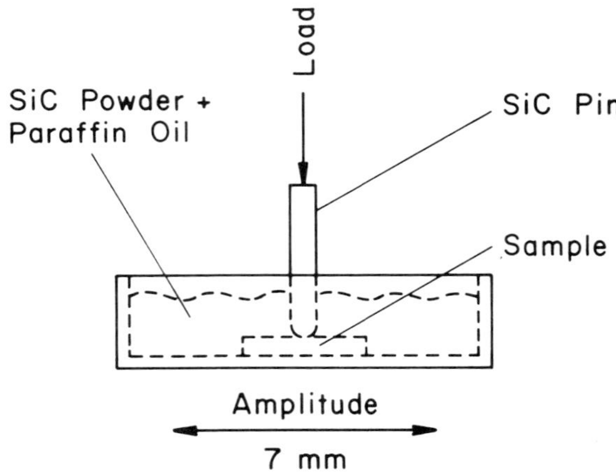

Fig. 1. A schematic picture of the abrasive wear test.

was extended simultaneously over both the substrate and the deposited layer. The scratch was repeated 500 times on the same track with a load of 10 gf normal to the sample surface.

The indentation traces as well as the wear scars and scratches were viewed in optical and scanning electron microscopes. The wear scars were also measured with the Alpha step profilometer.

3. RESULTS

Figure 2 presents load *vs.* indentation depth curves for i-C and the WC–Co substrate with maximum loads used during indentation of 50 gf and 25 gf respectively. As can be seen, the behavior of i-C differs significantly from that of pure WC–Co. The same indentation depth was reached for WC–Co with a load of only one-half of that used for i-C. The permanent plastic deformation for the substrate is considerably greater than that for the i-C deposition. In addition, the i-C reveals pronounced elastic recovery which has also been reported for hydrogenated i-C deposition [11]. When a maximum load of 10 gf was used, no indentation trace could be observed with either a scanning electron microscope or an optical microscope. The present measurements give values of 75 GPa and 18 GPa for the hardness of i-C and that of the pure WC–Co substrate respectively. The hardness of WC–Co cemented carbide corresponds to the maximum value obtained from a single grain.

On the basis of the hardness tests, excellent results from the abrasive wear measurements were expected. Figure 3 presents an optical micrograph of the wear scar at the WC–Co substrate–i-C boundary after 20×10^3 passes. The pure substrate has been severely damaged whereas the i-C deposition is almost undamaged. Also the SiC pin wore significantly during the test.

The wear scars were also studied with the Alpha step profilometer (Fig. 4). This also confirms that, although the pure substrate has worn very severely, the i-C deposition is still undamaged. The maximum wear depth of WC–Co is about 7.5 μm whereas no wear scar on the i-C deposition can be detected.

Scratch tests were used as an additional measure of hardness as well as to study the adhesion of the deposition to a substrate. Figure 5 presents the scratch at the pure substrate–i-C boundary. The scratch was repeated

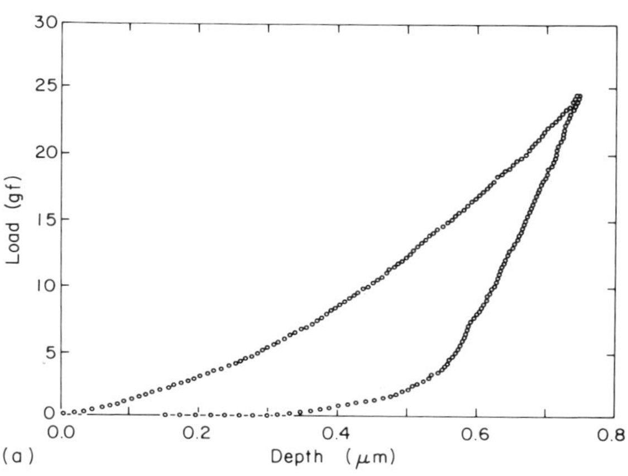

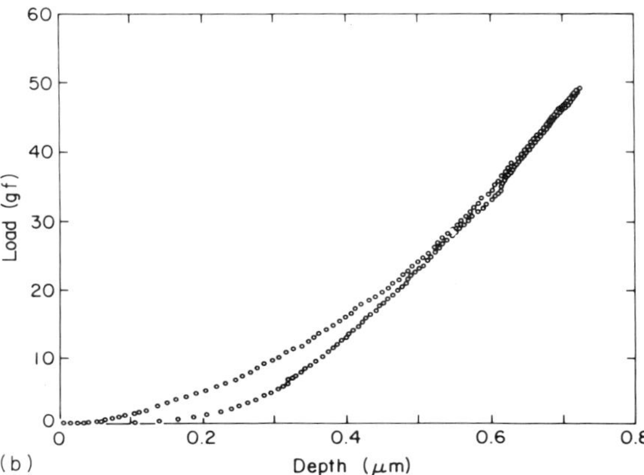

Fig. 2. Load *vs.* indentation depth for (a) WC–Co and (b) i-C.

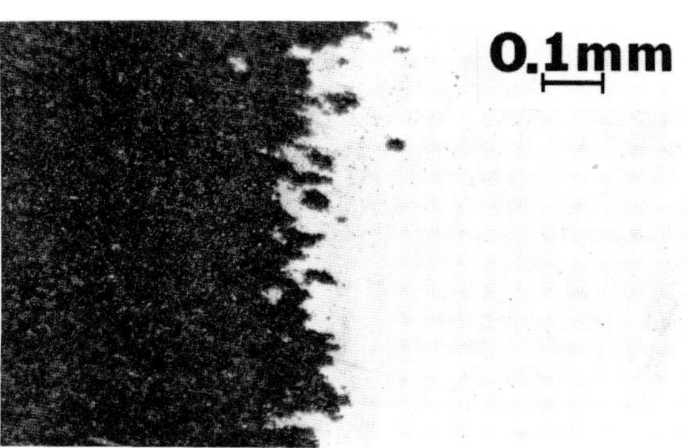

Fig. 3. Optical micrograph of the wear scar at the WC–Co–i-C boundary after 20×10^3 passes. The substrate is severely damaged, whereas the i-C deposition is unchanged.

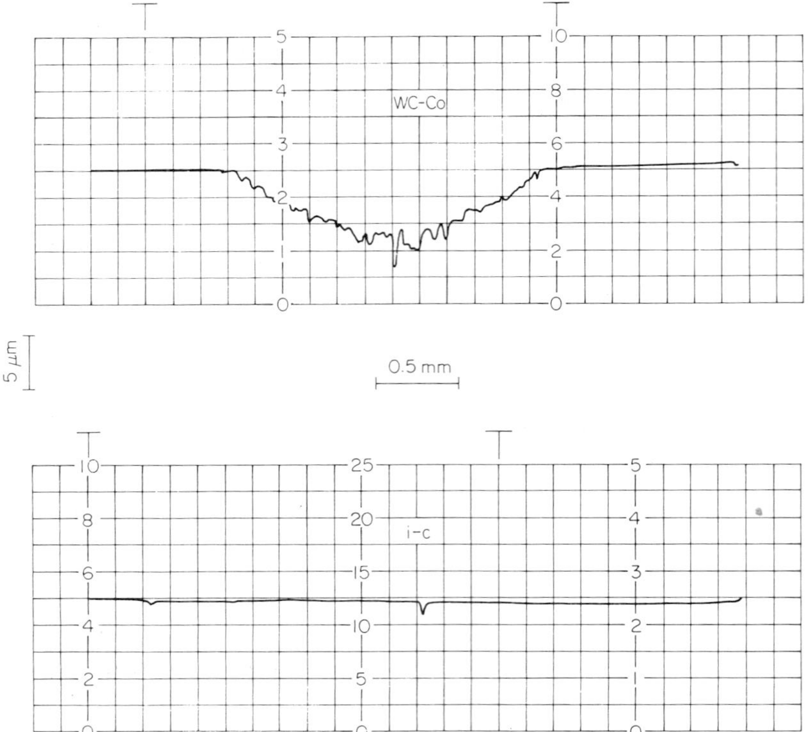

Fig. 4. Alpha step scans of the wear track of Fig. 3 on the WC–Co side and the i-C side.

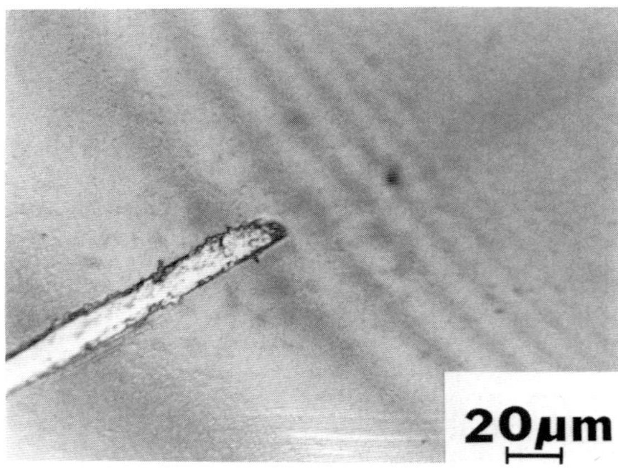

Fig. 5. Optical micrograph of the scratch caused by a diamond stylus with a load of 10 gf and after 500 repetitions on the same track. The Newton rings at the edge of the i-C deposition should be noted.

500 times on the same track with a load of 10 gf. Again the WC–Co side was damaged, whereas no scratch can be observed on the i-C side. This is consistent with the indentation hardness test where no indentation trace was found on i-C with a maximum load of 10 gf. In addition, no detachment was observed, indicating a good adhesion to the substrate.

4. DISCUSSION AND CONCLUSIONS

The microhardness, scratch and abrasive wear tests used in this work give consistent results. The scratch test confirms the results of the indentation tests about the extremely diamond-like hardness of the i-C depositions. Although strong elastic recovery was observed in the i-C case which may disturb interpretation, the loads corresponding to the maximum indentation depths in Fig. 2 can be unambiguously investigated. It can be seen that, in the i-C case, twice the load required to obtain an indentation depth of about 750 nm is needed as compared with the WC–Co case.

On this basis the excellent results of the abrasive wear tests are also easily understandable because the i-C depositions possess a higher hardness than the SiC abrasive used in this work [14]. In contrast, the hardness of the pure WC–Co substrate is lower than that of the abrasive, resulting in severe wear of WC–Co.

The structure of the i-C depositions is not yet known. Most often, transmission electron micrographs consist of diffusion rings indicating either the amorphous phase or the crystalline material with a very small grain size [15]. In some cases, distinct crystalline reflections have been observed of which all but one match those of crystalline diamond [16]. It has also been proposed that the structure consists of puckered and tilted carbon rings [17]. These rings would further be strongly interlinked via tetrahedral bonds. The broken bonds in such structures, it is explained, are satisfied by hydrogen. Justifications for such an interpretation have been based on the simulations of the electron diffraction data and the hydrogen concentration–hardness correlations for these films [17].

It is clear that the structure of i-C depositions without hydrogen differs from that explained above. Also a comparison of mass densities indicates the same. The density of hydrogenated i-C films has been reported to be in the range 1.2–2.25 g cm^{-3} [17]. We have found that the density of our i-C depositions is above 3 g cm^{-3} [18]. It is interesting to compare these values with those of graphite, amorphous carbon and crystalline diamond. The density of graphite is 2.25 g cm^{-3}, of amorphous carbon 1.8–2.1 g cm^{-3} and of crystalline diamond 3.51 g cm^{-3} [19].

On the basis of mass density, our depositions seem to be much closer to crystalline diamond than to other i-C films. Also the indentation hardness reflects this. We have also experimental evidence that a high hydrogen concentration decreases the hardness [20]. In fact, the hardness of our i-C depositions is considerably higher than that reported for i-C films involving hydrogen [17]. From a tribological viewpoint the i-C depositions without hydrogen thus seem to be most promising. Compared with crystalline diamonds, these types of deposition may also have some advantages. Many properties of crystalline diamonds are anisotropic. For example, wear resistance depends on crystal orientation [14]. We can suppose that the properties of the ion-beam-induced i-C depositions are isotropic. In addition, films prepared using ion-beam-enhanced deposition are in general dense and free from void and porosity [21], factors which are important in tribological and other applications.

In summary, in order to avoid the clear weakness caused by the large accumulation of hydrogen in i-C films produced by most common techniques such as ion plating and r.f. plasma deposition, we have used a pure ^{12}C beam to prepare diamond-like i-C layers on the WC–Co cemented carbide. These depositions reveal superior mechanical properties such as an extremely high hardness and an excellent resistance against scratching and abrasive wear.

REFERENCES

1 M. W. Guinan and J. H. Kinney, *J. Nucl. Mater.*, *103–104* (1981) 1319.
2 F. Seitz and J. S. Koehler, *Solid State Phys.*, *2* (1956) 305.
3 G. H. Vineyard, *Radiat. Eff.*, *29* (1976) 245.
4 D. A. Thompson, *Radiat. Eff.*, *56* (1981) 105.
5 R. Kelly, *Radiat. Eff.*, *32* (1977) 91.
6 J. Koskinen, J.-P. Hirvonen and A. Anttila, *Appl. Phys. Lett.*, *47* (1985) 941.
7 S. Aisenberg and R. Chabot, *J. Appl. Phys.*, *42* (1971) 2953.
8 M. Satou and F. Fujimoto, *Jpn. J. Appl. Phys.*, *22* (1983) L171.
9 K. Kitahama, K. Hirata, H. Nakamatsu and S. Kawai, *Appl. Phys. Lett.*, *49* (1986) 634.
10 M. J. Mirtich, D. M. Swec and J. C. Angus, *Thin Solid Films*, *131* (1985) 245.
11 C. Weissmantel, C. Schürer, F. Fröhlich, P. Grau and H. Lehmann, *Thin Solid Films*, *61* (1979) L5.
12 T. Miyazawa, S. Misawa, S. Yoshida and S. Gondo, *J. Appl. Phys.*, *55* (1984) 188.
13 S.-P. Hannula, D. Stone and C.-Y. Li, in E. A. Giess, K.-N. Tu and D. R. Uhlmann (eds.), *Electronic Packaging Materials Science, Materials Research Society Symp. Proc., Boston, MA, November 27–29, 1984*, Vol. 40, Materials Research Society, Pittsburgh, PA, 1985, p. 217.
14 J. Taeyaerts, *Proc. Int. Industrial Diamond Conf., Chicago, IL, 1969*, Industrial Diamond Association of America Inc., Moorestown, 1970, p. 1.
15 T. J. Moravec and T. W. Orent, *J. Vac. Sci. Technol.*, *18* (1981) 226.
16 E. G. Spencer, P. H. Schmidt, D. J. Joy and F. J. Sanssalone, *Appl. Phys. Lett.*, *29* (1976) 118.
17 C. Weissmantel, in K. J. Klabune (ed.), *Thin Films From Free Atoms and Particles*, Academic Press, Orlando, FL, 1985, p. 153.
18 A. Anttila, J. Koskinen, M. Bister and J.-P. Hirvonen, *Thin Solid Films*, *136* (1986) 129.
19 *CRC Handbook of Chemistry and Physics*, 51st edn., Chemical Rubber Co., Cleveland, 1970, p. B-79.
20 A. Anttila, J. Koskinen, R. Lappalainen, J.-P. Hirvonen, D. Stone and C. Paszkiet, *J. Appl. Phys.*, to be published.
21 T. Takagi, *Thin Solid Films*, *92* (1982) 1.

Materials Science and Engineering, 90 (1987) 349-355

Thick and Homogeneous Surface Layers Obtained by Reactive Ion-beam-enhanced Deposition*

L. GUZMAN[a], F. GIACOMOZZI[a], B. MARGESIN[a], L. CALLIARI[a], L. FEDRIZZI[a], P. M. OSSI[a] and M. SCOTONI[b]

[a]*Istituto per la Ricerca Scientifica e Tecnologica, 38050 Povo (Trento) (Italy)*

[b]*Facoltá di Scienze, Università di Trento, 38100 Trento (Italy)*

(Received July 10, 1986)

ABSTRACT

A recently developed method, reactive ion-beam-enhanced deposition (RIBED), which consists of simultaneous or sequential deposition and implantation steps, has proved to be very effective in the production of surface compounds with interesting surface properties. This hybrid technique allows the depth of the treated region to be nearly independent of the projected range of the implanted ions and thereby thicker layers to be obtained than by conventional ion implantation. Moreover, with respect to unimplanted deposited layers, the ion beam increases the uniformity of the films and their adhesion to the substrate.

In this work, we consider surface nitrides of chromium and boron, which are known to be hard and corrosion resistant.

The surface compounds were characterized using various different techniques: Auger electron spectroscopy, secondary ion mass spectrometry, X-ray diffractometry and scanning electron microscopy. Electrochemical measurements were carried out on the chromium and boron nitride layers, obtained by RIBED, on iron.

1. INTRODUCTION

There are many different methods for surface modification or, alternatively, coating of materials. Ion implantation has proved to be very effective in the production of compounds with predetermined (possibly non-equilibrium) composition, whose characteristics are interesting with respect to the improvement of surface properties (wear, corrosion and thermal oxidation resistance) [1]. However, the shallow thickness of the implanted layers constitutes a major limitation. In addition, sputtering from the surface drastically limits the concentration of implanted atoms.

Hard and protective thin film coatings are often used as an alternative treatment. Such films are normally applied by chemical vapour deposition or by such physical vapour deposition methods as ion plating or other ion-assisted processes. The idea of enhancing the properties of a physically deposited film by ion bombardment is not new and has been discussed, for example, by Weissmantel [2, 3] and more recently by Dearnaley *et al.* [4]. In order to obtain a thicker modified layer and higher implanted atom concentrations, new processes which involve both film deposition and ion implantation are being developed. One such process is ordinary ion mixing, in which a coating is first applied and subsequently bombarded with an ion beam. This process can be further extended by multiple sequential deposition and implantation steps, thus making the depth of the treated region nearly independent of the projected range of the implanted ions. Moreover, the ion beam plays a role in increasing the adhesion and homogeneity of the deposited film, and eventually in converting it into a chemically different material, *e.g.* a nitride.

Nowadays, coatings such as TiN, TiC, Cr(C, N) and Al_2O_3 deposited by sputtering or ion plating are used on a large scale in industry. New nitrides and carbides are being investigated as better alternatives to the present ones. Ion-assisted deposition techniques can be used

*Paper presented at the International Conference on Surface Modification of Metals by Ion Beams, Kingston, Canada, July 7-11, 1986.

0025-5416/87/$3.50

here advantageously as a fundamental tool for the development of new materials.

In the present paper, ion-enhanced deposition of hard coatings such as CrN, Cr_2N, BN and composites containing these materials with incorporated metal (iron) or excess boron is dealt with.

2. EXPERIMENTAL TECHNIQUES

2.1. Apparatus

An electron gun was placed under the implantation chamber of a low energy (30 keV) horizontal ion implanter, in order to perform reactive ion-beam-enhanced deposition (RIBED) experiments under controlled conditions in a high vacuum environment (base pressure, 3×10^{-6} Pa). The samples could be rotated around a horizontal axis (perpendicular to the ion beam) in order (i) to expose their surface to deposition and implantation sequentially or (ii) to allow simultaneous deposition and implantation. More details about the apparatus can be found in ref. 5.

Precalibrated film thicknesses and evaporation rates (kept fixed at about 0.1 nm s^{-1}) were monitored with a quartz crystal oscillator. The bulk temperature of the specimens was measured with a thermocouple and never exceeded 373 K.

2.2. Sample preparation

All samples were mechanically polished to a mirror-like finish using 1 μm diamond powder in the final step. In a typical RIBED experiment the first step is a 30 keV nitrogen ion preimplantation at a fluence between 1×10^{16} and 4×10^{16} N_2^+ cm^{-2}, intended to sputter clean the substrate surface and to create many favourable sites for deposition. In some cases, as in iron and some steels, it is well known that the substrate can further be strengthened by nitrogen ion implantation [4].

CrN films obtained by RIBED were produced by a sequence of between four and 25 steps, each one consisting of 10 nm chromium deposition, followed by a nitrogen implantation to a fluence of 8.4×10^{16} N_2^+ cm^{-2}, in order to achieve a stoichiometric 1-to-1 composition. Various substrates were used: iron, high speed steel, copper and glass. BN films obtained by RIBED on copper, iron and steel substrates were fabricated using the same technique with four sequential steps of 27 nm boron deposition, followed by a nitrogen ion implantation to a fluence of 1.4×10^{17} N_2^+ cm^{-2}, using a nominal boron-to-nitrogen ratio of 3 to 1. By partially screening the substrate, some unimplanted evaporated samples were obtained in every run; we shall refer to these as the reference samples.

2.3. Surface microanalyses and technological tests

The surface microstructure of the samples was examined by scanning electron microscopy (SEM); the depth composition profiles were obtained by Auger electron spectroscopy (AES) (the electron gun was coaxial to a single-pass cylindrical mirror analyzer with 0.6% intrinsic relative resolution), combined with 2 keV argon ion sputter etching. The AES depth profiles were evaluated using the sensitivity factor method described in ref. 6. Also, secondary ion mass spectrometry (SIMS) was used to obtain qualitative sputter profiles as well as evidence of the RIBED-induced phases. The structure of the samples was studied by X-ray diffractometry (XRD), using Cu Kα radiation.

The electrochemical behaviour and corrosion resistance of the Cr_2N and BN coatings obtained by RIBED on iron were tested in various aqueous solutions (H_2SO_4 and NaCl). SEM and AES analyses of the corroded surfaces were used to confirm the presence of protective layers.

3. RESULTS AND DISCUSSION

3.1. CrN

Figure 1(a) shows the AES sputter profiles for as-evaporated chromium on a copper substrate, whereas Fig. 1(b) refers to sequential chromium evaporations and nitrogen implantations (seven steps) on an analogous substrate. From Fig. 1(b) the nearly parallel chromium and nitrogen profiles give good evidence of the formation of a quasi-stoichiometric compound; also an indication of the presence of an f.c.c. phase (a = 4.15 Å) was found by XRD. This finding is further supported by the chromium Auger line shape, which is in reasonable agreement with that reported in the literature [7] for chromium

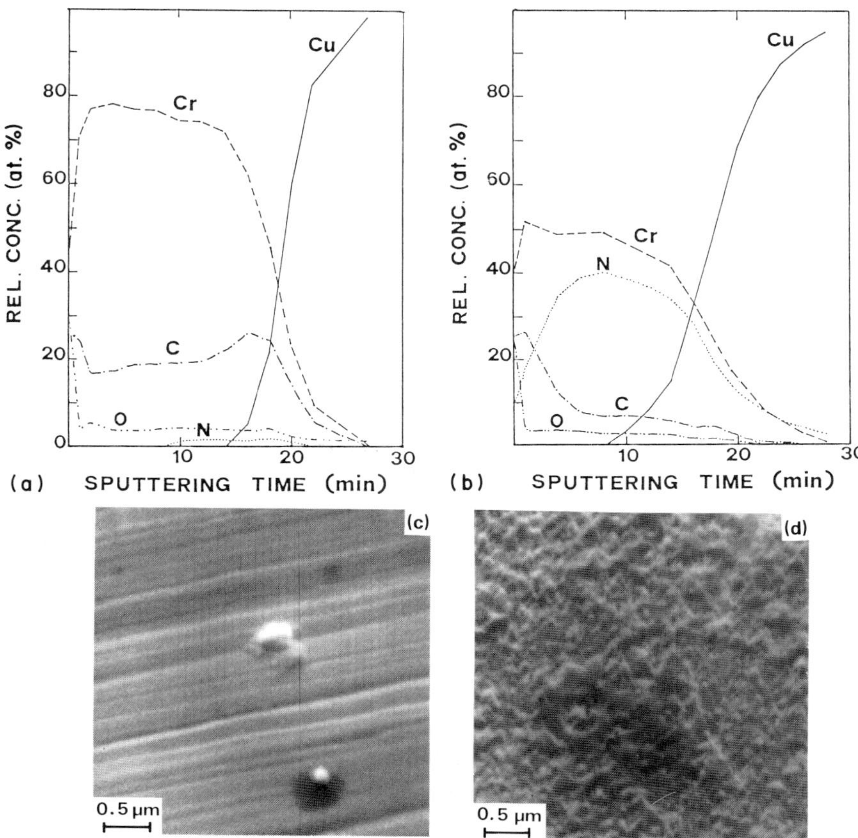

Fig. 1. (a) Auger depth concentration profiles for a chromium-evaporated copper substrate (reference sample); (b) Auger depth concentration profiles for a sequentially chromium-evaporated and nitrogen-implanted copper substrate; (c) SEM micrograph showing the morphology of the as-evaporated film; (d) SEM micrograph showing the morphology of the film obtained by RIBED.

in CrN as well as with that obtained from a reference CrN powder.

The as-evaporated chromium film contains oxygen as well as chromium because of the rather high reactivity of chromium with the residual gases in the chamber (H_2O(g), CO and CO_2) during evaporation. The relevant decrease in the contaminant concentrations after nitrogen ion implantation (see Figs. 1(a) and 1(b)) may be attributed to sputter cleaning of the surface. Comparison of Figs. 1(a) and 1(b) also gives strong evidence of ion-beam-induced interface broadening, which is definitely not an artefact due to sputter profiling.

As for the microstructure of the obtained films, the SEM micrograph for as-evaporated chromium (Fig. 1(c)) shows a structureless pattern following the surface morphology of the copper substrate with some defects; in contrast, the film obtained by RIBED (Fig. 1(d)) has a granular, less defective and more uniform morphology.

3.2. Fe–Cr–N surface alloys

Figure 2(a) shows the AES sputter profiles for a six-step sequentially chromium-evaporated and nitrogen-implanted iron substrate. The depth profiles indicate the formation of an Fe–Cr–N alloy surface with variable composition. Near the surface, β-Cr_2N seems to prevail; the presence of this compound (hexagonal structure; $a = 4.76$ Å and $c = 4.44$ Å) was confirmed by XRD, although the presence of CrN (cubic structure; $a = 4.15$ Å) could not be excluded.

Farther from the surface, Cr–Fe–N was first found and then, at a deeper level, Fe–N compositions. Ion-beam-induced interface broadening was seen, which may explain the enhanced film–substrate adhesion indicated by scratch testing [5].

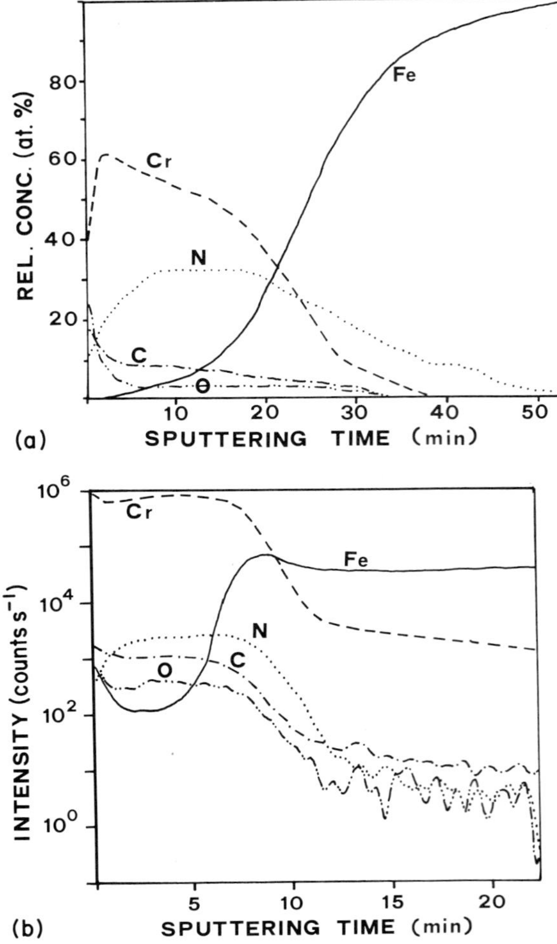

Fig. 2. (a) Auger depth concentration profiles for a sequentially chromium-evaporated and nitrogen-implanted iron substrate; (b) corresponding SIMS sputter profiles, which should be taken only as qualitative.

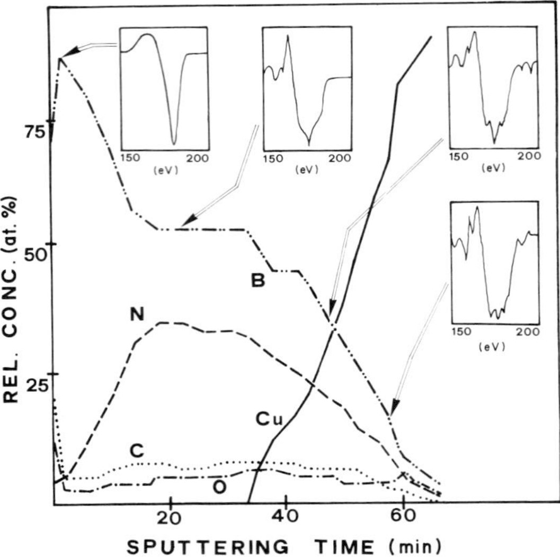

Fig. 3. Auger depth concentration profiles for sequentially boron-evaporated and nitrogen-implanted copper substrate. The B KLL Auger line shapes corresponding to different regions of the profile are shown in the insets.

The same specimens were investigated using SIMS. Figure 2(b) shows the corresponding sputter profiles, which are given in terms of the logarithm of the signal intensity. SIMS data cannot be converted easily to quantitative data but show nevertheless some interesting features concerning the chromium and nitrogen distributions as well as the broad Cr–Fe interface. Furthermore, secondary ion mass spectra were recorded, indicating the presence, near the surface, of chromium, iron, CrN, Cr_2N and Fe_2N, and, in the vicinity of the interface, in addition to the same species there is also CrFe, CrFeN and Fe_3N.

It should be noted that chromium nitrides are usually hard and resistant to corrosion and oxidation [8]. The structure of the chromium obtained by RIBED on steel, even with a much higher thickness (about 0.5 μm), is rather similar to that on iron, *i.e.* Cr_2N, whereas the coatings on copper and on glass consist mainly of CrN for the same nominal composition. This indicates that nitrogen migrates more into iron under ion bombardment.

3.3. *B–BN composites*

Figure 3 shows the AES sputter profiles for a four-step sequentially boron-evaporated and nitrogen-implanted copper substrate. Also the Auger line shapes of boron recorded throughout the profile are shown in the insets.

The multilayer coating obtained here shows features very similar to those observed after a single 100 keV nitrogen ion implantation step through a boron film 140 nm thick deposited on iron [9], *i.e.* the formation of a BN layer just below a surface layer of untransformed boron. Also the adhesion between the coating and the substrate appeared to be highly enhanced as a result of more effective ion mixing at the interface. Surprisingly a similar single 100 keV nitrogen ion implantation through a boron deposit on copper never succeeded in enhancing the adhesion, as can be seen in Fig. 4(a). On the contrary, the BN film deposited by RIBED and with the same thickness on

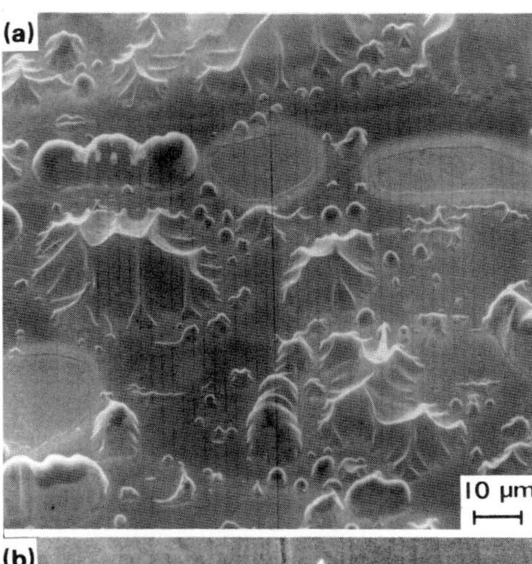

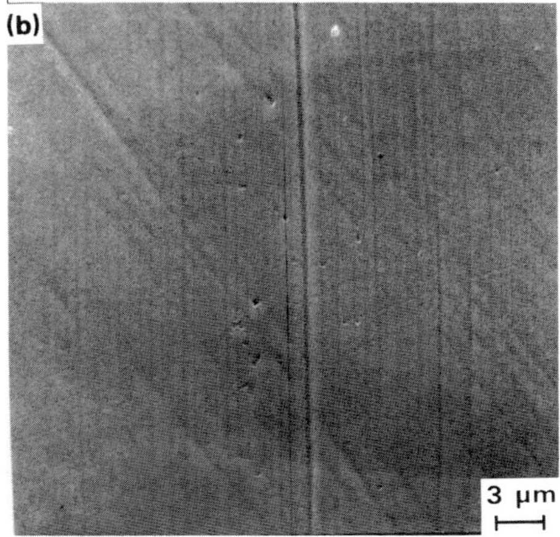

Fig. 4. Scanning electron micrographs of (a) a 100 keV nitrogen-ion-bombarded boron layer 140 nm thick on copper, showing extensive adherence failures, and (b) well-adherent homogeneous BN obtained by RIBED on copper.

copper appears to adhere well and to be homogeneous (Fig. 4(b)). The composition of the film determined against suitable reference samples is nitrogen deficient; the Auger line shape is representative of elemental boron in the near-surface region while under the surface it is typically that of BN even if the nitrogen-to-boron ratio attains a value of only 2 to 3 over an extended depth. The coating appears golden in colour, just as in other cubic nitrides, but at present we have no crystallographic evidence of its structure.

The corresponding SIMS sputter profiles have also shown the presence of a boron-rich BN layer. The secondary ion mass spectra indicate the presence of copper, BN, CuB and CuN in the interface region, although the last two compounds (CuB and CuN) are metastable and do not appear in the respective phase diagrams.

3.4. Electrochemical behaviour and corrosion resistance of coatings obtained by reactive ion-beam-enhanced deposition

The electrochemical behaviour of chromium and boron nitrides obtained by RIBED on iron was studied in 0.5 M H_2SO_4 and 1 M NaCl (pH 4) solutions. For this purpose, polarization resistance R_p measurements, potentiostatic $I(t)$ and potentiodynamic polarization curves were recorded.

The results obtained in H_2SO_4 are not highly significant, because the solution is very strong and corrosive; nevertheless a lower passivity current was observed with respect to pure Armco iron. The data from electrochemical measurements in 1 M NaCl solution are summarized in Table 1. The values of the free corrosion potential E_{corr} and corrosion current i_{corr} are given. The latter are obtained from Tafel lines (with the slopes β_a and β_c) and R_p [10]. The formula employed is

$$i_{corr} = \frac{\beta_a \beta_c}{2.3 R_p(\beta_a + \beta_c)}$$

For comparison, the values for pure iron and for BN-coated samples [11] and for Fe/B multilayers implanted with nitrogen ions [12] are also included. We first point out the noble free corrosion potential attained by the RIBED nitrides. In particular, Cr_2N shows an E_{corr} value which is about 100 mV more cathodic than iron.

The i_{corr} values indicate that the boron and chromium nitrides obtained by RIBED have very similar electrochemical behaviours. The values observed are interesting when compared with the other values reported in Table 1. Indeed, they are about two orders of magnitude lower than that of pure iron and about one order of magnitude lower than that of BN-coated iron samples [11].

The electrochemical parameters E_{corr} and i_{corr} clearly show the nobility and low reactivity of the nitride surface layers produced by RIBED. Observation of the surface morphology of the samples after anodic polarization tests indeed confirms the above observations.

The corrosive attack is localized and several pits are present in some zones where the coating is defective (Fig. 5). The low reactivity and the cathodic behaviour of the nitride layers

TABLE 1

Free corrosion potentials and instantaneous corrosion rates in 1 M NaCl (pH 4) solution (deaerated; area of the samples, 0.5 cm^2)

Sample	E_{corr} (mV)	R_p (kΩ)	β_a (V)	β_c (V)	i_{corr} (μA cm^{-2})
Armco Fe	−780	24.5	0.05	0.165	1.36
BN on Fe [11]	−810	178.7	0.08	0.14	0.25
Fe/B/Fe/B . . . on Fe implanted with N [12]	−820	20	0.04	0.16	1.39
BN obtained by RIBED on Fe	−720	650	0.065	0.16	0.062
Cr_2N obtained by RIBED on Fe	−670	720	0.07	0.17	0.06

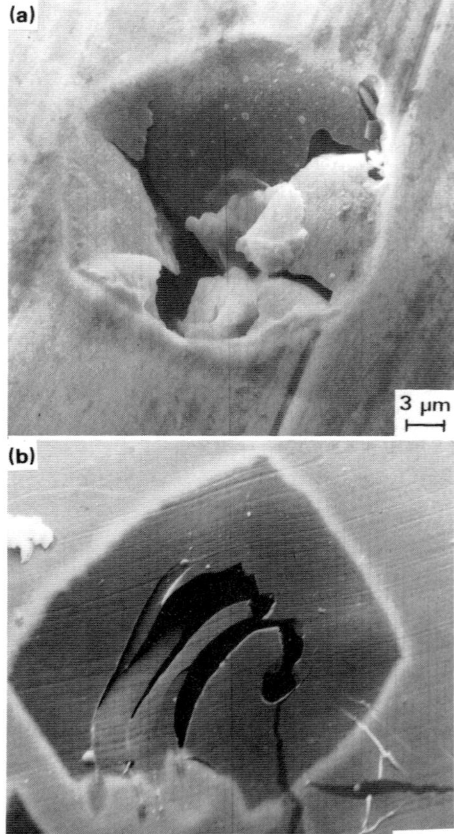

Fig. 5. Scanning electron micrographs of pits in otherwise chemically inert coatings obtained by RIBED: (a) Cr_2N; (b) BN. The different pit symmetries and mechanical behaviour in the two coatings should be noted.

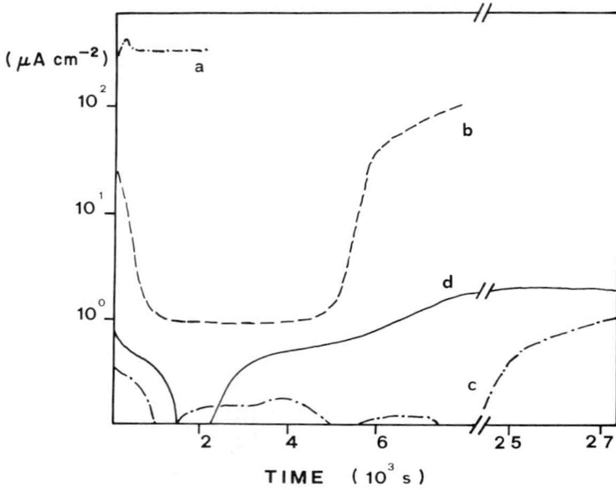

Fig. 6. Current *vs.* time plot for potentiostatic anodic polarization (at −600 mV (SCE)) of various samples in 1 M NaCl (pH 4) solution: curve a, Armco iron; curve b, Fe/B/Fe/B . . . on iron, implanted with nitrogen; (c) BN obtained by RIBED on iron; (d) Cr_2N obtained by RIBED on iron.

are highlighted by the presence of undamaged coating partially covering the holes. An interesting feature of the pits is their "cubic" symmetry, which may give further evidence of the crystallographic structure of the nitride coatings. Moreover, from a careful examination of the coating fracture in the pits, it is possible to learn something about the mechanical behaviour of the coatings. Both compounds appear to be hard, but BN seems to be much more brittle than Cr_2N. However, a direct measurement of the hardness was not performed.

A further interesting result appears in the potentiostatic $I(t)$ curves at −600 mV with respect to a saturated calomel electrode (SCE) (Fig. 6). The anodic current in the samples coated by RIBED remains very small (about two to three orders of magnitude lower than in Armco iron and about one to two orders of magnitude lower than in Fe/B multilayers implanted with nitrogen ions [12]) even after immersion for several hours in NaCl solution. The fact that the surface attained by the RIBED process is more homogeneous and compact than that obtained by single implantation or deposition confirms this.

4. CONCLUSIONS

A combined process such as ion beam mixing in which a coating is first applied and then ion bombarded can be further extended by multiple sequential deposition and implantation steps, which allow the following: (i) the depth of the treated region to be made nearly independent of the projected range of the implanted ions, thus enabling layers to be obtained which are thicker than those produced by conventional ion implantation: (ii) the substrate to be strengthened by a preimplantation step and the surface to be sputter cleaned, inducing a defective microstructure favouring adherence and uniformity of the film; (iii) better control of the composition, structure and microstructure of the coatings because of independent control of atom and ion beams. In this work, two different surface compounds were considered: chromium and boron nitrides, which were shown to be corrosion resistant. These ion-assisted coatings were deposited on different substrates, some of which can be hardened by a nitrogen preimplantation. Relatively thick surface layers can be grown, depending on the deposition and implantation rates. Since the surface region is selectively treated, these coatings can be used as finishing steps, allowing surfaces to be engineered for optimum properties independently from bulk property requirements.

ACKNOWLEDGMENTS

The authors are indebted to M. Anderle and R. Canteri for performing SIMS analysis and acknowledge technical help from Mr. E. Voltolini.

REFERENCES

1 G. Dearnaley, *Thin Solid Films, 107* (1983) 315.
2 C. Weissmantel, *Proc. 7th Int. Vacuum Congr. and 3rd Int. Conf. on Solid Surfaces, Vienna, 1977*, Berger, Vienna, 1977, p. 1533.
3 C. Weissmantel, *Vide, Couches Minces, 41* (1986) 45.
4 G. Dearnaley, P. D. Goode, F. J. Minter, A. T. Peacock and C. N. Waddell, *J. Vac. Sci. Technol. A, 3* (1985) 2684.
5 B. Margesin, F. Giacomozzi, L. Guzman, G. Lazzari, and V. Zanini, *Proc. 6th Int. Conf. on Ion Implantation Technology, Berkeley, CA, July 1986, Nucl. Instrum. Methods B.*, to be published.
6 L. E. Davies, N. C. MacDonald, P. W. Palmberg, G. E. Risch and R. E. Weber, *Handbook of Auger Electron Spectroscopy*, Physical Electronics Industries, Eden Prairie, MN, 1976.
7 I. L. Singer and J. S. Murday, *J. Vac. Sci. Technol., 17* (1980) 327.
8 S. Komiya, S. Ono, N Umezu and T. Narusawa, *Thin Solid Films, 45* (1977) 433.
9 L. Guzman, F. Marchetti, L. Calliari, I. Scotoni and F. Ferrari, *Thin Solid Films, 117* (1984) L63.
10 M. G. Fontana and N. D. Greene, *Corrosion Engineering*, McGraw-Hill, New York, 1978, p. 342.
11 F. Marchetti, L. Fedrizzi, F. Giacomozzi, L. Guzman and A. Borgese, in G. K. Wolf, W. A. Grant and R. P. M. Procter (eds.), *Proc. Int. Conf. on Surface Modification of Metals by Ion Beams, Heidelberg, September 17–21, 1984*, in *Mater. Sci. Eng., 69* (1985) 289.
12 M. Elena, L. Fedrizzi, V. Zanini, M. Sarkar, L. Guzman and P. L. Bonora, *Proc. 5th Int. Conf. on Ion Beam Modification of Materials, Cantania, June 9–13, 1986, Nucl. Instrum. Methods B.*, to be published.

Materials Science and Engineering, 90 (1987) 357-365

Ion Beam Modification of TiN Films During Vapor Deposition*

R. A. KANT and B. D. SARTWELL

Naval Research Laboratory, Washington, DC 20375 (U.S.A.)

(Received July 10, 1986)

ABSTRACT

Ion bombardment by titanium or nitrogen during reactive deposition of TiN using a titanium evaporant in a nitrogen gas is shown to have a strong influence on the composition, structure and properties of the resultant films. The implanted films were compared with unimplanted films produced under identical conditions and with a conventionally prepared sputter-deposited TiN coating. The results of the ion bombardment were that the TiN films had a reduced oxygen contamination, broader film-substrate interface, larger grains and increased lattice constants and were denser and considerably more adherent than those not ion bombarded. When compared with the conventionally prepared TiN coating, the implanted coatings were anomalously soft, behaving in a ductile rather than in the brittle fashion expected for a refractory material.

1. INTRODUCTION

Over the past few years, several reports have appeared in the literature which claimed that significant improvements in thin film properties are achieved if the deposition flux contains energetic ions [1]. One of the major hurdles to understanding the role of the ions has been an inability either to control independently or to characterize fully the various components of the flux of particles contributing to the growth of the film. While significant progress has been made by using low energy ion guns as sources of the ion flux [2-4], these guns are typically limited to use of gaseous materials and result in a relatively high chamber pressure which adds an additional uncontrolled flux of molecules at thermal energy to the surface. In addition, the ion flux from these guns is a combination of molecular and atomic species whose ratio often cannot be controlled. The study described in this paper presents an alternative experimental system based on the use of a moderate energy mass-analyzed ion implanter to study the role of bombardment by energetic ions during thin film deposition. With this system, virtually any ion species can be used and the uncertainties about its energy, charge state and current density are eliminated. This particular technique has been designated as ion-beam-assisted deposition (IAD) and is also referred to as ion-beam-enhanced deposition.

When viewed from the standpoint of a member of the ion implantation community, this technique of simultaneous ion implantation and thin film deposition is thought to offer many potential benefits. For example, the efforts to characterize ion-implanted layers adequately have been limited because until now only very thin layers have been available for study. The IAD process makes it possible to produce considerably thicker layers of similar composition and structure to those obtained by direct implantation. From the viewpoint of the coatings community, IAD may permit the fabrication of unique microstructures such as amorphous or metastable phases, extend the range of allowable processing conditions such as deposition at room temperature and provide for ion beam mixing of the film-substrate interface. It may also be possible to reduce or eliminate many of the problems often encountered in thin film deposition: poor adherence, high porosity and high internal stress.

Reported here is a summary of an extensive and ongoing investigation of the IAD process.

*Paper presented at the International Conference on Surface Modification of Metals by Ion Beams, Kingston, Canada, July 7-11, 1986.

0025-5416/87/$3.50

The work to date has principally involved implanting a TiN film during formation on room temperature substrates with titanium ions (Ti^+) or nitrogen ions (N_2^+). The structure and properties of these films were investigated as a function of the process parameters.

2. EXPERIMENTAL EQUIPMENT AND METHODS

Titanium vapor obtained from an electron beam evaporator was deposited on the entire front surface of a disk of mechanically polished AISI 52100 steel 1.25 cm in diameter. Prior to deposition, the substrates were bombarded with ions (typically 40 keV titanium ions or 30 keV nitrogen ions) to a fluence of 10^{16} ions cm^{-2} in order to sputter clean the substrate and to introduce titanium or nitrogen into the subsurface region of the substrate to help to produce a graded concentration at the substrate surface. During deposition, the vacuum chamber pressure was intentionally increased from a base level of 6.7×10^{-5} Pa to between 1.3×10^{-3} and 1.3×10^{-1} Pa with high purity nitrogen. In order to minimize contamination, the sample chamber was (1) isolated from other parts of the system using differential pumping, (2) equipped with a liquid nitrogen temperature cold wall which surrounded the specimen and (3) cryogenically pumped. The specimen was positioned 30 cm above the evaporator to improve lateral uniformity. This was necessary since the normal to the sample surface made an angle of 45° with the evaporant flux (Fig. 1). The ion beam entered the chamber horizontally and also made an angle of 45° with the sample normal. The film thickness or, more precisely, the mass per unit area was measured using a quartz crystal thickness monitor. The signal from this monitor was supplied to a computer which also monitored the ion beam current and used these data to control the evaporation rate such that the ratio of the ion bombardment rate to the deposition rate was held constant during the treatment. Some specimens were partially masked with a tantalum wire in order to allow an independent measure of film thickness to be made. Such measurements were made with a profilometer and with interference microscopy. Not all the surface was implanted. That portion of the specimen receiving both ions and vapor is referred to as the IAD region whereas the unimplanted region is referred to as the physical vapor deposition (PVD) region.

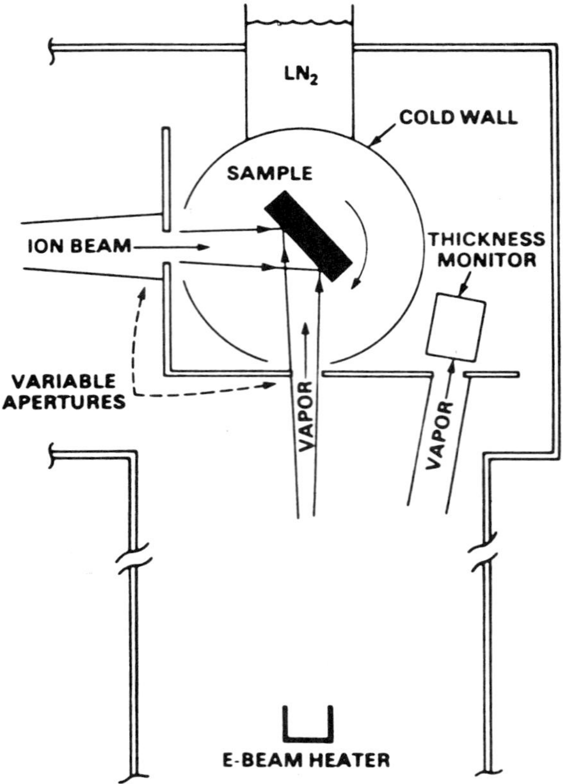

Fig. 1. Schematic representation of the IAD system: LN_2, liquid nitrogen; E-BEAM, electron beam.

Throughout this paper, the parameter R is used to characterize the extent of the ion irradiation treatment and is defined as the ratio of ion flux to vapor deposition rate. If both of these quantities are expressed in units of atoms per area, then R is dimensionless. Since the quartz crystal thickness monitor is shielded from the ion flux, the deposition rate used in calculating R is not influenced by ion beam effects such as sputtering and implantation processes. Strictly speaking, the quartz crystal thickness monitor measures only mass. Thus the actual value of R can only be determined if the composition of the PVD film is known. For the IAD coatings produced to date, $0.001 < R < 0.6$ and deposition rates ranged from 0.04 to 3.5 nm s^{-1}. Most of the films were 100–500 nm thick. However, thicker (1000–3000 nm) specimens were prepared for use in microhardness tests and X-ray diffraction analysis.

The steel substrates were mounted on a thermal sink to maintain the sample temperature near room temperature during processing; however, a series of IAD coatings was also produced on tantalum substrates which were not heat sunk but were suspended by wires. In this case the input power of the ion beam produced increased processing temperatures. The substrate temperatures ranged from 135 to 700 °C depending on the titanium ion beam energies which ranged from 40 to 190 keV. These temperatures were estimated by calculating the temperature at which the power dissipated into radiation equalled that delivered by the ion beam.

A number of surface-sensitive analytical techniques and mechanical tests were used to characterize the specimens. The composition as a function of depth was obtained using Auger electron spectroscopy (AES) with low energy argon ion sputtering. Rutherford backscattering and elastic recoil detection were also used in selected cases to provide additional compositional data. Transmission and scanning electron microscopies were used to investigate the microstructure, and X-ray diffraction was used to obtain crystal structure, orientation and lattice constants. Adhesion was investigated using scratch tests, cavitation erosion and thermal cycling; friction and wear were evaluated using dry (unlubricated) sliding; hardness was measured using a Knoop microhardness tester.

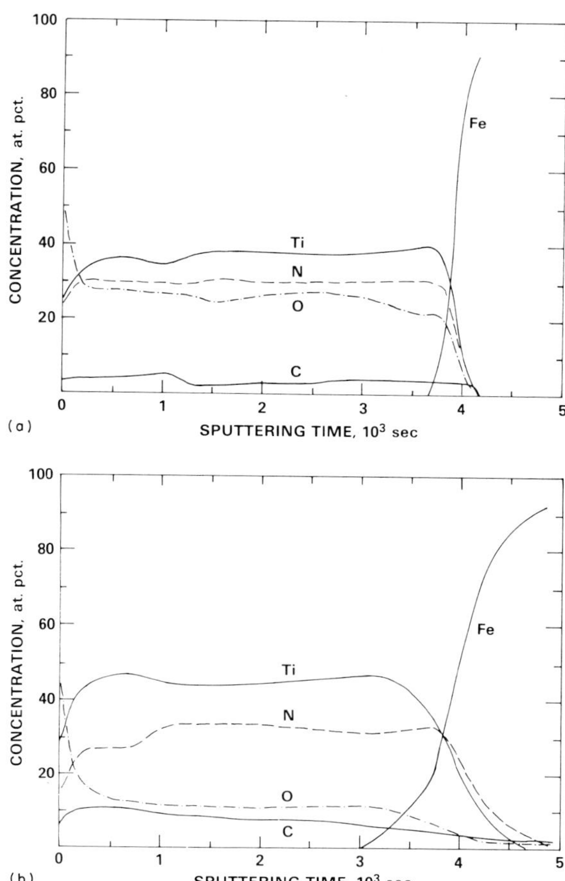

Fig. 2. Film composition measured by Auger electron spectroscopy as a function of sputtering time (thickness, 100 nm): (a) $R = 0.002$; (b) $R = 0.2$.

3. RESULTS

3.1. Film composition

Determination of the composition of TiN coatings by AES is complicated by the fact that the Ti(387 eV) peak overlaps the N(382 eV) peak. Handbook sensitivity factors [5] are also inappropriate for the elements titanium, nitrogen, oxygen and carbon detected in the coatings. The method used for determining the nitrogen-to-titanium ratio in the films has been described elsewhere [6]. Standards of bulk TiN, TiO_2 and TiC were used to establish the appropriate sensitivity factors.

Figure 2 shows a set of typical composition profiles for nitrogen-implanted films. It is easily seen that the slope of the titanium concentration at the interface is reduced as the arrival ratio R increases, indicating that the ion bombardment is producing a more graded composition at the interface. In addition, the level of oxygen contamination decreases with increasing R. In order to investigate the effects of the ion bombardment on the composition of the films more fully, these titanium-implanted IAD films were compared with the unimplanted portion of the same films (Table 1). These PVD films showed a strong trend toward incorporating a constant amount of gas; the sum of the oxygen and nitrogen concentrations was nearly constant at 62 at.%. For the IAD portions, this was reduced to closer to 50 at.% apparently by a reduction in the oxygen concentration. Nitrogen-to-titanium ratios ranged from 0.7 to 1.25 with a trend toward a larger uptake of nitrogen with increasing R.

TABLE 1

Film compositions *vs.* treatment parameters for 30 keV titanium ions (Ti^+)

R	*Deposition rate* (Å s^{-1})	*Amount* (at.%) *of the following elements*												
		IAD								*PVD*				
		N/Ti	*Ti*	*N*	*O*	*Ti, calculated*	*N, calculated*	*N, difference*	*O+N*	*N/Ti*	*Ti*	*N*	*O*	*O+N*
0.120	1.3	1.25	42	52	6	61	33	−20	58	0.53	48	25	27	52
0.120	1.8	1.00	48	47	5	67	28	−19	52	0.41	46	19	35	54
0.014	16.2	1.00	44	44	12	49	39	−5	56	0.81	38	31	31	62
0.005	25.3	1.00	45	45	11	45	45	0	56	1.00	38	38	24	62
0.010	12.8	1.00	45	45	10	45	45	0	55	1.00	38	38	24	62
0.050	2.6	0.91	47	43	10	45	45	2	53	1.00	38	38	24	62
0.050	2.4	0.77	50	39	11	47	42	4	50	0.91	38	35	27	62
0.050	2.4	0.77	52	40	8	48	44	4	48	0.91	38	34	28	62

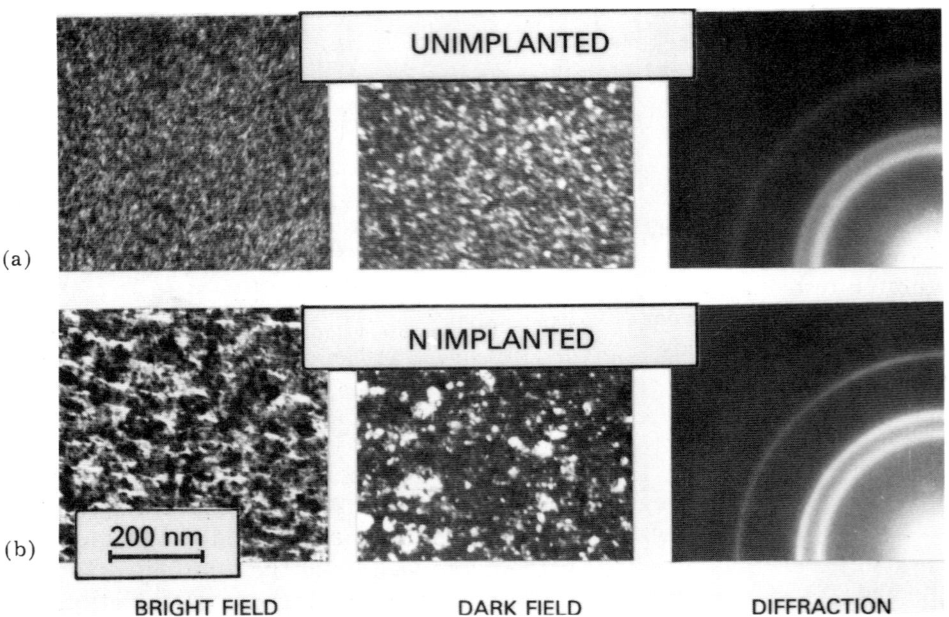

Fig. 3. Transmission electron micrographs and electron diffraction patterns of IAD and PVD films: (a) unimplanted; (b) nitrogen implanted.

3.2. Microstructure

Electron microscopy and X-ray diffraction were used to examine the microstructure of the films. High resolution scanning microscopy showed that the surfaces were all essentially flat, smooth and featureless with no evidence of porosity. Figure 3 consists of bright and dark field transmission electron micrographs of the implanted (IAD) and the unimplanted (PVD) structures. In both cases the grain size is quite small (only a few hundred ångströms). In the IAD case there is a clear trend toward larger grains and a bimodal size distribution is evident. The electron diffraction patterns indicated, and subsequent X-ray diffraction confirmed, that in both cases the crystal structure was that of TiN. The more precise X-ray measurements indicated that the PVD films had the smallest lattice constant (0.4221 nm), followed by the as-deposited IAD film (0.4227 nm) and an IAD film annealed to 850 °C for 15 min (0.4235 nm). All observed lattice constants were smaller than the value of 0.4240 nm given in the literature [7] for bulk TiN.

Preliminary estimates of the change in density due to the ion beam were made by determining the relative amounts of titanium in both the IAD and the PVD regions using energy-dispersive X-ray analysis and by measuring the film thickness. This analysis showed that the two regions had nearly equal amounts of titanium but that the IAD portion was considerably thinner. In addition, the densities were estimated using the mass deposited as measured by the quartz crystal thickness monitor and the thickness measured by interferometry or profilometry. The results indicate that the IAD films are as much as 62% denser than the PVD films and 19% denser than expected for conventionally prepared TiN. Because these measurements are subject to considerable uncertainty, the method has been refined and additional studies are under way.

3.3. Mechanical properties

Clearly, one of the most important properties for a thin film is its adherence to the substrate and yet adhesion remains one of the most difficult properties to test. Although all current adhesion tests are controversial, the results of scratch tests are widely reported; thus, this test was applied to our films. Figure 4 shows optical micrographs of the region of a scratch for both the IAD and the PVD films. The scratch is made by a Rockwell C indenter undergoing dry sliding with a load that increases from 10 to 50 N along the length of the scratch. The PVD film fails from the start and exhibits extensive peeling, cracking and buckling at great distances from the scratch. Such behavior provides convincing confirmation of the commonly held view that the deposition of PVD films at low temperature is not recommended. The IAD film, however, remains intact both outside and inside the scratch. Additional scratch tests were performed using a Knoop indenter and produced similar results (*i.e.* the films deformed with the substrate and remained intact) for the entire range of IAD parameters.

A potentially more aggressive adhesion test involved cavitation erosion whereby mass loss as a function of erosion time was measured for films deposited on M50 steel under a wide range of conditions ($0.002 < R < 0.4$ with 30 keV nitrogen ions). As indicated in Fig. 5, the onset of rapid mass loss is delayed by the presence of the IAD film. Scanning electron microscopy examination of the surface at various stages of the test revealed that large craters in the substrate were forming surrounded by regions in which the film was still intact. Because of this, and because in no instance was the film observed to have peeled off without substrate loss, it is believed that film loss occurred only when the substrate eroded and not because of adhesion failure at the substrate–film interface.

To evaluate adhesion further and also the ability of the IAD and PVD coatings to withstand thermal cycling, a coated AISI 52100 sample was annealed to 850 °C for 15 min in ultrahigh vacuum and then allowed to cool slowly to room temperature. As is shown in

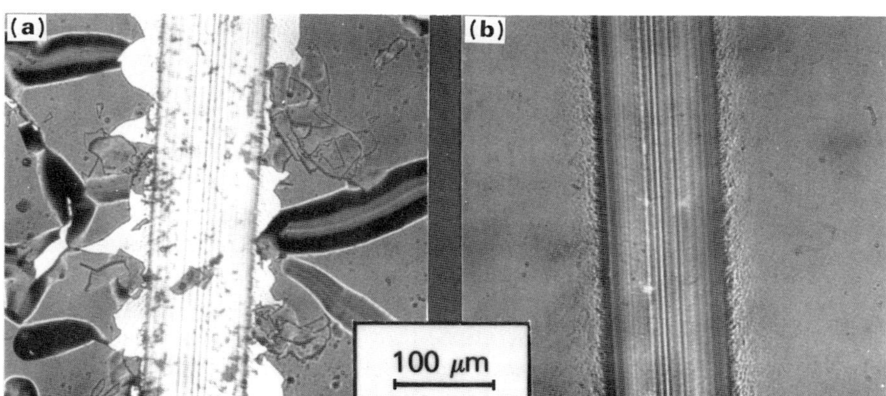

Fig. 4. Optical micrographs of (a) PVD and (b) IAD films following scratch testing for adhesion.

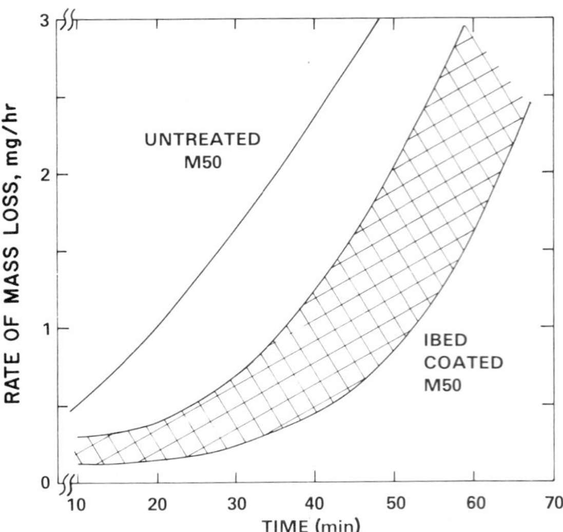

Fig. 5. Mass loss due to cavitation erosion of an IAD film.

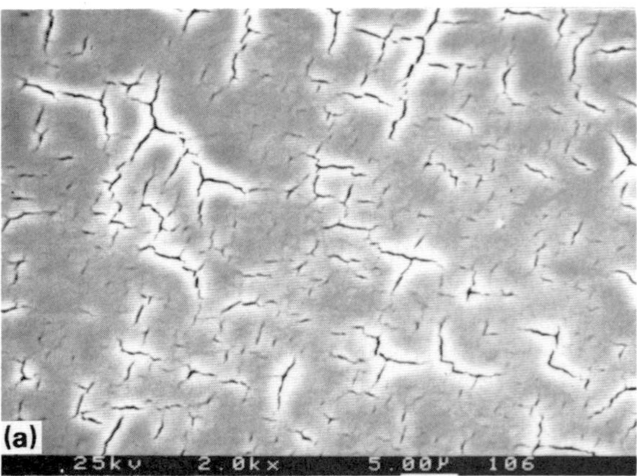

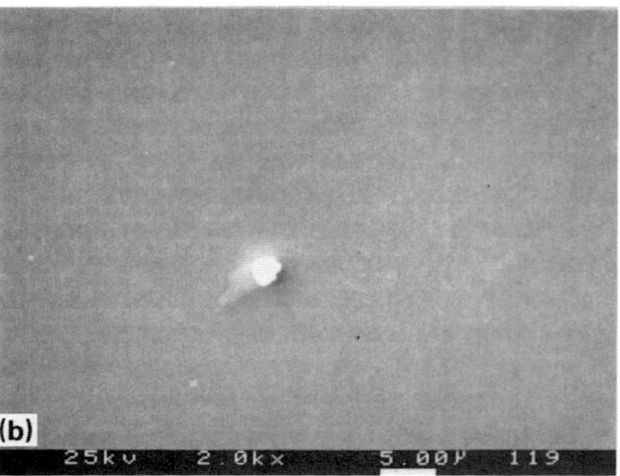

Fig. 6. Scanning electron micrographs following thermal cycling (850 °C; 15 min) (the contrast feature in the IAD case is the debris flake used for focusing since the surface is otherwise featureless): (a) PVD; (b) IAD.

Fig. 6, extensive cracking and spalling of the PVD film occurred, but the IAD film remained intact with an appearance identical with that observed prior to annealing.

Because the films deformed so readily in the scratch tests without exhibiting the brittle behavior expected for a hard refractory, it was suspected that the film hardness was less than expected for conventionally prepared TiN coatings. To confirm this, a Tucon microhardness tester was used with a Knoop indenter at loads ranging from 1 to 50 gf. At the heaviest load, the Knoop hardness was that of the AISI 52100 steel substrate (750 HK), which was expected since the depth of the indentation was greater than the thickness of the film. As the loads decreased, the hardness first tracked the normal increase in hardness expected for the substrate at lighter loads. Then, for the lightest loads and for the thickest films, the hardness decreased. This indicates that indeed the films have a hardness less than that of the substrate. Similar results were obtained for films deposited on beryllium, *i.e.* the hardness measured less than or equal to that of the substrate which is less than 500 HK.

Microhardness testing of IAD films on tantalum substrates heated by the ion beam determined that these coatings were also soft over the entire range of processing parameters. The microhardness of the IAD coating was also not affected by post-processing annealing to 850 °C for 15 min in ultrahigh vacuum.

The friction and wear behavior of the IAD TiN films also differed from that of conventionally prepared TiN. For these tests the IAD films are compared with a commercially prepared TiN coating made by reactive magnetron sputtering. A ball of AISI 52100 steel 2.5 cm in diameter underwent dry sliding with a normal force of 9.8 N at 0.1 mm s^{-1} and the frictional force was measured for each of 15 passes over the same region. The coefficient of friction was 0.6 for the commercial film and 0.2 for the IAD TiN. After a light polishing, the commercial film also had a value of 0.2 which is comparable with the published value of 0.225 for steel sliding on lubricated TiN [8]. The PVD films were stripped from the substrates either during ultrasonic cleaning prior to the tests or during the first pass. The IAD films, however, remained intact throughout the test. A measure of the wear was obtained by comparing the ratios of titanium to iron X-ray signals inside and outside the wear track using energy-dispersive X-ray analysis. While there was negligible loss of titanium for the commercial TiN, the IAD films exhibited about 8% loss which is consistent with a mild oxidative wear mechanism [9]. There was no significant difference between the behavior of films prepared with $0.001 < R < 0.2$.

4. DISCUSSION

The results presented above have shown that ion bombardment during reactive deposition of TiN has a strong influence on the composition, structure and properties of the resultant films. The IAD films were compared with films produced under identical conditions but without the ion bombardment (PVD regions). Ion irradiation of the IAD TiN films has reduced oxygen contamination, produced larger grains with increased lattice constants and has led to denser films which are tenaciously adherent. When compared with conventionally prepared TiN, the IAD coatings are anomalously soft, behaving in a ductile rather than in the brittle fashion expected for this refractory.

In general, film composition, structure and properties are influenced by the energy deposited by the ion beam and by ions that have come to rest within the film. For this work, both nitrogen and titanium ions were used in order to help to distinguish between effects due to energy deposition and those due to implanted ions. Since the ratio of the concentrations of nitrogen to titanium increased with increasing arrival ratio R when titanium ions were used, it is clear that the final character of these films is dominated by the energetics of the IAD process.

The oxygen contamination in these films is thought to be the result of the high reactivity between titanium and oxygen-containing molecules present in small amounts in the residual gases of the vacuum system. An investigation of these gases with a residual gas analyzer indicated that the largest contribution was water. In addition, an AES analysis of the titanium charge material used in the evaporator revealed that it was not a significant source of oxygen. It is interesting to note that the conventionally prepared (magnetron sputtered) TiN specimen had negligible oxygen despite the fact that it had been deposited in a vacuum system with a comparable or higher base pressure. The commercial process involves (1) 100 times higher partial pressures of nitrogen (13 Pa), which may overwhelm the available oxygen and may also lead to the formation of TiN in the vapor phase rather than at the film surface, (2) the presence of freshly deposited titanium adjacent to the workpiece, which is expected to provide additional pumping not available in the IAD system since there are apertures which restrict the deposition of titanium to the specimen, and (3) the presence of a plasma near the specimen which may produce more bombardment (*i.e.* a larger R) than has been used in the IAD treatments. Whatever the source of this oxygen contamination, its presence did serve to point out an interesting and potentially important process, namely the selective rejection of oxygen from IAD films due to the ion beam.

We conclude that the oxygen is being removed preferentially because the composition of the IAD film can, in many cases, be predicted from the composition of the PVD film by simply assuming that the IAD film composition is identical with that of the PVD film except that a fraction of the oxygen has been removed. For R values of about 0.05, this fraction is nearly two-thirds. Thus the IAD composition can be calculated from the PVD composition by simply reducing the oxygen content of the PVD film by a factor of 3 and renormalizing the composition to the remaining constituents. From the results of such calculations (Table 1), it is seen that there is excellent agreement with the observed compositions of the IAD films for intermediate values of the ratio R. For the higher values of R, the calculation indicates that the ion beam has also induced an increase in the uptake of nitrogen, *i.e.* these films contain an excess of nitrogen over that predicted by the calculation. Potential explanations for the ion-beam-induced processes include preferential sputtering of oxygen due to a smaller binding energy between the oxygen and titanium than between nitrogen and titanium, and/or an enhancement of the reactivity between the nitrogen and titanium due to an ion-beam-induced activation of the nitrogen. However, since an R value of the order of only 0.05 or less is required to reduce the oxygen level, the direct interaction between surface oxygen atoms and incoming ions required to remove them by sputtering before they are buried by the deposition flux seems unlikely. An alternative hypothesis is that there is radiation-induced segregation which is transporting buried oxygen to the surface where it is once again vulnerable to sputtering. Clearly, additional investigation is required to resolve these issues.

The evidence that the IAD films are significantly softer than conventionally prepared TiN is overwhelming and similar results have been reported by others [10, 11]. Moreover, this softness is accompanied by an anomalously ductile behavior. Since it is common practice to apply high compressive stresses in order to reduce brittle failure in the shaping of ceramic components, it may be that similar processes are acting here. From the buckling evident following scratch testing, the PVD films are clearly in a state of high compressive stress. In addition, hardness indentations in similarly prepared IAD TiN films deposited on silicon have been observed to decrease in size with increasing time [12] which also indicates a high compressive stress. Moreover, the IAD films were found to have a higher density and smaller lattice constant than the values in the literature, which is also consistent with a state of compressive stress. Preparations are under way to perform measurements of the intrinsic stress in the IAD films and the results are expected to produce an increased understanding of these phenomena.

The ion-beam-induced changes in microstructure include a broader film–substrate interface, modification of the grain size distribution and a larger lattice constant than for the PVD films. The grading of the interface concentrations is presumed to be the result of ion mixing. Since the IAD films were so adherent that they were not separated from their substrates during the scratch tests, by cavitation erosion or by thermal cycling, the improvement in film adhesion cannot yet be correlated with the extent of the ion bombardment. The presence of the bimodal grain size distribution is consistent with that expected for films deposited at considerably higher temperatures (0.3 of the melting point) [13] and is presumed to be due to ion-beam-enhanced diffusion. The fact that the IAD lattice constant is larger than that of the PVD case further suggests that the ion beam is producing an accelerated approach toward the equilibrium state.

5. SUMMARY

TiN films produced by IAD differ in structure and properties from films prepared under identical conditions but without the ion bombardment and from films prepared by conventional deposition techniques. Although the IAD films have nearly the same lattice structure, composition and grain size as conventionally prepared coatings, the IAD films exhibit significantly reduced hardness and a ductile behavior not normally associated with TiN. The observed beam-induced effects included a reduction in the rate of incorporating oxygen contamination by as much as a factor of 3, increased nitrogen incorporation with increasing R, dramatic increases in adhesion, and production of a microstructure consistent with films grown at higher temperatures. While a detailed understanding of the observed phenomena awaits further study, some processes have been suggested which serve as the basis for additional investigations. These processes include removal of oxygen by preferential sputtering, activated reaction between nitrogen and titanium, ion beam mixing of the interface region and an accelerated approach to the equilibrium state due to enhanced diffusion.

REFERENCES

1 R. F. Bunshah, J. M. Blocher, D. M. Mattox, T. D. Bonifield, G. McGuire, J. G. Fish, M. Schwartz, P. B. Ghate, J. A. Thornton, B. E. Jacobson and R. C. Tucker, *Deposition Technologies for Films and Coatings*, Noyes, Park Ridge, NJ, 1982.

2 Ch. Weissmantel, *Thin Solid Films, 58* (1979) 101.

3 L. Pranevicious, *Thin Solid Films, 63* (1979) 77.

4 J. M. E. Harper, J. J. Cuomo, R. J. Gambino and H. R. Kaufman, in O. Auciello and R. Kelly (eds.), *Ion Bombardment Modification of Surfaces: Fundamentals and Applications*, Elsevier, Amsterdam, 1984, pp. 127–162.

5 L. E. Davis (ed.), *Handbook of Auger Electron Spectroscopy*, Physical Electronics Industries, Eden Prairie, MN, 1976.

6 D. A. Baldwin, B. D. Sartwell and I. L. Singer, *Appl. Surf. Sci., 25* (1986) 364–379.

7 J. E. Sundgren, Formation and characterization of titanium nitride and titanium carbide films prepared by reactive sputtering, *Linkoping Studies in Science and Technology, Dissertation 79*, 1982.
8 A. K. Suri, R. Nimmagadda and R. F. Bunshah, *Thin Solid Films, 64* (1979) 191-203.
9 R. A. Kant, B. D. Sartwell, I. L. Singer and R. G. Vardiman, *Nucl. Instrum. Methods B, 7-8* (1985) 915-919.
10 G. Dearnaley, personal communication, 1986.
11 W. B. Nowak, R. Keukelaar and W. Wang, *J. Vac. Sci. Technol. A, 3* (6) (1985) 2242.
12 J. K. Hirvonen, personal communication, 1984.
13 B. A. Mouchan and A. V. Demchishin, *Phys. Met. Metallogr. (Engl. Transl.), 28* (1969) 83.

A New Rapid Technique for Characterizing Microstructures of Films Produced by Ion Beams in the Transmission Electron Microscope*

W. S. SAMPATH and P. J. WILBUR

Department of Mechanical Engineering, Colorado State University, Fort Collins, CO 80523 (U.S.A.)

(Received July 10, 1986)

ABSTRACT

A new technique that is suitable for studying the microstructures produced by ion-beam-processing techniques is outlined in this paper. The technique uses a thin plastic film on a transmission electron microscopy (TEM) grid as the substrate on which a film of the desired material is sputtered. The film can be used directly for TEM study of sputtered films or can be implanted with the desired element(s) and then characterized using TEM. Two examples of application of this technique are presented.

(i) Sputter deposition of aluminum in an oxygen atmosphere results in an amorphous oxide film.

(ii) Nitrogen (N_2^+) ion implantation in pure iron at high current densities (100 $\mu A\ cm^{-2}$) results in the formation of γ'-Fe_4N and ϵ-Fe_2N_{1-x} at a dose of $4 \times 10^{16}\ N_2^+\ cm^{-2}$ at an accelerating voltage of 20 kV. Radiation damage occurs at lower doses of nitrogen ion implantation and at all doses of argon (Ar^+) ion implantation up to $2.2 \times 10^{16}\ Ar^+\ cm^{-2}$.

1. INTRODUCTION

Ion-beam-processing technologies (sputter deposition, ion implantation and ion mixing) are gaining increasing attention for surface modification [1-3]. The number of possible microstructures obtainable by these processing techniques is enormous. The multiplicity points to the need for a quick technique to characterize the surfaces precisely. Transmission electron microscopy (TEM) is an ideal choice because of its widespread availability and good lateral resolution. A simple technique suitable for use by TEM is outlined in this paper.

*Paper presented at the International Conference on Surface Modifications of Metals by Ion Beams, Kingston, Canada, July 7-11, 1986.

2. EXPERIMENTAL TECHNIQUE

Conventional preparation of specimens suitable for TEM examination involve thinning from the bulk using electropolishing or other methods [4] or deposition of a thin film on cleaved crystals of rock salt or mica and the film is subsequently extracted by dissolving the substrate [5]. The technique presented in this paper uses a thin (200 Å) film of plastic (Formvar) on a TEM grid as the substrate on which a film of the desired material is sputtered. The casting of Formvar film was achieved by drawing a glass slide from a solution of 3% Formvar in ethylene dichloride and suspending it vertically to drain. When dry, the film is scored with a sharp razor edge and the film is easily removed by oblique immersion in water. TEM grids are placed on top of the floating film such that the rough side of the TEM grid faces the plastic film. A microscope slide covered with paper tape is used to gather the TEM grids covered with Formvar film by oblique immersion in water. The microscope slide is withdrawn from the water and suspended vertically to drain. When dry, the TEM grids covered with Formvar can be removed from the microscope slide with forceps and can then be used as the substrate for film deposition.

3. RESULTS OF CHARACTERIZATION OF SPUTTERED FILMS

The TEM grid is held firmly on a smooth metal block with the plastic facing the metal

0025-5416/87/$3.50

surface. The desired material can be sputtered on the plastic. After the film has been deposited, it is preferable to dissolve the plastic film by dipping in ethylene dichloride. The sputtered film can be characterized by TEM, electron diffraction and other analytical techniques. Sputter deposition of aluminum is considered here as an example.

Sputter deposition of aluminum in an argon atmosphere under a pressure of 5.5×10^{-4} Torr using argon ions of 1 kV results in a polycrystalline film of average grain size 200 Å, as shown by the micrograph in Fig. 1(a), and the electron diffraction pattern from this film is characteristic of polycrystalline film with randomly oriented grains, as shown in Fig. 1(b). Deposition of aluminum under the same conditions but under an oxygen partial pressure of 2×10^{-4} Torr leads to an amorphous oxide film. The loss of grain contrast can be seen in the micrograph in Fig. 1(c) and the electron diffraction pattern is characteristic of amorphous films, as shown in Fig. 1(d).

4. CHARACTERIZATION OF ION-IMPLANTED SURFACES

The films deposited on TEM grids by sputtering can be implanted with the desired element(s) and characterized by TEM. During ion implantation the TEM grid with the sputtered film is held firmly against a metal block, with the sputtered film in contact with the metal surface. The metal block is cooled by circulating water. The technique can be employed to study the effects of ion implantation on complex alloys, since the composition of the sputtered films is the same as that of the cathode [6] provided (i) that the cathode temperature is not too high and is chemically stable and (ii) that the sticking coefficients for

Fig. 1. Electron micrographs and diffraction patterns of sputtered aluminum films: (a), (b) no oxygen admitted (c), (d) oxygen pressure of 2×10^{-4} Torr.

the components on the substrate are the same. Evidence for the correspondence between sputtered films and the cathode compositions has been reported for various materials, such as type 347 stainless steel and brass [7], a number of aluminum alloys [8] and group III–V intermetallic films [9].

The results obtainable by this TEM technique because of ion implantation would be similar to results on bulk samples since the composition of the film can be made similar to that of bulk samples for most alloys. The small grain size in the film compared with that in bulk samples is not likely to have any effect in the formation of new phases, since the grain boundaries as crystal defects are

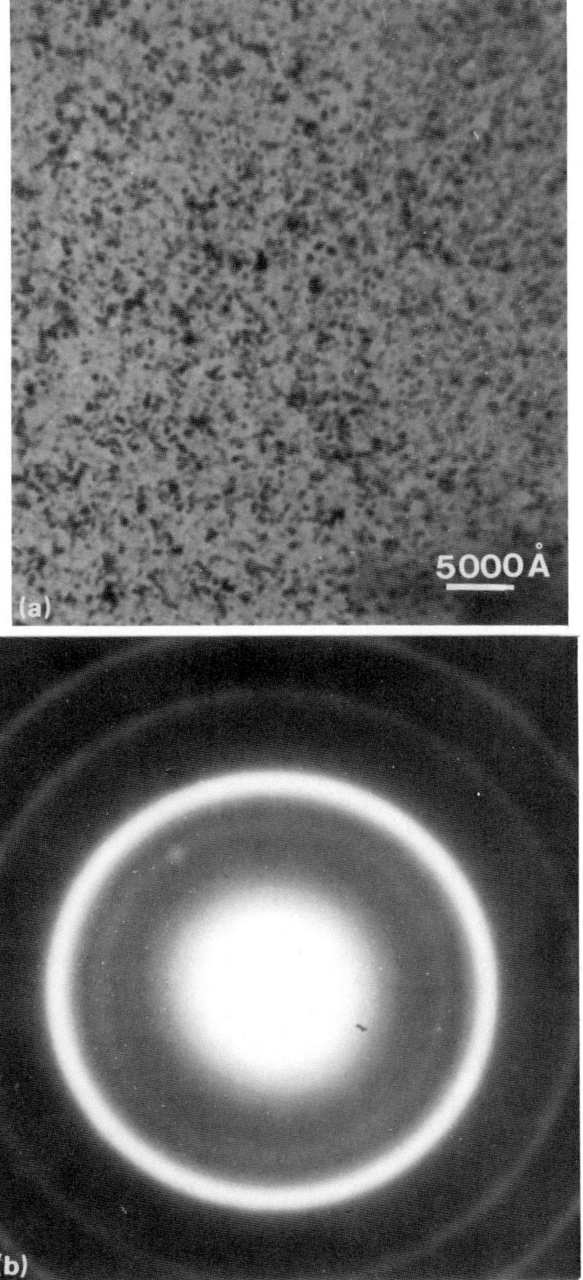

Fig. 2. (a) Electron micrograph and (b) diffraction pattern of a sputtered iron film.

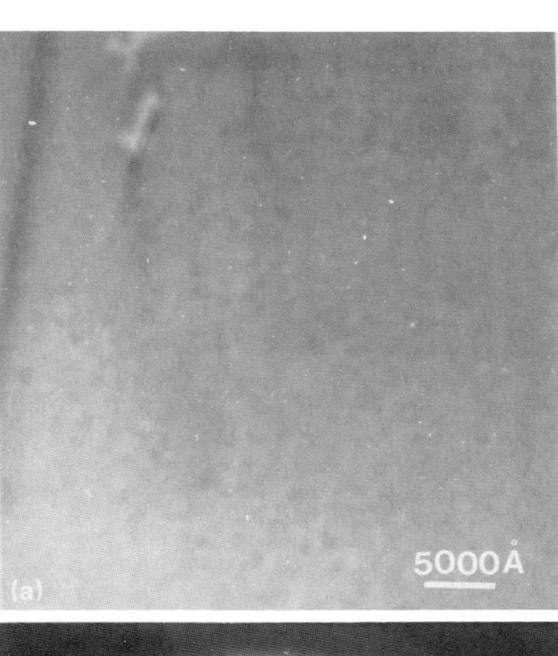

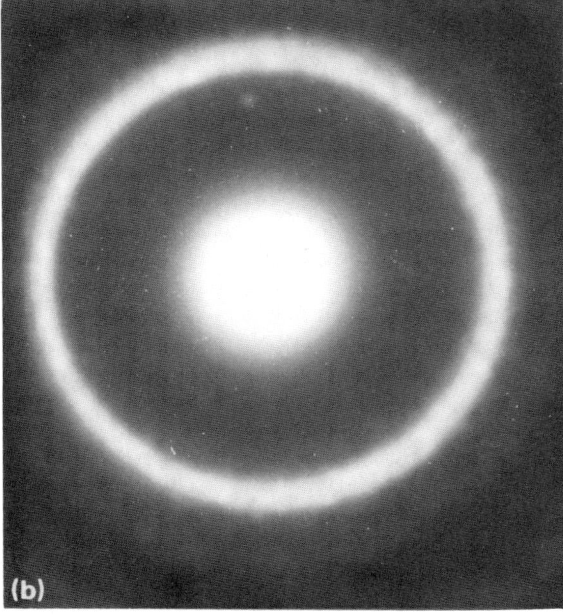

Fig. 3. (a) Electron micrograph and (b) diffraction and pattern of a sputtered iron film implanted with 4×10^{15} Ar^+ cm^{-2} at 20 kV and 100 μA cm^{-2}.

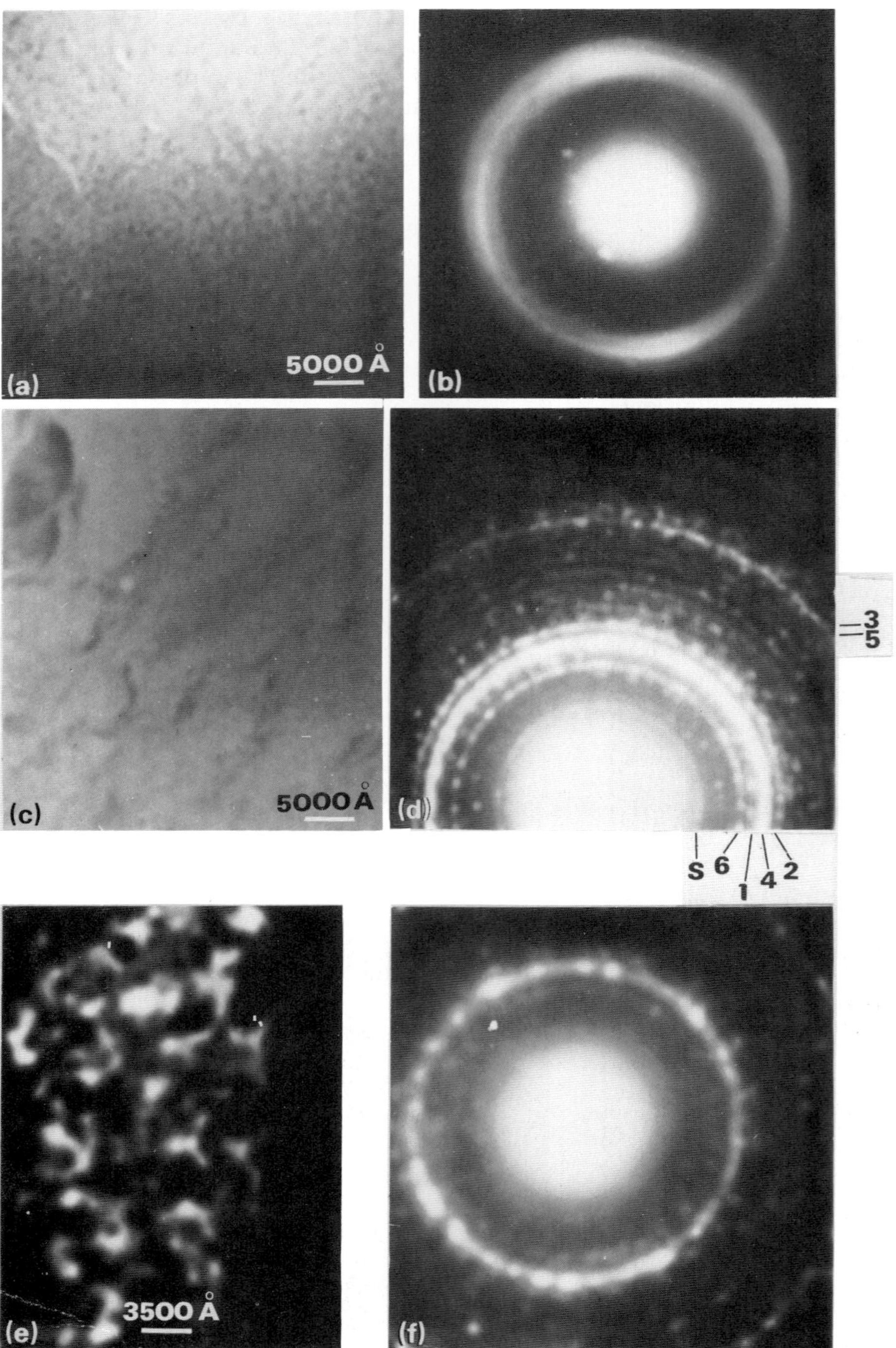

Fig. 4. Electron micrographs and diffraction patterns of sputtered iron films implanted with nitrogen at 20 kV and 100 μA cm^{-2}: (a), (b) dose of 4×10^{15} N_2^+ cm^{-2}; (c), (d) dose of 4×10^{16} N_2^+ cm^{-2} (s, superlattice reflection (γ'); 1, γ', (111); 2, γ', (200); 3, γ', (311); 4, α, (110); 5, α, (211); 6, ϵ, (11$\bar{2}$1)); (e), (f) dose of 1×10^{17} N_2^+ cm^{-2}.

insignificant compared with lattice damage due to even a small dose of energetic ions. The defects are important for mobility of species and formation of phases. The temperature in the film during implantation will not be significantly different from that of bulk samples based on a simple calculation. The thermal contact resistance between the sputtered film and metal block can be conservatively assumed as 0.0005 °C W^{-1} m^{-2} [10], which is the contact resistance between two shaped surfaces with perpendicular cuts. At a high current density of 100 μA cm^{-2} and an accelerating voltage of 40 kV, the rate of energy input is 40 000 W m^{-2}. This results in a temperature difference across the interface of 20 °C (40 000 × 0.0005). So the temperature of the film during ion implantation would be nearly the same as the temperature of bulk samples under identical conditions. The sputtered films need to be thicker than the range of implanted ions to allow diffusion. Films up to a thickness of 1500 Å are electron transparent for most materials in 100 kV electron microscopes and this is greater than the range of implanted ions under most conditions.

5. RESULTS OF CHARACTERIZATION OF ION-IMPLANTED FILMS

Iron films of thickness 200 Å were implanted with argon (Ar^+) and nitrogen (N_2^+) in a unique ion implanter at Colorado State University. The implanter produces large-area (10 cm in diameter) intense (up to 1500 μA cm^{-2}) unanalyzed beams and has been operated up to 80 kV [11]. The implanter has been designed to achieve low cost treatment of mechanically active surfaces. An electron micrograph and an electron diffraction pattern of an unimplanted as-sputtered iron film are shown in Fig. 2(a) and Fig. 2(b) respectively.

A TEM study of argon (Ar^+) ion implantation at doses of 4×10^{15}, 9×10^{15}, 1.3×10^{16}, 1.8×10^{16} and 2.2×10^{16} Ar^+ cm^{-2} yielded a selected area diffraction pattern with diffuse (110) reflection of α-Fe, and all the higher order rings were absent. This indicates a high defect density due to radiation damage. A typical result is shown in Fig. 3.

A TEM study of nitrogen ion (N_2^+) implantation in iron was done at doses of 4×10^{15}, 4×10^{16} and 1×10^{17} N_2^+ cm^{-2}. Nitrogen ion implantation at 4×10^{15} N_2^+ cm^{-2} yielded results similar to those of argon ion implantation. The selected area diffraction pattern at a dose of 4×10^{16} N_2^+ cm^{-2} contains diffraction rings of γ'-Fe_4N and ϵ-Fe_2N_{1-x}. Similar results were obtained by conversion electron Mössbauer spectroscopy at a dose of 2×10^{17} N_2^+ cm^{-2} and an accelerating voltage of 100 kV and were reported in ref. 12. The results in this study are in agreement with the results of ref. 12 if the dose in ref. 12 is scaled down by a factor of 5 to account for the lower accelerating voltage used in this study. At this dose, γ-Fe_4N was observed and reported in ref. 13. γ-Fe_4N is similar to γ'-Fe_4N, with the nitrogen atoms occupying an orderly position in γ'-Fe_4N. The dose rate used for the work reported in ref. 13 was 8 μA cm^{-2} compared with 100 μA cm^{-2} in this study. Enhanced diffusion occurs at high dose rates because of a high defect density during implantation and hence formation of γ'-Fe_4N is favored. At a dose of 1×10^{17} N_2^+ cm^{-2}, grain growth was observed. These results are shown in Fig. 4.

6. DISCUSSION AND CONCLUSION

A simple technique of specimen preparation for the TEM study of surfaces obtained by ion-beam-processing techniques has been presented. The technique yields specimens with large electron-transparent areas and the results of nitrogen ion implantation in iron have been shown to be in agreement with the results of other studies. The technique would be suitable for ion-beam-mixing studies.

REFERENCES

1 S. T. Picraux and L. E. Pope, *Science, 226* (4675) (1984) 615.
2 S. T. Picraux and P. S. Peercy, *Sci. Am., 252* (3) (1985) 102.
3 S. T. Picraux, *Phys. Today, 37* (1984) 38.
4 P. B. Hirsch, A. Howie, R. B. Nicholson, D. W. Pashley and M. J. Whelan, *Electron Microscopy of Thin Crystals*, Butterworths, London, 1965.
5 L. E. Murr, *Electron Optical Applications in Materials Science*, McGraw-Hill, New York, 1970.
6 K. L. Copra, *Thin Film Phenomena*, McGraw-Hill, New York, 1979.
7 T. F. Fisher and C. E. Weber, *J. Appl. Phys., 23* (1952) 181.
8 R. Hanau, *Phys. Rev., 76* (1949) 153.
9 C. Moulton, *Nature (London), 195* (1962) 793.
10 J. P. Todd and H. B. Ellis, *Applied Heat Transfer*, Harper and Row, New York, 1982.
11 P. J. Wilbur and L. O. Daniels, *Vacuum, 36* (1–3) (1986) 5.
12 G. Longworth and N. E. W. Hartley, *Thin Solid Films, 48* (1978) 95.
13 B. Rauschenbach and A. Kolitsch, *Phys. Status Solidi A, 80* (1983) 211.

Materials Science and Engineering, 90 (1987) 373-383

Surface Modification of Industrial Components by Ion Implantation*

PIRAN SIOSHANSI

Spire Corporation, Patriots Park, Bedford, MA 01730 (U.S.A.)

(Received July 10, 1986)

ABSTRACT

Ion implantation is an emerging technology that allows the selective modification of the surface properties of materials. Properties such as hardness, corrosion resistance and friction can be improved without adversely affecting the bulk properties of the material. This paper concentrates on the discussion of the industrial applications of the ion implantation process. The unique capabilities, outstanding features and overall potential of the technology are described. Based on the experiences of Spire Corporation, successful applications of the ion implantation process are presented and the criteria for choosing promising candidates are outlined. The technology is expected to have an increasing contribution in a great variety of applications from aerospace to orthopaedics.

1. INTRODUCTION

Even though the surface modification of materials by application of ion beams had been suggested over a decade ago, the approach has only gained increased interest in recent years. This is because there is typically a dormant period between the time of discovery of a new technology and the time that it finds a "niche" in the market-place. For most high technology processes, this dormant period lasts between 10 and 15 years. The acceptance of the new process in most instances is believed to go through various stages of growth following the so-called "learning curve" as shown in Fig. 1.

Following the discovery, there is usually a period of "grand expectations" when it is believed that the technology will solve many existing problems. This period is usually followed by a "disappointment" period. The disappointment is due to many reasons. The potential of a new technology is often exaggerated before it has any practical impact. There are also difficulties in the transfer of the technology from laboratory to production, but the lack of proper equipment to accommodate scale-up of the process for industrial components at acceptable cost is usually the overriding reason. Successful technologies can eventually resolve these problems. These technologies can find a way to cope with equipment design problems and to proceed slowly through a period of slow growth toward market maturity. A few of the more successful processes have had shorter dormant periods and have found universal acceptance in a broad spectrum of applications. In contrast, many new discoveries have never emerged from the dormancy period and have been temporarily or permanently abandoned.

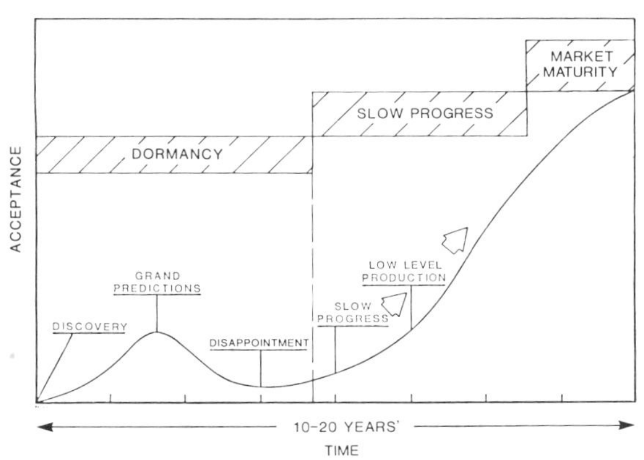

Fig. 1. Various stages of development for a new technology.

*Paper presented at the International Conference on Surface Modification of Metals by Ion Beams, Kingston, Canada, July 7-11, 1986.

0025-5416/87/$3.50

For example, ion implantation as a controlled process for doping semiconductors was first proposed [1] in 1954. The technology found widespread acceptance in the mid-1970s [2] and is now regarded as a universal process for fabricating semiconductor devices. The time span of 20 years from discovery to market maturity for ion implantation of semiconductors is almost average for high technology applications. For more traditional industries, the time span from discovery to market maturity is typically longer and may easily exceed 20 years. The medical device industry is perceived to be the most aggressive industry in recognizing the potential of new solutions and approaches to existing problems and has adopted high technology solutions in as little as 10 years.

Ion implantation as a process for improving the surface properties of non-semiconductors appears to be an emerging technology that is slowly finding its niche in the market-place. While it is becoming increasingly clear that ion implantation is not the universal treatment, as once thought, for changing the wear resistance of inexpensive tools and disposable items, the technology is finding a reputation for being the surface treatment of choice for ultraprecision products and tools. It is highly recommended for components where dimensional integrity is critical and where the danger of delamination of thin films and coatings rules out competing coating technologies. Thus, it is generally believed that there exists a growing market for ion implantation in applications such as medical tools and implantable devices, critical aerospace components, sensitive components of sophisticated reciprocating or rotating machinery, and certain types of ultraprecise and expensive toolings. Ion implantation has proved particularly successful in the treatment of beryllium-, titanium- and zirconium-based alloys, and various types of stainless, bearing grade and tool steels.

Ion implantation of ceramics and plastics is still in its infancy. In spite of some initial applications with exciting results, the process requires a much longer time to reach market maturity.

In terms of the time of development for the process to reach the market-place, ion implantation of non-semiconductors has been quite typical. The discovery of the process is generally acknowledged to be in 1973 with the publication of the first paper [3] in the field. The first industrial use of the process on the production level was in 1985 for the treatment of titanium-based orthopaedic implants [4]. The success of ion implantation is in part due to the availability of prototype ion implanters from the semiconductor industry but is also due to persistent efforts of a group of dedicated scientists in national laboratories, industry and universities who focused their attention on finding answers to some of the underlying questions involving the physics, chemistry and materials science aspects of the process.

With the current state of the technology and present prototype processing equipment, ion implantation is a valuable process for the treatment of precision products and tools. It is particularly suited to high technology, high value-added samples for which there is no suitable competing surface modification process. The aim of this paper is to review the status of ion implantation technology and to enumerate the successful applications of the process in various industries such as medical implantology, aerospace bearings and gears, and precision components and toolings.

Spire Corporation has been offering ion implantation services to various industrial customers for the past 7 years. The majority of applications have been performed on a small number of samples on a trial evaluation basis. A number of these applications have been highly successful and have prompted further close consideration of the process. At the present time, a few of these applications are in large-scale or pilot-scale production for ion implantation processing. From the point of view of organization, these components are broken down into two different categories (Table 1): (1) finished products that require a surface protection such as orthopaedic implants; (2) precision toolings (molds, dies, punches etc.) that are used to make other products.

The finished products are typically critical components of a sophisticated working machinery and as such have a highly complex geometry. They are made out of specialized alloys, have very tight dimensional tolerances and, as a rule, are expensive components. In many instances, failure of these devices leads to serious consequences and replacements are

TABLE 1

Successful examples of products and toolings for the ion implantation process

Type	*Field of application*	*Component*	*Material*	*Benefits*
Products	Medical	Orthopaedic implants	Ti–6Al–4V	Increased hardness and wear resistance
Products	Medical	Orthopaedic implants	Co–Cr alloys	Reduced corrosion current and ion release
Products	Aerospace	Gears or bearings	M50, AISI 52100, AISI 9310, type 440C steel and beryllium	Increased corrosion resistance and wear improvement
Products	Precision rotating or reciprocating machineries	Cryogenic refrigeration piston	Alloy steel	Low friction and improved wear properties
Products	Synthetic fibers; fiberglass	Extrusion die	Alloy steel	Corrosion and wear resistance
Toolings	Plastics; optics	Molds and dies	Steel	Increased life
Toolings	Metal forming	Stamping and punching dies	High speed steel	Increased life
Toolings	Ceramic	Punches	High speed steel	Increased life

quite costly. Often it is not the cost of the product that determines whether or not the added cost for the ion implantation is justified, but rather the (in-field) replacement costs of the critical component. Precision toolings are also expensive to manufacture and have ultratight tolerances. When these components are ion implanted, they typically last up to four to six times longer and reduce the unavailability of production equipment and hence can justify the added cost for ion implantation. Examples of the former and the latter groups that have emerged as prime candidates for ion implantation are presented in Table 1.

2. ORTHOPAEDIC IMPLANTS

Titanium and its alloys are generally recognized as one of the most biocompatible materials and, as such, titanium is the top contender to replace the Co–Cr alloys that have been traditionally used as orthopaedic implants. The questionable wear performance of titanium in abrasive and adhesive wearing conditions has been an impediment to more widespread use of the alloy up to the present time.

The ion implantation process has been shown to be extremely effective in enhancing the wear performance of titanium surfaces. At the same time, it reduces the wear of the ultrahigh molecular weight polyethylene mating part in contact with the alloy in total joint replacements. The latter point turns out to be just as important an issue as the wear of titanium and has attracted considerable attention.

The ion implantation process has many desirable features for processing of orthopaedic prostheses.

(1) The addition of benign elements such as carbon and nitrogen to titanium-based alloys is not expected to present any problem from the point of view of biocompatibility. The devices ion implanted with carbon and nitrogen can be marketed as "substantially equivalent" to unimplanted titanium.

(2) Ion implantation, under carefully controlled conditions, will process titanium devices without changing their surface finish or general appearance. The aesthetics and general appearance of orthopaedic devices are of great importance in this application;

the surgeons and prospective recipients are reluctant to accept devices that are discolored or non-uniform in appearance as this implies a lack of process control.

(3) The process is a clean vacuum process that can be readily applied to finished devices without introducing contaminants. Ion implantation is the last step in processing the orthopaedic devices prior to packaging and sterilization.

The ion implantation of carbon, nitrogen and similar ion species has been shown to be extremely effective in enhancing the overall properties of titanium-based alloys. Williams and coworkers [5, 6] have shown a very significant (up to a factor of 10 000) reduction in the corrosive wear of titanium-based alloys in laboratory experiments as a result of ion implantation. Zhang *et al.* [7] have performed a very careful study of nitrogen ion implantation on titanium alloys. They have compared the wear performance of nitrogen-implanted titanium not only with regular non-implanted samples but also with the traditional Co–Cr alloy used in orthopaedic prostheses. Ion-implanted titanium samples showed an even lower wear than Co–Cr-based alloys under similar conditions and showed a significant improvement over non-treated titanium. Zhang *et al.* reported that prototype titanium alloy implants have already been used in China on an experimental basis. Significant improvements in corrosion [8] and high cycle fatigue [9] resistance of the carbon- and nitrogen-ion-implanted Ti–6Al–4V (where the composition is in approximate weight per cent) have been measured and reported.

Spire Corporation has performed extensive independent tests on the wear performance of ion-implanted Ti–6Al–4V [4, 10]. Many different ion species (carbon, nitrogen, oxygen, boron, argon etc.) have been used in this evaluation. Spire's research has verified that the improved wear performance is related to the increased microhardness of titanium. As such, microhardness has been chosen as the main criterion for obtaining optimum ion implantation parameters for best wear results. Figure 2 shows the increased microhardness of Ti–6Al–4V as a function of nitrogen ion implantation. The Ionguard® process, as Spire refers to its proprietary ion beam process used for the optimal treatment of orthopaedic implants, has been successfully developed for the processing of Ti–6Al–4V total joints. The Ionguard process is currently used for the processing of a few thousand total joints per year on a production schedule started in mid-1985. Figure 3 shows the example of orthopaedic samples that are currently being processed by the Ionguard process.

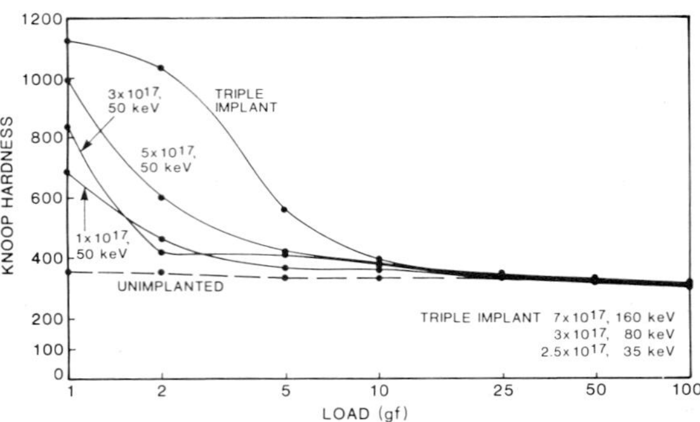

Fig. 2. Knoop microhardness *vs.* load for a variety of nitrogen ion implantation parameters.

3. Co–Cr-BASED ALLOY DEVICES

The Co–Cr-based alloys have traditionally been used for orthopaedic prostheses. This is due to the extreme hardness and wear resistance of the material in addition to the excellent corrosion resistance of the alloy. Some clinical studies have indicated that high concentrations of cobalt, chromium and other ingredients of the alloy have been identified in particular organs of patients who have received these devices for a prolonged time. The toxicity and carcinogenic nature of these ingredients (notably cobalt, chromium and nickel) have been a concern for the orthopaedic community. This phenomenon is aggravated in the newer generation of devices that make use of porous surfaces for bone ingrowth. This porous coated device represents a much larger surface area and accelerates attack by body fluids. The ion release concern is therefore proportionally higher for the new generation devices.

Ion implantation of nitrogen has been proven to be extremely effective in enhancing the corrosion resistance of Co–Cr-based alloys. This is not surprising. Chromium is an element well known for retaining nitrogen. The resultant chromium nitride surface has been

Fig. 3. Orthopaedic knee prostheses ready to be treated with the Ionguard process.

shown in corrosion experiments to achieve self-passivity at the corrosion potential and to reduce significantly the corrosion current. Spire has pursued an extensive study in this area under the sponsorship of the National Institutes of Health. The experimental results have been truly remarkable. The contribution of nitrogen ion implantation into Co–Cr alloy to reduce the ion release of the alloy may equal or surpass its contribution for the treatment of titanium-based alloys.

Ion beam mixing of noble metals such as platinum and gold into Co–Cr alloy has also been tried with varying degrees of success. The platinum ion beam mixing in particular is quite noteworthy. The polarization corrosion tests clearly show that platinum ions mixed into Co–Cr alloy not only change the corrosion potential and significantly reduce corrosion current but also create a hydrated calcium phosphate on the surface which may represent a biocompatible layer that will have a significant impact on the use of the process.

Other biomedical samples treated by ion implantation include hip compression plates and bone plates for use in patients with fractured hips or *long* bones (Fig. 4). These devices are made from stainless steel (type 316L) and suffer from fatigue failure in patients who subject these devices to excessive stress and strain cycles. Implantation of light ion species such as boron, carbon and particularly nitrogen has been very effective in enhancing the high cycle fatigue life of these alloys. The ion implantation of these devices is under preliminary testing and may prove to be a potentially large market in the future.

Dental implants and dental posts are typically made out of pure titanium or titanium-based alloys. Ion implantation has been used to improve the fatigue resistance of the dental posts. Dental implants, however, can benefit from added corrosion resistance with the superior biocompatibility of the titanium alloys as a result of ion implantation.

Various surgical tools have been processed by the ion implantation process. The applications include dental burrs, bone burrs, surgical scissors and scalpels, and titanium-based surgical tools. The main interest has been to prevent corrosive attack on the tools as well as to maintain the sharp cutting edge of these devices.

4. AEROSPACE BEARINGS AND GEARS

The ion implantation process has proved to be an ideal surface treatment for high

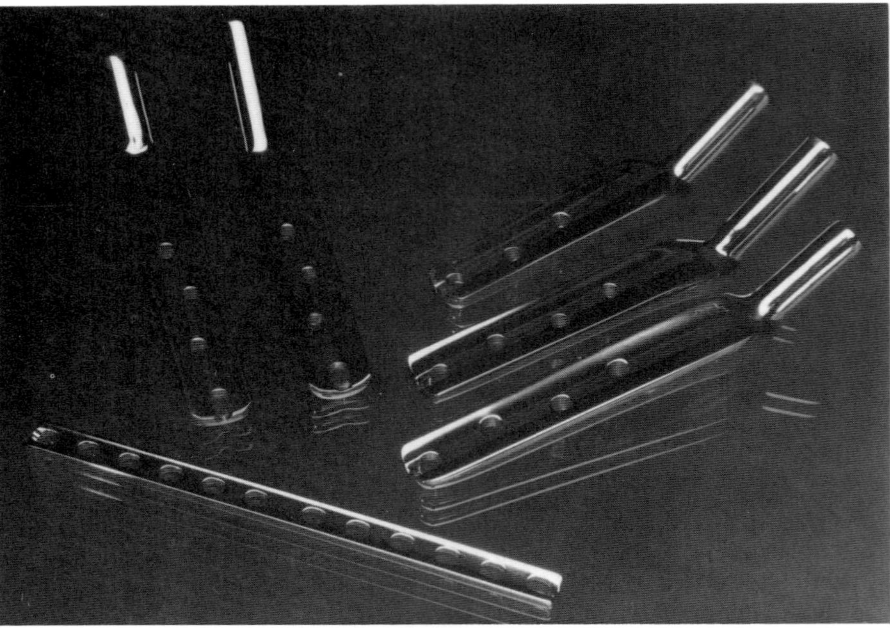

Fig. 4. Hip compression plates (right) and bone plate treated with the Ionguard process.

precision aerospace bearings. The process has been used to impart corrosion resistance to high grade bearing steels. Ion implantation has also been employed for increasing the wear resistance of bearings and gears and for lowering the coefficient of friction of these components. Aerospace bearings are a good example of a component that has a tight dimensional tolerance for which no other conventional coating procedure is suitable. Furthermore, the threat of delamination makes the application of coatings highly risky. Ion implantation is uniquely suited to this application. The preservation of dimensional integrity and surface finish and the absence of a sharp interface between the implanted layer and substrate material as a result of ion implantation remove the concern that exists with the delamination of hard coatings. Thus, ion implantation appears to be ideally suited to the processing of bearings.

The majority of pioneering work in this area has been performed at the Naval Research Laboratory (NRL). The primary objective has been to improve the corrosion resistance of the bearings for carrier-based aircraft. It has been observed that salt particles from the ocean environment accumulate in the aircraft engine's oil and cause corrosive attack on bearing components. Various ion implantation recipes have been developed by the NRL researchers [11-14]. The ion implantation of chromium has been shown to be extremely effective in increasing the resistance of the M50 bearings to generalized corrosion. Other ion species such as chromium, phosphorus, molybdenum and tantalum have also been tried with good results.

A Spire-General Electric team has implanted M50 NIL bearings with chromium and phosphorus. Corrosion tests have shown significant improvement in the resistance of the alloy to corrosive attack under laboratory conditions [15].

The ion implantation of titanium ions followed by carbon implantation has been used by a number of other groups for improving the wear resistance of bearing grade (AISI 52100 and type 440C) steel. Sioshansi and Au [11] and Au and Sioshansi [12] have reported a significant improvement in the wear resistance of mating surfaces that wear against AISI 52100 steel that has been ion implanted with titanium and carbon in lubricated conditions. The work of Singer and Jeffries [13] in this area also shows a much decreased wear as a result of titanium implantation into AISI 52100 steel. The work of Singer and Jeffries indicates that, under certain ion implantation conditions of titanium, there is no need for a subsequent carbon ion implantation. Their data support the formation of a carburized layer during titanium ion implantation as a result of latent hydrocarbons in the vacuum system. Pope *et al.* [14] have reported the formation of an amorphous layer on AISI 52100 steel as a result of dual-ion implantation of titanium and carbon.

Ion implantation of bearing and gear samples with tantalum has been performed at NRL [16] and independently at Spire Corporation. Tantalum is a particularly interesting ion species for the treatment of steel bearings and gears because tantalum can impart both corrosion and wear resistance to ferrous alloys. The NRL results have centered on evaluating the performance of tantalum ion implantation into AISI 9310 steel gears in sliding and scuffing wear situations. The results show a significant reduction in friction and wear of the materials. Spire Corporation has ion implanted tantalum in gears of rocket as well as helicopter engines. Preliminary laboratory results indicate that the tantalum-ion-implanted gears have performed significantly better than the standard.

Boron ion implantation into beryllium was reported in 1979 [17]. When implanted with boron, the beryllium gas bearings used in highly sophisticated aerospace applications showed a much reduced wear and a distinctly lower coefficient of friction. Similar investigation of beryllium or beryllium-coated mirrors proved that ion implantation is a powerful process for hardening the surface and providing resistance against abrasive wearing conditions.

The overall success of ion-implanted bearings has prompted the NRL to fund a manufacturing technology program at Spire Corporation.

A large ion implantation facility has been designed and manufactured that allows for prototype processing of large samples in a 1 m^3 chamber. The ion implanter can produce metallic ion beams with currents up to 10 mA. The other unique feature of the ion implantation facility is its capability to expand the beam and to use a wide-area uniform beam for the process. The beam is expanded either through a magnetic field or through a double raster-scanned electric field to expose an area as large as 0.2 m^2 with uniformity of better than 5%. The use of a widespread beam allows control of the temperature of the workpiece that is exposed to the beam. With proper heat conduction from the workpieces, the temperature can be maintained well below the tempering temperature of all ferrous alloys.

The facility incorporates an advanced workpiece-handling workstation. The mechanical system allows for manipulation of balls and roller elements for uniform exposure to the beam, for planetary systems for treatment of the inner and outer races of bearings and for most other samples with highly complicated geometries.

The design of the mechanical movement is such that it keeps the incident ion beams always within $\pm 30°$ of normal. This feature is necessary to allow a higher concentration of incoming ions in the substrate material. Figure 5 shows the general appearance of the ManTech ion implanter. Figure 6 shows the fixtures used for the treatment of roller elements.

As a result of this program, sizable quantities of hinge pin bearings and roller and ball bearings are being processed with chromium ions to make them impervious to corrosion. The program also entails processing of cobalt-bound tungsten carbide samples for improvements in wear performance.

The critical function of precision bearings in aerospace and other sophisticated operations, issues surrounding their indefinite shelf life and their tight dimensional tolerance make them ideal candidates for the ion implantation process. Furthermore, when the enormous cost of replacement of these bearings in aircraft engines is considered, the catastrophic consequences of their failure and the concomitant need to improve the longevity of these bearings have persuaded bearing manufacturers and bearing users to investigate the ion implantation process. It is expected that ion implantation will become an important process in aerospace and similar applications.

5. ULTRAPRECISE COMPONENTS OF SOPHISTICATED EQUIPMENT

The third category of components that are on the verge of market maturity for ion implantation treatment is sophisticated

Fig. 5. High current ion implantation system used for the NRL ManTech program.

Fig. 6. Fixturing used for uniform implantation of bearing roller elements.

components of machineries for which the ion implantation process has proved extremely useful.

For example, the refrigeration pistons in reciprocating cryogenic pumps have been ion implanted with titanium to a dose of 3×10^{17} ions cm^{-2} at 120 keV followed by carbon at a dose of 1.5×10^{17} ions cm^{-2} at 60 keV. After ion implantation, these components manifest much lower friction and operate satisfactorily in non-lubricated light load conditions. These components are processed by batches. The fixture drum is capable of processing 100 components at one time and can be loaded

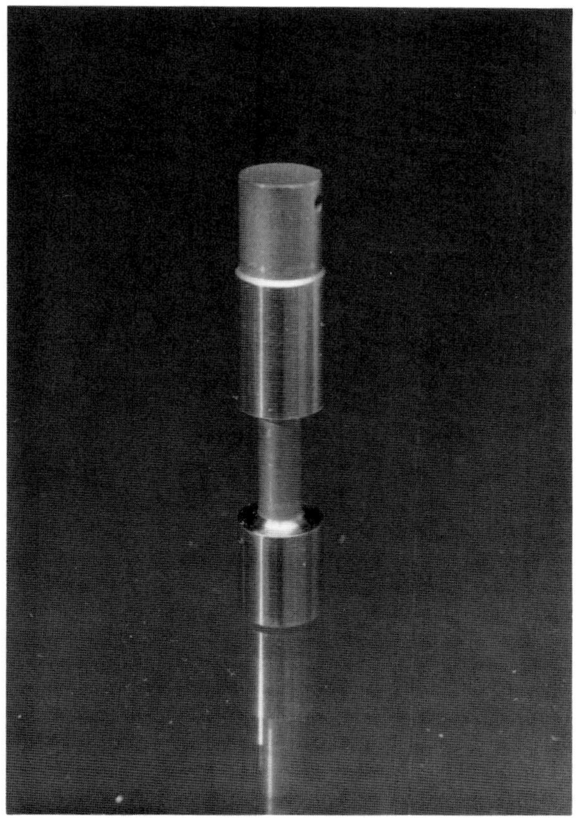

Fig. 7. Cryogenic refrigerator valve piston.

into the vacuum system. Figure 7 shows the cryogenic refrigeration piston and the fixture for processing these components.

Other components in this category include zirconium electrodes which monitor the flow of chemicals in highly corrosive environments. Zirconium is excellent for withstanding the corrosive attack of these alloys. However, the soft surface of zirconium introduces noise in the electrical signal of the electrode system. Ion implantation of nitrogen at a fluence of 1×10^{18} ions cm^{-2} at various energies up to 200 keV has proven to be extremely effective for improving these electrodes. To date, in excess of 400 of these electrodes have been processed with the ion implantation technique and this work is expected to continue and grow as the time goes on.

Ion implantation has proven to be very effective for reducing corrosion attack on Zircaloy (a zirconium alloy containing 1 wt.% Sn and small amounts of other impurities). Zircaloy is used extensively as the cladding for fuels in nuclear reactors. Ion implantation of tin, chromium and other elements has been very effective in reducing the high temperature aqueous corrosive attack on these alloys. Ion implantation of chromium and carbon has been effective in minimizing the fretting wear as well as the corrosion problem that these alloys are subjected to in a nuclear environment. Ion implantation processing of Zircaloy on a commercial basis appears to be a strong candidate for the near future.

6. PRECISION TOOLINGS

Precision toolings are components that are used for the manufacturing of products. These toolings are typically expensive items and the cost of ion implantation can be easily justified by the increase in lifetime. However, the major impact of ion implantation is due to the reduction of downtime and unavailability of a production line because of mechanical or chemical failure of the precision toolings. Several examples of these toolings are specified here.

6.1. Extrusion dies

Extrusion dies typically consist of a series of fine capillaries on a die. The objective in the treatment of these dies is to protect the walls of the holes against wear, erosion and corrosion. Spinnerettes for the extrusion of synthetic fibers is a case in point. The capillaries in these dies (typically stainless steel) are subject to erosive attack by the titanium dioxide particles that are added to synthetic material to minimize reflection of the fiber. They are also attacked by corrosion during the frequent cleaning which is necessary because of the build-up of synthetic material. When the sharp edges of the capillaries are removed, the hydrodynamics of the flow are changed and the die needs to be replaced. Spire has implanted a combination of chromium, titanium, yttrium and carbon on these spinnerettes with good results, especially from the point of view of corrosion resistance. Similarly, spinner dies for the extrusion of fiberglass have been treated by the ion implantation process with excellent success.

6.2. Injection molds

Injection molds are a good example of tools that have received the ion implantation process. Many of these molds are for plastics. Delves [18] has reported a distinct improvement in the life of these molds as well as of the injection screws used in this application. Our experiences generally agree with this assessment on injection molds. Spire's experience with injection molds has focused on those used in the semiconductor industry. In addition, Spire has treated very sensitive molds for fiber optic connectors. These molds are extremely delicate and have intricate details and structures. Titanium and carbon ion implantation has proven very effective for improving the longevity of these molds.

6.3. *Piercing, stamping and punching tools*

The literature on the industrial use of ion implantation gives extensive accounts of tools used in the metal stamping industry. Spire's focus in this area has again been on tools of high speed steel. Among the various tools, the improvements seen on the ion-implanted set of progressive dies used for piercing and punching holes in the semiconductor industry is noteworthy. Spire has also treated high speed steel stamping punches and dies for the aluminum can industry. The ion-implanted dies outlast untreated dies by a factor of 3–5 and there is a marked reduction in aluminum pick-up by the dies. The combination of titanium and carbon has been the most effective recipe for prolonged tool life. Pellet punches for forming fuel pellets in the nuclear industry are another tool that was successfully treated by ion implantation.

Cutting tools for synthetic fibers and material have been ion implanted with titanium and carbon with great success. The computer-controlled machinery for cutting several layers of material requires blades that stay sharp much longer. Unimplanted blades need to be resharpened after cutting through only a few meters of material. The ion-implanted blades last four to six times longer between sharpenings.

7. CONCLUSION

The industrial application of the ion implantation process has been reviewed and the potential of the technology and its main contribution have been emphasized. It is pointed out that the ion implantation process is clearly not the universal approach for treatment of inexpensive tooling where other conventional surface treatments are used. However, the ion implantation process is acknowledged to be ideally suited to processing of highly sophisticated, high value-added, high technology components with tight dimensional tolerances where other surface treatments have failed. As such, the ion implantation process is finding a niche in improving the wear and corrosion performance of "life-limiting" components in medical, aerospace and similar applications.

The ion implantation process is gradually making inroads in the industrial marketplace. The process is already used for production processing of orthopaedic implants. There are a number of other components that are processed on a pilot-scale production at this time. The ion implantation technology appears to be well positioned to find a more significant market for highly specialized applications in the coming years.

REFERENCES

1 J. Gale, personal communication, 1985.

2 S. Namba (ed.), *Ion Implantation in Semiconductors*, Plenum, New York, 1974.
G. Carter, J. S. Colligon and W. A. Grant (eds.), *Proc. Int. Conf. on Applications of Ion Beams to Materials, Warwick, 1975*, in *Inst. Phys. Conf. Ser., 28* (1976).

3 N. E. W. Hartley, W. E. Swindlehurst, G. Dearnaley and J. F. Turner, *J. Mater. Sci., 8* (1973) 900.

4 P. Sioshansi, R. W. Oliver and F. D. Matthews, *J. Vac. Sci. Technol. A, 3* (6) (1985) 2670.

5 J. M. Williams and R. Buchanan, in G. K. Wolf, W. A. Grant and R. P. M. Procter (eds.), *Proc. Int. Conf. on Surface Modification of Metals by Ion Beams, Heidelberg, September 17–21, 1984*, in *Mater. Sci. Eng., 69* (1985) 237.

6 J. M. Williams, G. M. Beardsley, R. Buchanan and R. K. Bacon, in G. K. Hubler, O. W. Holland, C. R. Clayton and C. W. White (eds.), *Ion Implantation and Ion Beam Processing of Materials, Materials Research Society Symp. Proc., Boston, MA, November 14–17, 1983*, Vol. 27, Elsevier, New York, 1984, p. 735.

7 Z. Jianqiang, Z. Xioazhang, G. Zintang and L. Hengde, in J. M. Williams, M. F. Nichols and W. Zingg (eds.), *Biomedical Materials, Materials Research Society Symp. Proc.*, Vol. 55, Materials Research Society, Pittsburgh, PA, 1986, p. 229.

8 E. D. Rigney, R. A. Buchanan and J. M. Williams, *11th Annu. Meet. Society for Biomaterials, San Diego, CA, 1985.*

9 R. G. Vardiman and R. A. Kant, *J. Appl. Phys., 53* (1) (1982) 690.

10 P. Sioshansi and R. W. Oliver, in J. M. Williams, M. F. Nichols and W. Zingg (eds.), *Biomedical Materials, Materials Research Society Symp. Proc.*, Vol. 55, Materials Research Society, Pittsburgh, PA, 1986, p. 237.
11 P. Sioshansi and J. J. Au, in G. K. Wolf, W. A. Grant and R. P. M. Procter (eds.), *Proc. Int. Conf. on Surface Modification of Metals by Ion Beams, Heidelberg, September 17-21, 1984*, in *Mater. Sci. Eng., 69* (1985) 161.
12 J. J. Au and P. Sioshansi, in G. K. Hubler, O. W. Holland, C. R. Clayton and C. W. White (eds.), *Ion Implantation and Ion Beam Processing of Materials, Materials Research Society Symp. Proc., Boston, MA, November 14-17, 1983*, Vol. 27, Elsevier, New York, 1984, p. 679.
13 I. L. Singer and R. A. Jeffries, in G. K. Hubler, O. W. Holland, C. R. Clayton and C. W. White (eds.), *Ion Implantation and Ion Beam Processing of Materials, Materials Research Society Symp. Proc., Boston, MA, November 14-17, 1983*, Vol. 27, Elsevier, New York, 1984, p. 673.
14 L. E. Pope, F. G. Yost, D. M. Follstaedt, S. T. Picraux and J. A. Knapp, in G. K. Hubler, O. W. Holland, C. R. Clayton and C. W. White (eds.), *Ion Implantation and Ion Beam Processing of Materials, Materials Research Society Symp. Proc., Boston, MA, November 14-17, 1983*, Vol. 27, Elsevier, New York, 1984, p. 661.
15 E. Bamberger, personal communication, 1986.
16 N. E. W. Hartley and J. K. Hirvonen, *Nucl. Instrum. Methods, 209-210* (1983) 933.
17 R. A. Kant, J. K. Hirvonen, A. R. Knudson and J. S. Wollam, *Thin Solid Films, 63* (1979) 28.
18 B. G. Delves, in V. Ashworth, W. A. Grant and R. P. M. Procter (eds.), *Proc. Int. Conf. on Ion Implantation into Metals*, Pergamon, Oxford, 1982, p.126.

Materials Science and Engineering, 90 (1987) 385–397

U.S. Navy Manufacturing Technology Program on Ion Implantation*

F. A. SMIDT[a], B. D. SARTWELL[a] and S. N. BUNKER[b]

[a]*Naval Research Laboratory, Washington, DC 20375-5000 (U.S.A.)*
[b]*Spire Corporation, Bedford, MA 01730 (U.S.A.)*

(Received July 10, 1986)

ABSTRACT

The U.S. Navy supported a manufacturing technology program to evaluate the potential of ion implantation for industrial-scale treatment of components for wear and corrosion protection. In this paper the overall project is described, the experiments conducted to define constraints imposed by heat removal, sputtering at non-normal angles of incidence, and contamination of the surface by residual gases in the chamber atmosphere are reviewed and the manner in which these problems were solved in the MANTECH facility is illustrated. Finally an economic assessment of ion implantation under batch processing conditions will be presented.

1. INTRODUCTION

Research and development on a large number of potential applications of ion implantation for materials processing have been reported in the literature over the past 10 years and a service industry has begun to develop. Several years ago, it was recognized that the major barriers to the acceptance of ion implantation as a practical materials-processing technique included the absence of a capability to process industrial quantities of actual parts with complex geometries, and a relatively high cost which was closely tied to the low throughput of facilities devoted primarily to research. The manufacturing technology program described in this paper was instituted to address these problems.

The U.S. Navy (and other components of the U.S. Department of Defense (DOD)) support manufacturing technology (MANTECH) programs to strengthen the industrial base of the U.S.A. by improving industrial capability through the introduction of new processes, techniques and equipment, by reducing acquisition, production and life cycle costs through improved processing methods and by bridging the gap between research and development advances and full-scale production, particularly where industry may not be capable of assuming the risk. Criteria for selection of MANTECH programs include a defined U.S. Navy need, demonstration of technical feasibility, a generic technology, an effort beyond the normal risk of industry and implementation within a 3–5 year time frame with a favorable cost payback. Research conducted at the Naval Research Laboratory (NRL) and other laboratories had demonstrated the efficacy of ion implantation in solving many wear and corrosion problems but, as previously noted, it had not been demonstrated on an industrial scale. A detailed analysis conducted in 1981 indicated that ion implantation met all the MANTECH criteria and, therefore, a proposal by NRL to develop a high throughput prototype industrial facility was accepted and funded in 1982 [1].

The program had several subprojects which included (a) definition of system requirements and processing parameters, (b) design, construction and operation of the facility, (c) demonstration of the successful treatment of representative components for corrosion and wear protection and (d) an economic benefit analysis for the process. Each of the subprojects in turn had several tasks which will be described in more detail below. NRL was the program manager and performed part of the research. Spire Corporation, Bedford, MA, was awarded the major subcontract for design, construction and operation of the facility

*Paper presented at the International Conference on Surface Modification of Metals by Ion Beams, Kingston, Canada, July 7–11, 1986.

0025-5416/87/$3.50

after a design competition. Eaton Corporation, Danvers, MA, built the implanter for the facility. The State University of New York at Stony Brook, University of Minnesota, and Virginia Polytechnic Institute and State University had research contracts to address specific issues concerning optimum implantation parameters for achieving specific material properties. The Naval Air Propulsion Center, Trenton, NJ, North Island Naval Air Station, San Diego, CA, and Draper Laboratories, Cambridge, MA, provided important links to operational components of the Navy and DOD. In this paper the major issues addressed in the Navy MANTECH program for the scale-up of ion implantation processing of materials for corrosion and wear protection from laboratory to industrial scale will be discussed.

2. DEFINITION OF SYSTEM REQUIREMENTS

2.1. Program objectives

The initial definition of system requirements proceeded from programmatic decisions on how large a scale-up would be representative of industrial practice. NRL experience indicated that a 0.5 mA beam attained with a Freeman source in a Varian–Extrion model 200A2F was adequate for research activities but required excessive implantation times for treatment of large numbers of parts. Commercial implanters were available in 1982 with beam currents up to 10 mA; this was considered a large enough increment to provide reasonable throughput and was representative of the problems of industrial production. This decision thus deferred the task of source development until the more practical problems of heat removal and manipulation of parts in a vacuum system could be shown to the tenable.

Other system requirements were defined by the components selected for demonstration implantation which included gas turbine bearings, instrument bearings and tooling. These choices dictated the requirement to handle flats, cylinders and spheres of various sizes and no limit the temperature rise to a level which would not degrade the material properties. The use of beam currents of 5–10 mA at acceleration voltages in the 25–160 kV range further defined system requirements for the means of uniformly distributing the ions across the surface of the component.

High current ion beams require space charge neutralization to propagate the beam in a controlled manner. Low energy electrons produced by interaction of the high energy ions with residual gas molecules normally provides this neutralization. However, if electrostatic rastering of the ion beam is attempted, the low energy electrons are removed from the beam envelope and the beam expands rapidly because of the positive space charge. A high current ion beam system therefore may require manipulation of the workpiece in a stationary beam rather than electrostatic scanning of the beam.

2.2. Heat removal

Heat removal is a major design constraint on high beam current systems. A 10 mA beam at 100 keV represents a heat input of 1 kW which must be dissipated to prevent overheating and degradation of the sample being implanted. Grabowski and Kant [2] analyzed the effectiveness of various methods to control specimen heating during ion implantation including radiation cooling and various approaches to conduction cooling. Radiation cooling was found to be totally inadequate for the power densities of interest in a commercial implanter. Three regimes of conductivity were treated for the case of one-dimensional heat transfer: (a) heat diffusion limited where the deposited energy has not diffused to a boundary, (b) heat capacity limited where the sample is isolated and the temperature rise is determined by the heat capacity of the sample and (c) steady state heat transfer to a sink.

The transition point between case (a) and cases (b) and (c) occurs at a value of

$$\tau = 1 = \frac{K}{\rho c}\left(\frac{\phi E}{J}\right)^{1/2} l \qquad (1)$$

where K is the thermal conductivity, ρ the density, c the heat capacity, ϕ the fluence, E the energy, J the power density and l the sample thickness. For values of τ less than unity the temperature increase is given by

$$\Delta T = \frac{2}{\pi^{1/2}}\left(\frac{\phi E}{K}\right)^{1/2} J^{1/2} \qquad (2)$$

where ΔT is the temperature rise of the surface above ambient. Figure 1 shows the tem-

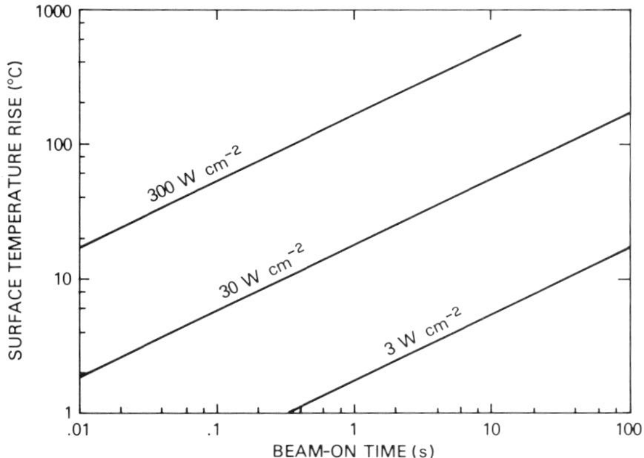

Fig. 1. Surface temperature rise for a carbon steel target with an ion beam incident on the surface for the case where diffusion of heat has not reached the boundary. The temperature rise as a function of dwell time for three power densities is shown.

perature rise as a function of beam residence time on a carbon steel sample 1 cm thick for selected power densities. Thus a 3 mA beam at 100 keV incident on a 1 cm^2 target would cause a 200 °C temperature rise in 1 s. This should be compared with a typical metallurgical implant of 3×10^{17} ions cm^{-2} which would require 16 s. It therefore is necessary to reduce the average current density or the time that the beam is on target by manipulation of the workpiece or rastering the ion beam.

Calculations [2] indicated that heat capacity could not be used to limit the temperature increase for most metallurgical implantations even if the actual workpieces were attached to large metal heat sinks in the vacuum chamber. Therefore, heat conduction to an infinite sink (*e.g.* water cooling) outside the chamber is required to maintain a surface temperature at a controlled value. The temperature at which AISI 52100 steel begins to soften is 250 °C; so a maximum temperature rise of 200 °C was specified as the heat control requirement for this project. Under steady state conditions the heat flux out of a plate in thermal contact with an ideal heat sink is

$$H_0 = \frac{K = (T_w - T_s)}{l} \tag{3}$$

where T_w is the surface temperature of the workpiece and T_s is the heat sink temperature. If $K = 0.21$ W cm^{-1} K^{-1} (for M2 tool steel), $l = 1$ cm and $\Delta T = 200$ °C, then the maximum heat flux would be 37 W cm^{-2}. In actual practice the situation is considerably more difficult since there an ideal heat sink never exists. The temperature drop across an interface is given by

$$\Delta T = \frac{H_{0c}}{h_c} \tag{4}$$

where h_c is the interface conductance. Experimental measurements of h_c for steel against steel in vacuum gave values of 0.05 W cm^{-2} K^{-1} [2]. A heat flux of 37 W cm^{-2} with an h_c of 0.05 would produce a temperature drop of 740 °C across the interface. Thus the vacuum interface is by far the most difficult problem to be solved.

Empirical studies of contact resistance show that it is directly related to the flatness of the contacting surfaces and the applied pressure and inversely dependent on the hardness [3]. Clamping a thin ductile film of high conductivity material such as indium between the two stainless steel surfaces raised h_c to 0.6 and a silver-filled conducting paint raised the value to 6 W cm^{-2} K^{-1}. The problem of heat removal is even more difficult with complex parts where there is only line or point contact between the component and the heat sink.

The preceding discussion makes clear that heat removal from the workpiece is the major engineering problem to be solved in high beam current ion implantation processing. The problem is so acute that it in fact eliminates the feasibility of using a high current density scanned beam to do piece-by-piece processing. A large-area beam with longer exposures was found to be necessary to reduce the heat load to a manageable level and this in turn dictates that the facility must be operated in a batch mode.

2.3. *Effects of sputtering*

Sputtering is an effect which must be taken into consideration for high fluence ion implantation. The sputtering yield depends on the energy deposition function (which in turn depends on the masses of the beam and target atoms and the energy of the incident ion), the escape depth for a sputtered ion and the binding energy to the surface. Sputtering yields for metal ions in the energy range 25–200 keV on metallic substrates are typically in the range 2–10. Ion implantation into a planar surface at normal incidence results in a roughly gaussian profile with a mean range R_p and straggling σ [4]. As the fluence increases, the surface is eroded and the profile gradually changes toward a flat profile from the surface to the depth of the mean range. The maximum concentration at steady state is $1/S$ where S is the sputtering yield. Steady state is approached after erosion to a depth of $\frac{3}{2}R_p$.

The sputtering yield is a function of the angle of incidence of the beam on the sample with the yield increasing until the angle between the ion beam and normal to the sample surface is 80° and decreasing thereafter. The increase in yield is due to the deposition of energy nearer to the surface while the decrease at large angles results from reflection at grazing angles of incidence. One of the consequences of the angular dependence of sputtering is that the retained dose at steady state is a strong function of the geometry of the part being implanted.

Grabowski *et al.* [5] conducted experiments on a stationary cylinder implanted with various ions at an energy of 150 keV and fluences from 0.03×10^{17} to 3×10^{17} ions cm^{-2}. The retained dose was then measured by ion-induced X-ray emission or Rutherford backscattering spectroscopy. Results for low fluence implantations could be explained solely on the basis of the larger area intersected by the beam, *i.e.*

$$D_0 = \phi_0 \cos\theta \tag{5}$$

where D_0 is the retained dose, ϕ_0 is the fluence at normal incidence (0°) and θ is the deviation from normal incidence. Results for high fluence implantations approaching steady state showed a much more complex angular dependence as shown in Fig. 2. The results from argon, titanium, chromium and tantalum ions incident on AISI 52100, M50 or type 304 steel cylinders are shown normalized to a fraction of the retained dose relative to 1.0 at normal incidence. These results can be explained in the following manner. If the concentration profile at steady state is approximated by a step function with the concentration of implanted ions equal to $1/S$ to a depth of $X = R_p$, then the retained dose D is given by $D = NR_p/S$ where N is the density of the target. For the non-normal incidence the range is shortened to $R_p \cos\theta$ and S is a function of the angle of incidence. Following Sigmund's

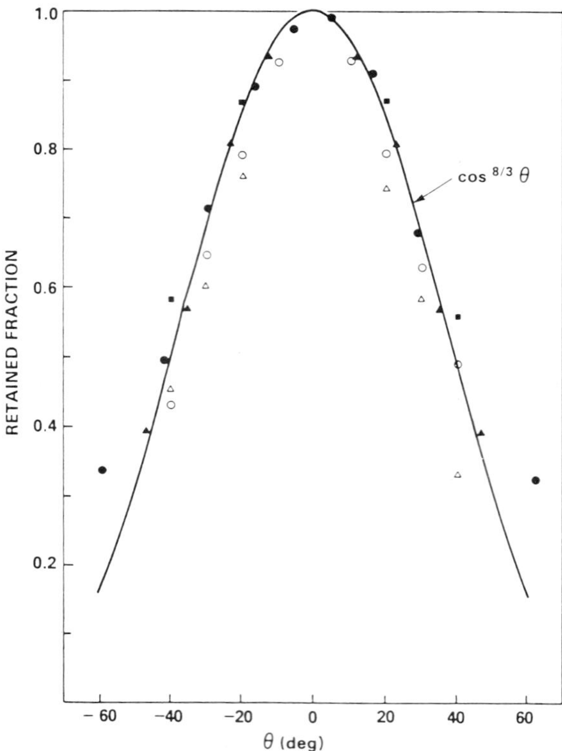

Fig. 2. Retained dose as a function of angle of incidence of the beam normalized to steady state concentrations at high fluences. The data obtained from the implantation of the ions indicated at an accelerating voltage of 150 kV into a stationary cylinder.

Symbol	*Ion*	*Target*	ϕ_0 ($\times 10^{17}$ ions cm^{-2})
■	Ar^+	AISI 52100	3
●	Ti^+	Type 304	10
▲	Cr^+	M50	2
○	Ti^+	AISI 52100	3
△	Ta^+	M50	1

model, $S = S_0 \cos^f\theta$ where S_0 is the sputtering yield at normal incidence and $f = 5/3$ for ions of nearly equal masses in the intermediate mass range of the periodic table [6]. Substituting this value for S gives a relation for retained dose of the form

$$D = \frac{NR_p}{S_0} \quad (\cos^{8/3}\theta) \tag{6}$$

This is in good agreement with the experimental data shown in Fig. 2.

The property change to be effected by ion implantation is usually a function of composition and thus the rapid variation in retained dose with angle of incidence is a concern. A specific example is the inhibition of corrosion of iron by implanting chromium. Corrosion resistance is imparted to ferrous alloys by the addition of 15 at.% Cr. The sputtering yield of chromium ions (Cr^+) on iron at 120 keV is 4.7; so the steady state concentration is 21 wt.%. It therefore becomes necessary to mask the samples so that the beam does not strike a curved surface at angles of incidence greater than ±30°.

Subsequent experience with fixtures for cylinders and spheres showed that under most circumstances these components were rotated under the ion beam so that variation in range and sputtering yield was averaged over all angles of incidence allowed by the mask. Manning [7] has written computer codes to calculate the composition profiles for a rotating cylinder and a rotating sphere using range and straggling calculations from E-DEP, corrections in range for the rotating component and a $\cos^{5/3}\theta$ sputtering yield dependence from 0° to 80° and an inverse function from 80° to 90°. A summary of the calculations for a rotating cylinder is shown in Fig. 3 for a case typical of chromium implantation into iron. The effect of rotation is to lessen the strong dependence of retained dose on angle of incidence so that the angle of incidence can be increased from ±30° to ±45° and still obtain a steady state concentration of 15 at.% Cr or more. This has a strong influence on beam

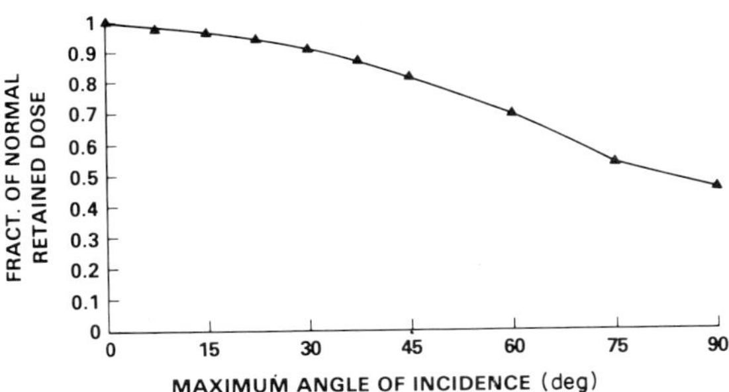

Fig. 3. Calculated values for the fraction of the maximum concentration of implanted atoms at steady state as a function of the maximum angle of incidence allowed by a mask over a rotating cylinder (iron bombarded with iron ions; $S_0 = 5.0$). The calculation assumes a sputtering yield of 5.0 at normal incidence and a $\cos^{5/3}\theta$ dependence for the sputtering yield from 0° to 80° and a linear decrease from 0° to 90°.

utilization and economics, as will be shown later in Section 4. Full exposure of the rotating cylinder without use of a mask results in a retained dose of 0.45 of that at normal incidence. The calculations for a rotating sphere show a slightly stronger angular dependence than the rotating cylinder, with a maximum concentration of 17.5 at.% at 30° *vs.* 18.2 at.% for the cylinder, 14.9 at.% at 45° *vs.* 16.3 at.% for the cylinder, 11.7 at.% at 60° *vs.* 13.8 at.% for the cylinder and 7.3 at.% at 90° *vs.* 9.1 at.% for the cylinder.

2.4. *Effects of residual gases in the target chamber*

Another processing variable which is often neglected in ion implantation is the residual gas content in the target chamber. Experience has shown that surface reactions can occur between reactive implant species such as titanium and carbonaceous molecules even when the chamber pressure is as low as 1×10^{-6} Torr [8]. Such reactions may be desirable if the modified surface exhibits low friction and wear as with the carburized surface layer formed during titanium implantation of steel or it may be undesirable as when second phases from near the surface that are less corrosion resistant than other areas of the surface.

Several experiments were conducted by Sartwell and coworkers [9-11] using an ultrahigh vacuum chamber with a cylindrical mirror Auger analyzer to perform *in situ* surface analysis during implantation. A controlled leak provided the capability to investigate the effect of various gases at low partial pressures on the resulting surface composition. The major observations are summarized in Fig. 4, which shows a comparison of the surface concentrations of carbon, oxygen and titanium as a function of titanium ion fluence implanted into high purity iron for a vacuum of 1×10^{-8} Torr and a partial pressure of CO of 1×10^{-6} Torr. The titanium concentration increases in both cases as the surface is sputtered away and previously implanted titanium atoms are exposed on the surface. In the ultrahigh vacuum environment, surface contaminants of carbon and the air-formed oxide are soon removed by sputtering. By contrast, implantation in a CO partial pressure of 1×10^{-6} Torr leads to increases in carbon and oxygen concentrations which closely parallel the increase in titanium concentration on the surface. Figure 5 shows a

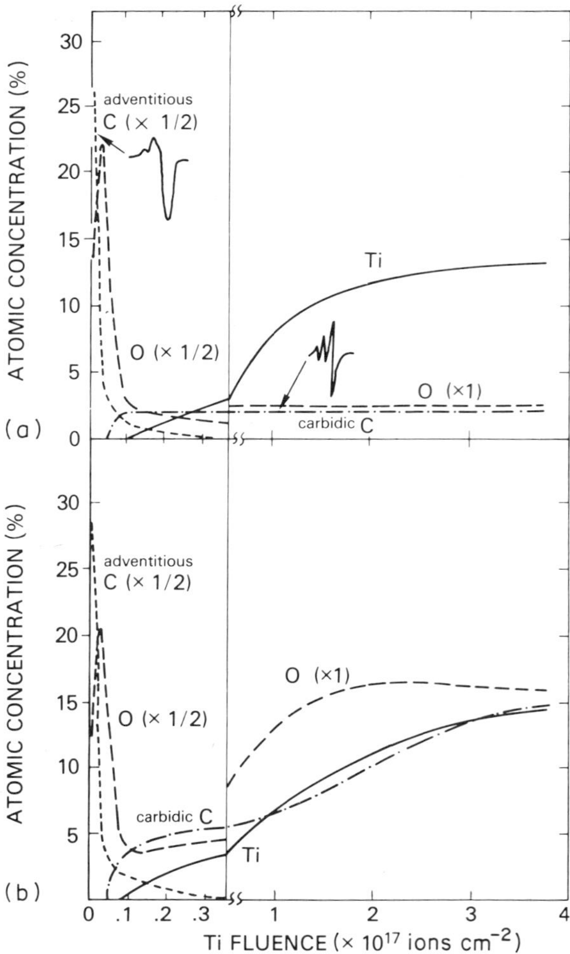

Fig. 4. The changes in the surface concentrations of carbon, oxygen and titanium during implantation of an AISI 52100 steel alloy as measured by Auger electron spectroscopy: (a) profile measured in an ultrahigh vacuum environment (1×10^{-8} Torr); (b) profile measured in the same chamber backfilled to a CO partial pressure of 1×10^{-6} Torr. A distinction between adventitious carbon and carbidic carbon was made on the basis of Auger line shape.

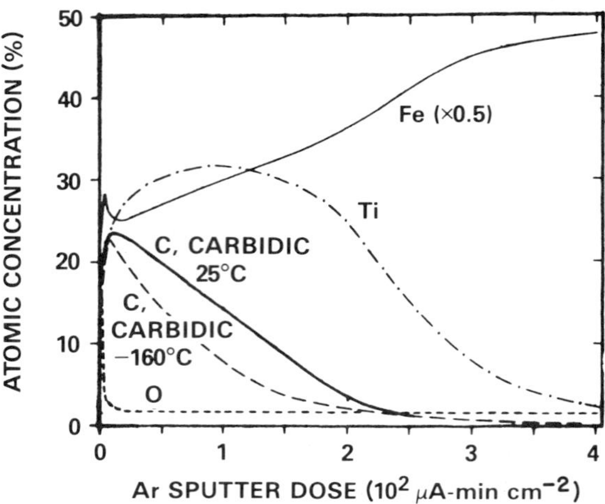

Fig. 5. Concentration depth profiles of titanium and carbon measured by ion profiling after implantation of 3×10^{17} Ti^+ cm^{-2} in a CO atmosphere after the conclusion of the experiment shown in Fig. 4 at 25 °C: - - -, the results of a similar experiment performed at −160 °C, showing that the carbon transport mechanism is not a normal thermally activated diffusion process.

depth profile at the end of the implantation experiment for the sample implanted in the CO atmosphere to a fluence of 3×10^{17} ions cm^{-2}. It is found that carbon has been transported from the surface in a diffusion-like profile. No oxygen transport has occurred under these same conditions. A similar experiment conducted with the sample maintained at −160 °C also shows substantial carbon transport, indicating that this is not a case of simple thermally activated diffusion.

Additional experiments [11] have shown that carburization occurs during titanium implantation in C_2H_3 at the same pressure as for the CO studies but not CH_4, that iron implantation into iron in a CO atmosphere does not produce carbon transport and that nitrogen transport occurs during titanium implantation into iron in an NH_3 atmosphere but not in an N_2 atmosphere. From these results, it can be concluded that chemisorption and dissociation of the gas molecule on the surface are critical steps since CH_4 and N_2 do not dissociate. The presence of an active species such as titanium is also necessary since iron ions did not produce transport even though carbon and oxygen chemisorbed on the surface. The transport of carbon and nitrogen but not oxygen is probably best explained by the interstitial site occupancy of carbon and nitrogen in iron. Additional modelling studies by Farkas *et al.* [12] indicate that the carbon transport mechanism involves both cascade mixing of the implanted titanium and radiation-induced segregation. Another observation from these experiments is that the incorporation of carbon into the alloy lowers the sputtering yield of titanium and thus higher retained doses are obtained in the CO and C_2H_2 atmospheres than in ultrahigh vacuum [11]. Although not all the mechanisms of these surface reactions are well understood, it is clear that residual gases even at partial pressures of 1×10^{-6}

Torr play an important role in implantation and therefore their concentration and composition must be controlled.

3. FACILITY DESIGN AND OPERATING EXPERIENCE

3.1. Implanter characteristics and beam optics

Facility design and operation were the responsibility of Spire Corporation. Armini and Bunker [13] have described the preliminary design in a previous publication which will only be summarized here. A cutaway schematic drawing of the facility is shown in Fig. 6. A quadrupole lens was designed to transform the 0.8 cm × 2.8 cm rectangular beam from the Eaton model NV 10-160 implanter into a variety of shapes including line, elliptical and focused spot beams. A beam deneutralizer was placed downstream from the quadrupole lens to disrupt the space charge neutralization intentionally so as to assist in beam expansion and to improve the homogeneity of the beam. A flood beam with a 40 cm diameter was envisaged for this mode of operation. Beam currents of 10 mA for nitrogen ions (N^+) and 3 mA for chromium ions (Cr^+) at 120 keV were specified in the implanter procurement. A uniformity in concentration of ±10% over the implanted area was a goal of the program.

Operational testing of the Eaton NV 10-160 implanter revealed a variation in beam profile with energy and mass. At low energies a diffuse spot with a large divergence was found. At high energies with beams of low mass elements, a low divergence core with a diffuse halo was observed while beams with higher mass elements produced less of a halo. The change in beam profile with operating conditions made it necessary to change the quadrupole lens setting with each change in operating conditions and this was found to be a time-consuming operation which required assistance from the professional staff.

A dual-axis raster scanner was designed using a modification of the original deneutralization plates [14]. This system has worked well especially at the higher voltages and has been found to require less set-up time, to need less supervision by the professional staff and to provide more efficient coverage of rectangular areas. A 1 cm × 2 cm spot can be obtained at voltages of 120–160 kV. Deflection up to 20 cm can be obtained by applying up to 20 kV to the deflection plates. The scans are driven by applying triangular pulses 180° out of phase to the parallel plates of the deflection system. A scan rate of 325 Hz on the horizontal plates and 23 Hz on the vertical plates is used. Implant uniformity for this system was evaluated using silicon wafers placed at various locations across the beam and was found to be ±2%.

A variety of beams have been obtained during evaluation of the implanter. A sum-

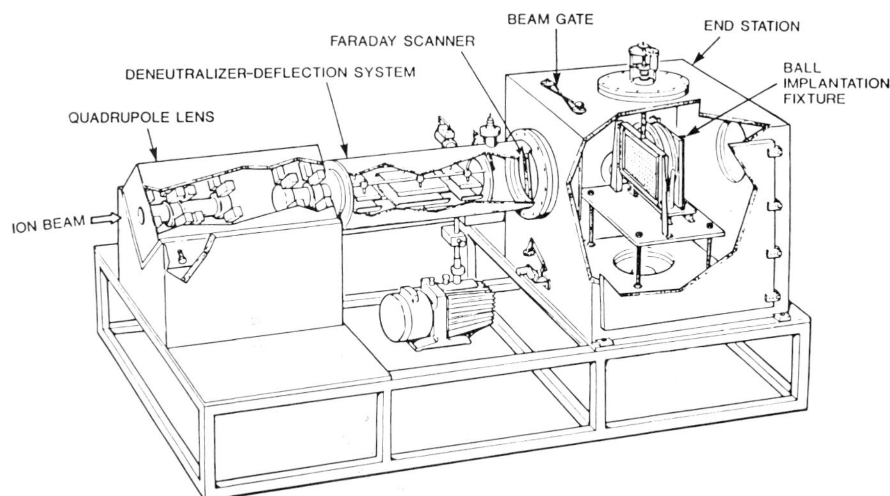

Fig. 6. Schematic diagram of the beam optics and end station designed by Spire Corporation for the MANTECH facility.

TABLE 1

Ion beams obtained from implanter

Ion species	*Current* (mA)	
	Typical	*Best*
Ti^+	5	8
C^+ (50 keV)	5	8
N (N^+ at 50–170 keV atom^{-1})	6	8
N (N_2^+ at 25–85 keV atom^{-1})	10	12
B^+	3	4
P^+		10
A^+		10
Cr^+	5	13
Mo^+	3	5

TABLE 2

Specialized fixtures for implantation

Fixture	*Uses*	*Implants*
Cooled shaft	Rotary drive	Inner races
Cooled platter	Rotary drive	Outer races
Multihole plate	Magnet platter	Flat ends of rollers
Multislot plate	Linear translator–magnet platter	Cylindrical sides of rollers
Multihole cage	Linear translator–magnet platter	Balls

mary of the results is given in Table 1 where typical and best values are shown. Typical indicates the best operating conditions for long-term stable operation of the implanter while best indicates the highest values obtained during short periods of operation. The results are in general better than the machine specifications. The most troublesome difficulty encountered to date has been deposition of source material in the arc chamber and on the extraction electrodes during routine shutdowns. This results in arcing and unstable operation and even an inability to restart the arc. This necessitates disassembly and cleaning of the arc chamber. The problem has been circumvented where possible by continuous three-shift operation but may ultimately require redesign of the arc chamber and extraction electrodes.

3.2. *Work chamber and fixtures*

The original work chamber and fixturing designs were also described in ref. 13 and are shown in Fig. 6. The work chamber consists of a 1 m^3 stainless steel chamber with a large vacuum door on one side. Various fixtures for handling different types of workpiece can be easily interchanged in the chamber. A quartz window for pyrometer measurements is located near the beam entrance. Vacuum pumping is provided by a 25 cm cryopump which can evacuate the chamber to 1×10^{-6} Torr in 30 min. Care has been taken to minimize the use of organics and other materials that outgas. A Faraday cup assembly can be placed in the beam to measure the incident ion flux

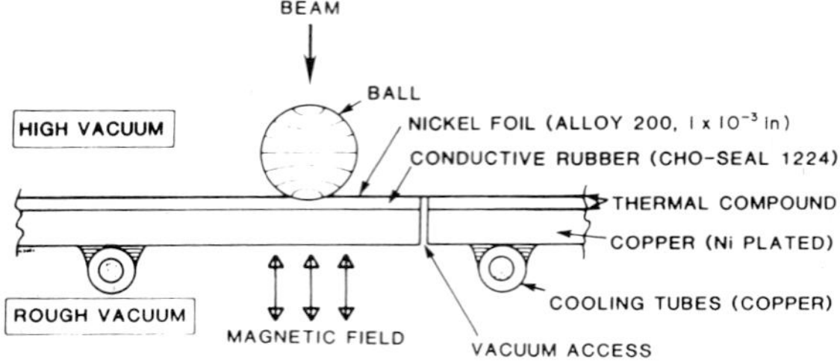

Fig. 7. Schematic diagram of the nickel foil cooling system for a ball bearing in the MANTECH facility. The ball is shown at 1000 G field pull-in depth (6×10^{-3} in or 10%). The cooling is limited by the rubber layer thickness and conductivity.

and a current integrator is used to determine implantation times.

MANTECH program requirements called for a demonstration of the capability to implant specimens of flat plate, roller bearing, ball bearing, inner race and outer race geometries with the ability to maintain steel specimens at a temperature of less than 250 °C and masking of the sample to avoid an angle of incidence for the beam greater than ±30° from normal. A number of fixtures were designed to accommodate the special requirements of each geometry. Table 2 lists the fixtures designed to date and their uses. Each of these fixtures can be inserted and removed from the work chamber in a short period of time. Solution of the design problems of heat removal and masking of the samples is similar on several of the fixtures although there may be differences in geometry. Figure 7 shows the approach used in the cooling of ball and roller elements. A water-cooled plate with a conductive rubber compound bonded to it is in turn covered by a thin nickel foil. A magnetic fixture applies

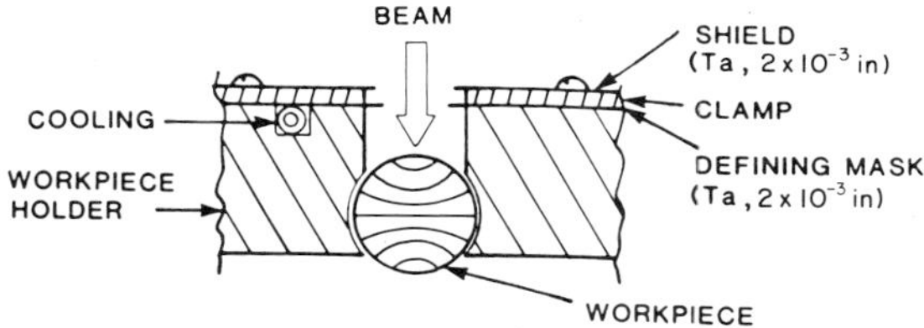

Fig. 8. Schematic drawing of the mask and cage assembly for moving the ball bearing in the MANTECH facility.

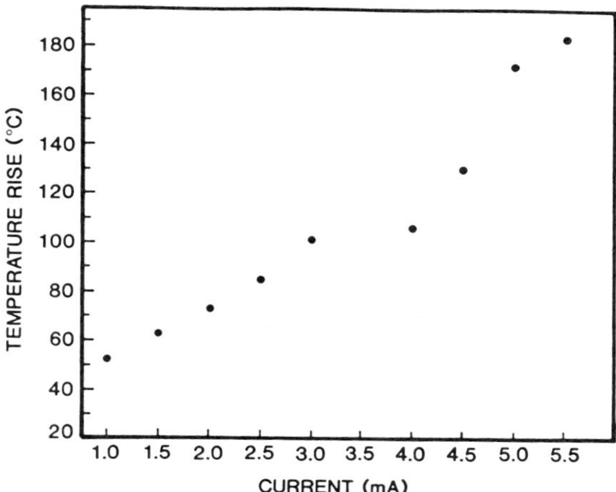

Fig. 9. Temperature rise for a static ball (thermocouple attached) as a function of beam current for the ball bearing cooled by the fixtures shown in Figs. 7 and 8.

a sufficient force on the ball to increase the contact area of the ball with the cooled surface. A cage assembly, shown in Fig. 8, is used for moving the ball and also holds a tantalum mask to limit the angle of incidence of the beam to ±30°. Motion of the ball is provided by two linear motion drives, one attached to the base and one to the cage, operating at right angles as illustrated in Fig. 6. The roller bearing fixture is similar in design but requires linear motion in only one direction. The roller and ball fixtures require a different cage for each size of rolling element to be implanted but use the same water-cooled base-magnet fixture. Figure 9 shows the equilibrium temperature rise for a static ball during implantation with a 120 kV beam of chromium ions at various beam currents.

Other fixtures listed in Table 2 with unique features include a water-cooled rotating shaft for implantation of bearing inner races and a rotating platter for outer races. The inner race achieves thermal contact to the race by an expandable D-block. The outer race fixture consists of a rotating tiltable water-cooled platter to which the outer races can be clamped. The contour of the raceway requires that each race be turned over and implanted from each side to achieve complete coverage of the raceway. Heating of components attached to these fixtures has been maintained within limits.

3.3. *Composition profiles and quality control*

The ability to predict and produce the composition profile for the implanted species is a necessary prerequisite for the routine use of ion implantation processing. While several computer codes such as E-DEP-1 and TRIM are available to calculate range and range straggling of ions in solids, other effects such as sputtering, target volume change and diffusion also influence the composition profile in high fluence implantations. A computer code called IMPLNT has been developed at NRL by Davisson [15] to include the effect of volume change and sputtering. IMPLNT uses E-DEP-1 to calculate the range and straggling for a single-element beam on a single-element target at normal incidence. The target is divided into layers and after each increment of fluence the composition of each layer is recalculated and appropriate changes are made in the range. Each increment of fluence is the fluence required to sputter off a layer. Provision is also made for the volume increase caused by the implanted beam atoms. The program is run iteratively until the goal fluence is reached. The number of sputtered atoms of each species is calculated by assuming the elemental sputtering yield weighted by the atom fraction of each species. Figure 10 shows profiles calculated for 100 keV chromium planted into iron at several fluences. The approach to steady state concentration at high fluences should be noted.

Experimental measurements of the implanted atom concentration profile can be made by several surface analysis methods. Rutherford backscattering analysis is widely used but can be difficult when the atomic numbers of the host and implanted elements are similar. Auger electron spectroscopy combined with low energy inert gas sputtering is also widely used for composition profile analysis but it is not a non-destructive technique, often requires calibration to yield absolute

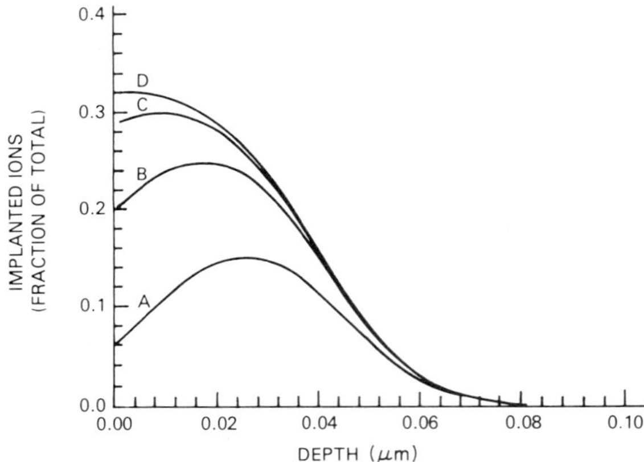

Fig. 10. Concentration profiles calculated for several fluences of 100 keV chromium ions implanted into iron using the computer code IMPLNT [14]: curve A, 0.6×10^{17} ions cm^{-2}; curve B, 1.2×10^{17} ions cm^{-2}; curve C, 1.8×10^{12} ions cm^{-2}; curve D, 2.4×10^{17} ions cm^{-2}.

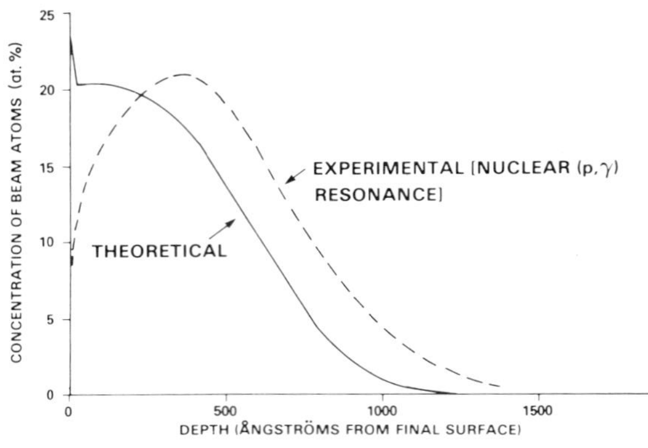

Fig. 11. Comparison of experimental measurements of the steady state concentration profile of chromium implanted into iron at 150 keV (fluence, 2×10^{17} ions cm^{-2}) and calculations performed by IMPLNT. The shift of the experimental curve to greater depths is believed to be due to cascade mixing using the (p, α) nuclear resonance technique [15].

concentrations and can be influenced by artifacts from radiation-enhanced diffusion or mixing. A nuclear reaction method based on the $^{52}Cr(p, \alpha)^{53}Mn$ resonant proton capture reaction has been developed by Gossett [16] for use as an absolute method for profiling chromium implanted into iron. The method uses the 1005 keV resonance to stimulate the reaction and measures the yield from α ray transitions to the ground state to measure concentration. Depth can be probed because of the energy loss of the proton in penetrating the solid. A depth resolution of about 10 nm can be achieved.

A comparison of an experimental profile determined by the (p, α) reaction and a prediction of the IMPLNT code is shown in Fig. 11 for 150 keV chromium implanted into an iron substrate for a fluence of 2×10^{17} ions cm^{-2}. It should be noted that the two curves are in agreement for the total number of ions implanted (area under the curves) and the maximum implant concentration, but the experimental curve is displaced by 15 nm to greater depths. A similar discrepancy has been noted for tantalum implants into iron [15]. The difference is attributed to cascade mixing and diffusional effects not incorporated in the IMPLNT code [12]. However, the calculations are considered to be sufficiently accurate to provide guidance for planning the implants.

Another requirement for an industrial process is a quality control procedure that is non-destructive, inexpensive and widely available. The accelerator-based analysis techniques and Auger analysis do not meet these criteria. An alternative method to measure total implanted concentration, although not the depth profile, is energy-dispersive X-ray analysis. This technique is widely available as an accessory on scanning electron microscopes, can be performed in a non-destructive manner if the component can be placed in the scanning electron microscope sample chamber and is relatively rapid and inexpensive. The implanted layer occupies only about one-tenth of the volume of material from which X-rays are excited by the electron beam probe at 10 kV; so a calibration sample or standard is required for a given implant configuration. Sprague [17] has demonstrated a reproducibility of approximately ±5% in the number of chromium atoms implanted into iron for fluences above 1×10^{16} atoms cm^{-2}.

4. COST ANALYSIS

One of the objectives of the MANTECH project was to establish a basis for cost analysis of ion implantation as a materials-processing method using the operating experience of the facility. The cost analysis consisted of three steps: (1) determining an hourly operating cost for the facility which was sufficient to defray costs, (2) determining actual times (and costs) to implant selected parts and (3) analyzing cost savings on these parts based on life cycle costs. Costs will vary somewhat depending on the number of shifts that the facility is operated and other assumptions. These are stated where relevant to allow recalculation and revision of the estimates as progress along the learning curve is made. Demonstration implantations were planned for approximately 700 aircraft bearings, 1500 printed-circuit board drills and beryllium alloy instrument bearings.

The operational costs for an implantation facility include capital investment, labor, materials, facilities and maintenance costs. These costs are summarized in Table 3 for a scenario of a fully operational facility. A capital investment of U.S. $\$1 \times 10^6$ is amortized over a 10 year lifetime to give an annual operating cost. Labor costs are assumed to include a full-time operator and half-time for a supervisory engineer. Additional labor costs for shift operation assume a 10% premium for the operator but no additional supervisory time. Materials and facilities costs are scaled to shift operations as shown in Table 3. Facility availability indicates the time that the facility is available for use exclusive of the time that the facility is not available because of unscheduled downtime and routine maintenance. Actual experience in the third quarter of operation indicated that availability of 90% was achieved on a three-shift basis and so this figure is considered attainable. Hourly operating costs range from U.S. $74 h^{-1} for three-shift operation to U.S. $144 h^{-1} for operation on only one shift.

The time (and cost) to process parts depends critically on the efficient utilization of the available beam current. The major factors which influence beam utilization include heat removal, efficiency of packing of components, and masking of the samples to reduce sputtering at non-normal angles of incidence. If heat removal is inadequate, then the beam current must be reduced or spread over a larger area. Experience has shown that heat removal is adequate to use a 5 mA beam if distributed over a 500 cm^2 area. The packing efficiency is the ratio of the projected area of the part to the planar area covered by the ion beam. Thus, squares can be packed at 100% efficiency while circles cover only 78% of the area in a linear array and 90.5% of the area in a hexagonal array. The mask for reduction of sputtering from curved surfaces further reduces the efficiency for implanting cylinders and balls since less area is exposed. The initial implantations were performed with an included angle of incidence of $\pm 30°$ which reduces the ideal total beam utilization efficiency to 33% for cylinders and 12% for spheres. Further calculations considering rotation of

TABLE 3

Implant facility operational costs (period: projected industrial operation)

	One shift	*Two shifts*	*Three shifts*
Assumptions			
Capital investment (U.S. $)	1000000	1000000	1000000
Years amortized	10	10	10
Work year (h)	2000	4000	6000
Availability (%)	85	85	90
Revenue (h)	1700	3400	5400
Operators	60000		
Engineers	84000		
Costs			
Capital equipment (U.S. $)	100000	100000	100000
Labor + overheads + G&A (U.S. $)	102000	168000	234000
Materials + G&A (U.S. $)	12000	18000	21000
Facilities, maintenance (U.S. $)	31000	37000	43000
Total annual cost (U.S. $)	245000	323000	398000
Hourly rate (U.S. $ h^{-1})	144.12	95.00	73.70

the component indicate that the mask angle requirement can be relaxed to 45°. This improves the efficiency by almost a factor of 2 to 50% for cylinders and 26% for spheres.

The processing time for a batch of parts is given by the following formula:

$$\text{batch hours} = \frac{\text{goal fluence (ions cm}^{-2}) \times \text{area (cm}^2) \times \text{number of implants} \times \text{factor} + \text{reload hours}}{6.2 \times 10^{18} \text{ ions A}^{-1}\text{ s}^{-1} \times 0.005 \text{ A} \times 3600 \text{ s h}^{-1}} \quad (7)$$

where the area includes the total surface area to be implanted as on cylinders and spheres, the number of implants considers multiple-energy implants or double implants for bearing races, and the factor includes additional time for rotating masked components. As a typical example, implantation of chromium ions into a flat component over a circle 35.5 cm in diameter would require 1.75 h, corresponding to a processing rate of 560 cm^2 h^{-1}. Run times will obviously vary depending on the particular component to be processed. Batch times typically are 2.5–4 h with a $\frac{1}{2}$ h reload time. The total number of parts implanted per batch ranges from 300 for roller bearings 7 mm in diameter to 14 for outer races with a 68 mm outside diameter. Calculated time to implant all parts of a single bearing was 1.10 h for a cost of U.S. $82.50 per bearing.

A life cycle cost analysis on this M50 gas turbine aircraft bearing indicates significant benefits from implantation. Analysis of replacement rates on these bearings shows that up to 70% of the bearings in forward locations in the engines are rejected at each inspection because of pitting corrosion and overall a rejection rate of 35% was found for all bearings inspected. Laboratory studies show that implantation with chromium to raise the surface chromium concentration of M50 steel from 4 to 17 at.% will increase the lifetime of the bearings by a factor of 2.5. Implantation will also reduce the incidence of maintenance overhauls resulting from bearing failure from fatigue cracks initiated at corrosion pits, estimated to be 1% of the total number of overhauls. Since an overhaul normally costs U.S. $100 000, a maintenance cost per bearing due to corrosion failures can be assigned a value of U.S. $380 per bearing. If each bearing costs U.S. $550, then the cost savings due to reduced replacement costs and reduced maintenance costs results in life cycle cost savings of U.S. $1693 per bearing or a 20 times return on the implantation cost.

5. CONCLUSIONS

This manufacturing technology project has provided the first comprehensive engineering analysis of the requirements for an industrial-scale ion implantation facility. Heat removal, high sputtering yields at non-normal angles of incidence and surface reactions with residual gases in the target chamber were shown to be significant problems to be resolved. A facility design employing a wide-area beam in a batch processing mode was shown to be the most feasible approach for an implanter with a 5–10 mA beam current output.

A prototype facility for both metal and nitrogen ion beams has been built by Spire Corporation, Boston, MA, and operated for a period of 9 months. Fixturing for implantation of all components of roller and ball bearings, as well as simpler component geometries, has been designed, built and tested. Several hundred bearings and representative quantities of tools and smaller instrument bearings have been implanted. Facility availability factors of 85%–90% have been achieved during the past quarter for operation at three shifts per day. Operating costs as low as U.S. $74 h^{-1} are projected for three-shift operations. Cost-benefit analysis on life cycle costs for a gas bearing subject to pitting corrosion indicates a 20 times return on the implantation cost. The project has successfully demonstrated the cost-effective use of ion implantation for surface treatment of large lots of parts under industrial production conditions.

REFERENCES

1 F. A. Smidt and B. D. Sartwell, *Nucl. Instrum. Methods B, 6* (1985) 70.
2 K. S. Grabowski and R. A. Kant, in H. Ryssel and H. Glawischnig (eds.), *Proc. 4th Int. Conf. on Ion*

Implantation: Equipment and Techniques, Berchtesgaden, September 13–17, 1982, Springer, New York, 1983, p. 364.

3 M. M. Yovanovich, *Prog. Astronaut. Aeronaut., 83* (1982) 83.

4 I. Manning and G. P. Mueller, *Comput. Phys. Commun., 7* (1974) 85.

5 K. S. Grabowski, N. E. W. Hartley, C. R. Gossett and I. Manning, in G. S. Hubler, O. W. Holland, C. R. Clayton and C. W. White (eds.), *Ion Implantation and Ion Beam Processing of Materials, Materials Research Society Symp. Proc., Boston, MA, November 14–17, 1983*, Vol. 27, Elsevier, New York, 1984, p. 615.

6 P. Sigmund, *Phys. Rev., 184* (1969) 383.

7 I. Manning, Naval Research Laboratory, unpublished research, 1986.

8 I. L. Singer, *J. Vac. Sci. Technol. A, 1* (1983) 419.

9 D. A. Baldwin, B. D. Sartwell and I. L. Singer, *Nucl. Instrum. Methods B, 7–8* (1985) 49.

10 B. D. Sartwell and D. A. Baldwin, in G. K. Wolf, W. A. Grant and R. P. M. Procter (eds.), *Proc. Int. Conf. on Surface Modification of Metals by Ion Beams, Heidelberg, September 17–21, 1984*, in *Mater. Sci. and Eng., 69* (1985) 539.

11 D. A. Baldwin, B. D. Sartwell and I. L. Singer, *Appl. Surf. Sci., 25* (1986) 364.

12 D. Farkas, M. Rangaswamy and I. L. Singer, in G. K. Hubler, O. W. Holland, C. R. Clayton and C. W. White (eds.), *Ion Implantation and Ion Beam Processing of Materials, Materials Research Society Symp. Proc., Boston, MA, November 14–17, 1983*, Vol. 27, Elsevier, New York, 1984, p. 609.

13 A. J. Armini and S. N. Bunker, *Nucl. Instrum. Methods B, 6* (1985) 214.

14 S. N. Bunker, Spire Corporation, Bedford, MA, unpublished research, 1986.

15 C. M. Davisson, *Nucl. Instrum. Methods B, 13* (1986) 421.

16 C. R. Gossett, *Nucl. Instrum. Methods, 168* (1980) 217.

17 J. A. Sprague, Naval Research Laboratory, unpublished research, 1986.

Materials Science and Engineering, *90* (1987) 399–405

Improving the Wear Resistance of Metal Forming Tools by Ion Implantation*

CHR. WEIST[a], G. K. WOLF[b] and P. BALLHAUSE[b]

[a]*Institut für Umformtechnik, Universität Stuttgart, D-7000 Stuttgart (F.R.G.)*
[b]*Physikalisch-Chemisches Institut, Universität Heidelberg, D-6900 Heidelberg (F.R.G.)*

(Received July 10, 1986)

ABSTRACT

A facility for wear tests under simulated production conditions was installed. These tests permit variations in essential parameters of the forming processes "backward can extrusion" and "upsetting between flat parallel dies", which were investigated in this work. For measuring sufficient wear, 10 000 to 20 000 workpieces had to be pressed. To reduce the time, in some cases a short-time testing procedure based on a radionuclide technique was used. In this, thin surface layer of the tool was activated by deuteron irradiation. During the wear process the activity of the tool decreased because of loss of material. This enabled the determination of wear after only 2000–3000 manufactured workpieces. By applying nitrogen-ion-implanted tools, wear decreased down to the level of conventionally nitrided tools. The great advantage of the former process is the low temperature of treatment; so the roughness and structure of the material are not affected and tools can be machined to their final dimensions before ion implantation. The results of nitrogen implantation are compared with those of nitrogen plus silver ion implantation and nitrogen plus tin ion implantation. A comparison of the two forming processes, different tool materials and various surface treatments should contribute to economic application of optimized wear-resistant tools in bulk metal forming processes.

1. INTRODUCTION

Bulk metal forming processes are capital-intensive manufacturing methods. Their efficiency depends on tool costs and indirect costs due to machine down-time caused by tool failure either by fracture or by wear. In cold-forging processes, tool fracture can nearly be excluded with the help of modern computational methods so that wear is the most important failure criterion. It affects the dimensional accuracy, shape and quality of the finished workpiece.

As well as constructional and tribological means, surface coatings or, in general, surface treatments offer the possibility of minimizing tool wear. This means that the tool material mainly bears the inner mechanical load, while the coating prevents the aggressive wear. In addition to well known coating processes, ion implantation is an alternative method of improving the mechanical surface properties.

2. ION IMPLANTATION OF METAL FORMING TOOLS

Figure 1 gives a survey of surface treatment processes, divided into reaction layers, in which the layer element diffuses into the base metal, and coating layers, which stick to the base material mainly by mechanical interlocking. The right-hand column shows the process modifications, distinguished either by the treatment medium or by the layer element. Ion implantation is in a special position because it is not a real surface coating process.

The implantation fluences, which are needed to influence the mechanical properties of metallic surfaces, are in the range 10^{16}–10^{18} ions cm^{-2}. This corresponds, for example, for 200 keV nitrogen ions (N^+) to a maximum concentration of about 11–12 at.%. Depending on the energy of the ions an elevated concentration up to a depth of about 0.4 μm is measured. The wear reduction by ion implantation is still detectable, however, if the wear reaches

*Paper presented at the International Conference on Surface Modification of Metals by Ion Beams, Kingston, Canada, July 7–11, 1986.

0025-5416/87/$3.50

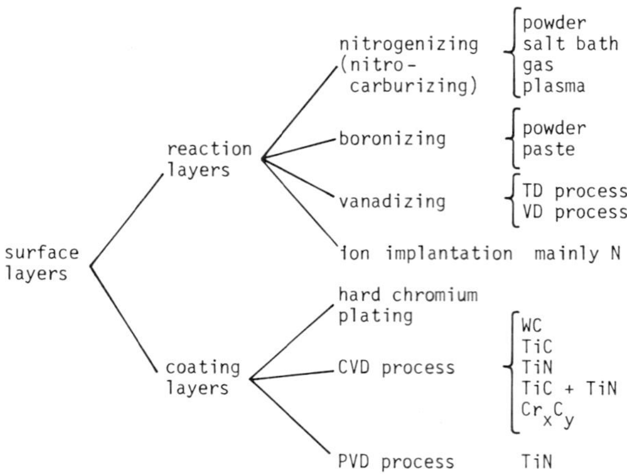

Fig. 1. Wear protection layers and coating processes in metal forming: TD, Toyota diffusion; VD, vanadium diffusion; CVD, chemical vapour deposition; PVD, physical vapour deposition.

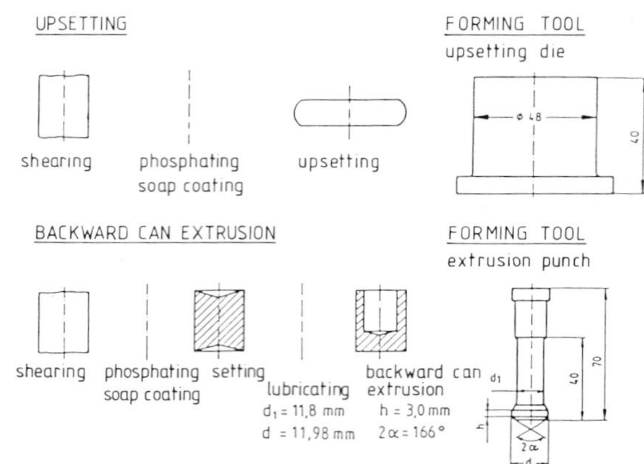

Fig. 2. Sequence of operation and tools for upsetting and backward can extrusion.

a multiple value of the initial implantation depth.

The main advantages of ion implantation in comparison with the common coating procedures are the low temperature of treatment (below about 300 °C), the fact that no (macroscopic detectable) variation in geometry and surface roughness occurs and the fact that the implantation process can be repeated at any time.

The emphasis of earlier investigations lay in surface hardening and wear reduction of bearing components and tools. In metal forming technology, ion-implanted tools were predominantly used in the U.S.A. and Great Britain. Nearly exclusively, nitrogen was implanted. Drozda [1] reported an increase in the tools life of 100%–500%. Similar results have been reported in refs. 2 and 3 for wire-drawing dies and cold-forging tools.

3. WEAR EXPERIMENTS

Wear is not a characteristic material value, but it is a system property. Therefore the transfer of results obtained in model experiments to practice is very difficult. Hence, there is a demand for utmost field-experienced simulation tests for metal forming technology.

3.1. *Metal forming processes and phases of operation*

The wear tests were carried out with the metal forming processes "backward can extrusion" and "upsetting between flat plane-parallel dies" (Fig. 2). In the upsetting process the pretreated workpiece is compressed between the lower and the upper die. In the backward can extrusion process the workpiece is put into a container or die. The forming process consists of the extrusion of the solid billet through the annulus formed by the container wall and the indenting extrusion punch to form a can. Upsetting provides an opportunity of isolating the influence of tool geometry from that of implantation (or coating in general), whereas in backward can extrusion both influences combine.

In both cases, billets of the steel 20 MnCr 5 with a diameter of 14 mm and a mass of 14.3 ± 0.1 g were used. In upsetting, the sheared parts were phosphated, lubricated and pressed with a geometrical strain (equivalent strain) φ of 1.1. For backward can extrusion with a relative reduction ϵ_A in area of 0.71 the pretreated billets were set so as to obtain good centring of the punch for the following extrusion process. After setting, the workpieces were pickled, phosphated and lubricated again since the effectiveness of soap as a lubricant is limited.

3.2. *Test conditions*

An experimental facility for wear tests under simulated production conditions was installed [4, 5]. Experiments were carried out using a C-frame knuckle-joint press with a 630 kN press rating and a 60 mm total height of stroke. The insertion of the billets and blowing-out of the pressed parts occurs automatically.

The extrusion punches were designed with a punch land diameter d of 12 mm and a total length l of 70 mm according to ref. 6. As tool material the cold-working steel X 155 CrVMo 12 1 (German material number 1,2379, similar to AISI D2) and the high speed working steel S 6-5-2 (1.3343) (similar to AISI M2) were used. Mostly the cold-working steel was hardened to a Rockwell C hardness of 62 ± 1 HRC and the high speed steel to 64 ± 1 HRC. The experiments were carried out with the case hardening steel 20 MnCr 5 (1.7147) (similar to AISI 5120) as the workpiece material. The specimens were first covered with a 12–15 μm zinc phosphate layer and subsequently treated with the soap lubricant Bondertube 236 (pure alkali soap).

3.3. Measuring methods

3.3.1. Backward can extrusion

3.3.1.1. Conventional method. In this case, some cans were taken at regular intervals from production and measured using an inside micrometer. The wear at the punch land leads (after a running-in period) to a proportional decrease in the inner can diameter. If no increase or decrease in the wear rate is detectable, the wear curve can be approximated by a straight regression line.

3.3.1.2. Thin layer difference method. Determining wear characteristics by conventional measurements involves the production of large batch sizes which is combined with large experimental effort involving time and costs. Therefore a suitable short-time testing procedure, the thin layer difference method was adapted for metal-forming processes by Nehl [7]. A thin surface layer of the punches was activated at the punch land by a (d, 2n) reaction in a cyclotron (Fig. 3). For the purpose of wear measurement, the reduction in activity of the tool is determined continuously, thus allowing an exact statement on the loss of material. The activated punches show in the activated area a high specific activity; nevertheless the total activity of the tool is relatively low (less than 3.7×10^5 Bq). The measurement of ionizing radiation is performed with an NaI-scintillation counter, which is positioned at a short distance from the tool and is protected from heat radiation. Count rates and corresponding times are measured and, subsequently, wear rates can be determined after a correction relating to the half-life of the radioactive isotopes. The thin layer difference method enables the wear behaviour after 2000–3000 pressed parts to be forecast after the running-in process.

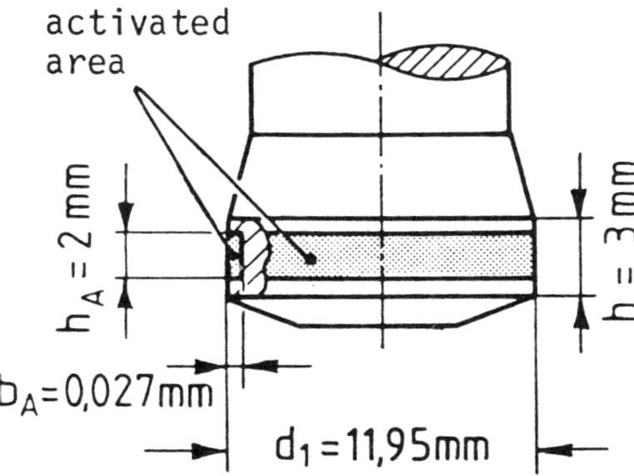

Fig. 3. Activated area at the punch land.

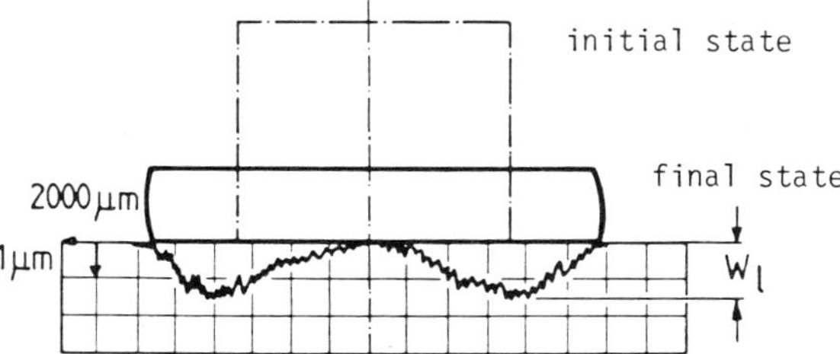

Fig. 4. Wear profile of upsetting dies.

3.3.2. Upsetting between flat plane-parallel dies

During the upsetting process the increase in specimen cross-section causes friction at the specimen die contact area which results in a characteristic profile of die wear (Fig. 4). These profiles were measured at certain intervals during the process with a roughness tester using a stylus with a reference plane, and thus the wear rate W_l was determined.

4. RESULTS

Tools wear firstly by adhesive cold welding of tool and workpiece material and secondly by pulling off these welds, which subsequently exist as hard particles in the active joint between the tool and workpiece and cause strong scoring as a consequence of abrasion.

In previous experiments [4] the high speed working steel S 6-5-2 as the tool material showed a better wear resistance than cold-working steel X 155 CrVMo 12 1, particularly at higher tool loadings.

4.1. Backward can extrusion

The standard experiments were carried out with the tool material S 6-5-2 and soap lubricant Bonderlube 236; the number n_c of strokes was 40 min^{-1} and the h_i/d_i 1.0. The depth of the can plays a decisive part in the experiments because, on the one hand, it determines the wear path and on the other hand, the soap lubricant becomes unstable above an h_i/d_i ratio of 1.0 and the lubricant film breaks so that an intensified solid contact between tool and workpiece takes place.

Figure 5 shows the wear course of nitrogen-ion-implanted extrusion punches (10^{18} N$^+$ cm^{-2}; 150 keV) compared with unimplanted punches. Although the wear depth of the implanted punches of 11 μm (10 000 cans) is substantially higher than the original implantation depth of about 0.2 μm, there was no progressive increase detectable in the wear process. This result proves that the effect of wear reduction goes far beyond the original implantation depth, which has been reported by several research groups (see for example refs. 8 and 9).

The surface of a punch is not affected by ion implantation. After 10 000 cans had been produced, the surface shows a roughness comparable with those of unimplanted tools. Figure 6 shows a comparison of ion-implanted punches with nitrocarburized punches [10]. For all nitrided and nitrocarburized ones the brittle compound layer was polished away because it chipped off with the first strokes, as experiments showed. Nevertheless, circumferential cracks arose perpendicular to the longitudinal axis in the stem area of the nitrided punches between 1200 and 12 500 manufactured parts. In one punch, which was used up to 12 500 pressed parts, several cracks appeared. Microscopic examinations of the cracks showed that cracks penetrated the diffusion layer and stopped a short distance after reaching the tougher base material. The reason for the formation of these cracks is probably the different behaviours of the diffusion layer and the parent material in the pulsating load which mainly acts in a compressive mode. For higher numbers of pieces the cracks will surely lead to fracture of the punches by the notch effect.

The wear rate of ion-implanted punches is only slightly above the upper limit of the wear rate of the nitrocarburized punches. In contrast with the nitrocarburized punches, none of the nitrogen-implanted punches showed cracks, although in both cases nitrogen exists as the alloying element in the surface. The reason for this difference is the relatively thin implanted region, which compared with the total cross-section bears nearly no inner load. In two cases, high speed working steel punches were used; the Rockwell C hardness of these, 60 ± 1 HRC, was distinctly below the usual value of 64 ± 1 HRC. In both cases the punches were implanted with nitrogen ions and subsequently with metal ions. Figure 7 shows the result for a punch implanted with nitrogen

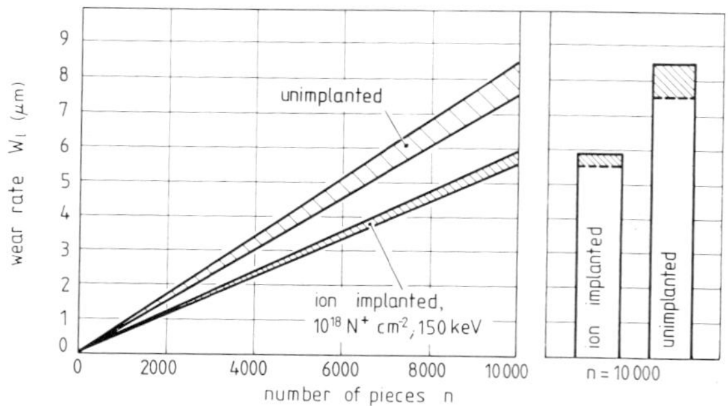

Fig. 5. Wear of nitrogen-ion-implanted and unimplanted extrusion punches (punch material, S 6-5-2 (similar to AISI M2); workpiece material 20 MnCr 5 (similar to AISI 5120); lubricant, Bondertube 236; $h_i/d_i = 1.0; n_c = 40$ min^{-1}).

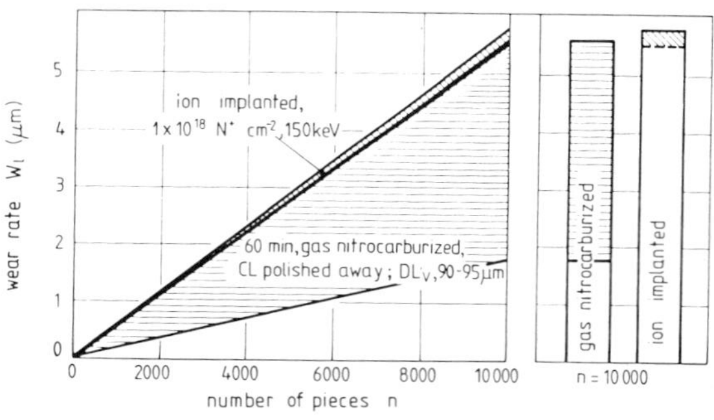

Fig. 6. Wear of nitrogen-ion-implanted and gas nitrocarburized extrusion punches (punch, material, S 6-5-2 (similar to AISI M2); workpiece material, 20 MnCr 5 (similar to AISI 5120); lubricant, Bondertube 236; $h_i/d_i = 1.0$; $n_c = 40$ min^{-1}): CL, compound layer; DL_V, visible diffusion layer.

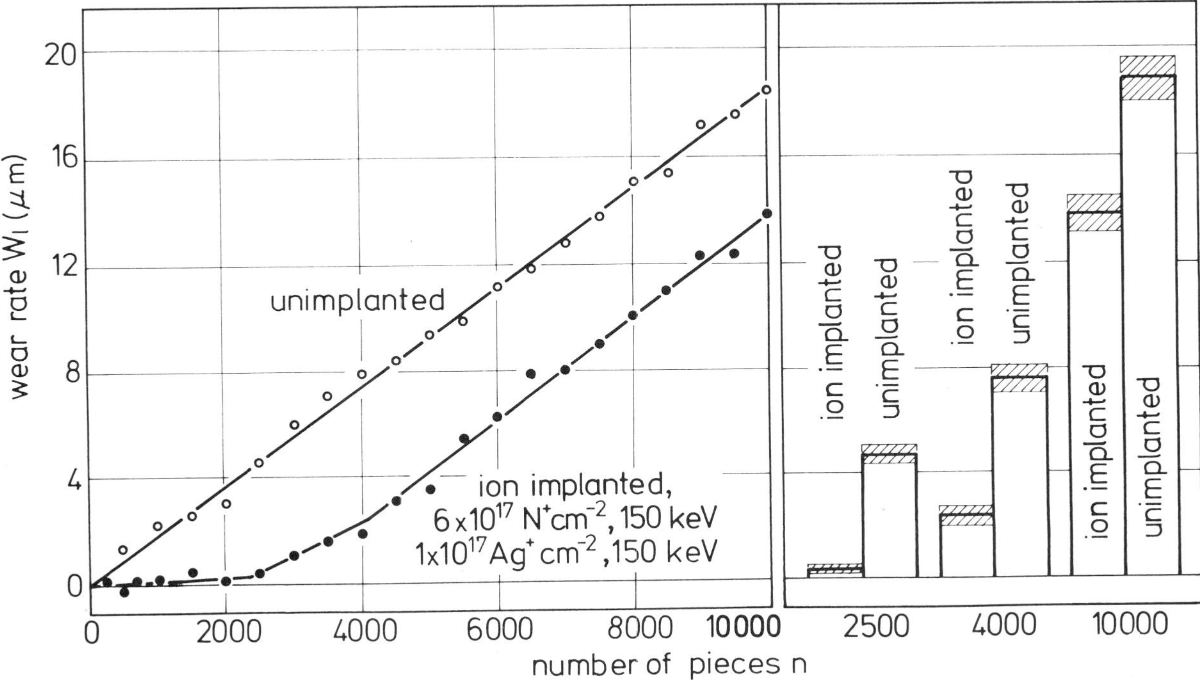

Fig. 7. Wear of nitrogen-and-silver-ion-implanted and unimplanted extrusion punches (punch material, S-6-5-2 (similar to AISI M2); workpiece material, 20 MnCr 5 (similar to AISI 5120); lubricant, Bondertube 236; $h_i/d_i = 1.0$; $n_c = 40$ min^{-1}).

and silver ions. Because of the lower hardness of the tool the wear of the unimplanted tool is higher than for the tools with a higher hardness. Obviously the wear-reducing effect of ion implantation is greater than the effect of a higher hardness of the material. This agrees with previous results [8].

Up to 2500 manufactured pieces the wear-reducing effect of ion implantation was more than 90%. Between 2500 and 4000 cans the slope of the wear curve is about the same as the slope of the curve for nitrogen-implanted punches. The initial strong wear-reducing effect can obviously be attributed to the implantation of metal ions. After 4000 pieces the wear curve corresponds nearly to that of unimplanted punches, *i.e.* the effect of ion implantation has become negligible. Although the wear-reducing effect is exhausted already after about 4000 manufactured pieces, the percentage of wear reduction is still comparable with the values for the nitrogen-implanted punches with a higher hardness.

Figure 8 shows a similar result for bombardment with nitrogen and tin ions. Up to 800 pressed pieces the wear rate remains very low. Between 800 and 3000 pieces, it increases in a similar way to the wear rate of nitrogen-implanted tools. Above 3000 pieces the wear curve looks like that of untreated tools.

The results prove that, for low and medium tool loads, ion implantation would lead to excellent results whereas, for higher loads, it reaches the wear-reducing effect of conventionally nitrided tools.

4.1.1. Comparison of ion implantation with other surface coatings in backward can extrusion

Figure 9 shows the wear reduction obtained by means of several surface coatings in backward can extrusion [10]. The best results were achieved with physically vapour-deposited (TiN) coated and with vanadized (Toyota diffusion process) punches.

With the exception of plasma nitriding, a substantial wear reduction can be achieved by several nitriding processes. Ion-implanted punches show a wear rate which is at the upper limit of the wear rate of nitrocarburized punches. In metal-forming processes with a lower tool load, ion implantation will show essentially better results.

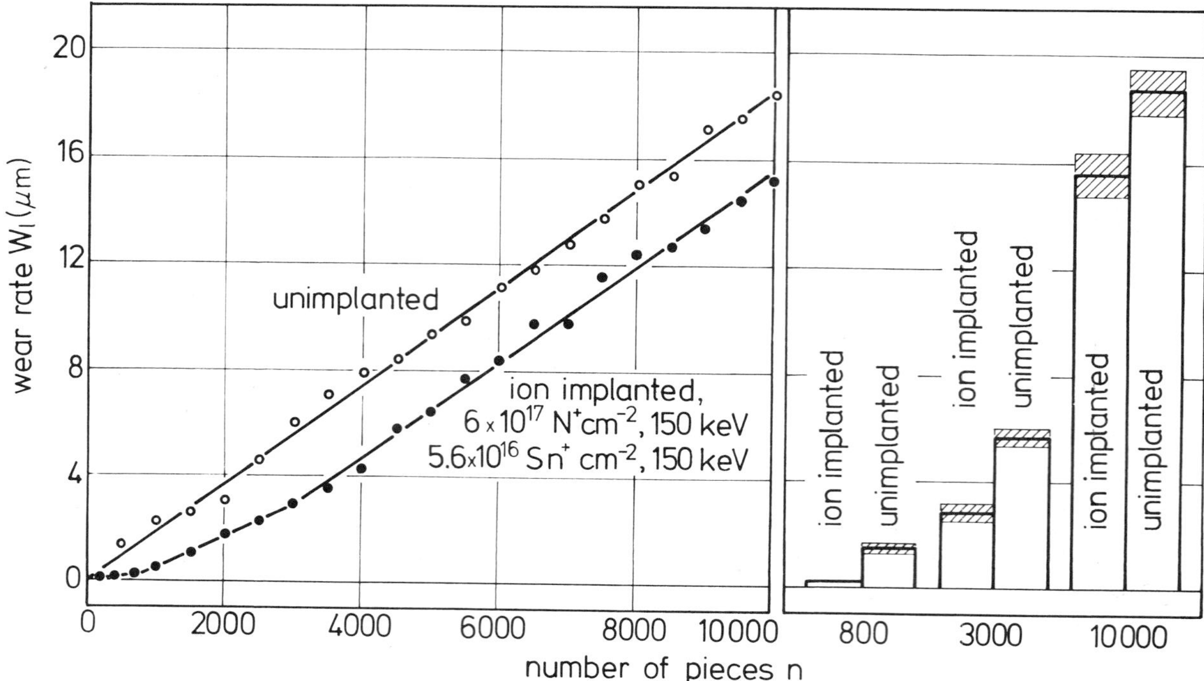

Fig. 8. Wear of nitrogen-and-tin-ion-implanted and unimplanted extrusion punches (punch material, S-6-5-2 (similar to AISI M2); workpiece material, 20 MnCr 5 (similar to AISI 5120); lubricant, Bondertube 236; $h_i/d_i = 1.0$; $n_c = 40\ \text{min}^{-1}$).

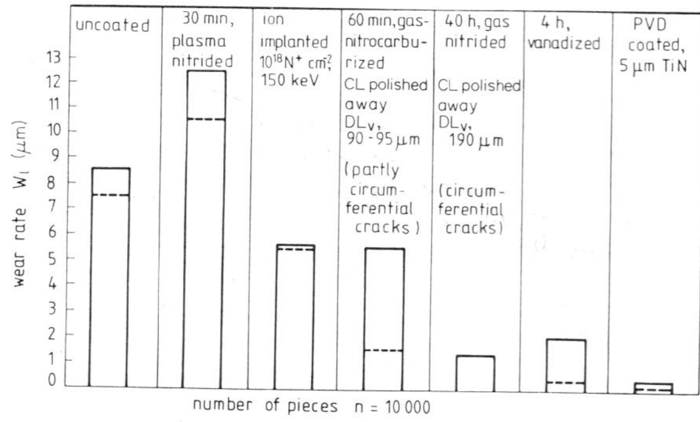

Fig. 9. Comparison of various coatings in backward can extrusion (punch material, S-6-5-2 (similar to AISI M2); workpiece material, 20 MnCr 5 (similar to AISI 5120); lubricant, Bondertube 236; $h_i/d_i = 1.0$; $n_c = 40\ \text{min}^{-1}$): CL, compound layer; DL_v, visible diffusion layer; PVD, physical vapour deposition.

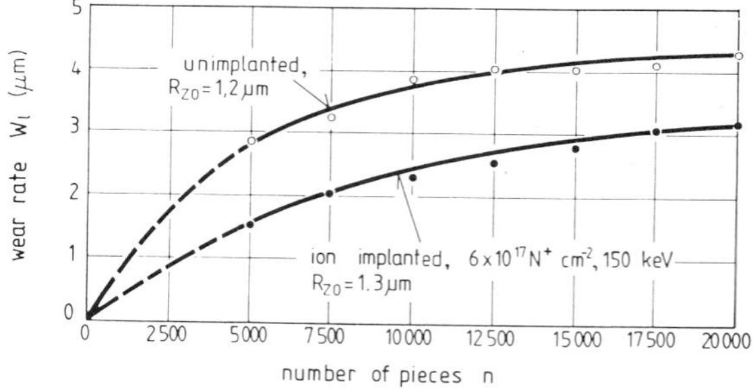

Fig. 10. Wear of nitrogen-ion-implanted and unimplanted upsetting dies (die material, X155 CrVMo 12 1 (similar to AISI D2); workpiece material 20 MnCr 5 (similar to AISI 5120); lubricant, Bondertube 236; $\varphi = 1.1$; $n_c = 45\ \text{min}^{-1}$).

4.2. *Upsetting between flat parallel dies*

In the upsetting process, implanted and unimplanted dies were tested simultaneously. The dies were ground before use. After 20 000 pressed parts, unimplanted dies of cold-working steel have a wear rate of between 2.0 and 4.3 μm and those of high speed working steel a wear rate of between 0.5 and 2.8 μm. The reason for this difference is the surface roughness, the hardness and the microstructure in decreasing importance.

Figure 10 shows the wear curve of nitrogen-ion-implanted and unimplanted cold working steel. The wear reduction due to ion implantation is to 20%–25%. The wear curve does not show the same linear slope as in backward can extrusion. Obviously, the higher surface roughness has the effect that the roughness

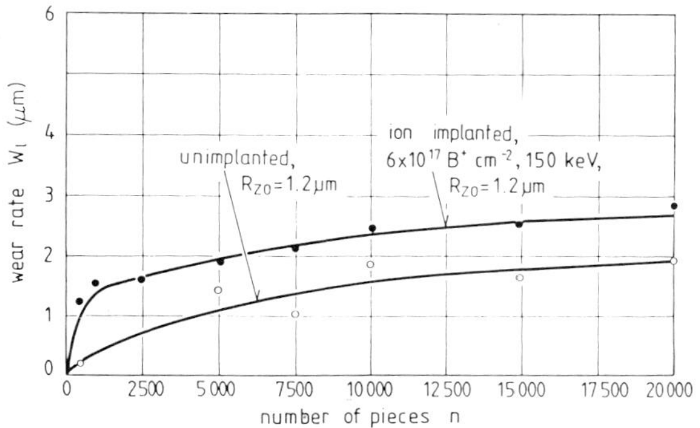

Fig. 11. Wear of boron-ion implanted and unimplanted upsetting dies (die material, X 155 Cr V Mo 12 1 (similar to AISI D2); workpiece material, 20 MnCr 5 (similar to AISI 5120); lubricant, Bondertube 236; $\varphi = 1.1$; $n_c = 45\ \text{min}^{-1}$).

peaks are worn out first. With increasing number of pieces the supporting surface area is increased and hence the slope of the wear curve decreases.

Figure 11 shows the wear curve of a boron-implanted upsetting die. The boron implantation does not cause any wear reduction. After the running-in period the wear rate shows the same behaviour as that of the unimplanted tool. Further experiments gave similar results. Because the wear-reducing effect also depends on the ion fluence [11], a change in ion fluence would be interesting in future experiments.

Surface roughness measurements of the upsetting dies before and after ion implantation proved that the surface does not roughen as a consequence of ion implantation.

5. CONCLUSION

The wear experiments carried out show that wear reduction in bulk metal forming can be achieved by using ion-implanted tools. Wear can be reduced by the same amount as by the use of nitrided tools. However, ion-implanted tools are much more resistant to crack formation. In the region of low and medium loaded tools, excellent results seem to be obtainable.

The great advantage of the implantation process is the low temperature of treatment (below 300 °C) and the fact that the roughness and structure of material are not affected. This offers the possiblity of treating steels which have a low tempering temperature. In addition, tools can be machined to their final dimensions before ion implantation.

ACKNOWLEDGMENT

We thank the Bundesministerium für Forschung und Technologie, Bonn, for supporting this work under Contract 03 T 0002 C 0 "Oberflächenvergütung durch Ionenimplantation".

REFERENCES

1 T. J. Drozda, Ion implantation, *Manuf. Eng., 64* (1985) 51-56.
2 R. E. Fromson and R. Kossowsky, Preliminary results with ion implantated tools, *Proc. 9th North Am. Manufacturing Research Conf., Philadelphia, PA, May 19-22, 1981*, Society of Manufacturing Engineers, Dearborn, MI, 1981, pp. 403-409.
3 G. Dearnaley, The effect of ion implantation upon the mechanical properties of metals and cemented carbides, *Radiat. Eff., 63* (1982) 1-15.
4 M. Weiergräber, Werkzeugverschleiβ in der Massivumformung, *Ber. Inst. Umformtech., Univ. Stuttgart*, Springer, Berlin, *73* (1983).
5 K. Lange and M. Weiergräber, Investigation of tool wear in bulk metal forming by model experiments, *CIRP Ann., 33* (1) (1984) 179-184.
6 International Cold Forging Group (ICFG), ICFG Data Sheet No. 5171, 1971.
7 E. Nehl, Wear test for bulk metal forming using radionuclide technique, *Proc. 7th Int. Congr. on Cold Forging, Birmingham, April 24-26, 1985*, ISME, Redhill, 1985, pp. 144-150.
8 H. Dimigen, K. Kobs, R. Leutenecker, H. Ryssel and P. Eichinger, Wear resistance of nitrogen-implanted steels, in G. K. Wolf, W. A. Grant and R. P. M. Procter (eds.), *Proc. Int. Conf. on Surface Modification of Metals by Ion Beams, Heidelberg, September 17-21, 1984*, in *Mater. Sci. Eng., 69* (1985) 181-190.
9 G. H. Feller and R. Klinger, Tribologisches Verhalten von Stählen nach Stickstoffionenimplantation, *Z. Metallkd., 76* (1985) 214-218.
10 Chr. Weist and H. Westheide, Application of chemical and physical methods for the reduction of tool wear in bulk metal forming processes, *CIRP Ann., 35* (1) (1986) 199-204.
11 K. Hohmuth, E. Richter, B. Rauschenbach and C. Blochwitz, Fatigue and wear of metalloid-ion-implanted metals, in G. K. Wolf, W. A. Grant and R. P. M. Procter (eds.), *Proc. Int. Conf. on Surface Modification of Metals by Ion Beams, Heidelberg, September 7-21, 1984*, in *Mater. Sci. Eng., 69* (1985) 191-201.

Materials Science and Engineering, 90 (1987) 407–416

Nitrogen Implantation of Type 303 Stainless Steel Gears for Improved Wear and Fatigue Resistance*

F. M. KUSTAS, M. S. MISRA and W. T. TACK

Martin Marietta Denver Aerospace, P.O. Box 179, Denver, CO 80201 (U.S.A.)

(Received July 10, 1986)

ABSTRACT

Fine-positioning mechanisms are responsible for accurate and reproducible control of aerospace system devices, i.e. filter grading wheels. Low wear and fatigue resistance of mechanism components, such as pinions and gears, can reduce system performance and reliability. Surface modification using ion implantation with nitrogen was used on type 303 stainless steel pinions and gears to increase tribological performance.

Wear–life tests of untreated, nitrogen-implanted and nitrogen-implanted-and-annealed gears were performed in a fine-positioning mechanism under controlled environmental conditions. Wear and fatigue resistance were monitored at selected time intervals which were a percentage of the predicted failure life as determined by a numerical stress analysis. Surface analyses including scanning electron microscopy and Auger electron spectroscopy were performed to establish the wear and fatigue mechanisms and the nitrogen concentration–depth distributions respectively.

Nitrogen implantation resulted in a significant improvement in both surface wear and fatigue spalling resistance over those of untreated gears. A 40% reduction in surface wear and a 44% reduction in dedendum spalling was observed. In contrast, the nitrogen-implanted-and-annealed gears showed a 46% increase in sliding wear area and an 11% increase in spall density compared with those of untreated gears, indicating that the post-implantation anneal was detrimental to wear and fatigue resistance.

*Paper presented at the International Conference on Surface Modification of Metals by Ion Beams, Kingston, Canada, July 7–11, 1986.

1. INTRODUCTION

Fine-positioning mechanisms are responsible for accurate and reproducible control of aerospace system devices, *i.e.* filter grading wheels. Low wear and fatigue resistance of mechanism components, such as type 303 stainless steel pinions and gears, reduce system performance and reliability. For example, misalignment by as little as ±20 μm over a projected life of 0.6×10^6 operating cycles is unacceptable for the effective positioning of filter grading wheels of the faint object spectrograph [1]. Conventional surface hardening techniques such as Malcomizing (proprietary processes licensed to Lindberg Heat Treating Corporation) and Tufftriding (proprietary processes licensed to Lindberg Heat Treating Corporation) suffer from several limitations, including (1) the potential contamination of Teflon-containing lubricants and critical flight hardware surfaces with hydrocarbon surface residue, (2) the required final finishing to remove this potentially brittle surface residue and nitride white layer and (3) the high process temperatures which restrict substrate selection.

An emerging surface modification technique, ion implantation, alleviates all the above concerns while producing wear- and fatigue-resistant surfaces. Research has shown significant improvements in the tribological properties of nitrogen-implanted austenitic stainless steels. Specifically, a reduction in wear rates of a factor of 25–50 has been reported for nitrogen-implanted type 304 stainless steel rotated against both unimplanted and nitrogen-implanted type 416 stainless steel [2]. Other investigators, *i.e.* Oliver *et al.* [3] and Singer and Jeffries [4], have also shown significantly reduced wear rates for nitrogen-implanted type 304 stainless steel, although

0025-5416/87/$3.50

the corresponding coefficients of friction were unchanged. In addition to wear benefits, fatigue life improvements of a factor of 8–10 have been reported for nitrogen-implanted stainless steels [5]. Therefore, the benefits of nitrogen implantation for improved tribological properties of austenitic stainless steel have been established by laboratory-scale testing.

The transfer of these laboratory results to actual component testing has been slowly progressing, being mainly confined to the metal-forming industry [6]. Recent developments, however, have been made for application to aircraft bearing and gear applications [7]. In order to extend this technology to precision mechanism components, in the study reported in this paper the nitrogen implantation of fine-dimensional type 303 stainless steel pinions and gears for subsequent wear-life evaluations was examined. Testing of gears provided a unique opportunity to examine both wear and fatigue resistance due to the complex stress states experienced from the combined sliding and rolling movement of gear tooth surfaces.

2. EXPERIMENTAL DETAILS

2.1. *Mechanism components*

Fine-dimension pinions and gears (Fig. 1), fabricated from type 303 stainless steel (nominal composition, 18wt.%Cr–9wt.%Ni–2-wt.%Mn–1wt.%Si–0.60wt.%Mo–0.15wt.%S-(minimum)–0.15wt.%C), were obtained for subsequent surface modification. Included were input (ten teeth, 96 pitch) and output (32 teeth, 64 pitch) pinions and two antibacklash gears (input gear, 180 teeth, 96 pitch; output gear, 160 teeth, 64 pitch).

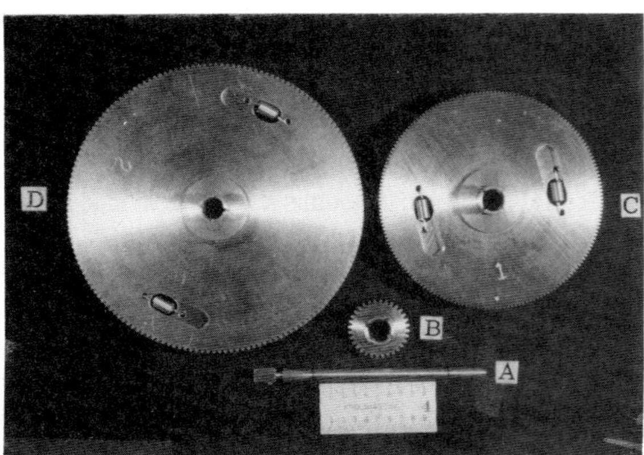

Fig. 1. Macrograph of fine-dimension type 303 stainless steel pinions and gears: A, input pinion; B, output pinion; C, input antibacklash gear; D, output antibacklash gear.

2.2. *Ion implantation*

Two pinion and gear sets were ion implanted with nitrogen using the following parameters: 3×10^{17} N_2^+–N^+ cm^{-2} at 93 keV. To prevent excessive temperature excursions, gears were mounted in a low melting point metal eutectic bath [8] and rotated until the desired fluence of 3×10^{17} N_2^+–N^+ cm^{-2} was achieved. This took 25 h for the small-diameter pinion and 4 h for the antibacklash gear. The chamber vacuum pressure was maintained at about 10^{-4} Torr and the implantation current density was 15 $\mu A\ cm^{-2}$ for the pinions and 10 $\mu A\ cm^{-2}$ for the gears.

2.3. *Heat treatment*

One set of nitrogen-implanted pinions and gears was given a post-implantation anneal of 300 °C for 4 h in an inert gas (argon) environment to stimulate Fe_2N and CrN formation. These parameters were selected from the results of a Mössbauer study of nitrogen-implanted type 304 stainless steel, which showed an increased percentage of Fe_2N after a post-implantation anneal [9]. It was anticipated that this anneal would not modify the nitrogen distribution because of the large percentage of nitride formers, *i.e.* iron and chromium, in the type 303 stainless steel matrix but produce a larger volume percentage of wear-resistant nitride compounds.

2.4. *Wear–life tests*

Prior to wear–life tests, all the mechanism components were ultrasonically cleaned in a Freon TF solvent, with subsequent assembly into a fine-positioning mechanism (Fig. 2) in a class 100 000 clean room bench. A space-qualified Teflon-containing lubricant, Brayco 899, was applied to all the pinion and gear tooth surfaces prior to test initiation. Subsequent wear–life tests were performed on the clean room bench to prevent air-borne particle contamination of the gear tooth surfaces.

The experimental parameters used included (1) an input pinion rotational frequency of 700 rev min^{-1}, (2) an input pinion contact

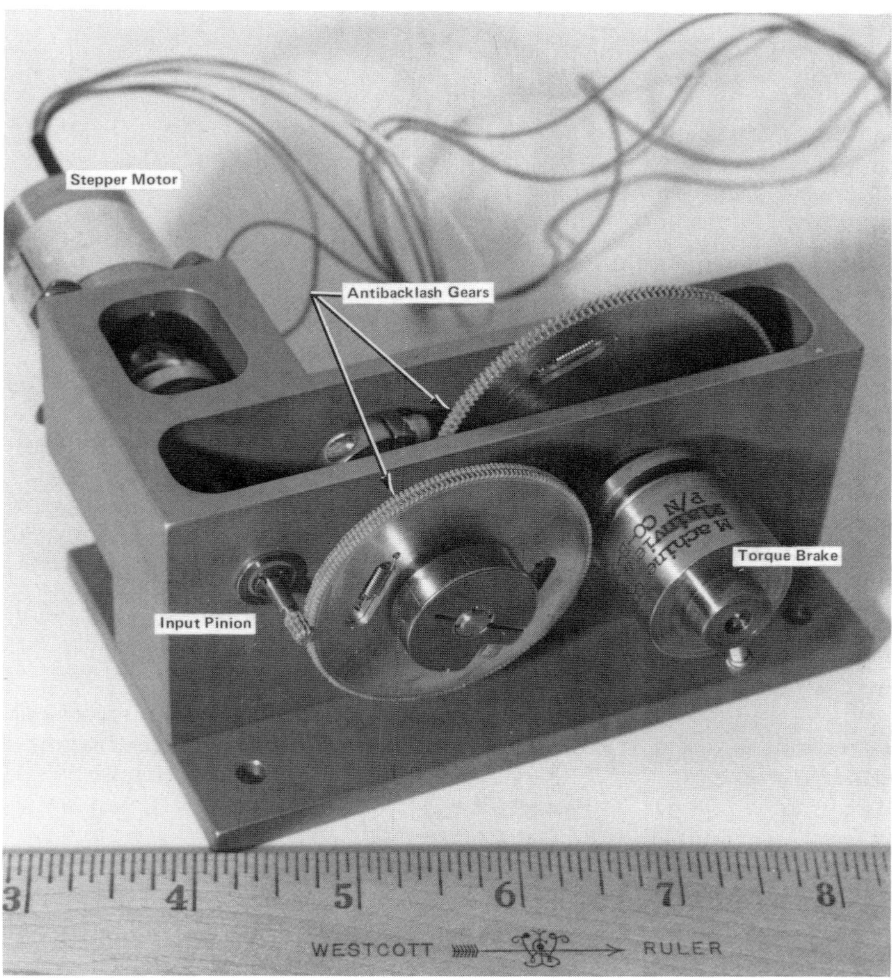

Fig. 2. Fine-positioning mechanism used for gear wear-life tests.

stress of 72.6 klbf in^{-2} and (3) a test duration of 1.68×10^6 cycles (40 h).

Tooth surfaces were examined using optical microscopy (at a magnification of 70×) at selected time intervals which were a percentage of the predicted life of the type 303 stainless steel gears, as determined from a numerical stress analysis. The results of this stress analysis predicted surface contact fatigue failure of the input type 303 stainless steel pinion in 1.24 h or 5.58×10^4 cycles.

The wear and fatigue resistance of the surface-modified gears was determined on a comparative basis with the untreated type 303 stainless steel gears. This was performed through the use of schematic sketches and micrographs of the gear tooth surface. Both surface wear and dedendum spalling were quantified during the test. Surface wear was characterized as a progression of an advancing sliding wear area. Dedendum spalling was also quantified since this type of fatigue damage is common for tooth bending fatigue which is the most frequent mode of gear failures [10].

2.5. *Surface analysis*

Characterization of the gear tooth surfaces after the wear-life tests was performed using electron microscopy to examine wear and fatigue degradation. In addition, chemical analysis was performed using Auger electron spectroscopy (AES) with ion milling to establish the nitrogen concentration (atomic per cent)-depth profiles. Quantification of the elements was made using the Auger electron sensitivity factors shown in Table 1 for a beam energy of 5 kV [11]. Elemental depth profiles

TABLE 1

Sensitivity factors used for calculation of atomic concentrations from Auger electron spectroscopy analysis [11]

Element	*Auger electron energy*	*Sensitivity factor*
Fe	650	0.191
Cr	529	0.310
Ni	848	0.270
C	272	0.140
O	503	0.400
N	379	0.230

were obtained using Ta_2O_5 as the sputter standard material, which has shown good agreement (within about 10%) to depths from previous measurements of sputter craters of implanted iron alloys [12].

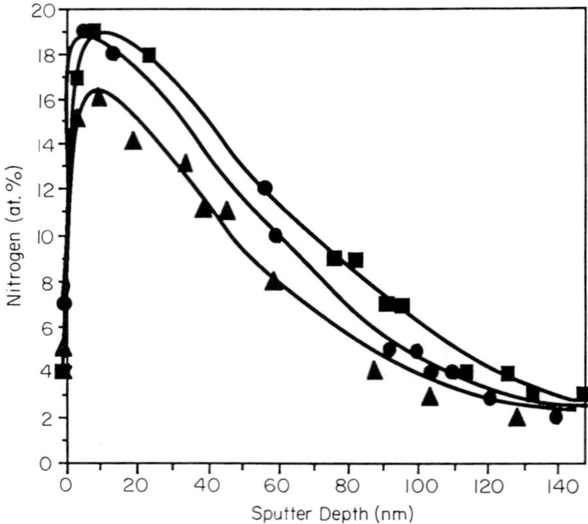

Fig. 3. AES profiles of nitrogen-implanted type 303 stainless steel pinion tooth showing a consistent mean range but a reduced peak nitrogen concentration: ■, top of face; ●, middle of face; ▲, bottom of face.

3. RESULTS AND DISCUSSION

3.1. Pre-implantation characterization

3.1.1. Unimplanted gear

The surface finish of a typical type 303 stainless steel gear tooth was representative of a machined or ground surface with no exterior debris [13]. The microstructure, consisted of fine-grained (ASTM size 8, 0.0224 mm) austenite with no indication of carbide sensitization. In addition, apparent sulfide inclusions with an orientation dependence were observed. Microhardness measurements (a Brinell hardness of about 150 HB) confirmed that the type 303 stainless steel was in the annealed condition.

3.1.2. Nitrogen-implanted gear

Nitrogen implantation did not modify the surface morphology of the type 303 stainless steel gears [13]. Therefore, no final finishing or chemical cleaning operations were needed for the nitrogen-implanted gear components. However, small discolored areas were observed at positions close to the pinion shaft, suggesting that localized oxidation may have occurred during implantation processing.

3.2. Surface analysis

3.2.1. Unimplanted gear

AES analysis of the unprocessed pinion showed a thin (about 3 nm) oxide layer which is characteristic of most polished metals [14]. The oxide stoichiometry, determined from the surface AES compositions [15], was nearly Fe_2O_3.

3.2.2. Nitrogen-implanted gear

Three AES profiles were performed along a nitrogen-implanted pinion tooth face to examine implantation profile uniformity. Profiles were performed along a diagonal line to prevent overlapping of the sputter craters. As shown in Fig. 3, the three profiles were all skewed toward the surface, indicating that sputter-limited conditions were controlling. Good agreement is seen for the mean ranges, although the peak concentrations were reduced slightly with distance down the tooth contour. In a similar manner, the retained nitrogen concentration was reduced by about 18% (middle location) and 32% (bottom location) down the tooth contour. These latter two effects are apparently due to increased off-normal beam incidence in the tooth root areas.

Analysis of the top location AES profile showed an oxide layer thickness increased to about 13 nm and the apparent formation of a duplex Fe_2O_3–Fe_3O_4 surface film. The latter oxide stoichiometry has been identified by Mössbauer analysis of implanted ferrous alloys [16]. In addition to the increased oxide layer thickness, carbon contamination extending to about 40 nm was observed, presumably as a result of the relatively poor vacuum conditions used during implantation.

3.2.3. Nitrogen-implanted-and-annealed gear

The post-implantation anneal altered the near-surface composition of the nitrogen-implanted layer in two major ways (Fig. 4): (1) an increased and broadened mean range and (2) a reduced peak nitrogen concentration. Because of the increased nitrogen penetration for the heat-treated sample, no correlations can be made between the two samples with respect to retained nitrogen concentrations.

These three effects are presumably due to enhanced nitrogen diffusion into the bulk as shown by the displaced gaussian mean. The reduced retained nitrogen concentration suggests back diffusion and loss of nitrogen at the free surface, which has been reported for nitrogen-implanted pure iron [17] but was not expected for nitrogen-implanted type 303 stainless steel because of the large concentration of nitride formers, such as chromium.

Uniformity of the nitrogen-implanted-and-annealed layer was observed (Fig. 5) after a nitrogen Auger electron map was taken after a sputter removal of 60 nm of material, roughly corresponding to the mean range of nitrogen atoms. The areas depleted in nitrogen shown in Fig. 5 correspond to the scratches on the tooth surface produced during specimen preparation.

The chemical state of the implanted and annealed nitrogen was investigated by analysis of the Auger electron energy spectra. It was found that the nitrogen line exhibited characteristic peak shapes that have been attributed to CrN formation by Baron *et al.* [18]. Additional AES profile modifications include a thickened oxide layer (about 80 nm) with a surface oxide stoichiometry between Fe_2O_3 and Fe_3O_4 and removal of the prior carbon contamination. It appears that the argon heat treatment atmosphere provided conditions for decarburization.

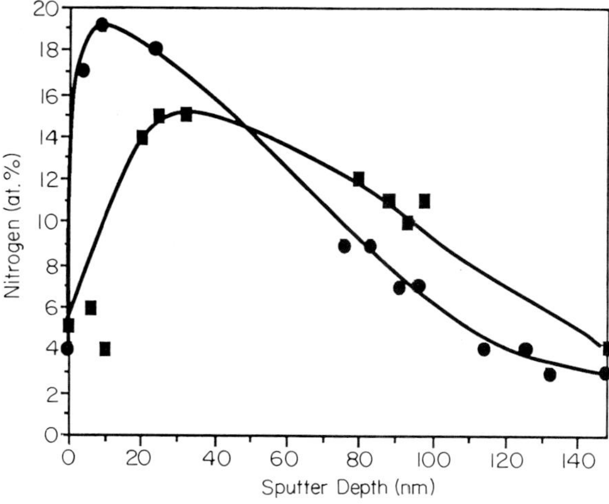

Fig. 4. AES profile of a nitrogen-implanted-and-annealed pinion (■) showing an increased and broadened mean range and a reduced peak nitrogen concentration compared with those for an as-implanted pinion (●).

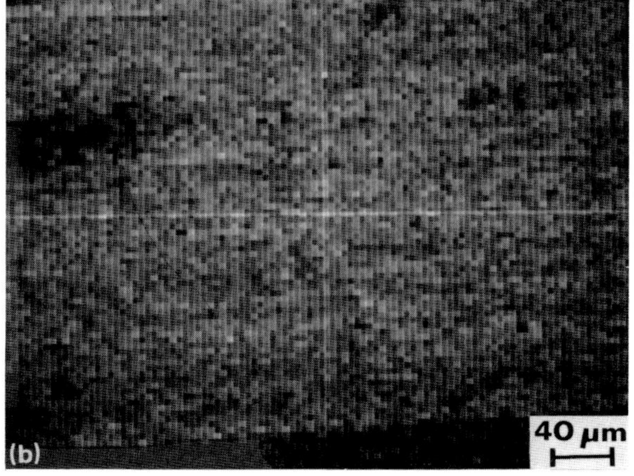

Fig. 5. Electron micrographs of a nitrogen-implanted-and-annealed gear tooth showing (a) the surface morphology and (b) a uniform nitrogen distribution by an AES nitrogen map.

3.3. *Surface wear*

Percentage surface wear as a function of number of wear cycles for the unimplanted, nitrogen-implanted and nitrogen-implanted-and-annealed input pinion gears are shown in Fig. 6. Uniquely different wear behaviors are shown for the three conditions.

Untreated type 303 stainless steel showed a duplex wear behavior as witnessed by the change in slope of the surface wear curve at about 10^5 cycles. It appears that the high initial wear rates (below 10^5 cycles) are a run-in phenomenon which changes to more uniform wear behavior at longer test durations.

The nitrogen-implanted gear appeared to reach uniform wear conditions much sooner than for either the untreated or the nitrogen-implanted-and-annealed gears. This treatment significantly reduced the surface wear area, by up to 40% compared with the untreated condition after 1.68×10^6 cycles. The lower slope of the surface wear cycle line for the nitrogen-implanted gear suggests that significant wear improvements may result for extended life requirements.

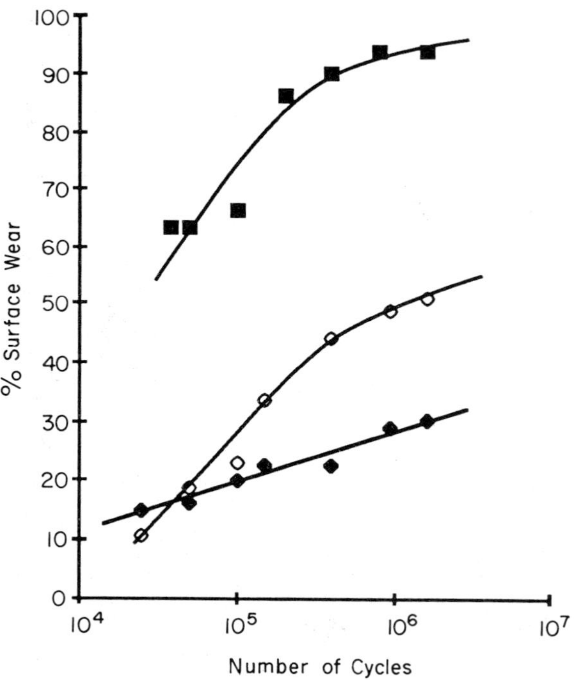

Fig. 6. Surface wear area as a function of the number of wear cycles showing the increased wear resistance from nitrogen implantation: ◇, baseline; ◆, nitrogen implanted; ■, nitrogen implanted and heat treated.

In contrast with the reduced surface wear for the nitrogen-implanted pinion, the nitrogen-implanted-and-annealed gear showed a much larger surface wear area than did the untreated gear throughout the wear test. As for the untreated gear, the wear behavior for the annealed gear exhibited two distinct wear regions, with evidence for a change in wear mode at about 10^5 cycles. An increase of about 46% in wear area was observed at the completion of the wear test.

In summary, nitrogen implantation produced the most wear-resistant surfaces, an effect which may extend to longer mechanism life requirements because of the lower slope of the surface wear–cycle line.

3.4. *Dedendum spalling*

Fatigue spalling was also quantified by visual examinations at selected time intervals during the wear tests. As shown in Fig. 7, spalling progression exhibited similar trends as for the surface wear. For example, nitrogen-implanted gears exhibited a superior fatigue spalling resistance to those of both the unimplanted and the nitrogen-implanted-and-annealed gears. Compared with the unimplanted gear, nitrogen implantation reduced dedendum spalling by 49%, while the nitrogen-implanted-and-annealed component showed an increase in spall density of 11%. In addition, both the

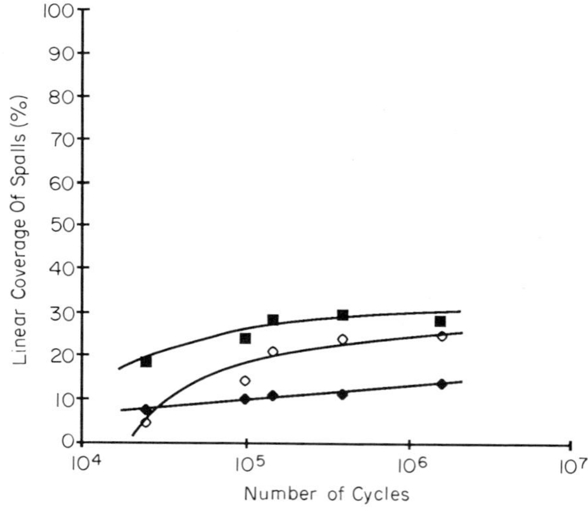

Fig. 7. Dedendum spalling as a function of the number of wear cycles showing the improved spalling resistance from nitrogen implantation: ◇, baseline; ◆, nitrogen implanted; ■, nitrogen implanted and heat treated.

unimplanted and the nitrogen-implanted-and-annealed gears again showed an anomaly in spalling resistance at about 10^5 cycles. This was consistent with the change in surface wear rate for the untreated and nitrogen-implanted-and-annealed gears, as shown in Fig. 6.

3.5. Wear characterization

3.5.1. Unimplanted gear

The unimplanted wear-tested pinion gear showed evidence of extensive sliding wear contact (Fig. 8). Sliding wear zones covered almost the entire addendum face (Fig. 8(a)), thus constituting about 50% of the total contact area. Pitch line areas were devoid of the sliding wear component across approximately 50% of the contact area. Sliding wear damage was primarily polishing of surface asperities; however, localized areas immediately below the pitch line did show more severe grooving with appreciable material removal (Fig. 8(b)). The predominant wear mode appeared to be of a "micromachining" variety as evidence of adhesive galling was not found. Adherent black deposits, immune from ultrasonic cleaning, were repeatedly found on all tooth surfaces and attributed to the Teflon-containing lubricant.

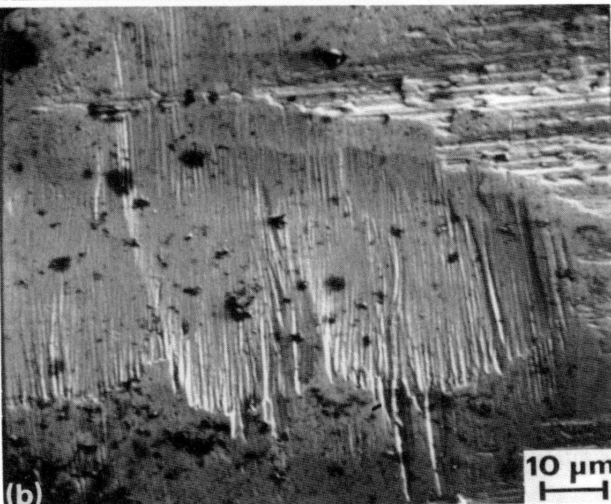

Fig. 8. Electron micrographs of an untreated wear-tested pinion tooth showing (a) a wear area coverage of greater than 50% and (b) dedendum area sliding wear damage.

3.5.2. Nitrogen-implanted gear

A much smaller sliding wear zone than that for the untreated gear was observed for the nitrogen-implanted pinion. As shown in Fig. 9(a), only about 30% of the total contact area was covered by this polished zone. The wear zone–machined surface interface (Fig. 9(b)) shows minor polishing of machined finish lines which was much less severe than observed for the untreated gear. However, this wear mode did effectively remove the implanted nitrogen distribution as shown in the AES profiles in Fig. 9(c).

Beneficial surface modifications that may have occurred as a result of nitrogen implantation include (1) precipitation of iron, chromium and complex (Cr, Fe) nitrides (*e.g.* Fe_2N [9], CrN [18, 19], $(Cr, Fe)_2N_{1-x}$ [2, 20]), (2) stabilization of the parent austenite phase [21] and (3) oxide modifications and enhanced oxide adhesion [22]. All these effects will reduce the extent of sliding wear damage.

3.5.3. Nitrogen-implanted-and-annealed gear

Post-implantation annealing drastically reduced the wear resistance of the type 303 stainless steel gear. As seen in Fig. 10(a), the sliding wear zone nearly covered the entire tooth contact area. Severe wear damage was found along the top land area, as shown in Fig. 10(b). It appears that localized plastic deformation has occurred at the tooth tip, resulting in accelerated sliding wear damage and significant material loss. In addition to the sliding wear damage, fatigue spalling was observed in the pitch line area (Fig. 11), apparently as a result of excessive rolling contact fatigue loads.

The significantly reduced wear and fatigue resistance for the nitrogen-implanted-and-annealed gear was unexpected, since previous investigators [23] have shown enhanced

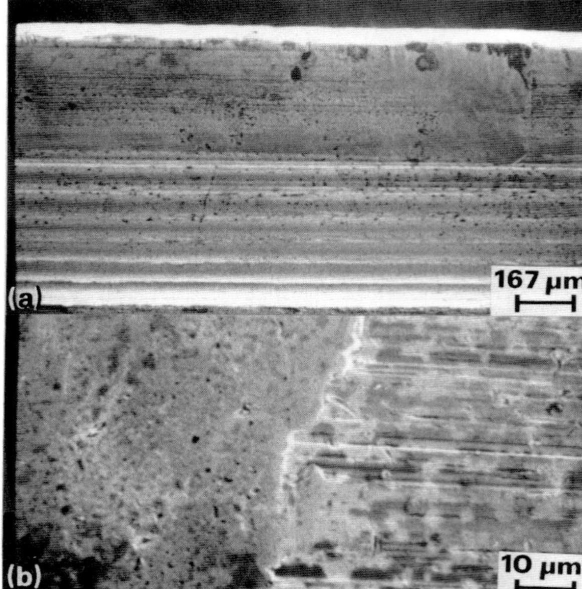

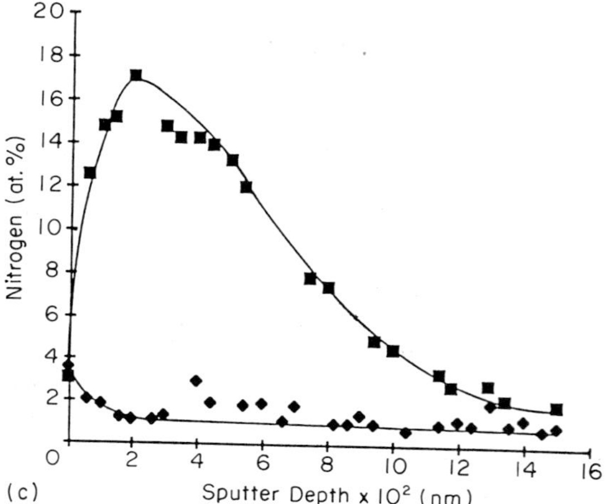

Fig. 9. (a) Electron micrograph of a wear-tested nitrogen-implanted pinion tooth showing wear area coverage of about 30%; (b) electron micrograph of a wear-tested nitrogen-implanted pinion tooth showing the polishing–wear area interface; (c) removal of nitrogen in the polished wear area (♦) and cut of the wear area (■).

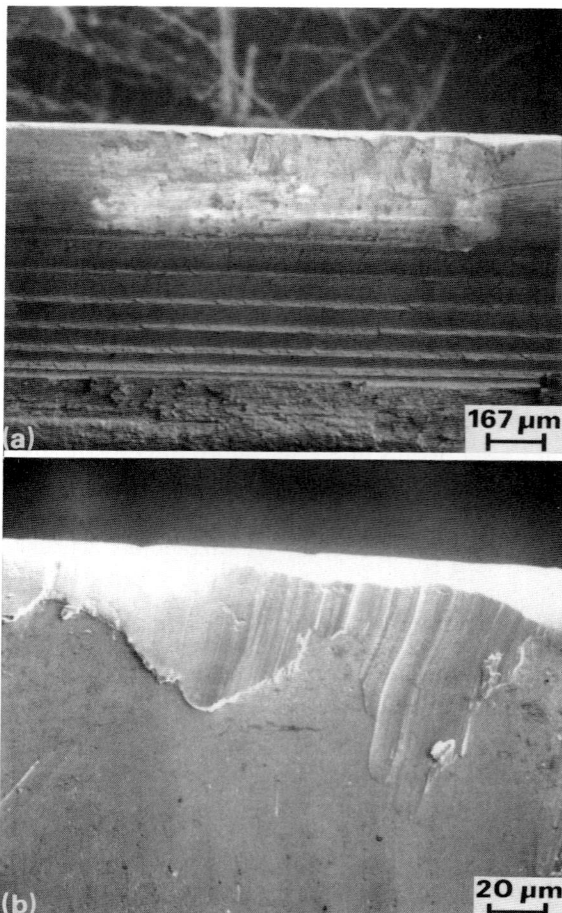

Fig. 10. Electron micrographs of a wear-tested nitrogen-implanted-and-annealed gear showing (a) nearly complete coverage of the wear area and (b) severe wear damage along the top land of the gear tooth.

fatigue lifes for nitrogen-implanted-and-heat-treated AISI 1018 steel.

The post-implantation anneal should not have affected the iron nitride compounds (*e.g.* ϵ-Fe_3N) which are stable up to about 400 °C after which they start to decompose [9]. In addition, the final anneal did not soften the type 303 stainless steel substrate as shown by nearly equivalent microhardness values for both untreated and nitrogen-implanted-and-annealed pinion shafts. Potential reasons for the reduced tribological properties include (1) a reduced near-surface nitrogen concentration (Fig. 4) and thus a reduced quantity of near-surface nitride compounds, (2) fracture of a potentially brittle thickened iron oxide layer (3–80 nm thick) during wear cycling and (3) thermal decomposition or degradation of CrN.

In order to establish the controlling mechanism for the degradation of wear resistance of the implanted-and-annealed material, a wear test of unimplanted and annealed type 303 stainless steel gears should be performed. Results of this wear test would indicate whether the thickened oxide from the thermal anneal or the effects of annealing on the im-

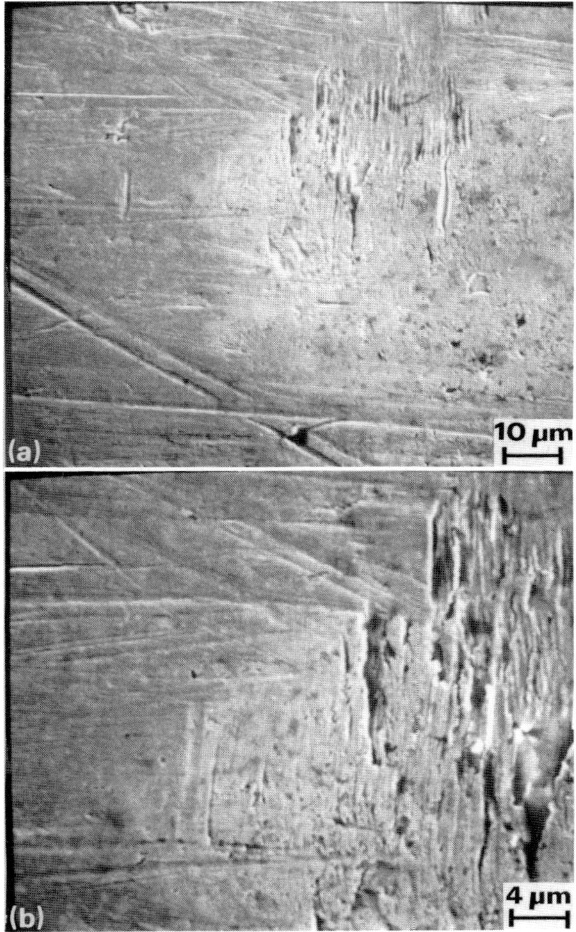

Fig. 11. Electron micrographs of a wear-tested nitrogen-implanted-and-annealed gear showing (a) pitch line area fatigue damage and (b) deep fatigue spalls.

planted nitrogen distribution are most detrimental to wear resistance.

4. SUMMARY

The increased wear and fatigue resistance for nitrogen implantation of type 303 stainless steel gear material confirms earlier research as to the benefits of this surface modification technique [2–5]. Significant implications for precision gear applications lie in the removal of final finishing requirements and the extension of wear resistance to longer cyclic durations. The latter advantage is important for long-life requirements and items where refurbishment is either costly or impractical (*i.e.* space system mechanisms).

Post-implantation annealing to increase the quantity of hard wear-resistant nitride compounds was ineffective for tribological property improvement, presumably as a result of detrimental modifications of the oxide character and thermal degradation of the CrN precipitates.

ACKNOWLEDGMENTS

The authors would like to thank Dr. J. K. Hirvonen of the Spire Corporation for supervising the implantation and Dr. S. R. Smith of the Denver Aerospace Failure Analysis Section for performing the AES analyses.

REFERENCES

1 F. M. Kustas, Preliminary failure analysis of faint object spectrograph mechanism drive motor PD-9500021-001, *Failure Anal. Rep. MeME84/150R*, August 1984 (Martin Marietta Denver Aerospace, Denver, CO).
2 J. K. Hirvonen, Ion implantation in tribology and corrosion science, *J. Vac. Sci. Technol., 15* (5) (1978) 1662–1668.
3 W. C. Oliver, R. Hutchings and J. B. Pethica, The wear behavior of nitrogen-implanted metals, *Metall. Trans. A, 15* (1984) 2221–2229.
4 I. L. Singer and R. A. Jeffries, Friction wear and deformation of soft steels implanted with Ti and N, in G. K. Hubler, O. W. Holland, C. R. Clayton and C. W. White (eds.), *Ion Implantation and Ion Beam Processing of Materials, Materials Research Society Symp. Proc., Boston, MA, November 14–17, 1983*, Vol. 27, Elsevier, New York, 1984, pp. 667–672.
5 N. E. W. Hartley, Tribological effects in ion implanted metals, in G. Carter, J. S. Colligon and W. A. Grant (eds.), *Proc. Int. Conf. on Applications of Ion Beams to Metals, Warwick, 1975*, in *Inst. Phys. Conf. Ser., 28* (1976) 210.
6 G. Dearnaley, Practical applications of ion implantation, *J. Met.* (September 1982) 18–28.
7 K. S. Grabowski, G. K. Hubler, E. T. Hodge, R. A. Jeffries, C. R. Clayton, Y. F. Wang and G. Kuhlman, Ion beam processing of bearing steels for corrosion and wear resistance, *NRL Rep. 5592*, June 1985 (Naval Research Laboratory).
8 D. J. Blak, Eutectic bath hardens to hold work in place, *Design News*, (June 1984) 97.
9 C. A. Dos Santos, M. Behar, J. P. DeSouza and I. J. R. Baumvol, Composition and thermal evaluation of nitrogen implanted steels: a systematic study, *Nucl. Instrum. Methods, 209–210* (1983) 907–912.
10 L. E. Alban, *Systematic Analysis of Gear Failures*, American Society for Metals, Metals Park, OH, 1985, pp. 85–86.

11 L. E. Davis, M. C. MacDonald, P. W. Palmberg, G. E. Riach and R. E. Weber, *Handbook of Auger Electron Spectroscopy*, Physical Electronics Division, Perkin-Elmer Corporation, Eden Prairie, MN, 2nd edn., 1978.

12 F. M. Kustas and M. S. Misra, Ion implantation of 440C steel or improved tribological properties, *Prog. Rep. R85-48681-002*, September 1985, p. 4 (Martin Marietta Denver Aerospace).

13 F. M. Kustas and M. S. Misra, Surface Modification of 303 SS gears, *Prog. Rep. R85-48681-003*, September 1985 (Martin Marietta Denver Aerospace).

14 K. Asami, K. Hashimoto and S. Shimodaira, *Corros. Sci., 17* (1977) 713.

15 M. Seo, J. B. Lumsden and R. W. Staehle, An AES analysis of oxide films of iron, *Surf. Sci., 50* (1975) 541.

16 D. L. Williamson, F. M. Kustas, D. F. Fobare and M. S. Misra, Mössbauer study of Ti-implanted 52100 steel, *J. Appl. Phys.*, to be published.

17 P. D. Goode and I. J. R. Baumvol, The influence of ion implantation parameters on the surface modification of steels, *Nucl. Instrum. Methods, 189* (1981) 161-168.

18 M. Baron, A. L. Chang, J. Schreurs and R. Kossowsky, Nitrogen distribution and nitride precipitation in $^{14}N^{+}$ ion implanted 304 and 316 steels, *Nucl. Instrum. Methods, 182-183* (1981) 531-538.

19 I. L. Singer and J. S. Murday, Chemical state of ion-implanted nitrogen in Fe18Cr8Ni steel, *J. Vac. Sci. Technol., 17* (1) (1980) 327-329.

20 F. G. Yost, S. T. Picraux, D. M. Follstaedt, L. E. Pope and J. A. Knapp, *Thin Solid Films, 107* (1983) 287.

21 R. G. Vardiman and I. L. Singer, *Mater. Lett., 2* (1983) 150.

22 J. T. Pollock, M. J. Kenny and P. J. Paterson, Enhancement of ferrous alloy surface mechanical properties by nitrogen implantation, in G. K. Hubler, O. W. Holland, C. R. Clayton and C. W. White (eds.), *Ion Implantation and Ion Beam Processing of Materials, Materials Research Society Proc. Symp., Boston, MA, November 14-17, 1983*, Vol. 27, Elsevier, New York, 1984, pp. 691-697.

23 W. W. Hu, H. Herman, C. R. Clayton, J. Kozubowski, R. A. Kant, J. K. Hirvonen and R. K. MacCrone, Surface-related mechanical properties of nitrogen implanted 1018 steel, in C. M. Preece and J. K. Hirvonen (eds.), *Ion Implantation Metallurgy, Proc. Conf.*, Metallurgical Society of AIME, Warrendale, PA, 1980, pp. 92-101.

Ion Beam Smoothing of Metallic Mirrors*

DAVID C. INGRAM[a], ANTONY W. McCORMICK[a], PETER P. PRONKO[a], JOHN A. WOOLLAM[b], PAUL G. SNYDER[b] and DAVID B. POKER[c]

[a] *Universal Energy Systems Inc., 4401 Dayton-Xenia Road, Dayton, OH 45432 (U.S.A.)*

[b] *University of Nebraska, Lincoln, NE 68588 (U.S.A.)*

[c] *Oak Ridge National Laboratories, Oak Ridge, TN 37830 (U.S.A.)*

(Received July 10, 1986)

ABSTRACT

The implantation of 150 keV molybdenum ions into molybdenum mirrors and 400 keV silver or copper ions into copper mirrors is found to increase the reflectivity of these mirrors. Analysis of ellipsometric data from these samples using the Bruggeman effective medium approximation demonstrates that the change is due to smoothing of the surface. The improvement reaches a maximum as a function of fluence and then declines as the surface is roughened through sputter-induced topography. The smoothing effect appears to be dependent on elastic energy deposition in the vicinity of the surface. However, if the energy density at the surface is too high, such as with 1 MeV gold ions, then spike effects appears to roughen the surface before significant smoothing can take place. Rotating the samples in the ion beam does not affect the initial smoothing, or its decline, but delays the development of the visible roughening.

1. INTRODUCTION

The possibility of using ions beams to smooth surfaces has been discussed and examined for many years [1, 2]. There are two approaches which may be tried. The first might be to attempt to smooth the surface by sputtering. This approach, as has been demonstrated many times, when carried to sufficiently high fluences, always results in a rougher rather than a smoother surface. This occurs through the development of structural features on the surface that are a consequence of various effects [3]. The second approach is to use defects created in the selvage to smooth the surface by preferential trapping of interstitials at sites of high energy such as steps in the surface [2]. In order to measure the roughness, electron microscopy [1, 2, 4] has been used as well as optical Nomarski micrographs and surface profilometry [5]. However, ellipsometry was also used as reported in ref. 5 and, although it did agree with the other techniques which demonstrated that surfaces may be smoothed during ion bombardment, it was found that ellipsometry is a much more sensitive tool since it could measure roughness on the nanometer scale rather than on the micrometer scale common to the other techniques. In particular, it was shown that, while the addition of heating during ion bombardment helped to smooth micron-sized features, the same treatment was deleterious to the nanometer-sized features and that the best "optically" smoothed surfaces, which also had a higher reflectivity, were achieved at room temperature.

In order to explore the limits of the smoothing effect, a study was undertaken with metallic laser mirror substrates. These substrates have much better surface finishes on them than the samples previously studied. Also, the ion irradiation was performed at room temperature to avoid oxidation or warpage of the substrates which might occur through heating during ion implantation.

2. EXPERIMENTAL DETAILS

Laser mirror substrates were obtained from various sources. Molybdenum mirrors 38 mm

*Paper presented at the International Conference on Surface Modification of Metals by Ion Beams, Kingston, Canada, July 7-11, 1986.

0025-5416/87/$3.50

in diameter and 6 mm thick, which had been polished mechanically, were obtained from Laser Power Optics. Mechanically polished copper mirrors 38 mm in diameter and 10 mm thick were obtained from CVI Laser Inc. Diamond-turned copper mirrors 38 mm in diameter and 15 mm thick were obtained from Peumo Precision Inc. The basic specification was that the mirrors should have a surface finish better than 10 nm r.m.s. and a flatness better than $\lambda/10$ where λ is the wavelength of the light used in the measurement, typically about 1 μm. Each mirror was implanted with 400 keV copper or silver ions using the 1.7 MV Tandetron at Universal Energy Systems or with 150 keV molybdenum ions from a 200 keV Extrian ion implanter at Oak Ridge National Laboratories. The implants, except where noted, were carried out with the beam at normal incidence to the target with a 10 mm $\times$ 5 mm aperture in front of the target. The targets were implanted at various fluences by moving them laterally with respect to the beam. Typically six different fluence implants were performed on all targets this way. All implants were carried out without additional target heating and thus the target temperature for all implants was close to room temperature. Some implants at Universal Energy Systems were also carried out with the beam at 45° to the target normal with and without the target being rotated. This was to determine the effect of rotating the target in the beam. For these implants an aperture was used which would produce a 1 cm $\times$ 1 cm area on the tilted target. The rotated implants were performed with the beam centered on the point of rotation. Because the beam spot on target was square, the implanted area on the rotated samples was larger than 1 cm^2. This meant that the fluence at the edge of the implant was less than at the center but that the fluence at the center was the same as for the stationary target. Thus, all optical measurements were carried out in the center of the rotated area.

Optical Nomarski micrographs were taken with a Nikon Optiphot microscope. The ellipsometry was performed with a multiwavelength and multiple-angle-of-incidence instrument at the University of Nebraska [5, 6]. In order to minimize contamination effects, the molybdenum mirrors were cleaned with distilled water, methanol and distilled water to remove a hydrocarbon layer which tended to build up on these samples during storage in their boxes and while they were exposed to laboratory air. The measurements were made with dry filtered nitrogen flowing over the surface. Contamination of this type was not found on the copper mirrors. The surface roughness on the molybdenum samples was estimated through the use of a Bruggeman effective medium approximation to the pseudodielectric function [6–8]. The real part $\langle\epsilon_1\rangle$ and the imaginary part $\langle\epsilon_2\rangle$ of the pseudodielectric function are defined from the ellipsometric parameters (ψ, Δ) and the angle (ϕ) of incidence as

$$\langle\epsilon_1\rangle + \mathrm{i}\,\langle\epsilon_2\rangle = \sin^2\phi \left\{1 + \left(\frac{1-\tan\psi\, e^{i\Delta}}{1+\tan\psi\, e^{i\Delta}}\right)^2 \tan^2\phi\right\} \quad (1)$$

An increase in the magnitude of $\langle\epsilon\rangle$ may indicate a smoother surface or a surface with a lower void fraction. To differentiate between roughness on the surface and voids near the surface, changes in the magnitude and sign of the components of $\langle\epsilon\rangle$ are used. If $\langle\epsilon_2\rangle$ increases and $\langle\epsilon_1\rangle$ changes very little, then the surface is probably becoming smoother. However, if $\langle\epsilon_2\rangle$ increases while $\langle\epsilon_1\rangle$ becomes more negative, then a decrease in the subsurface void fraction is probably occurring [9]. In addition to modeling the surface, it is also possible to measure changes in reflectivity R since

$$\langle R\rangle = \frac{(\langle n\rangle - 1)^2 + \langle K\rangle^2}{(\langle n\rangle + 1)^2 + \langle K\rangle^2} \quad (2)$$

where

$$\langle n\rangle - \mathrm{i}\,\langle K\rangle = \langle\epsilon\rangle^{1/2} \quad (3)$$

3. RESULTS

In Figs. 1 and 2 are shown the $\langle\epsilon_1\rangle$ and $\langle\epsilon_2\rangle$ components of the pseudodielectric function obtained from ellipsometric measurements on a molybdenum mirror implanted with 150 keV molybdenum ions at various fluences. The $\langle\epsilon_2\rangle$ curves show an increase with increasing fluence. The $\langle\epsilon_1\rangle$ curves fluctuate, showing no net change. The reflectivity as a function of wavelength obtained from these data is shown in Fig. 3. There is a sharp increase in reflectivity at a fluence of 2×10^{15} ions cm^{-2}

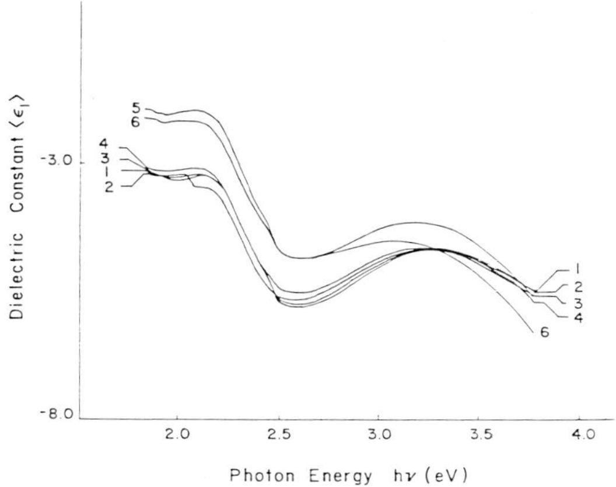

Fig. 1. Pseudodielectric function component $\langle\epsilon_1\rangle$ for a molybdenum mirror implanted with 150 keV molybdenum ions at various fluences: curve 1, 1×10^{14} ions cm^{-2}; curve 2, 2×10^{14} ions cm^{-2}; curve 3, 5×10^{14} ions cm^{-2}; curve 4, 1×10^{15} ions cm^{-2}; curve 5, 2×10^{15} ions cm^{-2}; curve 6, 5×10^{15} ions cm^{-2}.

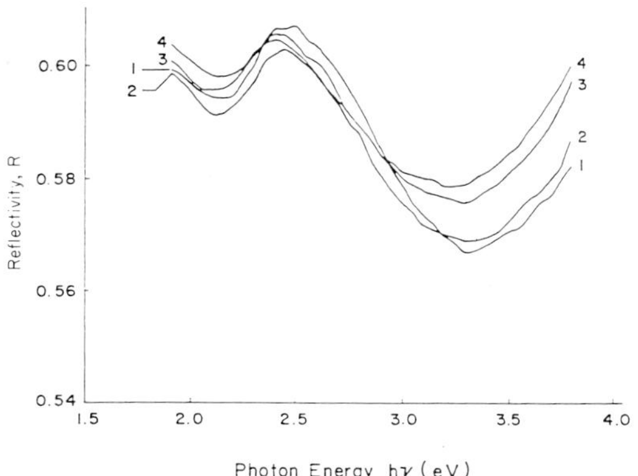

Fig. 3. Reflectivity as a function of incident photon energy for a molybdenum mirror implanted with 150 keV molybdenum ions at various fluences (cleaned with distilled H_2O–methanol and then with distilled H_2O): curve 1, unimplanted; curve 2, 1×10^{15} ions cm^{-2}; curve 3, 2×10^{15} ions cm^{-2}; curve 4, 5×10^{15} ions cm^{-2}.

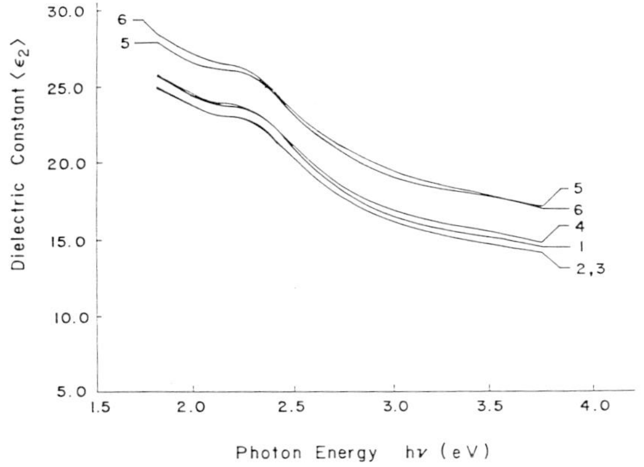

Fig. 2. Pseudodielectric function component $\langle\epsilon_2\rangle$ for a molybdenum mirror implanted with 150 keV molybdenum ions at various fluences: curve 1, 1×10^{14} ions cm^{-2}; curve 2, 2×10^{14} ions cm^{-2}; curve 3, 5×10^{14} ions cm^{-2}; curve 4, 1×10^{15} ions cm^{-2}; curve 5, 2×10^{15} ions cm^{-2}; curve 6, 5×10^{15} ions cm^{-2}.

commensurate with the increase in $\langle\epsilon_2\rangle$. The $\langle\epsilon_1\rangle$ and $\langle\epsilon_2\rangle$ data are consistent with local smoothing of the surface and, through the use of a Bruggeman [7] effective medium approximation, the samples have been modeled for fluences below 2×10^{15} ions cm^{-2} with a rough surface layer 1.4 nm thick with a 50% void fraction. This may be visualized as a surface which contains features no more than 1.4 nm high and with a length less than the wavelength of light used in the measurement. There may be progressive steps which cause the surface height to vary by much greater amounts although the specification of flatness to better than $\lambda/10$ means that across the entire surface of the sample the height varies less than 100 nm. Changes at these levels are not detectable by conventional optical measurements because they rely on the scattering of light from features which are of the order of the wavelength of the scattered light.

Further experiments on molybdenum mirrors have been conducted with 3 MeV nickel ions and 1 MeV gold ions. The nickel ions failed to produce any change up to the highest dose studied, which was 5×10^{15} ions cm^{-2}. The gold ions even at the lowest fluence used (1×10^{14} ions cm^{-2}) produced a rougher surface, with the roughening increasing with increasing fluence. The nickel presumably went too deep and created defects so far from the surface that they became trapped before reaching the surface. The gold ions, incident at this energy on a target as heavy as molybdenum, are liable to produce cascade spikes because of the tendency for non-linear cascade formation in these systems [10]. These events, if they intersect the surface would lead to very high sputtering on a local level, resulting in roughening on a local scale [11]. Such pits in the surface have been seen by similar ion target combinations with transmission electron microscopy [12].

Some of the diamond-turned copper mirrors were implanted with 400 keV silver ions at doses ranging from 5×10^{14} to 5×10^{16} ions cm^{-2}. The pseudodielectric functions of these mirrors are much less affected by implantation than the pseudodielectric functions of the molybdenum mirrors. The data show a reduction in roughness and improvement in reflectivity up to a fluence of 1×10^{16} ions cm^{-2}. Above this fluence the surface changes and the reflectivity is reduced (Fig. 4). Preliminary modeling of the ellipsometric data indicates that the subsurface void fraction is increasing as the reflectivity is decreasing. Further work is envisaged on modeling the results as well as cross-sectional transmission microscopy measurements to determine the void fraction [13].

A mechanically polished copper mirror was also implanted with 400 keV silver ions. The results are similar to those obtained from the diamond-turned mirrors except that the smoothing is relatively greater because of the slightly poorer starting surface of the mechanically polished mirror.

Diamond-turned copper mirrors have also been implanted with 400 keV copper ions. The results again indicate smoothing up to a fluence of 5×10^{15} ions cm^{-2} and roughening thereafter. As for the silver implants, the roughening takes place below the surface with the formation of subsurface voids. In an attempt to understand the roughening process better, some samples were also implanted at 45° to the beam. Figure 5 consists of optical Nomarski micrographs taken from a copper laser mirror which had received two implants. Both implants were performed using 400 keV silver ions to a fluence of 2×10^{16} ions cm^{-2} with the target at 45° to the beam. In one case the target was rotated and in the other the target was stationary. The marks left on the sample by the diamond-tipped tool are clearly visible on the unimplanted sample as is a mosaic pattern from the grain structure. The area which was rotated during implantation is smoother in appearance than the unimplanted region; both the tool marks and

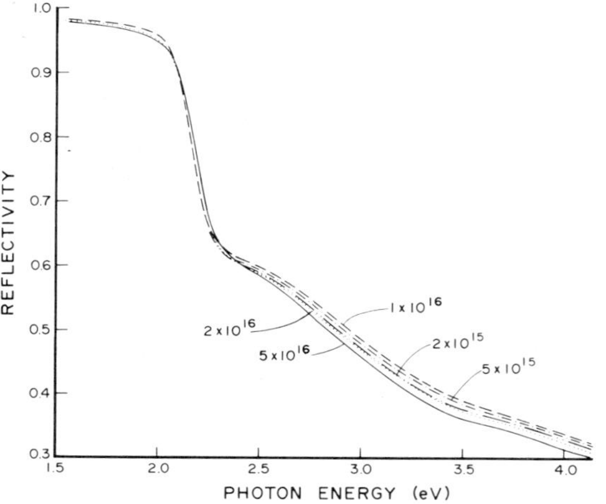

Fig. 4. Reflectivity as a function of incident photon energy for a copper mirror implanted with silver ions. The numerals on the curves indicate the fluences in ions per square centimeter.

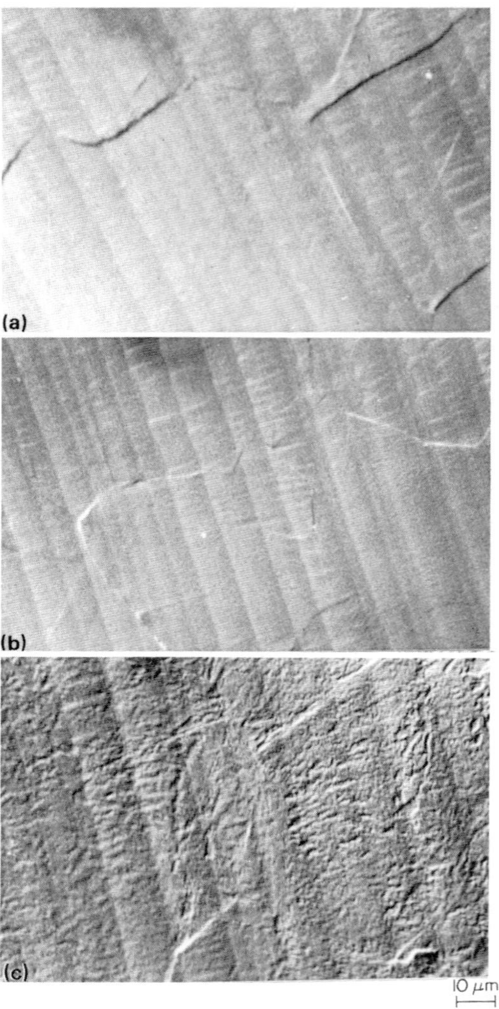

Fig. 5. Nomarski optical micrographs of a diamond-turned copper mirror: (a) unimplanted; (b) implanted with 400 keV silver ions to a fluence of 2×10^{16} ions cm^{-2} with the beam at 45° to the surface and while the sample was rotating; (c) same as (b) except that the sample was stationary during the implant.

the grain structure are less evident. The area which was not rotating has developed a topography which always occurs on these samples and which is presumably due to local variations in the sputtering rate. Ellipsometric measurements made on this target indicate that the area which was rotating was marginally rougher than the area that was stationary during implantation. It is only when the fluence is in excess of 1×10^{16} ions cm^{-2} at normal incidence for 400 keV silver ions or 2×10^{16} ions cm^{-2} for 400 keV copper ions that the sputter-induced roughening becomes evident, *i.e.* above the fluence for the maximum reflectivity as measured by ellipsometry. No visible roughening was seen on the molybdenum mirrors even though more energy deposition took place in the surface with both the gold and the molybdenum implants.

4. CONCLUSIONS

Energetic ion beams are shown, through ellipsometric measurements, to smooth metallic surfaces on an atomic scale. The process reaches a peak as a function of fluence and then declines just below the dose at which sputter-induced roughening of the surface begins to appear. Through the use of rotating targets to reduce the effect of the sputter-induced damage, it has been shown that this sputter-induced damage or roughening in itself is not the limitation to smoothing on the atomic scale. Since ellipsometry uses light which penetrates the surface a few tens of nanometers, it is sensitive, as was mentioned earlier, to voids just below the surface [8]. These voids also affect the reflectivity and it is possible that the loss of reflectivity detected in the copper mirrors at high fluences is due to the development of radiation-induced voids.

ACKNOWLEDGMENTS

Financial support from the National Science Foundation through Small Business Innovative Research (SBIR) Grant PHY-8318798 is gratefully acknowledged.

Discussions with George Carter and David Aspnes are also gratefully acknowledged.

REFERENCES

1 W. Marth, *Z. Angew, Phys., 13* (1961) 224.
2 R. Sizmann, *J. Nucl. Mater., 69–70* (1968) 386.
3 G. Carter and M. J. Nobes, in O. Auciello and R. Kelly (eds.), *Ion Bombardment Modification of Surfaces*, Elsevier, Amsterdam, 1984.
4 P. P. Pronko, A. W. McCormick, D. C. Ingram, A. K. Rai, J. A. Woollam, B. R. Appleton and D. B. Poker, in G. K. Hubler, O. W. Holland, C. R. Clayton and C. W. White (eds.), *Ion Implantation and Ion Beam Processing of Materials, Materials Research Society Symp. Proc., Boston, MA, November 14–17, 1983*, Vol. 27, Elsevier, New York, 1984, p. 559.
5 G. H. Bu-Abbud, D. L. Mathine, P. Snyder, J. A. Woollam, D. B. Poker, J. Bennett, D. C. Ingram and P. P. Pronko, *J. Appl. Phys., 59* (1986) 257.
6 P. G. Snyder, M. C. Rost, G. H. Bu-Abbud, J. Oh, J. A. Woollam, D. B. Poker, D. E. Aspnes, D. C. Ingram and P. P. Pronko, *J. Appl. Phys.*, to be published.
7 D. A. G. Bruggeman, *Ann. Phys. (Paris), 24* (1935) 636.
8 D. E. Aspnes, E. Kinsbron and D. D. Bacon, *Phys. Rev. B, 21* (1980) 3290.
9 D. E. Aspnes, J. B. Theeten and F. Hottier, *Phys. Rev. B, 20* (1979) 3292.
10 P. Sigmund, in R. Behrisch (ed.), *Sputtering by Particle Bombardment I, Top. Appl. Phys., 47*, Springer, Berlin, 1981.
11 J. E. Westmoreland and P. Sigmund, *Radiat. Eff., 6* (1971) 187.
12 K. L. Merkle and W. Jager, *Philos. Mag. A, 44* (1981) 741.
13 P. G. Snyder, A. Massengale, K. Memarzadeh, J. A. Woollam, D. C. Ingram and P. P. Pronko, submitted to *Symp. on Beam-Solid Interactions, Fall Meet. of the Materials Research Society, 1986.*

Materials Science and Engineering, 90 (1987) 423-432

Ion Sources and Implantation Systems*

JOHN H. KELLER[a] and JAMES W. ROBINSON[b†]

[a]General Technology Division, IBM, East Fishkill, Hopewell Junction, NY 12533 (U.S.A.)

[b]The Pennsylvania State University, Electrical Engineering, University Park, PA 16802 (U.S.A.)

(Received July 10, 1986)

ABSTRACT

Ion implantation makes possible unique metal surface characteristics that have found widespread applications. The still developing implanter technology offers a variety of systems suited to different purposes. The types of ion source are described in terms of attributes which influence the choice of a source for a particular application. Extractors are discussed to a lesser extent, and other system parts are mentioned briefly. Although present semiconductor ion implanters deliver 10–20 mA, the technology can be extended easily to over 100 mA, especially when mass analysis is not required. For large-area applications, the ion source can be one of several designs based on magnetic cusp confinement of a plasma containing the ion of interest.

1. INTRODUCTION

The important parts of an ion implantation system are the ion source and the extraction system. We review sources by comparing them as to type, by describing the historical development of sources used in high current ion implanters and by noting some of the newer developments. To a lesser extent, the properties of ion extractors are presented, and the issues of mass analysis and post-acceleration are treated briefly. The references are representative rather than all inclusive. Features related to the selection of a particular subsystem are described. Numerous excellent equipment surveys exist, *e.g.* those by Ryssel and Glawischnig [1] and Ziegler and Brown [2]. At this conference 2 years ago, the diversity of ion beam techniques and metallurgical applications was well documented. Although nitrogen implanting was a common topic, the range of papers included one by Goktepe [3] who was implanting 10 keV copper into gold. At another related conference, Brimhall *et al.* [4] irradiated NiTi with 2.5 MeV nickel ions. The wide range of applications requires that a variety of complementary implanting systems be available.

2. REVIEW OF HIGH CURRENT ION IMPLANTATION SOURCES

Many distinct types of ion source have been developed during the last 30 years, although much overlap also occurs. These were developed for isotope separation as well as for ion implantation. More recent applications include neutral beam injection as well as industrial ion beam sputtering. Several of these types are summarized in Table 1 which outlines the remainder of this section. The requirements that might be placed on an ion source are listed, and the various types are reviewed in turn, emphasizing the desirable properties of each. For many applications the multipole source is a prime contender.

2.1. *Source requirements*

When choosing a source, the system designer must consider the type of ion that is to be generated, the required ion current, the gas efficiency, the operating life of each expendable component, the current fluctuations, the beam shape and the adaptability to automatic control. Secondary considerations are the system weight, the system bulk, the power

*Paper presented at the International Conference on Surface Modification of Metals by Ion Beams, Kingston, Canada, July 7-11, 1986.

†On sabbatical leave at IBM, East Fishkill, Hopewell Junction, NY 12533, U.S.A.

0025-5416/87/$3.50

TABLE 1

Types of ion source and their characteristics

Source	*Electron confinement*	*Ion extraction*	*Special features*
Calutron	Magnetically confined arc	Side (cross-field)	Separate filament chamber
Bernas	Magnetically confined arc	Side	Single arc chamber
Freeman	Magnetron with end containment	Side	Simple structure; good gas efficiency
Multipole	Cusp magnetic fields	Field-free multiple aperture	Quiet with a high gas efficiency
Solenoid (PIG)	Magnetic with electrostatic end confinement	End (parallel to field)	Filament as large as extraction area
Duopigatron	Solenoid or cusp	Parallel to field or field free	Strong electromagnetic field layer separates anode and cathode chambers
Microwave (ECR)	Strong magnetic field	End	No filament; long operating life

consumption and the cost. Often a researcher's description of a source does not address all these issues.

Implanter sources should minimize beam noise so that, with background gas present, the beam can be space charge neutralized of the order of 99% [5]. Also with a quiescent plasma and a corresponding low level of high frequency noise, the extracted beam current will be steady. Consequently, beam divergence, which depends on the current as well as the extractor design, can be made small or at least constant. The gas efficiency should be high so that the extraction system can run at a high current density without excess charge exchange and gas scattering that may cause arcing.

Implanters that use magnetic mass analysis usually require an extraction slit which generates a ribbon-like beam compatible with analyzing slits. Yet, for fast processing, the target may be exposed to a diffuse unanalyzed beam extracted through multiple apertures [6, 7]. Beam energies may range from 500 eV [7] to 120 keV [8].

2.2. *Calutrons and Bernas sources*

The calutron and later the Bernas source were among the first developed for the related field of electromagnetic isotope separation as well as for ion implantation, which has almost the same requirements.

The calutron [9] ion source, which is sketched in Fig. 1, was developed for the Oak Ridge separators. This source resides inside a 180° analyzing magnet. The axial magnetic field in the source is that of the analyzing magnet, and ions are extracted through a slit which is perpendicular to the field. The filament is in a separate chamber and an arc is produced near the extraction slit. A later version used a 255° magnet with double focusing and produced 30 mA of tin ions (Sn^+) in one of its applications.

The Bernas source [10, 11] is a simplified calutron in which the filament is in the arc chamber located just beyond the end of the extraction slit. It is important that the filament be close to the plane of the slit (about 1 mm in the back of it [12]). This source is capable of producing 20 mA cm^{-2} of arsenic ions (As^+) which is extracted from a strong arc, again centered about 1–2 mm behind the extraction slit. The magnetic field is produced by a separate magnet. A version with a curved slit 10 cm long has been run by Lempert *et al.* [13] with an output current in excess of 90 mA. The curved slit provides source focusing in the axial plane of the analyzing magnet; it has also been used to give higher beam transmission through the magnet [14].

2.3. *Freeman sources*

Freeman [15] developed a different type of source which is the dominant source used to-

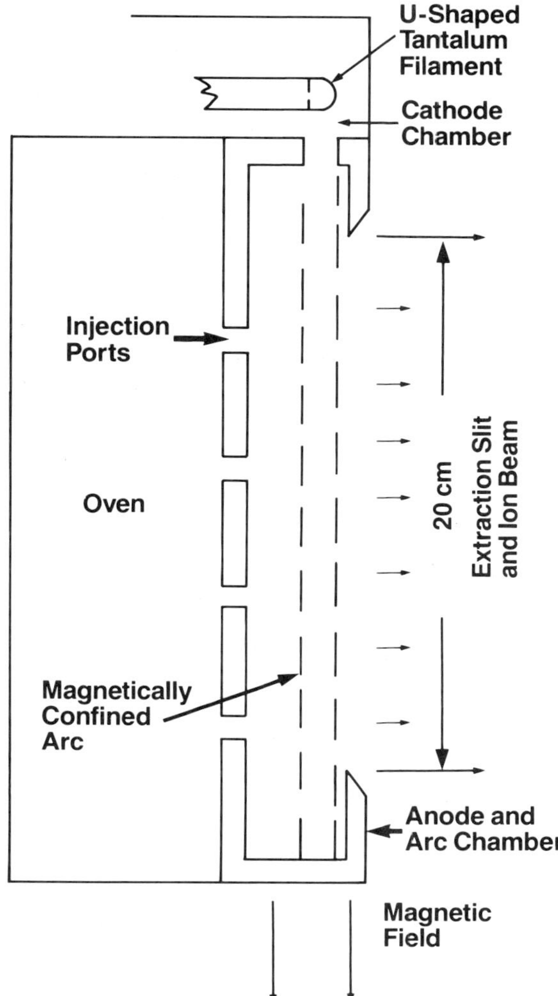

Fig. 1. Essential elements of a calutron source. Its features include a separate filament chamber and a magnetically confined arc close to the extraction slit. The gas to be ionized enters from the left and ions are extracted through a slit up to 20 cm long.

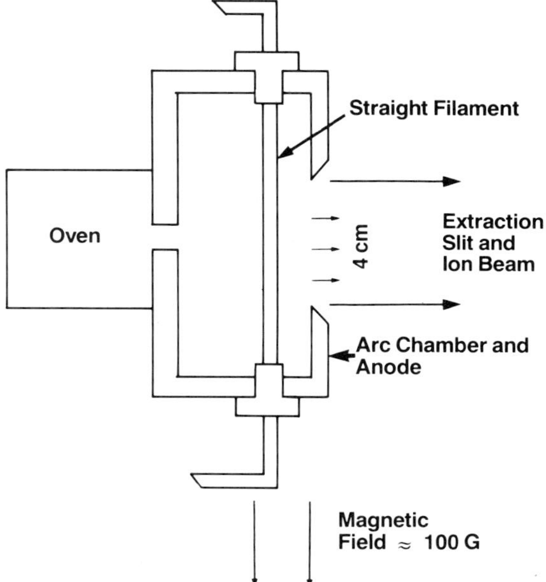

Fig. 2. The Freeman source replaces the arc column of the calutron with an emitting filament. The source operates as a magnetron. Reflex action was later achieved by adding reflectors at both ends of the filament.

day in high current ion implanters. Originally used in the Harwell separator, it was later adapted to the Lintott ion implanter [16]. This source, illustrated in Fig. 2, offers moderate current extraction through a slit in the side of a cylindrical anode. The cathode is a filament carefully positioned parallel to the axis but off-center toward the slit. A parallel magnetic field, combined with the field from the filament current, confines the emitted electrons to orbits which concentrate the ionization in the vicinity of the slit. The source thus runs in a magnetron mode. The plasma, which is stable and quiet, produces a stable extracted beam.

Williams [17] and Aiken [18] have made significant improvements to the Freeman source. These consisted mainly in providing end confinement for the energetic electrons. Williams used a magnetic mirror field and Aiken attached end disks to the filament which caused electrostatic confinement of the electrons. In its present form, the Freeman source is simple yet quite efficient. The present authors estimate that in normal operation it is two to three times as gas efficient as the Bernas source. That is, for the same arc current and extraction current, it will run at one-half to one-third of the pressure.

Some recent development work has been undertaken to extend filament lifetime. Careful design and the use of rhenium filaments have allowed the production of 10 mA of oxygen ions (O^+) for up to 40 h of operation between filament replacements [19]. Another alternative [20] is to place a filament in a separate chamber at the end of the cylinder. When this chamber is fed with a non-corrosive gas which is different from the gas in the ionizing chamber, the filament may be operated under conditions which optimize its lifetime. The plasma from this discharge is injected into the Freeman source through a small hole located where the conventional filament would normally terminate. The injected plasma jet spreads along field lines and replaces the usual filament as a source of ionizing electrons. In many ways, this modified Freeman source resembles a calutron.

2.4. Multipole sources

Multipole sources, also called magnetic cusp and bucket sources, have become an important source for neutral beam injection into fusion plasmas. They are characterized by a five-sided box (or cylinder) which supports rows of permanent magnets, with each row polarized opposite from its neighbors as illustrated in Fig. 3. The box is the anode of a d.c. discharge which is sustained by a hot-filament cathode more or less in the center of the box. The magnetic fields prevent the primary filament electrons from flowing directly to the anode. Thus, they have paths several times longer than the box dimensions and they are efficient ionizers of the feed gas. The sixth side of the box is a set of multiaperture extraction grids or extraction slits which can create well-focused beams of hundreds of milliampères.

The first multipole source was built by Limpaecher and MacKenzie [21]. This type of source, with a pole strength generally exceeding 800 G, has been studied extensively by Sterling *et al.* [22], Ehlers and Leung [23], Goede *et al.* [24], and more recently by Martin [25], Crecelius and Holzle [26], van Bommel *et al.* [27] and Keller *et al.* [28]. Besides the primary electron confinement, some plasma confinement also occurs and, after attempts by a number of researchers, Ehlers and Leung [29] achieved electrostatic confinement of the ions and thus the plasma by segmenting the anode structure. The segments between the magnetic cusps were biased at a potential above the potentials of the plasma and of the anode segments at the magnetic cusps.

In a related effort, multipole sources with lower magnetic fields of the order of 100 G have been developed for the National Aeronautics and Space Administration, notably by Kaufman [30], for industrial use and as ion propulsion sources. In this latter use the designs have emphasized light weight and efficiency. The relatively weak magnets at floating potential are interspersed with anode segments, as shown in Fig. 4, so that only the area between the poles is normally used as an anode. Thus the electron losses at the cusp or pole are reduced.

Both high and low field types can produce several hundred milliampères of ions and they use gas efficiently. These types of source are particularly applicable to surface modification of metals if mass analysis can be eliminated. Both types are normally broad-beam multiaperture sources typically about 15 cm in diameter, although they may be much larger. The sources may be designed to implant several hundred milliampères of ions at a few

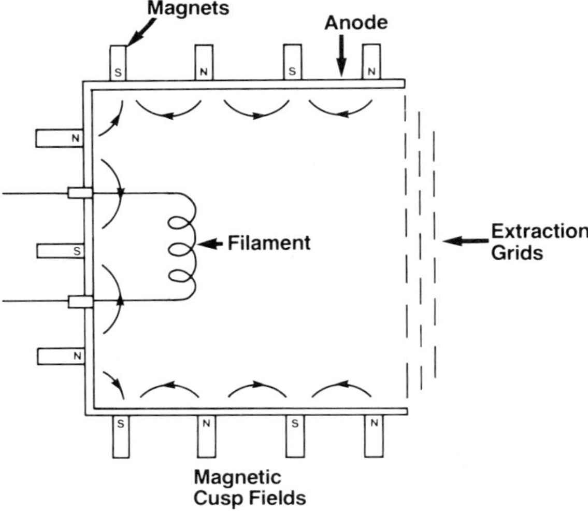

Fig. 3. Typical high field cusp or multipole source. The magnets, arranged in rows of opposite polarities, typically produce cusps of 1000 G while a quiet plasma is produced in the inner region where the field is less than 30 G.

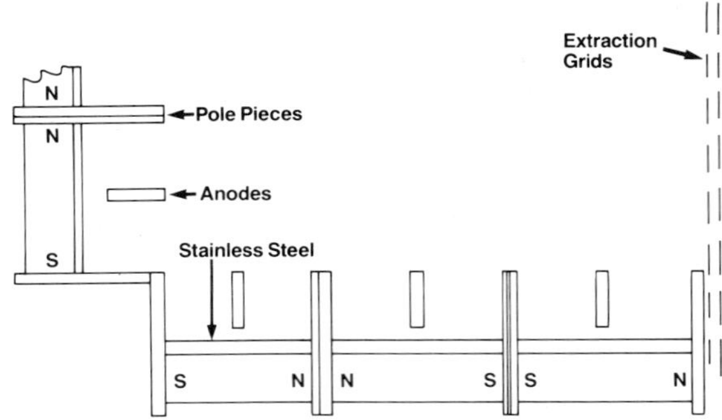

Fig. 4. Low field cusp or multipole design. Anode segments are placed between the cusps; thus, magnetic fields do not need to be as large as when the anode includes the cusp area.

milliampères per square centimeter or they may be used at lower current densities. Both high and low field types will probably be used in future commercial ion implantation systems.

Keller [31] and Cope and Keller [32] have adapted a multipole source to a single-slit extractor of the type that is used with a mass analyzer. For this source, which is shown in Fig. 5, Keller reported 55% gas efficiency for a 17.5 mA beam of argon. In normal operating conditions, this source was six times as gas efficient as a comparable Bernas source. Quadrupole, hexapole and decapole configurations were studied. The quadrupole configuration similar to that described by Brainard and O'Hagan [33] was the best for argon. However, for a dissociating molecule, the higher order cusps are considered to be better because the higher plasma volume for a given chamber size leads to greater dissociation.

The concave nature of the magnetic cusps corresponds to a stable configuration where the field is stronger as the distance away from the plasma increases. Cusp-contained plasmas have been found to be very uniform and quiescent [21]. Nicholson [34] has recently summarized the relevant theory. This stable mode of operation assures that the plasma density will be uniform throughout the region behind multiple extraction apertures.

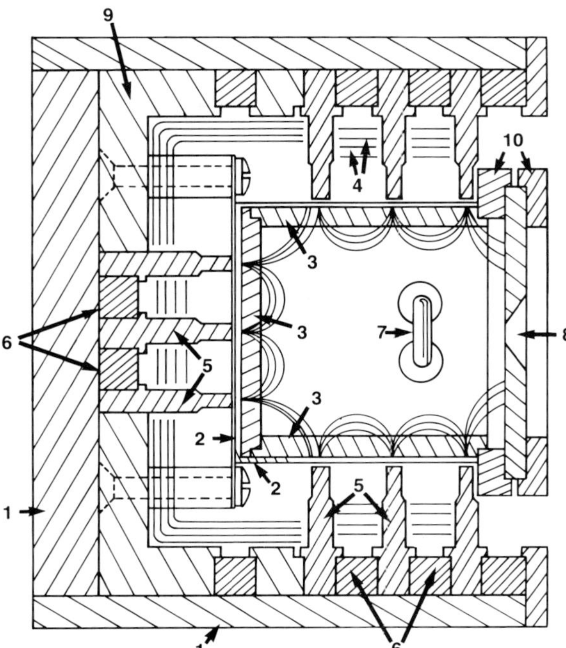

Fig. 5. The multipole source of Cope and Keller [32] showing the following: 1, copper cooling plates; 2, molybdenum heat shield box; 3, graphite anode; 4, heat shields; 5, pole pieces; 6, magnets; 7, U-shaped filament; 8, extraction slit; 9, magnet yoke; 10, insulator.

Multipole sources produce mainly singly charged ions of non-reactive gases, but other gases may be used. A major limiting factor is the operating life of the filament which is a function of the type of gas as well as the operating temperature. Oxygen causes rapid deterioration of tungsten, for example. Metal halides or hydrides may be used for a metal ion source, but halides especially may shorten filament life and metal coatings will form. Precautions must be taken to shadow insulators to prevent shorting.

Kaufman [35] and his associates were among many who have developed hot hollow cathodes that can be used in multipoles. Although hollow cathodes have much longer lifetimes than filaments, they entail added complexity and higher initial costs. Some starting mechanism is necessary. Alessi and Prelec [36] describe their recent experience with hollow cathodes and refer to still other work.

2.5. Solenoidal (PIG) sources

A predecessor of the multipole design, the solenoidal source, is still in use today [6, 7, 37]. This source uses a single large solenoid surrounding the box instead of the numerous small magnets of the multipole source. The magnetic field thus runs perpendicular to the extraction grids and causes electrons to orbit without easy access to the anode side walls. In such systems the field extends throughout the discharge chamber. This is contrasted with the nearly field-free region in the center of the multipole source which is achieved through the use of the closely spaced alternating poles. The solenoidal field configuration provides a less uniform plasma. However, the field strength can be adjusted easily, thereby making it possible to control the ionizing efficiency of the filament electrons. One design allows continuous adjustment of current extracted from a slit from 100 μA to 8 mA with no change other than the solenoid current [37]. In most other sources the output current is regulated by adjusting the amount of primary electrons produced in the source. The optimum low divergence beam is obtained by designing the extractor for use at a specific value of ion current. Some compensation for varying currents may be obtained by making the acceleration gap adjustable.

2.6. *Duopigatrons*

The name "duopigatron" has come to be associated with a variety of sources which have dual arc chambers. Electrons are produced from hot cathodes in one chamber and are drawn through a region of strong electric and magnetic fields where they are accelerated and multiply. These energetic electrons pass into a second reflex chamber where they produce ions that are extracted. In one form of duopigatron [38–41], electrons are confined in the ionizing chamber radially by the magnetic field and axially by electrostatic forces. The anode is in the form of a ring surrounding the discharge chamber. Another form of duopigatron [42, 43] injects electrons through the high field region into the second chamber which is virtually free of fields except for magnetic cusp fields around the periphery, much like the confinement fields described for multipole sources. This latter type of duopigatron is much like a multipole source except that the electrons are injected rather than being produced in the multipole chamber itself.

Duopigatrons have many characteristics similar to multipole systems. They can produce hundreds of milliampères with good gas efficiency and they are most suitable for nonreactive gases. An important advantage of duopigatrons is that the filament is in a separate chamber and may be protected from gases such as oxygen. Shubaly *et al.* [39] operated the filament in argon with both argon and electrons being injected into the ionizing chamber which contained oxygen. This system produced 138 mA of oxygen ions with 3.3 standard cm^3 of oxygen and argon combined and a filament life of 10–15 h. Menon *et al.* [43], using cusp confinement for a neutral beam injector, obtained a 48 A hydrogen beam with 900 standard cm^3. Shubaly and Hamm [44] reported 100 mA xenon beams with a gas flow of 1.4 standard cm^3. Bickes and O'Hagan [40] reported developing duopigatrons for neutral beam injection and Ghanbari *et al.* [41] are considering them for implanters.

2.7. *Other sources*

Hollow cathodes may replace filaments in most types of source [35, 36] or they may be used as sources themselves. Operation is at an elevated temperature such that thermionic emission occurs from the inside surface of a heated tube. The cathode fall supplies energy to heat the cathode which may also be heated with resistance wires wound around the tube. Insulation around the tube retains heat and improves the efficiency of operation. When used with some source such as a multipole, several ampères of electrons are extracted from the open end of the tube and the gas to be ionized, such as a rare gas or nitrogen, is fed at the other end. A recent source design [45] uses the hollow cathode as the source itself where ions are extracted from a slit in the side of the hollow cathode cylinder. The end caps are anodes and gas is fed through an orifice opposite the slit. The plasma potential is stabilized relative to the end cap anodes and the extracted beam is stable and uniform. At the highest reported discharge current of 3 A, the extracted ion current was 1 mA and contained significant numbers of cathode ions in addition to the feed gas ions. The source is simple and cheap with a long-life expectancy and the cathode is easily replaced when necessary. The gas efficiency is probably low.

A cathodeless discharge, such as achieved in microwave cavities or r.f. discharges, offers an alternative means of producing ions. One promising system is described by Sakudo *et al.* [46, 47] where power up to 2 kW at 2.45 GHz is fed into a ridged waveguide encircled by a solenoid. Ions are extracted from a slit at the end of the waveguide and between the ridges. The system, which will generate 10 mA of oxygen ions with a long operating life, may be fed with a variety of gases or vapors generated from ovens. It may be driven with a combination of oxide powders and CCl_4. For example it has been used to generate 3.5 mA of titanium ions (Ti^+). The source is useful with metal halides which do not attack the dielectric walls of the arc chamber. However, the halides prevent metal deposits from forming.

Liquid metal sources used mainly for focused ion beams have been reviewed by Prewett and Kellogg [48], and Adler and Picraux [49] have applied the well-developed vacuum arc technology to implanters.

3. EXTRACTOR DESIGNS

In the ion sources mentioned, the ions are formed at a low energy and they collect in a quiescent plasma near the apertures of an ion extraction system. From this state the ions are extracted and accelerated to the desired energy by a series of biased aperture plates or electrodes. Although it could be imagined that fine screens could be used rather than apertures, screens are not suitable because they erode and must be replaced and because the erosion releases contaminants. However, the use of apertures requires careful design of the extraction system. Computer programs involving two- and sometimes three-dimensional modeling have been used. The designer must account for space charge spreading of the beam as well as the focal properties of the apertures.

In a typical triode extractor, ions are first drawn through holes or slits in the first electrode that borders the plasma. It is usually at the plasma potential, although a negative bias from 50 to 100 V (limited by sputtering and thermal loading) can reduce aberrations [50, 51]. Whealton and Whitson [52] discuss both positive and negative biasing. The second electrode is biased negatively to attract ions which pass through it at energies somewhat greater than the desired final energy. The third electrode is biased positive relative to the second mainly to prevent neutralizing electrons beyond the extraction system from streaming back into the plasma. This system is commonly called an accel–decel system because the ions experience both acceleration and deceleration as they pass through the system.

Extractor systems may have only two electrodes [30], as is common on propulsion systems, or up to five [44]. With five, Shubaly and Hamm accelerated ions up to 250 keV in three stages. The usual choice is between three or four electrodes, where four allow more options for controlling aberration and distributing the stress at higher voltages up to 120 kV [8, 53–55].

A given extractor design may be scaled with reference to well-known dimensionless parameters [54]. If, for a cylindrically symmetric extractor, all dimensions are changed by the same factor, the optimum current does not change. Consequently, for high current applications, multiaperture designs are used.

Numerous modeling codes have been written for extraction systems. Shubaly *et al.* [56] provide a convenient introduction to desirable code features and references to various codes. Many extractor designs have been published and may be emulated for non-critical applications. However, a system designer can use simulation to gain information about the effective area of an aperture, the extent to which ions impact electrode structures, the beam focusing and the aberrations in the beam. A commonly used parameter to indicate beam quality is the brightness in ampères per square centimeter per steradian which can be deduced from the phase diagrams commonly produced by simulations. When ions can be extracted with a low angular divergence, the beam can be readily controlled and the target illuminated with well-defined angles of incidence. To achieve this controllability requires that the system be designed to operate at a specific current and that current fluctuations be low.

In well-designed systems the ions follow no sharp bends or excursions perpendicular to the line of travel. This objective can be met only by contouring the first, or plasma, electrode where ion energies are low and trajectories are strongly influenced by the fields near that electrode. Whealton [54] presented an instructive survey of the effects of various electrode shapes. Numerous other papers [27, 28, 55, 57] confirm the importance of using a tapered hole as have simulations done by the present authors [51]. Yet the narrow edge itself, which arises from the taper, does not imply that a thinner electrode will suffice. A rule of thumb is that the thickness of the plasma electrode should be of the order of 0.3–0.5 times the distance between the aperture and the accelerating electrode.

4. MASS ANALYSIS AND ACCELERATION

Mass analysis serves several functions which may or may not be pertinent for large-area implanters. Keller [58] has treated this topic, and Hanley [59] has also addressed it in his survey of implantation equipment.

(1) Where multiple species of ions are extracted from the source and only one is desired on the target, mass analysis is necessary.

(2) Focusing is inherent in mass analysis with a bending magnet. Magnets are preferred to other analyzing systems at high currents because the beam can be neutralized during the analysis.

(3) With mass analysis, ions pass through a slit which can serve as a barrier in a differential pumping scheme. This is useful if properties of a deposited film are influenced by background gas contamination.

(4) Post-acceleration schemes require well-focused beams typical of the mass analysis systems.

Several methods are in use for producing beams with energies higher than extractors can provide. The traditional Van de Graaff generator appears in many forms associated with a linear accelerator tube, one recently being described by Rathmell and Sundquist [60] who also describe a tandem generator. Whereas these accelerators provide steady currents, the r.f. quadrupole, which produces a pulsed beam, has become very prominent in recent years. Klein [61] provides a summary of r.f. quadrupole designs.

5. CONCLUSIONS

Steady developments in ion beam technology since about 1970 have provided a great variety of sources and beam-handling systems that have been adapted to many practical applications. Multipole or cusp confinement of the source plasma has the advantages of a high gas efficiency, a low noise and uniformity over a broad extraction area. When the plasma is so well defined, the extraction systems may be designed to produce highly directional and uniform beams that are easily analyzed and directed either to targets or into post-acceleration systems. When analysis is not needed, multiple beams can provide uniform exposure over a broad target. Much recent source development is concerned with extending component lifetimes for commercially available systems and broadening the range of usable feed materials.

REFERENCES

1 H. Ryssel and H. Glawischnig (eds.), *Proc. 4th Int. Conf. on Ion Implantation: Equipment and Techniques, Berchtesgaden, September 13–17, 1982*, Springer, New York, 1983.

2 J. F. Ziegler and R. L. Brown (eds.), *Proc. 5th Int. Conf. on Ion Implantation: Equipment and Techniques, Smugglers' Notch, VT, July 23–27, 1984*, in *Nucl. Instrum. Methods B, 6* (1985).

3 O. F. Goktepe, in G. K. Wolf, W. A. Grant and R. P. M. Procter (eds.), *Proc. Int. Conf. on Surface Modification of Metals by Ion Beams, Heidelberg, September 17–21, 1984*, in *Mater. Sci. Eng., 69* (1985) 13–20.

4 J. L. Brimhall, H. E. Kissinger and A. R. Pelton, in G. K. Hubler, O. W. Holland, C. R. Clayton and C. W. White (eds.), *Ion Implantation and Ion Beam Processing of Materials, Materials Research Society Symp. Proc., Boston, MA, November 14–17, 1983*, Vol. 27, Elsevier, New York, 1984, p. 163.

5 A. J. T. Holmes, *Proc. Int. Conf. On Ion Implantation Equipment, Povo, August 28–31, 1978*, in *Radiat. Eff., 44* (1979) 47–58.

6 J. C. Muller, E. Courcelle, D. Salles and P. Siffert, in J. F. Ziegler and R. L. Brown (eds.), *Proc. 5th Int. Conf. on Ion Implantation: Equipment and Techniques, Smugglers' Notch, VT, July 23–27, 1984*, in *Nucl. Instrum. Methods B, 6* (1985) 394–398.

7 Z. Yuchun, W. Yuming, R. Congxin, F. Xinding and C. Guoming, in J. F. Ziegler and R. L. Brown (eds.), *Proc. 5th Int. Conf. on Ion Implantation: Equipment and Techniques, Smugglers' Notch, VT, July 23–27, 1984*, in *Nucl. Instrum. Methods B, 6* (1985) 447–451.

8 M. A. Bell, J. H. Whealton, R. J. Raridon, D. E. Wooten and R. W. McGaffey, *Bull. Am. Phys. Soc., 29* (1984) 1369.

9 E. Newman, W. A. Bell, Jr., W. C. Davis, L. O. Love, W. K. Prater, K. A. Spainhour, J. G. Tracy and A. M. Veach, in S. Amiel and G. Engler (eds.), *Proc. 9th Int. Conf. on Electromagnetic Isotope Separators and Related Ion Accelerators, Kiryat Anavim, May 10–13, 1976*, in *Nucl. Instrum. Methods, 139* (1976) 87–93.

10 R. Bernas, in M. J. Higalsberger and F. P. Viehbock (eds.), *Electromagnetic Separation of Radioactive Isotopes*, Springer, Vienna, 1961, p. 300.

11 I. Chavet and R. Bernas, *Nucl. Instrum. Methods, 51* (1967) 77–86.

12 J. Camplan, personal communication, 1974.

13 G. Lempert, I. Chavet and M. Kanter, in S. Amiel and G. Engler (eds.) *Proc. 9th Int. Conf. on Electromagnetic Isotope Separators and Related Ion Accelerators, Kiryat Anavim, May 10–13, 1976*, in *Nucl. Instrum. Methods, 139* (1976) 57–63.

14 R. Meunier, J. Camplan, J.-L. Bonneval, J.-L. Daban-Haurou, D. Deboffle, D. Leclercq, M. Ligonniere and G. Moroy, in S. Amiel and G. Engler (eds.), *Proc. 9th Int. Conf. on Electromagnetic Isotope Separators and Related Ion Accelerators, Kiryat Anavim, May 10–13, 1976*, in *Nucl. Instrum. Methods, 139* (1976) 101–104.

15 J. H. Freeman, *Nucl. Instrum. Methods, 22* (1963) 306.

16 D. Aiken, in S. Amiel and G. Engler (eds.), *Proc. 9th Int. Conf. on Electromagnetic Isotope Separators and Related Ion Accelerators, Kiryat Anavim, May 10–13, 1976*, in *Nucl. Instrum. Methods, 139* (1976) 125–134.

17 N. Williams, in K. G. Stephens, I. H. Wilson and J. L. Moruzzi (eds.), *Proc. Int. Conf. on Low Energy Ion Beams, Salford, September 5–8, 1977*, in *Inst. Phys. Conf. Ser., 38* (1978) 70–77.

18 D. Aiken, *Proc. Int. Conf. on Ion Implantation Equipment, Povo, August 28–31, 1978*, in *Radiat. Eff., 44* (1979) 159–167.

19 M. Guerra, V. Benveniste, G. Ryding, D. H. Douglas-Hamilton, M. Reed, G. Gagne, A. Armstrong and M. Mack, in J. F. Ziegler and R. L. Brown (eds.), *Proc. 5th Int. Conf. on Ion Implantation: Equipment and Techniques, Smugglers' Notch, VT, July 23–27, 1984*, in *Nucl. Instrum. Methods B, 6* (1985) 63–69.

20 E. Yabe, S. Takeshiro, K. Sunako, K. Takayama, R. Fukui, K. Takagi, K. Okamoto and S. Komiya, in J. F. Ziegler and R. L. Brown (eds.), *Proc. 5th Int. Conf. on Ion Implantation: Equipment and Techniques, Smugglers' Notch, VT, July 23–27, 1984*, in *Nucl. Instrum. Methods B, 6* (1985) 119–122.

21 R. Limpaecher and K. R. MacKenzie, *Rev. Sci. Instrum., 44* (1973) 726–731.

22 W. L. Stirling, P. M. Ryan, C. C. Tsai and K. N. Leung, *Rev. Sci. Instrum., 50* (1979) 102–108.

23 K. W. Ehlers and K. N. Leung, *Rev. Sci. Instrum., 50* (1979) 1353–1361.

24 A. P. H. Goede, T. S. Green and B. Singh, *J. Appl. Phys., 51* (1980) 1896–1899.

25 A. R. Martin, *Proc. 3rd Int. Conf. on Low Energy Ion Beams, Loughborough, March 28–31, 1983*, in *Vacuum, 34* (1984) 17–24.

26 G. Crecelius and R. Holzle, *IEEE Trans. Nucl. Sci., 32* (1985) 1785–1787.

27 P. J. M. van Bommel, P. Massmann, E. H. A. Granneman, H. J. Hopman and J. Los, *Proc. 3rd Int. Conf. on Low Energy Ion Beams, Loughborough, March 28–31, 1983*, in *Vacuum, 34* (1984) 25–29.

28 R. Keller, F. Nohmayer, P. Spadtke and M.-H. Schonenberg, *Proc. 3rd Int. Conf. on Low Energy Ion Beams, Loughborough, March 28–31, 1983*, in *Vacuum, 34* (1984) 31–35.

29 K. W. Ehlers and K. N. Leung, *Rev. Sci. Instrum., 53* (1982) 1429–1433.

30 H. R. Kaufman, *NASA Contract. Rep. CR 134845*, 1975 (National Aeronautics and Space Administration).

31 J. H. Keller, in H. Ryssel and H. Glawischnig (eds.), *Proc. 4th Int. Conf. on Ion Implantation: Equipment and Techniques, Berchtesgaden, September 13–17, 1982*, Springer, New York, 1983, pp. 97–105.

32 D. Cope and J. H. Keller, *J. Appl. Phys, 56* (1984) 96–100.

33 J. P. Brainard and J. B. O'Hagan, *Rev. Sci. Instrum., 54* (1983) 1497–1505.

34 D. R. Nicholson, *Introduction to Plasma Theory*, Wiley, New York, 1983, pp. 200–208.

35 H. R. Kaufman, *NASA Contract, Rep. CR 159527*, 1978; *NASA Contract. Rep. CR 135226*, 1977; *NASA Contract. Rep. CR 135100*, 1976 (National Aeronautics and Space Administration).

36 J. G. Alessi and K. Prelec, *Proc. Workshop on Polarized ^{3}He Beams and Targets, Princeton, NJ, October 22–24, 1984*, in *AIP Conf. Proc., 131* (1985) 18–24.

37 W. Scaife, D. Wagner and W. Faul, in J. F. Ziegler and R. L. Brown (eds.), *Proc. 5th Int. Conf. on Ion Implantation: Equipment and Techniques, Smugglers' Notch, VT, July 23–27, 1984*, in *Nucl. Instrum. Methods B, 6* (1985) 39–45.

38 M. R. Shubaly and M. S. deJong, *IEEE Trans. Nucl. Sci., 30* (1983) 1399–1401.

39 M. R. Schubaly, R. G. Maggs and A. E. Weeden, *IEEE Trans. Nucl. Sci., 32* (1985) 1751–1753.

40 R. W. Bickes, Jr., and J. B. O'Hagan, *Rev. Sci. Instrum., 53* (1982) 585–591.

41 E. Ghanbari, J. Boers, R. Liebert, L. Ayers and P. Bazeley, *Nucl. Instrum. Methods B, 10* (1985) 767–770.

42 W. L. Stirling, C. C. Tsai and P. M. Ryan, *Rev. Sci. Instrum., 48* (1977) 533–536.

43 M. M. Menon, C. C. Tsai, J. H. Whealton, D. E. Schechter, G. C. Barber, S. K. Combs, W. K. Dagenhart, W. L. Gardner, H. H. Haselton, N. S. Ponte, P. M. Ryan, W. L. Stirling and R. E. Wright, *Rev. Sci. Instrum., 56* (1985) 242–248.

44 M. R. Shubaly and R. W. Hamm, *IEEE Trans. Nucl. Sci., 28* (1981) 1316–1318.

45 A. Tonegawa, E. Yabe, D. Satoh, K. Sunako, K. Takayama, K. Takagi, R. Fukui, K. Okamoto and S. Komiya, in J. F. Ziegler and R. L. Brown (eds.), *Proc. 5th Int. Conf. on Ion Implantation: Equipment and Techniques, Smugglers' Notch, VT, July 23–27, 1984*, in *Nucl. Instrum. Methods B, 6* (1985) 129–132.

46 N. Sakudo, K. Tokiguchi, H. Koike and I. Kanomata, *Rev. Sci. Instrum., 49* (1978) 940–943.

47 N. Sakudo, K. Tokiguchi and H. Koike, *Proc. 3rd Int. Conf. on Low Energy Ion Beams, Loughborough, March 28–31, 1983*, in *Vacuum, 34* (1984) 245–249.

48 P. D. Prewett and E. M. Kellogg, in J. F. Ziegler and R. L. Brown (eds.), *Proc. 5th Int. Conf. on Ion Implantation: Equipment and Techniques, Smugglers' Notch, VT, July 23–27, 1984*, in *Nucl. Instrum. Methods B, 6* (1985) 135–142.

49 R. J. Adler and S. T. Picraux, in J. F. Ziegler and R. L. Brown (eds.), *Proc. 5th Int. Conf. on Ion Implantation: Equipment and Techniques, Smugglers' Notch, VT, July 23–27, 1984*, in *Nucl. Instrum. Methods B, 6* (1985) 123–128.

50 J. H. Whealton, C. C. Tsai, W. K. Dagenhart, W. L. Gardner, H. H. Haselton, J. Kim, M. M. Menon, P. M. Ryan, D. E. Schechter and W. L. Stirling, *Appl. Phys. Lett., 33* (1978) 278–279.

51 J. W. Robinson and J. H. Keller, unpublished work, 1985.

52 J. H. Whealton and J. C. Whitson, *J. Appl. Phys., 50* (1979) 3964–3966.

53 C. N. Meixner, M. M. Menon, C. C. Tsai and J. H. Whealton, *J. Appl. Phys., 52* (1981) 1167–1174.

54 J. H. Whealton, *J. Appl. Phys.*, *53* (1982) 2811–2817.
55 A. J. T. Holmes, E. Thompson and F. Watters, *J. Phys. E*, *14* (1981) 856–859.
56 M. R. Shubaly, R. A. Judd and R. W. Hamm, *IEEE Trans. Nucl. Sci.*, *28* (1981) 2655–2657.
57 B. Piosczyk and G. Dammertz, *Rev. Sci. Instrum.*, *55* (1984) 1421–1424.
58 J. H. Keller, Beam optics design for ion implantation, *Nucl. Instrum. Methods, Part II*, *189* (1981) 7–14.
59 P. R. Hanley, in H. Ryssel and H. Glawischnig (eds.), *Proc. 4th Int. Conf. on Ion Implantation: Equipment and Techniques, Berchtesgaden, September 13–17, 1982*, Springer, New York, 1983, pp. 2–24.
60 R. D. Rathmell and M. L. Sundquist, in J. F. Ziegler and R. L. Brown (eds.), *Proc. 5th Int. Conf. on Ion Implantation: Equipment and Techniques, Smugglers' Notch, VT, July 23–27, 1984*, in *Nucl. Instrum. Methods B*, *6* (1985) 56–62.
61 H. Klein, *IEEE Trans. Nucl. Sci.*, *30* (1983) 3313–3322.

Author Index

0025-5416/87/$3.50

Materials Science and Engineering, 90 (1987) 434-438

Subject Index

0025-5416/87/$3.50

ANDERSONIAN LIBRARY
WITHDRAWN
FROM
LIBRARY
STOCK
UNIVERSITY OF STRATHCLYDE